Indian Ocean Geology and Biostratigraphy

Studies Following Deep-Sea Drilling Legs 22-29

Edited by
J. R. Heirtzler, H. M. Bolli,
T. A. Davies, J. B. Saunders,
and J. G. Sclater

American Geophysical Union
Washington, D.C. 20006

Published under the aegis of the AGU Geophysical Monograph Board; Bruce Bolt, Chairman; Thomas E. Graedel, Rolland L. Hardy, Pearn P. Niiler, Barry E. Parsons, George R. Tilton, and William R. Winkler, members.

Sponsored by the Joint Oceanographic Institutions for Deep Earth Sampling (JOIDES) and Supported by the National Science Foundation.

Copyright © 1977 by the American Geophysical Union, 1909 K Street, N.W., Washington, D.C. 20006

ISBN 0-87590-208-1

Library of Congress Card Number 77-88320

Printed in the United States of America by LithoCrafters, Inc., Chelsea, Michigan 48118

Preface

The Drilling Vessel GLOMAR CHALLENGER entered the Indian Ocean when she left Darwin, Australia on January 13, 1972. She traveled around that Ocean in a counterclockwise fashion drilling 72 holes at 64 sites during 7 cruises over all the major ridges and basins from the Red Sea to the Antarctic continent. She finished her work there when she eventually arrived at Christchurch, New Zealand on February 27, 1973. During this time nearly 30 km of seafloor was penetrated and much of that material was recovered and studied and is archived at the Deep Sea Drilling Project core depository.

In the Indian Ocean CHALLENGER had several notable achievements. On Leg 22 she drilled in a water depth of 6243 meters - the deepest at that time. On Leg 23 she had a single hole penetration of 1300 meters and at another hole had a penetration into basaltic basement of 81 meters for two new records at that time. She drilled in the Red Sea, which is a new spreading center, and in the Argo Abyssal Plain where some of the oldest seafloor in the world is found. She spent 69 days at sea traveling 7400 miles on Leg 28 working into the ice south of Australia. On Ninetyeast Ridge lignite was found indicating major subsidence. There, and at other sites, the northward drift of India and the opening of the Indian Ocean was in evidence in the recovered seafloor materials.

The plans of the IPOD Project do not call for any further drilling in the Indian Ocean in the forseeable future and so the work completed four years ago may not be rivalled for some time. This fact prompted the editors to organize this synthesis of all the work published in the Initial Reports and in other related publications. Within a manageable sized single book it is not possible to give sufficient coverage to all geographic areas or to all disciplines or subdisciplines. Some of these shortcomings will be evident here.

Each of the articles is a synthesis and then in two introductions - one to the geology and one to the paleontology and biostratigraphy - a brief further synthesis of the articles is attempted. We feel that as a result of these articles a complete history of the Indian Ocean is established and rests upon a solid factual basis.

We wish to thank the authors for their patience with us during the editing and publication of this work and to thank J. Beckman, D. Bukry, C. Caron, J. Curray, H. Dick, R. Douglas, R. Heath, J. Kennett, R. Kidd, B. Luyendyk, E. Martini, D. Ninkovitch, J. Peirce, K. Perch-Nielson, I. Primola-Silva, E. Schreibner, J. Thiede, G. Thompson, T. Vallier, and Tj. van Andel for reviewing manuscripts. We also thank Mrs. Florence Mellor who typed and edited the entire manuscript.

The Editors

Foreword

Certain human efforts produce a curious mixture of humility and pride; humility -- in that truly important matters are addressed; pride -- in that we are capable, collectively, of addressing them. Now in its ninth year of operation, the Deep Sea Drilling Project is an excellent example of such an effort. The Project is a uniquely successful attempt to increase man's knowledge of the Earth, the age, history and processes of development of the ocean basins and the structure and composition of the oceanic crust. An equally important aspect is the significant information DSDP has made available to the world's scientific community -- and, ultimately, to the public at large -- concerning the resource potential either directly or indirectly related to oceanic realms.

Begun as an 18-month program of ocean drilling in August, 1968, its levels of high achievement earned four successive extensions; thus, the Project will now operationally continue until at least the Fall of 1979.

Its current phase is called IPOD, the International Phase of Ocean Drilling. Although initially the Project was financed entirely by the United States, scientists from more than two dozen nations participated in its work. In 1970, the first tentative discussions were begun with the USSR Academy of Sciences. These led to the signing of a Memorandum of Understanding between our two countries by which the Soviet Union became a participating member of DSDP for a period of five years, beginning on 1 January 1974.

Since then, similar Memorandums of Understanding have been executed between the United States and Germany, England, France and Japan.

Each of these nations, together with nine U. S. oceanographic institutions, is represented on the Executive and Planning Committees of the Joint Oceanographic Institutions for Deep Earth Sampling (JOIDES); it is JOIDES which provides the scientific guidance for DSDP. Indeed, it was JOIDES, formed in mid-1964 by four U. S. oceanographic institutions, which can truly be regarded as the scientific wellspring from which the Project evolved.

Principal operational tool of the Deep Sea Drilling Project is the research vessel GLOMAR CHALLENGER. The University of California, in its capacity as principal contractor with the National Science Foundation for the Project, sub-contracts with Global Marine Inc. of Los Angeles to accomplish the actual drilling and coring operations using the CHALLENGER.

CHALLENGER is a scientific ship created especially to meet the goals of the Project and has never been used by anyone else. Launched in Galveston, Texas, in March 1968, she is 400 feet long, displaces 10,500 tons and bears a 142-foot high drilling tower amidships. In her service to science, she has traveled over a quarter-of-a-million miles across the world's oceans and marginal seas -- with the exception of the ice-bound Arctic Ocean.

In these past nine years, the Project has developed a priceless legacy of information, principal among which is a library of over 30 miles of sediment cores collected from drill holes in every major ocean basin. These cores are kept in two repositories; at Scripps Institution of Oceanography in California and at Lamont-Doherty Geological Observatory in New York. This carefully preserved resource provides the basis for study of the past history of the oceans, the rifting and separation of continents, generation and destruction of ocean basins, interaction of changing ocean currents and climate and the evolution of flora and fauna. The cumulative effect of the Project's findings has been to provide the basis for a continuing revolution in thinking concerning the history of the entire Earth, the origins of every resource for which we mine or drill, and our self appreciation as inhabitants of this planet.

DSDP successes have not been entirely scientific in nature. Among breakthroughs achieved is computer-controlled dynamic positioning which system keeps the CHALLENGER over the bore hole in full oceanic water depths. By the end of 1970, a re-entry system had been developed by which worn bits could be brought to the surface, replaced and then re-entry gained to the same hole for yet deeper penetration into the seabed.

The Deep Sea Drilling Project stands at the doorstep of bold and vital exploration. The results of what its scientists will learn can have significant impact on the future of all who live on this planet.

Those of us who have stood central to the management of the Deep Sea Drilling Project have been privileged to serve that community of scholars, engineers, seamen and drillers who have truly forged its purposes into realities.

                                             M. N. A. Peterson
                                             Project Manager
                                             Principal Investigator
                                             Deep Sea Drilling Project

Contents

Preface		iii
Foreword		iv
Chapter 1.	An Introduction to Deep Sea Drilling in the Indian Ocean by John G. Sclater and James R. Heirtzler	1
Chapter 2.	Paleobathymetry and sediments of the Indian Ocean by John G. Sclater, Dallas Abbott and Jörn Thiede	25
Chapter 3.	Sedimentation in the Indian Ocean through time by Thomas A. Davies and Robert B. Kidd	61
Chapter 4.	Volcanogenic sediments in the Indian Ocean by Tracy L. Vallier and Robert B. Kidd	87
Chapter 5.	Mesozoic-Cenozoic sediments of the Eastern Indian Ocean by Peter J. Cook	119
Chapter 6.	Models of the Evolution of the Eastern Indian Ocean by J. J. Veevers	151
Chapter 7.	Deep sea drilling on the Ninetyeast Ridge: Synthesis and a tectonic model by Bruce P. Luyendyk	165
Chapter 8.	Eastern Indian Ocean DSDP sites: Correlations between petrography, geochemistry and tectonic setting by Fred A. Frey, John S. Dickey, Jr., Geoffrey Thompson and Wilfred B. Bryan	189
Chapter 9.	Large ion lithophile elements and Sr and Pb isotopic variations in volcanic rocks from the Indian Ocean by K. V. Subbarao, R. Hekinian and D. Chandresekharam	259
Chapter 10.	Seismic velocities and elastic moduli of igneous and metamorphic rocks from the Indian Ocean by Nikolas I. Christensen	279
Chapter 11.	The magnetic properties of Indian Ocean basalts by M. W. McElhinny	301
Chapter 12.	Introduction to stratigraphy and paleontology by Hans M. Bolli and John B. Saunders	311

Chapter 13. Paleontological-biostratigraphical investi-   325
             gations, Indian Ocean sites 211-269 and 280-
             282, DSDP Legs 22-29 by Hans M. Bolli

Chapter 14. Mesozoic calcareous nannofossils from the   339
             Indian Ocean, DSDP legs 22 to 27 by Hans R.
             Thierstein

Chapter 15. Paleocene to Eocene calcareous nannoplankton   353
             of the Indian Ocean by Franca Proto Decima

Chapter 16. Distribution of calcareous nannoplankton in   371
             Oligocene to Holocene sediments of the Red
             Sea and the Indian Ocean reflecting paleo-
             environment by Carla Müller

Chapter 17. Synopsis of cretaceous planktonic foraminifera 399
             from the Indian Ocean by Rene Herb

Chapter 18. Maastrichtian to Eocene foraminiferal assem-   417
             blages in the northern and eastern Indian
             Ocean region: Correlations and historical
             patterns by Brian McGowran

Chapter 19. Oligocene plankton foraminiferal assemblages   459
             from Deep Sea Drilling Project sites in the
             Indian Ocean by Robert L. Fleisher

Chapter 20. Indian Ocean Neogene planktonic foraminiferal 469
             biostratigraphy and its paleoceanographic impli-
             cations by Edith Vincent

Chapter 21. Synthesis of the cretaceous benthonic foramini-585
             fera recovered by the Deep Sea Drilling Project
             in the Indian Ocean by Viera Scheibnerova

Chapter 22. Neogene deep water benthonic foraminifera of   599
             the Indian Ocean by E. Boltovskoy

CHAPTER 1. AN INTRODUCTION TO DEEP SEA DRILLING IN THE INDIAN OCEAN

John G. Sclater

Department of Earth and Planetary Sciences
Massachusetts Institute of Technology
Cambridge, Massachusetts 02139

James R. Heirtzler

Department of Geology and Geophysics
Woods Hole Oceanographic Institution
Woods Hole, Massachusetts 02543

Introduction

The Indian Ocean is the most complex of the three major oceans. It has structures which are peculiar to this ocean as well as a large assortment of features which are found elsewhere. The early work in the ocean was carried out aboard the DANA, SNELLIUS, MABAHISS, CHALLENGER, ALBATROSS, and OB (Yentsch, 1962) and the results were published by Wiseman and Seymour-Sewell (1939), Seymour-Sewell (1925) and Fairbridge (1948, 1955). The first really substantial investigation of the ocean came with the International Indian Ocean Expedition which extended from 1959 to 1966. During the course of this expedition ten countries and forty-six ships took part in loosely coordinated cruises to the ocean. A compilation of the data from these cruises has been completed in Atlas form (Udintsev et al., 1975) and we present a chart of the general bathymetry from this Atlas as Figure 1. The International Indian Ocean Expedition resulted in the publication of much scientific work and led directly to a series of expeditions to undertake specific objectives. Perhaps the most successful of these was the drilling program carried out between 1971 through 1972 by the D/V GLOMAR CHALLENGER. The purpose of this paper is to present a simple description of the morphology and structure of the Indian Ocean, to review the problems presented to the D/V CHALLENGER before drilling commenced and to present, within this general context, the highlights of the findings of the drilling program.

Morphology of the Indian Ocean

Many bathymetric charts of the Indian Ocean have been published. The most detailed of these, Laughton et al. (1970) for the Arabian Sea and Carlsberg Ridge, Fisher et al. (1968) for the Somali Basin, Fisher et al. (1971) for the Central Indian Ridge and portions of Sclater and Fisher (1974) have been compiled by Udintsev et al. (1975) to produce a bathymetric chart of the whole ocean. Our modification of this chart (Figure 1) shows the principal

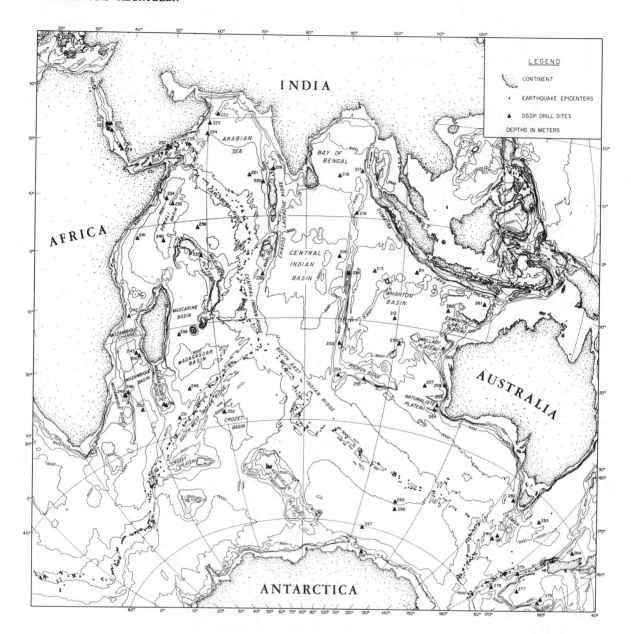

Fig. 1. Principal bathymetric features of the Indian Ocean (Udintsev, et al., 1975). Triangles mark the Deep Sea Drilling Sites and dots the epicenters occurring between 1963 and 1973 taken from the ESSA earthquake tape. To contrast the spreading center with the aseismic ridges, we have shown earthquakes along the active mid-ocean ridges, but not in the Java-Sumatra Trench or other off axis areas.

morphological features on which are superimposed the earthquake epicenters from the mid-ocean ridges and the sites of the Deep Sea Drilling holes in the Indian Ocean. The nomenclature of the major features has often given rise to confusion especially in comparisons of Russian work with that of western countries. In this study we have attempted whenever possible to follow the nomenclature of Laughton et al. (1971).

Mid Ocean Ridge System

The major topographic feature in the Indian Ocean is the broad active mid-ocean ridge system which starts in the Gulf of Aden (Laughton et al., 1970) and trends southwest as the Carlsberg Ridge into an en echelon pattern of transform faults and spreading centers (Fisher et al., 1971) called the Central Indian Ridge. At 25°S, 70°E this ridge bifurcates into the Southeast Indian and Southwest Indian Ridges. The Southwest Indian Ridge eventually joins the mid-Atlantic Ridge at the Bouvet Triple Junction. The Southeast Indian Ridge, a broad swell with less rough topography than the other ridges, trends southeast to the Amsterdam and St. Paul Islands and then almost due east between Australia and Antarctica joining up with East Pacific Rise after some major offsets at the Macquarie Triple Junction.

The Aseismic Ridges and Plateaus

Characteristic of the Indian Ocean is the abundance of relatively shallow ridges and plateaus that are free from earthquake activity. The most prominent of these features are the Chagos-Laccadive and Ninetyeast Ridges both of which are thought to be oceanic and the Seychelles Bank which is attached to the Mascarene Plateau. The Seychelles Bank is known to be continental (Matthews and Davies, 1966) and is thought to be a micro-continent split from India during the northward movement of that continent. The origin of the Mascarene Plateau is unknown. Other aseismic ridges of considerable morphological importance are the Mozambique and Madagascar Ridges, the Crozet and Kerguelen Plateaus, Broken Ridge and the Wallaby and Naturaliste Plateaus (Figure 1).

Further aseismic ridges and islands which show up as morphological features but are less extensive in size are the Rodriquez Ridge, the Ob Seamount Province and the Cocos-Keeling, Christmas group of islands in the Wharton Basin. Because all these features are relatively small and appear of secondary importance they are not discussed either in this introduction or in the papers presented in the volume.

The Ocean Basins

The mid-ocean ridges, the aseismic ridges and the continental shelves divide the Indian Ocean into a number of more or less isolated basins (Figure 1). These ridges and the continental edges have a very large effect upon the distribution of sediments. Seismic reflection data have enabled an isopach map of sediment thickness (Figure 2) to be constructed which shows that 40 percent of the total sediment present is to be found in the Arabian Sea and in the Bay of Bengal, probably from Himalayan erosion (Ewing et al., 1969). More recent seismic studies in the Bay of Bengal by Curray and Moore (1971) suggest that sediment thickness may exceed 12 km and that Himalayan denudation may be proceeding at an average rate of 70 cm/1000 years. Other thick terrigenous sediments are found off East Africa and in the Mozambique Channel. South of the polar front, high biological productivity has given rise to thick accumulations of siliceous ooze. Elsewhere the sediments are relatively thin and on the crestal area, 100 km on either side of the ridge axis, they are virtually absent.

Scattered seismic refraction stations from a variety of sources (Laughton et al., 1971) illustrate that, except under the micro-continents and the ridge axes, normal oceanic crust is found. The only seismic data available on a mid-ocean ridge is in the complex

Fig. 2. An Isopach map of unconsolidated sediments contoured in two-way reflection time in seconds (at 0.1 sec contour intervals). 1 sec is equivalent approximately to 1000 m (from Ewing et al., 1969).

area of the Central Indian Ridge and this data does not yield much information about the deep structure.

Tectonic Synthesis of the Indian Ocean

It is impossible in a short review to do anything other than highlight the work and theories that have led to our present understanding of this very complex ocean. Perhaps the most important contributions that came out of the International Indian Ocean Expedition were those of Vine and Matthews (1963) and Heezen and Tharp (1965). Vine and Matthews (1963) developed the concept of sea floor spreading by an analysis of magnetic anomaly observations across the Carlsberg Ridge. They presented strong evidence that this active ridge was a spreading center. Heezen and Tharp (1965) made the first major attempt to describe the physiography of the Indian Ocean. They emphasized the entire mid-ocean ridge system and the large fracture zones offsetting the ridge axis. They contrasted the broad ridge system with the long

linear smooth topped aseismic ridges such as the Chagos-Laccadive and
Ninetyeast Ridges. Later work by LePichon and Heirtzler (1968) and
Fisher et al. (1971) and Ewing et al. (1969) supported many of their
suggestions. In particular LePichon and Heirtzler (1968) showed that
the rate of separation on the active ridge axis increases from near
1 cm/yr (half rate) on the Carlsberg Ridge to greater than 3 cm/yr
(half rate) on the Southeast Indian Ridge. At the same time as this
analysis of sea floor data was underway McElhinney (1970) published
an analysis of paleomagnetic data from the surrounding continents.
He concluded that initial movement within Gondwanaland began during
the Upper Permian, when Africa and Antarctica together made a northward and counterclockwise movement away from the other continents,
which remained stationary. The final break-up occurred during the
Upper Cretaceous, with Indian and Australia separating from Antarctica
and drifting to their present positions during the Tertiary. There
was little movement between either Africa, South America, or Antarctica
and the present South Pole during the Tertiary, but India must have
moved 59° or 5500 km, between earliest Eocene and the Miocene (the
latter is the time of formation of the Himalaya). This corresponds
to a rate of drift of 11 cm/yr. Similarly, Australia moved northward
at 8 cm/yr between the Oligocene and the present.

During the International Indian Ocean Expedition a large quantity of
residual magnetic anomaly data was collected along track by various
research vessels. Two compilations of this data have been published
(Sclater et al., 1971 and Udintsev et al., 1975). Most of this and
much other magnetic data collected during and after the International
Indian Ocean Expedition was first used by Fisher et al. (1971) and
then McKenzie and Sclater (1971) to examine the evolution of the ocean
from the late Cretaceous to present. Fisher et al. (1971) showed in
a topographic chart that rather than being a simple north-south ridge
with a central deep, as suggested by Heezen and Tharp (1965), the
Central Indian Ridge consisted of a complex suite of en echelon fracture zones and active spreading centers trending N60°E. Further, they
and McKenzie and Sclater (1971) assuming a constant spreading rate about
the present pole of motion showed that the Mascarene plateau fit snuggly
into the Chagos-Laccadive Ridge in the earliest Oligocene. This new
feature, stretching from Mauritius to the Maldives, was perpendicular
to the older anomaly sequences in the Arabian Sea and Central Indian
Basin. McKenzie and Sclater (1971) made this discovery the basis of a
tectonic evolution in the whole ocean (Figure 3). They recognized
only four plates, Africa, India, Australia, and Antarctica, and from
the Eocene to present assumed only three plates. Prior to the Eocene,
Australia and India were separated by an active transform fault running
parallel to the Ninetyeast Ridge.

Working backwards in time and terminating in the Late Cretaceous,
McKenzie and Sclater (1971) identified three phases of spreading:

(a) Present to Oligocene (present to 36 Ma). This period was dominated by the closing of the Red Sea and the Gulf of Aden, slow spreading
on the Carlsberg (1.2 cm/yr half rate) and Southwest Indian Ridges
(1.0 cm/yr half rate), and faster rates on the Central Indian (2.0
cm/yr half rate) and Southeast Indian Ridges (Figure 3b). The uplift
and denudation of the Himalaya occurred during this time and the phase
terminated with the Mascarene Plateau fitting into the Chagos Laccadive
Ridge.

For the rest of this introduction, we do not reference each report
whenever a site is mentioned. However, if we cite something from
a paper in the Initial Report, then we reference it.

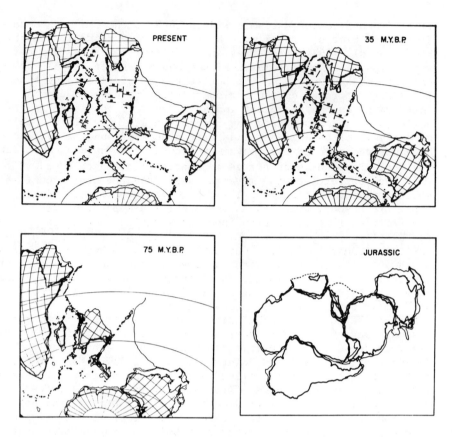

Fig. 3. Stages in the evolution of the Indian Ocean (after McKenzie and Sclater, 1971): a) present; b) at 35 Ma (anomaly 13); c) at 75 Ma (anomaly 32) and d) prior to the breakup of Gondwana (after Smith and Hallam, 1971).

(b) Eocene to Paleocene (36 Ma to 53 Ma). The main movement during this phase is the continued separation of the Australian part of the Indian plate from Antarctica at 2.5 cm/yr (half rate). Relatively little movement is apparent on the Carlsberg and Central Indian Ridges in the northwest Indian Ocean. On the Southwest Indian Ridge the anomalies from the Mascarene basin abutt those on the Crozet basin and thus during this time span the ridge has changed from an en echelon set of ridges and transform faults into one long continuous transform fault. This phase terminates with the connection of Australia to Antarctica and the juxtaposition of the south end of the Ninetyeast and Broken Ridges against the Kerguelen Plateau.

(c) Paleocene to Late Cretaceous (53 Ma to 70 Ma). This period is dominated by the southward movement of the Indian plate, separated from Australia by the Ninetyeast transform fault. Two spreading centers, south of India and north of the Seychelles (Figure 3c), generated sea floor that is now found in the Crozet basin and the Central Indian basin on one hand, and in the Arabian basin and the basin north of the Seychelles on the other. These spreading centers were linked by the Southwest Indian transform fault, which continued south of Africa to meet with the Africa/Antarctic ridge. The Seychelles spreading center was terminated in the west by the Owen Fracture Zone. The northern side of the Indian plate must have lain in a subduction zone in the Tethys Sea. The southward movement of

India started slowly 53 Ma but accelerated between 65 and 70 Ma to a total rate of 17 cm/year. The slow start is probably related to the collision of India with the subduction zone to the north and the onset of the uplift of the Himalaya. It is possible that the post 70 Ma movements could be extrapolated back in time (perhaps to 100 Ma), bringing India farther south until its southern tip met the spreading center and its northeast section reached the southern end of the Ninetyeast Ridge. In this model, the seafloor south of the Seychelles and Madagascar should be older than 75 Ma, although the northward movement of the Seychelles and the growth of the Amirante Island Arc may have obscured any traces of the old crust.

(d) Late Cretaceous to Jurassic (75 Ma to 140 Ma). Owing to the absence of clearly identified magnetic lineations McKenzie and Sclater (1971) could not extend the tectonic history back beyond the Late Cretaceous. They were unable to account for either the ocean crust generated between Africa and Antarctica or for the apparent displacement of Madagascar from East Africa. These motions must have occurred before 75 Ma and were related to the breakup of the original Gondwana continents.

(e) Gondwanaland Reconstructions. Many possible reconstructions of Gondwanaland have been suggested. Holmes (1965) includes an excellent short review of the early attempts. Magnetic anomaly information presented by Weissel and Hays (1972) between Australia and Antarctica and paleomagnetic evidence have tended to support simple variations on the DuToit (1937) position of the continents. This fit has received much support by the computer analysis of continental coastlines by Smith and Hallam (1970). Their preferred fit was that most generally accepted for the continents before the D/V CHALLENGER entered the Indian Ocean (Figure 3d). Another fit considered was a modification of that of Wilson (1963) suggested by Crawford (1969) where Indian fits into the west coast of Australia. However, as will be seen in the next section, this fit is not compatible with magnetic anomaly information and the basement ages inferred from holes in the Wharton Basin (Sclater and Fisher, 1974; Luyendyk, 1974).

Major Problems Undertaken by DSDP in the Indian Ocean

Before the D/V GLOMAR CHALLENGER entered the ocean the J.O.I.D.E.S. Indian Ocean Advisory Panel consisting of E. Bunce, R. Fisher, J. Heirtzler, M. Langseth, R. Schlich and M. Talwani (Chairman) set up a suite of major tectonic and sedimentological objectives. In this section we outline these objectives and the attempts made by the scientists on board the D/V GLOMAR CHALLENGER to meet them.

In our topographic map of the Indian Ocean we present the position of all the sites drilled in this ocean (Figure 1). Compilations of the lithofacies sections for Leg 22 (Von der Borch et al., 1973), Leg 23a (Whitmarsh et al., 1974a), Leg 23b (Whitmarsh, et al., 1974b), Leg 24 (Fisher et al., 1974), Leg 25 (Schlich et al., 1974), Leg 26 (Davies et al., 1974), Leg 27 (Veevers et al., 1975) and Leg 28 (Hayes et al., 1975) are presented as Figure 4 through 11.

NOTE: The dates used in this review differ slightly from those in the original paper by McKenzie and Sclater (1971). We have chosen to use the new dates because they are better established and conform well with the Deep Sea Drilling information presented later in this volume (Sclater et al., 1974).

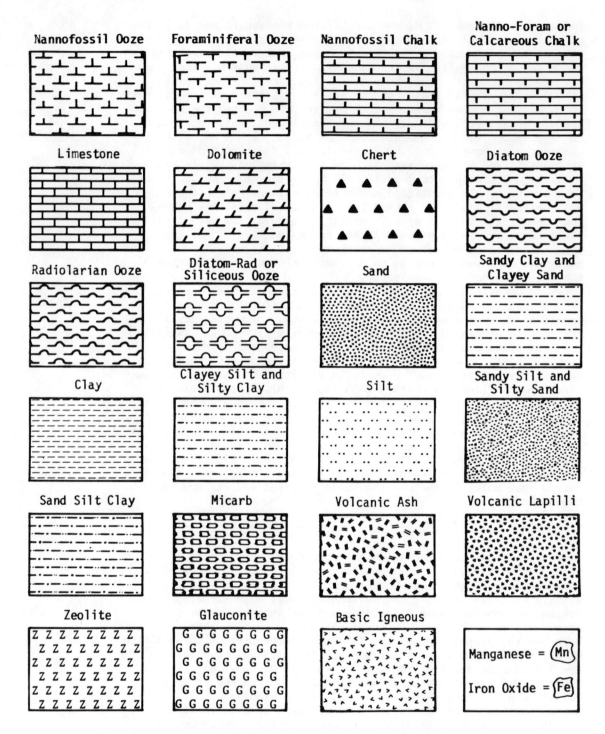

Fig. 4. Standard symbols used to illustrate lithology.

holes on legs 22, 26 and 27 was placed in the Wharton Basin and two holes were located in the Somali Basin on Leg 25 and one in the Mozambique Basin on Leg 25 and 26. The holes in the Wharton Basin were most successful. They showed that the basin aged to the south

We discuss the three sites around and to the south of the Tasman Rise from Leg 29 (Kennett et al., 1975) but we do not show any sections.

1) The Reconstructed Position and Early History of the Gondwana Continents

From the tectonic history set out by McKenzie and Sclater (1971) outlined in the previous section it was clear that back to 70 Ma the tectonic history of the Indian Ocean was well known in all except the Wharton and Mozambique Basins. Prior to this time very little was known about this history. As a consequence, a suite of in the Late Cretaceous (sites 211, 212, 213 and 256) and that in the Early Cretaceous and Jurassic this basin and the Argo Abyssal Plain became older to the southwest (sites 260, 261, 263). An excellent review of the implications of these sites is presented by Veevers (1977) in a later paper in this volume. The holes in the Somali Basin did not reach basement but indicated that west of the Chain Ridge the basin is at least as old as Late Cretaceous (site 241). The two holes in the Mozambique Basin are quite confusing. They imply that the basin ages to the south. This direction is opposite to that proposed by Bergh and Norton (1976) who surveyed a well identified magnetic anomaly sequence southeast at site 250. A possible explanation is that the basalt recovered at site 248 is a sill and is much younger than the crust on which it is located. Apart from those in the Wharton Basin, the sites in the basins were not as useful in helping elucidate the reconstructed position of the Gondwana continents and the early history of the ocean as had been hoped.

2) The Post 70 Ma Tectonic History of the Indian Ocean

Though there was a consensus as to the general features of the tectonic history of the Indian Ocean as presented by McKenzie and Sclater (1971) significant problems still remained. These included (a) the exact timing of the break apart of the Mascarene Plateau and the Chagos-Laccadive Ridge (site 238); (b) the dating of the ocean crust either side of the Ninetyeast Ridge (sites 213 and 215); (c) the age of the crust to the west Chagos-Laccadive Ridge (site 220); (d) the age of the Arabian Sea south of the Asian continent (sites 221 and 222) and just north of the Seychelles (site 236); and finally, (e) the age of the Mascarene Basin south of Mauritius (site 245). In general these sites tended to confirm the tectonic history outlined in the previous section with minor modifications.

3) The Origin and Tectonic History of the Aseismic Ridges

From the point of view of the tectonics this project and the drilling between Australia and Antarctic were two of the major successes of the project in the Indian Ocean. First a suite of five holes from Legs 22 and 26 along the Ninetyeast Ridge (sites 254, 253, 217, 216, 214) demonstrated unequivocally that this ridge was attached to the Indian Plate, was formed at sea level at a spreading center, and sunk at the same rate as the plate to which it was attached. A discussion of the implications of these results is presented by Luyendyk (1977) later in this volume. Drilling on the Chagos-Laccadive Ridge (site 219) and on the Mascarene Plateau (site 237) also showed that these ridges had once been at sea level and had subsided to the depth predicted at the same rate as the oceanic crust to which they were attached. Drilling on the Madagascar Ridge (sites 246, 264) was inconclusive as the sites did not reach basement, drilling on the Mozambique Ridge (site 249) showed it was early Cretaceous and gave a very important control date on the age of the breakup of Gondwanaland (Luyendyk, 1974). Site 255 on Broken Ridge

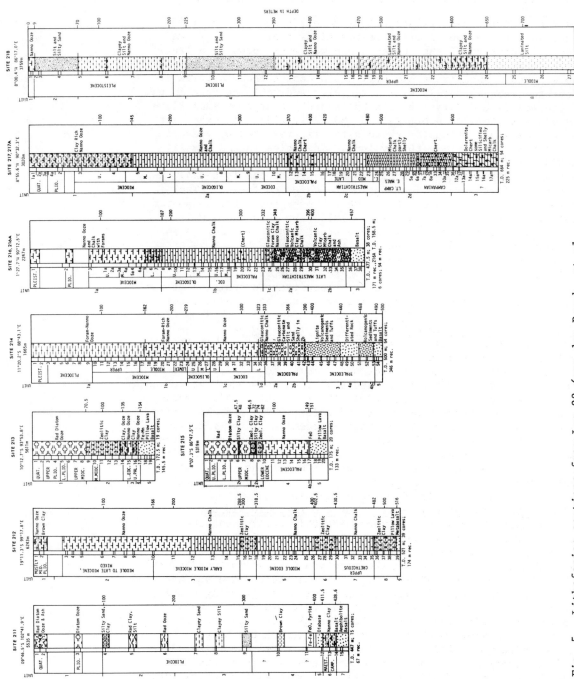

Fig. 5. Lithofacies sections from Leg 22 (von der Borch, et al., 1974). See figure 4 for an explanation of the symbols.

INTRODUCTION 11

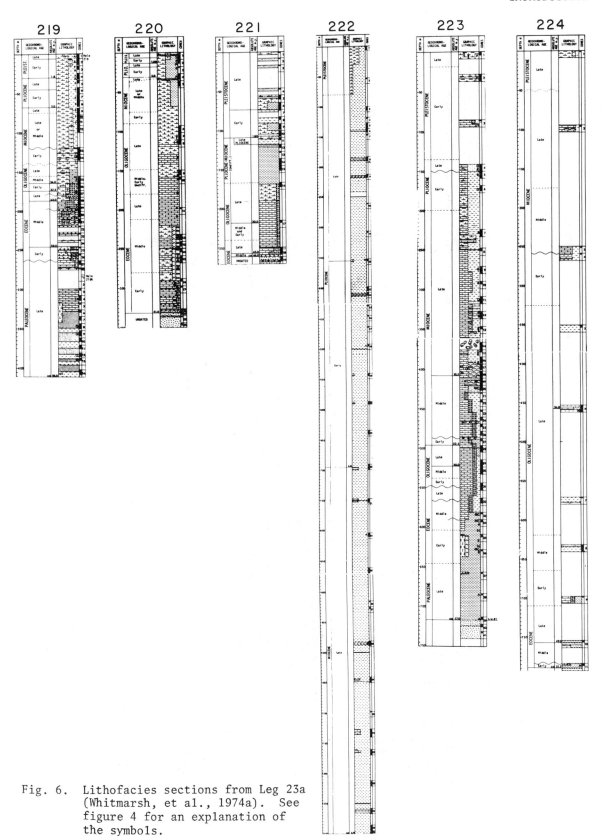

Fig. 6. Lithofacies sections from Leg 23a (Whitmarsh, et al., 1974a). See figure 4 for an explanation of the symbols.

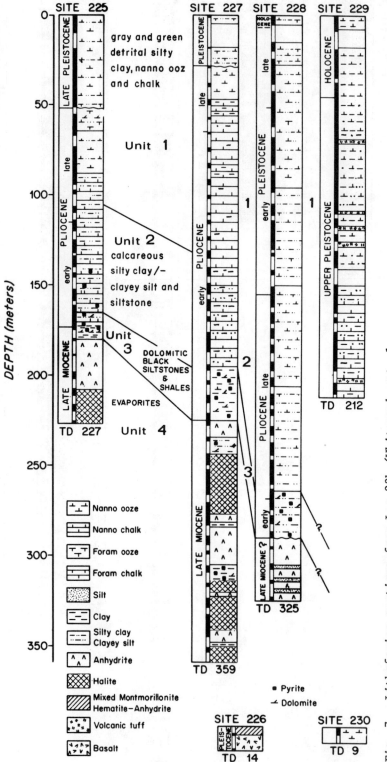

Fig. 7. Lithofacies sections from Leg 23b (Whitmarsh, et al., 1974a). See figure 4 for an explanation of the symbols.

showed that it has a complex history of subsidence and uplift terminating with uplift in the Eocene and then a slow subsidence to the present depth. The drilling on the aseismic ridges had two other implications. First their basement ages provided estimates of the age of the regional oceanic crust and second the detailed sections enabled estimates to be made of their subsidence history. In the case of the Ninetyeast Ridge this first point helped constrain the position of India within Gondwanaland (Sclater and Fisher, 1974) and the second was invaluable in the construction of the paleobathymetric charts. Sclater et al. (1977) and Davies and Kidd (1977) later in this volume make considerable use of these histories and charts in their synthesis of the sediments recovered in the Indian Ocean.

4) The History of Antarctic Glaciation

The distribution of the facies boundaries and of the first ice-rafted detritus in Leg 28 (sites 268, 267, 266) off the coast of Antarctica present major evidence concerning glaciation on Antarctica. These studies and oxygen isotope work on site 284 from DSDP Leg 29 (Shackleton and Kennett, 1975) were presented as evidence that the major glaciation in Antarctica did not begin until the Late Oligocene. During the Miocene the sites close to Antarctica indicate that the climate gradually deteriorated and cool waters pushed further north until the end of the Miocene when a major glacial pulse occurred involving a rapid building and subsequent retreat of the Antarctica ice sheet (Hayes and Frakes, 1975).

5) The Opening of Australia and Antarctica and Circumpolar Circulation

Much of the latter part of Leg 29 was devoted to the problem of timing and the nature of the breakup between Australia and Antarctica. Site 281 on the southern extension of the south Tasman Rise bottomed is continental granite indicating that this whole feature was continental. Further, Kennett et al. (1975) using the reconstructions of Weissel and Hayes (1972) pointed out that this Rise which originally abutted Antarctica did not pass the edge of the Antarctic continent until the late Oligocene. They suggest that just after this happened a shallow and deep water circumpolar current became activated and that this current caused the pronounced Oligocene hiatus in the holes just south of the Australian continent (sites 282, 280). This is one of the few cases where a sediment hiatus can be related to changes in possible circulation patterns. It has major implications for the study of paleo-oceanography.

6) The History of the Large Sedimentary Basins in the Indian Ocean

Clearly the D/V CHALLENGER is not capable of drilling deep enough in the large sedimentary basins to get a complete stratigraphic sequence. Thus a limited objective involved drilling in the distal sections of four of these basins. On Leg 22 the end of the Bengal fan was drilled (site 218) and in the sediments recovered there was evidence that there had been at least four major pulses of sedimentation across this fan since the middle Miocene. On Leg 23 a widespread lower Middle Eocene chart reflector was discovered in the southeast Arabian Sea. On Leg 25, site 241 in the Somali abyssal plain terminated in a Turonian (middle Cretaceous) claystone and sandstone. It appears that the relatively high sedimentation rates and bore rates in the sedimentary record in the 1174 m hole can be both due to epeirogenic movements of Africa and to the initiation of a current system that increased productivity in the surface waters. Site 250 in the Mozambique Basin shows that since the Miocene sediments have been under the control of active bottom circulation flowing clockwise.

Other objectives of the sedimentary drilling program were the in-

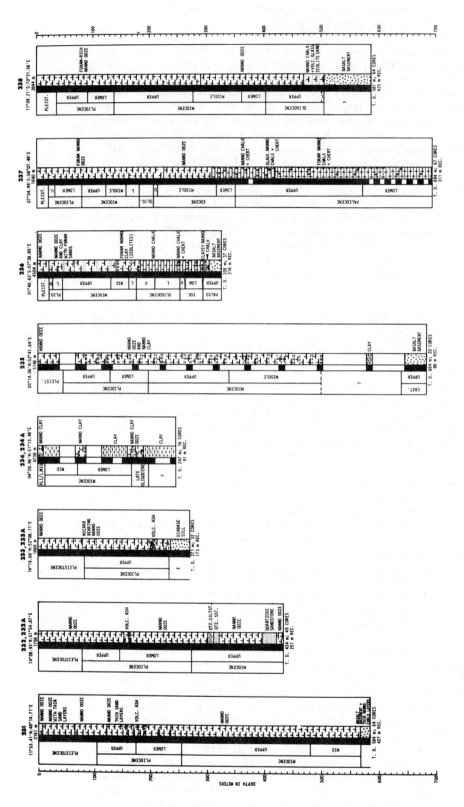

Fig. 8. Lithofacies sections from Leg 24 (Fisher, et al., 1974). See figure 4 for an explanation of the symbols.

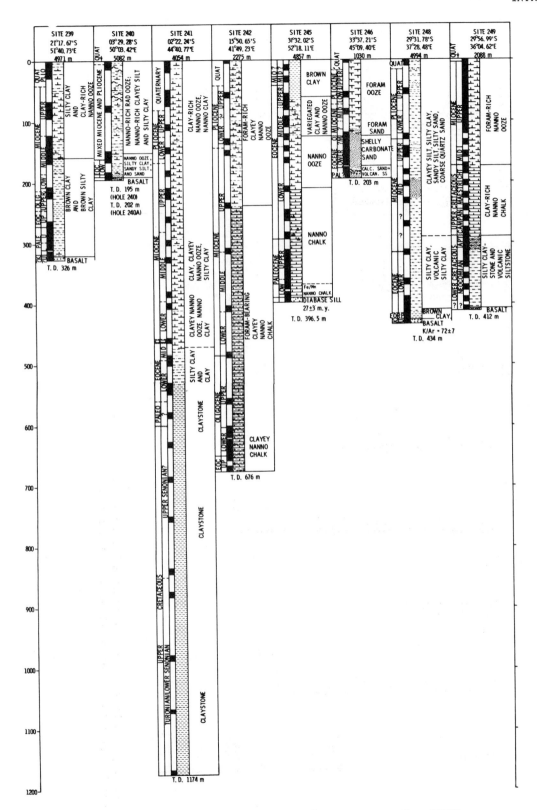

Fig. 9. Lithofacies sections from Leg 25 (Simpson, et al., 1974).
See figure 4 for an explanation of the symbols.

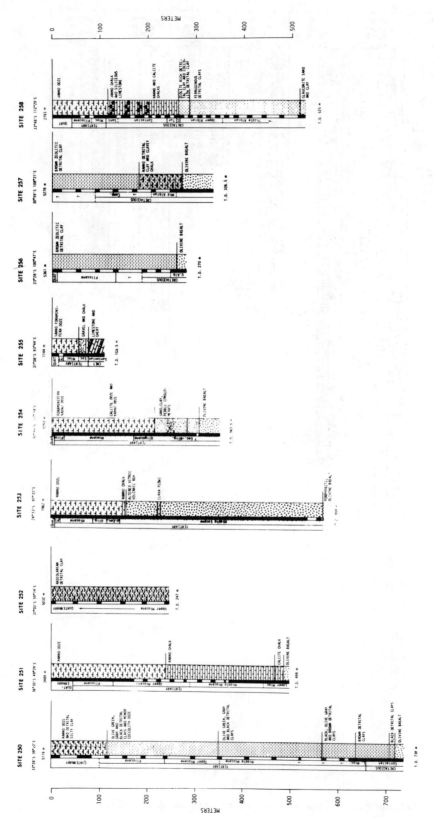

Fig. 10. Lithofacies sections from Leg 26 (Davies, et al., 1974). See figure 4 for an explanation of the symbols.

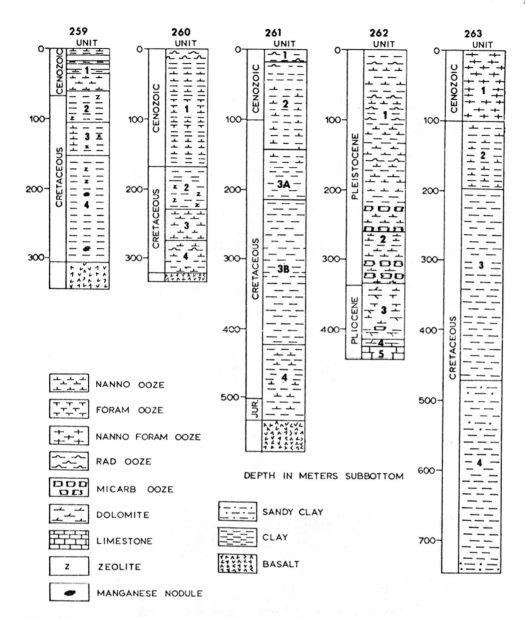

Fig. 11. Lithofacies sections from Leg 27 (Veevers, et al., 1975).

vestigation of near coast sediments off the west coast of Australia (sites 257, 260, 261, 263) and the general examination of volcanogenic sediments. In this volume Cook (1977) examined the sediments of the eastern Indian Ocean and Vallier and Kidd (1977) review the volcanogenic content of the cores.

7) <u>Magnetic Time Scale</u>

A series of holes, 214, 234, 239, 245, 265, 266 and 267, were drilled on well identified magnetic anomalies. All those on crust younger than anomaly 13 seemed to agree well with the Heirtzler et al. (1968) time scale. However, those older than Eocene gave ages considerably younger than that predicted by this time scale. This led Sclater et al. (1974) and Schlich (1975) to revise the Heirtzler et al. (1968) geomagnetic time scale. There is still some controversy over site 239. But the latest analysis of sediments above

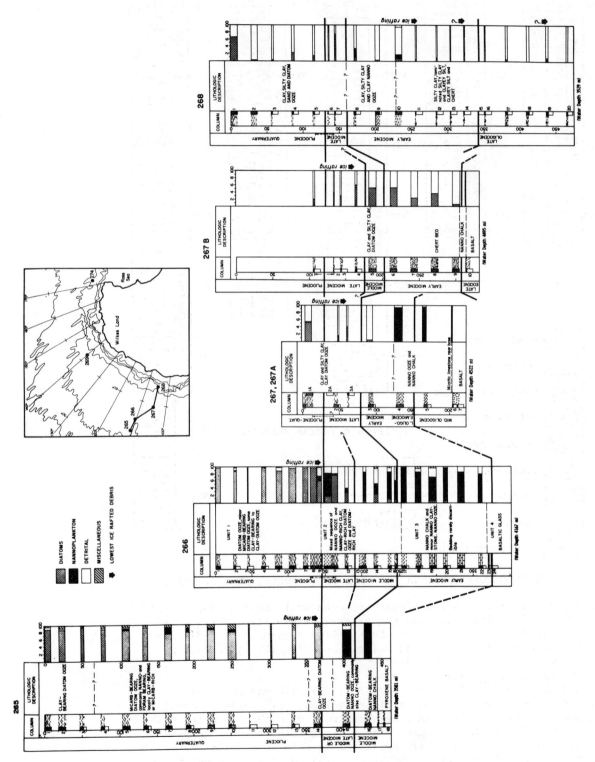

Fig. 12. Lithofacies sections from Leg 28 (Hayes, et al., 1975).

INTRODUCTION 19

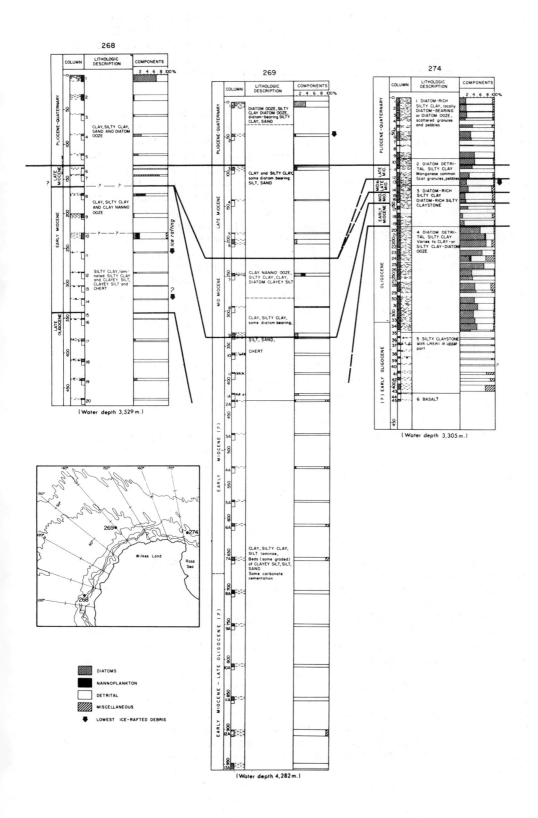

basement is evidence that these are younger than those initially reported (see Sclater et al., this volume). If substantiated, it is probable that the modification presented by Sclater et al. (1974) of the Heirtzler et al. (1968) time scale is the better justified of the two suggested revisions. Site 261 on anomaly M22 has been used to calibrate the Mesozoic magnetic time scale.

8) <u>Gulf of Aden and Red Sea</u>

The first part of the Leg 24 was devoted to the Gulf of Aden. The drilling in the Red Sea established as expected that the central basin was very young. Further, it was established that the prominent reflector at 500 m marks the top of a Miocene evaporite. No evaporites were found at site 231 in the Gulf of Aden suggesting it was not closed by the Owen Fracture Zone during the initial opening. Also from the age of the sediments above basement at this site the Gulf of Aden started to open in the Middle Miocene.

9) <u>Brines and Metalliferous Deposits in the Red Sea</u>

The second part of Leg 23 was devoted to examining the brines and metalliferous deposits in the Red Sea. It was found that the mineralization in the ATLANTIS II deep is restricted to the area under the brines. Further, evaporite shales enriched in Cu, Zn and B were also discovered. Stoffers and Ross (1975) suggested on the basis of this evidence that the brines were probably derived by a leaching process from the local evaporite strata and not from the cooling of newly created oceanic crust.

10) <u>Fresh Unweathered Basalt</u>

One of the secondary objectives of drilling in the Indian Ocean was to recover as much basalt from beneath the sediment as possible. This basement drilling was most successful and fresh olivine basalt was recovered at the base of the sediment column on each leg (Figures 5 through 12). The most basalt drilled was 80.5 m at site 238. This site also had the longest recovery of 40.6 meters. Two papers in this volume, (Subbarao, et al., 1977 and Frey et al., 1977) discuss the results of this basement drilling program. Further, Christensen (1977) and McElhinny (1977) respectively discuss the seismic and magnetic properties of these rocks at the end of the first section.

Problems Still Left for Drilling in the Indian Ocean

Because of serious weather constraints no serious drilling was attempted in the Indian Ocean south of 40°S and west of 110°E. As a result of this we know very little about the Crozet and Kerguelan Plateaus and have no sites in the Crozet Basin to compare with the sites in the Australia/Antarctic Basin further to the east. This absence of sites leads to serious problems in interpreting the pre-Oligocene circulation patterns in the ocean. Also, there is an absence of high latitude sites in the southern ocean in either shallow or deep water. Thus there is a lack of a continuous shallow water carbonate reference core close to Antarctica and an absence near the Antarctic coast of sites that would help elucidate the glacial history of Antarctica and the westward flow of bottom water from the Weddell Sea into the Indian and Pacific Basins. Further problems that are of interest to resolve are the carbonate sedimentary history on either side of the major ridge systems, deep drilling on the Madagascar Ridge to check if it is continental or not, and finally some deep penetration drilling on the aseismic ridges to examine their origin. It is hoped that in a later phase of the Deep Sea Drilling Project an attempt will be made to undertake each of these problems.

Acknowledgements. John Sclater would like to thank the Office of Naval Research (Contract N00014-75-C-0291) for the support that enabled the completion of this review.

References

Bergh, H. W., and I. O. Norton, Prince Edward Fracture Zone and the evolution of the Mozambique Ridge, J. Geophys. Res., 81, no. 29, 5221-5239, 1976.

Christensen, N. I., Seismic velocities and elastic moduli of igneous and metamorphic rocks from the Indian Ocean, in Indian Ocean Geology and Biostratigraphy, AGU, Washington, this volume, 1977.

Cook, P. J., Mesozoic-Cenozoic sediments of the eastern Indian Ocean, in Indian Ocean Geology and Biostratigraphy, AGU, Washington, this volume, 1977.

Crawford, A. R., India, Ceylon, and Pakistan: New age data and comparisons with Australia, Nature, 223, 380-384, 1969.

Curray, J. R., and D. G. Moore, Growth of the Bengal deep-sea fan and denudation of the Himalayas, Bull. Geol. Soc. Amer., 82, 563-572, 1971.

Davies, T. A., and R. B. Kidd, Sedimentation in the Indian Ocean through time, in Indian Ocean Geology and Biostratigraphy, AGU, Washington, this volume, 1977.

Davies, T. A., B. P. Luyendyk, et al., Initial Reports of the Deep Sea Drilling Project, Washington (U. S. Government Printing Office), 26, 1974.

Du Toit, A. L., Our wandering continents, Oliver and Boyd, Edinburgh and London, 1937.

Ewing, M., S. Eittreim, M. Truchan, and J. I. Ewing, Sediment distribution in the Indian Ocean, Deep Sea Res., 16, 231-248, 1969.

Fairbairn, R. W., The juvenility of the Indian Ocean, Scope, 1, 29-35, 1948.

Fairbridge, R. W., Some bathymetric and geotectonic features of the eastern part of the Indian Ocean, Deep Sea Res., 2, 161-171, 1955.

Fisher, R. L., C. G. Engel, and T. W. C. Hilde, Basalts dredged from the Amirante Ridge, Western Indian Ocean, Deep Sea Res., 15, 521-524, 1968.

Fisher, R. L., E. T. Bunce, et al., Initial Reports of the Deep Sea Drilling Project, Washington (U. S. Government Printing Office), 24, 1974.

Fisher, R. L., J. G. Sclater, and D. P. McKenzie, Evolution of the Central Indian Ridge, Western Indian Ocean, Bull. Geol. Soc. Am., 82, 553-562, 1971.

Frey, F. A., J. S. Dickey, G. Thompson, and W. B. Bryan, Eastern Indian Ocean DSDP sites: Correlations between Petrography, Geochemistry and Tectonic Setting, in Indian Ocean Geology and Biostratigraphy, AGU, Washington, this volume, 1977.

Hayes, D. E., and L. A. Frakes, General synthesis, Deep Sea Drilling Project, Leg 28, in Hayes, D. E., Frakes, L. A., et al., Initial Reports of the Deep Sea Drilling Project, 28, Washington (U. S. Government Printing Office), 919-942, 1975.

Hayes, D. E., Frakes, L. A., P. J. Barrett, D. A. Burns, P. Chen, A. B. Ford, A. G. Kaneps, E. M. Kemp, D. W. McCollum, D. J. W. Piper, R. E. Wall and P. N. Webb, Initial Reports of the Deep Sea Drilling Project, Washington (U. S. Government Printing Office), 28, pp. 1017, 1975.

Heezen, B. C., and M. Tharp, Physiographic diagram of the Indian Ocean, the Red Sea, the South China Sea, the Sulu Sea and the Celebes Sea, Geol. Soc. Amer. Inc., New York, 1965.

Heirtzler, J. R., G. O. Dickson, E. M. Herron, W. C. Pitman and X. LePichon, Marine magnetic anomalies, geomagnetic field reversals, and motions of the ocean floor and continents, J. Geophys. Res., 73, 2119-2136, 1968.

Holmes, A., Principles of physical geology, second edition, Ronald Press, New York, 1288 p., 1965.

Kanaev, V. P., Relief of the Indian Ocean, in Relief of the Earth (Morphostructure and Morphosculpture), Nauka, Moscow, 276-286, 1967.

Kennett, J. P., R. E. Houtz, P. B. Andrews, A. R. Edwards, V. A. Gostin, M. Hajor, M. A. Hampton, D. G. Jenkins, S. V. Margolis, A. T. Ovenshine, and K. Perch-Nielsen, Development of the Circum-Antarctic Current, Science, 186, 144-147, 1974.

Kennett, J. P., R. E. Houtz, P. B. Andrews, A. R. Edwards, V. A. Gostin, M. Hayes, M. A. Hampton, D. Graham Jenkins, S. V. Margolis, A. T. Overshine, and K. Perch-Nielsen, Cenozoic paleoceanography in the southwest Pacific Ocean, Antarctic glaciation and the development of the circumantarctic current, in Kennett, J. P., R. E. Houtz, et al., Initial Reports of the Deep Sea Drilling Project, Washington (U. S. Government Printing Office), 29, 1155-1170, 1975.

Laughton, A. S., R. B. Whitmarsh, and M. T. Jones, The evolution of the Gulf of Aden, Phil. Trans. Roy. Soc. London, A, 267, 227-266, 1970.

Laughton, A. S., D. H. Matthews, and R. L. Fisher, The structure of the Indian Ocean, in The Sea, Part II, 4, edited by A. E. Maxwell, John Wiley and Sons, New York, 543-586, 1971.

Laughton, A. S., D. P. McKenzie, and J. G. Sclater, The structure and evolution of the Indian Ocean, 24th International Geological Congress, section 8, 65-73, 1972.

LePichon, X., and J. R. Heirtzler, Magnetic anomalies in the Indian Ocean and sea floor spreading, J. Geophys. Res., 73, 2101-2117, 1968.

Luyendyk, B. P., Geophysical measurements along the track of D/V GLOMAR CHALLENGER, in Davies, T. A., B. P. Luyendyk, et al., Initial Reports of the Deep Sea Drilling Project, Washington (U. S. Government Printing Office), 26, 1974.

Luyendyk, B. P., Deep sea drilling on the Ninetyeast Ridge: Synthesis and Tectonic Model, in Indian Ocean Geology and Biostratigraphy, AGU, Washington, this volume, 1977.

Luyendyk, B. P., and T. A. Davies, Results of DSDP Leg 26 and the geologic history of the southern Indian Ocean, in Initial Reports of the Deep Sea Drilling Project, in Davies, T. A., and Luyendyk, B. P., Washington (U. S. Government Printing Office), 26, 1974.

Matthews, D. H., and D. Davies, Geophysical studies of the Seychelles Bank, Phil. Trans. Roy. Soc., A, 259, 227-239, 1966.

McElhinny, M. W., Formation of the Indian Ocean, Nature, 228, 977, 1970.

McElhinny, M. W., B. J. J. Embleton, L. Daly and J. P. Pozzi, Paleomagnetic evidence for the location of Madagascar in Gondwanaland, Geology, 4, 455-457, 1976.

McElhinny, M. W., The magnetic properties of Indian Ocean basalts, in Indian Ocean Geology and Biostratigraphy, AGU, Washington, this volume, 1977.

McKenzie, D. P., and J. G. Sclater, The evolution of the Indian Ocean since the Late Cretaceous, Geophys. J. Roy. Astron. Soc., 25, 437-528, 1971.

Schlich, R., Bathymetric, magnetic and seismic reflection data, Deep Sea Drilling Project, Leg 25, in Simpson, E. S. W., Schlich, R., et al., *Initial Reports of the Deep Sea Drilling Project*, Washington (U. S. Government Printing Office), 25, 763-830, 1974.

Schlich, R., Structure et age de l'ocean Indien Occidental, *Memoire Hors-Sene* No. 6, *de la Societe Geologique de France*, Paris, 103, 1975.

Schlich, R., E. W. W. Simpson, and T. L. Vallier, Regional aspects of Deep Sea Drilling in the western Indian Ocean, Leg 25, DSDP, in Simpson, E. S. W., R. Schlich, et al., *Initial Reports of the Deep Sea Drilling Project*, Washington (U. S. Government Printing Office), 25, 743-760, 1974.

Sclater, J. G., and R. L. Fisher, The evolution of the east central Indian Ocean, *Bull. Geol. Soc. Amer.*, 85, 683-702, 1974.

Sclater, J. G., R. Jarrard, B. McGowran, and S. Gartner, Comparison of the magnetic and biostratigraphic time scales since the late Cretaceous, in von der Borch, C. C., J. G. Sclater et al., *Initial Reports of the Deep Sea Drilling Project*, Washington (U. S. Government Printing Office), 22, 1974.

Sclater, J. G., U. Ritter, W. Hilton, L. Meinke and R. L. Fisher, Charts of residual magnetic field plotted along track for the Indian Ocean, M. P. L., *Scripps Institution of Oceanography*, SIO Report 71/7, 1971.

Sclater, J. G., D. Abbott, and J. Thiede, Paleobathymetry and sediments of the Indian Ocean, in *Indian Ocean Geology and Biostratigraphy*, AGU, Washington, this volume, 1977.

Seymour-Sewell, R. B., Part 1, The geography of the Andaman Sea Basin, 1-26, in Geographic and oceanographic research in *Indian waters*, *Mem. Asiatic Soc. Bengal*, 9, 550 p., 1925.

Shackleton, N. J., and J. P. Kennett, Paleotemperature history of the Cenozoic and the initiation of Antarctic glaciation: oxygen and carbon isotope analysis in DSDP sites 277, 279 and 281, in Kennett, J. P. and R. E. Houtz, et al., *Initial Reports of the Deep Sea Drilling Project*, Washington (U. S. Government Printing Office) 29, 1975.

Simpson, E. S. W., Schlich, R., J. M. Gieskes, W. Girdley, A. L. Leclaire, B. V. Marshall, C. Moore, C. Müller, J. Sigal, T. L. Vallier, S. M. White, B. Zobel, *Initial Reports of the Deep Sea Drilling Project*, Washington (U. S. Government Printing Office) 25, pp. 884, 1974.

Smith, A. G., and A. Hallam, The fit of the southern continents, *Nature*, 225, 139-144, 1970.

Stoffers, P., and D. A. Ross, Sedimentary history of the Red Sea, in R. B. Whitmarsh, D. A. Ross, et al., *Initial Reports of the Deep Sea Drilling Project*, Washington (U. S. Government Printing Office), 23b, 849-866, 1975.

Subbarao, K. V., R. Hekinian, and D. Chandrasekharam, Large ion lithophile elements and Sr and Pb isotopic variation in volcanic rocks from the Indian Ocean, in *Indian Ocean Geology and Biostratigraphy*, AGU, Washington, this volume, 1977.

Udintsev, G. G., R. L. Fisher, V. F. Kanaev, A. S. Laughton, E.S.W., Simpson and D. I. Zhiv, Geological-geophysical atlas of the Indian Ocean, *Academy of Sciences of the U.S.S.R.*, Moscow, 1975.

Vallier, T. L., and R. B. Kidd, Volcanogenic sediments in the Indian Ocean in *Indian Ocean Geology and Biostratigraphy*, AGU, Washington, this volume, 1977.

Veevers, J. J., Models of the evolution of the Eastern Indian Ocean, in *Indian Ocean Geology and Biostratigraphy*, AGU, Washington, this volume, 1977.

Veevers, J. J., J. R. Heirtzler, et al., Initial Reports of the Deep Sea Drilling Project, Washington (U. S. Government Printing Office), 27, 1975.

Vine, F. J., and D. H. Matthews, Magnetic anomalies over oceanic ridges, Nature, 199, 947-949, 1963.

von der Borch, C. C., J. G. Sclater, S. Gartner, R. Hekinian, D. A. Johnson, B. McGowran, A. C. Pimm, D. Thompson, J. Veevers, and L. Waterman, Initial Reports of the Deep Sea Drilling Project, Washington (U. S. Government Printing Office), 8, 1973.

Weissel, J. K., and D. E. Hayes, Magnetic anomalies in the southeast Indian Ocean, in Antarctic Oceanology, II, the Australian-New Zealand sector, ed. by D. E. Hayes, A.G.U., Washington, 1972.

Whitmarsh, R. B., D. Ross, S. Ali, J. E. Boudreaux, R. Coleman, R. L. Fleisher, R. W. Girdler, F. T. Manheim, A. Matter, C. Nigrini, P. Stoffers, P. R. Supko, Initial Reports of the Deep Sea Drilling Project, Washington (U. S. Government Printing Office), 23, 528-975, 1974a.

Whitmarsh, R. B., O. E. Weser, S. Ali, J. E. Boudreaux, R. L. Fleisher, D. Jipa, R. B. Kidd, T. K. Mallik, A. Matter, C. Nigrini, H. N. Siddiquie, and P. Stoffers, Arabian Sea, Initial Reports of the Deep Sea Drilling Project, Washington (U. S. Government Printing Office), 23, 1-527, 1974b.

Wilson, J. I., Hypothesis of earth's behavior, Nature, 198, 925-929, 1963.

Wiseman, J. D. H., and R. B. Seymour-Sewell, The floor of the Arabian Sea, Geol. Mag., 74, 219-230, 1937.

Yentsch, A. E., Submarine geology, geophysics and geochemistry, in a partial bibliography of the Indian Ocean, A. E. Yentsch, ed., Woods Hole Oceano. Inst., Woods Hole, Mass., 45-69, 1962.

CHAPTER 2. PALEOBATHYMETRY AND SEDIMENTS OF THE INDIAN OCEAN

John G. Sclater

Department of Earth and Planetary Sciences
Massachusetts Institute of Technology
Cambridge, Massachusetts 02139

Dallas Abbott[1]

Department of Earth and Planetary Sciences
Massachusetts Institute of Technology
Cambridge, Massachusetts 02139

Jörn Thiede
School of Oceanography, Oregon State University
Corvallis, Oregon 97331

Abstract. We establish a simple relation between subsidence and age for both normal ocean floor and the aseismic ridges in the Indian Ocean. This subsidence is accounted for by the cooling and contraction of the lithospheric plate as it moves away from a center of spreading. We use the relation between subsidence and age to construct paleobathymetric charts of the ocean for the early Oligocene (36 m.y.b.p.), the early Eocene (53 m.y.b.p.) and the late Cretaceous (70 m.y.b.p.). We conclude from these charts that the Indian Ocean between the middle Cretaceous and the Oligocene may have been separated by the Ninetyeast Ridge/Kerguelen Plateau complex and the Madagascar, Amirantes, Mascarene, Chagos complex into three basins which were not connected at depths below 2,000 m. We discuss the implications these complexes and the active mid-ocean ridge axis may have had for deep water circulation patterns in the Indian Ocean.

As an example of the application of these charts we use 19 drill sites to reconstruct the past history of the Calcite Compensation Depth (CCD) in the Indian Ocean. The average depth of this boundary shallows from more than 4,500 m at present to 4,000 m in the Oligocene and remains approximately constant till the Campanian. In the Wharton Basin it shallows in the Albian and Aptian. Major differences from the average in the Northern Arabian Sea and the Australia and Antarctic Basins can be accounted for by proximity to the continental shelf and the presence of bottom water in the Antarctic Circumpolar Current. We assume a constant CCD of 4,000 m for the rest of the Indian Ocean and compute the surface distribution of carbonate and clay sediments in the Oligocene. The shallowing of the CCD results in a marked reduction in the surface distribution of calcareous sediments between the present and

---

[1]Currently at Lamont-Doherty Geological Observatory, Columbia University, Palisades, New York 10964.

the Oligocene. The reason for such a dramatic diminution of carbonate sediments is not known.

Introduction

In the plate theory of tectonics, oceanic crust is created at a spreading center by the intrusion of hot magma. As this magma moves away from the spreading center it loses heat, cools and contracts. Plate models based on these simple concepts give an excellent match to the observed increase in depth with age of the oceanic crust (Parsons and Sclater, 1977). In general, the depth of the ocean floor increases from $2,700 \pm 300$ m at a spreading center to $5,800 \pm 300$ m in oceanic crust of early Cretaceous to late Jurassic age. The depth versus age relationship is a simple consequence of plate creation and there is no reason to believe that plate creation was any different in the late Cretaceous than now. Thus, if the past motions of the plates are known from mapping fracture zones, identifying magnetic lineations and DSDP site information, it is possible to predict the past bathymetry of the ocean floor as a function of time (Sclater and McKenzie, 1973; Berger and von Rad, 1972).

Unfortunately not all of the ocean floor has been created by the simple sea floor spreading process. There are major structural units called aseismic ridges some of which have not been created in this fashion. Further, many of these ridges have been found to have been topped by shallow water sediments early in their history. These structural units, of which the Chagos-Laccadive and Ninetyeast Ridges are the most prominent, abound in the Indian Ocean. A model for their subsidence history has to be developed before quantitative paleobathymetric charts can be constructed. The analysis of sediments recovered in DSDP holes has been invaluable in resolving the past history of these ridges. For example, sediments recovered from the Ninetyeast Ridge in the Indian Ocean are evidence that this feature was formed at sea level close to an active spreading center and that it has subsided at exactly the same rate as the oceanic crust to which it is attached (Sclater and Fisher, 1974; Pimm et al., 1974). This also holds true for ancient ridges from other oceans (Thiede, 1977; Detrick et al., 1977). If information is available, it is possible to predict in a quantitative fashion the past subsidence history of aseismic ridges.

To construct a series of paleobathymetric charts and then to use them to examine sediment facies changes through time it is necessary first to establish a consistent magnetic and biostratigraphic time scale. We start this paper by establishing such a time scale and follow by justifying the depth versus age relation we propose to use for normal ocean crust and the aseismic ridges. Then we present the tectonic history we propose to follow and from this history and the depth versus age relation we produce paleobathymetric charts for the Indian Ocean at 36, 53 and 70 m.y.b.p. As an example of the possible application of the relation between depth and age and these charts we examine the past history of the Calcite Compensation Depth (CCD) in the Indian Ocean as a function of time. We use this curve and the paleobathymetry to examine the variation with time of the surface distribution of calcareous and clayey sediments.

The tectonic history of the Indian Ocean is not well understood (see also Johnson et al., 1976). Many of the aseismic features of the ocean floor are as yet undrilled. Thus, the present study is mainly a semi-quantitative attempt to determine the depth of the major topographic features of the Indian Ocean with time followed by a speculative use of the CCD and these charts to predict past surface sediment distributions.

As the paper is mainly speculative we have concentrated in the text on the major points and have left the details of how the various charts were constructed to the appendices.

There are other major sedimentary problems in the Indian Ocean upon which the paleobathymetry will have some effect. Such problems include the explanation of the deep and shallow water Oligocene hiatus (Davies, et al., 1974) and the influence the past topography has upon the terrigenous input to the major basins. These specific questions are tackled by Davies and Kidd (this volume) in the following paper in this volume and only briefly alluded to in this manuscript.

Justification of the Chosen Time Scale

Sediments recovered in DSDP holes give information concerning the past history of the site. To relate this information to the general tectonics of the area it is necessary to have a biostratigraphic time scale and to know the relationship between this time scale and the ages predicted by the magnetic anomalies on the sea floor. For the biostratigraphic information we have used the Berggren and van Couvering (1974) time scale for the Tertiary and the preliminary time scale of Thierstein (1977) for the Cretaceous/Jurassic. Relating the biostratigraphic time scale to the magnetic anomaly information is more difficult. It is clear from deep sea drilling data in the Indian Ocean that the original Heirtzler et al. (1968) scale as modified by Larson and Pitman (1972) is inadequate due to large discrepancy in the Eocene and Paleocene (Site 213 from Sclater et al., 1974; Site 236 from Vincent, 1974; Vincent et al., 1974; and Sites 239 and 245 from Schlich et al., 1974a).

Sclater et al. (1974) have proposed a modification to the Heirtzler et al. (1968) magnetic scale which brings the age of the distinctive magnetic anomalies into better agreement with the sediment information from the DSDP holes. This modification has been questioned by both Larson and Pitman (1975) and Schlich (1975). Larson and Pitman (1975) argue in particular that Sites 239 and 10 in the South Atlantic indicate an older age for anomalies 31 through 34 than given by Sclater et al. (1974). However, at Site 239 the lowermost distinctive nannofossil horizon is Micula mura (Latest Maastrichtian) and not Tetralithus aculeus as cited in the original report (Larson and Thierstein, personal communication). Also we consider the location of Site 10 with respect to magnetic anomalies to be too poorly known to be considered as a basis for a time scale. Further, Site 211, which may well be on the quiet zone attributed to anomaly 33, has a basal sediment age of early to middle Campanian. This is significantly younger than that predicted by Larson and Pitman (1975, Figure 1). Clearly the relation between the magnetic and the late Cretaceous biostratigraphic time scales is still controversial. For the sake of consistency we have decided to use the Sclater et al. (1974) modification of the Heirtzler et al. (1968) scale because it involves no dramatic spreading rate change in the South Atlantic in the late Cretaceous.

The Depth of the Ocean Floor as a Function of Age

To determine the past bathymetry of the ocean floor it is necessary to establish a relation between depth and age for all portions of the ocean crust. To a first approximation the floor of the Indian Ocean can be separated into typical ocean crust and aseismic ridges (Figure 1). In this section we compare the depth versus age data for the Indian Ocean with that found in other oceans and we develop a subsidence model for the aseismic ridges.

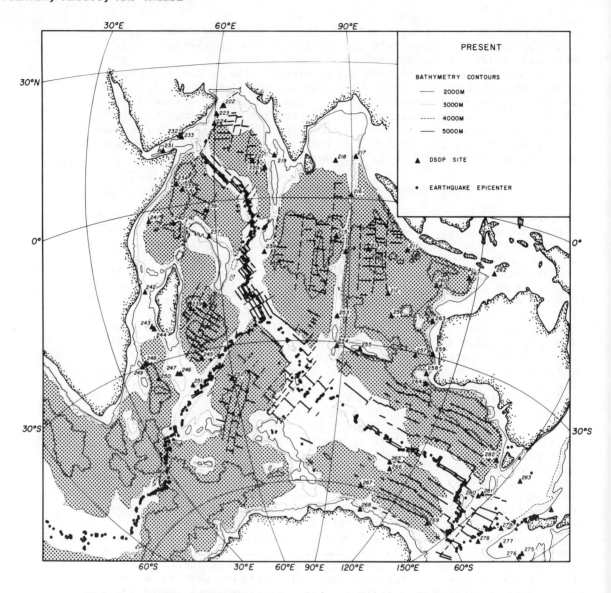

Fig. 1. A plot of JOIDES Deep Sea Drilling Sites on a bathymetric and tectonic chart of the Indian Ocean. Bathymetric contours from the new Russian Atlas of the Indian Ocean. Magnetic lineations are from McKenzie and Sclater (1971); Weissel and Hayes (1972); Sclater and Fisher (1974); and Schlich (1975). The shaded areas represent depths greater than 4000 m.

Typical Ocean Crust

We have defined typical oceanic crust as that produced by the normal sea floor spreading process. In the Indian Ocean we contrast it with the aseismic ridges which show up as distinct regions of shallow topography on the bathymetric chart (Figure 1). We have listed all the sites on normal ocean floor which penetrated basement (Table 1) and computed the isostatically corrected depth and biostratigraphic age of the basal sediments. We plotted the depths as a function of age and compared these depths with that expected for normal ocean crust (Figure

PALEOBATHYMETRY AND SEDIMENTS 29

TABLE 1. Depth and Age Relation for Sites on Normal Ocean Crust

Site	Latitude	Longitude	Depth (m)	Sediment Thickness	Sediment Correction* (m)	Topographic Relief Correction	Mean corrected Depth+ (m)	Contact	Fossil Zones
211	09°46.53'S	102°41.95'S	5528	428.5	259	—	5787	excellent	Eiffellithus augustus zone
212	19°11.34'S	99°17.84'E	6240	516	287	-240	~6287	excellent	Nephrolithus frequens zone [0] in lowermost chalk
213	10°12.71'S	93°53.77'E	5609	154	97	-100	5606	excellent	Discoaster mohleri zone [0], not older than Planorotalites pseudomenardii zone (P4)
215	8°07.30'S	86°74.50'E	5319	155.5	99	—	5428	excellent	Fasciculithus tympaniformis zone; Morozovella pusilla, M. angulata (P3), Planorotalites pseudomenardii (P4) zone
220	6°30.97'N	70°59.02'E	4036	329	210	—	4246	excellent	Morozovella aragonensis (P8) zone, Discoaster lodoensis zone
221	7°58.18'N	68°24.37'E	4650	261	170	—	4820	excellent	Chiasmolithus grandis (13-P14) ? zone
231	11°53.41'N	48°14.71'E	2152	566.5	357	—	2509	excellent	Discoaster exilis-Sphenolithus heteromorphus zone, Globorotalia fohsi (N12) zone
235	03°14.06'N	52°41.64'E	5130	651.5	450	—	5580	poor but sediment in basalt	Micula mura zone
236	01°40.62'S	57°38.85'E	4487	306	198	—	4685	excellent	Discoaster mohleri zone, Planorotalites pseudomenardii (P4) zone
238	11°09.21'S	70°31.56'E	2832	506	310	+700	3842	excellent	Sphenolithus predistentus zone, Pseudohastigerina barbadoensis (P19) zone
239	21°17.67'S	51°40.73'E	4971	320	223	—	5194	excellent contact poor fossils	Tetralithus aculeus zone
240	03°29.28'S	50°03.42'E	5082	190	126	—	5208	fair	Discoaster multiradiatus zone, Morozovella subbotinae - M. wilcoxensis (P16) zone
245	31°32.02'S	52°18.11'E	4857	389	245	—	5102	excellent	Chiasmolithus danicus (NP3) zone, Globoconusa daubjergensis (P1) zone
248	29°31.78'S	37°28.48'E	4994	422	273	—	5267	excellent	Cyclococcolithus formosus, probably somewhat older than NP12
250	33°27.74'S	39°22.15'E	5119	725	498	—	5617	excellent contact but fossils 15m above basalt	Marthasterites furcatus zone, Inoceramus sp.
251	36°30.26'S	49°29.08'E	3489	487	296	—	3785	excellent contact	Sphenolithus heteromorphus (NN5) zone, Globorotalia peripheroroda (N4-N8) zone
256	23°27.35'S	100°46.46'E	5361	251	172	-350m*	~5700	excellent contact	Eiffellithus turriseiffelli zone
257	30°59.16'S	108°20.00'E	5278	262	186	—	5464	excellent contact but fossils 15m above basalt	Prediscosphaera cretacea zone
259	29°37.05'S	112°41.78'E	4696	304.3	200	—	4896	excellent contact	Crybelosphirites stylosus zone
260	16°08.67'S	110°17.92'E	5702	323.0	219	—	5921	excellent contact	Prediscosphaera cretacea zone
261	12°56.83'S	117°53.56'E	5667	542.0	353	—	6020	excellent contact	Stephanolithion bigoti zone
265	53°32.45'S	109°56.74'E	3582	444.5	324.5	—	3906	excellent contact	Denticula hustedtii/C. lanta zone
266	56°24.13'S	110°06.70'E	4167	370	291	—	4358	excellent contact	Discoaster deflandrei, Sphenolithus moriformis zone, Catapsydrax unicavus zone
267	59°14.55'S	104.29.94'E	4495	314	229	—	4724	excellent contact	Globigerina linaperta (P15-P16) zone

* Topography correction dominant.

+ The isostatic loading of the sediments is corrected for by the method of Sclater et al, 1971, Appendix III, $\rho s = 1.8$ g/cm^3

TABLE 1. Depth and Age Relation for Sites on Normal Ocean Crust (Cont.)

Fossil Zones	Extrapolated age (Relative)	Numerical m.y.b.p.	Magnetics Anomaly No.	Magnetics Age m.y.b.p.	Comments
Eiffellithus augustus zone	Early-Middle Campanian	74-82	>33b	?	Wharton Basin 120 km southeast of anomaly 33b
Nephrolithus frequens zone [0] in lowermost chalk	Late Cretaceous	84-110	>33b	?	On extension of Investigator Fracture Zone
Discoaster mohleri zone [0], not older than Planorotalites pseudomenardii zone (P4)	Late Paleocene	55-58	25-26	59	East of Ninetyeast Ridge
Fasciculithus tympaniformis zone; Morozovella pusilla, M. angulata (P3), Planorotalites pseudomenardii (P4) zone	Late Paleocene	58-60	–	–	West of Ninetyeast Ridge
Morozovella aragonensis (P8) zone, Discoaster lodoensis zone	Early Eocene	50-52	–	–	Western Flank of Chagos Laccadive Ridge
Chiasmolithus grandis (13-P14) ? zone	Middle Eocene	43-51	19-20?	44-46?	Arabian Abyssal Plain
Discoaster exilis-Sphenolithus heteromorphus zone, Globorotalia fohsi (N12) zone	Middle Miocene	14-17	–	–	Arabian Sea
Micula mura zone	Late Maastrichtian	65-68	–	–	50 km east of Chain Ridge
Discoaster mohleri zone, Planorotalites pseudomenardii (P4) zone	Late Paleocene	56-58	26-27	–	North of Seychelles
Sphenolithus predistentus zone, Pseudohastigerina barbadoensis (P19) zone	Early Oligocene	30-35	–	–	Northeast end of Argo Fracture Zone
Tetralithus aculeus zone	Maastrichtian to Late Campanian	65-76	end of 31	69	Mascarene Basin
Discoaster multiradiatus zone, Morozovella subbotinae – M. wilcoxensis (P16) zone	Late Paleocene	53-57	–	–	Somali Basin
Chiasmolithus danicus (NP3) zone, Globoconusa daubjergensis (P1) zone	Early Paleocene	59-64	29-30	64-66	Mascarene Basin
Cyclococcolithus formosus, probably somewhat older than NP12	Late Paleocene to Early Eocene	50-60	–	–	In Madagascar Basin 50 km east of Mozambique Ridge
Marthasterites furcatus zone, Inoceramus sp.	Santonian-Coniacian	>82-87	–	–	In Madagascar Basin
Sphenolithus heteromorphus (NN5) zone, Globorotalia peripheroroda (N4-N8) zone	Early Miocene	16-22	–	–	North flank of Southwest Indian Ridge
Eiffellithus turriseiffeli zone	Late Albian	95-98	–	–	South Wharton Basin near tongue of Broken Ridge
Prediscosphaera cretacea zone	Middle Albian	94-102	–	–	Southeast Wharton Basin, N.G. of Naturaliste Plateau
Crybelosphirites stylosus zone	Aptian	102-107	–	–	Off westcoast of Australia
Prediscosphaera cretacea zone	Albian	94-102	–	–	Gascoyne Abyssal Plain
Stephanolithion bigoti zone	Late Oxfordian	151-155	M23	150	Argo Abyssal Plain
Denticula hustedtii/C. lanta zone	Late to Middle Miocene	10-22-5	5-5b	10-15	Australia Antarctic Ridge (Area C)
Discoaster deflandrei, Sphenolithus moriformis zone, Catapsydrax unicavus zone	Early Miocene	16-23	6-7	20-5-26	Australia Antarctic Ridge (Area C)
Globigerina linaperta (P15-P16) zone	Late Eocene	38-44	?	?	Australia Antarctic Ridge (Area C)

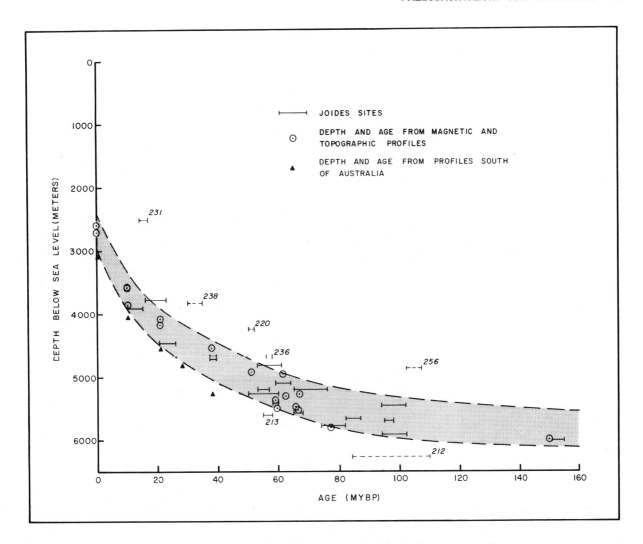

Fig. 2. The relationship between depth and age in the Indian Ocean for topographic profiles over oceanic crust with identified magnetic lineations and for JOIDES Sites on normal ocean crust. The two dashed lines are 600 m apart and represent the scatter in the data for the depth versus age relation shown by most oceanic crust (Parsons and Sclater, 1977). Sites which do not fall within these lines have been identified.

2). We also superimposed upon this plot the depth versus age data for the spreading center crests and anomalies 5, 6, and 25 from Sclater et al., 1971. Further, on the basis of topographic and seismic information collected along the track of the GLOMAR CHALLENGER and published topographic, magnetic, and seismic surveys, we added fourteen depth/age points to those originally computed by Sclater et al. (1971) (Table 2). These depths, except for one point south of Australia, all fall within the depth versus age relationship expected for normal ocean crust. Only 7 out of 43 DSDP holes do not fall within the expected relationship for depth and age. Of these 7 there are excellent reasons why 5 fall out of the range. Starting with the younger of these anomalous sites, Site 231 is in the Gulf of Aden and close to the African continental margin. Site 238 is virtually on a fracture zone close to the foot of the Chagos-

TABLE 2. Depth and Age from Magnetic, Topographic and Sediment Thickness Surveys.

Region	Anom. No.	Age m.y.	Depth m	Sediment Thickness	Corrected + Depth	References
Southeast Indian Ridge	14-16	36-40	4465	115 (76)	4541	Sclater et al. (1976) Luyendyk et al. (1974)
Wharton Basin (22-3)	27	62.5	5200	150	5300	Sclater and Fisher (1974)
Wharton Basin (22-4)	30	66.5	5400	200	5533	Sclater and Fisher (1974) Veevers (1974)
Wharton Basin (22-3a)	25-26	59-60	5300	300	5500	Sclater and Fisher (1974) Ewing et al. (1969)
Wharton Basin	33	76	5750	100	5816	Sclater and Fisher (1974) Carpenter and Ewing (1974) Veevers (1974)
Central Indian Basin	31	67	5168	500	5498	Eittreim and Ewing (1972)
Argo Abyssal Plain	M-23	150	5675	550	6005	Falvey (1972), Larson (1975), Veevers (1974)
Arabian Sea	21	51	4635	450 (300)	4935	Whitmarsh (1974a, 1974b)
North of the Seychelles	26-27	60-63	4751	300	4951	Sclater et al. (1971) Bunce (1974)
Mascarene Basin	30	67	4983	450	5283	Schlich (1975)
Australia-Antarctic Ridge	0	0	3100	–	3100	Weissel and Hayes (1972)
	5	9-10	4050	–	4050	Weissel and Hayes (1972)
	8	27-29	4835	–	4835	Weissel and Hayes (1972)
	13	37-39	5280*	–*	5280	Weissel and Hayes (1972)

* Sediment thickness from figure 5 of Houtz and Markl (1972) included in original depth estimate.
+ Depth adjusted for isostatic loading of sediments (see Sclater et al., 1971; Appendix III), ρs assumed = 1.8 g/cm³.

Laccadive Ridge. Site 220 is on the western flank of the same ridge. Site 212 is too deep because it is located in an extension of the Investigator Fracture Zone (Sclater and Fisher, 1974; Figure 1). Site 256 lies on a topographic spur which trends in a northeasterly direction away from Broken Ridge. It is not surprising that it is too shallow (Sclater and Fisher, 1974; Figure 1). Support for the anomalous nature of this site comes from the chemical composition and concentration of rare earth elements which suggest the basalt has more affinities to Sites 214 and 216 on the Ninetyeast Ridge than other sites in the Wharton Basin (Frey et al., this volume). There are two sites which are unexpectedly different from normal ocean crust. Site 236 is on a topographic bulge north of the Seychelles, which shows up very clearly on the 1964 Russian chart of the Indian Ocean (Anonymous, 1964). Site 213 which is located in a wide northeast-southwest trough between the Ninetyeast Ridge and the Cocos-Keeling Island complex (Figure 1 from Sclater and Fisher, 1974) is too deep. The explanation of these troughs which trend in a different direction to the other prominent features in the Wharton Basin is unknown.

In general the depth versus age data from the Indian Ocean is similar to that observed in the Pacific and Atlantic. Thus we have confidence from this data that we can construct paleobathymetric charts with the depth versus age relation of the Pacific and Atlantic.

DSDP Holes on Aseismic Ridges

One of the dominant features of the Indian Ocean which contrasts it with the central and southern Atlantic and the central and eastern Pacific is the large number of aseismic ridges. Their past tectonic history is critically important to determining the paleobathymetry of this ocean. The major aseismic features are Broken Ridge, the Ninetyeast and Chagos-Laccadive Ridges, the Mascarene Plateau, the Madagascar Ridge and Crozet and Kerguelen Plateaus (Figure 1). Five sites (214, 216, 217, 253, and 254) were drilled on the Ninetyeast Ridge and one each on the Chagos-Laccadive Ridge (219), Mascarene Plateau (237), Madagascar Ridge (246), Mozambique Ridge (249), and Broken Ridge (255). Two sites were drilled on the Naturaliste Plateau (258, 264) and although neither of them reached basement they gave important information concerning the past history of this feature.

From paleontological information it is possible to determine whether the environment of deposition was shallow water marine, shelf or deep ocean facies. This information was used with considerable success by Pimm et al. (1974), Luyendyk and Davies (1974) and Luyendyk (this volume) to analyze the past history of the Ninetyeast Ridge. This is the best known aseismic ridge in the Indian Ocean and forms the type example to which the limited information gathered on the other aseismic ridges is compared. Sclater and Fisher (1974) have argued that the Ninetyeast Ridge is the same age as the oceanic crust to the west and is attached to the Indian plate. Further there is evidence (Table 3) that the depth of the Ninetyeast Ridge increases with age. In fact, this depth increase is close to that predicted by assuming that the ridge was created at sea level and that since the onset of shallow water marine facies it has subsided at the same rate as the oceanic crust to which it is attached (Table 3; Figure 3). With this assumption, it is possible from the DSDP holes, to predict the subsidence history at every point along the ridge.

All other sites on aseismic ridges in the Indian Ocean have evidence from the sediments for subsidence from a depth close to sea level to their present depth (Table 3). The most striking is Site 219 on the

TABLE 3. Depth and Age from Sites on Aseismic Ridges

Site	Latitude	Longitude	Age Basal Sediment m.y.b.p.	Age Non-Shelf* Facies m.y.b.p.	Depth m	Sediment Thickness	Corrected+ Depth	Expected Depth from North Pacific subsidence curve.**
	Description of Subsidence History							

A) NINETYEAST RIDGE

214	11°20.21'S	88°43.08E	57-59	49-54	1671	490	1980	2300
	Shallow water Paleocene basalt sediments with the onset of oceanic facies starting in early Eocene							
216	01°27.73'N	90°12.48'E	69-65	65-60	2262	457	2530	2600
	Shallow water late Maastrichtian basal sediments with onset of oceanic facies in the earliest Paleocene							
217	98°55.57'N	90°32.33'E	76-82	68-72	3030	~663	3442	2800
	Shallow water early Campanian basal sediments with the onset of oceanic facies in the early Maastrichtian							
253	24°52.65'S	87.21.97'E	43-49	43-45	1962	558	2294	2200
	Shallow water mid-Eocene basal sediments with the onset of oceanic facies in the uppermost mid-Eocene							
254	30°58.15'S	87.53.72'E	36.39	20-26	1253	~301	1435	1400
	Shallow water Oligocene/Eocene boundary basal sediments with the onset of oceanic facies in the early Miocene/late Oligocene							

B) OTHER

219	09°01.75N	72°52.67'E	54-60	49-54	1764	~600	2137	2300
	Shallow water late Paleocene basal sediments with the onset of oceanic facies in early Eocene							
237	07°04.99'S	58°07.48'E	60-65	49-60	1623	~694	2054	2250-2500
	Shallow water early Paleocene basal sediments with the possible onset of oceanic facies in the early Eocene/late Paleocene							
246	33°37.21'S	45°09.60'E	49-54	20-23	1030	~194	1151	1400
	Shallow water early Eocene basal sediment with the onset of oceanic facies in the early Miocene							
249	29°56.99'S	36°04.62'E	121-134		2088	408	2352	
	Shallow water early Cretaceous basal sediments - subsidence history unknown?							
255	31°07.87'S	93°43.72'E		20-23	1144	~109	1211	1400
	Cretaceous limestone at bottom possibly uplifed in the Paleocene with onset of oceanic facies in the early Miocene							
258	33°47.69'S	112°28.42'E	93-102	93-102	2793	~525	3119	3000
	Shallow water mid-Albian to Cenomanian basal sediments with the onset of oceanic facies in the mid-Albion with rapid shoaling through the Cretaceous							
264	34°58.13'S	112°02.68'E	70-100		2873	171	3003	2800
	Volcanoclastic conglomerate as basal sediments subsidence to moderately great depths by Santonian							

* By non-shelf we mean the onset of clearly shallow water marine facies (see text).
+ Depth adjusted for isostatic loading of the sediments (see Sclater et al., 1971, Appendix 3), $\rho s = 1.8$ g/cm^3.
** From Sclater et al. (1971).

Chagos-Laccadive Ridge, where the onset of oceanic facies occurs in the early Eocene (Whitmarsh, 1974b). Assuming that the ridge was formed at sea level close to a spreading center and since then has sunk at the same rate as the adjacent crust gives an expected depth very close to that observed. This and the general morphological similarities to the Ninetyeast Ridge argue strongly that the Chagos-Laccadive Ridge can be treated in the same fashion as this ridge. Site 237 lies on the Mascarene Plateau. There is no direct evidence from the sediments for subsidence from shallow to abyssal depths. However, the Mascarene Plateau and the Chagos-Laccadive Ridge have a similar subsidence history as is clear from the benthonic foraminifera and sedimentation rates at both Sites 237 and 219 (Vincent et al., 1974). For the purposes of constructing the paleobathymetric charts we have assumed that the plateau and the ridge have had the same history of subsidence.

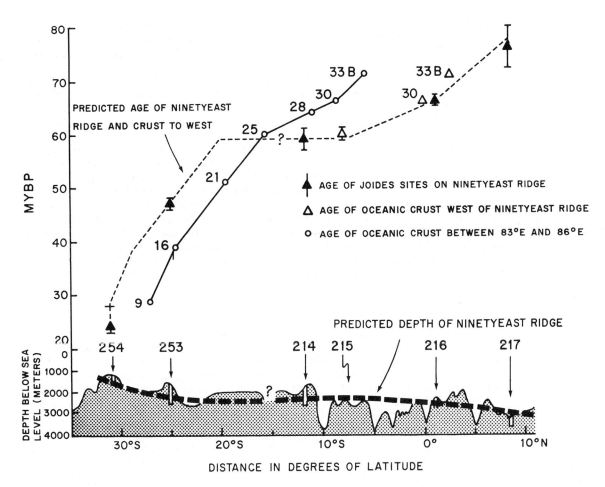

Fig. 3. Plot of the ages of the sediment 'basement' contacts on the Ninetyeast Ridge, the oceanic crust between 83°E and 86°E, and between 86°E and the Ninetyeast Ridge, all against latitude. Also shown is the actual depth of the crest of the Ninetyeast Ridge and the predicted depth (dashed line) assuming that it was formed at sea level and has the same age as the oceanic crust immediately to the west (after Sclater and Fisher, 1974).

Drilling at Site 246 on the Madagascar Ridge was terminated prematurely owing to technical problems. There is evidence from the basal sediments of a shallow water environment in the early Eocene. However, by the early Miocene the onset of deep water facies is observed. This history of subsidence from a shallow to deep water environment suggests that the ridge behaves like oceanic basement, and that prior to the Oligocene it was close to or above sea level. Unfortunately, we know nothing about its history prior to the Eocene. Site 249 is on the Mozambique Ridge. Though the oldest sediments were clearly formed near sea level the subsidence history is unknown. The same is true for Site 258 on the Naturaliste Plateau. However, this plateau subsided quite fast after formation (Table 3). Site 255 on Broken Ridge has the most complex history of the sites on the aseismic ridges. Luyendyk and Davies (1974) believe that the basal Cretaceous limestone was formed at depths less than 600 m but that the Ridge was uplifted in the Paleocene and then started to sink in the late Eocene. The exact rate of sinking is not known but it was clearly close to that of normal ocean crust in order for the site to have reached a depth of 1,144 m by the present (Luyendyk and Davies, 1974, Table 3).

For our reconstructions we have assumed that the Ninetyeast and Chagos-Laccadive Ridges subsided from sea level at the same rate as normal ocean crust and that this subsidence started at the time of onset of marine shallow water facies. We have assumed the same for the other ridges. There is much less justification for this but it is probably acceptable as a first approximation. However, not this, but the absence of drill sites on the Crozet and Kerguelen Plateaus is the most important problem with our interpretation of the aseismic ridges. We know nothing about the age of these Plateaus and nothing about their subsidence history. Clearly both are critical to our understanding of past bathymetry of the Indian Ocean and, hence, the past flow of bottom water around Antarctica. In fact, as we will show in our reconstruction, the absence of information from Kerguelen renders it impossible for us to make quantitative predictions about past bottom water circulation around Antarctica prior to the Miocene.

Tectonic History

In order to construct quantitative paleobathymetric charts it is necessary to know the past tectonic history of the oceans. We have used as the basis of our history the plate tectonic analysis of McKenzie and Sclater (1971). Though there has been a significant upgrading of data especially between Australia and Antarctica (Weissel and Hayes, 1972), in the Wharton Basin (Sclater and Fisher, 1974), the Arabian Sea (Whitmarsh, 1974a) and the Crozet and Mascarene Basins (Schlich et al., 1974b) these data have not radically changed the reconstruction history worked out by McKenzie and Sclater (1971). However, there are three modifications that are important. Weissel and Hayes (1972) have shown that Australia and Antarctica separated approximately 53 m.y.b.p. (anomaly 22). Also Sclater and Fisher (1974) have argued that India started to move with respect to Australia roughly 33 m.y.b.p. (anomaly 11). These same authors have shown that the Ninetyeast Ridge was initiated at this time and has remained fixed to the Indian plate. As a consequence of these changes, we have chosen to reconstruct the paleobathymetry at 36, 53 and 70 m.y.b.p. We present a 36 m.y.b.p. reconstruction because this is close to the time of the major change in direction of all the spreading centers in the central and western Indian Ocean. It is also close to the time at which the Chagos Fracture Zone separated into the Mascarene Plateau and the Chagos-Laccadive Ridge. The 53 m.y.b.p. reconstruction is the time we believe Australia started to separate from Antarctica. The last reconstruction at 70 m.y.b.p. is the time of anomaly 32 and is the furthest we can go back in time using magnetic anomalies. We have considerable confidence in the 36 m.y.b.p. reconstruction and some confidence in the position of the continents for 70 m.y.b.p. but not in the exact position of the ridge axes. However, the 53 m.y.b.p. reconstruction is speculative because there are serious problems with aligning fracture zones and lineation trends.

The basic magnetic lineations superimposed upon a contour chart of the Indian Ocean are presented in Figure 1. The present positions of all the JOIDES deep sea drilling sites are also shown on this chart. Throughout the reconstructions that follow we shall assume only four major plates for this ocean (Figure 4a). These are (1) the African plate bounded by the Carlsberg, Central Indian and Southwest Indian Ridges, (2) the Indian plate bounded by the Carlsberg, Central Indian, Southeast Indian

PALEOBATHYMETRY AND SEDIMENTS 37

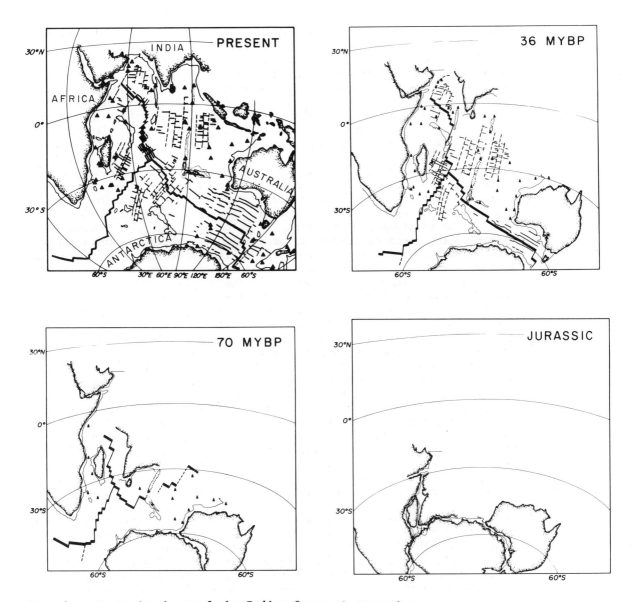

Fig. 4a. Tectonic chart of the Indian Ocean at present.

Fig. 4b. Tectonic chart of the Indian Ocean 36 m.y.b.p. (after McKenzie and Sclater, 1971).

Fig. 4c. Tectonic chart of the Indian Ocean 70 m.y.b.p. (after McKenzie and Sclater, 1971).

Fig. 4d. Reconstructed position of the Indian Ocean (after Smith and Hallam, 1970). The paleolatitude was determined using the paleomagnetic data for Australia.

and Ninetyeast Ridges, (3) the Australian plate bounded by the Ninetyeast and Australia/Antarctic Ridges and (4) the Antarctic plate bounded by the Southwest and Southeast Indian Ridges and the Australia/Antarctica Ridge. For the 36 m.y.b.p. reconstruction (Figure 4b) we have closed Arabia onto Africa, assumed the Indian and Australian plates are rigidly connected, and used the poles given by McKenzie and Sclater (1971) to close India/Australia to Antarctica and Africa to India/Antarctica. For Africa with respect to India/Antarctica the Chagos-Laccadive Ridge was assumed to close onto the Saya de Malha bank at the time of anomaly 11 (McKenzie and Sclater, 1971). With our modified time scale this date is now close to 36 m.y.b.p. and marks the time of anomaly 13. This is also close to, although probably a little earlier than, the time of the sharp change in spreading direction on the Southeast Indian Ridge (Sclater et al., 1976). Thus the change in direction of spreading and the split up of the old Chagos Fracture Zone may be contemporaneous. There are clearly errors in detail in the parameters chosen by McKenzie and Sclater (1971). However, the errors are small and to a first approximation this chart is a good estimate of the position of the continents 36 m.y.b.p. It is of interest to note that by this time the Southwest Indian Ridge is almost completely closed.

The 45 m.y.b.p. reconstruction of McKenzie and Sclater (1971) has more serious errors. We constructed the tectonic chart for this time and found that the lineations in the Crozet Basin overlapped those in the Southern Mascarene Basin. As a consequence, we decided to follow Weissel and Hayes (1972) and place the rifting apart of Australia and Antarctica at 53 m.y.b.p. Further, we split India and Australia before 36 m.y.b.p. (Sclater and Fisher, 1974) and followed Molnar and Francheteau (1975) in using the Chagos trench and the fracture zones in the Mascarene Plateau and anomalies in Arabian and Somali Basins to determine the early Tertiary history of the motion of Africa and India. We constructed a tectonic chart of the Indian Ocean for 53 m.y.b.p. using the parameters given in the above papers (Table 4a). The lineations in the Crozet Basin still overlap those in the Mascarene Basin but the overlap is not very large. This suggests that by this time the Southwest Indian Ridge had completely disappeared as a spreading center between the two basins. We follow McKenzie and Sclater (1971) and suggest that it was dominantly a fracture zone. However, if this is the case the lineations in the Crozet and Mascarene Basins should be parallel. They are close to being parallel but do not match exactly. In the present qualitative treatment this is not considered a severe restriction since the data from the Crozet Basin and the preliminary poles of Molnar and Francheteau (1975) may have significant error. Another flaw of this reconstruction is that the fracture zones in the Ceylon Basin do not align with those in the Crozet Basin and the lineations do not appear to coalesce on the same pole as the Ninetyeast Ridge. Clearly more detailed work with all the available magnetic data is needed to resolve these problems. However, even given all these objections, we feel that it is unlikely that the positions of the continents and the magnetic lineations have been significantly misplaced in our proposed reconstruction and we used it to construct a preliminary bathymetric chart of the Indian Ocean 53 m.y.b.p. However, we do not show the tectonic reconstruction for fear that it might be believed quantitatively reliable.

For the 70 m.y.b.p. reconstruction we have used the anomaly 32 poles of McKenzie and Sclater (1971) but corrected the misprint in their pole for the earthquakes between Africa and Antarctica (Table 4b). As there is little information on the floor of the Indian Ocean older than anomaly 32 it is difficult to check this reconstruction in as much detail as that for 53 m.y.b.p. On the other hand, the positions of the continents and of the Ninetyeast Ridge are compatible with the sites on this Ridge and Site 250 between Africa and Antarctica. Also the overlap of anomaly 32 on the Indian and Antarctic plates severely con-

PALEOBATHYMETRY AND SEDIMENTS   39

TABLE 4a. Rotation Poles and Angles for the 53 m.y.b.p. Reconstruction

	Latitude	Longitude	Angle	Reference
South America to Africa →	58.0	−37.0	26.14	Le Pichon and Hayes (1971)
Atlantic Ridge to Africa →	58.0	−37.0	13.07	Le Pichon and Hayes (1971)
Arabia to Africa →	36.5	18.0	−6.14	McKenzie and Sclater (1971)
Africa and above				
Carlsberg Ridge to India →	13.63	40.91	29.91	Molnar and Francheteau (1975)
to India →	14.35	40.75	14.96	
Southwest Indian Ridge to India →	14.35	40.75	14.96	
India and above				
to Antarctica →	−5.3	49.4	27.6	Sclater and Fisher (1974)
Southeast Indian Ridge to Antarctica →	−5.3	49.4	13.8	
Southwest Indian Ridge to Antarctica →	−5.3	49.4	13.8	
Antarctica and above				
to Australia →	−6.0	40.5	31.6	McKenzie and Sclater (1971)
Southeast Indian Ridge to Australia →	−6.5	37.8	15.8	
Australia and above				
to South-Pole →	0.0	216.0	19.0	Sclater and Fisher (1974)

TABLE 4b. Rotation Pole for the Indian Ridge to the South Pole for 70 m.y.b.p.

	Latitude	Longitude	Angle
Southwest Indian Ridge	7.18	87.97	−10.91

All the rest of the rotations are the same as those given in Table 10 of McKenzie and Sclater (1971).

strains the possible geometry of the plates. However, neither the position of the ridge axis between India and Africa nor that between Africa and Antarctica is well known. We have assumed that the spreading center between India and Africa jumped south sometime prior to anomaly 32 to account for the extra crust in the northern Mascarene Basin. For the ridge between Africa and Antarctica we have assumed that it consisted of two very long fracture zones, the westerly being the Prince Edward and the easterly separating the Crozet and Mascarene Basins and offset by three short ridge sections south of the Madagascar Basin. The position of the spreading center in the Wharton and Central Indian Basins and the history of the Ninetyeast Ridge and Broken Ridge are reasonably well known. However, the spreading centers in the western portion of the basin and the exact position of Africa with respect to Antarctica is still uncertain. It is likely a better understanding of these two continents will have to wait until the lineations observed in the Madagascar Basin (Bergh and Norton, 1976) have also been recognized and identified in the Crozet Basin south of the southwest Indian Ridge. However, for a preliminary paleobathymetric chart it is unlikely that the reconstructed positions of the continents and ocean floor are in serious error and since we have considerable confidence in the position of the continents we present this tectonic reconstruction as Figure 4c.

In order to produce a paleobathymetric chart for 70 m.y.b.p. it is necessary to know the positions of 4,000 m and 5,000 m isobaths. This in turn requires the determination of the position of the 90 m.y.b.p. and 120 m.y.b.p. isochrons on the 70 m.y.b.p. tectonic reconstruction (Appendix I). By 120 m.y.b.p. all the continents except Africa and Antarctica were probably closed onto each other and even in the case of Africa and Antarctica there was little sea floor between them. To determine the position of the 90 m.y.b.p. and 120 m.y.b.p. isochrons we have chosen to start with a tentative position for the continents all closed into one continent called Gondwana. From this position of the continents we work forward in time. For our speculative 70 m.y.b.p. paleobathymetry we have assumed that the Gondwana continents started in positions close to those given by Smith and Hallam (1970) (Figure 4d). In the early Cretaceous, Africa and South America split away from Madagascar, India, Australia, and Antarctica. Sometime in the mid-Cretaceous Africa picks up Madagascar which then becomes attached to Africa about the same time India separates from Antarctica and by the early Cretaceous we have the configuration shown in Figure 4c. In this reconstruction the dominant features are spreading centers completely surrounding the south tip of India with large fracture zones connecting them to the Tethys, a spreading center in the Wharton Basin and the mid-Atlantic Ridge. This configuration continues until 53 m.y.b.p. when Australia separates from Antarctica and the ridge axis west of the Ninetyeast Ridge jumps south. The next major change occurs in the late Eocene-early Oligocene when there is a reorientation in spreading direction in the central Indian Ocean, the Chagos Fracture Zone opens, the Ninetyeast Ridge moves away from the Southeast Indian Ridge and Southwest Indian Ridge is created. It has been noticed by many that this major change in morphology occurs at the onset of rapid uplift in the Himalayas. It is probable that these two features are causally related. After this major change, the development of the ocean essentially continues in the same fashion until the present.

The Paleobathymetric Charts

Once the relation between depth and age has been determined, the paleobathymetric charts follow directly from the tectonic history.

PALEOBATHYMETRY AND SEDIMENTS   41

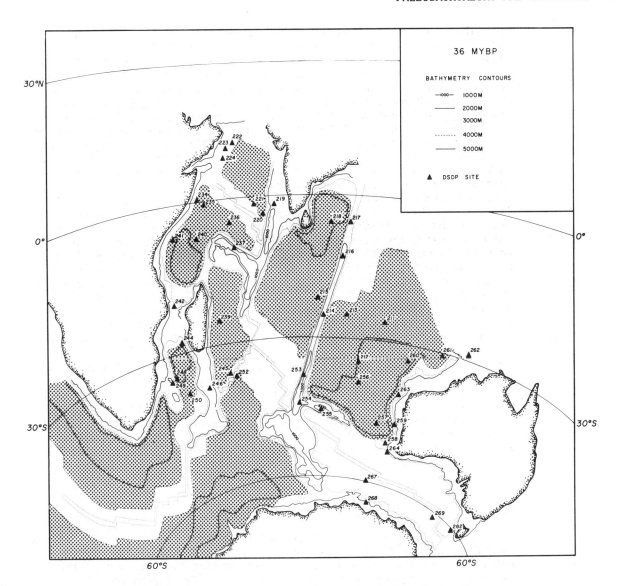

Fig. 5. The predicted bathymetry of the Indian Ocean for the early Oligocene (36 m.y.b.p; anomaly 13). The shaded area represents depths greater than 4000 m.

The method we used to construct these charts is discussed in detail in Appendix I. In this section we discuss the problems found in constructing each chart and the implications the contours might have upon the flow of deep and bottom water in the ocean.

The 36 m.y.b.p. (Early Oligocene, Anomaly 13) Chart

For the 36 m.y.b.p. reconstruction (Figure 5) we determined the relative depth of the Ninetyeast Ridge by assuming it was created at or above sea level and has sunk at the same rate as the crust to which it is attached. At this time much of the Ninetyeast Ridge was shallower than 2,000 m and the southern portion and the section near Broken Ridge were clearly close to or above sea level (Luyendyk et al., 1974). Almost all of the Ninetyeast Ridge was probably above 3,000 m and if it was connected to India to the north, the Ridge might have acted as a major barrier to the eastward migration of water masses and sediments. To the west we have less control on the subsidence history of the Chagos-Laccadive Ridge and the Mascarene Plateau. The sediments in holes 219 and 237 are evidence that by 36 m.y.b.p. both these regions probably had a depth of close to 1,000 m or less (Vincent et al., 1974). Some of the Chagos-Laccadive Ridge was probably close to sea level. Further, the Madagascar Plateau, on the basis of Site 246, was also close to sea level at this time and this plateau and the southwest Indian Ridge may have acted as a major barrier to the flow of bottom water from the Madagascar Basin to the Mascarene Basin.

Our knowledge of the past history of the Crozet and Kerguelen Plateaus is very sketchy. The Crozet Islands are young, but the Kerguelen-Heard Islands may be as old as Oligocene (Watkins et al., 1974). On the other hand, the bulk of both plateaus may be considerably older than the islands which show above sea level. As both features lie on old crust, we have assumed that both were in existence 36 m.y.b.p. and in their present geometry with probably only the southeastern end of Crozet significantly shallower then than now. It is clear from the 36 m.y.b.p. chart (Figure 5) that Kerguelen and its possible extension into Antarctica are very important as the subsidence history of this feature must have controlled the possible flow of bottom water around Antarctica.

Other features of some importance to deep water circulation patterns are the Amirantes Islands and the Amirantes Trench. There is a major hiatus in the shallow water carbonate sediments which extends across the entire central and eastern Indian Ocean (Luyendyk and Davies, 1974) for a short time span in the Oligocene. However, it does not extend in either deep or shallow water into the Arabian Sea or the Somali Basin. We tentatively suggest on this evidence that in the Oligocene this trench and island system was shallow and formed a barrier to deep water flow.

The 53 m.y.b.p. (Early Eocene, Anomaly 22) Chart

For the 53 m.y.b.p. chart (Figure 6) the reconstructed position of the end of anomaly 32 (70 m.y.b.p.) was taken to mark the 4,000 m contour. The 5,000 m contour was estimated by determining the position of the 100 m.y.b.p. isochron on this reconstruction. The depths for the Ninetyeast Ridge, the Chagos-Laccadive Ridge, the Mascarene Plateau and Broken Ridge were determined from the DSDP sites on these features. Long portions of all of these Ridges were probably close to sea level at this time. Unfortunately we know very little about the Madagascar Ridge and the Crozet and Kerguelen Plateaus. We have assumed for our 53 m.y.b.p. reconstruction that the Madagascar Ridge was close to sea level and that the Kerguelen Plateau was already in existence. We have omitted the Crozet Plateau.

The dominant features of this reconstruction are the closure of Australia and Antarctica and the segmentation at the 4,000 m level of the ocean into a series of small isolated basins; the Madagascar, Somali, Mascarene, Ceylon and Wharton Basins, all lying to the north of the Southwest and Southeast Indian Ridges. Another important feature is the possible existence of major land bridges along the Ninetyeast and Broken Ridges between Australia and India and along the Chagos Fracture Zone between Madagascar and India.

Between 53 m.y.b.p. and 36 m.y.b.p. Broken Ridge is above sea level and we have postulated that the Kerguelen Plateau was very shallow.

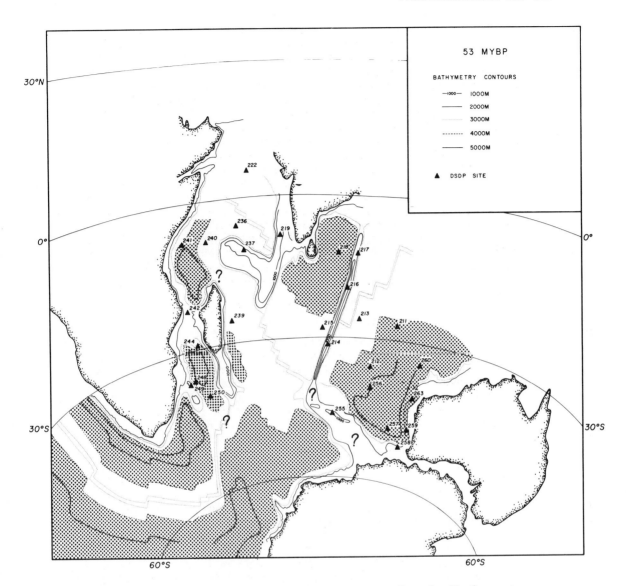

Fig. 6. The estimated bathymetry of the Indian Ocean for the Early Eocene (53 m.y.b.p.; anomaly 22). The shaded area represents depths greater than 4000 m.

During this time span Tasmania and its extension to the south was always close to Antarctica. Thus, except for the small region between Broken Ridge and Naturaliste Plateau, the basins between Antarctica and Australia were totally closed to deep water either from the east or west. It is possible that this geometry could have led to major changes in the deep sea sedimentary record. The absence of an Oligocene hiatus at Site 267 may be evidence of such a change.

The 70 m.y.b.p. (Late Cretaceous, Anomaly 32) Chart

Apart from the position of the ridge axis between India and Australia-Antarctica, the 4,000 m contour in the Wharton Basin, and the contours off the coast of Africa, all other contours and ridge axes on the 70 m.y.b.p. reconstruction are speculative (Figure 7). The position of spreading center between Madagascar and India is also speculative as is that between Africa and Antarctica. However, the contours in the Wharton Basin and either side of the Southeast Indian Ridge determined by estimating the 90 m.y.b.p. isochron are probably reasonable. It is likely that at this time all of the existing Ninetyeast Ridge and most of Broken Ridge were above 2,000 m and that the portions of both ridges around Sites 255 and 217 were above sea level. The exact paleobathymetry of the basin between India and Madagascar is much less certain. It is probable that almost all of this basin was above 4,000 m. The only really deep portions of the Indian Ocean occur in the Somali Basin and in the Madagascar and South Enderby Basins. This is to be expected as in our speculative history of early opening these are the first basins to be formed. As with our other reconstructions Kerguelen is a major problem. Kaharoeddin et al. (1973) have reported on some Cretaceous shallow water fossils from this ridge. If correct these fossils are evidence that the Kerguelen Plateau was elevated and in its position against Broken Ridge 70 m.y.b.p. However, much more data is needed before we can accept with confidence such an important conclusion.

Use of Paleobathymetric Charts

It has been shown in the Pacific that there is a relation between the circulation of Antarctic bottom water and the level of dissolution of calcareous sediments. Paleobathymetric charts would be useful in the determination of the paths of such currents in the past. To do this, however, other information not now available is required. First, we do not know how bottom water was formed prior to the Oligocene. Second, we have no understanding of the exact structure of the Kerguelen/Antarctica contact and the relative closure of South America and Antarctica beyond the Miocene. Consequently, we cannot predict with any confidence the flow of such a water mass around Antarctica.

It is thus obvious that we cannot use the paleobathymetric charts at present to obtain a quantitative model of how and why the deep water sediments are formed in the Indian Ocean. However, we can backtrack the DSDP sites (Berger and von Rad, 1972) and estimate the depth of the CCD through time for the Indian Ocean. These depths can then be superimposed upon the paleobathymetric charts to map the surface distribution of calcareous and non-calcareous sediments at different geologic times. In the next section we compute the CCD as a function of time and use this and the paleobathymetric charts to investigate the surface distribution of calcareous sediments in the Indian Ocean.

Distribution of Sedimentary Facies in the Indian Ocean

The Calcite Compensation Depth

The most readily observable boundary between distinct sedimentary facies in the oceans is the Calcite Compensation Depth (CCD), the level where the dissolution rate of calcite equals the supply rate in the deep ocean (Peterson, 1966). The CCD is found today at approximately 3.5 to 4.5 km water depth (Berger and Winterer, 1974) sloping upwards close to continental margins and in the polar to subpolar regions of the world oceans. The depth variation of the CCD with time is still under considerable discussion (van Andel, 1975). First, the coverage of the world ocean with DSDP sites is geographically uneven and second, only a certain portion of the drill sites are suitable for determining the age/depth position of the CCD. This unevenness and scarcity of

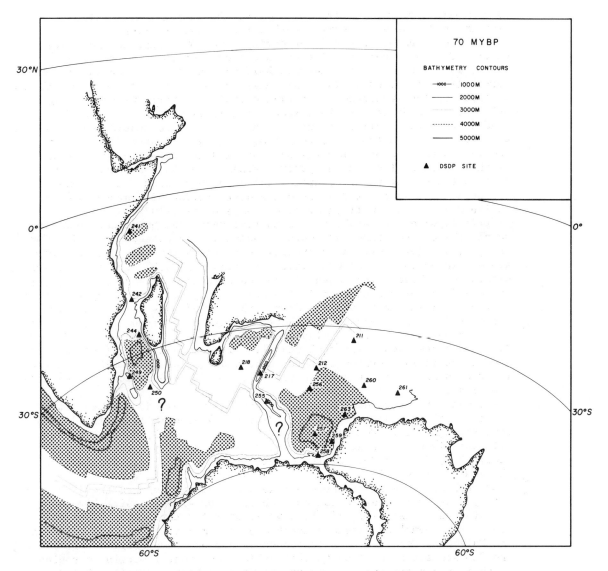

Fig. 7. A possible bathymetry of the Indian Ocean for the Late Cretaceous (70 m.y.b.p.; anomaly 32). The shaded area represents depths greater than 4000 m.

data have resulted in a considerable delay in compiling the geological history of this boundary between major pelagic sediment facies. The first attempt to describe the history of the CCD on a world wide scale has been completed only recently by van Andel (1975).

The Indian Ocean poses a special problem for reconstructing the CCD as a function of time. There are no deep sites in the central Indian Basin and none south of the Southeast or Southwest Indian Ridges. Of the sites that are available many did not reach basement and of the few that did only nineteen have an adequate sediment record. Certain sites, especially those of DSDP legs 23, 27 and 28, were drilled close to the Arabian and Australian continental margins or in the polar regions south of Australia where the CCD is known to be situated much shallower than in the open ocean. Thus these sites cannot easily be compared with others from the central portions of the Indian Ocean. This problem is discussed at some length later in this section.

Fifty-three holes were drilled during seven DSDP legs in the Indian Ocean. For the nineteen holes which showed a crossing of the CCD, twelve were located in the eastern half and the other seven in the western half of the ocean. The past water depth of all sites was calculated by applying the subsidence of normal oceanic basement and correcting only for the sediment overload (see Appendix II for the exact method used). We denoted all crossings of the CCD on these subsidence curves by a point and plotted these points against age and depth (Figure 8a). The time scales adopted for the Cenozoic and the Mesozoic are the same as those used earlier in the paper. Finally, we have avoided using gaps in the coring records, artificial or natural ones, because they are negative evidence and because the distribution of hiatuses in the Indian Ocean is not well understood (Moore et al., 1977).

It is important to evaluate the initial data from holes carefully to obtain the correct time and depth at which a hole crosses the CCD. For example, a major question is the degree to which the carbonate sediments are displaced to water depths which are below the CCD. Indications for displacement have to be sought in the descriptions of sedimentary structures and textures contained in the initial reports. If a major portion of the total sediment consists of reworked calcareous fossils of older zones as in Sites 260 and 261, there is excellent evidence that much of the carbonate sediments is displaced. Thin beds of calcareous sediments have been found intercalated with clays at a number of sites. In some cases it was obvious from the core photographs (Site 212) that the lower boundary of these beds is very sharp while they grade upwards into the clays.

Although the cases discussed above have been avoided, the crossings still reveal considerable scatter (Figure 8a) especially during Paleocene through Middle Eocene times and during late early Miocene through Pliocene times. However, this scatter is due to only a few sites, namely 212, 223 and 239 for the older interval, 222, 261, and 265 for the younger interval.

At each of these sites there are good reasons to believe the data are not typical of the deep ocean as a whole. For example, Site 212 is situated on an extension of the Investigator Fracture Zone. This explains the anomalous basement depth and also makes it highly likely that the Paleocene-Eocene carbonate sequences are of slump origin. Sites 222 and 223 are situated close to the Arabian continental margin, a region which is today and was throughout the Neocene highly fertile (Thiede, 1974). The age of the oceanic basement at Site 239 in the southern Mascarene Basin is only known to be pre-Campanian (Simpson et al., 1974) the oldest sediments consisting of brown clay interbedded with nanno ooze and thus suggesting a hiatus between the basement and the oldest sediment (van Andel and Bukry, 1973). The Neogene cores of Site 261 contain partly calcareous sediments interbedded with clays (Veevers and Heirtzler, 1974); the nanno oozes were probably displaced to water depths below the CCD. Site 265 lies north of Antarctica, and has a much shallower CCD in the Miocene than the other sites. The reason for this is unknown but it could possibly be related to the proximity of the site to the deep circumpolar current. We return to a discussion of this site and Sites 222, 223 and 239 in the next section.

If the above six sites are ignored then this considerably simplifies the history of the CCD. During the Tithonian the CCD was close to 3.5 km. It shallowed during the Albian and Aptian to 2.5 to 3.0 km and then dropped to 4.0 km for the Maastrichtian. From then until the Oligocene it remained fairly flat before another drop to approximately

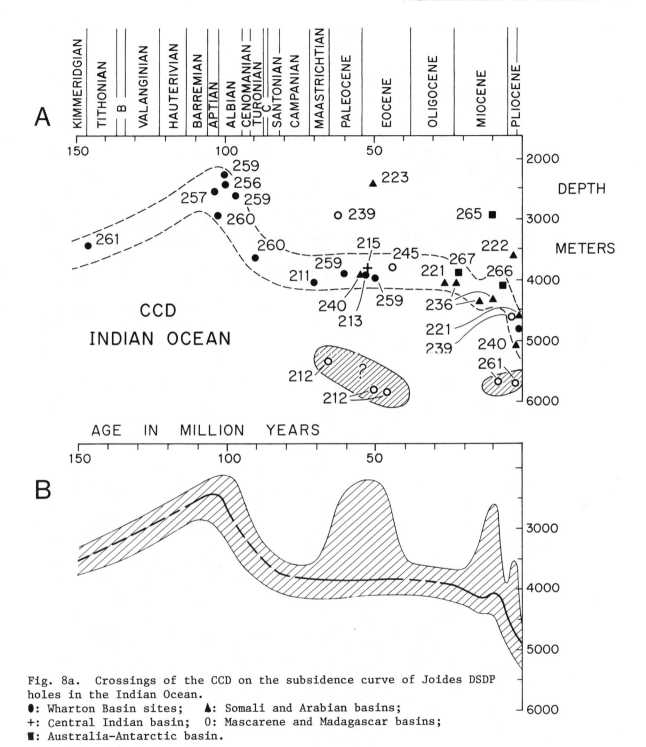

Fig. 8a. Crossings of the CCD on the subsidence curve of Joides DSDP holes in the Indian Ocean.
●: Wharton Basin sites;   ▲: Somali and Arabian basins;
+: Central Indian basin;   O: Mascarene and Madagascar basins;
■: Australia-Antarctic basin.

Fig. 8b. The range of distribution of the Indian Ocean Calcium Carbonate Compensation Depth (CCD) since Jurassic times. The solid line has been fitted by hand through the majority of the crossings on figure 8a, the stippled lines and the hatched area circumscribe the scatter of all variable points except sites 212 and 261.

4.5 to 5.0 km occurred during the later part of the Neogene (heavy line on Figure 8b).

The CCD curve which we obtain for the Indian Ocean is similar to that presented by van Andel (1975) except for regions older than the Eocene. The differences all result from the use by van Andel (1975) of a subsidence curve for younger crust which starts at 3,100 m rather than 2,700 m. Such a curve will tend to deepen sediment found very close to basement by up to 400 m and results in the deeper CCD presented by van Andel (1975) for the pre-Eocene sites.

Having subtracted six sites from the original nineteen it is important to realize that the data base for this curve is not only regionally inadequate but also very poorly controlled in time. For example, there is only one crossing available for the time span from 150 to 110 million years (Site 261), there is only one site from the Turonian to Maastrichtian (Site 260) and no sites from the middle Eocene to late Oligocene. However, certain tentative conclusions can be drawn from the data. First, the CCD in the Tithonian was close to 3,500 m. Second, though poorly distributed, the data from the Indian Ocean sites do not support the idea of a CCD very close to the sea surface at the Cretaceous/Tertiary boundary as proposed by Worsley (1974). Third, the simplified curve, with all the caveats mentioned above, can be combined with the paleobathymetric charts (Figures 5, 6 and 7) to provide the surface distribution of calcareous and non-calcareous sediments at selected time slices (Figures 9a, b, c, and d).

Distribution and Interpretation of Major Sediment Facies Changes

The calcite compensation depth in any present ocean is not horizontal but rather it is an undulating surface usually at a depth between 5.0 and 3.5 kms which shallows towards the continental margins and polar regions (Berger and Winterer, 1974). Though there is a large difference in the depth of the CCD between oceans, if only individual basins are concerned, the differences in depth reduce to about 500 m if the near coastal and polar areas are neglected. If it were possible to map the surface of the CCD at a given epoch back in time for a given basin then in principle interfacing the CCD surface with the paleobathymetry would define the distribution of surface sediments in the basin at that time. All ocean floor which lay above the CCD would have calcareous sediment deposited upon it and conversely all ocean floor which lay below the surface would have clay and radiolarian ooze deposited upon it. However, with the present distribution of JOIDES samples and the absence of continuous coring on the early legs it is impossible except in a few isolated areas to determine the surface of the CCD at a given point in time for any basin. As a result of this, to compute these sediment facies maps, it is necessary to make a further assumption beyond those which have enabled us to compute the CCD as a function of age. The assumption is that within ±300 m from the present back to 70 m.y.b.p. the CCD remains horizontal and is given by the mean point of the CCD curve through time (Figure 8a). If we neglect the Australia/Antarctic Basin then all the sites back to the Paleocene appear to lie within ±300 m of the mean. Before the Paleocene this assumption appears less valid. We applied this assumption of a horizontal CCD to the paleobathymetry to produce the sediment facies maps. We believe only the 35 m.y.b.p. chart to be quantitative. The other two charts for 53 and 70 m.y.b.p. have merely been included for the sake of completeness and to speculate as to how far our assumptions would lead us.

The errors in the charts are difficult to evaluate. We believe that apart from the Australian-Antarctic Ridge and between South Africa and

PALEOBATHYMETRY AND SEDIMENTS 49

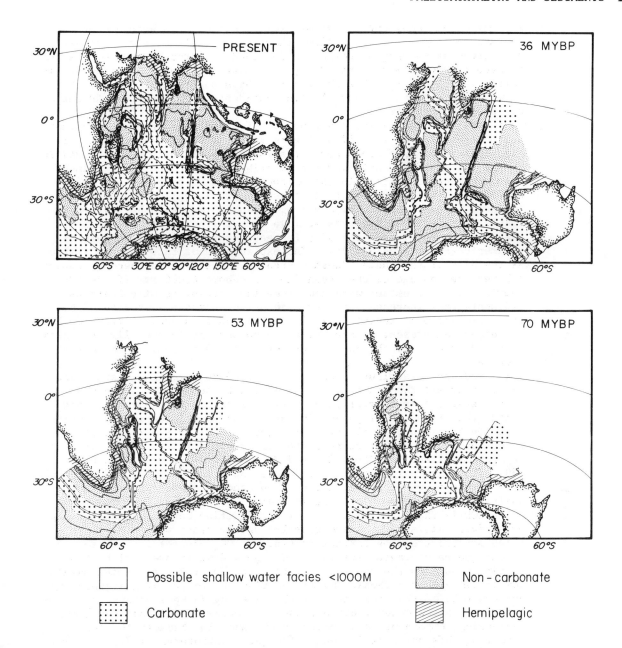

Fig. 9a. Present distribution of carbonate, clay, hemipelagic and shallow water marine sediments (Kolla et al., 1976).

Fig. 9b. Probable distribution of surface sediments in the early Oligocene, (36 m.y.b.p.).

Fig. 9c. Speculative distribution of the surface sediments in the early Eocene, (53 m.y.b.p.).

Fig. 9d. Highly speculative distribution of the surface sediments in the Late Cretaceous, (70 m.y.b.p.).

Antarctica our sediment facies map for 36 m.y.b.p. is accurate on the average to ±300 m. For the older charts it is difficult to estimate the errors as so many unknowns are involved. A probable maximum estimate is around 400-500 m for the 53 m.y.b.p. chart as the 4,000 m contour is fairly well known for this chart but perhaps more than 500 m for the 70 m.y.b.p. chart as we have little confidence in either the CCD curve or the paleobathymetry.

We compiled the present distribution of calcareous and non-calcareous sediments (Kolla et al., 1976; Figure 9a) and compared this with our distribution of calcareous sediments for the early Oligocene (Figure 9b). As the CCD was much shallower the distribution of calcareous sediment was much more restricted in the Oligocene than at present. We are confident of the above conclusions as the CCD is well controlled back to the Oligocene and the 36 m.y.b.p. paleobathymetry is believed reliable. It is clear that, as we move back further in time, the major basins are covered by a steadily increasing proportion of calcareous sediments (Figures 9c and 9d). The explanation of this is very simple. As the size of the Indian Ocean decreases and the basins become younger, a larger percentage of the ocean floor is above 4,000 m. If the CCD remains fairly constant with time then the percentage of calcareous sediments will increase. An estimate of the surface area covered by carbonate sediment reveals that in this ocean it decreases from roughly 60 percent at present to approximately 40 percent in the Oligocene back up to 50 percent by the Eocene and more than 60 percent in the early Cretaceous.

The sediment distribution charts are based on three key assumptions. First, that all the basins have the same CCD; second, that the CCD, when averaged over 10 million year time slices, changes smoothly; and third, that from the Oligocene to the Campanian this depth has remained at a constant depth fairly close to 4,000 m. Clearly our assumptions do not hold for the Oligocene hiatus (Pimm and Sclater, 1974; Davies and Luyendyk,1974) and for Sites 222 and 223, Sites 265 and 266 and Site 239. It is our belief that these sites and Sites 256, 257, 259 and 260 give information concerning the past changes of environment of the basins in which these sites are located.

The CCD is the level where calcite dissolution equals carbonate supply by surface waters. Carbonate dissolution increases with decreasing temperature, increasing partial pressure of CO_2 and total pressure. As a consequence of these factors the CCD goes down in areas of high productivity such as the equator. It goes up near the continental margin due to removal of carbonate by dislocation due to the high organic carbon controls of the sediment. It also goes up in regions over which the Antarctic bottom water flows because such water is undersaturated due to decreasing temperature and increased CO_2 and also because the rate of supply of this water is increased (Berger and Winterer, 1974). We now examine the anomalous sites within the framework of this model.

Sites 222 and 223 are close to the coast of Arabia and thus it is not surprising that they show a shallow CCD. Sites 265 and 266 in the Miocene lie under the path of the deep Antarctic Circumpolar Current. This may be the explanation of why these sites should show an anomalously shallow CCD. Site 239 is an anomaly. However, it is possible that bottom water passed from the Enderby basin off Antarctica to the Tethys. If such a current was undersaturated in carbonate it could explain the shallow CCD at Site 239. The explanation of the shallow Sites 256, 257, 259 and 260 in the Wharton Basin is much simpler. Sclater and Fisher (1974) and Markl (1974) predict that India and its Tibetan extension closed against western Australia between 100 and 120 m.y.b.p. Thus from initial opening at 120 m.y.b.p. until at least 100 m.y.b.p., the Tibetan extension of India, the Exmouth Plateau and the Kerguelen Ridge would have made the Wharton a totally enclosed basin with nearby continental shelves and consequent removal of carbonate.

In this study we have concentrated upon questions to which we think we have plausible answers. However, there are problems which have arisen which we cannot answer. For example, in the previous section we have attributed the shallowness of the CCD in the Miocene at Sites 265 and 266 to circumpolar currents. Further Kennett et al., (1975) have attributed the Oligocene hiatus in Sites 280, 281 and 282 which range in depth from 4200 m to 1600 m to the opening of the gap between the Tasman Rise and Antarctica and the onset of a vigorous shallow and deep circumpolar current. We feel both our own interpretation of Sites 265 and 266 and that of Kennett et al. (1975) of Sites 280, 281 and 282 very plausible. The problem with both interpretations lies in the fact that the late Oligocene is complete in both holes 267 and 269. However, at this time, both sites were close to the ridge axis (Figure 5) and directly under the proposed axis of the supposedly very vigorous Antarctic Circumpolar Current. At present we do not have a convincing answer to this problem which yet again illustrated the difficulties of trying to speculate about global circulation with an inadequate data base.

A second problem occurs with the Oligocene hiatus in the Indian Ocean sediments. If it is defined solely as an absence of calcareous sediments where they would otherwise be expected then the evidence for its occurrence is limited only to shallow water sites. Also these sites (249, 246, 214, 216, 217, 264, 258, and 261) all lie east of a line joining the east coast of Madagascar, the Seychelles, the Chagos-Laccadive Ridge and India. Thus the hiatus is evidence only for (1) a possible shallow and deep water barrier between the Amirantes and the Seychelles and (2) some shallow water relation between the sediments in the sites east of our proposed shallow water line joining Africa to India. At present there is no data supporting a deep water hiatus in the eastern Indian Ocean.

Conclusions on Sediment Distributions

The major conclusions from our study of the variation of CCD with time and the examination of the influence of this on the distribution of surface sediments with time are (a) that the mean CCD shallows from around 5,000 m at present to close to 4,000 m for the Oligocene through the Campanian; (b) that the variations of the CCD away from this mean can be attributed to deep water currents and proximity to continental margins; (c) that the surface distribution of calcareous sediments in the Oligocene is considerably less than what is observed at present; and (d) that the presence of calcareous sediments throughout the history of almost all the aseismic ridges can be explained by the fact that except during the Albian and Aptian the CCD was never shallower than 2,500 m.

There are some outstanding problems which our study has not resolved. These include the exact vertical distribution of the Oligocene hiatus in the Indian Ocean, the past sedimentary history of the Kerguelen and Crozet Basins and the general history of bottom water circulation south of the southwest and southeast Indian Ridges. We feel strongly that these and many other paleooceanographic problems of major order might well be rendered soluble by a carefully designed program of deep sea drilling in the southern Indian Ocean.

Acknowledgements. Much of the speculation discussed in this paper arose out of spirited discussions with R. Schlich, H. Bergh, I. Norton and P. Molnar. We are grateful for their interest and, at first, often unheeded advice. The paper was reviewed by J. van Andel, E. L. Winterer and E. Vincent and we have benefited significantly from their comments. The work of Sclater and Abbott was supported by contract number N00014-75-C-0291 from the Office of Naval Research. The assistance of N. A. Brewster in compiling the sediment data is gratefully acknowledged.

References

Anonymous, Russian Chart of the Indian Ocean, 1:15,000,000, Committee of Management for Geodesy and Cartography of the Soviet Ministry of the U.S.S.R., Moscow, 1964.

Anderson, R. N., D. McKenzie, and J. G. Sclater, Gravity, bathymetry and convection in the earth: Earth Planet. Sci. Lett., 18, 391-407, 1973.

Berger, W. H., and U. von Rad, Cretaceous and Cenozoic sediments from the Atlantic Ocean, in Haues, D. E., Pimm, A. C. et al., Initial Reports of the Deep Sea Drilling Project, Washington, D. C. (U. S. Government Printing Office), 14, 787-954, 1972.

Berger, W. H., and E. L. Winterer, Plate stratigraphy and the fluctuating carbonate line, International Assoc. of Sed., Spec. Publ. 1, 11-48, 1974.

Berggren, W. A., and J. A. van Couvering, The late Neogene-biostratigraphy, geochronology and paleoclimatology of the last 15 million years in marine and continental sequences, Palaeogeogr. Palaeoclimatol., Palaeoecol., 16, 1-216, 1974.

Bergh, H. W., and I. O. Norton, Prince Edward Fracture Zone and the evolution of the Mozambique Ridge, J. Geophys. Res., 81, no. 29, 5221-5239, 1976.

Bunce, E. T., Djibouti to Seychelles Islands, in Fisher, R. L., Bunce, E. T. et al., Initial Reports of the Deep Sea Drilling Project, Washington, D. C. (U. S. Government Printing Office), 24, 591-605, 1974.

Carpenter, G., and M. Ewing, Crustal deformation in the Wharton Basin, J. Geophys. Res., 78, no. 5, 846-850, 1974.

Davies, T. A., and R. B. Kidd, Sedimentation in the Indian Ocean through time in Indian Ocean Geology and Biostratigraphy, AGU, Washington, this volume, 1977.

Davies, T. A., and B. P. Luyendyk, in Davies, T. A., Luyendyk, B. P., et al., Initial Reports of the Deep Sea Drilling Project, Washington (U. S. Government Printing Office) 26, 1192, 1974.

Davis, E. E., and C.R.B. Lister, Fundamentals of ridge crest topography, Earth and Planet. Sci. Letts., 21, 405-413, 1974.

Detrick, R., J. G. Sclater, and J. Thiede, Subsidence of aseismic ridges, Earth Planet. Sci. Letts., 34, 185-196, 1977.

Eittreim, S. L., and J. Ewing, Mid-plate tectonics in the Indian Ocean, J. Geophys. Res., 77, no. 32, 5413-6421, 1972.

Ewing, M., S. Eittreim, M. Trauchan, and J. I. Ewing, Sediment distribution in the Indian Ocean, Deep-Sea Res., 16, 231-248, 1969.

Falvey, D. A., Sea floor spreading in the Wharton Basin (Northeast Indian Ocean) and the breakup of eastern Gondwanaland, Australian Petroleum Exploration Association, 12, no. 2, 1972.

Frey, F. A., J. S. Dickey, G. Thompson, and W. B. Bryan, Eastern Indian Ocean DSDP sites: correlations between petrography, geochemistry and tectonic setting, in Indian Ocean Geology and Biostratigraphy, AGU Washington, this volume, 1977.

Hayes, D. E., A. C. Pimm, J. P. Beckmann, W. E. Benson, W. H. Berger, P. H. Roth, P. R. Supko and U. von Rad, Initial Reports of the

Deep Sea Drilling Project, 14, Washington (U. S. Government Printing Office) 975, 1972.

Heirtzler, J. R., G. O. Dickson, E. M. Herron, W. C. Pitman and X. Le Pichon, Marine magnetic anomalies, geomagnetic field reversals, and motions of the ocean floor and continents, J. Geophys. Res., 73, 2119-2136, 1968.

Houtz, R. E., and R. G. Markl, Seismic profiler data between Antarctica and Australia, in Antarctic Oceanology II, The Australian-New Zealand Sector, edited by D. E. Hayes, Amer. Geophys. Un., Washington, 147-164, 1972.

Johnson, B. D., C. McA., Powell, and J. J. Veevers, Spreading history of the eastern Indian Ocean and greater India's northward flight from Antarctica and Australia, Bull. Geol. Soc. Amer., 87, 1560-1566, 1976.

Kaharoeddin, A., F. M. Weaver, and S. W. Wise, Cretaceous and Paleogene cores from the Kerguelen Plateau, Antarct. J. U. S., 8, 297-298, 1973.

Kahle, H. G., and M. Talwani, Gravimetric Indian Ocean Geoid, Zeitschrift für Geophysik, 39, 491-499, 1973.

Kennett, J. P., R. E. Houtz, P. B. Andrews, A. R. Edwards, V. A. Gostin, M. Hajos, M. A. Hampton, D. Graham Jenkins, S. V. Margolis, A. T. Overshine, and K. Perch-Nielsen, Cenozoic paleooceanography in the southwest Pacific Ocean, Antarctic glaciation and the development of the circum-antarctic current, in Kennett, J. P., Houtz, R. E., et al., Initial Reports of the Deep Sea Drilling Project, 29, Washington, (U. S. Government Printing Office) 1155-1170, 1975.

Kolla, V., A.W.H. Be, and P. E. Biscaye, Calcium carbonate distribution in the surface sediments of the Indian Ocean, J. Geophys. Res., 81, no. 15, 2605-2616, 1976.

Larson, R. L., Late Jurassic sea floor spreading in the Eastern Indian Ocean, Geology, 3, 69-71, 1975.

Larson, R. L., and W. C. Pitman, World wide correlation of Mesozoic magnetic anomalies and its implications, Bull. Geol. Soc. Amer., 83, 3645-3662, 1972.

Larson, R. L., and W. C. Pitman, World wide correlation of Mesozoic magnetic anomalies and its implications: discussion and reply, Geol. Soc. Annual Bull., 86, 267-272, 1975.

Le Pichon, X., and D. E. Hayes, Marginal offsets, fracture zones and the early opening of the South Atlantic, J. Geophys. Res., 76, no. 26, 6283-62-93, 1971.

Luyendyk, B. P., and T. A. Davies, Results of DSDP Leg 26 and the geologic history of the southern Indian Ocean, in Davies, T. A., Luyendyk, B. P., et al., Initial Reports of the Deep Sea Drilling Project, Washington, D. C. (U. S. Government Printing Office) 26, 909-943, 1974.

Luyendyk, B., T. A. Davies, et al., Report on Site 253, in Davies, T. A., Luyendyk, B. P., Initial Reports of the Deep Sea Drilling Project, Washington, D. C. (U. S. Government Printing Office) 26, 153-232, 1974.

Luyendyk, B. P., Deep sea drilling on the Ninetyeast Ridge: synthesis and tectonic model, in Indian Ocean Geology and Biostratigraphy, AGU, Washington, this volume, 1977.

Markl, R. G., Evidence for the breakup of eastern Gondwanaland by the early Cretaceous, Nature, 251, no. 5471, 196-200, 1974.

McKenzie, D., and C. Bowin, The relationship between bathymetry and gravity in the Atlantic Ocean, J. Geophys. Res., 81, no. 11, 1903-1915, 1976.

McKenzie, D. P., and J. G. Sclater, The evolution of the Indian Ocean since the Late Cretaceous, Geophys. J. Roy. astr. Soc., 25, 437-528, 1971.

Menard, H. W., Gravity anomalies and vertical tectonics as indicators of plate motions, *Amer. Geophys. Un. Trans.*, 54, 4, 239, 1973.

Molnar, P., and J. Francheteau, The relative motion of "hot-spots" in the Atlantic and Indian Oceans during the Cenozoic, *Geophys. J. Roy. astr. Soc.*, 43, no. 3, 763-774, 1975.

Moore, T. C., T. H. van Andel, C. Sancetta, and N. Pisias, Cenozoic hiatuses in pelagic sediments, *Proc. 3rd Internat. Plankt. Conf.* (Kiel, 1974), 1977, in press.

Parsons, B., and J. G. Sclater, An analysis of the variation of ocean floor heat flow and depth with age, *J. Geophys. Res.*, 1977, in press.

Peterson, M.N.A., Calcite: rates of dissolution in a vertical profile in the central Pacific, *Science*, 154, 1542-1544, 1966.

Pimm, A. C., B. McGowran, and S. Gartner, Early sinking history of Ninetyeast Ridge, Northeastern Indian Ocean, *Bull. Geol. Soc. Amer.*, 85, no. 8, 1219-1224, 1974.

Pimm, A. C., and J. G. Sclater, Early Tertiary hiatuses in the sediment record of Deep Sea Drilling Sites in the northeastern Indian Ocean, *Nature*, 252, 362-365, 1974.

Schlich, R., Structure et age de l'ocean Indien Occidental, *Memoire Hors-Serie* 6, Paris, France, 103, 1975.

Schlich, R., E.S.W. Simpson, and T. L. Vallier, Introduction, in Simpson, E.S.W., Schlich, R. et al., *Initial Reports of the Deep Sea Drilling Project*, Washington, D. C. (U. S. Government Printing Office), 25, 5-24, 1974a.

Schlich, R., E.S.W. Simpson, and T. L. Vallier, Regional aspects of deep sea drilling in the western Indian Ocean, Leg 25, DSDP, in Simpson, E.S.S., Schlich, R., et al., *Initial Reports of the Deep Sea Drilling Project*, Washington, D. C., (U. S. Government Printing Office), 25, 743-760, 1974b.

Sclater, J. G., R. N. Anderson, and M. L. Bell, Elevation of ridges and evolution of the central eastern Pacific, *J. Geophys. Res.*, 76, 7888-7915, 1971.

Sclater, J. G., C. Bowin, R. Hey, H. Hoskins, J. Peirce, J. Phillips, and C. Tapscott, The Bouvet Triple Junction, *J. Geophys. Res.*, 81, no. 11, 1857-1869, 1976.

Sclater, J. G., and R. L. Fisher, The evolution of the east central Indian Ocean, *Bull. Geol. Soc. Amer.*, 85, 683-702, 1974.

Sclater, J. G., R. Jarrard, B. McGowran, and S. Gartner, Comparison of the magnetic and biostratigraphic time scales since the late Cretaceous, in von der Borch, C., Sclater, J. G., et al., *Initial Reports of the Deep Sea Drilling Project*, Washington, D. C. (U. S. Government Printing Office), 22, 381-386, 1974.

Sclater, J. G., S. Hellinger, and C. Tapscott, The paleobathymetry of the Atlantic Ocean (in press), 1977.

Sclater, J. G., L. A. Lawver, and B. Parsons, Comparison of long wavelength residual elevation and free air gravity anomalies in the North Atlantic and possible implications for the thickness of the lithospheric plate, J. Geophys. Res., 80, 1031-1052, 1975.

Sclater, J. G., and D. P. McKenzie, The paleobathymetry of the south Atlantic, *Bull. Geol. Soc. Amer.*, 84, 3203-3216, 1973.

Simpson, E.S.W., and R. Schlich et al., *Initial Reports of the Deep Sea Drilling Project*, Washington, D. C. (U. S. Government Printing Office), 25, p. 884, 1974.

Smith, A. G., and A. Hallam, The fit of the southern continents, *Nature*, 225, 139-144, 1970.

Thiede, J., Sediment coarse fractions from the western Indian Ocean and the Gulf of Aden (Deep Sea Drilling Project Leg 24), in Fisher, R. L., Bunce, E. T., et al., *Initial Reports of the Deep Sea Drilling Project*, Washington, D. C. (U. S. Government Printing Office), 24, 651-765, 1974.

Thiede, J., Evidence from sediments on the Rio Grande Rise (SW Atlantic Ocean): Subsidence of an aseismic ridge, Amer. Assoc. Petro. Geol. Bull., (in press), 1977.

Thierstein, H. R., Mesozoic biostratigraphy of marine sediments by calcareous nannoplankton, Proc. 3rd Internat. Plankt. Conf. (Kiel, 1975), (in press), 1977.

van Andel, T. H., Mesozoic-Cenozoic calcite compensation depth and the global distribution of calcareous sediments, Earth Planet. Sci. Lett., 26, 187-194, 1975.

van Andel, T. H., and D. Bukry, Basement ages and basement depths in the Eastern Equatorial Pacific from Deep Sea Drilling Project Legs 5, 8, 9 and 16, Bull. Geol. Soc. Amer., 84, 2361-2370, 1973.

Veevers, J. J., Seismic profiles made underway on Leg 22, in von der Borch, C., Sclater, J. G., et al., Initial Reports of the Deep Sea Drilling Project, Washington, D. C., (U. S. Government Printing Office), 22, 351-368, 1974.

Veevers, J. J., and J. R. Heirtzler, Bathymetric, seismic profiles, and magnetic anomaly profiles, in Veevers, J. J., Heirtzler, J. R., et al., Initial Reports of the Deep Sea Drilling Project, Washington, D. C. (U. S. Government Printing Office), 27, 339-382, 1974.

Vincent, E., Cenozoic planktonic biostratigraphy and paleooceanography of the tropical western Indian Ocean, Deep Sea Drilling Project, Leg 24, in Fisher, R. A., and Bunce, E. T., et al., Initial Reports of the Deep Sea Drilling Project, Washington, D. C., (U. S. Government Printing Office) 24, 1111-1150, 1974.

Vincent, E., J. M. Gibson, and L. Brun, Paleocene and early Eocene microfacies, benthonic foraminifera, and paleobathymetry of deep sea drilling project sites 236 and 237, Western Indian Ocean, Deep Sea Drilling Project, Leg 24, in Fisher, R. L., and Bunce, E. T., et al., Initial Reports of the Deep Sea Drilling Project, Washington, D. C., (U. S. Government Printing Office), 24, 859-886, 1974.

Vogt, P. R., and O. E. Avery, Detailed magnetic surveys in the northeast Atlantic and Labrador Sea, J. Geophys. Res., 79, no. 2, 363-389, 1974.

Watkins, N. D., B. M. Gunn, J. Nougier, and A. K. Baksi, Kerguelen: continental fragment or oceanic island?, Bull. Geol. Soc. Amer., 85, 201-212, 1974.

Weissel, J. K., and D. E. Hayes, Magnetic anomalies in the southeast Indian Ocean, in Antarctic Oceanology II, the Australian-New Zealand sector, edited by D. E. Hayes, A.G.U., Washington, D. C., 1972.

Weissel, J. K., and D. E. Hayes, The Australian-Antarctic discordance: new results and implications, J. Geophys. Res., 79, no. 17, 2579-2587, 1974.

Whitmarsh, R. B., Some aspects of plate tectonics in the Arabian Sea, in Whitmarsh, R. B., Weser, O. E., Ross, D. A., et al., Initial Reports of the Deep Sea Drilling Project, Washington, D. C., (U. S. Government Printing Office), 23, 527-536, 1974a.

Whitmarsh, R. B., Summary of general features of Arabian Sea and Red Sea Cenozoic history based on Leg 23 cores, in Whitmarsh, R. B., Weser, O. E., Ross, D. A., et al., Initial Reports of the Deep Sea Drilling Project, Washington, D. C. (U. S. Government Printing Office), 23, 527-536, 1974b.

Worsley, T., The Cretaceous-Tertiary boundary crust in the ocean, Soc. Econ. Paleont. Mineral.. Spec. Publ., 20, 94-125, 1974.

Appendix I

The Construction of the Paleobathymetric Charts

In this section we outline in detail the basis and process by which we computed the paleobathymetric charts.

Sclater et al. (1971) have shown that for much of the ocean floor younger than 80 m.y.b.p. there is a simple relation between depth and age. Many authors have shown that this relation is compatible with a simple cooling model where new crust is intruded at the ridge axis which cools and contracts as the crust increases in age. Parsons and Sclater (1977) have demonstrated that for crust younger than 60 - 80 m.y.b.p. and older than 1 m.y.b.p. to the relative subsidence of the ocean crust, ΔH, follows within ±300 m the following simple empirical relation

$$\Delta H = 350 \sqrt{t} - 200$$

where t is in millions of years. For older ocean floor the depth follows a relation predicted by the plate model (Parsons and Sclater, et al., 1977). The closeness of the fit to this simple relation is a very strong agreement that the ridge crest topography at least in the range 0-70 m.y.b.p. or 2700 to 5500 m is dominated by simple conductive cooling of the originally hot crust.

The simple depth versus age relation has been obtained by averaging many closely spaced topographic profiles and with some modification probably holds for between 70 and 80 percent of the ocean floor. However, the absolute depth of the relation does not hold for four specific types of ocean crust (1) local topographic relief, (2) aseismic ridges, (3) localized regions of young crust that are anomalously shallow (for example, the Azores High or the Reykjanes Ridge) and (4) areas of crust which appear to have long wavelength topographic anomalies which may correlate with the long wavelength gravity field.

Fortunately, the first two and possibly the third region of anomalous crust may have simple explanations which make them readily adaptable to the normal subsidence relation. McKenzie and Bowin (1976) have demonstrated that most local topographic relief originates at the ridge axis by the construction of excess crust. This relief is fixed into the crust at the spreading center and remains with the crust throughout its history. Thus localized offsets from the subsidence curve for calculating CCD's with time can simply be accounted for by assuming the observed offset has occurred throughout the entire history of the crust. For the aseismic ridges, Detrick et al. (1977) and ourselves in an earlier section have shown that a similar model accounts for the subsidence of these features. However, in this case subsidence normally starts with the aseismic close to or above sea level. The third and fourth type of anomalous regions are much more difficult to model. Vogt and Avery (1974) and Sclater et al. (1977) have argued that the Azores high and the Reykjanes Ridges are localized topographic anomalies with their origin totally within the oceanic plate. If this is the case then these local regions will probably show the same subsidence curve as normal ocean floor except offset by a certain amount. The residual depth anomalies that correlate with the gravity field are much more difficult to model as they are thought to result from sub-lithospheric processes (Anderson et al., 1973; Menard, 1973 and Sclater et al., 1975) and it is not clear how the process responsible for the anomalies will vary with time. Fortunately apart from an area near Crozet Island another between Australia and Antarctica (Weissel and

Hayes, 1974) and the Bengal fan gravity low (Kahle and Talwani, 1973), these features are not widespread in the Indian Ocean and no allowance has been made for them in constructing the paleobathymetric charts.

For the paleobathymetric charts we have assumed that the floor of the Indian Ocean consists of only typical ocean floor and aseismic ridges. From the relation given by Parsons and Sclater (1977) it can be seen that the 3,000 m isobath corresponds to 2 m.y. old ocean crust and the 4,000 m and 5,000 m isobaths to approximately 20 m.y. and 50 m.y. old crust respectively. For the aseismic ridges we have assumed they were created close to sea level at a spreading center and have sunk with the crust to which they are attached.

Having determined the relation between depth and age we assume that this relation has been the same for the past 100 m.y. and use the histories of the relevant plate motions to determine the age of the ocean floor. First the number of plates to be considered in the history are determined and their relative histories designated by a series of rotation parameters which describe their history through time. In the Indian Ocean we assume four plates, Africa, India, Australia and Antarctica. From the rotation parameters there are two simple ways to construct charts of the age of the ocean floor. The first is to digitize the position of the ridge crest or the earthquakes which define the ridge axis and then to rotate these coordinates through the parameters given for the relation rotation of the plates. In this method it is necessary to rotate one plate with respect to another by the rotation prescribed for the given age. The ridge axis, between the two plates, is then rotated by half the angle of rotation and then by the rotations necessary to define the 2, 20, and 50 m.y. isochrons either side of the ridge axis. These rotations give the positions of the ridge crest and the 3,000, 4,000 and 5,000 m isobaths on the given reconstruction (Sclater and McKenzie, 1973; Sclater et al., 1977).

Unfortunately this simple method cannot be applied to the Indian Ocean because of the dramatic change in spreading direction around 36 m.y.b.p. For this paper, we digitized the magnetic lineations in the Indian Ocean and for each reconstruction we attached them to their relevant plate and rotated the plates by the necessary amount to overlap anomalies 13, 22 and 32 (36, 53 and 70 m.y.b.p.). The parameters for the 36 and 70 m.y.b.p. reconstructions are given in McKenzie and Sclater (1971) and for the 53 m.y.b.p. reconstruction they are presented in Table 4a. To determine the paleolatitude we rotated the plates with respect to a fixed Australia and then rotated Australia the distance necesssary to superimpose the paleomagnetic poles for Australia on the North Pole.

We then determined by eye, using the rotated lineations as a guide, the positions of 2, 20 and 50 m.y.b.p. isochrons on the given tectonic chart. For example, on the 36 m.y.b.p. chart the position of the 3,000 m contour was drawn on crust 38 m.y. old (anomaly 15). Anomaly 24 which has an age of 56 m.y.b.p. is 20 m.y. older than the crust at the ridge axis. Thus, the position of anomaly 24 on the 36 m.y.b.p. chart marks the 20 million year old isochron and hence the 4,000 m isobath. Similarly, the 5,000 m isobath on the 36 m.y.b.p. reconstruction is marked by the position of the 85 m.y.b.p. isochron. Anomaly 34 is approximately 80 m.y. old. Thus by extrapolating the age of the ocean floor a little beyond this anomaly it is possible to estimate the position of this isochron.

On the 53 m.y.b.p. reconstruction a position just beyond anomaly 32 (70 m.y.b.p.) was used to obtain the 4,000 m contour. The 5,000 m contour is given by the position of the 103 m.y.b.p. isochron. This was extrapolated from the tectonic history. For the 70 m.y.b.p. recon-

58 SCLATER, ABBOTT, AND THIEDE

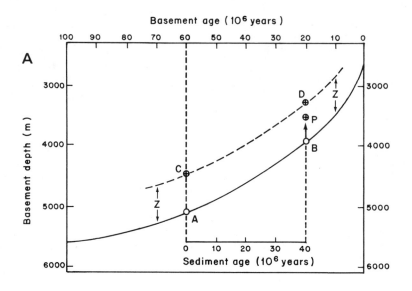

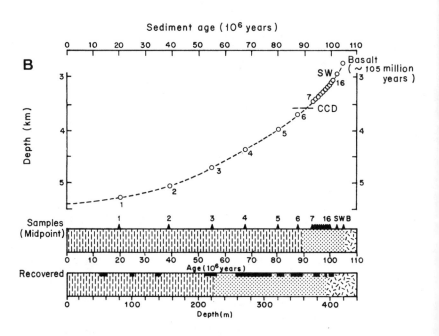

Fig. 10a. Paleodepth determination by vertical backtracking parallel to idealized subsidence track.
A and B: present depth and paleodepth (40 m.y.b.p.) on idealized curve;
C: actual size;
D: analogue to B on parallel curve;
Z: distance between A and C;
P: final paleodepth after correction for isostatic loading (see text) (after Berger and Winterer, 1974).

Fig. 10b. Backtracking of site 137, leg 14 (Hayes et al., 1972). Symbols from left to right: clay, calcareous ooze, basalt. The sample numbers represent core numbers from site 137 (after Berger and Winterer, 1974).

struction the 90 and 120 m.y.b.p. isochrons extrapolated from the tectonic histories were used to determine the 4,000 m and 5,000 m contours.

It is clear from the above analysis that only the 3,000, 4,000 and 5,000 m depths on the 36 m.y.b.p. reconstruction, the 3,000 and 4,000 m depth on the 70 m.y.b.p. reconstruction are known with any certainty.

Appendix II

Backtracking

In order to explain precisely what we did to compute the past depths of the calcium carbonate line we present here a precis of our method. A more complete description can be found in the excellent paper by Berger and Winterer (1974).

To determine the past depth of the carbonate line from Deep Sea Drilling data, we need to know its depth at the time the sediment column passes through the carbonate clay boundary. In order to determine this depth we make two basic assumptions, first, that sea level has not changed by more than 100 to 200 m since the Jurassic and second, that there is a simple relation between depth and age for oceanic crust that also has not changed since the Jurassic. Having made these assumptions we can immediately obtain an estimate of the depth of deposition from the standard subsidence curve if we know the age of basement and the age of the sediment in question.

We assumed the subsidence curve of Sclater et al. (1971) and Parsons and Sclater (1977) starting with 2,700 m as the depth for newly formed sea floor. The age of the basement defines the starting point (A) on the idealized subsidence curve, whereas the age of the sediment determines the distance we have to back track on this curve to find the depth (B) at which the sediment was being deposited (Figure 10a; basement age = 60 m.y.b.p.; sediment age = 40 m.y.b.p.; A = 5,200 m; B = 4,000 m. These depths are slightly shallower than those used by Berger and Winterer (1974). This results from the fact that we have used a slightly shallower subsidence curve). However, for two reasons this method is too simple. First, a particular piece of the sea floor does not follow the absolute paleodepth curve rather it subsides parallel to the curve. Second, we have neglected to correct for the isostatic adjustment of the sediment. To correct for these features we use actual site depth and depth from the bottom of the hole. We relocate the curve parallel to itself by the vertical distance between the actual site depth (C) and the expected depth (A). The subsidence curve now goes through the actual depth (C) and we find a preliminary paleo-depth (D) as before. To compensate for isostatic loading we go downward from point (D) by the total depth of sediment in the hole and then up again by two-thirds of the depth from the bottom of the hole of the sediment in question to obtain point (P). The correction for sediment loading was calculated from the relation given by Sclater, et al. (1971, Appendix III) assuming a mean density of 1.8 g/cm^3 for the sediments. For example for Z = 700 m, a total sediment thickness of 600 m and a depth from the bottom of the hole for 40 m.y.b.p. sediment of 500 m the final paleodepth is 3,550 m (Figure 10a).

CHAPTER 3. SEDIMENTATION IN THE INDIAN OCEAN THROUGH TIME

Thomas A. Davies

Department of Geology, Middlebury College
Middlebury, Vermont

Robert B. Kidd

Institute of Oceanographic Sciences
Wormley, Godalming, Surrey, United Kingdom

Abstract. The nature and distribution of the Mesozoic and Cenozoic sediments of the Indian Ocean, as revealed by deep-sea drilling, are summarized and paleobathymetric reconstructions are used to show probable patterns of sedimentation in Late Cretaceous, Early Eocene and Early Oligocene times. Mesozoic sedimentation was characterized by relatively rapid accumulation of clays in the restricted Wharton and Mozambique basins. Calcareous and terrigenous sediments accumulated around the margins of these basins. In the early Tertiary more open conditions prevailed, but tectonic activity and changing bottom circulation have resulted in a patchy and discontinuous record, especially in the Eocene and early Oligocene. Oligocene to Recent patterns of sedimentation closely resemble those of the present day.

Accumulations of terrigenous sediment in the western and northern parts of the Indian Ocean are clearly related to tectonic activity in the neighboring land areas, and there is a sharp distinction between ocean basins with abundant terrigenous sediment input (Mozambique, Mascarene, Arabian, Somali, northern Central Indian) and those without (South Central Indian, Wharton). Hiatuses in the early Tertiary sedimentary record can be related to changing patterns of bottom circulation associate with glaciation in Antarctica and the subsequent development of strong bottom circulation. The pattern of bottom circulation changed with the development of the present circumpolar circulation, permitting sedimentation to resume in many places.

Introduction

The purpose of this paper is to describe briefly the major features of the sediments of the Indian Ocean, and of the changing patterns of sedimentation through the late Mesozoic and Cenozoic, as revealed by deep-sea drilling at more than 50 locations (Figure 1). We will pay particular attention to some aspects of the sediments which may yield information concerning environmental conditions in the ocean, or in the surrounding regions, at various times in the past. These are: (1) the history of terrigenous sedimentation; (2) the development of sedimentation in different individual basins; and (3) the gross distribution in space and time of hiatuses, or periods of reduced sedimentation, in the sedimentary record. The history of the changing carbonate

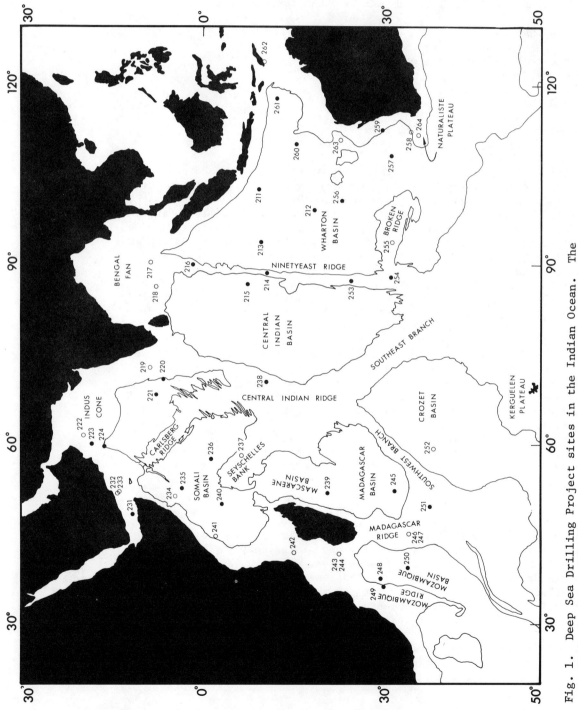

Fig. 1. Deep Sea Drilling Project sites in the Indian Ocean. The principal physiographic features are defined by the 4000 m contour. Solid circles represent sites where basement was reached.

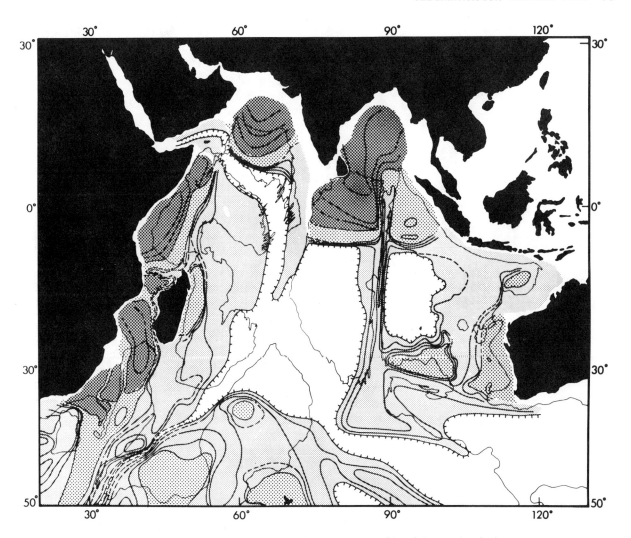

Fig. 2. Sediment thickness in the Indian Ocean as defined by seismic reflection records (modified from Ewing, et al., 1969). Blank = less than 0.1 sec. of sediment; light stipple = 0.1 - 0.5 sec; coarse stipple = 0.5 - 1.0 sec; heavy stipple = greater than 1.0 sec.

compensation depth (CCD), which also has paleo-environmental significance, is considered in the paper on paleo-bathymetry by Sclater et al. (this volume, 1977), along with a discussion of sedimentation on the aseismic ridges and plateaus. Similarly the volcanogenic sediments are the subject of a separate paper by Vallier and Kidd (this volume, 1977).

The broad features of the sediment distribution have been inferred from seismic reflection profiles (Ewing, et al., 1969). The major thicknesses of sediment are found in the Bengal and Indus fans and in the basins off Africa (Figure 2). The controlling influence of physiography on sediment distribution is clearly shown. The southern part of the Central Indian basin and the western Wharton basin, for example, are virtually devoid of sediment because of their isolated positions. The young, active mid-ocean ridge crests are similarly barren on account of their relative youth. It might be noted in

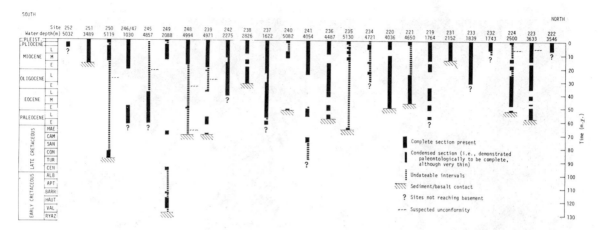

3. Stratigraphic section sampled at sites in the western Indian Ocean. Sites 243 and 244 are omitted since no cores were recovered from these sites. Basement ages at Sites 248 and 249 were determined radiometrically. (From Nature, 253, no. 5486, 15-19).

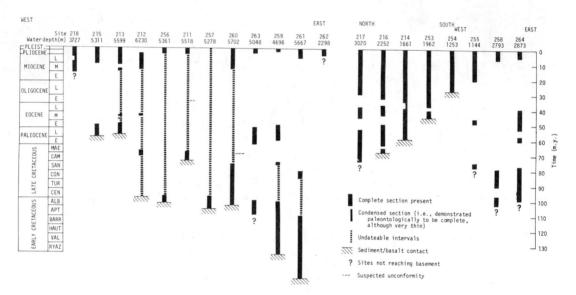

Fig. 4. Stratigraphic section sampled at sites in the eastern Indian Ocean. Ridge and plateau sites (right) and basin sites (left) are separated. The basement age for Site 212 is derived from geophysical evidence. (From Nature, 253, no. 5486, 15-19).

passing that the thicknesses of sediment measured by deep-sea drilling agree well with those inferred from seismic profiles, the obvious differences usually being attributable to the selection of drill sites in atypical locations (Luyendyk and Davies, 1974).

Drilling has shown that, in common with the other oceans, the sediments of the Indian Ocean are patchy and discontinuous in their distribution. In Figures 3 and 4, we have summarized the stratigraphic sections sampled at 48 Indian Ocean sites. Coring at these sites was near continuous in most cases; clearly the record is incomplete at many locations. Gaps in the sedimentary record were encountered in all types of sediment and at sites on a wide variety of topographic features over a broad range of depths. Major hiatuses of ocean wide extent appear to show temporal groupings centered in the Oligocene, Early Tertiary, and Late Cretaceous, and may be interpreted in terms of developing oceanic circulation patterns (Davies, et al., 1975).

Present Day Distribution of Sediments

Figures 5 through 10 show the known distribution of sediments in the Indian Ocean at the present time. For clarity, data from different time periods are shown on different maps. A Mercator projection is used for convenience, and to facilitate comparison with Figure 1 and with other presentations of Indian Ocean data. Data for the maps were plotted by computer, a digital code being assigned to each major sediment type. Provision is made in the coding for mixed lithologies and for intervals represented by several different sediment types. Also indicated on the maps are the partial outlines of the continents, sites with unconformities of appropriate age, and the approximate extent of the ocean crust which had not been formed as of the end of the time interval under consideration. Data for the latter are taken from Larson and Pitman (1972), McKenzie and Sclater (1971), Pitman et al. (1974) and Sclater and Fisher (1974). In assigning ages to the sediments we have accepted the ages stated in the Initial Reports of the Deep Sea Drilling Project.

The sediments are grouped into five broad classes on the basis of composition: terrigenous (clastic detrital), volcanogenic, calcareous, siliceous, and clays. Volcanogenic sediments include both volcanic ashes and other sediments which, according to the criteria of Vallier and Kidd (1977) based on the nature and relative abundances of clay minerals, might be considered volcanic in origin. The clays are subdivided into pelagic ("normal") clays and other ("abnormal") types. This distinction is based primarily on accumulation rate, but color, composition, organic carbon content, and bedding features are also taken into account. What we consider pelagic clays accumulated at rates of less than 5.0 m/m.y., have no bedding features or significant coarse fraction, and have compositions indicative of extremely slow accumulation, including negligible amounts of organic carbon and pyrite. Distinguishing these from other types of clays helps to draw attention to some features of sedimentation in the Indian Ocean which might otherwise be masked.

The data summarized in Figures 5 through 10 are taken from the Initial Reports, and from the lithologic data files at the Deep Sea Drilling Project (DSDP). We will, therefore, make no attempt to describe the sections sampled at the different sites in any detail, but simply draw attention to the main features of the present distribution of sediments in the Indian Ocean.

Mesozoic Sediments

Data concerning the Mesozoic sediments of the Indian Ocean are relatively sparse, since these sediments are found at only a few localities, principally around the ocean margins. The Jurassic is represented by only one site (261), northwest of Australia (Figure 5), where the sediments are predominantly clays with significant amounts of calcareous material near the base. These clays are probably of volcanic origin (Vallier and Kidd, 1977). Although not sampled, we infer from land sections (Brown et al., 1968) that Jurassic carbonates or detrital sediments, and volcanogenic sediments exist along the northwest Australian continental margin.

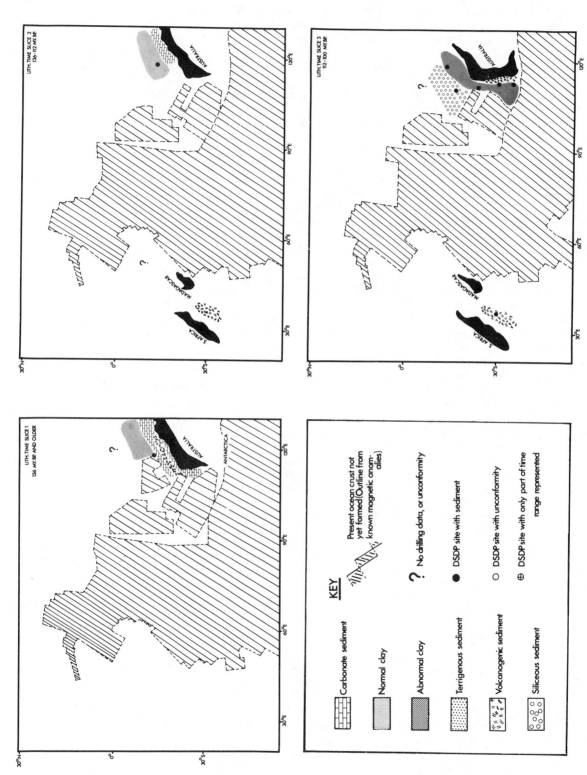

Fig. 5. Present distribution of sediments in the Indian Ocean. Top left: sediments 136 m.y. old and older; top right: 136 – 112 m.y.b.p; bottom right: 112–100 m.y.b.p.

The Early Cretaceous is little better represented than the Jurassic. Volcanogenic sediments are found on the Mozambique Ridge, while off Australia terrigenous detrital clays are seen (Figure 5). These clays contain significant amounts of pyrite, which may indicate that they accumulated under relatively anoxic conditions, although pyrite is a common diagenetic mineral. Farther north (Site 260) siliceous sediments were sampled. Radiolarites of similar age are known from the Carnarvon Basin in northwest Australia (Brown et al., 1968). On Naturaliste Plateau, southwest of Australia, Aptian-Albian detrital sediments derived from the weathering of basaltic rocks occur. These extend upwards into the Late Cretaceous and are probably related to the Cretaceous volcanic episode in the Perth Basin, described by Brown et al. (1968).

Late Cretaceous sediments are more widespread (Figure 6). Around the margins of the Wharton Basin the Late Cretaceous is represented by terrigenous detrital clays, but in the center of the basin these give way to true pelagic clays (see earlier definitions). Late Cretaceous limestones are found on Broken Ridge, and limestones and dolomites at the northern end of Ninetyeast Ridge. Less is known about the Late Cretaceous sediments of the western Indian Ocean. Volcanogenic sediments of early Late Cretaceous age are found on Mozambique Ridge, overlain by calcareous sediments. In the western Somali Basin Late Cretaceous terrigenous sediments, presumably representing the progradation of the African margin, are inferred to exist, on the basis of results from site 242 and from seismic profiles in the region. Detrital clays, similar to the Wharton Basin clays, are found in the Mozambique Basin and off Madagascar. The clays found off Madagascar (Site 239) are rich in montmorillonite implying nearby volcanism (Vallier and Kidd, 1977). Along the margins of the Indian Ocean gaps in the Late Cretaceous record are common (Figure 6). It seems likely that these are related to Early Tertiary tectonic activity and erosion, but the timing of the relationship is difficult, if not impossible, to determine because of the incomplete nature of the overlying Tertiary record (see Davies et al., 1975).

Early Tertiary Sediments

Paleocene calcareous sediments are found on many of the ridges and shallow plateaus of the Indian Ocean, while in the deep basins this time period is represented by pelagic clays (Figures 6 and 7). Off East Africa and the west coast of India Late Paleocene coarse terrigenous sediments and silty detrital clays are found. Volcanogenic sediments have been sampled in the southwest Indian Ocean, on Ninetyeast Ridge, and on Owen Ridge in the northwest. Paleocene sediments are missing from parts of the Australian continental margin, Broken Ridge, and the basins off Africa. This is probably the result of erosion, following late Eocene - Oligocene tectonic activity, and/or changes in bottom circulation patterns.

For the same reason, the Eocene sedimentary record, especially in the western Indian Ocean, is not well preserved. Drilling on the Naturaliste Plateau, for example, suggests that Eocene sediments were deposited over the whole region and subsequently removed by erosion, since only isolated pockets of sediment remain (Luyendyk and Davies, 1974). Even so, a pattern of sediment distribution broadly similar to that for the Paleocene may be discerned. Eocene volcanic ashes on Ninetyeast Ridge are found farther south than their Paleocene counterparts; while to the north the volcanogenic sediments pass upwards into pelagic carbonates. On Broken Ridge, the oldest sediments overlying

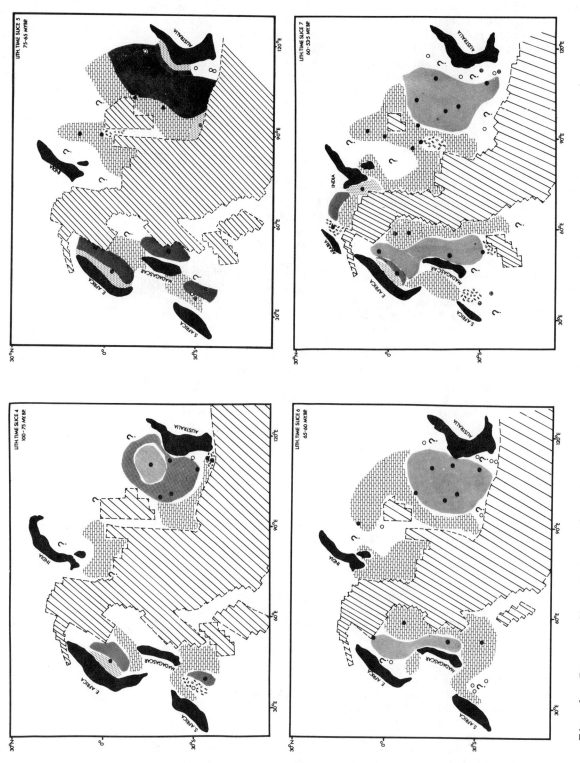

Fig. 6. Present distribution of sediments in the Indian Ocean. Top left: 100-75 m.y.b.p; top right: 75-65 m.y.b.p; bottom left: 65-60 m.y.b.p; bottom right: 60-53.5 m.y.b.p.

SEDIMENTATION THROUGH TIME 69

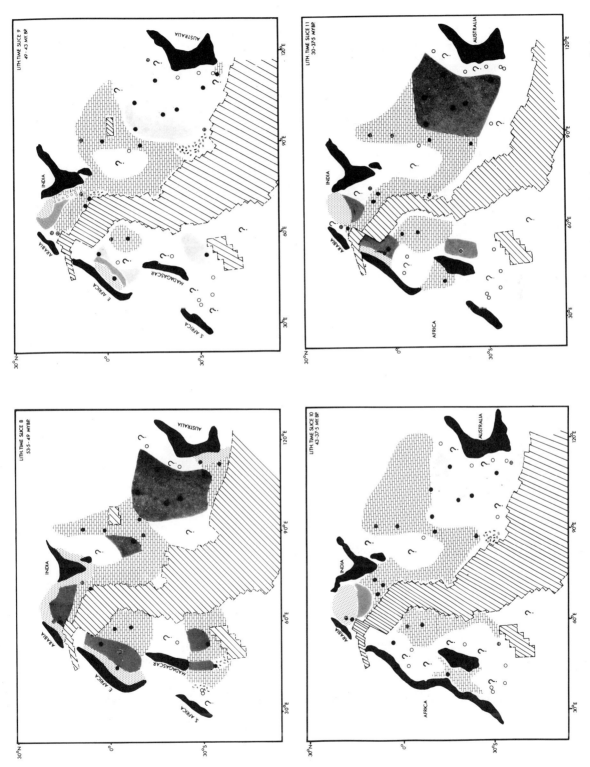

Fig. 7. Present distribution of sediments in the Indian Ocean. Top left: 53.5-49 m.y.b.p; top right: 49-43 m.y.b.p; bottom left: 43-37.5 m.y.b.p; bottom right: 37.5-30 m.y.b.p.

the Cretaceous limestones are very late Eocene littoral gravels. Eocene volcaniclastic sediments are found mixed with carbonates on the Madagascar Ridge and in the Mozambique Basin (Vallier and Kidd, 1977, this volume), and on Owen Ridge. Areas of Early Eocene siliceous sediment flank the coastal regions of India and Arabia. These have given rise to cherts, recognizable as a seismic reflector of regional extent. The siliceous sediments are gradually replaced upwards by Mid to Late Eocene terrigenous sediments. It is probable that these terrigenous sediments are locally derived, and not from the Indus (Kidd, 1975). Possible sources are the river systems of southwest Arabia, or a more extensive Tigris/Euphrates system debouching from the present Persian Gulf area. The Indus must, however, have begun to influence sedimentation in the Arabian Sea by Late Eocene times, since Indus derived materials are a firmly established feature of sedimentation in the area by Oligocene time (Weser, 1974).

The Early Oligocene is perhaps the most poorly represented part of the Early Tertiary. Oligocene sediments are absent from all around the margins of the Indian Ocean and from the shallow ridges and plateaus, with the exception of the Mascarene and Chagos-Laccadive platforms (Davies et al., 1975). Despite the incompleteness of the Oligocene record, some observations can be made. At the south end of Ninetyeast Ridge (Site 254) detrital volcanic sands and silts, which accumulated in a lagoonal or littoral environment are found. Other evidence of volcanic activity is recorded at Sites 234 and 238, associated with Carlsberg Ridge volcanism. Northwards on Ninetyeast Ridge, and probably on Kerguelen Plateau and Broken Ridge also, patchy accumulations of pelagic carbonates are found. Oligocene carbonates have also been sampled from the strait between Madagascar and Africa, on the Chagos-Laccadive and Mascarene plateaus, and in the region north of them. Terrigenous sediment is found off the north coast of East Africa and in the northern Arabian Sea. The first Indus-derived sediments appear in the Lower Oligocene section at site 221 (Weser, 1974). Also, it is certain that the build-up of the Bengal Fan, east of India, had commenced by this time. Currary and Moore (1974) have suggested that much of the early sediment supply for this feature came from the Godvari River in southeastern India, which, they believe, began depositing terrigenous sediment in deep basins off the coast as early as the Late Cretaceous. In the deep basins, the Oligocene is represented only by thin, unfossiliferous clays which accumulated at a very slow rate.

Miocene to Recent Sediments

Since the Oligocene the basic pattern of sediment distribution in the Indian Ocean has remained essentially unchanged (Figures 8 through 10). Unconformities raise problems in determining sediment distributions for the Early and Middle Miocene. Two notable developments are the terrigenous accumulations off the Somali coast and off the Zambesi River. In the Lower Miocene, detrital clays extend into the eastern Somali Basin, but pass upwards into carbonate sediments. Since the prograding margin of the East African rise continued to infill the western Somali Basin, it is perhaps surprising that clay sedimentation is not sustained in the east. The terrigenous influence of the Zambesi fan is first recorded in the Middle Miocene, at Site 248, in the northern Mozambique Basin. It is probable, however, that the detrital clays reaching Site 250 (further south) in the Early Miocene are distal products from this same source.

In the northeast, drilling results provide a record of sedimentation on the Bengal fan itself only since the Middle Miocene. How-

SEDIMENTATION THROUGH TIME 71

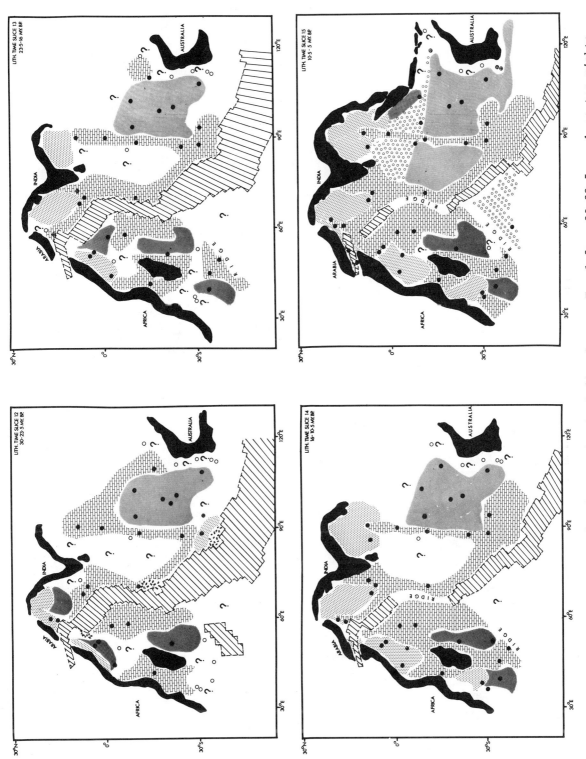

Fig. 8. Present distribution of sediments in the Indian Ocean. Top left: 30-23.5 m.y.b.p; top right: 23.5-16 m.y.b.p; bottom left: 16-10.5 m.y.b.p; bottom right: 10.5-5 m.y.b.p.

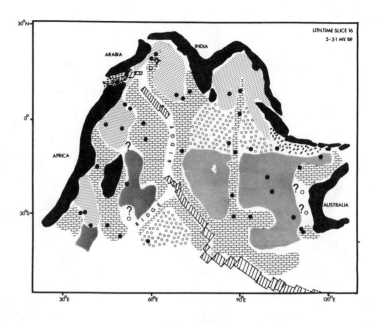
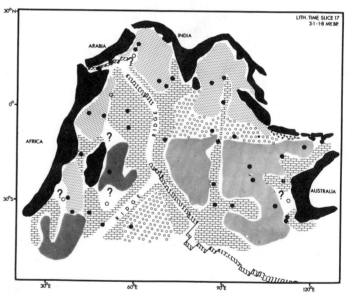

Fig. 9. Present distribution of sediments in the Indian Ocean. Top: 5-3.1 m.y.b.p; bottom: 3.1-1.8 m.y.b.p.

ever, as noted previously, terrigenous sediments have been accumulating in that region since at least the Late Cretaceous. In Miocene time, the influence of the Ganges-Brahmaputra system became dominant west of the northern Ninetyeast Ridge. To the east of the ridge terrigenous sediments from this source continued to accumulate (Site 211), passing through the Nicobar fan, at least until the beginning of the Pleistocene (Pimm, 1974). Active sedimentation on the Nicobar fan probably ceased in the middle Pleistocene. Any terrigenous sediments which subsequently passed between the Ninetyeast and Andaman-Nicobar ridges have accumulated in the northern part of the Sunda-Java Trench.

Drilling in the Gulf of Aden sampled only Middle Miocene to Recent

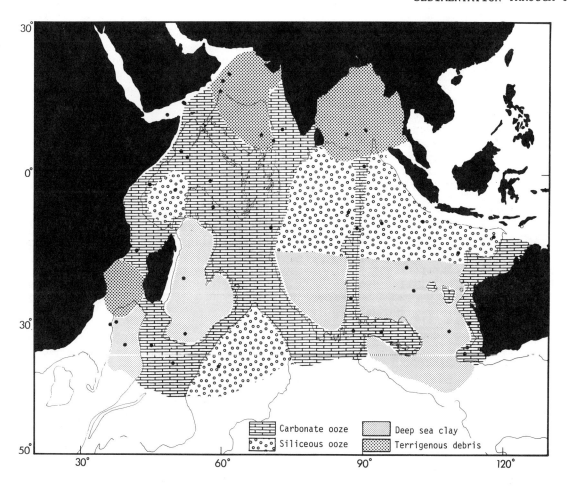

Fig. 10. Distribution of Pleistocene and Recent sediments in the Indian Ocean.

sediments. These were of mixed calcareous and detrital type, and were invariably slumped. The detrital component is predominantly terrigenous, but there is a significant volcanogenic component, especially in the Lower Pliocene.

The present day pattern of sedimentation is shown in Figure 10. Carbonate sediments are accumulating in the shallow areas along the African and Australian coasts, on the shallow ridges and platforms of the western Indian Ocean, in the shallower parts of the Somali Basin, and on Ninetyeast and Broken ridges in the east. On the basis of drilling results the CCD lies deeper than 4000 m, probably around 4500 m, the CCD for much of the present ocean (Berger and Winterer, 1974), and may attain 5000 m in the equatorial zone, especially in the western Indian Ocean. These observations compare favorably with those of Kolla, et al., 1977) who find, on the basis of piston cores, that the carbonate critical depth (CCrD), the depth below which less than 10% carbonate is present, is deepest (5100 m) in the equatorial region and gradually becomes shallower southwards, reaching 3900 m in the area south of 50°S. Below the CCD, the deep basins of the Indian Ocean are receiving deep-sea clays and siliceous sediments, except in the central Wharton Basin (Site 212) and southern Mascarene Basin (Site 235), where it appears that no sediment is accumulating, probably because bottom circulation in these regions is sufficiently strong to inhibit sedimentation. Siliceous sediments are confined to the equatorial high productivity regions and to the sub polar regions of the Crozet Basin.

Thick accumulations of terrigenous sediment are associated with areas where major rivers debouch into the ocean. The erosion of the youthful mountain ranges north of the Indian subcontinent has resulted in the build-up of the Indus and Bengal fans. Other areas of rapid sedimentation are the Zambesi fan, which receives sediments from a wide area of southern Africa, and the western Somali Basin, where the prograding continental margin has built up an immense thickness of sediment. Terrigenous detrital debris is a common component of the sediments of the Mascarene and Mozambique Basins. This is a consequence of the strong currents which sweep the narrow southeast African continent shelf and move sediment to the deep sea (Vincent, 1976). By contrast, the sediments of the Central Indian and Wharton basins consist only of very slowly accumulating pelagic clays, as might be expected from the comparative remoteness of the regions and the fact that there is very little runoff from western Australia to bring terrigenous debris from that continent.

South of the Indonesian island arc is the major development of present day volcanogenic sediment. The arc contains 14% of the world's active volcanoes and, because of their silicic nature, their products are almost entirely pyroclastic. These volcanic ashes are distributed over a wide area of the northeast Indian Ocean. Elsewhere, near regions of active volcanism, volcanogenic materials form a minor, but significant, component of the sediments.

The Evolving Pattern of Sedimentation

The paleobathymetric reconstructions of Sclater, et al. (this volume, 1977) permit us to view the developing pattern of sedimentation in the Indian Ocean in the context of the changing shape and location of the ocean basin, and to make some statements concerning the environmental conditions in the ocean at various stages in its development. The addition of paleobathymetry permits significant refinement over earlier attempts (e.g., Luyendyk and Davies, 1974) to reconstruct the past distribution of ocean sediments.

The Indian Ocean is the youngest of the three major ocean basins, being formed by the separation of Africa, Antarctica, India and Australia. The separation between Africa and Antartica commenced during, or prior to the Early Cretaceous (Luyendyk, 1974), to be followed sometime in Albian-Aptian times by the separation of India from Australia/Antarctica (Sclater and Fisher, 1974). These events may be recorded in the Early Cretaceous volcanogenic sediments of the the Mozambique Ridge and the volcanic detrital sediments of the Naturaliste Plateau, respectively. The separation of India and Australia produced a narrow basin, or series of basins, in which fine-grained terrigenous debris accumulated. The circulation in this basin must have been quite restricted, since the sediments are relatively rich in organic debris and pyrite (although, as noted earlier, pyrite is a common diagenetic product). North of Australia, more open ocean conditions prevailed.

The somewhat meager evidence we have suggests that this pattern of sedimentation persisted throughout the Cretaceous. The pattern of sedimentation at the end of the Cretaceous (Campanian-Maastrichtian) is shown in Figure 11. India lies south of the equator. its southern tip extending south of 30°S, and an almost continuous topographic barrier, formed by Ninetyeast and Broken ridges, extends south and east from the

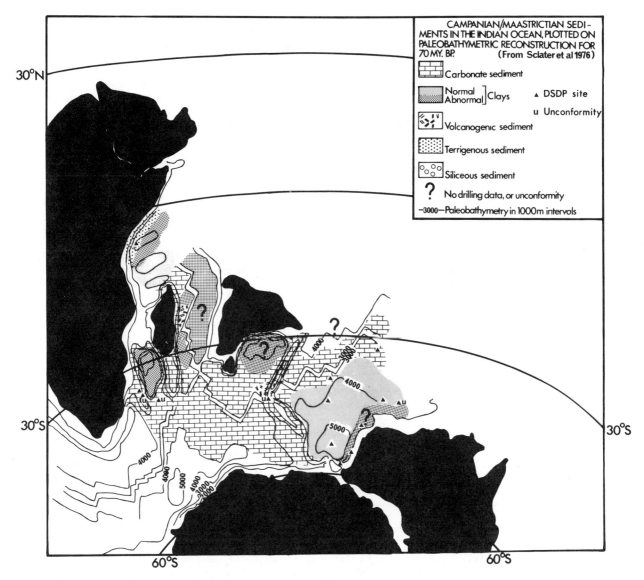

Fig. 11. Campanian-Maastrichtian sediments plotted on a paleobathymetric reconstruction for 70 m.y.b.p. (Reconstruction from Sclater, et al., this volume).

eastern corner of India towards the Australian-Antarctic continent. To the east of this barrier lies the Wharton Basin, now a wide embayment open to the north and reaching depths of over 5000 m in the southeast. In the west of Ninetyeast Ridge is a wide, relatively shallow area bounded by India, Madagascar, Africa and Antarctica. In only one area, between Madagascar and south Africa, are depths greater than 5000 m attained, most of the region lying between 3000 and 4000 m. To the southwest the region opens to the Cape Basin and the embrionic South Atlantic, to the north a narrow strait between India and Madagascar leads to Tethys.

The eastern basin was a region of quiet, deep-water clay accumulation. In the deep parts of this early Wharton Basin, the sedimentary record comprises only very thin, unfossiliferous clays. A zone of siliceous sedimentation may have been present north of the eastern branch of the spreading Indian Ocean Ridge, since Late Cretaceous radiolarites are known from northern India and the Andaman Islands (Stoneley, 1974). This would presumably reflect low latitude (equatorial) upwelling in Southern Tethys. However, there is no evidence from the drilling results for such a zone of siliceous sediments and so it is omitted from Figure 11. On the basis of drilling results, the CCD appears to have been considerably shallower than at present, probably lying between 3000 and 4000 m, since the deepest site at which in situ carbonates were found was Site 211 (Pimm, 1974). (The carbonates at Site 212, an anomalously deep site in a fracture zone, are not believed to be in situ (Pimm, 1974). This figure agrees reasonably well with that given by Sclater et al. (this volume,1977), derived by a slightly different method. It seems likely that volcanic activity was occurring along the line of incipient rifting marked by Broken Ridge and Naturaliste Plateau, although unfortunately Maastrichtian sediments are not preserved at sites drilled on these features. The southern end of Ninetyeast Ridge was the scene of volcanic activity, perhaps volcanic islands, as recorded by the accumulation of volcanic ash and shallow water sediments at site 216. The northern part of Ninetyeast Ridge, and presumably Broken Ridge and the Naturaliste Plateau, formed shallow platforms, a few hundred meters deep, where pelagic carbonates quietly accumulated.

Little is known of Campanian-Maastrichtian sedimentation in the western Indian Ocean. The sediments sampled at site 239, off Madagascar, indicate proximity to volcanic activity. The clays of the Mozambique Basin are (relatively) rich in pyrite and organic carbon indicating restricted circulation. And we speculate that significant amounts of coarse terrigenous detrital sediments were deposited on the East African rise as the margin continued to build out into the western Somali Basin.

By Early Eocene time the division of the Indian Ocean into separate east and west basins was most pronounced (Figure 12). India by now lay athwart the equator and the Ninetyeast Ridge extended southwards over almost 30 degrees of latitutde to the point where it joined the northwest-southeast trending Broken Ridge-Kerguelen Plateau-Naturaliste Plateau complex. Since the oldest sediments overlying the Cretaceous limestones on Broken Ridge are Late Eocene littoral gravels, we must assume that in Early Eocene time Broken Ridge was above sea level, forming a substantial island (Luyendyk and Davies, 1974). East of Ninetyeast Ridge, the Wharton Basin was steadily deepening and enlarging, while remaining open to Tethys and the western Pacific to the north and east. West of Ninetyeast Ridge deep basins had formed off Africa and the East African rise continued its progradation oceanwards. Generally, however, the western Indian Ocean remained shallow, being dominated by the mid ocean ridge and the extensive shallow seas of the Chagos-Laccadive Ridge and Mascarene Plateau. The northern connection between the western basin and Tethys must have been becoming more constricted at this time, due to the steady northward movement of India.

Description of the Eocene pattern of sedimentation is complicated by the occurrence of gaps in the record at sites near the coasts and in the Mascarene Basin. The gaps in the record at sites close to the coast are likely to be attributable, directly or indirectly, to tectonic activity (Davies et al., 1975). The unconformities on Broken Ridge and in the north near Owen Ridge, for example, are clearly the result of tectonic uplift and erosion (Luyendyk and Davies, 1974; Whitmarsh et al., 1974). Similarly, as pointed out earlier, drilling on the Naturaliste Plateau suggests that Eocene sediments were deposited over

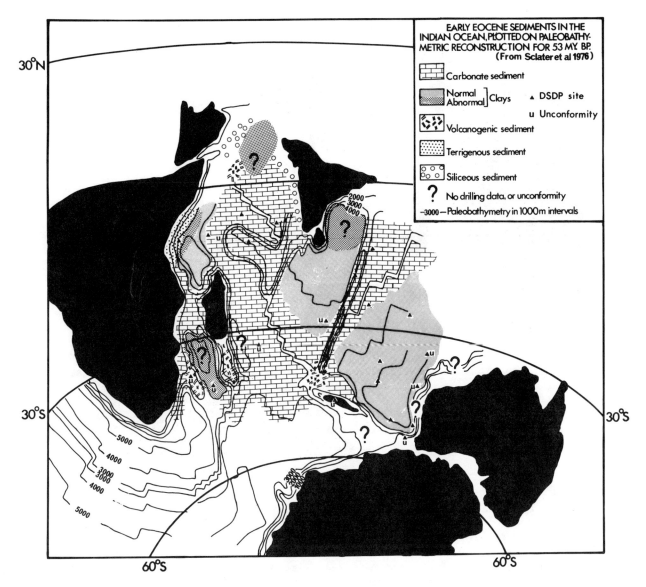

Fig. 12. Early Eocene sediments plotted on a paleobathymetric reconstruction for 53 m.y.b.p. (Reconstruction from Sclater, et al., this volume).

the whole region but subsequently removed by pre-Late Miocene erosion (Luyendyk and Davies, 1974). The gaps in the sedimentary record of the Mascarene Basin could be the result of Late Eocene erosion (see later discussion).

The broad pattern of Early Eocene sedimentation suggests that the CCD was probably still between 3000 and 4000 m, hence, below these depths unfossiliferous clays were slowly accumulting in the Wharton and Central Indian basins. Sediments in the deep basins off Africa differ in having a substantial silty terrigenous component and significant amounts of pyrite, which may suggest some restriction to circulation. On the northern part of the Ninetyeast Ridge, the submerged margins of Broken Ridge and on Naturaliste Plateau and the continental margins

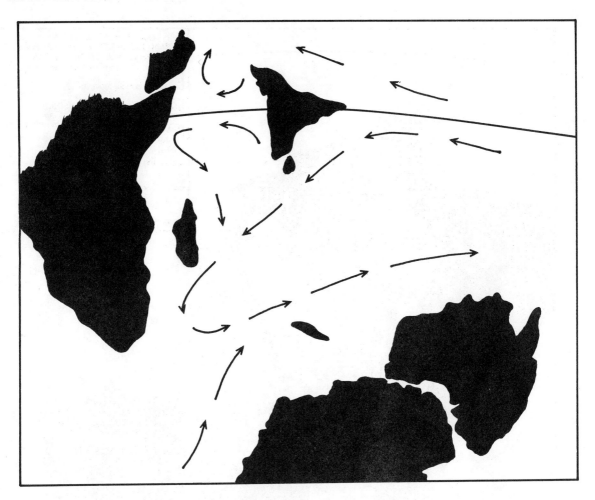

Fig. 13. Possible surface current patterns in the Early Eocene.

of Australia, pelagic carbonates were being deposited. Carbonates were also being laid down on the extensive shallow platform formed by Chagos-Laccadive Ridge and Mascarene Plateau, and in the shallow region to the north. Volcanic activity continued at the southern end of Ninetyeast Ridge, with a great thickness of volcanic ashes accumulating in very shallow water. Volcanic activity was also extensive in the southwestern Indian Ocean and on Owen Ridge. Areas of siliceous sediment flank the coastal regions of Arabia and India, suggesting the establishment of gyral circulation in the Arabian Sea with upwelling off the Indian and Arabian coasts. Figure 13 indicates possible surface current directions. Preservation of siliceous fossils may have been enhanced by the introduction of pyroclastic material derived from the Deccan traps (Vallier and Kidd, this volume, 1977).

Middle and Late Eocene patterns of sedimentation are essentially the same as the Early Eocene, although the record is patchy and discontinuous. Sometime in the Middle to Late Eocene Broken Ridge must have subsided below sea level once more, to permit the deposition of Late Eocene littoral gravels. During the Middle to Late Eocene we see the first major influx of terrigenous debris into the northern Arabian Sea, presumably the result of the closing of the Indus Trough, north of India (Stoneley, 1974) and the uplift of the Himalayas.

The Early Oligocene geography of the Indian Ocean more closely resembled its present day configuration (Figure 14). Three distinct regions can be recognized: northwestern, central, and eastern. The northwestern region forms an almost totally enclosed basin bounded on the west by Africa and Arabia, on the south and east by India and the Chagos-Laccadive Ridge-Mascarene Plateau complex and with limited (if any) access to the Tethyan Region to the north. The dominant feature of this area is the northwest-southeast trending Carlsberg Ridge which effectively divides the basin into two parts. In the south a sill, shallower than 4000 m, separates the Somali Basin from the Mascarene Basin. The central region is bounded on the west by the Chagos-Laccadive Ridge, Mascarene Plateau, Madagascar and the Madagascar Ridge, on the north by India, and on the south and southeast by Ninetyeast Ridge and Kerguelen Plateau. The central area is split by the inverted-Y of the active spreading ridges. East of Ninetyeast Ridge and north of Broken Ridge the Wharton Basin forms a wide, deep basin with a rapidly developing seaway extending to the southeast between Australia and Antarctica. To the north the Wharton Basin appears still to have extensive communication with the western Pacific region. Separation between Australia and Antarctica had commenced at the beginning of the Eocene, and by Early Oligocene time Antarctica had moved south to its present position over the south pole.

Figure 14 shows the pattern of sedimentation in the Early Oligocene. Apart from the extensive unconformities, the most noticeable features are the large amounts of terrigenous debris entering the northern parts of the ocean. These are clearly the consequence of the collision between Asia and India. Volcanic activity was continuing at the southern end of Ninetyeast Ridge, associated with the separation between the ridge and Kerguelen Plateau. The CCD at this time was probably still at about 4000 m. Carbonates accumulated at deeper levels in the Somali and Arabian basins, but these appear to have been displaced from shallower depths (Thiede, 1974).

Since the Oligocene the basic elements of the Indian Ocean paleogeography have remained essentially unchanged. The major new developments were: 1) rapid spreading along the Southeast Indian Ridge causing Australia to move northwards through about twenty degrees of latitude; 2) creation of the Sunda Arc brought about by subduction of the Indian Plate under the Eurasion Plate; and 3) continued movement of India northwards and opening of the Gulf of Aden, resulting in the destruction of eastern Tethys (Stoneley, 1974). As a consequence of the northward movement of Australia, wide areas of carbonate sediment formed on the flanks of the Southeast Indian Ridge, and deep basins formed off Australia and Antarctica. Carbonate accumulation was further enhanced by a general lowering of the CCD (Sclater et al., 1977).
This would, perhaps, explain why the Early Miocene terrigenous clays of the eastern Somali Basin appear to give way upwards to carbonate sediments. Subduction south of Indonesia caused the Plio-Pleistocene build-up of volcanogenic sediments in that area. Closure of the northern Arabian Sea and the progressive development of the Indus Fan meant that terrigenous sediments entirely dominated Miocene to Recent sedimentation north of the Carlsberg Ridge.

Siliceous sediments in the Wharton and Crozet basins are an obvious feature of the Late Miocene and Pliocene distributions, as at the present day. From this one can suppose that oceanographic conditions, and in particular circulation, must have been similar to the present over the past 10 m.y. Figure 15 shows the present pattern of sedimentation on the same projection as the earlier paleogeographic reconstructions, for ease of comparison.

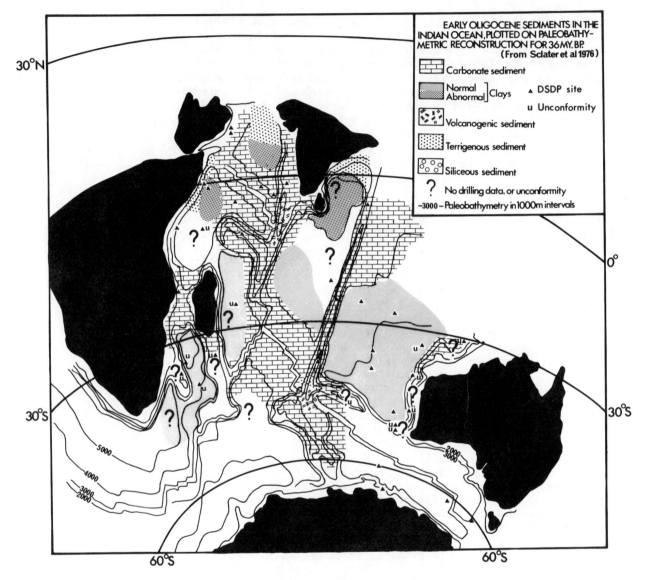

Fig. 14. Early Oligocene sediments plotted on a paleobathymetric reconstruction for 36 m.y.b.p. (Reconstruction from Sclater et al., this volume).

Discussion

In the foregoing sections of this paper we have summarized what is known of the present distribution of Mesozoic and Cenozoic sediments in the Indian Ocean, and used paleobathymetric reconstructions to show the evolving pattern of sedimentation in the Indian Ocean. We now return to some features of the sediments which seem of particular interest from the point of view of paleo-environmental studies.

Terrigenous Sedimentation

Terrigenous (clastic detrital) sediments represent the mechanical load of rivers. Since there is a correlation between mechanical load

SEDIMENTATION THROUGH TIME 81

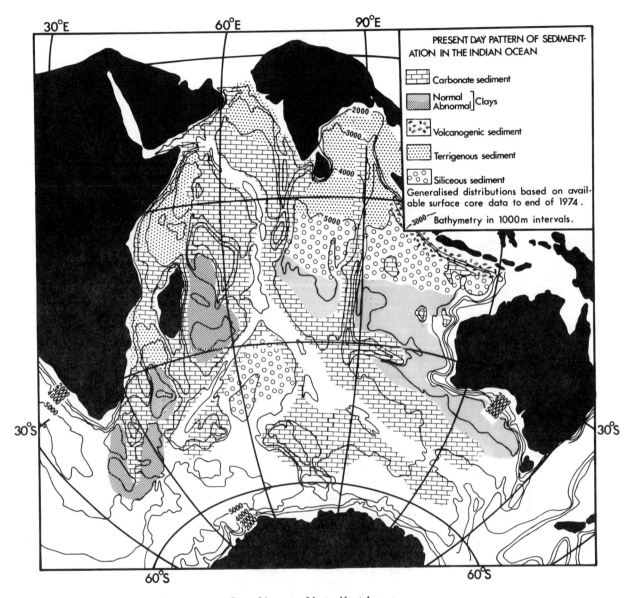

Fig. 15. The present pattern of sediment distribution.

and mean continental elevation (Garrels and Mackenzie, 1971), thick accumulations of terrigenous sediment can be considered indicative of tectonic uplift and erosion of neighboring land areas. This is well demonstrated in the history of terrigenous sedimentation in the Indian Ocean.

The principal accumulations of terrigenous sediment in the Indian Ocean lie along the western and northern margins. With the exception of some volcanic detrital sediments on the Naturaliste Plateau, which can be related to Cretaceous volcanism in southwest Australia, comparatively little terrigenous sediment is found in the eastern part of the ocean, as might be expected since Australia has been a flat, low lying landmass since early in the Mesozoic (King, 1962). The four major accumulations of terrigenous sediment are clearly associated with major tectonic events. Thus terrigenous sediments first appear in abundance in the Arabian Sea during the Eocene, soon after the first contact of India and Pakistan (Stoneley, 1974), and floods of sediment begin rapidly building the Indus and later Bengal Fans during the Oligocene and Miocene, following the closure of the Indus trough and subsequent uplift of the Himalayas. The thick terrigenous sediments of the prograding continental margin off the Somali coast are associated with a long history of tectonic activity in central East Africa, while the rapid growth of the Zambesi Fan from mid Miocene times onwards is surely related to Neogene epirogenic uplift and tilting of southern and central Africa (King, 1962). The close connection between tectonic events in the East African hinterland and rates of sediment accumulation in the Somali and Mozambique basins has been discussed in some detail by Girdley, et al. (1974).

Basin Sedimentation

The supply of terrigenous sediment has clearly been the major influence on sedimentation in the different basins. Figure 2 shows a clear contrast between basins with an abundant supply of terrigenous sediment (Mozambique, Mascarene, Somali, Arabian, northern Central Indian) and basins which are relatively starved (southern Central Indian, Wharton). Most of the basins are deep enough that the pelagic components have been only pelagic clays since early in their histories. Exceptions are the Crozet Basin, and equatorial parts of the Wharton, Central Indian, and southern Somali basins, where siliceous components associated with the equatorial and subantarctic high productivity regions appear.

During the early history of many of the deep basins conditions of restricted circulation accompanied by rapid accumulation of fine terrigenous debris ("abnormal" clays) prevailed. These conditions eventually gave way to more open circulation and reduced rates of clay accumulation.

Hiatuses and Ocean Circulation

Perhaps the most interesting aspect of the history of sedimentation in the Indian Ocean is the widespread occurrence of periods of greatly reduced or total lack of sedimentation over areas which are ocean wide in extent. Whether erosion, nondeposition or sediment accumulation occurs is determined by the balance between the rate at which sediment is supplied and the rate at which it is removed (van Andel, et al., 1975). Both supply and removal are brought about by a variety of processes. Certainly gaps in the sedimentary record in many places, especially around the margins of the ocean, can be attributed to tectonic activity, either directly (i.e uplift and erosion), or indirectly as a result of tectonic events interrupting the supply of sediment (Kent, 1974). In regions remote from land, however, where the sediment is predominantly biogenic, the rate of supply is largely a function of surface productivity, while the rate of removal is determined by the strength of bottom currents and by the propensity of cold deep water to dissolve calcite and silica. The distribution in space and time of hiatuses and periods of reduced sedimentation therefore yields valuable information concerning the development of oceanic circulation.

Gaps in the sedimentary record are most prevalent in the Early Oligocene. They are found all around the margins of the Indian Ocean and on the shallow ridges and plateaus, with the notable exception of the Mascarene and Chagos-Laccadive platforms (Davies, et al., 1975). At times corresponding to these gaps the deep basin sediments frequently

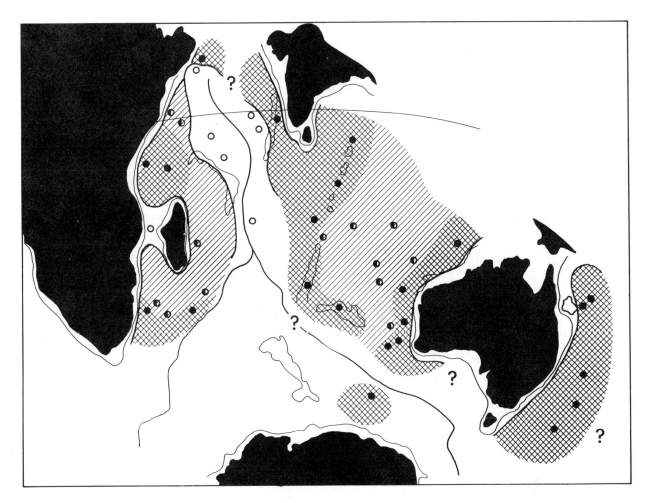

Fig. 16. DSDP sites plotted on an Early Oligocene reconstruction. Open circles represent sites where the sedimentary section is complete; solid circles where there is a proven unconformity; and half circles where there is an undated interval or an inferred unconformity. The shading indicates the extent of the Oligocene unconformity, crosshatching showing the extent of the proven unconformity. (From Nature, 253, no. 5486, 15-19).

show evidence of reduced sedimentation rates. This has been ascribed to the activities of ocean currents (Leclaire, 1974; Davies, et al., 1975).

At the present time cold 'aggressive' bottom water forms in the Antarctic shelf regions and spreads north into the southwest Indian Ocean, where it behaves as a western boundary undercurrent. It also drifts north through fracture zones in the Southeast Indian Ridge and passes through the gap between Broken Ridge and the Naturaliste Plateau into the Wharton Basin. This pattern of distribution is mirrored by the regional development of the Oligocene hiatus (Figure 16) and suggests a causal relationship. There is evidence that as Antarctica moved to its present polar position during the Eocene there was a general deterioration of climate, ultimately resulting, near the end of the

Eocene, or beginning of the Oligocene, in glaciation, and the production of copious amounts of cold 'aggressive' bottom water (Kennett, et al., 1974). Until deep circumpolar circulation was established in the Late Oligocene (Kennett, et al., 1974), this bottom water must have moved northwards into the Indian Ocean, establishing vigorous bottom circulation. This inhibited sedimentation and caused erosion in many places. Once circumpolar circulation was established the intensity of bottom circulation in the Indian Ocean probably decreased somewhat, permitting sedimentation to resume in most places.

A similar scenario could probably be developed to explain the occurrence of extensive Paleogene and Late Mesozoic hiatuses (Davies, et al., 1975). However, the evidence for the regional extent of these is too ambiguous at present to merit a detailed exposition.

Acknowledgements. We would like to emphasize that in preparing this short summary we have drawn heavily upon the observations of our colleagues, the scientific staffs of the GLOMAR CHALLENGER. We have attempted to give appropriate recognition in every case and apologize for any omissions which may have crept in. We would like to thank J. G. Sclater and his co-workers for kindly making available paleobathymetric reconstructions of the region and P. B. Woodbury and our colleagues of the DSDP Computer Group for assistance in plotting the lithologic data. R. B. Kidd acknowledges the help of Dr. M.N.A. Peterson (DSDP) and Professor H. Charnock (IOS) in arranging his secondment to the DSDP Headquarters and West Coast repository, La Jolla, in order to study core materials stored there.

References

Berger, W. H. and E. L. Winterer, Plate stratigraphy and the fluctuating carbonate line, in Pelagic sediments on land and under the sea edited by K. J. Hsü and H. C. Jenkyns, Intern. Assoc. Sed. Spec. Publ. (Blackwell Scientific Publications) 1, 11-48, 1974.

Brown, D. A., K. S. W. Campbell, and K. A. W. Crook, The geological evolution of Australia and New Zealand, London (Pergamon Press) 409 pages, 1968.

Curray, J. R., and D. G. Moore, Sedimentary and tectonic processes in the Bengal deep-sea fan and geosyncline, in Geology of Continental Margins, edited by C. A. Burk, C. L. Drake, New York (Springer-Verlag), 617-628, 1974.

Davies, T. A., O. E. Weser, B. P. Luyendyk and R. B. Kidd, Unconformities in the sediments of the Indian Ocean, Nature, 253, 15-19, 1975.

Ewing, M., S. Eittreim, M. Truchan, and J. I. Ewing, Sediment distribution in the Indian Ocean, Deep Sea Res., 16, 231-248, 1969.

Garrels, R., and F. T. Mackenzie, Evolution of the sedimentary rocks, New York (Norton), 397 pages, 1971.

Girdley, W. A., L. Leclaire, C. Moore, T. L. Vallier, and S. M. White, Lithologic summary, leg 25, Deep Sea Project, in Simpson, et al., Initial Reports of the Deep Sea Drilling Project, Washington (U. S. Government Printing Office), 25, 725-759, 1974.

Kennett, J. P., et al., Development of the Circum-Antarctic Current, Science, 186, 144-147, 1974.

Kent, P. E., Leg 25 results in relation to East African coastal stratigraphy, in Simpson, et al., Initial Reports of the Deep Sea Drilling Project, Washington (U. S. Government Printing Office), 25, 679-684, 1974.

Kidd, R. B., Sedimentary processes in the development of the Indus submarine fan, in *Reports Volume of the IX International Congress of Sedimentology*, Nice, France, 1975.

King, L., *The morphology of the earth*, New York (Hafner), 699 pages, 1962.

Kolla, V., A. Be, and P. E. Biscaye, Calcium carbonate distribution in the surface sediments of the Indian Ocean, *J. Geophys. Res.*, 1977.

Larson, R. L., and W. C. Pitman III, World wide correlation of Mesozoic magnetic anomalies and its implications, *Geol. Soc. Amer. Bull.*, 83, 3645-3662, 1972.

Leclaire, L., Late Cretaceous and Cenozoic pelagic deposits--paleoenvironment and paleo-oceanography of the central western Indian Ocean, in Simpson, et al., *Initial Reports of the Deep Sea Drilling Project*, Washington (U. S. Government Printing Office), 25, 481-512, 1974.

Luyendyk, B. P., Gondwanaland dispersal and the early formation of the Indian Ocean, in Davies, et al., *Initial Reports of the Deep Sea Drilling Project*, Washington (U. S. Government Printing Office), 26, 945-952, 1974.

Luyendyk, B. P., and T. A. Davies, Results of DSDP leg 26 and the geologic history of the southern Indian Ocean, in Davies, *et al.*, *Initial Reports of the Deep Sea Drilling Project*, Washington (U. S. Government Printing Office), 26, 909-943, 1974.

McKenzie, D. P., and J. G. Sclater, The evolution of the Indian Ocean since the Late Cretaceous, *Geophys. J. Roy. Astr. Soc.*, 25, 437-528, 1971.

Pimm, A. C., Sedimentology and history of the northeastern Indian Ocean from Late Cretaceous to Recent, in von der Borch, et al., *Initial Reports of the Deep Sea Drilling Project*, Washington (U. S. Government Printing Office), 22, 717-803, 1974.

Pitman, W. C. III, R. L. Larson, and E. M. Herron, *The age of the ocean basins*, Boulder (Geol. Soc. Amer.), (Map), 1974.

Sclater, J. G., and R. L. Fisher, The evolution of the east central Indian Ocean, *Geol. Soc. Amer. Bull.*, 85, 683-702, 1974.

Sclater, J. G., D. Abbott, and J. Thiede, Paleobathymetry and sediments of the Indian Ocean, in *Indian Ocean Geology and Biostratigraphy*, Washington, (AGU), 1977.

Stoneley, R., Evolution of the continental margins bounding a former southern Tethys, in *Geology of Continental Margins*, edited by C. A. Burk and C. L. Drake, New York, (Springer-Verlag), 889-906, 1974.

Thiede, J., Sediment coarse fractions from the western Indian Ocean and Gulf of Aden (Deep Sea Drilling Project Leg 24), in Fisher, et al., *Initial Reports of the Deep Sea Drilling Project*, Washington (U. S. Government Printing Office), 24, 651-765, 1974.

Vallier, T. L., and R. B. Kidd, Volcanogenic sediments in the Indian Ocean, in *Indian Ocean Geology and Biostratigraphy*, Washington, (AGU), 1977.

van Andel, Tj. H., G. R. Heath, and T. C. Moore, Cenozoic history and paleooceanography of the central equatorial Pacific Ocean, *Geol. Soc. Amer. Memoir* 143, 1975.

Vincent, E., Planktonic foraminifera sediments and oceanography of the Late Quaternary, Southwest Indian Ocean, *Allen Hancock Foundation Monograph 9*, Los Angeles, (University of Southern California Press), 235 pages, 1976.

Weser, O. E., Sedimentological aspects of strata encountered on leg 23 in the northern Arabian Sea, in Whitmarsh, R. B., et al., *Initial Reports of the Deep Sea Drilling Project*, Washington (U. S. Government Printing Office), 23, 503-519, 1974.

Whitmarsh, R. B., O. E. Weser, D. A. Ross, et al., *Initial Reports of the Deep Sea Drilling Project*, Washington (U. S. Government Printing Office), 23, 1180 pages, 1974.

CHAPTER 4. VOLCANOGENIC SEDIMENTS IN THE INDIAN OCEAN

Tracy L. Vallier

U. S. Geological Survey
Menlo Park, California

Robert B. Kidd

Institute of Oceanographic Sciences
Wormley, Godalming, Surrey, United Kingdom

Abstract. Volcanogenic contributions to sediments in the Indian Ocean are significant not only volumetrically but also as indicators of tectonic events. Major accumulations are related to ocean basin tectonic events such as initial stages of rifting, high spreading rates, the formation of presently aseismic ridges, and to landmass volcanism, some of which is associated with subduction. The composition of volcanism changed with time from basaltic during late Mesozoic and early Cenozoic tensional activity to silicic during the Late Cenozoic as subduction occurred beneath the Indonesian Arc. There was a causal relationship between early stages of sea floor spreading and quantities of volcanogenic sediments; thick basal piles of volcanogenic sediments accumulated during early parts of ocean basin evolution. Volcanogenic sediments on presently aseismic ridges occur both in thick basal piles and in beds higher within the sediment columns. On aseismic ridges, basal sediments probably are related to the initial formation which occurred at the intersection of a mid-ocean ridge and a transform fault, and beds higher in the sediment columns may be related to uplift which formed the present ridge topography.

Introduction

As a result of the Deep Sea Drilling Project, scientists have had better opportunities to study time and space relationships of oceanic sediments and to construct tectonic and sedimentologic models that help explain geologic histories of the ocean basins. Volcanogenic sediments are important contributors to marine sediment accumulations, particularly near island arcs where volcanic sediment wedges can be several thousand meters thick. They also can be significant contributors to oceanic sediments far from volcanic centers where their abundance, however, is generally much less than biogenic and other non-biogenic sediment contributors. Their importance as contributors to the sedimentary column and as probable indicators of tectonic events has long been suspected; even so, it is often difficult to determine the amount of volcanic detritus in marine sediments because (1) it frequently is not recognized because of masking by other sediments, and (2) it is easily affected by diagenetic processes. Consequently, it is probable that the importance of volcanogenic contributions has been underestimated.

Attempts have been made to relate the history of volcanism to local or regional tectonics. For example, McBirney (1971) noted that volcanism occurred at about the same time over large areas throughout the Mesozoic and Cenozoic irrespective of the local setting. Kennett and Thunell (1975) and Ninkovich (1977) are among the latest authors who related volcanism to tectonic events. Kennett and Thunell (op. cit.) concluded that explosive volcanism has increased dramatically on a global basis during the last two million years which correlates with widespread synchronism in increased Quaternary orogenesis. They further believed that episodes of volcanism require episodes in sea floor spreading, and thereby implied that rapid spreading and changes in spreading rates and directions can trigger volcanic events.

This paper describes the time and space relationships of volcanogenic sediments in the Indian Ocean and attempts correlations of major accumulations with tectonic events such as the initial opening of the ocean, the formation of aseismic ridges, and rates of sea floor spreading. Also, a relationship with landmass volcanism is attempted.

The term volcanogenic component is given to those materials, both primary and secondary, that are the result of volcanic activity. The products can be pyroclastic, epiclastic, and authigenic in origin and can consist of primary volcanic materials such as volcanic glass, feldspar, pyroxene, etc. and/or of secondary minerals such as smectite, palagonite, palygorskite, zeolites, and silica minerals.

Most data presented were collected during the six legs (22-27) of the Deep Sea Drilling Project in the Indian Ocean, plus site 264 from Leg 28 on the Naturaliste Plateau (Figure 1). Site report chapters in the Initial Report volumes have been useful during this compilation and the following articles from those volumes have been particularly helpful: Coleman (1974), Cook et al. (1975), Cronan (1974), Fleet and Kempe (1974), Matti et al. (1974), McKelvey and Fleet (1974), Moore et al. (1974) Pimm (1974), Robinson et al. (1974), Schlich (1974), Thiede (1974), Vallier (1974), Venkatarathnam (1974), Warner and Gieskes (1974) and Weser (1974). Data were updated by reinvestigations of smear slide and X-ray results. Smear slide data are semiquantitative at best and the reader should be cautious about some investigations based entirely on smear slide results. However, the combination of these with X-ray data can be extremely useful.

The accuracy of the quantitative X-ray diffraction data is primarily a function of the precision of the method used and the degree of similarity between the minerals in the samples and the calibration standards (Cook et al., 1975, p. 1004-1006). The reliability of the identification of the minerals reported is considered to be very good. Actual percentage concentrations of the minerals are actually ratios of quantifiable portion of the sample where the total is normalized to 100%. The nonquantifiable part of the sample consists of amorphous matter, any mineral group for which adequate standards are not available, and the failure to identify the mineral. Thus, a sample which actually contains 5% quartz, 2% mica, 43% an unidentified mineral, and 50% amorphous silica is reported as 71.8% quartz, 28.2% mica, an unidentified mineral as a major component, and an amorphous value of 50%. The actual percentage figures are subject to several inherent inaccuracies mainly due to the differences in crystal structures and the degree of crystallinity between minerals in sediments and mineral calibration standards. This is discussed in more detail by Cook et al. (1975).

The concentration trends, however, are regarded as being highly reliable especially within a limited geologic setting. Thus, the data are valuable for preliminary studied and synthesis. Percentages calculated from relative amounts of clay minerals are used in this report.

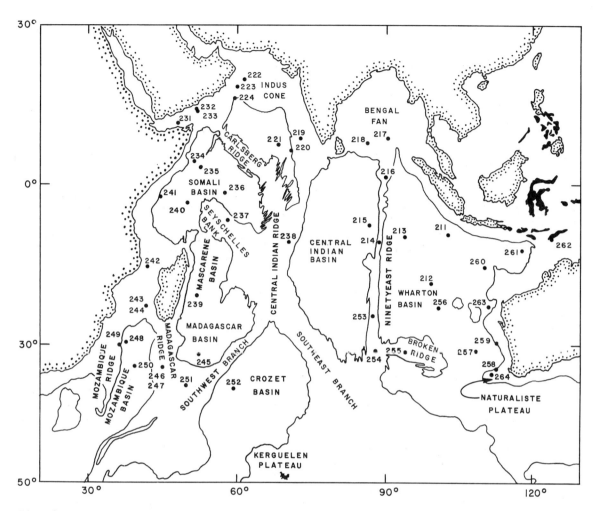

Fig. 1 Deep Sea Drilling Project sites in the Indian Ocean.

Volcanic glass and primary volcanic minerals that are diagnostic of a specific volcanic rock type, e.g. olivine, augite, and calcic plagioclase for basalt, give the most useful information for determining volcanic contributions to the non-carbonate parts of the sediments. In altered volcanogenic sediments, major, minor, and trace mineral chemistries can assist interpretations (Donnelly, personal comm.). Clays, zeolites, and members of the silica mineral group, when combined with other mineralogic, stratigraphic and chemical data, also can be helpful. In this study we have relied heavily on clay mineral content to evaluate volcanic contributions in sediment columns that have undergone mineralogic changes during diagenesis. To assist in evaluating the clay mineral contributions, we have used two percentages that are based on data from the <2µ carbonate-free X-ray data. The first of these is the montmorillonite percentage which is calculated from the clay percentages as follows:

$$\frac{\text{montmorillonite(M)}}{\text{montmorillonite(M) + palygorskite (P) + kaolinite(K) + illite(I) + chlorite(C)}} \times 100 = \%M$$

the second is the montmorillonite plus palygorskite percentage which is calculated as follows:

$$\frac{\text{montmorillonite(M)} + \text{palygorskite(P)}}{\text{montmorillonite(M)} + \text{palygorskite(P)} + \text{kaolinite(K)} + \text{illite(I)} + \text{chlorite(C)}} \times 100 = \%(M+P)$$

Generally, when these percentages exceed 50% and 75% respectively, they signify probable volcanic contributions to the sample. If they exceed 90%, a strong volcanic input is suggested. Other mineralogic and stratigraphic indicators are combined with these percentages to interpret the origin of the non-carbonate parts of the sediments columns.

No account has been taken of sedimentation rates in these calculations. A high montmorillonite content in sediments deposited at 10 m/m.y., for instance, is relatively more significant than a high content in sediments deposited at a rate of 1 m/m.y. Such a technique is useful in sequences that can be adequately age dated, but is of questionable value in non-fossiliferous columns, some of which contain hiatuses.

Volcanogenic Components in Marine Sediments

The amount of volcanogenic contributions in marine sediments probably has been underestimated because they are difficult to distinguish from other sedimentary commmponents unless the volcanic components are in discrete beds. For example, if terrigenous and/or biogenous debris are abundant, the volcanogenic contributions can be masked by the relative flood of these other components. Furthermore, fresh volcanic sediments are particularly unstable in the marine realm and are susceptible to diagenetic processes which result in the physical breakdown of pumice and shard shapes, an obliteration of the original euhedral forms of minerals, and the formation of authigenic minerals such as smectites, zeolites, and silica minerals. Additional quantities of such minerals can be added to the marine realm through erosion of landmass outcrops and by diagenesis of nonvolcanic components in the marine environment; therefore, their presence in a sample does not necessarily mean that there was a direct volcanic contribution to the sediment. It is for these reasons that a volcanogenic origin of a sample that contains smectites, zeolites, and silica minerals should be assigned only after the stratigraphic relationships, such as discrete beds of glass or altered ash, and the abundances of other components are evaluated.

Primary volcanogenic components are the direct result of volcanic activity. The major primary components are glass, lava fragments, feldspar, pyroxene, amphibole, olivine, quartz, iron oxide minerals, mica (biotite and muscovite), and rock fragments. Secondary volcanogenic components include such authigenic components as palagonite, members of the smectite group (e.g., montmorillonite), zeolites (predominantly phillipsite and clinoptilolite), palygorskite, and the silica minerals cristobalite, tridymite, quartz, and opal.

Primary volcanogenic components are not discussed further, but secondary ones warrant additional attention because a volcanic origin is given to some sequences in DSDP holes of the Indian Ocean that are composed almost entirely of them. Montmorillonite (smectite) is believed to form from the alteration of volcanic materials, particularly magnesium-rich volcanic glass and palagonite (Peterson and Griffin, 1964; Nayudu, 1964; von Rad and Rosch, 1972) and Bradshaw (1975) concluded that Jurassic montmorillonite in eastern England is the result of alteration of air-fall volcanic ash. Montmorillonite can form in many environments (Millot, 1970), but we believe that in deep marine sediments where montmorillonite is the only clay mineral present in a sample, a volcanic origin can be assigned to the clay-sized portion of that sample. If the sample is rich in montmorillonite but contains abundant detrital clays such as kaolinite, chlorite, and illite, then all or part of the montmorillonite also may be detrital.

The clay mineral palygorskite apparently can form either as a diagenetic product of volcanic materials in the marine realm or in non-volcanic continental areas such as northern Africa and Arabia (Hathaway and Sachs, 1965, Siever and Kastner, 1967; Bonatti and Joensuu, 1968; Millot et al., 1969; Hathaway et al., 1970; and von Rad and Rosch, 1972). A dual origin of this mineral is recognized in sediments of the Indian Ocean. Some is eroded from non-volcanic terranes in Africa and Arabia, whereas other palygorskite probably forms diagenetically. It is thought that marine diagenetic palygorskite is derived from degraded montmorillonite clays by precipitation from mangesium-rich solutions containing excess silica that was derived from the devitrification of volcanic ash and/or the dissolution of opaline fossils (von Rad and Rosch, 1972). If palygorskite is mostly associated with montmorillonite, and/or zeolites, and/or primary volcanic components, we believe that it was derived from the diagenesis of volcanic material and, as such, is considered to be a secondary volcanogenic component. However, if it is associated with abundant detrital minerals, then it may not be a secondary volcanogenic component.

Zeolite minerals, particularly phillipsite and clinoptilolite, are frequently specified as authigenic products of altered volcanic material. There is very little doubt that phillipsite is of volcanic origin (Arrhenius, 1963), and clinoptilolite is believed to form in the marine environment by the reaction of magnesium-rich solutions with volcanic glass or with minerals of the smectite family (Bonatti, 1972). Stonecipher (1974) drew attention to the association of zeolites and volcanic materials. However, the presence of clinoptilolite in carbonate and siliceous oozes suggests that it can form in sediments which apparently lack volcanic detritus (von Rad and Rosch, 1972). In this paper we consider zeolites important volcanogenic components only if other evidence of volcanic input is present in the sample (e.g., abundant montmorillonite, primary volcanic components, etc.).

Volcanic glass has been recognized as a possible source of silica for chert because its devitrification can place silica into solution. For example, Gibson and Towe (1971) and Mattson and Pessagno (1971) proposed a volcanic origin for the Eocene cherts in the North Atlantic and Caribbean. However, Wise and Weaver (1974) reviewed evidence for the volcanic origin of silica minerals and opal in chert and concluded that the vast majority of oceanic chert deposited since the mid-Paleozoic is biogenic in origin. It is apparent, therefore, that the silica minerals cristobalite, tridymite, and quartz, plus opaline silica, form in sediments where there is an abundance of excess silica which can be generated by the alteration of volcanic glass but more likely by the dissolution of siliceous microorganisms (e.g. Radiolaria, diatoms, sponge spicules, and silicoflagellates). Consequently, these silica minerals and opal may not be diagnostic of volcanic origin and are used as evidence only if associated with other components that have a probable volcanic source.

In summary, volcanogenic origins for sediments are assigned after analyses of stratigraphy, sedimentology and mineralogy. Sediments can be mainly volcanogenic if they are composed of a mixture of primary and secondary components, or if they are wholly made up of either primary or secondary components. For example, if other stratigraphic and sedimentologic data are supportive, a volcanogenic origin can be assigned for the following: (1) primary volcanogenic components; (2) primary components and montmorillonite; (3) montmorillonite as the only clay mineral; (4) montmorillonite and palygorskite as the only clay minerals; and (5) mixtures of the above with zeolites and/or silica minerals which combined comprise most of the sample.

Where detrital materials are abundant, the sample is assigned a detrital origin. For example, if the terrigenous detrital clays kaolinite, illite-mica, and chlorite comprise more than 50 percent of the clays minerals, the sample is considered to be detrital (J. Hein, pers. commun., 1975). Clay minerals ratios in the <2μ size fraction (noncarbonate) are used in this paper as a semi-quantitative method of determining the relative importance of volcanic contributions in a sample. For instance, if the percentage of montmorillonite in the total clay assemblage (montmorillonite, palygorskite, kaolinite, chlorite and illite) in the <2μ size X-ray sample is greater than 90% (i.e., M/M+P+K+I+C x 100 > 90%), then that size fraction is thought to be of volcanic origin. If that percentage is greater than 75% but less than 90%, the clay fraction is considered to be mostly of volcanic origin and if that percentage is between 50% and 75%, some volcanic input is suspected. Furthermore, if palygorskite is a significant contributor along with montmorillonite (i.e., M+P/M+P+K+I+C x 100 >75%), the clay fraction also is assigned a volcanic origin. These assignments are strengthened if primary components and/or zeolites and/or silica minerals are present.

Dispersal of Volcanogenic Sediments

It is not surprising that volcanogenic components can reach all parts of the world oceans considering the available methods of dispersal. In fact, the volcanogenic components probably are more effectively dispersed than any other component because some extremely explosive volcanic eruptions eject ash into all levels of the atmosphere where it is transported by the various wind systems. Besides wind, dispersal also occurs by: (1) subaerial and submarine pyroclastic flows and falls; (2) subaerial erosion and subsequent transport by streams, wind, and ocean currents, including those of surface, deep geostrophic, and turbidity origins; and (3) in polar regions, even on sea ice. Not much is known about deep submarine volcanism and subsequent transport of components by ocean currents, but certainly this method of dispersal should not be overlooked.

Winds are important mechanisms for the dispersal of fine-grained ash. For example, Nelson et al. (1968) traced ash from one volcano over part of the northeastern Pacific Ocean, Huang et al. (1975) traced the "Eltanin ash" over a large part of the south Pacific and Hays and Ninkovich (1970) correlated ashes in the north Pacific. Direct input of coarser-grained pyroclastic debris from explosive activity, both subaerial and subaqueous, also is an effective method of providing volcanic components to the marine realm. This method not only is important near explosive volcanoes that surround the Pacific Ocean, for example, but also may be significant around oceanic islands and seamounts. Erosion of volcanic rocks and sediments on landmasses play a strong role in providing volcanogenic sediments to the world oceans where they are subsequently transported by ocean currents.

Contributions from basaltic volcanoes have a local, or at most a regional effect on the sediment columns, whereas the more silicic volcanoes can disperse ash world-wide. The amount of pyroclastic material erupted from seamounts is not known, but the explosivity of the magma and the depth of the eruption below sea level are two important variables that govern the quantity (McBirney, 1963). Lava flows are subject to fragmentation (e.g. hyaloclastites) when erupted into water. These fragments subsequently can either undergo gravity transport which may extend to the base of the seamount slope or, if fine-grained, can be transported long distances by ocean currents. Furthermore, reaction of hot lavas and hyaloclastites with sea water will hasten the formation of smectites which subsequently can then be transported by bottom currents.

Volcanic Contributions to Indian Ocean Sediments

The volcanic contributions to Indian Ocean sediments at DSDP sites are summarized in Figures 2-6. Relative importances are assigned based on stratigraphic, X-ray, smear slide, thin section, and chemical studies.

Hole 213, in the northeastern Indian Ocean, is selected to show how primary and secondary volcanogenic components are used to determine the volcanic contributions in a sediment column that has several different lithologies (Figure 2). Volcanic glass, montmorillonite, palygorskite, phillipsite, and iron oxide minerals are considered volcanogenic components. Site 213 is located in the northwestern Wharton Basin in 5113 meters of water; the hole was continuously cored to a depth of 172 meters of which 154 meters are sediments and the remainder is basalt (see Figure 7 from Site Report, Chapter 4, in von der Borch, Sclater et al., 1974). Stratigraphic evidence at this site is consistent with it having subsided according to the Sclater depth curve, dropping below the calcium carbonate compensation depth (CCD) in the middle Tertiary, probably the early Eocene. Through sea floor spreading, the site migrated into an area of high siliceous plankton productivity in the later Tertiary (Pimm, 1974). It was influenced by terrigenous sedimentation throughout most of its history, but because of the presence of volcanic glass in the upper part of the hole and the relatively high quantities of diagenetic minerals in the bottom, volcanic contributions in the column can be evaluated. Silicic volcanic glass, as determined by smear slide analyses, is common in the top eight cores which range in age from late Miocene to Recent. In Cores 8 through part of 15, the zeolitic brown clay unit contains montmorillonite, phillipsite, and palygorskite with some kaolinite, illite, feldspar, and quartz. Below the brown clay unit is the nanno ooze of Cores 15 and 16 where montmorillonite and palygorskite are important contributors to the clay-sized fraction and, in the basal layers of unit 4, goethite also is a contributor.

We interpret the volcanic contributions to sediments in Hole 213 to be the following. The basal iron-manganese oxide facies is related to mid-ocean ridge volcanism which provided both volcanic ash and hydrothermal products from vents along the crest. Deep sea weathering of basalt and subsequent dispersal of authigenic minerals by currents also contributed materials to these basal sediments. In the overlying nanno ooze, montmorillonite and palygorskite abundances indicate a continued volcanic input. The brown clay unit is an undated interval and probably contains at least one major hiatus. Here, volcanic input is shown by the abundance of montmorillonite, calcic plagioclase, palygorskite, and phillipsite; some terrigenous input is recognized by the relatively high kaolinite content. Much of the volcanic debris was probably derived from early activity along the Indonesia Arc (van Bemmelen, 1949) and from island volcanoes and seamounts, although the relatively high kaolinite content suggests that some montmorillonite may have been eroded from a subaerial volcanic terrane. In the upper Miocene to Recent radiolarian-diatom ooze, the large admixture of silicic volcanic glass indicates volcanism in the Indonesian Arc (van Bemmelen, 1949; Katili,

Figure 2. Volcanic Contributions to Site 213, Northeastern Indian Ocean

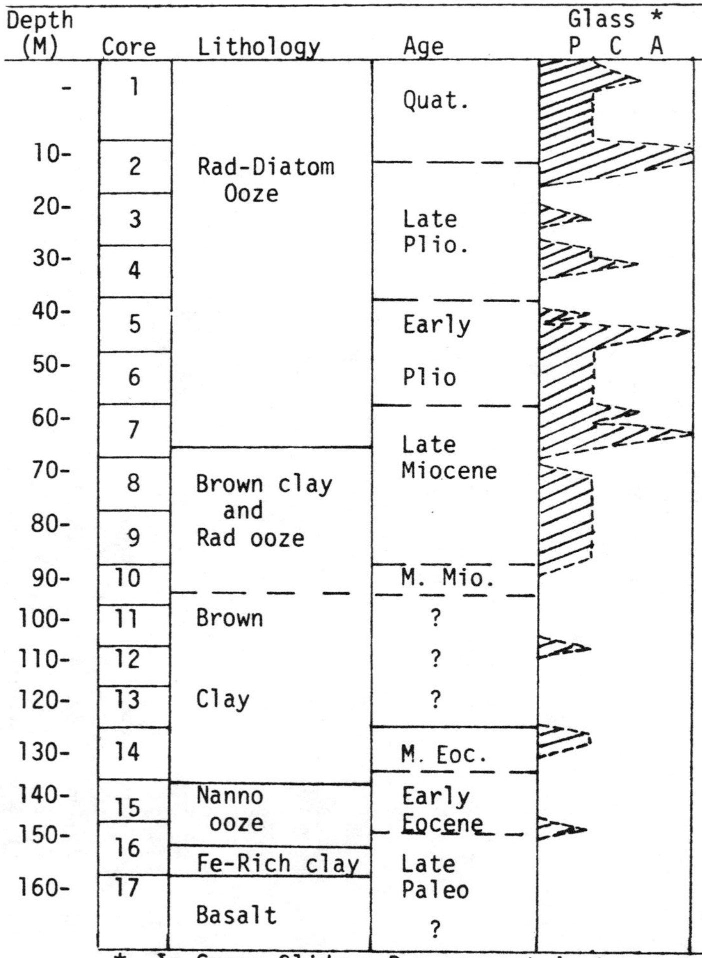

Depth (M)	Core	Lithology	Age	Glass* P C A	Amor.	Mont.	Paly.	Phil.	Kaol.	Ill.	Quar.	Plag.	K-Feld	Chl
-	1	Rad-Diatom Ooze	Quat.		86	33.6	-	-	17.7	19.2	22.0	8.1	-	-
10-					88	25.9	-	-	26.4	14.8	27.3	5.6	-	-
20-	2				83	35.1	-	-	19.9	16.8	18.6	5.9	3.7	-
30-	3		Late Plio.		82	49.2	-	-	19.0	12.4	11.8	3.7	4.0	-
40-	4													
50-	5		Early Plio											
60-	6				81	39.7	-	-	22.4	13.9	14.9	4.0	5.0	-
70-	7	Brown clay and Rad ooze	Late Miocene		79	53.6	-	-	19.0	7.8	13.5	6.0	-	-
80-	8				75	67.9	-	-	11.6	-	12.0	3.6	5.0	-
90-	9		M. Mio.		75	64.8	5.7	3.5	10.2	3.7	6.5	1.4	4.1	-
100-	10		?		74	73.4	-	-	8.6	3.8	9.3	4.9	-	-
110-	11	Brown	?		81	41.3	12.5	3.5	12.2	8.1	12.5	-	9.9	-
120-	12		?		79	18.1	11.4	36.4	-	10.5	12.0	-	111.7	-
130-	13	Clay	M. Eoc.		85	38.1	29.6	-	1.4	7.1	9.1	-	12.5	2.1
140-	14	Nanno ooze	Early Eocene											
150-	15	Fe-Rich clay	Late Paleo		87	12.0	57.3	-	6.0	-	7.7	-	17.0	-
160-	16	Basalt	?											
	17													

* In Smear Slides: P = present in trace amounts (Tr-2%), c = Common (2-10%) A = Abundant (>10%)

+ X-ray Mineralogy (<2µ Carbonate Free)
In smear slides, P = Present in trace amounts (Trace – 2%); C = Common (2-10%); and A = Abundant (>10%); 2) For discussions of X-ray methods, see text and Cook et al., 1975.

Fig. 2. Site 213, DSDP Indian Ocean drillsite, showing amounts of volcanogenic components in smear slide and X-ray results. 1) X-ray mineralogy from Matti and others (1974).

EXPLANATION FOR FIGURES 3-6

Abbreviations

I = Interval (presence of hiatuses, condensed sequences, etc.)
G = Percentage of volcanic glass and vitric tuff in smear slides
M = Percentage of montmorillonite (see text) in clay-sized X-ray results
P = Percentage of montmorillonite plus palygorskite (see text) in clay-sized X-ray results
Z = Percentage of zeolites in silt-sized X-ray results
R - Relative importance of all volcanogenic components in an interval

Symbols for Relative Percentages

	■	▨	⦀
G	>10%	1-10%	trace-1%
M	>90%	75-90%	50-75%
P	>95%	85-95%	75-85%
Z	>25%	5-25%	trace-5%
R	mostly volcanogenic	strong contribution	moderate contribution

Symbols For Intervals

- - -   undated Interval   ⋀⋀⋀ Condensed sequence

-u-u   suspected unconformity   ⋀⋀   Hiatus

1975; Ninkovich, 1977). From these data, we can conclude that there has been some volcanic input to the sediments throughout most of the time interval. At times, it was masked by the large influx of biogenous debris. Basaltic volcanism probably was dominant until the Miocene, after which silicic glass began accumulating from Indonesian volcanoes.

A desired result from this method of investigation would be a correlation of volcanic activity as shown by volcanogenic sediments in each hole drilled by DSDP in the Indian Ocean. However, although some correlations are possible, optimum results cannot be achieved. Some of the reasons, which are true for DSDP data in all oceans, are as follows: (1) all sites were chosen to solve a specific local or regional problem; therefore, in one specific region there might be sites (e.g. on a ridge and a basin) where very different tectonic and oceanographic conditions governed sedimentation; (2) there was an absence of adequate fossil data in many holes due to nonfossiliferous sequences and unconformities; (3) there was a lack of continuous coring; and (4) there was poor sediment recovery at some sites. The Indian Ocean, on the other hand, is probably the most likely of the major oceanic areas drilled by DSDP to yield good results for the following reasons: (1) meaningful stratigraphic penetrations were made at 50 sites located in all major basins as well as on most of the structural highs; (2) the percentage of stratigraphic columns actually cored is 55% compared to 37% in the other oceans and 15 sites were continuously cored; and (3) the major hiatuses in the sediment sequences have been identified (Davies et al., 1975) which permits some age controls where sediments are non-fossiliferous. Therefore, when specific regions of the Indian Ocean are evaluated (Figures 3-6), major times of volcanism become apparent. With these considerations in mind, we have divided the discussion of the Indian Ocean into five parts. The first four are geographic areas which closely correspond to the four quadrants of the Indian Ocean and include all meaningful sites drilled; the fifth part is a discussion of volcanic sediments from columns on presently aseismic ridges. These sites are 214, 216, and 253 on Ninetyeast Ridge, 254 and 255 on Broken Ridge, 223 and 224 on Owen Ridge, 219 and 238 on Chagos-Laccadive Ridge, 237 on the Mascarene Plateau, 246 on Madagascar Ridge, 249 on Mozambique Ridge, and 264 on the Naturaliste Plateau.

Northeastern Indian Ocean

Volcanogenic contributions to sediments in holes of the northeastern Indian Ocean are summarized in Figure 3. Unfortunately, the many hiatuses and condensed sequences in the sediment columns do not allow thorough analyses of volcanic input through time and some intervals lack diagnostic fossils for age control. For example, Holes 213-215 have brown clay units which probably represent long time intervals and at site 211, the brown silty clay and brown clay unit represents about 60 m.y. The relatively abundant montmorillonite and palygorskite in the brown clay units probably represent altered volcanogenic material, but specific times of major additions are not known.

Noteworthy of sediments in almost all holes that penetrated basalt in the Indian Ocean, and particularly in the northeastern quadrant, is the presence of a basal sediment unit which contains volcanogenic detritus. These sediments indicate volcanic activity coincident with an early development of the sites, which in turn probably is related to magma generation, rifting, and uplift along a spreading center. Although most volcanogenic components in the basal sediments are fine-grained, coarse basaltic tuffs are common at some aseismic ridge sites (e.g. Ninetyeast Ridge). Coarse-grained pyroclastic sediments should be expected in areas where nearby volcanoes approached and grew above sea level and explosive eruptions occurred when magma gas pressures exceeded hydrostatic pressures. Basal enrichments of metalliferous particles and some metal-rich clays probably are related to hydrothermal activity along the spreading ridge (Bostrom and Peterson, 1966; Dymond et al., 1973). Other volcanogenic clays may have formed during submarine weathering of basalt flows and hyaloclastites and from the alteration of fine-grained volcanic ash that was transported to the sites by winds and ocean currents from distant volcanoes.

Volcanic activity was noteworthy during the Late Cretaceous and Paleocene in the northeastern Indian Ocean. This might be related to rapid rates of sea floor spreading (Schlich, 1974; Vallier, 1974). However, because evidence of volcanism is mostly in the basal parts of the columns, this apparent increase also may be a consequence of many sites being in basinal locations; a large part of the sea floor in the northeastern Indian Ocean was formed during that time interval. Data from

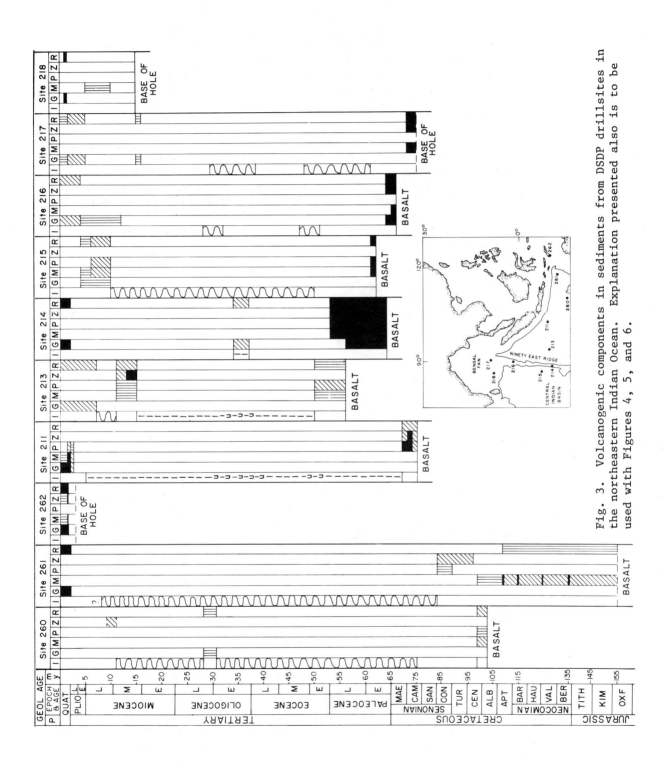

Fig. 3. Volcanogenic components in sediments from DSDP drillsites in the northeastern Indian Ocean. Explanation presented also is to be used with Figures 4, 5, and 6.

field studies on land suggest that the Late Cretaceous marks the initial collision of India and Asia and eruption of the Deccan Traps followed in the Paleocene (McElhinney, 1970).

Besides the Late Cretaceous-Paleocene pulse of volcanic activity, high montmorillonite contents in Upper Jurassic and Lower Cretaceous sediments suggest that some volcanism was associated with the earliest opening of the Indian Ocean. Particularly noteworthy in this regard are the individual beds of nearly pure montmorillonite in the Upper Jurassic and Lower Cretaceous sediments at site 261.

Other times of volcanism, as recorded by the volcanogenic components, were in the early and middle Oligocene (sites 214 and 260) and in the middle Miocene to Recent intervals (all sites). Silicic volcanic glass is particularly abundant in the Pliocene and Pleistocene. Strong input of silicic glass did not begin until the late Miocene, about 10 m.y. ago, as indicated by abundant glass at sites 215 and 216. The middle Miocene to Recent volcanism is related to the development of volcanic events along the Indonesian Arc which in turn marks subduction of the Indian Ocean Plate beneath the arc (van Bemmelen, 1949; Pimm, 1974; Katili, 1975).

The significance of the Oligocene volcanism will be discussed more completely in a later section. In this region of the Indian Ocean it probably marks early subduction along the Indonesian Arc (van Bemmelen, 1949, p. 230). During the Oligocene, present day velocities and directions of plate motions were established in the Indian Ocean (Moore et al., 1974, p. 411).

Northwestern Indian Ocean.

Relative amounts of volcanic contributions to sediments at DSDP sites in the northwestern Indian Ocean are shown in Figure 4. Of importance are the concentrations of volcanic debris near basement basalt, a possible Paleocene-Eocene pulse, an Oligocene event, and an apparent late Miocene to Recent increase in activity. A Turonian-Campanian event is recorded by sediments at Site 241 which lies along the east African continental rise.

Volcanic-rich basal units occur at sites 219, 220, 221, 223, 231, 232, 236, and 237. Some are coarse-grained tuffs, whereas others contain only diagenetic products of volcanic debris. There were significant amounts of volcanic sediments deposited during the Paleocene-Eocene interval, some of which probably are related to mid-ocean ridge rifting. However, some of the fine-grained volcanogenic sediments may be related to the eruption and subsequent erosion of the Deccan Traps (e.g., volcanic sediments in the Eocene at sites 219 and 223) and to volcanism coincident with the separation of India and the Seychelles as outlined by Davies (1968).

The Oligocene pulse is more apparent in sediments from sites in the northwestern Indian Ocean than elsewhere. Disseminated silicic glass occurs in the Oligocene in holes 223, 224, and 241 and in discrete beds and zones in holes 236 and 237. The cause of this volcanism is not known, but it may be related to explosive volcanic activity in Yemen, Ethiopia, and Arabia (Coleman, 1974) or to the development of volcanic eruptive centers that formed during an early phase of subduction along the Indonesian Arc (van Bemmelen, 1949). A third possibility is that island volcanoes on the Mascarene Plateau or other ridges in the northwestern Indian Ocean erupted silicic phyroclastics during that time interval.

Beds of nearly pure (>90%) montmorillonite in the Turonian-Campanian at site 241 indicate a Late Cretaceous pulse of volcanic activity. The

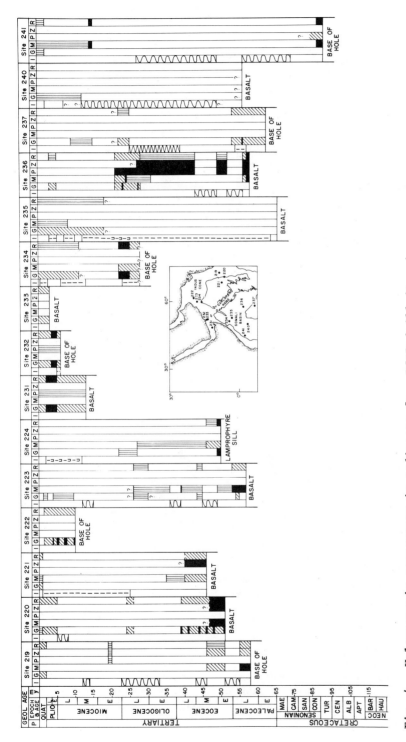

Fig. 4. Volcanogenic components in sediments from DSDP drillsites in the northwestern Indian Ocean. See Figure 3 for explanation.

cause and source of the volcanism are not known, but it may be related to Late Cretaceous volcanism on Madagascar (Besairie, 1972).

Early Miocene to Recent silicic volcanism is evident at sites 219 to 223, 231 to 236, 240, and 241. Glasses in this age range from Leg 24 sites (231-236) have a wide diversity of chemistries as suggested by their refractive indices (Cronan et al., 1974) which indicates a variety of source volcanoes, perhaps as far apart as the Indonesian Arc and the Red Sea areas.

Palygorskite is a common clay mineral in most cores from DSDP holes in the northwestern Indian Ocean. Although some may be authigenic, through generation of extra silica by the alteration of volcanic ash or the solution of opaline fossil tests (Hathaway and Sachs, 1965), more likely it is eolian in origin and derived from desert regions of North Africa (Goldberg and Griffin, 1970).

Southwestern Indian Ocean

The nature, distribution and possible reasons for volcanogenic sediments in sites of Leg 25 in the southwestern Indian Ocean are reported by Vallier (1974). Figure 5 summarizes the Leg 25 data plus data from site 238 (Leg 24) and sites 250-252 (Leg 26).

Basal volcanogenic sediments are common at most sites in the southwestern Indian Ocean. They are particularly abundant in the Upper Cretaceous and Paleocene at site 239, in the Paleocene and Lower Eocene at site 245, in Oligocene sediments at site 238, and in Lower Miocene sediments at site 251. Sediments that overlie basalt at site 248 may be significantly younger than the basalt; however, they have high metal contents (Marchig and Vallier, 1974) which suggest that they represent the hydrothermal products that normally develop early in a site's history. At site 245, Paleocene and lower Eocene sediments contain beds of nearly pure (>90%) montmorillonite and some dispersed silicic volcanic glass (Vallier, 1974; Warner and Gieskes, 1974).

Lower Eocene coarse-grained basaltic tuffs at site 246 on Madagascar Ridge are coeval with fine-grained, partly volcanogenic sediments at site 248 in the adjacent Mozambique Basin. Oligocene volcanism is recorded in sediments at sites 238, 239, and 242, all of which are in the northern part of the southwestern Indian Ocean region. At site 238, near the junction of Chagos-Laccadive Ridge and the Central Indian Ridge, coarse-grained basaltic tuffs overlie basalt. This basaltic volcanism correlates with silicic volcanism in sites of the northwestern Indian Ocean, but causes probably are not directly related.

Early Miocene to Recent volcanism is recorded in sediments from sites 238 and 239 and 250 through 252. At sites 238 and 239 these volcanic components may be related to the volcanism along the Mascarene Plateau, but in sites of the Mozambique and Crozet basins, the components likely are related to volcanism along the Southwest Indian Ridge and the Crozet Islands (Vallier, 1974).

Southeastern Indian Ocean

Concentrations of basaltic volcanogenic sediments in basal sediments also are common in holes drilled in the southeastern Indian Ocean, such as those at sites 212, 253, 256, 257, and 259 (Figure 6). Late Cretaceous volcanism contributed small amounts of volcanic sediments to sites 212, 257, and 258, and in the Aptian-Albian to sites 257 and 258. Whereas the volcanogenic sediments on the Australian margin can be related to landward activity (Brown et al., 1968), those at site 212 are probably derived from the mid-ocean ridge, formerly positioned in this area (Sclater, et al., 1977).

VOLCANOGENIC SEDIMENTS 101

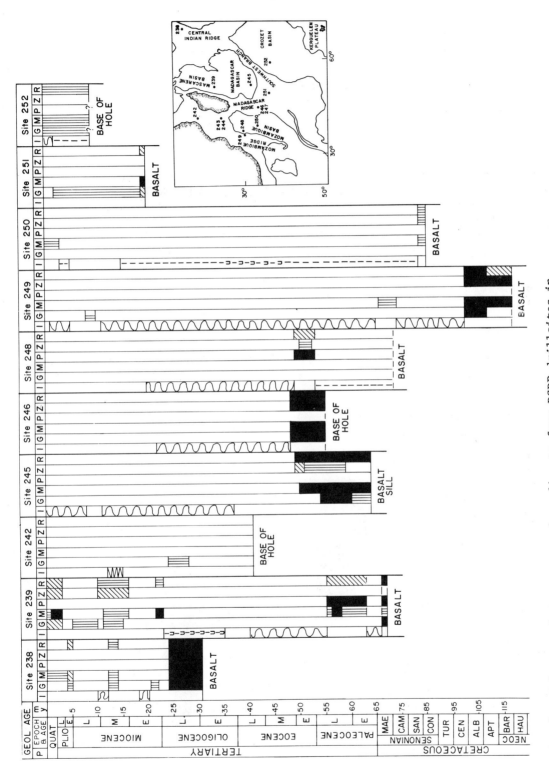

Fig. 5. Volcanogenic components in sediments from DSDP drillsites in the southwestern Indian Ocean. See Figure 3 for explanation.

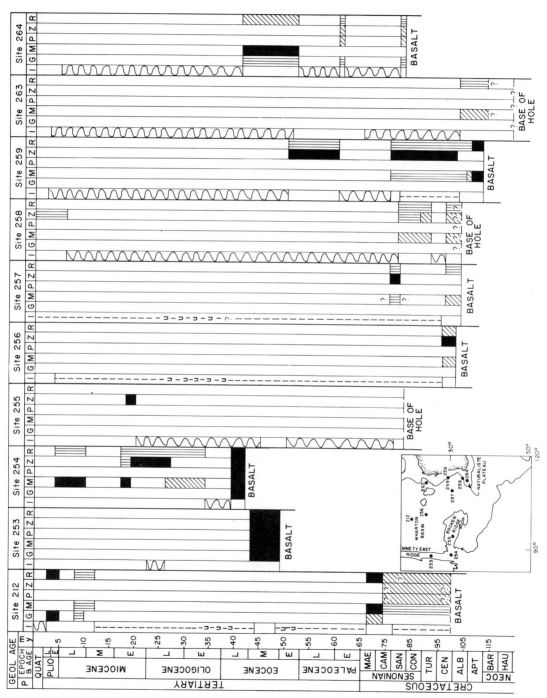

Fig. 6. Volcanogenic components in sediments from DSDP drillsites in the southeastern Indian Ocean. See Figure 3 for explanation.

Coarse basaltic tuffs of Eocene age overlie basalt at site 253 (Ninetyeast Ridge) and site 254 (Broken Ridge) and indicate shallow water or subaerial eruptions during the early stages of ridge formation (see discussion below). Eocene volcanism also contributed sediments to the Naturaliste Plateau (site 264).

Pliocene-Pleistocene silicic glass is dispersed in sediments at site 212 which lies in the southern Wharton Basin. This glass probably was derived from volcanoes in the Indonesia Arc (Pimm, 1974).

Discussion

Reviewing volcanism in the four quadrants of the Indian Ocean as shown by the volcanogenic sediments, it is clear that our methods do delineate periods of relatively important input of volcanogenic debris to the ocean. Furthermore, it is obvious that the amount of volcanogenic materials in many columns has been underestimated in the past when the presence of volcanic glass and other primary components was taken as the only indicators. It is apparent that at any given time since the early formation of the Indian Ocean, volcanism has occurred somewhere in the region. Nevertheless, several volcanic pulses are recognizable (Table 1).

We believe that the Oligocene volcanic event was a very important pulse in the Indian Ocean. No doubt it extended to the southeastern quadrant and it would have been recognized at most sites were it not for the presence of hiatuses. Davies et al. (1975) reviewed the occurrences of unconformities in the Indian Ocean and explained the presence of oceanwide hiatuses in the late Eocene to early Oligocene and early Tertiary as a consequence of climatic events in Antarctica and their effect on circulation patterns in the world oceans. Another major hiatus occurs in the Upper Cretaceous. Figure 7 is constructed from their map of the extent of the hiatuses in early Oligocene time. Volcanogenic sediments recorded in our columns (Figures 3-6) occur only at those sites in the western and northwestern parts where the hiatuses do not occur. Therefore, it is probable that the Oligocene volcanic pulse was oceanwide in extent but its products either were eroded off or were redistributed by intensified oceanic circulation. Similarly, it can be argued that volcanic input to Eocene sediment sequences could have been more extensive (see Figure 8).

Aseismic Ridges

Almost all presently aseismic ridges in the Indian Ocean have basaltic basement rocks and many of the basal sediment units are strongly volcanogenic. In several holes, basaltic pyroclastics indicate subaerial or shallow water explosive eruptions. In other holes, the coarse volcaniclastic sediments occur in units above the basal sediments, thereby recording local volcanic activity at a later time in the evolution of the ridge. We believe that the early volcanic events mark times of formation at the spreading center and that the later event marks uplift of the ridge or another tectonic event in the ridge's evolution.

The major ridges and plateaus are discussed individually in the sections that follow. Some conclusions are presented that link volcanic and tectonic events along each ridge.

Ninetyeast Ridge

Sites 214, 216, and 253, drilled on Ninetyeast Ridge, have fairly thick sequences of volcanogenic sediments resting on basalt (Pimm, 1974;

TABLE 1. Volcanogenic Input to Indian Ocean Sediments

Northeast	Northwest	Southwest	Southeast
Recent to Late Miocene	Recent to Early Miocene	Recent to Early Miocene	Pleistocene and Pliocene
Middle Oligocene	Oligocene	Oligocene	————
Paleocene to Late Cretaceous	Eocene to Late Paleocene	Eocene to Late Paleocene	Eocene
————	Late Cretaceous	————	Late Cretaceous
Early Cretaceous to Late Jurassic	————	Early Cretaceous	————

McKelvey and Fleet, 1974). Although basaltic basement was not penetrated at site 217, high montmorillonite and zeolite contents in the upper Cretaceous sediments near the base of the hole, indicate a volcanic influence during that time interval. In Hole 214, a volcanic sediment sequence (390 to 490 meter depth interval) is interbedded with lignite and intermediate differentiated flow rocks. This sediment sequence probably is mostly pyroclastic except for a few thin beds of conglomerate which indicate an epiclastic input. Because of the absence of marine fossils and the presence of lignite beds, it was concluded that many of the volcanic sediments were deposited subaerially or in shallow, possibly fresh water (Pimm, 1974). Site 216 basal sediments, however, contain predominantly epiclastic debris with some basaltic ash beds, all of which are mixed with shallow water marine sediments. Furthermore, at site 253 on the southern part of the ridge, basaltic volcaniclastic sediments comprise about 388 meters of the section. Some are hyaloclastites that were formed by the fragmentation of basaltic lavas as they were quenched by sea water. Many of the originally holohyaline tuffs are largely replaced by smectites and some contain analcime and calcite cements. Their chemistries (Table 2) show a wide range of values which emphasizes the altered nature of the samples. McKelvey and Fleet (1974, p. 563) suggest that the elemental concentrations on a carbonate-free basis represent variations due to hydrothermal alteration, which formed the two smectites and analcime, and to postdepositional ion migration and other effects.

These accumulations of basaltic sediments above basement emphasize the explosive nature of the magmas during formation of the Ninetyeast Ridge and indicate that eruptions occurred in shallow water and/or that volcanoes formed islands. Luyendyk and Davies (1974) modelled the origin of the ridge as a combination of a point source and a leaky transform fault. From the evidence of volcanogenic sediments there is no doubt that the Ninetyeast Ridge was a high relief feature during the early part of its formation.

Broken Ridge

Site 254 lies near the intersection of Ninetyeast and Broken ridges and, as such, could be assigned to either or both of those ridges. The site has an early volcanic history similar to the Ninetyeast Ridge sites; basal volcanic sediments were deposited in the late Eocene. The vol-

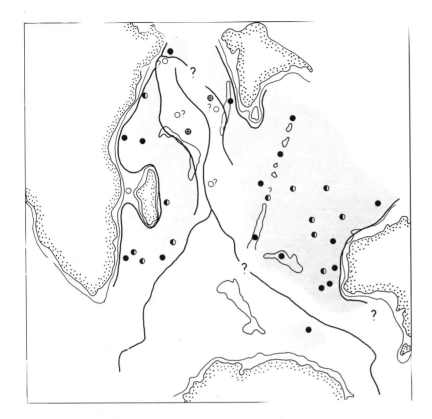

Fig. 7. Significance of volcanogenic sediments with regards to the extent of early Oligocene hiatuses (from Figure 4 of Davies et al., 1975). Open circles represent sites where the sedimentary section is complete; solid circles show where there is a proven early Tertiary unconformity; half filled circles indicate an undated interval or an inferred unconformity; crosses on the open circles indicate sites where volcanogenic sediments occur in the Eocene sections. Shading shows the extent of the Eocene unconformity, both proven and suspected.

canic sediments are predominantly basaltic in composition and were deposited in shallow water (Davies, Luyendyk et al., 1974, Chapter 7). High montmorillonite contents and some zeolites in overlying strata indicate a continued volcanic influence, particularly during the Oligocene and Miocene.

Owen Ridge

Sites 223 and 224 lie on Owen Ridge, whose topographic prominence is due to movement along the Owen Fracture Zone. Their stratigraphic records show differing structural histories (Weser, 1974). Both sites had strong volcanic inputs during their early development which occurred before the formation of Owen Ridge. At site 223, the oldest sediments are upper Paleocene and lower Eocene brown clays which consist mostly of montmorillonite, zeolites, cristobalite, and palygorskite, thereby implying a volcanic origin for most of that unit. Early in its history, site 223 probably had a basinal setting and then, in about the middle Eocene, uplift began in connection with movement along the Owen Fracture Zone. This fracture zone is thought to have been active as a transform fault between the Indian and Somalia-Arabia plates until the early

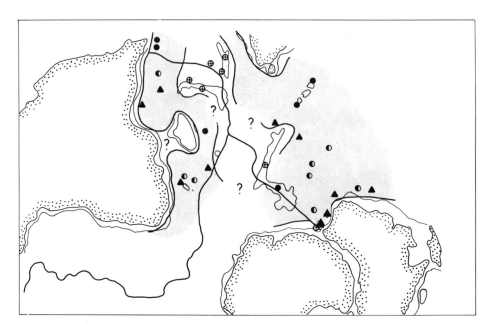

Fig. 8. Significance of volcanogenic sediments with regards to the extent of middle Eocene hiatuses (from Figure 5 of Davies et al., 1975). Open circles represent sites where the sedimentary section is complete; solid circles show sites that have a proven early Tertiary unconformity; half filled circles indicate an undated interval or an inferred unconformity; solid triangles show sites where an early Tertiary unconformity cannot be separated from the overlying Oligocene unconformity; crosses on the open circles indicate sites where volcanogenic sediments occur in the Eocene sections. Shading shows the extent of the Eocene unconformity, both proven and suspected.

Miocene or late Oligocene (Whitmarsh et al., 1974). At site 224 initial volcanic activity in the early and middle Eocene is suggested by a basal layer of red montmorillonite clay, containing abundant palygorskite and cristobalite. Clay minerals are predominantly montmorillonite until the middle Oligocene. Uplift apparently occurred in the early or middle Miocene (Weser, 1974).

Volcanism during the early histories of these sites probably is not associated with the formation of the ridge as was the case with the Ninetyeast Ridge. Instead, the topographic edifice formed subsequent to volcanism related to the mid-ocean ridge.

Naturaliste Plateau

Naturaliste Plateau (site 264) experienced significant basaltic volcanic activity during the early and middle Eocene, which is much younger than the Late Cretaceous age of acoustic basalt basement. As such, it marks an Eocene event which is associated with the opening of the southeastern Indian Ocean between Australia and Antarctica and possibly also to the uplift of that plateau.

Chagos-Laccadive Ridge

Sites 219 and 238 lie on the Chagos-Laccadive Ridge. At site 219, the lowest unit penetrated has a shallow water origin and sediments are

TABLE 2. Chemical analyses of samples from the altered vitric ash sequence, DSDP Site 253. Results were recalculated on a carbonate-free basis (from McKelvey and Fleet, 1974)

SiO_2^a (%)	TiO_2^a (%)	$Al_2O_3^a$ (%)	FeO^{*b} (%)	MnO (%)	MgO (%)	CaO (%)	Na_2O (%)	K_2O (%)	H_2O^+ (%)	Cr (ppm)	Co (ppm)	Ni (ppm)	Cu (ppm)	Zn (ppm)	Pb (ppm)
34.1	1.1	2.7	14.54	1.14	3.47	5.37	3.87	3.15	7.82	30	76	23	77	134	4
(11.3)	(0.9)	(-)	19.03	0.44	6.30	34.40	5.15	3.96	n.d.	160	73	40	126	273	9
(-)	(0.5)	(-)	12.33	0.37	6.40	40.04	5.61	1.66	n.d.	109	126	67	148	183	10
40.7	0.1	3.0	14.31	0.03	7.58	3.15	3.60	1.06	n.d.	10	16	-	22	16	2
44.8	0.9	5.8	10.79	0.08	5.83	8.15	3.53	1.83	n.d.	38	56	-	92	129	2
39.8	0.3	7.6	8.69	0.18	5.59	13.41	3.51	2.25	n.d.	139	65	105	116	70	3
56.1	0.6	8.4	8.46	0.05	7.20	5.30	3.46	1.04	8.94	92	66	61	185	87	3
(16.3)	(0.2)	(-)	8.47	0.62	8.10	13.73	3.61	1.16	n.d.	106	59	24	197	79	3
(19.0)	(0.4)	(-)	15.06	0.24	5.97	28.68	2.46	3.29	n.d.	132	71	20	52	175	7
53.9	0.2	5.8	5.61	0.01	6.19	2.71	3.51	0.86	n.d.	20	16	-	61	107	3
45.1	2.1	6.6	14.58	0.13	5.23	2.59	2.42	1.92	6.67	29	71	9	93	238	4
(13.1)	(0.3)	(0.5)	14.08	0.13	10.50	4.19	2.49	2.83	n.d.	128	95	40	127	127	11
40.9	0.9	3.7	13.97	0.07	7.74	1.56	4.48	1.59	n.d.	70	64	26	154	116	2
(28.0)	(0.7)	(3.4)	13.02	0.14	8.48	6.39	4.40	3.21	n.d.	108	83	18	164	148	4
38.1	0.3	4.6	11.34	0.09	2.66	8.24	4.80	1.27	n.d.	111	72	125	178	81	3
40.0	0.7	8.4	10.36	0.10	7.73	2.49	3.82	2.16	7.83	133	66	35	120	116	4
37.9	0.2	5.2	8.84	0.14	7.99	1.90	5.90	0.25	n.d.	165	62	83	161	90	4
(38.6)	(0.2)	(-)	9.94	0.11	9.28	1.43	3.55	0.38	n.d.	161	57	88	135	69	2
44.4	0.3	7.4	6.37	0.09	3.33	1.38	1.58	0.83	n.d.	43	47	2	74	63	2
33.7	1.7	6.5	13.06	0.27	9.00	4.15	2.18	1.97	6.36	81	62	25	145	141	3

Note: n.d. = Not determined; Not detected.
aElements determined by direct-reading spectrography, values in parenthesis are unreliable.
b* = Total iron assuming it is all in the ferrous state.

detrital and of late Paleocene age. High montmorillonite and zeolite contents indicate volcanic inputs, some of which probably is related to erosion of the nearby Deccan Traps of India. High montmorillonite and zeolite contents continue into the middle Eocene chalks. It is interesting to note that palygorskite does not appear in the samples until the late Eocene, coincident with a decrease in montmorillonite. This may reflect a change of source rocks rather than diagenesis of volcanic components. At site 238, on the southern part of the ridge, volcaniclastic sediments make up a significant proportion of the bottom 35 meters of the Oligocene nanno chalk. These volcanic sediments have abundant zeolites and montmorillonite with some palagonite, a high plagioclase (anorthite) content, pyroxene, and basaltic rock fragments which indicate a basaltic source. The volcanic activity evident in sediments from the basal unit at site 238 probably is related to early formation of the Chagos-Laccadive Ridge.

Mascarene Plateau

Basaltic basement rocks were not penetrated at site 237 on the Mascarene Plateau. However, the presence of small amounts of basaltic ash at several horizons in the upper Eocene and lower Paleocene sediments indicates significant volcanism in the area. The association of shallow water sediments and basaltic ash found in older sediments at this site suggests that the Mascarene Plateau was a shallow feature during the early stages of its development and that basaltic ash was being expelled from nearby volcanoes which may have been built up as islands along the plateau, similar to the present day Mascarene Islands of Mauritius and Réunion.

Madagascar Ridge

Acoustic basement was not penetrated at site 246 on the Madagascar Ridge. Beds of upper Paleocene (?) and lower Eocene coarse volcaniclastic sediments are of basaltic composition and were erupted in shallow water (Vallier, 1974). These accumulations are approximately the same age as the volcanogenic sediments in the Madagascar (site 245) and Mozambique (site 248) basins. Paleocene-Eocene volcanism in the southwestern Indian Ocean is undoubtedly related to activity along the Southwest Indian Ridge which at that time formed a continuation of the Madagascar Ridge (Sclater et al., 1977).

Mozambique Ridge

Significant basaltic volcanism is recorded in Lower Cretaceous (Aptian-Albian) sediments at site 249 on Mozambique Ridge. This volcanism is much younger than the early Neocomian basaltic basement rocks. The basement basalt is highly vesicular, with vesicles having diameters of one centimeter, which indicates subaerial or shallow submarine eruptions. The dark brown carbonaceous detrital sediments and limestones which overlie the basalt probably were deposited in a relatively deep euxinic basin. These sediments contain abundant montmorillonite which may be related to the subaerial erosion and transport of clays from the flood basalts in southeastern Africa. The tuffaceous basaltic sediments that overlie the carbonaceous unit indicate shallow water or subaerial explosive eruptions. This event may mark the uplift of the Mozambique Ridge, an interpretation that is strengthened by the fact that the volcanic sediments are overlain by a major unconformity caused either by current scouring along a topographic high or by subaerial erosion.

Kerguelen Plateau

Kerguelen Plateau was not drilled during the Deep Sea Drilling Project cruises in the Indian Ocean. However, some data from that plateau may add to an understanding of aseismic ridges and to the volcanism in the Indian Ocean. Watkins et al. (1974) concluded that the oldest rocks exposed on the island are late Oligocene in age and that the plateau migrated southwestward from the mid-ocean ridge system. Kerguelen, according to them, is a simple oceanic island located over a region of large-scale magma generation and this is supported by the paleobathymetric maps of Sclater et al., 1977. We agree with these authors that the oldest marine sediments should contain early Tertiary or possibly even Late Cretaceous fossils. No doubt, the history of Kerguelen Plateau in the early Tertiary is the same or very similar to the history of Broken and Ninetyeast ridges. However, volcanism continued during the late Tertiary, which differs from the volcanic history of Broken and Ninetyeast ridges.

Implications of Aseismic Ridge Volcanism

From the occurrences of volcanic sediments on the major aseismic ridges of the Indian Ocean, we can make several observations and suggestions: (1) all coarse-grained volcaniclastic sediments on the ridges are basaltic in composition. (2) Some aseismic ridges developed along the mid-ocean rift during early stages of sea floor spreading and have persisted as topographic highs throughout most of their histories. At these sites the volcanogenic sediments directly overlie the igneous basement rocks. (3) Other presently aseismic ridges have coarse basaltic sediments high in the sediment columns, and this suggests that the ridge topography formed much later than the creation of basaltic basement. Sea floor was generated at a spreading ridge, subsequently subsided, and then was uplifted at the same time as basaltic pyroclastic materials were erupted. (4) The silicic volcanogenic sediments that occur in the younger, generally upper Tertiary, strata, probably were transported to the sites by wind from distant volcanoes.

Ninetyeast Ridge has coarse basaltic tuffs directly above basalt basement at all drill sites that penetrated basalt. The age of that ridge increases towards the north which is consistent with the complex origin suggested by Luyendyk and Davies (1974) where a volcanic source interacted with variously offset spreading ridges and created a volcanic rise that subsequently moved northward. Evidence from the volcanogenic sediments confirms that shallow water and subaerial explosive activity was nearly continuous when this activity is traced in a southern direction. The Chagos-Laccadive and Broken ridges also have pyroclastic basaltic sediments directly above basaltic basement which strengthens our contention that basaltic pyroclastic activity is an important part of the early formation of those ridges. Volcanoes apparently develop a ridge and contribute most of their volcaniclastic sediments locally; however, some of the finer-grained components must be distributed into surrounding basins as well. This volcanic activity is an important source of basaltic debris in the Indian Ocean.

Not to be overlooked are two ridges in the southwestern Indian Ocean and one in the southeastern part that have basaltic pyroclastic sediments higher in the stratigraphic columns which apparently are not related to the early development of the ridges. Mozambique Ridge (site 249) has Aptian and Albian basaltic sediments overlying Neocomian carbonaceous marine sediments. Because of this, and because of the possibility of a subaerial unconformity above the basaltic sediment, we believe that the basaltic pyroclastic activity is related to the uplift of that ridge. The geologic history of the Mozambique Ridge began with the extrusion of tholeiitic basalt in shallow water at a spreading center, followed by subsidence and the deposition of euxinic sediments in a restricted marginal basin. Subsequently, in the Aptian and Albian, basaltic volcanism occurred contemporaneously with rifting of the sea floor and uplift of the ridge. Current erosion did not allow additional sediment accumulation until the Late Cretaceous. At site 246 on the Madagascar Ridge, basaltic pyroclastics accumulated in the Paleocene (?) and early Eocene when water was shallow. Basement was not penetrated and we think that these basaltic pyroclastic sediments record the activity of a former Southwest Indian Ridge or the uplift of Madagascar Ridge. The Naturaliste Plateau (site 264) also shows Eocene volcanism which probably records the separation of Australia from Antarctica by the Southeast Indian Ridge.

Volcanogenic Sediments and Sea Floor Spreading

A natural consequence of sea floor spreading is volcanism. Along spreading centers where sea floor is being added, basaltic volcanism is dominant, and where it is being subducted more silicic (e.g. andesitic) volcanism generally is prevalent. At other plate boundaries, such as along transform faults, volcanic activity also takes place. Oceanic islands and seamounts that are younger than the sea floor upon which they form indicate that volcanic activity takes place within plates as well as along their borders. No doubt, other types of intraplate volcanism occur which are generally not observed but probably result in the formation of hypabyssal intrusives both within the basaltic basement and in the over-lying sediments. Vallier (1974) attempted a correlation between the rate of sea floor spreading and the times of volcanic sedimentation in the southwestern Indian Ocean and was able to establish some relations, particularly between high rates of spreading in the Paleocene and Eocene and volcaniclastic sedimentation along the Madagascar Ridge and within the Madagascar and Mozambique basins.

The abundance of volcanic sediments in most of the Upper Cretaceous and lower Tertiary basal units is a consequence of the formation of new crust at a spreading ridge and attendant pyroclastic activity associated with the formation of volcanic islands and seamounts along the crest. Therefore, volcanogenic sediments should be common in basal units. Modern analogies occur all along the mid-ocean ridge system. For example, in the Indian Ocean, volcanoes on Réunion and Mauritius distribute volcanic sediments locally and even small volcanoes like Surtsey in the Atlantic can contribute abundant sediments to the sea floor.

Some sites, however, have volcanogenic sediments, most often pyroclastic, in the column far above the basal sediment units. Examples are in the mid-Cretaceous at sites 249, 259, and 263, in the Paleocene-early Eocene at sites 215, 220 (?), 236, 241, 246, and 248, and in the Oligocene at sites 214, 217, 220, 234, 236, 237 and 241 (?). These volcanic pulses probably can be related to other events such as changes in directions and rates of plate motion or to the creation (i.e. uplift) of aseismic ridges. Of course, where the sediments are silicic, they are most likely related to an arc environment or to the eruption of highly differentiated magmas in other geologic settings.

Landmass Volcanism

Continental and Island Volcanism

Vallier (1974) and Coleman (1974) reviewed the volcanic activity on landmasses that both border on and lie within the western Indian Ocean. In the northeastern Indian Ocean, most volcanism is related to the development of the Indonesian Arc (van Bemmelen, 1949, Katili, 1975; Ninkovich, 1977). In western Australia only the Neocomian Bunbury Basalt testifies to volcanism (Quilty, pers. comm., 1974); that unit probably is related to the initial breakup of Gondwanaland and the early formation of the Indian Ocean basin. Major landmass volcanism is reviewed in Figure 9.

A review of Karroo and younger volcanism in southeastern Africa is given by Flores (1970), in South Africa by Truter (1949), and in southeastern Africa and Madagascar by Blant (1973). Flores (1970) related volcanism in southeastern African and Madagascar to the widening split between Africa and Madagascar which occurred when major fractures reached subcrustal depths and caused extensive outpourings along linear fissures, first in late Karroo (Jurassic) times (e.g., Lebombo Range) and then in the Cretaceous and Tertiary (e.g., Madagascar and mid-channel volcanism). During the Jurassic-Early Cretaceous time interval, volcanism was extensive in southeastern Africa, but in the Late Cretaceous-Recent interval, volcanic activity was limited to only a few alkalic volcanic centers (Haughton, 1963, p. 292-295).

VOLCANOGENIC SEDIMENTS 111

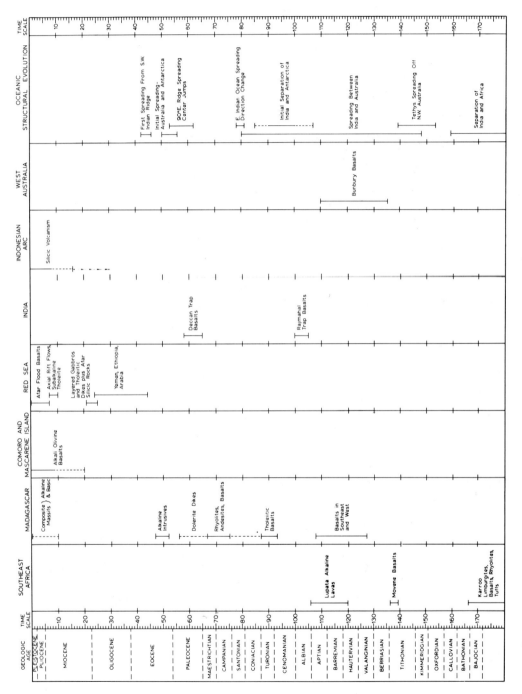

Fig. 9. Major landmass volcanism around the Indian Ocean and major structural events related to the opening of the Indian Ocean.

Besairie (1972) described the Cretaceous-Recent volcanism in Madagascar. Major volcanism occurred in the Early Cretaceous (Valanginian-Aptian), Late Cretaceous (Turonian and Campanian), Paleocene (tuffs in the north), Eocene, Miocene, and Pliocene-Pleistocene. The island of Nosy Bé, along the northwestern coast of Madagascar, has an older Pliocene phase of basaltic volcanism, and the latest phase (Recent) includes outpourings of basalt and ankaratrite.

No attempt has been made here to completely review the volcanic history of India. However, the two recorded major basaltic trap rock series should be mentioned. The oldest, the Rajmahal Trap Basalts, have been dated at about 100-105 m.y. and the youngest, the Deccan Trap Basalts, are dated at about 59-65 m.y. (McElhinny, 1970).

Late Tertiary to Recent volcanism is recorded in the Comoro Islands. The archipelago apparently began forming in the early Miocene (Aquitanian), and Khartala Volcano still is active. The three islands which form the Mascarene Island group are Mauritius, Réunion, and Rodriguez. Volcanism on these islands (McDougall and Chamalaun, 1969, p. 1438) occurred during the last 8 m.y. Mauritius is the oldest and consists mostly of alkali olivine basalt and some of its differentiates, known as the "Older Volcanic Series", which were erupted between 8 and 5 m.y. ago. Réunion has one active volcano, Piton de la Fournaise, which erupts olivine basalt lava. However, according to Fisher et al., (1971), older volcanism along the "Chagos Fracture Zone" may have contributed some volcanic products in the Mascarene Plateau area from as far back as the Late Cretaceous. According to Watkins et al. (1974), Kerguelen basaltic volcanic activity was particularly noteworthy in the late Oligocene and early Miocene with subsequent cone building associated with intrusions of dikes and stocks and still later eruptions of basanite flows.

The Indonesian Arc has been distributing abundant silicic volcanic glass to the Indian Ocean at least from the late Miocene to present times (Venkatarathnam, 1974) and some since the late Oligocene (van Bemmelen, 1949). The Oligocene silicic glass is most evident in sediments from the western Indian Ocean and probably is related to volcanism along the Indonesian Arc. It should be recognized that strong monsoonal winds may have carried ash that distance. If so, the presence of this glass in the sediments substantiates the fact that Indonesian Arc silicic volcanism began in the Oligocene.

Coleman (1974) reported extensive volcanism in the Eocene-Oligocene interval in Yemen, Ethiopia, and Arabia. In the Miocene (about 22 m.y. ago) layered gabbros and tholeiitic dike swarms invaded the hinge line of monoclinal downwarps along the southern Arabian coast, and within the Afar Depression small granitic intrusive and silicic volcanic rocks (22-25 m.y. ago) are probably related to the downwarping. Thin lava flows spread out over large areas around the Red Sea in the Miocene, and during formation of the axial rift (+ 5 m.y. ago), subalkaline oceanic tholeiite began filling the axial trough with contemporaneous alkaline basalt eruption centers along the eastern margin, which has continued to the present. Late Miocene to Quaternary flood basalts have been erupted in the Afar Depression.

Landmass Volcanism and Indian Ocean Evolution

Landmass volcanism has been used previously in discussions of the Indian Ocean area to record times of major rifting and subduction. For example, Davies (1968) used evidence of synchronous early Tertiary volcanic activity in India and the Seychelles to demonstrate the time for their separation and Flores (1970) regarded synchronous volcanism in Madagascar and Africa as indicative of rifting between those two landmasses. The history and evolution of the Indian Ocean basin has been well documented by using magnetic anomaly patterns, which are quite adequate for about the last 80 m.y. (since anomaly 33). However, before that time, interpretations are based on poorly defined magnetic anomalies, the sediment record, and the surrounding landmass geologic histories. Therefore, conclusions based on landmass volcanism might hold a key to the evolution of the Indian Ocean Basin, particularly from Late Jurassic to Late Cretaceous times. Similar ages of volcanic activity on landmasses that are believed to have been close together in the geologic past might signify times of major tensional stress. Evidence from landmass volcanism would be strengthened if large parts of the ocean floor were formed by rapid spreading at the same time as the landmass volcanism.

The oldest volcanism, recorded in southeastern Africa, Madagascar, Australia, and India probably marks the initial breakup and rifting of Gondwanaland in the Late Jurassic and Early Cretaceous. A second pulse in Madagascar, India, and the Seychelles during the Late Cretaceous and early Tertiary may indicate additional tensional stresses and volcanism related to further movement of India from Madagascar, Africa, and the Seychelles. Coincidentally, this also was a time of rapid sea floor spreading (Schlich, 1974). Then, in the Eocene-Recent interval volcanism occurred in northeastern Africa and Arabia (Coleman, 1974), probably associated with rifting around the Red Sea. Oligocene-Recent volcanism marks the latest pulse along the Indonesian Arc. Miocene-Recent volcanism has occurred in Madagascar, East Africa, and on the oceanic islands such as the Mascarene, Comoro, and Crozet islands. The Oligocene-Recent volcanism is mostly associated with growth of oceanic islands on the mid-ocean ridge system and associated fracture zones, with subduction of the Indian Ocean Plate beneath Asia, and with rifting in East Africa.

Conclusions

Volcanogenic components are significant contributors to sediments in the world oceans both volumetrically and as indicators of ocean basin tectonism. Kennett and Thunnell (1975), working mostly with silicic volcanic glass, were able to recognize major explosive events in the Plio-Pleistocene which are related to compressive activity associated with subduction along island arcs. However, significant explosive activity apparently is related to tensional events also, particularly along the mid-ocean ridge system. The amount of volcanic detritus in marine sediments generally has been underestimated because it is difficult to recognize and is easily affected by diagenetic processes which destroy primary characteristics.

Clay mineral ratios seem particularly worthwhile in the evaluation of relative amounts of volcanic components in sediment columns that have undergone diagenesis. When these clay mineral ratios are combined with the presence of other components such as primary volcanic glass and minerals, zeolites, and silica minerals, and with stratigraphic studies which may indicate volcanic ash beds, the relative importance of volcanic components in a sediment column can be determined with some confidence.

Volcanic components probably are more widely and effectively dispersed in the world oceans than any other non-biogenic sediment contributor because of the available methods of transport. Wind is the most effective agent for dispersal of fine ash from explosive eruptions, but it can also be inferred that ocean currents carry volcanic components long distances.

The time and space relationships of volcanogenic sediments in the Indian Ocean are now fairly well established. Some of the major accumulations can be correlated with tectonic and other events such as the initial opening of the Indian Ocean, formation of presently aseismic ridges, rates of sea floor spreading, subduction of the Indian Plate along the Indonesian Arc, and landmass volcanism. Volcanic activity contributed sediments to at least some part of the Indian Ocean at any given time, but on a regional scale specific pulses can be recognized. The nature of volcanism has changed with time; basaltic volcanism was dominant before the late Cenozoic and silicic volcanism gained intensity during the late Cenozoic. Basaltic volcanism is related to tensional tectonism associated with early opening of the basin, mid-ocean ridge volcanism, and the formation of aseismic ridges, whereas the silicic volcanism is mostly related to the compressional tectonism along the Indonesian Arc.

There may be a causal relationship between rates of sea floor spreading, or changes in rates and directions of spreading, and amounts of volcanic sediments, but it is not proven in this study. Basal accumulations in some parts of the ocean basin were formed during early stages of basin formation when spreading rates were coincidentally high. For example, the Late Cretaceous pulse in the southeastern Indian Ocean and the Paleocene-Eocene pulse in the southwestern Indian Ocean probably correspond to early opening and rapid spreading rates respectively. The Oligocene pulse may be related to changes in rates and directions when the present day pattern was established and the silicic activity is related to the initial subduction of the Indian Plate under the Indonesian Arc. Landmass volcanism can be roughly correlated with sea floor volcanism and plate movements. In this regard, Cretaceous basaltic volcanism may relate to tensional events associated with opening of the ocean. Also, Indonesian Arc silicic volcanism can be correlated with the subduction of sea floor that apparently began in the Oligocene.

Aseismic ridges have basaltic volcanic detritus either directly above acoustic basement or concentrated higher in the sediment column. The basal sediments probably are related to volcanism that occurred during the initial formation of the ridges, which took place at an intersection of the mid-ocean ridge and a transform fault. Subaerial or shallow submarine volcanism was nearly continuous at these intersections as shown by the coarse accumulations on Ninetyeast, Chagos-Laccadive, and Broken ridges. Coarse basaltic sediments higher in the columns at some ridge sites (e.g. Mozambique Ridge) may be related to the final uplift that formed this present day topography.

Acknowledgements. We thank James Hein, Harry Cook, Oscar Weser, and Stan White who read early drafts of the manuscript and made several helpful suggestions. The assistance of colleagues and repository staff at the Deep Sea Drilling Project, Scripps Institution of Oceanography, is gratefully acknowledged. We are grateful for the use of illustrating, typing, and photographic services at the U. S. Geological Survey in Menlo Park, California. Samples and some data used in this report were provided by the National Science Foundation.

Robert Kidd acknowledges the help of Dr. M.N.A. Peterson, project manager of the Deep Sea Drilling Project, and Professor H. Charnock, I.O.S., in arranging his visit to the DSDP headquarters and west coast repository, La Jolla, California, in order to study the DSDP core materials.

References

Arrhenius, G., Pelagic sediments, in The Sea, edited by M. N. Hill, Interscience Publications, New York, 3, 655-727, 1963.

Besairie, Henri, Geologie de Madagascar, I. Les Terrains Sedimentaires, An. Geol. de Madagascar, 35, Tanarive, Madagascar, 463, 1972.

Blant, G., Structure et paleogeographie du littoral meridional et oriental de l'Afrique, in Sedimentary Basins of the African Coasts, 2, south and east coast, edited by G. Blant, Assoc. African Geol. Surveys, Paris, 193-233, 1973.

Bonatti, E., Authigenesis of minerals-Marine, in the Encyclopedia of Geochemistry and Environmental Sciences, edited by R. W. Fairbridge, Van Nostrand Reinhold Company, New York, 52, 1972.

Bonatti, E., and O. Joensuu, Palygorskite from Atlantic deep sea sediments, Am. Mineral., 53, 975-983, 1968.

Bostrom, K., and M.N.A. Peterson, Precipitates from hydrothermal exhalations in the East Pacific Rise, Econ. Geol., 61, 1258-1266, 1966.

Bradshaw, M. J., Origin of montmorillonite bands in the Middle Jurassic of eastern England, Earth and Planet. Sci. Lett., 26, 245-252, 1975.

Brown, D. A., K. S. Campbell, and K.A.W. Cook, The geological evolution of Australia and New Zealand, Pergamon Press, London, 409, 1968.

Coleman, R. G., Geologic background of the Red Sea, in Whitmarsh, R. B., Weser, O. E., Ross, D. A. et al., Initial Reports of the Deep Sea Drilling Project, 23, Washington (U. S. Government Printing Office), 813-819, 1974.

Cook, H. E., P. D. Johnson, J. C. Matti, and I. Zemmels, Methods of sample preparation and X-ray diffraction data analysis, X-ray mineralogy laboratory, Deep Sea Drilling Project, University of California, Riverside, in Hayes, D. E., Frakes, L. A. et al., Initial Reports of the Deep Sea Drilling Project, 28, Washington (U. S. Government Printing Office), 999-1007, 1975.

Cronan, D. S., V. V. Amiani, D.J.J. Kinsman, and J. Thiede, Sediments from the Gulf of Aden and Western Indian Ocean, in Fisher, R. L., Bunce, E. T., et al., Initial Reports of the Deep Sea Drilling Project, 24, Washington (U. S. Government Printing Office), 1047-1110, 1974.

Davies, D., When did the Seychelles leave India?, Nature, 220, 1225-1226, 1968.

Davies, T. A., B. P. Luyendyk, et al., Initial Reports of the Deep Sea Drilling Project, 26, Washington (U. S. Government Printing Office), 1129 pp., 1974.

Davies, T. A., O. E. Weser, B. P. Luyendyk, and R. B. Kidd, Unconformities in the sediments of the Indian Ocean, Nature, 253, 15-19, 1975.

Dymond, J., J. B. Corliss, G. R. Heath, C. W. Field, E. J. Dasch, and E. H. Veeh, Origin of metalliferous sediments from the Pacific Ocean Bull. Geol. Soc. Amer., 84, 3355-3372, 1973.

Fisher, R. L., J. G. Sclater, and D. McKenzie, Evolution of the Central Indian Ridge, western Indian Ocean, Bull. Geol. Soc. Amer., 82, 553-562, 1971.

Fleet, A. J., and D.R.C. Kempe, Preliminary geochemical studies of the sediments from DSDP Leg 26, southern Indian Ocean, in Davies, T. A., Luyendyk, B. P., et al., Initial Reports of the Deep Sea Drilling Project, 26, Washington (U. S. Government Printing Office), 541-551, 1974.

Flores, Giovanni, Suggested origin of the Mozambique Channel, Geol. Soc. Africa Trans., 73, pt. 1, 1-16, 1970.

Gibson, T. G., and K. M. Towe, Eocene volcanism and origin of Horizon A, Science, 172, 152-154, 1971.

Goldberg, E. D., and J. W. Griffin, The sediments of the northern Indian Ocean, Deep Sea Res., 17, 513-537, 1970.

Hathaway, J. C., and P. L. Sachs, Sepiolite and clinoptilolite from the Mid-Atlantic Ridge, Am. Mineral., 50, 852-867, 1965.

Hathaway, J. C., P. F. McFarlin, and D. A. Ross, Mineralogy and origin of sediments from drill holes on the continental margin off Florida, U. S. Geol. Surv. Prof. Paper 581-E, 26 pp., 1970.

Haughton, S. H., The stratigraphic history of Africa south of the Sahara, Edinburgh and London (Oliver and Boyd), 365 pp., 1963.

Hayes, James D., and D. Ninkovich, North Pacific deep-ash chronology and age of present Aleutian underthrusting, Geol. Soc. Amer. Mem. 126, 263-290, 1970.

Huang, T. C., N. D. Watkins, and D. M. Shaw, Atmospherically transported volcanic glass in deep-sea sediments: Volcanism in sub-Antarctic latitudes of the south Pacific during late Pliocene and Pleistocene time, Bull. Geol. Soc. Amer., 86, 1305-1315, 1975.

Katili, John A., Volcanism and plate tectonics in the Indonesian island arcs, Tectonophysics, 26, 165-188, 1975.

Kennett, J. P., and R. C. Thunell, Global increase in Quaternary explosive activity, Science, 187, 497-503, 1975.

Luyendyk, B. P., Gondwanaland dispersal and the early formation of the Indian Ocean, in Davies, T.A., Luyendyk, B.P., et al., Initial Reports of the Deep Sea Drilling Project, 26, Washington (U. S. Government Printing Office) 945-952, 1974.

Luyendyk, B. P., and T. A. Davies, Results of DSDP Leg 26 and the geologic history of the Southern Indian Ocean, in Davies, T.A., Luyendyk, B.P., et al., Initial Reports of the Deep Sea Drilling Project, 26, Washington (U. S. Government Printing Office), 909-943, 1974.

Marchig, V., and T. L. Vallier, Geochemical studies of sediment and interstitial water, sites 248 and 249, Leg 25, Deep Sea Drilling Project, in Simpson, E.S.W., Schlich, R., et al., Initial Reports of the Deep Sea Drilling Project, 25, Washington (U. S. Government Printing Office), 405-416, 1974.

Matter, Albert, Burial diagenesis of pelitic and carbonate deep-sea sediments from the Arabian Sea, in Whitmarsh, R.B., Weser, O.E., Ross, D.A., et al., Initial Reports of the Deep Sea Drilling Project, 23, Washington (U. S. Government Printing Office), 421-443, 1974.

Matti, J. C., I. Zemmels, and H. E. Cook, X-ray mineralogy data, northeastern part of the Indian Ocean, Leg 22, Deep Sea Drilling Project, in von der Borch, C.C., Sclater, J.G., et al., Initial Reports of the Deep Sea Drilling Project, 22, Washington (U. S. Government Printing Office), 693-710, 1974.

Mattson, P. H., and E. A. Pessagno, Caribbean Eocene volcanism and the extent of Horizon A, Science, 174, 138-139, 1971.

McBirney, A. R., Factors governing the nature of submarine volcanism Bull. Volcanologique, 26, 455, 1963.

McBirney, A. R., Comments, Earth Science, 2, 69, 1971.

McDougall, I., and F. H. Chamalaun, Isotopic dating and geomagnetic studies on volcanic rocks from Mauritius, Indian Ocean, Bull. Geol. Soc. Amer., 80, 1419-1442, 1969.

McElhinney, M. W., Formation of the Indian Ocean, Nature, 288, 977-979, 1970.

McKelvey, B. C., and A. J., Fleet, Eocene basaltic pyroclastics at site 253, Ninetyeast Ridge, in Davies, T.A., Luyendyk, B.P., et al., Initial Reports of the Deep Sea Drilling Project, 26, Washington (U. S. Government Printing Office), 553-565, 1974.

Millot, G., Geology of clays, Masson, Paris, France, 429 pp., 1970.

Millot, G., H. Paquet, H., and A. Ruellan, Neoformation d'attapulgite dans les sols a carapaces calcaires de la Basse Monlonya (Moroc Oriental), Acad. Sci. (Paris) Comptes rendus, 268, (ser. D), 2771-2773, 1969.

Moore, D. G., J. R. Curray, R. W. Raitt, and F. J. Emmel, Stratigraphic seismic section correlations and implications to Bengal Fan history, in von der Borch, C. C., Sclater, J. G., et al., Initial Reports of the Deep Sea Drilling Project, 22, Washington (U. S. Government Printing Office), 403-412, 1974.

Nayudu, Y. R., Palgonite tuffs (hyaloclastites) and the products of post-eruptive processes, Bull. Volcanologique, 27, 332-341, 1964.

Nelson, C. H., L. D. Kulm, P. R. Carlson, and J. R. Duncan, Mazama ash in the northeastern Pacific, Science, 161, 47, 1968.

Ninkovich, D., Late Cenozoic clockwise rotation of Sumatra, Earth Planet. Sci. Lett., 1977.

Peterson, M.N.A., and J. J. Griffin, Volcanism and clay minerals in the southeastern Pacific, Jour. Mar. Res., 22, 287-312, 1964.

Pimm, A. C., Sedimentology and history of the northeastern Indian Ocean from Late Cretaceous to Recent, in von der Borch, C. C., Sclater, J. G., et al., Initial Reports of the Deep Sea Drilling Project, 22, Washington (U. S. Government Printing Office), 717-804, 1974.

Robinson, P. T., P. A., Thayer, P. J. Cook, and B. K. McKnight, Lithology of Mesozoic and Cenozoic sediments of the eastern Indian Ocean, Leg 27, Deep Sea Drilling Project, in Veevers, J. J., Heirtzler, J. R., et al., Initial Reports of the Deep Sea Drilling Project, 14, Washington (U. S. Government Printing Office), 1001-1047, 1974.

Schlich, R., Sea floor spreading history and deep-sea drilling results in the Madagascar and Mascarene Basins, western Indian Ocean, in Simpson, E.S.W., R. Schlich, et al., Initial Reports of the Deep Sea Drilling Project, 25, Washington (U. S. Government Printing Office) 663-678, 1974.

Sclater, J. G., D. Abbott, and J. Thiede, Paleobathymetry and sediments of the Indian Ocean, in Indian Ocean Geology and Biostratigraphy, AGU, Washington, 1977 (this volume).

Siever, R., and M. Kastner, Mineralogy and petrology of some Mid-Atlantic Ridge sediments, Marine Res., 25, 263-278, 1967.

Stonecipher, S. A., Distribution of zeolites in marine sediments, Bull. Geol. Soc. Amer. Abstracts with Programs, Cordilleran Section, 1974 Annual Meeting, 262-263, 1974.

Thiede, J., Sediment coarse fractions from the western Indian Ocean and the Gulf of Aden, Deep Sea Drilling Project, in Fisher, R. L., Bunce, E.T., et al., Initial Reports of the Deep Sea Drilling Project, 24, Washington (U. S. Government Printing Office), 651-766, 1974.

Truter, F. C., A review of volcanism in the geological history of South Africa (anniversary address by the president), Geol. Soc. Africa Trans., 52, 29-88, 1949.

Vallier, T. L., Volcanogenic sediments and their telation to landmass volcanism and sea floor-continent movements, western Indian Ocean, Leg 25, Deep Sea Drilling Project, in Simpson, E.S.W., Schlich, R., et al., Initial Reports of the Deep Sea Drilling Project, 25, Washington (U. S. Government Printing Office) 515-542, 1974.

van Bemmelen, R. W., The geology of Indonesia, Government Printing Office, The Hague, 732, 1949.

Venkatarathnam, K., Mineralogical data from sites 211, 212, 213, 214, and 215 of the Deep Sea Drilling Project, Leg 22, and origin of non-carbonate sediments in the Equatorial Indian Ocean, in von der Borch, C.C., Sclater, J. G., et al., Initial Reports of the Deep Sea Drilling Project, 22, Washington (U.S. Government Printing Office), 489-502, 1974.

von der Borch, C. C., J. G. Sclater, et al., Initial Reports of the Deep Sea Drilling Project, 22, Washington (U. S. Government Printing Office), 890 pp., 1974.

von Rad, U., and H. Rosch, Mineralogy and origin of clay minerals, silica and authigenic silicates in Leg 14 sediments, in Hayes, D. E., Pimm, A.C., et al., Initial Reports of the Deep Sea Drilling Project, 14, Washington (U. S. Government Priting Office), 727-751, 1972.

Warner, T. B., and J. M. Gieskes, Iron-rich basal sediments from the Indian Ocean: Site 245, Deep Sea Drilling Project, in Simpson, E.S.W., Schlich, R., et al., Initial Reports of the Deep Sea Drilling Project, 25, Washington (U. S. Government Printing) 395-403, 1974.

Watkins, N. D., B. M., Gunn, J. Nougier, and A. K. Baksi, Kerguelen: continental fragment or oceanic island? Bull. Geol. Soc. of Amer., 85, 201-212, 1974.

Weser, O. E., Sedimentological aspects of strata encountered on Leg 23 in the northern Arabian Sea, in Whitmarsh, R. B., Weser, O. E., Ross, D. A., et al., Initial Reports of the Deep Sea Drilling Project, 23, Washington (U. S. Government Printing Office), 503-519, 1974.

Whitmarsh, R. B., N. Hamilton, and R. B. Kidd, Paleomagnetic results for the Indian and Arabian Plates from Arabian Sea Cores, in Whitmarsh, R. B., Weser, O.E., Ross, D.A., et al., Initial Reports of the Deep Sea Drilling Project, 23, Washington (U. S. Government Printing Office), 521-525, 1974.

Wise, S. W., and F. M. Weaver, Chertification of oceanic sediments, in Hsu, K.J., and Jenkyns, H. C., Pelagic Sediments: On Land and Under the Sea, Int. Assoc. of Sedimentologists Spec. Pub. No. 1, 301-326, 1974.

CHAPTER 5. MESOZOIC-CENOZOIC SEDIMENTS OF THE EASTERN INDIAN OCEAN

Peter J. Cook

The Australian National University
Box 4, P. O.
Canberra A.C.T., Australia 2600

Abstract: Nineteen sites have been drilled in the Eastern Indian Ocean. Sediments range in age from Oxfordian (155 m.y.) to Recent. In the abyssal zone they are composed primarily of pelagic clay, calcareous ooze, or siliceous ooze. Locally, the clays contain zeolites, and nannofossil remains. The clays are commonly siliceous, the silica is mainly terrigeneous, but some is biogenic, and a small amount may be volcanic. The abyssal zone calcareous oozes are composed mainly of foraminiferal and nannoplankton remains and have probably been deposited below the CCD by turbidity currents or slumping. The siliceous oozes appear to be associated with the equatorial zone of high productivity. On the ridge and plateau sites, calcareous oozes are abundant; other sediment types include volcanogenic mud, sand, and gravel, pyroclastics, and glauconitic calcareous sediments. Terrigeneous silts and clays associated with the late Cenozoic Bengal Fan form a thick sequence in the northwest corner of the Eastern Indian Ocean.

Compositionally, the Eastern Indian Ocean sediments are similar to oceanic sediments from elsewhere, although less rich in some metals. The $CaCO_3$ content of the sediments varies markedly, though the general trend is of poorly calcareous Mesozoic sediments (with a significant increase in the Albian) and richly calcareous Cenozoic sediments. The highest rates of sedimentation appear to correspond with times of maximum development of calcareous sediments. The pattern of sedimentation for the Mesozoic suggests that most of the sediment was derived from Australia, with a minor source from the south (Antarctica) and possibly also a small contribution from the northwest. In the Cenozoic the dominant sediment source was still from the east but was mainly calcareous material derived from the shelf and slope. Bengal fan sediments form a major contribution in the northwest. Overall, sedimentation was discontinuous at most sites, with numerous unconformities within the sequence. The northward movement of India was important as it opened a seaway to the west of Australia and enabled cool erosive currents to enter the region, thus modifying the sediment pattern and producing major hiatuses. Subsequently, the influx of Antarctic Bottom Water, and later the development of the Circum-Antarctic Current were important in influencing sedimentation in the Eastern Indian Ocean.

Introduction

The Eastern Indian Ocean is a geologically complex area of ridges, plateaus and basins. It has been investigated by Legs 22, 26 and 27 of the Deep Sea Drilling Project, and also to a limited extent by Leg 28. Nineteen sites were drilled within the study area (Figure 1) in a

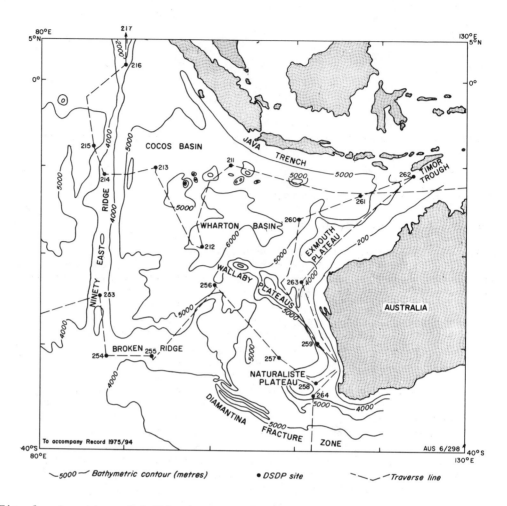

Fig. 1. Location of drill sites, cruise tracks and major physiographic features in the Eastern Indian Ocean.

variety of physiographic-geologic locations ranging in water depth from 1253 m to 6243 m, including the Naturaliste Plateau (sites 258, 264); on ocean ridges such as Broken Ridge (site 255), and the Ninetyeast Ridge (sites 214, 217, 253, 254); and on the Bengal Fan (site 218). Nine sites (211, 212, 213, 256, 257, 259, 260, 261, 263) were drilled in ocean basin locations (the Wharton Basin and its subsidiary Cocos Basin). The Bengal Fan sediments (site 218) are dealt with elsewhere in this publication and consequently will not be considered in detail here. Site 262 in the Timor Trough is regarded as being geologically outside the Indian Ocean. One site (215) was drilled in the central Indian Basin. The stratigraphic information obtained from the nineteen drill sites in the Eastern Indian Ocean is summarized in Figure 2. Up to 773 m of sediment was drilled in the Bengal Fan (site 218), 746 m in the Wharton Basin (site 263), and 664 m on the Ninetyeast Ridge (site 217). The oldest basement penetrated in the region was at site 261 (155 million years) in the northeast corner of the Wharton Basin. This paper will examine the regional stratigraphy, sedimentology and geochemistry, and will attempt to relate them to the history and paleo-environments of the Mesozoic-Cenozoic Eastern Indian Ocean.

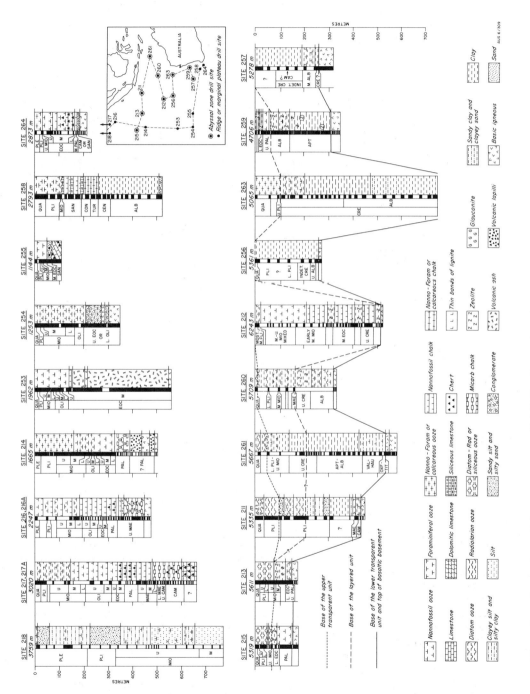

Fig. 2. Stratigraphic sequence of DSDP sites in the Eastern Indian Ocean.

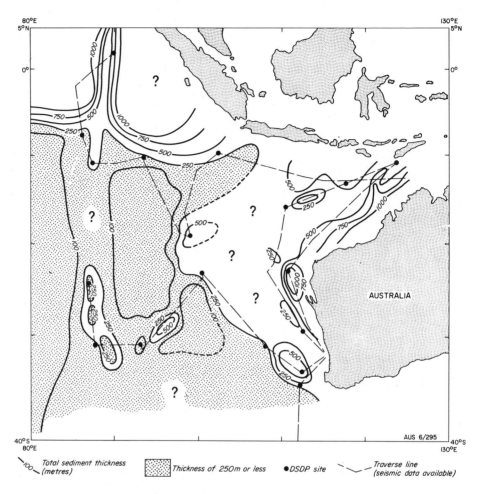

Fig. 3. Isopachous map of total sediment thickness in the Eastern Indian Ocean as determined from Ewing et al. (1969) and DSDP seismic profiles.

Sediment Thickness and Distribution

The sediments of the Eastern Indian Ocean range in age from Oxfordian (155 m.y.) to Recent, and in thickness from 100 m or less in the western portion of the region to more than 1000 m adjacent to the Australian continental shelf and within the Bengal Fan. The isopachous map obtained by incorporating the results of Ewing et al. (1969) and the DSDP seismic profiles (Figure 3) serves to illustrate the dominant sediment sources in the region; the Australian continent to the east and the Ganges and Brahmaputra Rivers to the northwest. The slight increase in sediment thickness near the eastern end of the Indonesian archipelago is thought to be the result of increased sedimentation associated with the equatorial zone of high productivity and proximity to the volcanic arc. Also apparent from figure 3 is the relatively thick sedimentary sequence on many ridges, plateaus and rises, where thick carbonate sequences have accumulated. Conversely, in the ocean basins the sedimentary sequence is comparatively thin in most areas.

Previous workers (Ewing et al., 1969; Veevers, 1974; Veevers and Heirtzler, 1974a) have shown that in many areas, particularly the abyssal

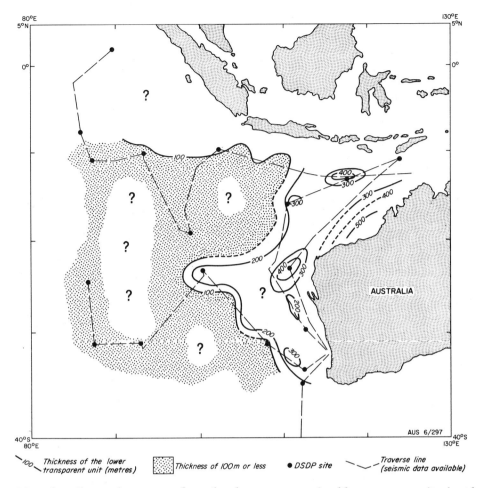

Fig. 4. Isopachous map for the lower acoustically opaque unit in the Eastern Indian Ocean as determined from DSDP seismic profiles.

zone, the sedimentary sequence consists of three acoustically distinct zones: a lower transparent unit, a middle layered unit, and an upper transparent unit. Deep Sea Drilling (Figure 2) has revealed that the lower transparent unit is a brown, grey and green clay, and claystone with minor nannofossil ooze. The layered unit consists predominantly of light-colored nannofossil or foraminiferal-nannofossil ooze with some graded beds. The upper transparent unit is siliceous diatom-radiolarian ooze. On the shallower ridges and plateaus, the three-fold acoustic division is generally absent. In the ocean basins, the boundary between the lower transparent and the layered unit is very sharp, possibly corresponding to an unconformity in many areas. Comparison of the acoustic record with the drilling results (Figure 2) indicates that in most areas this boundary may be correlated with a major Mesozoic-Tertiary break. The only instance where the boundary between these two acoustic units extends down into the Cretaceous is at site 212 (von der Borch et al., 1974). Conversely, to the south the lower transparent unit extends up into the Cenozoic (e.g. at site 257).

Looking briefly at the extent and thickness of the three acoustic units (Figures 4-6) it is apparent that they are geometrically quite distinct sedimentary bodies. The lower transparent unit (Figure 4) shows

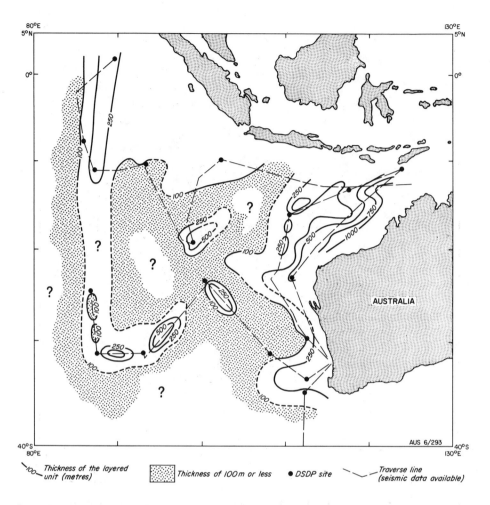

Fig. 5. Isopachous map for the layered unit in the Eastern Indian Ocean as determined from DSDP seismic profiles.

marked thinning to west, consistent with an easterly source. The layered unit has a more complex distribution pattern (Figure 5). Again the section is thickest to the east, particularly adjacent to the northwest Australian shelf, but in addition, it is also thick on the ridges and plateaus. In the abyssal zone it becomes much thinner and is absent in many places. Locally, it is ponded in deep basins (near site 212). A second type of acoustically layered sediment as exemplified by the seismic profile at site 218 (von der Borch et al., 1974, p. 331) is present near the Bengal-Nicobar Fan. However, unlike the layered sediments elsewhere, these sediments are non-calcareous or poorly calcareous, and represent the distal portion of the Bengal Fan turbidites. These sediments are discussed elsewhere in this volume. The upper transparent unit (Figure 6) is of very limited extent, occurring only at sites 211, 213, 215, 260 and 261. It is present as an east-west band across the northern part of the Eastern Indian Ocean, thinning rapidly to the south. Most of the sediment is siliceous biogenic material associated with equatorial high productivity. A minor amount of volcanic ash probably derived from the Indonesian archipelago is present in the sequence.

Seismic data from the ridges and plateaus are not abundant, although as mentioned earlier the layered unit (calcareous oozes) is commonly thicker

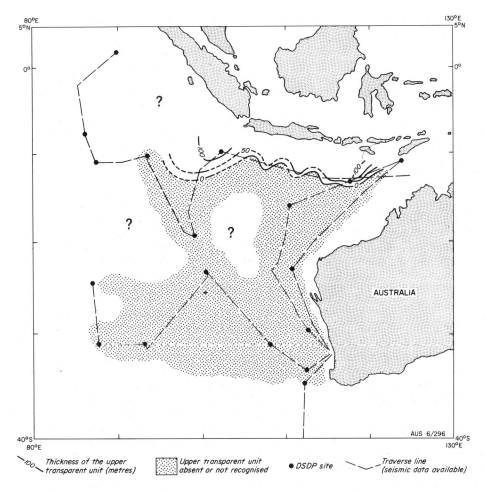

Fig. 6. Isopachous map for the upper acoustically opaque unit in the Eastern Indian Ocean as determined from DSDP seismic profiles.

on the shallow features than in the abyssal zone. An additional trend which might be expected is that as the Ninetyeast Ridge becomes progressively older and deeper to the north (Pimm, 1974a; Sclater and Fisher, 1974; Luyendyk and Davies, 1974), the sediments would also become thicker. This does not appear to be the case either from the isopachous map (Figure 3) or from the information provided by drill sites on the ridge (sites 217, 216, 214, 253 - Figure 2). In fact, sediment thickness is highly variable along this ridge; probably due to a combination of hiatuses in the sedimentary record, locally thick sequences of ponded sediments, and localized input of volcanogenic material, such as at site 253 (Figure 2). The sediment thickness across marginal plateaus is rather more uniform, although Petkovic (1975a, b) has shown that the sediment thickness on the Naturaliste Plateau ranges from 1 km or less on the western flank, to over 2 km on the eastern flank.

Lithology

The lightology of sediments at DSDP sites in the Eastern Indian Ocean has been previously discussed by Pimm (1974a), Luyendyk and Davies (1974), Robinson et al. (1974), and Bezrukov (1974), and this summary draws extensively upon their work. It has been indicated earlier that the threefold acoustic division of the sedimentary sequence can be related in a general way to lithology. However, the detailed picture (Figure 2) is complex with a wide range of lithologies present at both the deep-water and shallow-water drill sites. These would be grouped into a smaller number of sediment types. The lateral and vertical disposition of these sediment types in the abyssal zone and shallow-water sites is shown in Figure 7 which illustrates schematically the facies changes which occur between drill sites. In the abyssal zone the trend is again abundant siliceous oozes to the north and west, calcareous oozes in the center, and clays to the south. A feature now evident is the occurrence of clay containing nannofossils, at the base of the sequence.

The lateral changes in the shallower drill sites are rather different. Calcareous oozes are dominant and there is a general lack of clays on ridge and plateau sites, except at site 258 on the Naturaliste Plateau where clays are abundant at the base of section. Volcanogenic sediments are common in places. Cherty limestones are also unique to the ridges and plateaus. The various sediment types will now be discussed.

Detrital Sediments

Siliceous clay and claystone. Siliceous clay and claystone occur throughout the Mesozoic-Cenozoid of the abyssal zone drill sites but are particularly abundant in the Mesozoic. They contain the highest percentage of clay-size material (82% clay, see Table 1) and are also the most siliceous of the sediments, with an average SiO_2 content of 59.3 percent (Table 2). Most of the silica is in the form of microcrystalline or cryptocrystalline quartz and crystobalite. Robinson et al. (1974) consider that .. "The degree of induration (of the clays) generally increased with depth at any given site and correlates directly with the percent of microcrystalline quartz and cristobalite in the sediment." The clays range from brown, green and grey to black. The classical brown or redbrown pelagic clay is comparatively rare. Generally, the darker the clay the higher the percentage of organic carbon; the clays of site 263, contain up to 10 percent organic carbon, compared with an average of about 0.5 percent organic carbon in most deep sea clays. Fossils are generally absent or rare, but undoubted bioturbation or color mottling of probable biogenic origin is common in places. Where there has been little or no biogenic activity the sediments are finely laminate.

Mineralogically, siliceous clay and claystone are composed of approximately equal amounts of quartz/crystobalite and clay minerals, with minor to trace feldspar, iron oxides, zeolites, calcite (from nannofossils), detrital heavy minerals, and apatite (mainly from fish remains, though Cook (1974a) records a thin phosphorite at site 259). Some of the terrigenous silica is brought to the area by aeolian transport particularly from Australia. However, Cook (1974b) suggests on the basis of an SiO_2-Ba correlation that a significant proportion of the silica is of biogenic origin. The clay minerals appear to be predominantly of detrital origin, with kaolinite most common and illite, chlorite, montmorillonite and palygorskite present in varying amounts. Venkatarathnam (1974), has shown that three provinces may be delineated in the Eastern Indian Ocean on the basis of sediment source - an Australian Province, an Indonesian Province, and a Ganges Province. The abundance of kaolinite in the siliceous clay and claystone indicates their location was mainly within an ancestral Australian province. Kaolinite is particularly common at site 263 with the kaolinite occurring not only in the form of dispersed clay-size material but also as pellets, some of which may have formed in situ as kaolinite blebs or by the alteration of pre-existing pellets

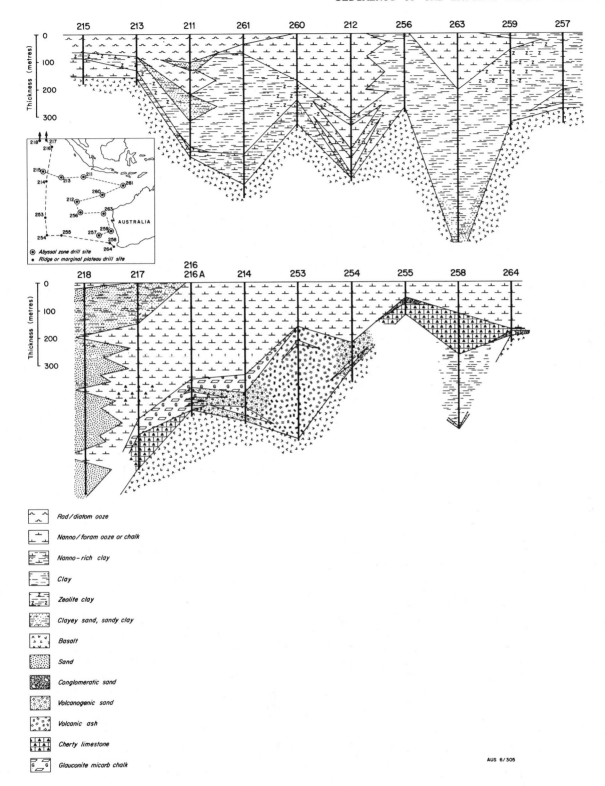

Fig. 7. Vertical and lateral changes in sedimentary facies as determined from deep water (A) and shallow water (B) drill sites in the Eastern Indian Ocean.

TABLE 1. The sand-silt-clay distribution of various sediments from DSDP sites in the Eastern Indian Ocean. Number in brackets indicates the number of analyses used for determining the mean.

		% Sand	% Silt	% Clay
Foram-nanno ooze	(80)	25.6	41.5	32.9
Rad-diatom ooze	(2)	5.9	33.0	61.1
Nanno ooze	(53)	3.1	51.1	45.8
Micarb chalk	(4)	0.1	35.3	64.6
Clay	(95)	0.8	17.0	82.2
Zeolitic clay	(19)	0.7	21.0	78.3
Volcanogenic sands & muds	(9)	12.9	40.2	46.9
Volcanic ash	(29)	29.9	30.1	40.1

(see Figure 8 and Plate 4, Figure 6 of Robinson et al., 1974). Others are probably of intraclastic origin. This abundance of kaolinite together with a high sedimentation rate suggests that site 263 was close to a fluviatile source. The occurrence of calcareous and arenaceous foraminifera, and well sorted, cross-bedded sandstones may indicate comparatively shallow (?bathyal rather than abyssal) depositional conditions. Elsewhere, siliceous clays were mainly deposited in deep water. Davies and Kidd (1977) point out that the presence of pyrite in some Aptian-Albian clays in the Wharton Basin may indicate restricted depositional conditions at that time. The volcanic component in the clays throughout much of the Eastern Indian Ocean seems to be comparatively minor, although at site 211 montmorillonite is as abundant as kaolinite. Whether the montmorillonite is detrital or has formed by in situ alteration of volcanic material is not certain. In addition to modifying the mineral assemblage, diagenesis in the siliceous clays also results in nodules of dolomite, barite, siderite and calcite. Manganese micro-nodules are present but comparatively rare.

Zeolitic clay. Zeolitic clay occurs sporadically throughout the post-Aptian/pre-Pliocene section of the abyssal zone drill sites (Figure 2). It is similar in appearance and composition to siliceous clay and claystone, though less abundant. It varies in color from brown to yellow-brown; bedding ranges from poor to moderate and from lenticular to laminate. Bioturbation is seldom present. The presence of zeolites (up to 10% of the total sediment) is reflected in the abundance of silt-size crystals (Table 1) and higher percentages of Na_2O and CaO (Table 2). Pimm (1974a) suggests on the basis of the Leg 22 data that clinoptilolite is the dominant zeolite in Eocene and older sediments whereas phillipsite is dominant in the younger sediments. Elsewhere, this trend is not evident. In general the zeolite crystals are euhedral; in some instances they are scattered throughout the clay; commonly they are present as clusters, in some cases infilling radiolarian molds (Robinson et al., 1974; Plate 1). This mode of occurrence suggests that the zeolite crystals have undergone little or have formed diagenetically. Although there is a notable absence of volcanic glass, a volcanic association of the zeolites is supported by the abundance of palygorskite and montmorillonite in the zeolitic clays.

Nannofossil clay. Nannofossil-rich clay represents an intermediate member in a continuous sediment sequence between the two end members,

TABLE 2. Average major element composition of various sediment types from DSDP sites in the Eastern Indian Ocean, and other deep-sea sediments

Oxide (Percent) [1]	Foram-nanno ooze (10)	Nanno clay (6)	Rad-diatom ooze (1)	Zeolotic clay (10)	Clay and Claystone (21)	Glauconitic silty clay (1)	Volcanic ash (20)	Average Pelagic Sediment[4] Calcareous	Average Pelagic Sediment[4] Clayey	Average Pelagic Sediment[4] Siliceous
SiO_2	20.4	50.3	56.0	56.2	59.5	26.8	29.4	27.0	55.5	64.0
Al_2O_3	3.0	13.0	14.1	10.7	11.9	12.8	3.6	8.0	17.6	13.4
Fe_2O_3 [2]	2.6	6.7	4.3	5.9	7.4	12.3	9.17	3.9	8.3	6.3
CaO	35.3	5.5	1.1	4.7	1.6	0.8	5.7	28.5	1.4	1.6
MgO	1.3	2.8	2.3	2.3	2.5	1.8	5.65	2.3	3.8	2.5
Na_2O	1.6	2.1	4.8	2.8	1.8	–	2.95	0.8	1.5	0.9
K_2O	0.9	3.9	2.2	2.7	2.15	1.0	1.4	1.5	3.3	1.9
MnO	0.21	0.18	1.2	0.4	0.25	0.05	0.17	0.32	0.47	0.41
P_2O_5	0.08	0.22	0.14	0.11	0.24	–	–	0.15	0.14	0.27
TiO_2	0.18	0.77	0.55	0.51	0.59	1.28	0.5	0.44	0.84	0.65
CO_2	28.1	2.0	2.2	2.2	1.2	–	7.4	23.3	0.77	0.93
H_2O [3]	4.1	11.4	12.8	9.9	8.3	–	–	3.9	6.5	7.1
Total	97.8	98.8	101.1	98.4	96.6			100.1	100.0	99.9

1. Number in brackets indicates the number of analyses used for determining the mean.
2. Total iron expressed as Fe_2O_3.
3. Total H_2O.
4. Average pelagic sediment recalculated from El Wakeel and Riley (1961).

siliceous clay and nannofossil ooze, and is not really a distinctive sediment type as are most of the other sediments discussed. This type of clay is found mainly in the abyssal zone drill holes, but also occurs at site 217, as a result of the mixing of nannofossil ooze with distal clays of the Bengal Fan. Nannofossil clay varies in composition from siliceous clay primarily in the abundance of calcium carbonate (from nannofossil tests), averaging 7.4 percent total CaO plus CO_2 (Table 2). It ranges in age from Valanginian to Recent, but is most abundant at the base of the sequence (Figure 7). This is particularly so at sites 213, 211, 261, 260, and 257. The implication of this preferred stratigraphic location is that this location corresponded with a time shortly after the formation of a mid-ocean ridge when the sea floor was elevated above the CCD.

Sandy mud and muddy sand. Detrital sandy mud and muddy sand are comparatively rare in the Eastern Indian Ocean drill holes. They occur at or near the base of the sequence (site 254), or within deep water turbidite sequence (sites 211, 215(?), and 218).

Those at the base of site 254 are typical poorly sorted and associated with conglomerates (including some containing volcanic clasts), and breccias. Bioturbation and macrofossil fragments are comparatively common, resulting in the destruction of most laminations. The basal sandy mud and muddy sand are poorly calcareous; they contain abundant glass shards and montmorillonite, with phillipsite commonly present. Pyrite is also relatively abundant perhaps indicating moderately restricted depositional conditions at times. Glauconite, kaolinite, clinoptilolite, analsite, and a variety of heavy minerals are present in trace amounts. Davies, Luyendyk et al. (1974) suggest that some of these sediments may have formed under very shallow water conditions, which at times were sufficiently vigorous to produce intraformational breccias. The source area for the coarse basaltic pebbles was evidently close by.

The sandy mud/muddy sand of the turbidite sequence contains up to 15 percent calcite derived from included microfossils, in contrast to the poorly calcareous basal sandy mud/muddy sand. It is, in addition, finer-grained, with no associated conglomerates. Feldspar is abundant, comprising up to 85 percent of the total sediment in places; heavy minerals and opaques each comprise up to 5 percent of the total sediment. Thompson (1974) concluded that the heavy mineral assemblage is consistent with a Himalayan source area. At site 218, there appear to have been four major pulses of relatively coarse terrigenous sedimentation, with the second pulse of sufficient magnitude to carry the distal portion of the sediment load as far as site 211, a distance of almost 4000 km from the head of the Bay of Bengal.

Sand and gravel. Sand and gravel composed primarily of detrital (non-calcareous material is rare at abyssal sites occurring only at site 263 where a few thin laminae composed of sand-size quartz and kaolinite grains are present near the base of the thick clay sequence. Sand and gravel is more common on the ridge and plateau sites, particularly sites 254 and 255 where the sands and gravels are of Eocene-Oligocene age, and site 264 with Campanian and Santonian sands and gravels. At all these sites coarse sediments rest directly on major unconformities (Figure 2) with the clasts derived primarily of the underlying unit.

Volcanogenic Sediments

Sediments of undoubted volcanic affinities are found only on the ridges and plateaus and are dealt with in detail elsewhere in this volume. Therefore they are only considered briefly here.

Volcanogenic gravels (possibly with some detrital affinities) occur, as previously mentioned, at sites 254 and 264. Hayes et al. (1975) report that a conglomerate 38 m thick at site 264, is composed of subround basaltic pebbles up to 6 cm diameter in a tuffaceous sandy matrix, cemented by sparry calcite. The gravel unit at site 254 is 91.5 m thick, with subrounded to angular basaltic pebbles, up to 10 cm diameter. Hydrothermal ferruginization and pyritization are common.

Volcanogenic sand and gravel are present in the lower part of the Ninety-east Ridge sequence. At site 253 the volcanogenic sequence is 388 m thick, and consists mainly of altered vitric ash and lapilli. Montmorillonite and calcite are the dominant minerals, with minor plagioclase and zeolites and traces of glauconite (probably an alteration produce of volcanic glass), cristobalite, pyrite, and other accessory minerals. McKelvey and Fleet (1974) consider that the site 253 pyroclastics were produced primarily by the fragmentation of basaltic lavas quenched by sea water. At sites 216 and 253, and the upper portion of the volcanogenic sequence at site 214, micarb fragments of microcrystalline calcite of uncertain origin is abundant. The micarb may be derived from nannoplankton, primary precipitation of calcium carbonate (unlikely under marine conditions) or by the in situ alterations of glass. A biogenic origin for the micarb is supported by the presence of thin laminae of nannofossil ooze or chark in places, the occurrence of scattered macrofossils within the sequence, and locally abundant bioturbation. These biogenic features also support for a submarine site of deposition. However the tuffs at site 214 (Unit 3 of von der Borch et al., 1974) do not contain any marine fossils, and include beds of lignite up to 80 cm thick (believed by Cook (1974) to be autochthonous), which supports a terrestrial (e.g. swamp) environment.

Pimm (1974a, Figure 18) indicates that a few thin beds (up to 15 cm) of rhyolitic ash ranging in age from Pliocene to Recent are present in the northern part of the Eastern Indian Ocean. These deposits appear to be derived from the nearby volcanic province of the Indonesian archipelago, whereas the thicker, more ancient volcanogenic deposits are related to tectonism and volcanism within the Indian-Australian plate.

Biogenic Sediments

The biogenic sediments of the Eastern Indian Ocean are overwhelmingly calcareous ooze and chalk, with only minor siliceous ooze. In many places they are admixed with varying amounts of detrital sediments, particularly clays, or more rarely with volcanogenic sediments, the degree of admixing producing differences in color, texture, and mineralogy. In some instances, the original biogenic nature of the sediments may be masked by diagenesis.

Foraminiferal-nannofossil and nannofossil ooze and chalk

Calcareous ooze is abundant at most drill sites in the Eastern Indian Ocean (Figure 7). It is mainly Cenozoic and comprises the majority of the acoustically layered sediments (Figure 2). Most calcareous oozes are composed of mixtures of foraminiferal and nannofossil remains; some are composed exclusively of nannofossils; few, if any, are entirely foraminiferal. It appears that nannofossil ooze is commoner in the lower part of the sedimentary sequence, particularly in the ocean basin sites, whereas foraminiferal-nannofossil ooze is more common in the upper part. Work by Pimm (1974a) and Berggren et al. (1974) on the species distribution of foraminifera in Leg 22 drill holes indicates that not only do foraminifera as a whole decrease in abundance down many of the drill holes, but also that the foraminiferal species least resistant to dissolution are the first to disappear.

Calcareous ooze ranges from pink to greyish orange, light brown, and green depending on the abundance of clay. Bedding varies from thin to laminate; graded beds are present in places. The foraminiferal-nannofossil ooze which contain 25.6 percent sand-size material is markedly coarser-grained than the nannofossil ooze, which contains an average of only 5.9 percent sand (Table 1). Calcareous ooze is composed predominantly of low-magnesian calcite (though aragonite is also abundant at the top of site 263), with minor amounts of clay, detrital quartz, feldspar, zeolites, pyrite (commonly found in the tests of the foraminifera) and accessory heavy minerals. The chemical composition of the calcareous ooze varies with the amount of included detrital and volcanogenic material, overall however, foraminiferal and nannofossil oozes are chemically similar (Table 2).

There appear to be three distinct deposition locations for calcareous ooze (Figure 7): (i) Near the base of the sedimentary sequence. This probably corresponds to the period shortly after the development of a spreading ridge when the sea floor in that area was above the CCD. In general, these calcareous oozes are nannofossil-rich and are not abundant. They occur only in the Mesozoic. (ii) Cenozoic calcareous oozes (ranging into the Recent) which are in ocean basin location well below the CCD, and yet are highly calcareous. This location, together with the presence of some graded beds, and the laminated nature of the deposits and the commuted and reworked nature of the fossils indicates that they have probably been deposited by turbidity flows or large-scale slumps. (iii) Cenozoic calcareous oozes (ranging into the Recent) which are on ocean ridge or plateau locations above the CCD.

Micarb ooze and chalk

Ooze and chalk composed mainly of fragments of calcite (micarb) of uncertain origin are referred to in the DSDP scheme for sedimentary rocks as a micarb ooze, or if lithified as a micarb chalk. Micarb fragments are a common component of many of the foraminiferal or nannofossil oozes, but micarb chalk is present only in the northern part of the Eastern Indian Ocean (Figure 7), particularly at sites 214, 216 and 217. It ranged from light grey to grey-green and olive-grey; some interbeds of mud are present but generally bedding is poorly developed. Silt-size calcite fragments form up to 80 percent of the total sediment, with a few recognizable foraminiferal, nannofossil or macrofossil (*Inoceramus* oysters) fragments in places. Other components include clay (up to 20%) and glauconite (up to 5%) with minor silica (mainly of biogenic origin), quartz, feldspar, pyrite, dolomite, and apatite, and trace amounts of various heavy minerals. The micarb chalk is very fine-grained containing on average only 0.1 percent of sand-size material (Table 1). Von der Borch and Sclater et al. (1974) regard the fragmentary nature of the micarb chalk, together with the presence of glauconite and the macrofossils as consistent with a vigorous shallow marine environment.

Cherty limestone and chalk

Cherty carbonates are of fairly limited distribution. They are found only in ridge or plateau locations (Figure 7), and are restricted to the Eocene-Paleocene and the Santonian-Campanian (Figure 2). Tertiary cherty limestone is composed predominantly of nannofossils with some foraminiferal and siliceous remains, minor amounts of clay, and traces of volcanic quartz. Chert is present as scattered light grey nodules and irregular patches. Chalk showing various degrees of silicification, as well as siliceous infillings of foraminifera, occurs sporadically. Most of the silica is secondary and of uncertain origin. These Tertiary siliceous carbonates appear to be biogenic calcareous deposits which have accumulated above the CCD.

The late Cretaceous cherty carbonates are somewhat thicker and coarser-grained. They are composed of abundant nannofossils, foraminifera and *Inoceramus* fragments. Micarb makes up 5-20 percent of the limestone; dolomite rhombs, glauconite, clay, mica, collophane, and pyrite are all present in minor to trace amounts. Silica occurs as nodules veins, and patches, and as a matrix. At site 217, von der Borch and Trueman (1974) found a positive correlation between the abundances of silica and dolomite rhombs; and concluded that both formed diagenetically in response to conditions prevailing below the sediment-water interface and invoke a refluxion mechanism of the type proposed by Adams and Rhodes (1960). The origin of the silica is not clear though von der Borch, Sclater et al. (1974) make the observation that at site 217 siliceous remains decrease in abundance as chert becomes more common. They also suggest of the basis of the fragmentary nature of the shells, the presence of abundant burrows, and the occurrence of cross-beds. The depositional environment for the cherty limestone was comparatively shallow.

Siliceous ooze

Siliceous ooze is present only in the northern part of the Eastern Indian Ocean (sites 211, 213, 215, 260 and 261). The distribution of the upper acoustically transparent unit (Figure 6) indicating its extent. With the exception of a thin bed of Albian age at site 260 (Figure 2), siliceous ooze is restricted to Miocene and younger sediments. Siliceous ooze, ranges from brown to green and grey; it is composed of radiolarians and diatoms, with minor sponge spicules and silicoflagellates. A few nannofossil fragments are present in some beds; iron and manganese oxides occur sporadically as micro-nodules or as partings. Volcanic glass (from the nearby Indonesian volcanic arc), is commonly present in trace amounts, and locally comprises up to 25 percent. Terrigenous clay is invariably present, ranging from minor to abundant. Texturally the siliceous ooze corresponds to a silty clay (Table 1); the sand-size material is probably composed of larger siliceous organisms. Compositionally it is characterized by high SiO_2, MnO, and Ba and low CaO + CO_2 (Tables 2 and 3). The appearance of these pelagic siliceous oozes in the late Tertiary is probably the result of the northward movement of the region into the zone of high equatorial productivity.

Chemical Composition

The chemical composition of sediments from the DSDP sites in the Eastern Indian Ocean is dealt with by Pimm (1974b), Fleet and Kempe (1974), Calder et al. (1974), and Cook (1974 a,b). Average compositions (Tables 2 and 3) were derived from these publications. In general the data are insufficient to establish vertical or synchronous lateral trends and it is possible to do nothing more than point to the compositional variation of the various sediment types, and to indicate some probable inter-element associations.

Major and minor oxides

Most Mesozoic-Cenozoic sediments of the Eastern Indian Ocean are either calcareous ooze (predominantly calcium carbonate) or siliceous clay (silica and alumino-silicates), with a scatter of sediments between the two end-members. In general the major and minor oxide content of these sediments is similar to the "average" deep-sea sediment (Table 2).

TABLE 3. Average trace element composition of various sediment types from DSDP sites in the Eastern Indian Ocean, and average pelagic sediments after (a) Goldberg and Arrhenius (1958) and (b) Turekian and Wedepohl (1961).

Trace Element	Foram-nanno ooze *(35)	Nanno clay (6)	Rad-diatom ooze (7)	Zeolitic clay (10)	Clay and Claystone (21)	Glauconitic silty clay (1)	Volcanic ash (20)	Average(a) Pacific pelagic sediment	Average(b) pelagic carbonate	Average(b) pelagic clay
Sr	950	113	189	180	154	22	224	710	2000	180
Ba	366	277	1150	880	860	45	165	390	190	2300
Li	25	40	34	30	48	192	40	59	–	2.6
Cu	66	82	544	196	200	111	54	740	30	250
Pb	10	32	74	19	35	–	3	150	9	80
Zn	54	139	129	116	160	27	46	–	35	165
Co	5	32	66	30	25	78	32	160	7	74
Ni	21	91	95	75	–	79	34	320	30	225

* Number in brackets indicates the number of analyses used for determining the mean.

SiO_2 occurs in the Eastern Indian Ocean sediments predominantly as quqrtz (including chalcedony-chert), tridymite and cristobalite (including opaline material), and alumino-silicates; minor amounts are present as feldspar, feldspathoid and zeolite, and traces occur as pyroxene and amphibole. Terrigenous SiO_2 is dominant, but biogenic silica is an important component of the Cenozoic siliceous oozes. It may also be important in siliceous clays and claystones despite the lack of direct evidence of a biogenic origin for the silica. Locally, particularly in ash and volcanoclastic sequences, the SiO_2 is obviously of direct volcanic origin; the siliceous volcanogenic input is rather less certain in the pelagic clays and zeolitic clays. Other terrigenous and/or volcanic oxides which show a positive correlation with SiO_2 include Al_2O_3, TiO_2, Fe_2O_3, Na_2O, K_2O and H_2O, predominantly the result of the quartz-clay mineral association. Concentrations of these elements are similar to those found in "average" pelagic sediments (Table 2), although the high TiO_2 content of the glauconitic silty clay (1.28%) and nannofossil clay (0.77%) is possibly an indication of titaniferous detrital grains derived from a nearby source area.

The Fe_2O_3 content of the sediments results in part from iron oxide associated with clays, or more rarely glauconite, as indicated by the high Fe_2O_3 content (12.3%; Table 2) of the glauconitic silty clay. However, abnormally high iron contents are also found in some thin basal iron oxide units overlying basalt. Pimm (1974b) studied the geochemistry of this facies at sites 211, 212, and 213 and found that the total iron content ranged up to 16 percent. This iron-rich facies is believed to be formed by volcanic hydrothermal exhalations. Similar facies in the Pacific contain relatively high metallic concentrations particularly of copper (von der Borch and Rex, 1970; Cronan, 1973) though the Eastern Indian Ocean occurrences do not. Manganese appears to be most abundant in the siliceous oozes where it occurs as bands and patches of manganese oxides containing up to 8 percent MnO. Overall, the siliceous oozes are high in manganese (1.2% MnO; Table 2). It is uncertain whether this is the result of a slow rate of sedimentation, submarine fumaroles, biochemical concentration, or all three.

CaO is an abundant oxide in many pelagic sediments, and is particularly common in calcareous ooze where CaO + CO_2, (occurring mainly as calcite), averages more than 60 percent of the total sediment. Minor amounts of CaO are also present as feldspar and zeolite in some sediments. The P_2O_5 content of the sediments varies markedly though the range and corresponds closely with average values from other areas (Table 2). Apatite is occasionally present as biogenic collophane or bone and teeth fragments in places, but in most pelagic sediments the phosphate is probably adsorbed on clays. A P_2O_5- Fe_2O_3 correlation reported by Cook (1974b) from Leg 27 results may indicate the presence of trace amounts of vivianite or alternatively adsorption of phosphate by iron oxides.

Trace elements

In general, the trace element content of the Mesozoic-Cenozoic Eastern Indian Ocean sediments is similar to average values for deep-sea sediments (Table 3). However, the Eastern Indian Ocean clays have a lower metal content than the average pelagic clay of Turekian and Wedepohl (1961). This may be ascribed to the greater input of terrigenous material into the Eastern Indian Ocean, and perhaps also the paucity of submarine fumarolic activity.

Strontium is associated with the biogenic carbonates, substituting for calcium in the calcite lattice. The highest strontium value (950 ppm) is found in foraminiferal-nannofossil ooze (Table 3); this is significantly

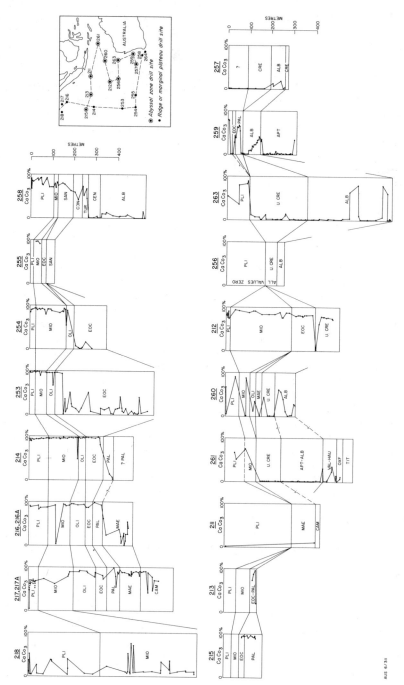

Fig. 8. Variation of CaCO₃ content at deep (A) and shallow (B) drill sites in the Eastern Indian Ocean. All available analyses were used to compile these curves but on average only 1 sample point in 5 is indicated in the diagrams.

below the value for average pelagic carbonates (2000 ppm). Barium is highest in the siliceous oozes and lowest in the glauconitic clays. Some of the barium is presnt as barite nodules and veins, such as those found at site 263. Brongersma-Sanders (1966) has speculated that much of the barium in marine sediments is associated with siliceous organisms, and factor analysis on the Indian Ocean sediments (Cook, 1974b) would seem to support a biogenic silica-barium correlation. A minor amount of the barium is also associated with iron and manganese-rich sediments. Copper, zinc, cobalt and nickel also appear to be associated with ferromanganese oxides, probably in an adsorbed state, whereas lead seems to be associated predominantly with the clay component.

In conclusion, the major minor and trace elements in the Eastern Indian Ocean sediments may be grouped into four groups: (i) the alumino silicate group which is derived from terrigenous and volcanic sources and includes SiO_2, Al_2O_3, TiO_2, Fe_2O_3, Na_2O, K_2O, H_2O, Zn and (?)Pb; (ii) the ferromanganese group, the product of syngenetic (or very early diagenetic) mineralization and including Fe_2O_3, MnO, Cu, Zn, Co, and Ni; (iii) the biogenic silica group derived mainly from diatoms and radiolaria and composed of SiO_2, Ba and possibly MnO; and (iv) the biogenic carbonate group, derived mainly from foraminifera and nannofossils and composed of CaO, CO_2, Sr and possible MgO.

Calcium carbonate content

The general picture which has emerged so far is of poorly calcareous Mezozoic sediments and highly calcareous Cenozoic sediments and this is again clearly evident in Figure 8. In addition the present-day sediment distribution pattern (see Davies and Kidd, 1977) in the Eastern Indian Ocean indicates calcareous sediments are abundant on continental margins and ocean ridges above the CCD, or locally as ponded sediments below the CCD. Elsewhere in the abyssal zone the sediments are either siliceous oozes or clays.

The general pattern of carbonate distribution at Ninetyeast Ridge sites (the data are insufficient to establish any trend for Broken Ridge) is one of low $CaCO_3$ content at the base of the section followed by a rapid increase and then stabilization at a fairly uniformly high $CaCO_3$ content for the remainder of the Cenozoic. The carbonate content at the base of the section may be the result of a high rate of volcanic input in the time shortly after the formation of the ridge (e.g. site 253). It seems likely from both the sedimentology (Luyendyk and Davies, 1974) and calcimetry that in the Cretaceous the Naturaliste Plateau was below the CCD but that it underwent rapid uplift during the late Cenomanian-early Turonian and continued to occupy a shallow location for the remainder of the Cretaceous and the Tertiary. Sites 256 and 257, appear to have below the CCD and outside the range of calcareous turbidity flows or large-scale slumps throughout most of their depositional history. At site 263 the low $CaCO_3$ content is in part the result of a very high rate of terrigenous sedimentation and the consequent swamping of the calcareous component. Conversely, at site 212, which is situated in a small abyssal depression where allochthonous carbonates have collected over a long period, the $CaCO_3$ content is uniformly high. Despite marked variations between sites it is nevertheless possible to recognize regional trends from Figure 8. There is a general tendency for the $CaCO_3$ content to be moderately high at the base of a sequence but to decrease rapidly to nearly zero. An increase in $CaCO_3$ occurs in the Albian at a number of sites (257, 259 and ?263), particularly those near the eastern margin of the study area, but by the late Albian the $CaCO_3$ content has again decreased and remained generally low throughout the rest of the Cretaceous, except for site 212.

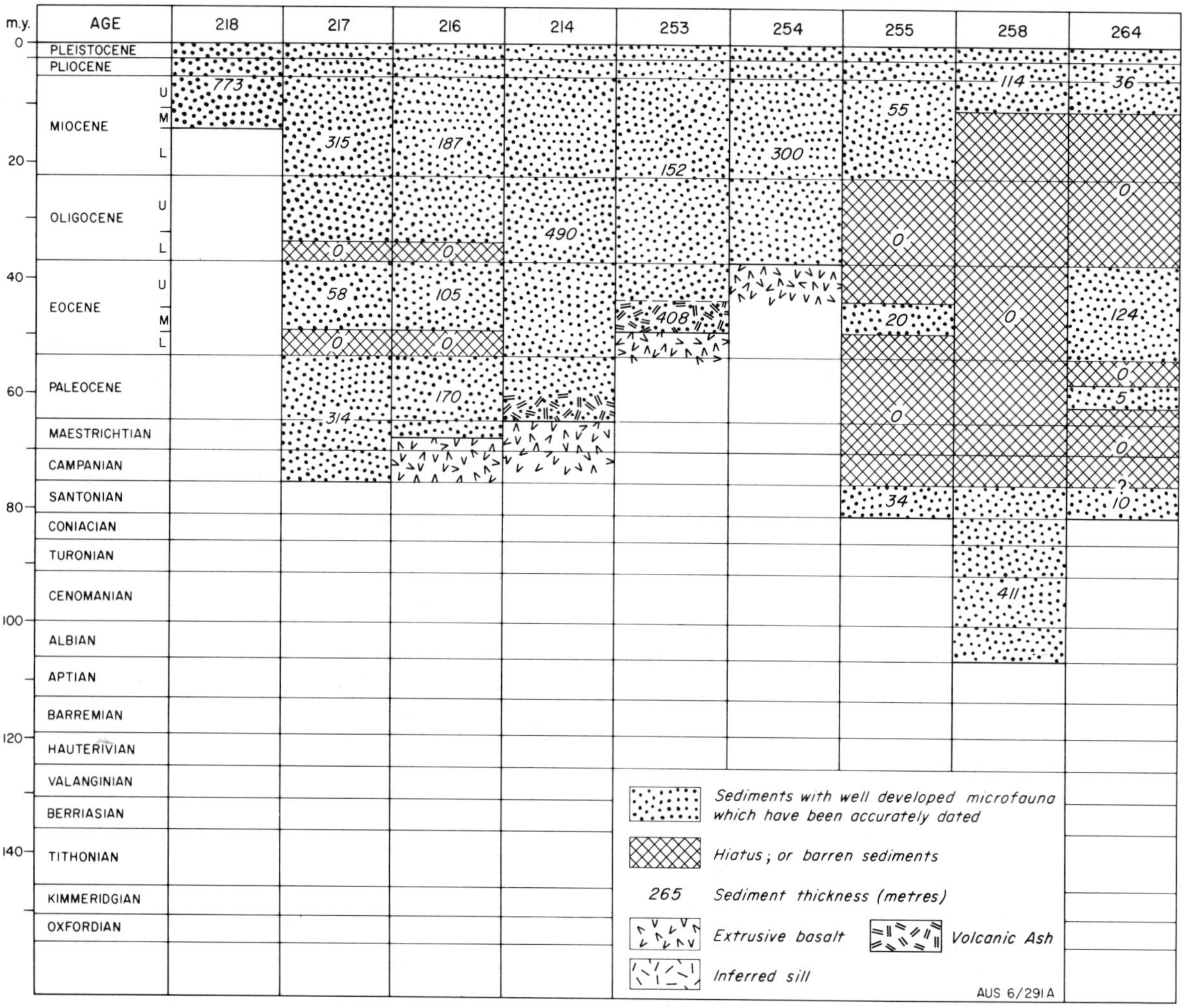

Fig. 9. Schematic representation of the distribution of hiatuses at shallow drill sites in the Eastern Indian Ocean.

A marked increase in $CaCO_3$ content had again decreased and remained generally low throughout the rest of the Cretaceous, except for site 212. A marked increase in $CaCO_3$ content then took place at about the Cretaceous-Tertiary boundary at most sites. Major fluctuations occur in the calcimetry of the Cenozoic sediments but the average carbonate content is much higher than that generally found in the Mesozoic.

The sequence of events at a deep-sea site is well established (Berger, 1972) and involves first the formation of a relatively shallow spreading center (commonly above the CCD) followed by a gradual sinking of the location to well below the CCD. However, superimposed upon this general sequence are variations in the CCD. Van Andel (1975) has suggested that from the late Jurassic to the early Cenozoic the CCD was shallow (above 4000 m) in all the world oceans. However the increase in carbonate content in the Albian at several Indian Ocean sites (Figure 8) suggests a temporary though short-lived deepening of the CCD at that time. Van Andel (1975) also indicates that there was a marked drop (down to 4500 m) in the CCD about 40 m.y. ago. This would seem to coincide with the thick Cenozoic carbonates in the Eastern Indian Ocean in general, however the picture is complicated by the massive influx of allochthonous calcareous ooze from the continental slope into the abyssal zone in the late Cretaceous in places (site 212). This event must be related to the onset of a major phase of carbonate deposition on the shelf and slope which is in turn likely to reflect changes in the CCD.

Sedimentation

Hiatuses and unconformities

Deep-sea drilling has shown large gaps in the sedimentary record of all of the oceans. Hiatuses and unconformities within the Indian Ocean sequence have been considered in some detail by Pimm and Sclater (1975) and Davies et al. (1975). They concluded that several major breaks are primarily the result of climatic events in Antarctica and variations in following the separation of Australia and Antarctica and the setting up of a circum-Antarctic circulation system.

Only at site 255 (Figure 2) is an angular unconformity (between Santonian limestone and chert and Eocene sands and gravels) recognizable in the cores. However, an angular discordance of regional magnitude between the lower acoustically transparent unit and the layered unit can be recognized from the seismic profiles. All other unconformities are less obvious and there is no angular discordance. The stratigraphy of both deep-water and shallow-water sites is summarized schematically in Figures 9 and 10. At some drill sites there is obviously a major hiatus. At site 258 (Figure 10) there is, for instance, a gap of about 65 m.y. during which no sediments were laid down or of which all trace has been removed. At other sites, however, the magnitude of the hiatus is less certain. At site 256 for example, there are no fossils representative of the entire interval from the Cenomanian to the Pliocene, an interval of about 100 m.y. However, 105 m of barren sediments was laid down at some time during this interval. If Figure 10 is comapred with Figure 11 it is apparent that the deep-sea record is far more incomplete than the record from shallower areas.

At shallow water sites (Figure 9) there appear to be well defined hiatuses at all sites except 214 and 264, in the early Eocene and at most sites (excluding 214, 253 and 254) in the early Oligocene. There are marked differences between the sites on the Ninetyeast Ridge where the sedimentary record is comparatively uninterrupted once sedimentation has been initiated, and the Broken Ridge-Naturaliste Plateau sites where there is a major break in the sequence extending from the base of the Campanian to the Paleocene (20 m.y. duration). This was followed by sporadic sedimentation in the Eocene and then a second break in the Oligocene and late Miocene (10-15 m.y. duration) on the Naturaliste Plateau. At site 255 there has obviously been uplift and subsequent erosion in the late Cretaceous-early Tertiary. This same phase of tectonic uplift similarly may be responsible for loss of section at sites 258 and 264, although Davies et al. (1975) consider that the gaps on the Naturaliste Plateau are more the result of variation in oceanic circulation and sediment supply than tectonic events. The Oligocene hiatus (extending into the early and middle Miocene in places) is also best developed on Broken

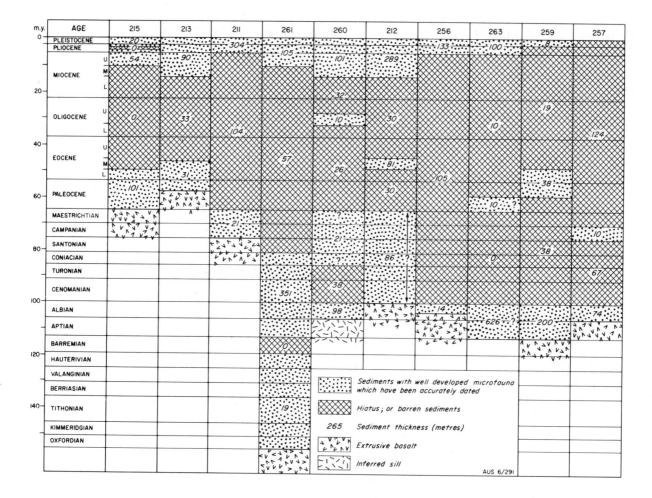

Fig. 10. Schematic representation of the distribution of hiatuses at the deepwater drill sites in the Eastern Indian Ocean.

Ridge-Naturaliste Plateau (Figure 9). There is no evidence of angular discordance associated with this gap in any of the sequences, nor is thre any sedimentological or paleontological evidence of a major change in water depth. Consequently the hiatus may be the result of the rate of sedimentation decreasing to zero, or the influx of erosive bottom current which prevented sedimentation, or eroded the sedimentary sequence, or both.

At the deep-sea sites the distribution of hiatuses is less clear, not because of their absence but because of their abundance. The almost total lack of datable late Cretaceous and Tertiary sediments is a marked feature of Figure 10. The late Cretaceous hiatus appears to be particularly marked in the southern and eastern portions of the ocean, whereas the general Tertiary hiatus occurs throughout the region. The various gaps in the sequence appear to be:

(i) Barremian (approximately 5 m.y. duration)

(ii) Cenomanian-Paleocene (approximately 40 m.y. duration). This hiatus occurs only in the southern and eastern parts of the abyssal zone.

(iii) Middle Eocene-early Miocene (approximately 35 m.y. duration). This hiatus occurs throughout the abyssal zone.

(iv) Middle and upper miocene (approximately 9 m.y. duration). This hiatus is present only in the southern and eastern parts of the abyssal zone.

The deep ocean hiatuses are synchronous with those of the shallower ridges and plateaus, but also extend beyond them. Hiatuses of similar are are also recognizable on the continental shelf (Veevers and Johnstone, 1974) and as far afield as Papua (Veevers and Evans, 1973). The late Cretaceous hiatus initially affected only the ocean basin (Cenomanian-Santonian) but later (Campanian-Paleocene), both ocean basin and some shallow sites (Broken Ridge and Naturaliste Plateau) were affected. However, the hiatus at both locations may not necessarily have resulted from the same immediate cause. At the shallow sites it appears to be the consequence of uplift and subsequent erosion whereas at the abyssal sites it probably resulted from strong bottom currents which may have both prevented sedimentation and produced some erosion. These strong bottom currents were probably ancestral Circum-Antarctic Current moving north through the narrow juvenile Indian Ocean. The late Cretaceous-early Tertiary uplift of Broken Ridge and perhaps also the Naturaliste Plateau may have restricted the path of such currents resulting in an intensification of the bottom currents in places.

Davies et al. (1975) propose that Tertiary hiatuses in both the Indian and Pacific Oceans may be correlated with periods of climatic deterioration which produced glaciation in Antarctica and the formation of large amounts of cold, erosive Antarctic Bottom Water which moved north into the Wharton Basin, and also periods of increased storminess and erosion and a shallowing of the CCD. However, major breaks also occur in the Cretaceous record, yet there is no evidence of climatic deterioration or of an Antarctic ice cap during this time. Quilty (1975) records uniformly warm to hot conditions throughout the late Cretaceous of the Western Australian region. But, if Antarctic Bottom Waters did not exist at that time there must nevertheless have been water movement associated with west wind drift after India started to move north, leaving open ocean to the west of Australia. The northward deflection of this westerly current may be responsible for the late Cretaceous sedimentary gaps at abyssal sites off the west coast of Australia. Once separation of Australia and Antarctica was complete and the Circum-Antarctic Current established (estimated by Kennett et al. (1975) to have occurred at about 30 m.y.), there was probably little direct input of Antarctic Bottom Water. In addition, by the Oligocene, Broken Ridge and the Naturaliste Plateau restricted the entrance of deep water into the Wharton Basin. On the West Australian shelf, the Oligocene to middle Miocene was marked by a progradation of carbonates across the shelf, leading to a massive build-up of carbonates on the upper slope some of it ultimately slumping down the slope and into the Wharton Basin.

Therefore it appears that no single mechanism can be used to account for all the Mesozoic-Cenozoic hiatuses which occur in the deep water of the Eastern Indian Ocean and also, in some instances, on the shallower flanking structures. The late Cretaceous-Paleocene hiatus may be ascribed to extensive west wind drift, and an associated deep current system which was forced north along the eastern side of the Indian Ocean. The Eocene hiatus may, as suggested by Davies et al. (1975), be the consequence of climatic deterioration, glaciation, and the development of large bodies of northward-moving Antarctic Bottom Water. Subsequent northward movement of Australia and the attendant development of the Circum-Antarctic Current precluded the entry of this deep current system by the late Oligocene.

Rates of sedimentation

Determination of the rate of sedimentation from the Eastern Indian Ocean

TABLE 4. Variation in the rate of sedimentation in the Mesozoic and the Cenozoic as determined from DSDP sites and from oil company drill holes on the Western Australian Margin

Age	Number of Values	Rate of Sedimentation (m/10^6yrs)		
		Minimum	Maximum	Average
Cenozoic				
Quaternary	11	3.5	21.5	11.1
Pliocene	16	2.5	88.5	14.4
Miocene	14	0.6	14.9	6.3
Oligocene	5	1.8	10.0	4.5
Eocene	10	3.3	21.5	6.1
Paleocene	5	3.4	6.8	4.7
Cretaceous				
Maestrictian	3	1.4	23.6	12.4
Campanian	2	1.7	1.7	1.7
Santonian	2	4.4	5.6	5.0
Coniacian	2	6.4	20.0	13.2
Turonian	1	-	-	3.1
Cenomanian	1	-	-	0.8
Albian	5	6.0	42.0	20.3
Aptian	1	-	-	19.6
Barremian	-	-	-	-
Valanginian	1	-	-	1.7
Hauterivian	1	-	-	1.7
Berriasian	-	-	-	-

drill sites is made difficult by the many gaps and the thick sequences of barren sediments in the stratigraphic record. Average rates of sedimentation were determined using only those parts of the sequence which could be confidently dated from their fossil assemblages. These average values show that there were quite marked fluctuations in the rate of sedimentation throughout the Mesozoic-Cenozoic, ranging from 0.8 to 20.3 m per million years (Table 4). There were also some well defined peaks of sedimentation rates in the Aptian-Albian, Coniacian, Maestrictian, and Pliocene-Quaternary. These maxima do not appear to coincide with those proposed by Veevers et al. (1974) from seismic information, but as their time intervals are greater, no direct comparison of their results and those given here can be made. Minima in the rate of sedimentation are commonly, thought not exclusively, found within periods which contain hiatuses. There is also a tendency for sedimentation maxima to be common at times when highly calcareous sediments were deposited in the abyssal zone. A third feature evident from Table 4 is that the sedimentation rate was far more variable in the Mesozoic than the Cenozoic, perhaps indicating the more unstable Mesozoic conditions (climatic, oceanographic and/or tectonic) in the Eastern Indian Ocean region. It should perhaps be noted that the high rate of sedimentation in the Aptian-Albian coincides with a time of considerable magmatic activity and tectonism in Australia (Veevers and Evans, 1973). Some increase in the rate of sedimentation in the Pliocene-Pleistocene might be attributable to the increase in volcanism which

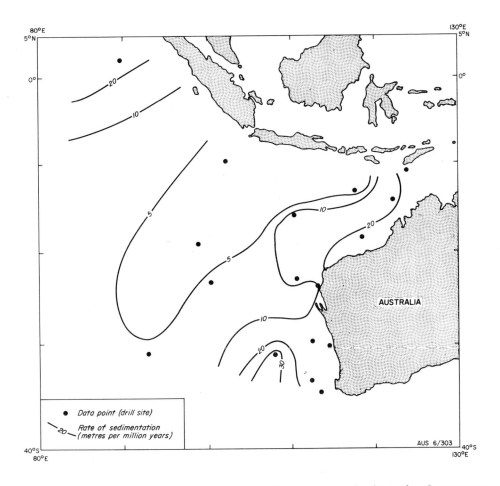

Fig. 11. Variation in the rate of sedimentation during the Cretaceous in the Eastern Indian Ocean.

is thought to have taken place in the past two million years (Kennett and Thunell, 1975), although most of the increase is probably the result of an increase in the rate of calcareous sedimentation at that time.

It is not possible to determine the regional variation in the sedimentation rate for each epoch because there are unsufficient data, consequently the regional change can only be determined for the Cretaceous and the Cenozoic. For the Cretaceous (Figure 11) there is a clear increase in the rate of sedimentation as the Australian continent is approached, which is consistent with it being a major source of sediment at that time. There is the suggestion of a source area to the south (?Antarctica) which would be feasible, as Australia and Antarctica did not separate until the Paleocene. There may also have been a source for Cretaceous sediments to the northwest as indicated by the increase in the rate of sedimentation in that direction. This would seem to add support to the proposal of Crawford (1974), Larson (1975) and others that a continental landmass (such as Tarim) lay to the northwest of Western Australia during the Jurassic-Cretaceous. The central portion of the Eastern Indian Ocean was evidently a considerable distance from land throughout the Cretaceous.

The pattern of variation in the rate of sedimentation is rather different in the Cenozoic (Figure 12). The rate still increases markedly towards Australia, a feature also noted by Veevers et al. (1974), but now the

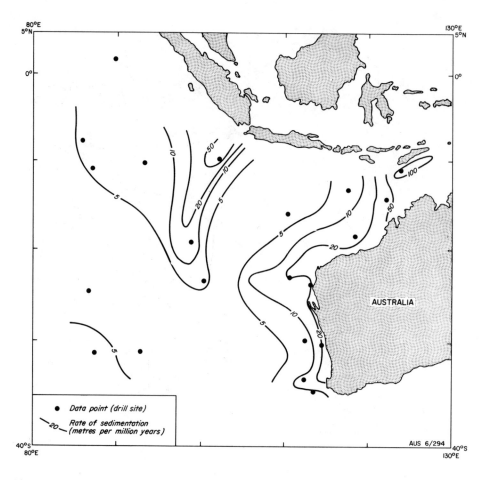

Fig. 12. Variation in the rate of sedimentation during the Cenozoic in the Eastern Indian Ocean.

sediment is primarily calcareous and it is more the Australian shelf and upper slope which is contributing the bulk of the sediment rather than the continent itself. There is no evidence of a source area to the south, but a marked increase in sedimentation rate occurs in the center of the basin and also to the northwest. The northwesterly trend is attributable to the Bengal Fan which constituted a major source of sediment supply to this area by the late Cenozoic (Curray and Moore; 1971). The maximum in the center of the ocean basin is probably the consequence of this area being a low point into which turbidity currents descended, and where they were able to deposit their suspended load.

Sedimentary History

On shore and shelf sedimentary sequence indicate that marine conditions persisted throughout much of the Phanerozoic on the northwest margin of Australia, with a northern connnection open to the Tethys (Veevers et al., 1971). A new phase of sea-floor spreading may have taken place as early as the Callovian (Veevers and Heirtzler, 1974b). At about this time a spreading ridge was initiated, running approximately east northeast. At first nannofossil-bearing or nannofossil-rich clays were deposited on the ridge in places. As spreading proceeded, and the sea floor previously on the ridge flanks descended below the CCD, brown pelagic clays were deposited. From then until the early Aptian, the rate of sedimentation was slow (1-2 m per million years). There were probably several hiatuses, although the only one recognized to date (from site 261) is the gap in the Barremian. This slow rate of sedimentation may also indicate a low arid hinterland from which there was little input of terrigenous sediment. Locally, there was rupturing during the period 155-113 m.y. which may have produced horsts and grabens and some semi-enclosed small ocean basins (Veevers and Heirtzler, 1974b). However, in general the period 155-113 m.y. was probably relatively quiet tectonically.

During the Aptian-Albian, conditions became more tectonically active; an accelerated phase of ocean-floor generation took place as indicated by the widespread oceanic basalts of Aptian-Albian age, and the occurrence of zeolitic clays. A number of isolated basins may have formed at this time (Davies and Kidd, 1977). There was continental tectonism, coupled with the development of extensive epicontinental seas. This tectonism resulted in a marked increase in the rate of sedimentation, particularly in the vicinity of site 263 where a thick sequence of organic-rich clays was deposited. Davies and Kidd (1977) suggest that conditions were quite restricted in the Wharton Basin at this time. In the Albian there was an increase in the carbonate content of the sediments, possibly associated with a warming phase. By the close of the Albian (100 m.y.) a broad seaway, opening to the north, now with only limited areas of restricted circulation (Davies and Kidd, 1977), existed off Western Australia (Figure 13B). Sediment supply (mainly clay) was predominantly from the east with some input from the south (Antarctica) and possible a minor amount from the northwest.

Sedimentary conditions changed markedly in the Cenomanian. The sediments became less calcareous, there was a major decrease in the rate of sedimentation, perhaps accompanied by extensive erosion of the existing sediments. These conditions appear to have prevailed throughout most of the late Cretaceous and extending into the early Tertiary in places. The most likely cause of this change was the northward movement of India commencing in the Cenomanian (95-100 m.y.), providing a seaway to the west of Australia so allowing the influx of cooler westerly currents which swept west, then north along the eastern margin of the ancestral Eastern Indian Ocean by the latter part of the Cretaceous (Figure 13C). By the Santonian or earlier, the uplift and volcanism associated with the early development of the nInetyeast Ridge had commenced in the vicinity of site 217, which was, at this time sufficiently shallow for coral reefs and in places coal to form on its crest. By the Paleocene (about 60 m.y.) the Ninetyeast Ridge existed as an elongate feature, shallowest at its southern (young) end (Figure 13D). Volcanogenic and calcareous sedimentation continued along its crest and flanks at this time. Elsewhere in the Eastern Indian Ocean, which was broader by this stage, there was little or no sedimentation. Deep currents from the west, deflected north, continued to sweep across the shallow features in the south and over the abyssal plain.

At about 50-55 m.y. the final separation of Australia and Antarctica commenced. Some calcareous sedimentation continued in shallower waters, but with little or no sedimentation in deeper waters. However in the early Eocene (about 50 m.y.) and early Oligocene (35-40 m.y.) the hiatuses extended onto the shallower feature (Figure 13E). By 30-35 m.y. the separation of Australia and Antarctica was complete and a strong Circum-Antarctic Current established. This, together with the development of the Ninetyeast Ridge resulted in a marked decrease in the influx of Antarctic Bottom Water. Conditions became warmer as the plate drifted north and this, together with the deepening of the CCD at about 40 m.y.

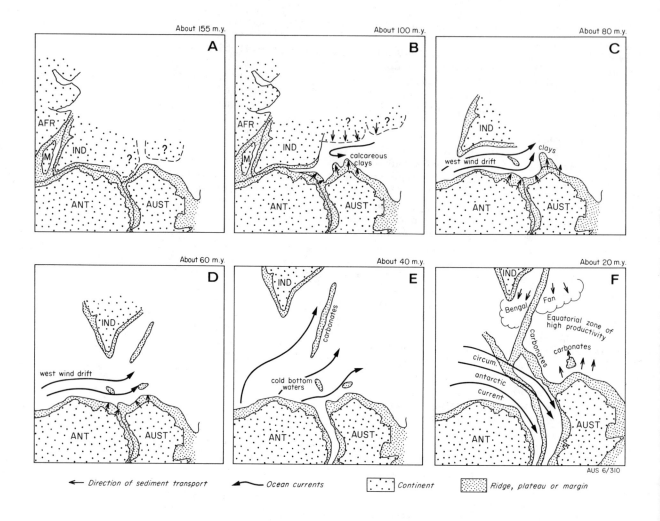

Fig. 13. The pattern of continental dispersal during the late Mesozoic-Cenozoic and the influence of this movement on ocean currents and sedimentation. Position of continents and the time intervals are only approximate, but are based on Smith and Hallan (1970), Luyendyk (1974), McElhinny (1970), and Embleton and McElhinny (1975). For more detailed paleogeographic reconstruction see Sclater et al. (this volume).

(van Andel, 1975) produced a massive development of carbonates especially in shallower waters. By the Miocene (10-20 m.y.), these carbonates had prograded across the shelf and periodically slumped down the slope and into the abyssal plain. At about this same time the Bengal Fan began adding large volumes of clayey and silty sediment to the northwest portion of the Eastern Indian Ocean (Figure 13F). A third feature, the movement of the northern margin of the plate into the equatorial zone of high productivity also produced an increase in the abundance of siliceous sediment (radiolarian-diatom ooze). All of these processes resulted in a marked increase in the rate of sedimentation throughout much of the Eastern Indian Ocean in the late Tertiary and Quaternary. They also produced the pattern of sedimentation that we now see at the present day.

References

Adams, J. E., and M. L. Rhodes, Dolomitization by seepage refluxion, Amer. Assoc. Petroleum Geol. Bull., 44, 1912-1920, 1960.

Berger, W. H., Deep sea carbonates: dissolution facies and age depth constancy, Nature, 236, 392-395, 1972.

Berggren, W. A., G. P. Lohmann, and R. Z. Poore, Shore laboratory report on Cenozoic planktonic foraminifera: Leg 22, in C. C. von der Borch, J. G. Sclater et al., Initial Reports of the Deep Sea Drilling Project, 22, Washington (U. S. Government Printing Office), 613-618, 1974.

Bezrukov, P. L., Sedimentary formation of the Indian Ocean and their relationship with tectonics, Geotectonics, 1, 1-7, 1974.

Brongersma Sanders, M., Barium in pelagic sediments and diatoms, Konihkl. Ned. Akad. Wetensch. Proc., 70, 93-99, 1966.

Calder, J. A., G. J. Horvath, D. J. Shultz, and J. W. Newman, Geochemistry of the stable carbon isotopes in some Indian Ocean sediments, in T. A. Davies, B. P. Luyendyk et al., Initial Reports of the Deep Sea Drilling Project, 26, Washington (U. S. Government Printing Office), 613-618, 1974.

Cook, A. C., Report on the petrography of a Paleocene brown coal sample from the Ninetyeast Ridge, Indian Ocean, in C. C. von der Borch, J. C. Sclater et al., Initial Reports of the Deep Sea Drilling Project, 22, Washington (U. S. Government Printing Office), 485-488, 1974.

Cook, P. J., Phosphate content of sediments from Deep Sea sites 259 to 263, Eastern Indian Ocean, in J. J. Veevers, J. R. Heirtzler et al., Initial Reports of the Deep Sea Drilling Project, 27, Washington (U. S. Government Printing Office), 455-462, 1974a.

Cook, P. J., Major and trace element geochemistry of sediments from Deep Sea Drilling Project, Leg 27, sites 259-263, Eastern Indian Ocean, in J. J. Veevers, J. R. Heirtzler, et al., Initial Reports of the Deep Sea Drilling Project, 27, Washington (U. S. Government Printing Office), 481-498, 1974b.

Crawford, A. R., A greater Gondwanaland, Science, 184, 1179-1181, 1974.

Cronan, D. S., Basal ferruginous sediments cored during leg 16, Deep Sea Drilling Project, in T. H. van Andel, G. R. Heath et al., Initial Reports of the Deep Sea Drilling Project, 16, Washington (U. S. Government Printing Office, 601-604, 1973.

Curray, J. R. and D. G. Moore, Growth of the Bengal deep-sea fan and denudation of the Himalayas, Geol. Soc. Amer. Bull., 82, 565-572, 1971.

Davies, T. A., and B. P. Luyendyk et al., Initial Reports of the Deep Sea Drilling Project, 26, Washington (U. S. Government Printing Office), 1129 p., 1974.

Davies, T. A., O. E. Weser, B. P. Luyendyk, and R. B. Kidd, Unconformities in the sediments of the Indian Ocean, Nature, 253, 15-19, 1975.

Davies, T. A., and R. B. Kidd, Sedimentation in the Indian Ocean through time in Indian Ocean Geology and Biostratigraphy, AGU, Washington, this volume, 1977.

El Wakeel, S. K., and J. P. Riley, Chemical and mineralogical studies of deep sea sediments, Geochim. Cosmochim. Acta., 25, 110-146, 1961.

Embleton, B. J. J., and M. W. McElhinny, The paleoposition of Madagascar: paleomagnetic evidence from the Isalo Group, Earth Planet. Sci. Letters, in press.

Ewing, M., S. Eittreim, M. Truchan, and J. I. Ewing, Sediment distribution in the Indian Ocean, Deep Sea Res., 16, 231-248, 1969.

Fleet, A. J., and D. R. C. Kempe, Preliminary geochemical studies of the sediments from DSDP Leg 26, Southern Indian Ocean, in T. A. Davies, B. P. Luyendyk et al., Initial Reports of the Deep Sea Drilling Project, 26, Washington (U. S. Government Printing Office), 541-552, 1974.

Goldberg, E. D., and G. Arrhenius, Chemistry of Pacific pelagic sediments, Geochim. Cosmochim. Acta., 13, 153-212, 1958.

Hayes, D. E., L. A. Frakes et al., Initial Reports of the Deep Sea Drilling Project, 28, Washington (U. S. Government Printing Office, 1975.

Kennett, J. P., Development of the circum-Antarctic current, Science, 186, 144-147, 1974.

Kennett, J. P., R. E. Houtz, et al., Initial Reports of the Deep Sea Drilling Project, 29, Washington (U. S. Government Printing Office), 1975.

Kennett, J. P., and R. C. Thunell, Global increase in Quaternary explosive volcanism, Science, 187 (4176), 497-503, 1975.

Larson, R. L., Late Jurassic sea-floor spreading in the Eastern Indian Ocean, Geology, 3 (2), 69-71, 1975.

Luyendyk, B. P., Gondwanaland dispersal and the early formation of the Indian Ocean, in T. A. Davies, B. P. Luyendyk et al., Initial Reports of the Deep Sea Drilling Project, 26, Washington (U. S. Government Printing Office, 2945-2952, 1974.

Luyendyk, B. P., and T. A. Davies, Results of DSDP Leg 26 and the geologic history of the southern Indian Ocean, in T. A. Davies, B. P. Luyendyk et al., Initial Reports of the Deep Sea Drilling Project, 26, Washington (U. S. Government Printing Office), 909-943, 1974.

McElhinny, M. W., Formation of the Indian Ocean, Nature, 228, 977-979, 1970.

McKelvey, B. C., and A. J. Fleet, Eocene basaltic pyroclastics at site 253, Ninetyeast Ridge, in T. A. Davies, B. P. Luyendyk et al., Initial Reports of the Deep Sea Drilling Project, 26, Washington (U. S. Government Printing Office), 553-565, 1974.

Petkovic, P., Origin of the Naturaliste Plateau, Nature, 253, 30-33, 1975a.

Petkovic, P., Naturaliste Plateau, in J. J. Veevers, ed., Deep Sea Drilling in Australasian Waters, Challenger Symposium, Sydney, Australia, 24-25, 1975b.

Pimm, A. C., Sedimentology and history of the northeast Indian Ocean from the late Cretaceous to Recent, in C. C. von der Borch, J. G. Sclater et al., Initial Reports of the Deep Sea Drilling Project, 22, Washington (U. S. Government Printing Office), 717-804, 1974a.

Pimm, A. C., Mineralization and trace element variations in deep-sea pelagic sediments of the Wharton Basin, Indian Ocean, in C. C. von der Borch, J. G. Sclater et al., Initial Reports of the Deep Sea Drilling Project, 22, Washington (U. S. Government Printing Office), 469-476, 1974b.

Pimm, A. C., and J. G. Sclater, Early Tertiary hiatuses in the northeastern Indian Ocean, Nature, 252, 362-365, 1975.

Quilty, P. G., Late Jurassic to Recent geology of the western margin of Australia, in J. J. Veevers, ed., Deep Sea Drilling in Australasian Waters, Challenger Symposium, Sydney, Australia, 15-23, 1975.

Robinson, P. T., P. A. Thayer, P. J. Cook and B. K. McKnight, Lithology of Mesozoic and Cenozoic sediments of the Eastern Indian Ocean, Leg 27, Deep Sea Drilling Project, in J. J. Veevers, J. R. Heirtzler et al, *Initial Reports of the Deep Sea Drilling Project*, 27, Washington (U. S. Government Printing Office), 1001-1047, 1974.

Sclater, J. G., and R. L. Fisher, Evolution of the East Central Indian Ocean, with emphasis on the tectonic setting of the Ninetyeast Ridge, *Geol. Soc. Amer. Bull.*, 85, 683-702, 1974.

Shepard, F. P., Nomenclature based on sand-silt-clay ratios, *Jour. Sediment. Petrol.*, 24, 151-158, 1954.

Smith, A. G., and A. Hallam, The fit of the southern continents, *Nature*, 225, 139-144, 1970.

Thompson, R. W., Mineralogy of sands from the Bengal and Nicobar Fans, sites 218 and 211, Eastern Indian Ocean, in C. C. von der Borch, J. G. Sclater et al., *Initial Reports of the Deep Sea Drilling Project*, 22, Washington (U. S. Government Printing Office), 711-714, 1974.

Turekian, K. K., and K. H. Wederpohl, Distribution of the elements in some major rock units in the earth's crust, *Geol. Soc. Amer. Bull.*, 172, 175-192, 1961.

Veevers, J. J., Seismic profiles made underway on Leg 22 in C. C. von der Borch, J. G. Sclater et al., *Initial Reports of the Deep Sea Drilling Project*, 22, Washington (U. S. Government Printing Office), 351-367, 1974.

Veevers, J. J., and P. R. Evans, Sedimentary and magnetic events in Australia and the mechanism of world-wide Cretaceous transgressions, Nature (Phys. Sci.), 245, 33-36, 1973.

Veevers, J. J., D. A. Falvey, L. V. Hawkins, and W. J. Ludwig, Seismic reflection measurements of northwest Australian margin and adjacent deeps, *Amer. Assoc. Petrol. Geol.*, 58 (9), 1731-1750, 1974.

Veevers, J. J., and J. R. Heirtzler, Bathymetry, seismic profiles, and magnetic anomaly profiles, in J. J. Veevers, J. R. Heirtzler et al., *Initial Reports of the Deep Sea Drilling Project*, 27, Washington (U. S. Government Printing Office), 339-381, 1974a.

Veevers, J. J., and J. R. Heirtzler, Tectonic and paleogeographic synthesis of Leg 27, in J. J. Veevers, J. R. Heirtzler et al., *Initial Reports of the Deep Sea Drilling Project*, 27, Washington (U. S. Government Printing Office), 1049-1054, 1974b.

Veevers, J. J., and M. H. Johnstone, Comparative stratigraphy and structure of the Western Australian margin and the adjacent deep ocean floor, in J. J. Veevers, J. R. Heirtzler et al., *Initial Reports of the Deep Sea Drilling Project*, 27, Washington (U. S. Government Printing Office) 571-585, 1974.

Veevers, J. J., J. G. Jones, and J. A. Talent, Indo-Australian stratigraphy and the configuration and dispersal of Gondwanaland, *Nature*, 229, 383-388, 1971.

Venkatarathnam, K., Mineralogical data from sites 211, 212, 213, 214 and 215 of the Deep Sea Drilling Project, Leg 22 and origin of non-carbonate sediments in the equatorial Indian Ocean, in C. C. von derBorch, J. G. Sclater et al., *Initial Reports of the Deep Sea Drilling Project*, 22, Washington (U. S. Government Printing Office), 489-502, 1974.

van Andel, Tj. H., Mesozoic/Cenozoic calcite compensation depth and the global distribution of calcareous sediments, *Earth and Planet. Sci. Lett.*, 26, 187-194, 1975.

von der Borch, C. C., and R. W. Rex, Amorphous iron oxide precipitates in sediments cored during leg 5, Deep Sea Drilling Project, in D. A. McManus et al., *Initial Reports of the Deep Sea Drilling Project*, 5, Washington (U. S. Government Printing Office), 541-544, 1970.

von der Borch, C. C., J. G. Sclater et al., *Initial Reports of the*

Deep Sea Drilling Project, 22, Washington (U. S. Government Printing Office), 890 pp., 1974.

von der Borch, C. C., and N. A. Trueman, Dolomitic basal sediments from the northern end of Ninetyeast Ridge, in C. C. von der Borch, J. G. Sclater *et al.*, *Initial Reports of the Deep Sea Drilling Project*, 22, Washington (U. S. Government Printing Office), 477-483, 1974.

CHAPTER 6. MODELS OF THE EVOLUTION OF THE EASTERN INDIAN OCEAN

J. J. Veevers

Macquarie University
Sydney, Australia

  Abstract. The recognition by Le Pichon and Heirtzler (1968) of sea floor spreading magnetic anomalies in the Indian Ocean started a phase of exploration that culminated in the drilling in 1972 and early 1973 of twenty-three sites in the eastern Indian Ocean. Several models of the evolution of the region were made during the course of this exploration. Augmented by subsequent work, the magnetic and deep sea drilling data can be linked in a model that entails the generation of the eastern Indian Ocean by four stages of spreading.

                              Introduction

  As described here, the eastern Indian Ocean lies within the area outlined by the equator and Lat. 50°S, and Long. 80° and 120°E (Figure 1). Several matters point to a complex history of this region:
  (a) only part of it is relatable to the present spreading ridge;
  (b) it contains many elevated features (Ninetyeast Ridge, Broken Ridge, Naturaliste Plateau, Wallaby Plateau) whose structure and origin are obscure; and
  (c) part of the record has been lost, presumably by plate consumption, at the Java Trench.
The first reported seafloor spreading magnetic anomalies (Le Pichon and Heirtzler, 1968) were found for a short distance on either side of the current spreading ridge. McKenzie and Sclater (1971) found magnetic anomalies south of Ceylon, but except Falvey (1972), since modified by Larson (1975), no recognizable magnetic anomalies from the Wharton Basin were reported until 1974 (Sclater and Fisher, 1974; Markl, 1974). Until deep sea drilling started in 1972, the only direct sampling of the deep seafloor was by piston-coring, which revealed Cretaceous sediments near Perth, Eocene sediment south of Java, and nothing older than Miocene elsewhere (Le Pichon and Heirtzler, 1968). The drilling of twenty-three oceanic sites in the region during 1972 and early 1973 has made good this deficiency. The sites of deep sea drilling and the location of selected magnetic anomalies are shown in Figure 1.

Some Models Published Since 1971

  1) Dietz and Holden (1971) (Figure 2) postulated that, in a generalized reconstruction of Pangaea, the Tethys Sea occupied the region of the Indian Ocean, and that the area between India and Australia was occupied by a bay of Tethys, which they called Sinus Australi. By postulating further a single spreading ridge from which India, and, subsequently, Australia dispersed from Antarctica, Dietz & Holden concluded

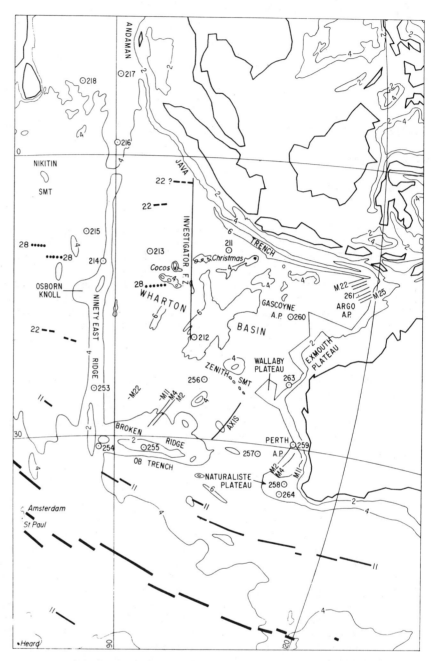

Fig. 1. Location of eastern Indian Ocean, showing isobaths (km), selected magnetic anomalies (nos. 11, 22, and 28 of the scale of Heirtzler, et al., 1968, and M2, etc. of the scale of Larson and Pitman, 1972), deep sea drilling sites (211, etc.), and the axis of the southeast Indian Ocean spreading ridge (heavy lines). Lambert equal-area projection.

that the Wharton Basin possibly contained pre-Mesozoic oceanic crust.
 2) In an almost simultaneous publication, Veevers, Jones, and Talent (1971) postulated that the entire Indian Ocean, except perhaps a small area off northwest Australia, was generated by seafloor spreading since

EVOLUTION OF THE EASTERN INDIAN OCEAN 153

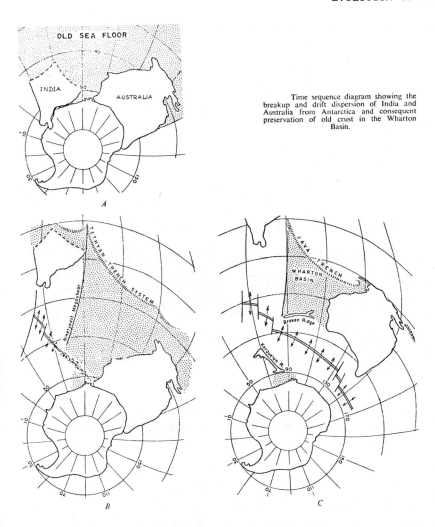

Fig. 2. Reproduced from Dietz and Holden (1971), by permission of the authors and of <u>Nature</u>.

the Cretaceous. In their model (Figure 3) the fragments of Gondwanaland were dispersed in two stages: (a) in the Cretaceous, India-Africa separated from Antarctica-Australia by spreading from ridges that subsequently became the Ninetyeast Ridge and the southwest branch of the Indian Ocean Ridge; and (b) in the Cenozoic, Africa-India and Antarctica-Australia each broke into the separate continents by spreading along the current northwest spreading ridge, with the southwest ridge and Ninetyeast Ridge acting as transform faults. In this model, the eastern margin of India was placed next to the margin of southwest Australia, in accord with the distribution of marine and non-marine facies along the western margin of Australia.

In the course of an account of the evolution of the Southern Oceans, Heirtzler (1971) independently derived a model similar to that of Veevers et al. (1971).

3) From the discovery of a set of seafloor spreading magnetic anomalies in the Wharton Basin south of Sumatra, and the incorporation of drilling results from DSDP Legs 22 and 26, Sclater and Fisher (1974) (Figure 4) made a quantitative model of the evolution back to 80 my BP. The dis-

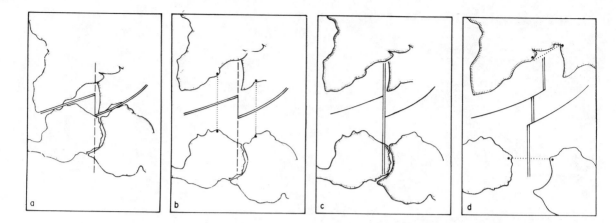

Fig. 3. Dispersal of Gondwanaland. *a* and *b*; Beginning and end of Cretaceous phase of dispersal. *c* and *d*, Beginning and present state of Cenozoic phase of dispersal. Heavy lines, crustal plate boundaries which may be passive (broken lines), or acting as faults (single lines) or spreading ridges (double lines). Lines of small dots indicate paths of displacement between once contiguous points on continents indicated by large dots. Central segment of Indian Ocean Ridge system taken as datum. Reproduced from Veevers, et al. (1971), by permission of the authors and of Nature.

covery in the Wharton Basin of magnetic anomalies that young northward, the use of the drilling results, and the construction of a rigorous quantitative model make this the first authoritative synthesis of the eastern Indian Ocean.

The model entails the location of India against Antarctica, as in the Smith and Hallam (1970) reconstruction of Gondwanaland, and the dispersal of India from Antarctica-Australia at 100 my BP. Several different plates, which we identify in terms of the associated continental fragment or fragments, were involved: 100-53 my BP -- India and Antarctica-Australia; 53-32 my BP -- India, Australia, and Antarctica; 32-0 my BP -- Antarctica, India-Australia.

4) Luyendyk (1974) compiled all the deep sea drilling and magnetic results except the subsequent work of Markl (1974) and Larson (1975) (Fig. 5) and produced an outline sketch of the early dispersal, in the late Jurassic and Cretaceous, of Gondwanaland (Figure 6). Luyendyk used the Smith & Hallam (1970) fit of India against Antarctica, but, citing Falvey (1972) and Dickinson (1971), extended the area of India to fill the Sinus Australi of Dietz and Holden (1971), as postulated at the same time by Crawford (1974), Curray and Moore (1974), McElhinny and Embleton (1974), and Veevers et al. (1975).

5) Curray and Moore (1974) postulated three hot spots (bight of India, tip of India, Naturaliste area of southwest Australia) in their Gondwanaland reconstruction, which they modified from Smith and Hallam (1970) 'by increasing the area of Indian continental crust believed to have underthrust Asia.' They postulated that the hot spot from the bight of India formed the Ninetyeast Ridge and lies today under St Paul and Amsterdam islands, and that the hot spot from the Naturaliste area was responsible for forming Broken Ridge and Kerguelen Plateau after separation of India from Antarctica and Australia (Figure 7A). In the initial breakup, in the early Cretaceous, India separated from combined

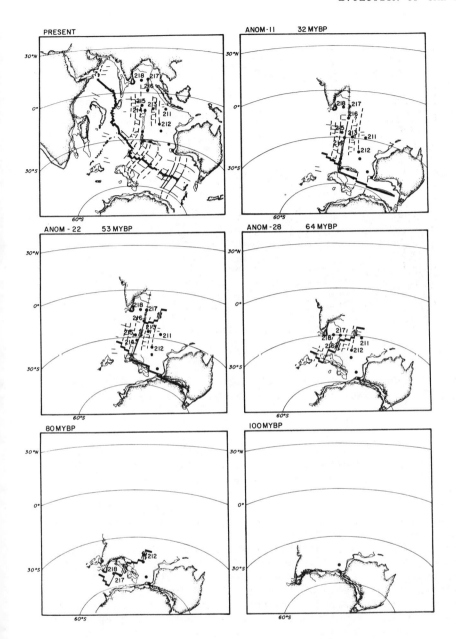

Fig. 4. Reconstruction of the relative positions of India, Australia, and Antarctica regressing from the present to the latest early Cretaceous (100 to 110 my BP). The small black dots represent seismic epicenters on the active mid-ocean ridges. The heavy black dots mark DSDP stations. Those from Leg 22 have been numbered to aid identification of points on the ocean floor. Appropriate information has been left on the reconstructions in diagrammatic form. They can be identified as follows: the thick, black lines, spreading centers; the heavy continuous lines, transform faults, the north-south and dashed lines, fracture zones; the east-west trending lines, identified magnetic anomalies. Reproduced from Sclater and Fisher (1974) by permission of the authors and of the Geological Society of America.

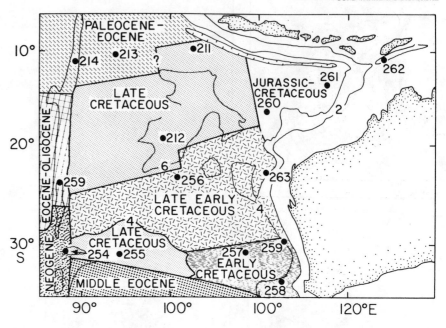

Fig. 5. Age provinces based on magnetic anomaly data and DSDP sites from Legs 22, 26, 27. Reproduced from Luyendyk (1974) by permission of the author.

Australia and Antarctica in a direction approximately perpendicular to the north-east trending continental margin of India, and this motion continued to about 75 m.y. ago. Then the direction of spreading changed to become approximately north-south, parallel to the present Ninetyeast Ridge (Figure 7B), until 35 m.y. ago. From 55 m.y. ago (initial collision of India and southern Asia) to 35 m.y. ago, the rate of seafloor spreading decreased. Finally, at approximately 35 m.y. ago, India converged toward Asia in a more northeasterly direction (Figure 7C).

6) Veevers and Heirtzler (1974) presented reconstructions of the late Jurassic and early Cretaceous episodes of spreading of the area off western Australia (Figure 8). An advance was made by Heirtzler's recognition of 'a group of northeast-trending magnetic anomalies from the region west of Perth that probably indicate the trend of spreading here'. This is part of the set of anomalies that Markl (1974) independently recognized as the Perth sequence.

7) Markl (1974) identified the Perth sequence as M1-M11 on the Larson and Pitman (1972) scale, and recognized an equivalent sequence (the Broken Ridge sequence) of magnetic anomalies to the west-north-west (Figure 1). With this discovery, the set of primary data for the eastern Indian Ocean became complete.

8) Using the concept of an enlarged India, referred to in sections 4 and 5 above, Johnson, Powell and Veevers (1976) derived a quantitative model (Figure 9) from the magnetic and drilling data that had accrued by 1975. In this model, the eastern Indian Ocean is generated in four stages of spreading during the Cretaceous and Cenozoic:

(i) 130-80 my BP -- Greater India is dispersed from Antarctica-Australia by seafloor spreading from an east-west ridge; (ii) 80-53 my BP -- a new axis of spreading transfers some of the ocean-backed part

EVOLUTION OF THE EASTERN INDIAN OCEAN 157

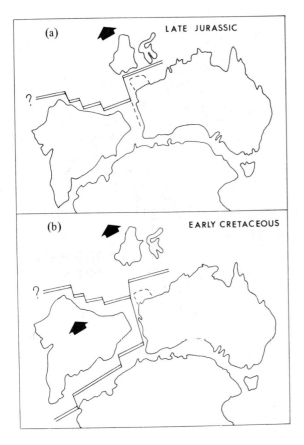

(a) Late Jurassic plate boundaries in the eastern Indian Ocean showing an east-west ridge system separating an ancient landmass from northern Gondwanaland. Mercator projection. (b) The plate system in the Early Cretaceous showing the separation of Indian from Antarctica and Australia. Arrows indicate plate motions relative to Antarctica.

Fig. 6. Reproduced from Luyendyk (1974) by permission of the author.

of the Indian plate to the Antarctica-Australia plate. India moves northward rapidly; (iii) 53-32 my BP -- with the spreading of seafloor between Australia and Antarctica, three plates existed; (iv) 32-0 my BP -- with the cessation of spreading between India and Australia, a new plate (India-Australia) came into being.

Note on Late Jurassic Spreading

The occurrence of late Jurassic fossils near the base of site 261 in the Argo Abyssal Plain (Veevers et al., 1974) allowed Larson (1975) to calibrate the magnetic anomalies identified as late Cretaceous to Eocene by Falvey (1972) as M22-M25 (Kimmeridgian and Oxfordian). Larson's (1975) identifications of these anomalies were made in face of the difficulties pointed out by Luyendyk (1974, p. 945), who noted that 'there is some doubt as to whether these anomalies are related to seafloor spreading processes.' Falvey's (1972) attempt to compute

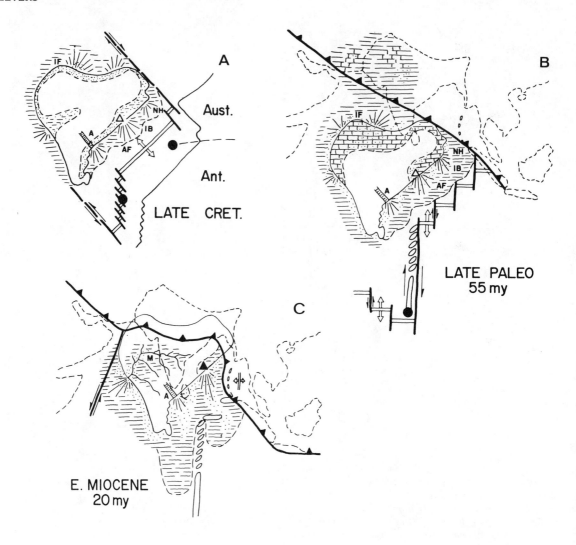

Fig. 7. Reproduced from Curray and Moore (1974) by permission of the authors and the publisher.

a pole of rotation was modified by Falvey and Veevers (1974), who showed that the data are too sparse and the positioning not precise enough (the navigation of the two sets of parallel lines across the anomalies was found by star-fixes only) to determine a precise pole of rotation.

This late Jurassic spreading predates the dispersal of at least the main elements of Gondwanaland.

Discussion

Dietz and Holden's (1971) and Veevers et al., (1971) models have been superseded by models that incorporate subsequent drilling and magnetic data in the eastern Indian Ocean.

Sclater and Fisher (1974) were the first to provide a quantitative reconstruction of the complex plate motions in the eastern Indian Ocean. They traced the history back to the beginning of the Heirtzler et al., (1968) time scale ($\sim$ 80 my BP), leaving the period from 80 m.y. ago to

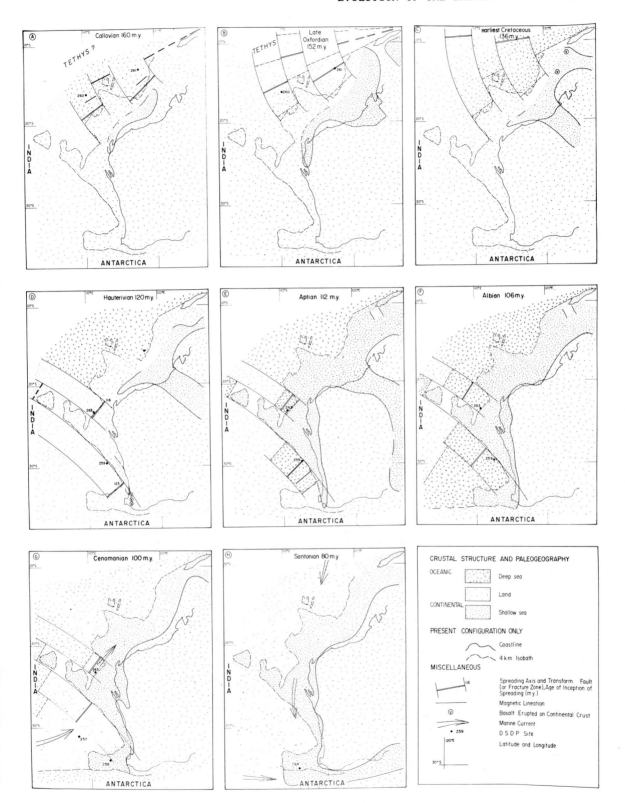

Fig. 8. Redrawn from Veevers and Heirtzler (1974).

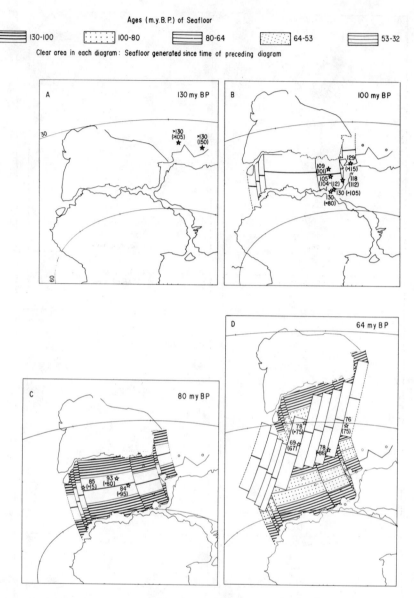

Fig. 9. Reproduced from Johnson et al. (1976) by permission of the authors and the Geological Society of America. Spreading history from 130 my BP through five intermediate ages to the present. Lambert equal-area projection. Deep sea drilling sites shown in newly generated areas of seafloor as stars with basement ages (in my BP) predicted in model and (in brackets) observed from drilling, and in older areas as circles. The part of the seafloor subsequently subducted along the Andaman-Java Trench is shown as if subduction had not taken place.

the initial breakup of Gondwanaland to be documented. This was done by the discovery of early Cretaceous oceanic crust by drilling at DSDP sites 256, 257, 259, and, at least suggested, at 263. Finally, Markl's (1974) confirmation of Heirtzler's discovery of northeast-trending magnetic anomalies near Perth, his discovery of the Broken Ridge set,

EVOLUTION OF THE EASTERN INDIAN OCEAN 161

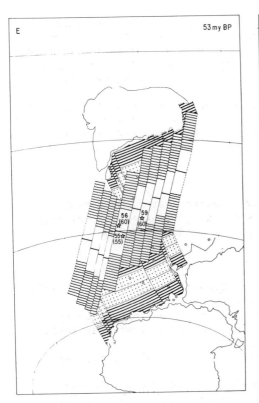
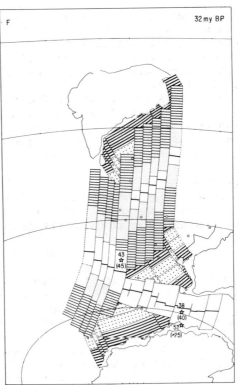

Fig. 9. Continued

and the dating of both sets as early Cretaceous filled in the last remaining first-order gap in knowledge. Preliminary attempts at putting together the information then available off western Australia (Veevers and Heirtzler, 1974) and off India (Curray and Moore, 1974) culminated in the quantitative synthesis of the eastern Indian Ocean by Johnson et al., (1976).

Future work in the eastern Indian Ocean should be aimed at detailing the Perth, Broken Ridge, and Argo Abyssal Plain sets of magnetic anomalies, at resolving the structure of the marginal and oceanic plateaus of the region, and at testing by quantitative methods the postulated motions of hot spots.

Postcript

Since writing the manuscript, I have learned that J. R. Curray and D. G. Moore developed in 1972 a series of sketch diagrams essentially the same as those given in Figure 9.

References

Crawford, A. R., A greater Gondwanaland, Science, 184, 1179-1181, 1974.
Curray, J. R., and D. G. Moore, Sedimentary and tectonic processes in the Bengal deep-sea fan and geosyncline, in The geology of continental margins, edited by C. A. Burk and C. L. Drake, Springer, New York, 617-627, 1974.
Dickinson, W. R., Plate tectonic models for orogeny at continental margins, Nature, 232, 41-42, 1971.
Dietz, R. S., and J. C. Holden, Pre-Mesozoic oceanic crust in the eastern Indian Ocean (Wharton Basin)?, Nature, 229, 309-312, 1971.
Falvey, D. A., Sea-floor spreading in the Wharton Basin (northeast Indian Ocean) and the break-up of eastern Gondwanaland, Jour. Aust. Pet. Expl. Assoc., 12, 86-88, 1972.
Falvey, D. A., and J. J. Veevers, Physiography of the Exmouth and Scott Plateaus, Western Australia, and adjacent northeast Wharton Basin, Marine Geology, 17, 21-59, 1974.
Heirtzler, J. R., The evolution of the Southern Oceans, in Research in the Antarctic, Amer. Assoc. Adv. Sci., 667-684, 1971.
Heirtzler, J. R., G. O. Dickson, E. M. Herron, W. C. Pitman, and X. Le Pichon, Marine magnetic anomalies, geomagnetic field reversals, and motions of the ocean floor and continents, Jour. Geophy. Res., 73, 2119-2136, 1968.
Johnson, B. D., C. M. Powell, and J. J. Veevers, Spreading history of the eastern Indian Ocean and Greater India's northward flight from Antarctica and Australia, Bull. Geol. Soc. Amer., 87, 1560-1566, 1976.
Larson, R. L., Late Jurassic seafloor spreading in the eastern Indian Ocean, Geology, 3, 69-71, 1975.
Larson, R. L., and W. C. Pitman, World-wide correlation of Mesozoic magnetic anomalies and its implications, Bull. Geol. Soc. Amer. Bull., 83, 3645-3662, 1972.
Le Pichon, X., and J. R. Heirtzler, Magnetic anomalies in the Indian Ocean and sea-floor spreading, Jour. Geophys. Res., 73, 2101-2117, 1968.
Luyendyk, B. P., Gondwanaland dispersal and the early formation of the Indian Ocean, in Davies, T.A., Luyendyk, B.P. et al., Initial Reports of the Deep Sea Drilling Project, 26, Washington, D. C. (U. S. Government Printing Office) 945-952, 1974.
Markl, R. G., Evidence for the breakup of eastern Gondwanaland by the early Cretaceous, Nature, 251, 196-200, 1974.
McElhinny, M. W., and Embleton, B.J.J., Australian paleomagnetism and

the Phanerozoic plate tectonics of eastern Gondwanaland, Tectonophysics, 22, 1-29, 1974.

McKenzie, D., and J. G. Sclater, The evolution of the Indian Ocean since the Late Cretaceous, Geophys. Jour. Roy. Astron. Soc., 25, 437-528, 1971.

Sclater, J. G., and R. L. Fisher, Evolution of the East Central Indian Ocean, with emphasis on the tectonic setting of the Ninetyeast Ridge, Bull. Geol. Soc. Amer., 85, 683-702, 1974.

Smith, A. G., and A. Hallam, The fit of the southern continents, Nature, 225, 139-144, 1970.

Veevers, J. J., and J. R. Heirtzler, Tectonic and paleogeographic synthesis of Leg 27, in Veevers, J.J., Heirtzler, J.R., et al., Initial Reports of the Deep Sea Drilling Project, 27, Washington (U. S. Government Printing Office) 1047-1054, 1974.

Veevers, J. J., J. G. Jones, and J. A. Talent, Indo-Australian stratigraphy and the configuration and dispersal of Gondwanaland, Nature, 229, 383-388, 1971.

Veevers, J. J., and J. R. Heirtzler, et al., Initial reports of the Deep Sea Drilling Project, 27, Washington (U. S. Government Printing Office), 1974.

Veevers, J. J., C. M. Powell, and B. D. Johnson, Greater India's place in Gondwanaland and in Asia, Earth Plan. Sci. Lett., 27, 383-387, 1975.

CHAPTER 7.   DEEP SEA DRILLING ON THE NINETYEAST RIDGE: SYNTHESIS
AND A TECTONIC MODEL

Bruce P. Luyendyk

Department of Geological Sciences, University of California
Santa Barbara, California   93106

Abstract. Legs 22 and 26 of the Deep Sea Drilling Project drilled five sites on the Ninetyeast Ridge. Four main observations emerged:
1) The Ridge has an extrusive volcanic basement with a distinctive petrochemistry.
2) The Ridge is older to the north, ranging from Campanian or older at 9°N to Eocene-Oligocene at 31°S. This age gradient is in the same sense as that for the Indian basin to the west, implying that the Ridge belongs, tectonically, to the Indian plate.
3) It was formed in shallow water, sometimes subaerially, and subsided with time in accordance with known age-depth curves.
4) The Ninetyeast Ridge was formed in more southerly latitudes and has since been transported northwards.
These observations, combined with other geophysical data, show that the Ridge is primarily a sunken oceanic island and seamount chain. Reconstructions based on a fixed hot spot assumption demonstrate strong support for this point source model. The model predicts that the Ridge is from 90 to 20 m.y. old and was formed primarily from volcanism associated with a source beneath the Amsterdam-St. Paul islands, but contributions from a source beneath Kerguelen are also probable. The Kerguelen source mainly generated the Broken Ridge and Naturaliste Plateau, and the Kerguelen Plateau between >100 and 80 mybp.

Introduction

This paper reviews and synthesizes the DSDP results from the Ninetyeast Ridge obtained during Leg 22 (von der Borch, Sclater et al., 1974) and Leg 26 (Davies, Luyendyk et al., 1974) during Phase II and III of Indian Ocean DSDP operations.

Topography

The Ninetyeast Ridge was unknown prior to the International Indian Ocean Expedition in the early 1960's. A good historical review of its discovery is given in Sclater and Fisher (1974). The Ridge extends from about 32°S to near 10°N, where it becomes buried by the sediments of the Bengal Fan (Moore et al., 1974). At 30°S, it intersects with the east-west trending Broken Ridge forming an impressive topographic bight. The Ridge trends slightly east of north and crosses 90°E near the equator. Its relief is about 2000 m and summit depths vary from less than 2000 m in the south to almost 3000 m in the north (Figure 1). Osborn Knoll, near 14°S, is a 200 km-diameter

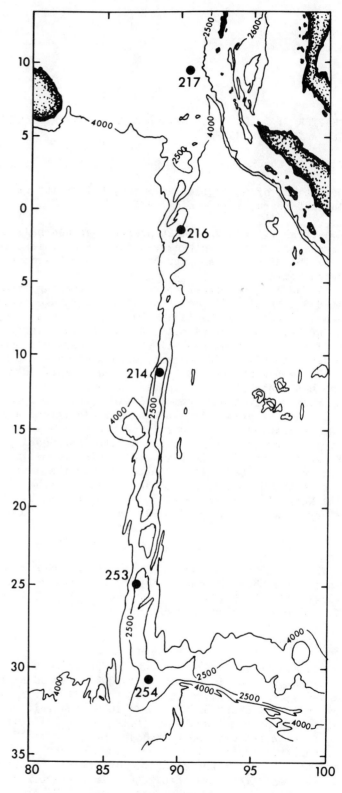

Fig. 1. Bathymetric chart of the Ninetyeast Ridge modified from Laughton et al. (1970). Contours are in corrected meters.

circular appendage to the west side of the Ridge. The Ninetyeast Ridge demonstrates three distinctive physiographies from north to south (Sclater and Fisher, 1974; Luyendyk and Davies, 1974). North of 7°S are found lobate structures, reminiscent of many of the Pacific type seamount chains, which have also been described as en-echelon in character (e.g., Bowin, 1973; Curray and Moore, 1974). South of here and down to Osborn Knoll, the Ridge is much narrower (100 km), has straight, steep sides and is symmetrical in cross section. South of Osborn Knoll the Ridge is much wider (200 km), shallower and asymmetric with a very steep eastern side. This multiple morphology is believed related to the tectonic origin of the Ridge as discussed below.

Tectonic Setting

Sclater and Fisher (1974) have found that Cenozoic magnetic anomalies become older to the northwest of the Ninetyeast Ridge and older to the southeast of the Ridge (Figure 2). This suggests that the Ridge lies on or near a fracture zone (left-offset transform fault). Both Bowin (1973) and Sclater and Fisher have noted a deep fracture valley immediately to the east of the Ridge which is probably this fault or part of the fault system (Figure 2). The Ninetyeast Ridge would then belong structurally to the Indian plate. Prior to anomaly 28 time (ca. 64 mybp, Sclater et al., 1974) this transform fault had a left lateral offset of about 10°. Sometime between anomaly 28 and anomaly 22 (ca. 53 mybp) time, the ridge west of the Ninetyeast Ridge jumped south 11° (Sclater and Fisher, 1974; Sclater et al., 1976) capturing part of the Antarctic plate to the Indian, and producing a greater left-lateral offset. Between these two times, it is possible that parts of the Ridge between 7°S and Osborn Knoll were formed by a leaking along the transform. The Indian and Australian plates remained separated along this transform fault up to about 32 mybp. The Ridge is evidently younger than 32 mybp suggesting that mechanisms other than transform leaking have formed it since this time.

Previously Proposed Origins

A long litany of hypotheses can be found for the origin of the Ninetyeast Ridge. Soon after its discovery, Heezen and Tharp (1965) proposed that this and other aseismic ridges in the Indian Ocean were fragments of continental crust. Francis and Raitt (1967) conducted seismic refraction studies which failed to reach the Moho but found velocities with continental affinities. They also suggested the Ridge was a horst, which was supported by Laughton et al. (1970), and which Bowin (1973) disproved later on the basis of gravity data. McKenzie and Sclater (1971) suggested it was uplifted from the edge of the Indian plate in response to subduction of the Wharton (Australian/Antarctic) plate. Veevers et al. (1971) thought the Ridge was an extinct spreading axis. Falvey (1972) thought its origin was due to combined spreading and subduction. DSDP results showed that the Ridge is volcanic in origin, not continental, and that it formed in shallow water and then subsided rather than being uplifted. More recent ideas which are compatible with the drilling results suggest that the Ridge formed from a hot spot or mantle plume (Morgan, 1972a, b; Bowin, 1973; Curray and Moore, 1974) or a combination of a hot spot and leaky transform mode (Sclater and Fisher, 1974; Luyendyk and Davies. 1974; Vogt and Johnson, 1975; Luyendyk and Rennick, (1977). These last hypotheses are covered in some detail below.

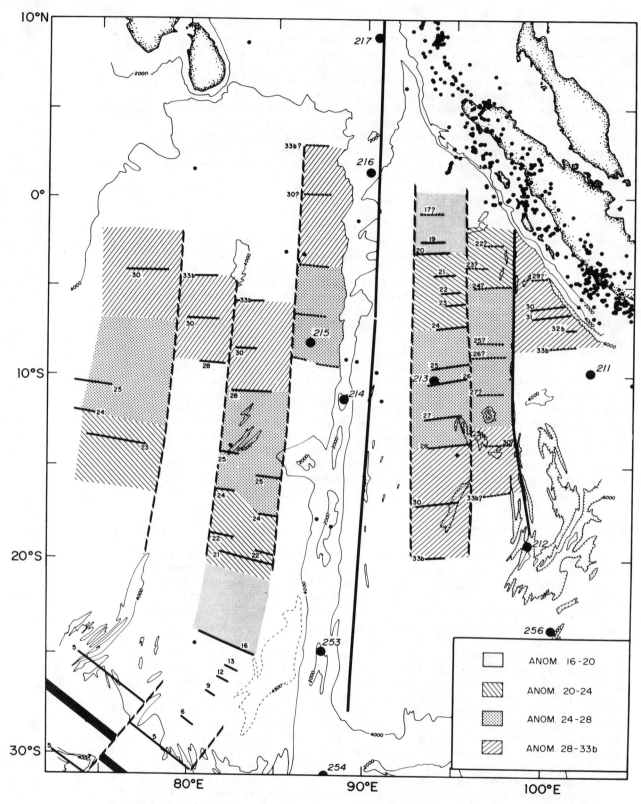

Fig. 2. Crustal ages in the vicinity of the Ninetyeast Ridge determined from Cenozoic magnetic anomalies; from Sclater and Fisher (1974), figure 9.

Drilling Results

The main observations from Legs 22 and 26 can be enumerated as known facts about the Ridge.

1) The Ridge has an extrusive volcanic basement, largely submarine, with a distinctive petrochemistry.

2) It is older to the north, ranging from Campanian or older at site 217, to Eocene-Oligocene at site 254. The age gradient is in the same sense as that for the Indian basin to the west implying that the Ridge belongs, tectonically, to the Indian plate. Just south of the equator the Ridge is the same age as the Indian plate.

3) The Ninetyeast Ridge was formed in shallow water, sometimes sub-aerially, and then subsided with time.

4) The Ridge was formed at more southerly latitudes and has moved steadily northward.

Stratigraphy

The stratigraphic section on the Ridge consists of basal extrusive volcanics overlain by pyroclastics or volcanic sediments which are in turn covered by calcareous ooze and chalk (Figure 3). The units immediately above the basalt have an extremely varied lithology. At site 254, volcanic sediments conformably overlie the basalt and contain a shallow water biofacies. These sediments are ferrugenous silty clays and fine sandstones and contain weathered fragments of the underlying basalt. The basal sedimentary unit at site 253 is 388 meters of volcanic ash and lapilli which was deposited in about 4 million years. This unit is altered and contains three lithic types; dark olive-green, micarb-bearing altered vitric ash, black altered vitric ash and lapilli (with mollusc fragments), and flows of black vesicular olivine basalt and scoria with carbonate fillings. The basal sediment units at site 214 are tuffs, tuffaceous sediments, lignite, and glauconitic carbonate silt and sand. The lignite occurs in beds up to 80 cm thick and contains clay aggregates and pyrite. Clay aggregates in the tuff are iron stained. Site 216 has a basal unit of volcanic clay and micarb chalk. The volcanic clay occurs in chloritized aggregates. Glauconitic foram clay and chalk occurs near the top of the unit. Igneous basement was not reached at 217 and volcanic constituents are not present in the basal sediments. However, the lower sediment section does indicate shallow water conditions with the presence of oysters, minor glauconite, and possible dessication cracks.

The age of the basalt-sediment contact increases steadily northward (Table 1). These ages are well determined as late middle Eocene at site 253, Maestrictian at site 216 and Campanian-Maestrictian at site 217 (Figure 3). The oldest sediment at site 214 is Paleocene lignite dated by means of palynomorphs (Harris, 1974). The age data at site 254 are very confusing. Nonconcordant dates are compounded by the possibility of reworking of foraminifera and nannoplankton. The basal volcanic unit contains a few benthonic foraminifera, no nannoplankton, but ostracods and molluscs. The macrofauna seem to indicate a mid-Tertiary (Eocene-Oligocene?) age, but molluscs are poorly preserved and the ostracod assemblages are rather unique, rendering age assignments uncertain. Palynology studies indicate an early Miocene (late Oligocene?) age but this date is uncertain due to a high proportion of new species (Kemp, 1974). The Leg 26 scientists seem to accept a late Eocene-early Oligocene age but this is possibly too old. Potassium argon age determinations are concordant with paleontologic estimates at sites 214 and 216 (McDougall, 1974). Determinations at sites 253 and 254 do not agree (Rundle et al., 1974). This is possibly because

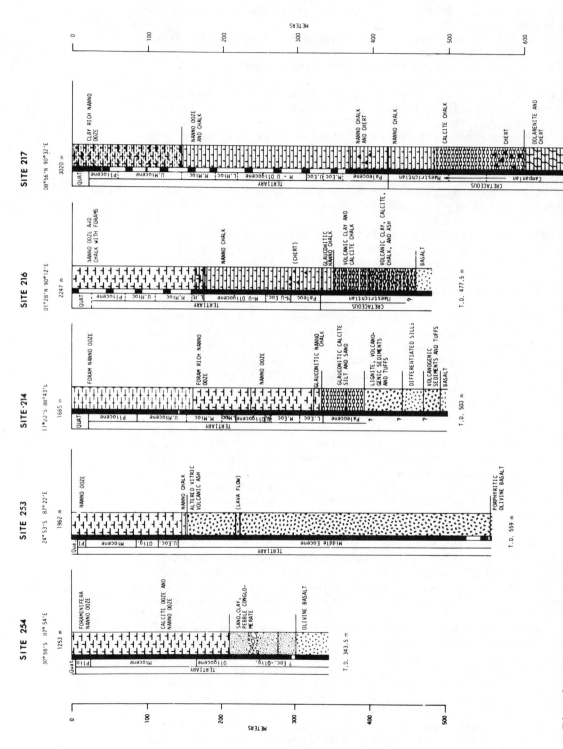

Fig. 3. Stratigraphy of sites drilled on the Ninetyeast Ridge. Black columns indicate cored intervals. Data from von der Borch, Sclater et al. (1974), and Davies, Luyendyk et al. (1974). Illustration is from Luyendyk and Davies (1974), figure 8.

K_2O concentrations at 214 and 216 (1.2-0.4 wt%) are 5 to 10 times greater than those at 253 and 254. Rundle et al. (1974) feel confident about a determination of 49 ± 5 at 254, which would make this site older than 253. Considering the available studies, a late Eocene-early Oligocene age is favored here.

The sedimentary section above the basal units is more uniform, being mainly carbonate ooze and chalk of biogenous origin. The chalks occur near the base of the carbonate section except at site 254 and are glauconitic at 214 and 216. Cherts are found at sites 216 and 217. Site 217 has the most varied stratigraphy. The base of this hole shows dolomite (dolarenite) of secondary origin which has replaced shallow-water bioclastic material. Carbonate oozed here of mid-late Miocene to Recent age contain a terrigenous clay admixture, which could represent turbidite contributions from the Bengal Fan.

Two hiatuses can be recognized in the calcareous section on the Ridge (Pimm, 1974; Luyendyk and Davies, 1974; Pimm and Sclater, 1974; Davies et al., 1975). These are centered on the early Eocene and on the late Eocene-early Oligocene, although paleontologically, the younger hiatus is not sharply developed (S. Gartner, personal comm., 1975). The youngest dates for these disconformities are found at 217 and are middle Oligocene and middle Eocene. The lower boundary of the upper hiatus has a maximum age of late Eocene at 214. The lower hiatus, near the Eocene-Oligocene boundary, occurs near the estimated age at site 254, pointing to a possible cause of the confusing basal stratigraphy at this site.

Two unconformities seen by Curray and Moore (1971) on the Bengal Fan have been correlated with those at site 217 (Moore et al., 1974). Pimm and Sclater (1974) and Davies et al. (1975) have indicated that these same two unconformities are found virtually throughout the entire Indian Basin at sites much deeper than the Ridge. The cause of the hiatuses is controversial (op. cit.) but evidently they mark a dramatic oceanographic change for these two times.

Subsidence History

Bowin (1973) and Sclater and Fisher (1974) have demonstrated that the summit depths of the Ninetyeast Ridge increase systematically northward in accordance with the Sclater-Menard age-depth curve (Sclater et al., 1971). Therefore, we can conclude that the Ridge subsided with time at the same rate as that of a section of ocean floor. The absolute depth would of course be offset by the elevation of the Ridge structure. Luyendyk and Davies (1974) have "backtracked" (Berger, 1973) the Ninety-east Ridge sites along the age-depth curves (Figure 4). This analysis shows that the two hiatuses discussed above formed over a wide depth range, from 3000 to 1000m. Sites 253 and 214 formed above sea level by this diagram. Also, at site 253, the subsidence rate matches the sedimentation rate for the thick volcanic ash sequence. In the manner which Luyendyk and Davies (1974) interpret the data, parts of the Ninetyeast Ridge were islands for much of the period between 75 and 25 mybp.

Pimm et al. (1974) have described the environmental facies for sites 214, 216 and 217 for the purpose of showing the Ridge's subsidence history. Sites 253 and 254 have been analyzed in a similar manner and the combined results are shown in Figure 5. The environmental facies vary from subaerial, lagoonal or shallow bank at the base of the section, to open oceanic (pelagic) near the top (Figure 5). About a dozen criteria separate the section into four or five subgroups. Major distinguishing criteria are the apparent diversity of calcareous microfossils, the presence of glauconite, presence of macrofossils, and vol-

TABLE 1. Minimum Age of DSDP Sites, Ninetyeast Ridge
(age of lowermost dateable material)

Site	Unit	Distance Above Basement Contact	Age	Method
217 (08°55.57'N)	1. micrite-dolomite	610.5 subbottom; basement not reached	late Campanian[a]	forams, nannos
216 (01°27.73'N)	1. micarb volcanic clay	+ 4.5 m	late Maestrictian[b]	forams, nannos
	2. basalt	-10.0 m	64.1 ± 1.0[c]	K-Ar
214 (11°20.21'S)	1. volcanic silty clay	+50.5 m	Paleocene[d]	forams, nannos
	2. volcanic clay	-28.3 m	middle Paleocene[e]	palynology
	3. basalt	- 0.78 m	53.4[c]	K-Ar
253 (24°52.65'S)	1. vitric volcanic ash	+ 3.5 m	late middle Eocene	nannos
	2. basalt	0	101 ± 3g	K-Ar
254 (30°58.15'S)	1. micarb ooze	+76 m	mid to upper Eocene ?[h]	b. forams
	2. silty clay	+15 m	Eocene-Oligocene[h]	ostracods
	3. silty clay	+15 m	mid-Tertiary[h]	molluscs
	4. silty clay-sand	+ 9 m	early Miocene[i]	palynology
	5. basalt	variable	18-49g	K-Ar

a – Ch. 8, von der Borch, Sclater et al. (1974)
b – Ch. 7, op cit.
c – McDougall (1974)
d – Ch. 5, von der Borch, Sclater et al.
e – Harris (1974)
f – Ch. 6 in Davies, Luyendyk et al. (1974)
g – Rundle et al. (1974)
h – Ch. 7, Davies, Luyendyk et al. (1974)
i – Kemp (1974)

canic material. Subaerial facies are indicated at site 254 by volcanogenic sediments from a nearby highly weathered basalt terrain, probably exposed. At site 214, this same facies is indicated by the presence of lignite and water-lain volcaniclastic debris. Subaerial exposure may also be indicated at 217 by a dessication crack in a lower section of claystone. Lagoonal and littoral facies are found at 254 and 217. This facies is indicated by a littoral assemblage of foraminifera, shallow-water ostracods, and pelecypods and gastropods at 254, and by dolomitization under hypersaline conditions, plus algal and reef structures at 217. A shallow bank facies is found at all sites. At 214, it is in direct contact with subaerial facies, and at 253 and at 216 it overlies volcanic basement. This facies is distinguished by low diversity of calcareous microfossils which are predominantly nonpelagic or hemipelagic. Macrofossil debris is common as is the presence of glauconite and volcanic debris. The transition of this facies to the oceanic type is shown by an increase in the pelagic constituents, rare glauconite and benthonic foraminifera, and the absence of volcanogenic material. The oceanic facies is indicated by a predominance of cosmopolitan nannoplankton and foraminifera.

The environmental facies are plotted on the paleodepth subsidence curves (Figure 4). In theory the facies boundaries should plot as horizontal lines because they are interpreted to represent changes in depth (vertical axis, Figure 4). The boundaries are actually more complex than that. The subaerial boundary is satisfactory but the shallow-oceanic boundary has much relief, particularly at site 216. This analysis points to the caution with which these paleodepth curves and the environmental bounds must be interpreted. For instance, these subsidence curves are not corrected for isostatic adjustment to sediment load. Although this can be done for normal oceanic crust to which a two-dimensional approximation can be applied (Berger, 1973), the Ninetyeast Ridge structure is essentially a point load in time near a plate edge and the load correction for this is not obvious or simple.

Evidence for Northward Motion

Two lines of evidence support northward motion for the Ninetyeast Ridge: the existence of cold water species and assemblages deep in the sediment section, and latitude determinations from paleomagnetic studies. Gartner et al. (1974) and Pimm et al. (1974) have discussed paleontologic indicators of northward motion from Leg 22 sites. One indicator is a diversity gradient in planktonic foraminifera from high to low, downsection at some sites. Higher diversity can indicate warmer environments. A diversity gradient is particularly well documented at 217 where extratropical assemblages occur below the Cretaceous/Tertiary boundary (Gartner et al., 1974) pointing to translation of this site from higher southern latitudes. At site 214 (11°20'S) pollen assemblages and benthonic foraminifera are similar to those temperate assemblages seen on southern Australia and New Zealand (Pimm et al., 1974). On the other hand, tropical or Tethyan foraminifera assemblages occur in the basal Maestrictian units at site 216 (Pimm et al., 1974). This site (1°28'N) presumably originated near 40°S as discussed below. However, northward movement was extremely rapid in the late Cretaceous (Figure 9) so that in only 10 my, 216 would have been in the tropical zone. At the Leg 26 sites, Paleogene nannoplankton assemblages at site 253 indicate transitional to subtropical conditions while the Miocene and Pliocene assemblages suggest subtropical to tropical (discoasters, sphenoliths, scyphospheres). Assemblages at site 254 are wholly subtropical or temperate.

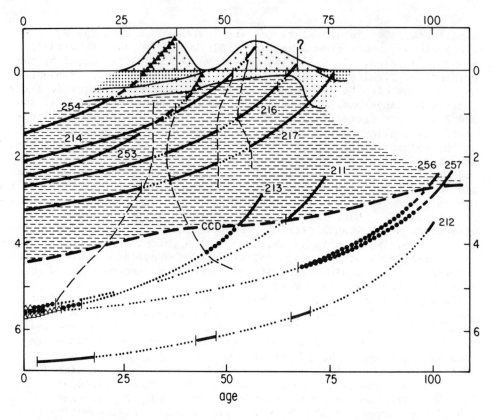

Fig. 4. Paleodepth and lithology of sites from the Ninetyeast Ridge and Wharton Basin modified from Luyendyk and Davies (1974), figure 9. Solid triangles are volcanic sediments, open triangles siliceous, solid line calcareous, heavy dots are clays, light dots are gaps or undated clays where dashed. Facies symbols are as with figure 5. The dashed lines outline the two unconformities on the Ridge and the CCD in the Wharton Basin.

Paleomagnetic studies have been made on the igneous rocks of 253 and 254 (Peirce et al., 1974) and work is in progress on the basal volcanic sequence of 253 (Cockerham et al., 1975; see Table 1). These results indicate paleolatitudes of near 50°S for the origin of the Ridge (southern portion). This result would fit the tectonic model of Luyendyk and Rennick (1977) discussed below. However, it also suggests a dramatic 20 of northward motion for 254 and 25° for 253. Paleontologic results do not express this suggested change in paleo-environment.

The evidence for northward movement of the Ridge supports interpretations that it has belonged tectonically to the Indian plate, because it is well known that India has traversed a considerable distance northward since the break up of Gondwanaland (McElhinney, 1973; Sclater and Fisher, 1974).

Petrology and Geochemistry

Basalt was retrieved at all sites on the Ridge except 217. Recovery at 253 was limited to only 35 cm, however. The petrology and chemistry of these rocks have received considerable attention and are discussed

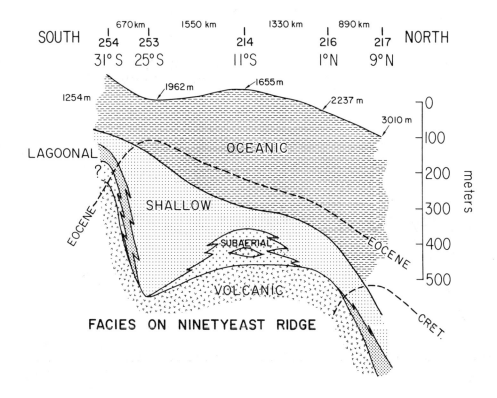

Fig. 5. Diachronous facies on the Ridge after Pimm et al. (1974), and this paper. The top of the Eocene and Cretaceous are indicated.

in Hekinian (1974a,b), Thompson et al. (1974), Frey and Sung (1974), Kempe (1974), and Frey et al. (1977). Three igneous rock types have been distinguished chemically: oceanic andesites (214), iron-rich tholeiites (214, 216) and tholeiitic basalts (253 and 254). The petrography of the rocks is highly varied. Amygdaloidal and vesicular basalt occurs at sites 214, 216, and 254. Site 214 and 216 basalts contain phenocrysts of plagioclase and magnetite and agglomerations of clinopyroxene. Olivine is absent in these basalts, but it is present in site 254 basalts and common (up to 15%) in site 253 basalts. Rocks classified as intermediate differentiates (oceanic andesites) occur at site 214. They are hypocrystalline and non-porphyritic and contain laths of oligoclase-andesine (Hekinian, 1974a, b).

Major element and trace element (including rare earth elements) analyses have been made on rocks from each site. In addition Fleet et al. (1977) have analyzed REE in volcanogenic sediments of 253 and 254. Compared to other Ninetyeast Ridge basalt, site 214 oceanic andesites have a high SiO_2 and alkali-metal content (Figure 6). Compared to oceanic spreading ridge basalts, the Ninetyeast Ridge basalts at sites 214, 216 and 254 are enriched in large ion lithophile elements (LIL) such as Sr, Ba and Zr and in light REE (La, Ce). Fleet et al. (1977) found light REE enrichment in site 254 volcaniclastics. These LIL element and REE abundances are the most distinguishing characteristics of Ninetyeast Ridge basalts.

The results from site 253 are not so clear. Frey and Sung (1974) found that the LIL element and REE are relatively enriched in samples taken from both a thin inter-ash flow and the apparent basement.

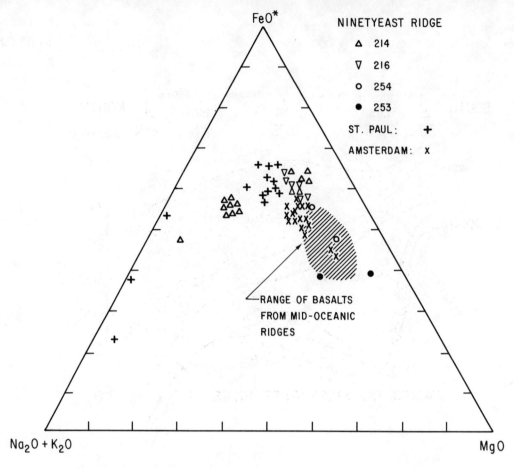

Fig. 6. FMA diagram (FeO* is total iron as FeO, units are wt %) of rocks from the Ninetyeast Ridge and the Amsterdam-St. Paul islands. From Frey and Sung (1974), figure 2.

Fleet et al. (1976) also found the volcanic ashes here to be REE enriched. However, Frey et al. (1977) studied further samples from the basement and found them relatively depleted in LIL elements and REE contrary to their previous result. Also they argue that the inter-ash flow does not compare in major element composition to the 214, 216 and 254 results (e.g., lower FeO/MgO). They conclude that the lower basalts of 253 were derived from a different mantle composition than that for 214 and 216. In view of the limited basalt recovery at 253 (35 cm), it is not clear whether this statement could be generally applied to the entire Ridge structure in the vicinity of 253.

Much discussion has been devoted to comparing the petrochemistry of spreading ridge rocks to basalts from Iceland, aseismic ridges, and eastern Indian Ocean islands like Amsterdam-St. Paul and Kerguelen (Hekinian, 1974b; Thompson et al., 1974; Frey and Sung, 1974). Briefly, the chemistry of the Ninetyeast Ridge rocks bears strong affinity to oceanic islands and other aseismic ridges. This distinctive chemistry may result from a mantle plume source in line with the models of Schilling (1971, 1973). For example, Frey and Sung (1974) and Luyendyk and Rennick (1977) have proposed that hot spots under Amsterdam-St. Paul and/or Kerguelen have constructed the Ridge and point out that geochemical data from these islands compare favorably with the results of the Ninetyeast Ridge analyses (Gunn et al., 1971; Watkins et al., 1974; Stephenson, 1972).

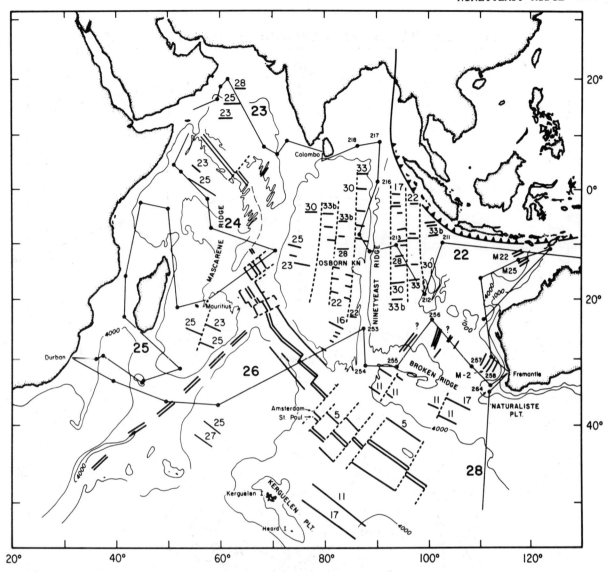

Fig. 7. Magnetic anomalies and DSDP sites from the Indian Ocean (legs 22 and 26 numbered). Data from McKenzie and Sclater (1971) and Sclater and Fisher (1974).

A Tectonic Model

Luyendyk and Rennick (1977) have produced computerized reconstructions of the plates in the eastern Indian Ocean since about the early Cretaceous in order to determine possible origins of the Ninetyeast Ridge, Broken Ridge-Naturaliste Plateau, and the Kerguelen Plateau (Figure 7). The reconstructions employ a fixed "hot-spot" reference frame (Morgan, 1972a, b) and conclude that all of the features were formed from two fixed hot spots which are presently under the Amsterdam-St. Paul islands and the Kerguelen Islands. The Ninetyeast Ridge was formed from both the Amsterdam-St. Paul (AMSP) and the Kerguelen (KER) hot spots and is from 90 to about 20 million years old. The other aseismic ridges were all formed from the KER spot. Broken Ridge plus Naturaliste Plateau is from 100 to zero m.y. old.

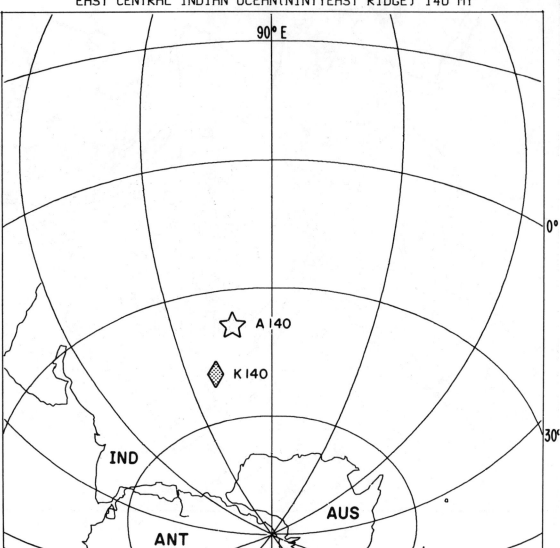

Fig. 8. Reconstructed positions of the Indian, Australian and Antarctic plates in a reference frame attached to the Amsterdam-St. Paul (AMSP, star) and Kerguelen (KER, diamond) hot spots. After Luyendyk and Rennick (1977) figure 2. Arrow heads show track of volcanic source relative to the plate which passes over it. Successive diagrams show predicted ages along these tracks as A140, K140, etc. Spreading boundaries are shown as double lines; transforms, single. Other notes: a) 140 mybp; b) 64 mybp, OKN = Osborn Knoll, I28 = anomaly 28; c) 32 mybp, dotted outline of India shows position of India at 17 mybp, d) present.

To perform the reconstructions Luyendyk and Rennick determined the relative motions of the Indian, Australian and Antarctic plates from marine magnetic anomalies and then adjusted the paleolatitudes by rotating the paleomagnetic poles for Australia to the spin axis. In

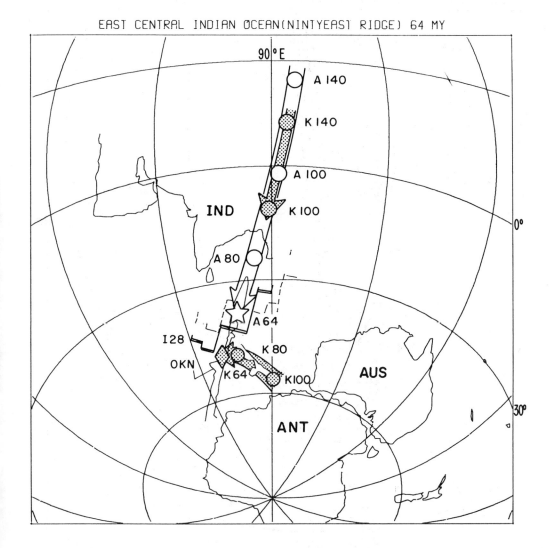

addition, they constrained the AMSP spot to remain under the Ridge for the period 100-53 mybp. Under the guiding assumptions then, this experiment involves predicting both the trends and the age gradients of aseismic ridges laid down by the fixed hot spots using marine magnetic anomaly data and paleomagnetic results as constraints. These predicted ages compare favorably with Deep Sea Drilling Project dates and other observations on the aseismic ridges, which are an independent geologic data set.

Reconstructions

The east Gondwanaland continents are shown reassembled in Figure 8a, from Smith and Hallam (1970). The age of this reconstruction is taken as early Cretaceous. Several lines of evidence suggest sea floor of almost this age off west Australia (Wharton Basin). DSDP sites 256, 257, 258, 259 (Figures 1, 7) all indicate early Cretaceous sediments in the region (Davies, Luyendyk et al., 1974; Veevers, Heirtzler et al., 1975). In addition, Markl (1974) claims to have found early Cretaceous-late

EAST CENTRAL INDIAN OCEAN (NINTYEAST RIDGE) 32 MY

Jurassic magnetic anomalies in the Perth Basin. Rifting in east Gondwanaland was then probably in the early Cretaceous and was between India, and Australia joined with Antarctica. In the period 140 to 64 mybp (Figure 8b), India moved northward while Australia/Antarctica moved generally eastward. The movements of the plates over the fixed sources have created the northernmost Ninetyeast Ridge, almost the entire Naturaliste Plateau-Broken Ridge structure, and the majority of the Kerguelen Plateau. During the period 140 to 100 mybp, KER was under the Indian plate and generated a volcanic line which was later overprinted by the AMSP source.

By 64 m.y. (anomaly 28, Figure 8b), the Indian-Antarctic plate boundary has moved north and away from the AMSP spot. In fact, it can be predicted that this hot spot may soon be under the Antarctic plate. Because the spreading center is close to AMSP, here the Ninetyeast Ridge should be almost the same age as the Indian plate upon which it rests. In this diagram, Osborn Knoll, a circular construction presently at

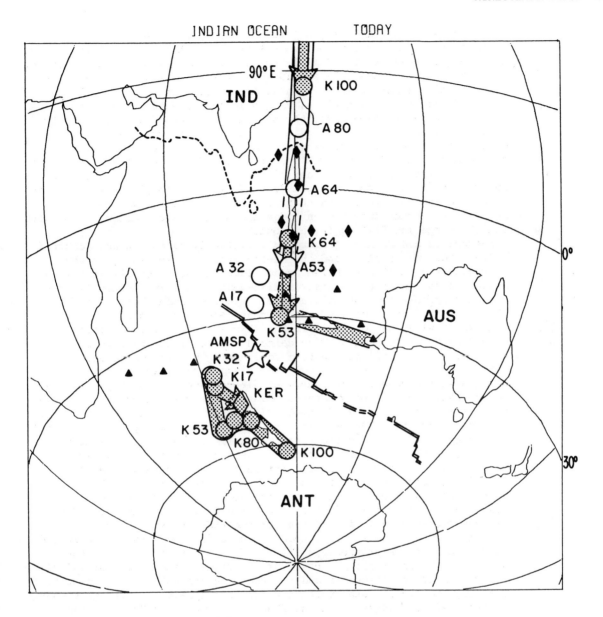

about 14°S is indicated (OKN). It is possible that the Osborn Knoll was originally laid down on the Antarctic plate by the KER spot as is suggested by Figure 8b.

A major event sometime between 64 and 53 mybp (anomaly 22 time) is the jump of the Indian-Antarctic spreading center immediately west of the Ninetyeast Ridge, 11 degrees to the south (Sclater and Fisher, 1974; Sclater et al., 1976). The exact timing of this jump is unknown due to lack of survey data west of the Ridge (between the 86°E fracture zone and the Ninetyeast Ridge). It seems likely that the jump was in response to the plate boundary moving too far north from the AMSP and/or the KER hot spots. If the jump south occurred at 64 mybp (Figure 8b) the KER spot was captured under the Indian plate. The jump also would have captured Osborn Knoll from the Australia/Antarctic plate. By 53 mybp this would result in a continuously generated Ninetyeast Ridge, part of which would be produced by overlapping contributions from the AMSP and

KER sources. Alternatively, a jump at 53 mybp positions the plate boundary squarely between the two hot spots, much like the situation of 64 mybp. Between 64 and 53 mybp, then, both spots would have been under the Australia/Antarctic plate. In this case, Osborn Knoll could have been formed via AMSP-Australia/Antarctic relative motion and then was captured onto the Indian plate. A gap in the Ninetyeast Ridge would result after AMSP moved under the Antarctic plate. This gap could be filled by leaking along the India-Australia transform fault along the Ninetyeast Ridge. This possibility may explain the narrow linear morphology of the Ninetyeast Ridge south of 7°S and north of Osborn Knoll.

Another significant event near 53 mybp is the rifting of Australia and Antarctica (Weissel and Hayes, 1972). This may also play a role in the southward jump of the Indian-Antarctic ridge. The rifting splits the KER volcanic trace into Broken Ridge and Naturaliste Plateau on the north and the Kerguelen Plateau in the south.

Between 53 and 32 m.y. (Figure 8c), virtually no motion has occurred between the Indian plate and the mantle while the spreading on the Southeast Branch of the Indian Ocean Ridge is due to mainly southward motion of the Antarctic plate. Indian-mantle relative motion is constrained by only two Australian pole positions in this case (McElhinney, 1973), so different motions must be considered a possibility. The Ninetyeast Ridge has drifted off the AMSP in this reconstruction. To constrain the Ridge and AMSP to coincide would necessitate rotating all the plates due west. We are free to do this, but it seemed more satisfying to allow India to continue along a meridian via rotational poles near the equatorial plane. At first glance it then appears that the southern end of the Ninetyeast Ridge is between 64 and 53 m.y. old and was laid down by the KER spot. It is also possible that construction continued from the AMSP spot via the lithosphere dam hypothesis of Vogt and Johnson (1975). The Ninetyeast Ridge in the south occurs near the boundary of the older and thicker lithosphere of the Wharton Basin on the east and the younger and thinner India lithosphere on the west. Therefore, according to Vogt and Johnson, plume activity in the mantle on the west would partly flow east and be deflected upwards upon impinging on the thicker lithosphere. The Ninetyeast Ridge could still be constructed even if it drifted off the AMSP and KER spots.

A chart of the Indian Ocean today (Figure 8d) shows the aseismic ridges, DSDP sites and the progressive tracts of the spots relative to the plates. AMSP is at the Southeast Branch ridge crest. The volcanically active islands of Amsterdam and St. Paul are presently over the spot. The trajectories of both spots show striking agreement with the trends of the aseismic ridges. Although the trajectory of AMSP along the Ninetyeast Ridge is fixed, the trajectory of KER is a variable dependent for its geologic reality upon the correctness of the India-AMSP relative motion.

Discussion

Three independent data sets are compared in Figure 9: crustal ages immediately west of the Ninetyeast Ridge determined from marine magnetic anomalies (Sclater and Fisher, 1974; Sclater et al., 1976); ages of Deep Sea Drilling Project sites on the Ridge (Leg 22: sites 214, 215 off to the west, 216, 217; Leg 26: sites 253, 254) and the age gradients predicted by the model. To test the models, the predicted ages must be the same as the DSDP ages, and both cannot be older than the magnetic anomaly ages (age of Indian plate). This test is satisfied very well by the AMSP trace prior to about 50 m.y., which is the age near 25°S. The KER spot may have contributed to the Ridge between 64 and 53 mybp

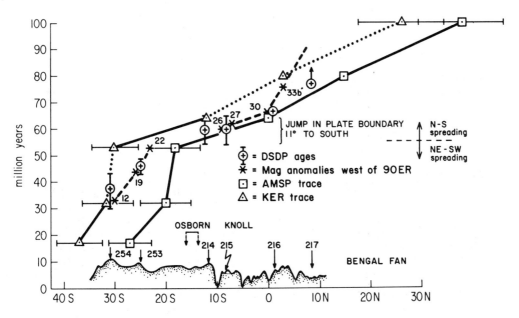

Fig. 9. A north-south topographic profile along the Ninetyeast Ridge showing ages on the Ridge from DSDP results (circles), ages immediately west of the Ridge determined from magnetic anomalies (*) and predicted ages according to the hot spot model. The triangles are the predicted ages from the KER spot. This trend is dotted for portions when the KER spot was under the Antarctic plate. Squares are the AMSP trace. Error bars are the α_{95} for the VGP's used (Luyendyk and Rennick, 1977).

Figure 9 shows that if the DSDP ages are to be believed, KER could have been under the Indian plate for only a short time after 64 mybp. Since 53 m.y., and south of about 25°S, the age patterns are much less satisfying. According to the model, AMSP contributed little to this portion of the Ridge as all DSDP ages are older than predicted. However, the possibility that KER contributed to the Ridge between 32 and 17 mybp looks strong.

Paleomagnetic data support the hypothesis that these aseismic ridges were formed from fixed hot spots (Table 2). These measurements show two things: 1) the Ninetyeast Ridge sites and Naturaliste Plateau have moved north, 2) the Kerguelen Plateau has moved south (Watkins et al., 1974). The results from the southern Ninetyeast Ridge (253, 254) and the Naturaliste Plateau (258) show an origin consistent with a spot fixed near 50°S (KER). Results from the northern sites (214, 216, 217) would be expected to yield paleolatitudes near 40° at AMSP.

Conclusions

The Deep Sea Drilling results on the Ninetyeast Ridge have contributed a great deal to understanding the origin and history of this feature. Most obviously, these observations show that the Ridge is oceanic in origin and is not a continental fragment. The geology of the Ridge is more comparable to that of an island-seamount chain such as the Hawaiian-Emperor, than of any other oceanic feature. The thin pile of sediments on the Ninetyeast Ridge show a history which is supportive of this comparison, including periods of subaerial exposure, probably to form islands, and of subaerial and submarine erosion and gradual subsidence

TABLE 2. Paleolatitudes of basal rock units

Ninetyeast Ridge (-38.5 and 49.5)	Present Latitude	Paleolatitude
Site *253 M. Eocene volcanoclastics[a]	-24.9	-50.5 ± 11 (49)°
Site 253 M. Eocene basalt[b]	-24.9	-51.8 (1)°
Site 254 Oligocene (?) basalt[b]	-31.0	-48.2 ± 4.6 (9)°
Naturaliste Plateau (-49.5)		
Site *258 M. Albian silt	-33.75	-50.4 ± 13 (17)°
Kerguelen Island (-40.5)		
Early Miocene lavas[c]	-49.0 to -49.25	-44.0 ± 8 (22)+
Heard Island (-49.5)		
Pleistocene lavas	-53.1	-43.2 ± 8 (48)°
Early Tertiary lavas[d]	-53.05	-47.0 ± ? (6)°

* work in progress
° number of samples
a = Cockerham et al. (1975)
b = Peirce et al. (1974)
c = Watkins et al. (1974)
d = Irving et al. (1974)

in accordance with established age-depth curves. Paleontologic and paleomagnetic data show northward motion for the Ridge which ties it clearly to the continental drift of India.

Petrochemical data, and a rigorous tectonic model are perhaps the strongest links to a hot spot, or fixed volcanic source origin for the Ninetyeast Ridge. Other formative modes may have operated, like leaks along a transform fault, or lithosphere damming, but the fixed point source is also central to these ideas. An attractive possibility is that the Ridge was constructed from a point source conduited through a transform fault system, now buried. This could explain why there is no volcanic trace connecting the end of the Ridge to the Amsterdam-St. Paul islands. The strength of the hot spot model lies in the demonstration that reconstructions constrained by marine magnetic anomalies and Australian VGP data successfully combine to satisfy independent geologic data from the Ridge.

Future Work

In the best of all possible worlds, one would like to have more drill sites on the Ninetyeast Ridge. Sites 254 and 217 are not completely satisfying in that 254 has weak dating control and 217 did not reach basement. Redrills should be attempted in the vicinity of these sites. More sites are needed in the region of suspected plate boundary jump-between 253 and 214, and also between 214 and 216. Osborn Knoll should be separately drilled to see whether its geology is separate from the Ninetyeast Ridge.

The exact timing of the ridge crest jump west of the Ninetyeast Ridge is critical. To this end geophysical profiles between the Ridge and 86°E are needed. The paleolatitudes of basal units on the aseismic ridges need further paleomagnetic study. Sites 214, 216, and 217 on the Ninety-east Ridge should show a paleolatitude of 38.5°S consistent with an AMSP origin and in contrast to the 50°S result for 253 and 254. Measurements at sites 258 and 264, plus from the older rocks on Kerguelen and Heard

Islands, are also needed. Finally, geochemical comparisons should be strengthened by REE analyses on the Amsterdam-St. Paul and Kerguelen Islands.

Acknowledgments. This research was funded by the Academic Senate of the University of California and by National Science Foundation grant DES 75-03709. Fred Frey, Stefan Gartner, John Sclater, Tony Pimm and John Peirce provided useful comments on this work and on the manuscript.

References

Berger, W. H., Cenozoic sedimentation in the eastern Tropical Pacific, Bull. Geol. Soc. Amer., 84, 1941-1954, 1973.

Bowin, C. O., Origin of Ninetyeast Ridge from studies near the equator, Jour. Geophys. Res., 78, 6029-6043, 1973.

Cockerham, R., B. Luyendyk, and R. Jarrard, Paleomagnetic study of sediments from site 253 DSDP, Ninetyeast Ridge, Trans. Amer. Geophys. Un., 56, 978, 1975.

Curray, J. R., and D. G. Moore, Growth of the Bengal Deep Sea Fan and denudation in the Himalayas, Bull. Geol. Soc. Amer., 82, 563-572, 1971.

Curray, J. R., and D. G. Moore, Sedimentary and tectonic processes in the Bengal Deep Sea Fan and geosyncline, in, Burke, C. A., and C. L. eds., The geology of continental margins, Springer-Verlag, New York, 617-627, 1974.

Davies, T. A., and B. P. Luyendyk, in Initial Reports of the Deep Sea Drilling Project, XXVI, Washington (U. S. Govern. Printing Office) 1129 pp., 1974.

Davies, T. A., O. E. Weser, B. P. Luyendyk, and R. B. Kidd, Unconformities in the sediments of the Indian Ocean, Nature, 253, 15-19, 1975.

Falvey, D. A., Sea-floor spreading in the Wharton Basin (northeast Indian Ocean) and the breakup of eastern Gondwanaland, Aust. Petrol. Expl. Assoc. J., 12, 86-88, 1972.

Fleet, A. J., P. Henderson, and D. Kempe, Rare earth element and related chemistry of some drilled southern Indian Ocean basalts and volcanogenic sediments, Jour. Geophys. Res., 81, 4257-4268, 1976.

Francis, T.J.G., and R. W. Raitt, Seismic refraction measurements in the southern Indian Ocean, Jour. Geophys. Res., 72, 3015-3041, 1967.

Frey, F. A., and C. M. Sung, Geochemical results for basalts from sites 253 and 254, Ch. 23 in Davies, T. A. and Luyendyk, B. P., eds., Initial Reports of the Deep Sea Drilling Project, XXVI, Washington, (U. S. Govern. Printing Office) 567-572, 1974.

Frey, F. A., J. S. Dickey, G. Thompson, and W. B. Bryan, Eastern Indian Ocean DSDP Sites: Correlations between petrography, geochemistry and tectonic setting, in Indian Ocean Geology and Biostratigraphy, AGU, Washington (this volume) 1977.

Gartner, S., D. A. Johnson, and B. McGowan, Paleontology synthesis of Deep Sea Drilling results from Leg 22 in the northeastern Indian Ocean Ch. 40 in von der Borch, C., and Sclater, J., eds., Initial Reports of the Deep Sea Drilling Project, XXII, Washington, (U. S. Govern. Printing Office), 805-814, 1974.

Gunn, B. M., L. E. Abranson, J. Nougier, N. D. Watkins, and A. Hajash, Amsterdam island, an isolated volcano in the southern Indian Ocean, Contrib. Miner. Petrol., 32, 79-97, 1971.

Harris, W. K., Palynology of Paleocene sediments at site 214, Ninetyeast Ridge, Ch. 24 in von der Borch, C., and Sclater, J., eds., Initial Reports of the Deep Sea Drilling Project, XXII, Washington, (U. S. Govern. Printing Office) 503-520, 1974.

Heezen, B. C., and M. Tharp, Tectonic fabric of the Atlantic and Indian Oceans and continental drift, Phil. Trans. Rpy. Soc. Series A, 258, 90-106, 1965.

Hekinian, R., Petrology of igneous rocks from Leg 22 in the northeastern Indian Ocean, Ch. 17 in von der Borch, C., Sclater, J., eds., Initial Reports of the Deep Sea Drilling Project, XXII, Washington, (U. S. Govern. Printing Office) 413-448, 1974a.

Hekinian, R., Petrology of the Ninetyeast Ridge (Indian Ocean) compared to the other aseismic ridges, Contr. Mineral. and Petrol., 43, 125-147, 1974b.

Irving, E., P. J. Stephenson, and A. Major, Magnetism in Heard Island rocks, Jour. Geophys. Res., 70, 3421-3427, 1965.

Kemp, E., Preliminary palynology of samples from site 254, Ninetyeast Ridge, Ch. 34 in Davies, T. A., and Luyendyk, B. P., eds., Initial Reports of the Deep Sea Drilling Project, XXVI, Washington, (U. S. Govern. Printing Office) 815-818, 1974.

Kempe, D.R.C., The petrology of the basalts, Leg 26, Ch. 14 in Davies, T.A., and Luyendyk, B. P., eds., Initial Reports of the Deep Sea Drilling Project, XXVI, Washington (U. S. Govern. Printing Office, 465-504, 1974.

Laughton, A. A., D. H. Matthews, and R. L. Fisher, The structure of the Indian Ocean, in Maxwell, A. E., ed., The Sea: Ideas and Observations, 4, pt. II, New York, John Wiley and Sons, Inc., 543-586, 1970.

Luyendyk, B. P., and T. A. Davies, Results of DSDP Leg 26 and the geologic history of the southern Indian Ocean, Ch. 36, in Davies, T.A. and Luyendyk, B. P., eds., Initial Reports of the Deep Sea Drilling Project, XXVI, Washington, (U. S. Govern. Printing Office, 909-943, 1974.

Luyendyk, B. P., and W. Rennick, Tectonic origin of aseismic ridges in the eastern Indian Ocean, Bull. Geol. Soc. Amer., in press, 1977.

Markl, R., Evidence for the breakup of eastern Gondwanaland by the early Cretaceous, Nature, 251, 196-200, 1974.

McDougall, I., Potassium-Argon ages on basaltic rocks recovered from DSDP, Leg 22, Indian Ocean, Ch. 12 in von der Borch, C., and Sclater, J., eds., Initial Reports of the Deep Sea Drilling Project, XXII, Washington (U. S. Govern. Printing Office) 377-380, 1974.

McElhinney, M. W., Paleomagnetism and plate tectonics, Cambridge Univ. Press, 357 pp., 1973.

McKenzie, D. P., and J. G. Sclater, The evolution of the Indian Ocean since the late Cretaceous, Royal Astron. Soc. Geophys. Jour., 25, 437-528, 1971.

Morgan, W. J., Plate motions and deep mantle convection, in Shagam, R., et al., eds., Studies in earth and space sciences (Hess volume), Geol. Soc. Amer. Mem., 132, 7-22, 1972a.

Morgan, W. J., Deep mantle convection plumes and plate motions, Amer. Assoc. Petrol. Geol. Bull., 56, 203-213, 1972b.

Moore, D. G., J. R. Curray, R. W. Raitt, and F. J. Emmel, Stratigraphic-seismic section correlations and implications to Bengal Fan history, Ch. 16, in von der Borch, C., Sclater, J., et al., eds., Initial Reports of the Deep Sea Drilling Project, XXII, Washington (U. S. Gov. Printing Office) 403-412, 1974.

Peirce, J. W., C. R. Denham, and B. P. Luyendyk, Paleomagnetic results of basalt samples from DSDP Leg 26 southern Indian Ocean, Ch. 18, in Davies, T. A. and Luyendyk, B. P., eds., Initial Reports of the Deep Sea Drilling Project, XXVI, Washington (U. S. Gov. Printing Office) 517-527, 1974.

Pimm, A. C., Sedimentology and history of the northeastern Indian Ocean from late Cretaceous to Recent, Ch. 39, in von der Borch, C., Sclater,

J., et al., eds., Initial Reports of the Deep Sea Drilling Project, XXII, Washington (U. S. Gov. Printing Office) 717-804, 1974.

Pimm, A. C., B. McGowan, and S. Gartner, Early sinking history of the Ninetyeast Ridge, northeastern Indian Ocean, Bull. Geol. Soc. Amer., 85, 1219-1224, 1974.

Pimm, A. C., and J. G. Sclater, Early Tertiary hiatuses in the northeastern Indian Ocean, Nature, 252, 362, 1974.

Rundle, C., M. Brook, N. Snelling, P. Reynolds, and S. Barr, Radiometric age determinations, Ch. 17, in Davies, T. A., and Luyendyk, B. P., eds., Initial Reports of the Deep Sea Drilling Project, XXVI, Washington (U. S. Gov. Printing Office), 513-516, 1974.

Schilling, J-G., Sea-floor evolution: rare earth evidence, Phil. Trans. Roy. Soc., London, Series A, 258, 663-703, 1971.

Schilling, J-G., Iceland mantle plume: geochemical study of Reykjanes Ridge, Nature, 242, 565-571, 1973.

Sclater, J. G., R. Anderson, and M. L. Bell, The elevation of ridges and the evolution of the central eastern Pacific, Jour. Geophys. Res., 76, 7888-7915, 1971.

Sclater, J. G., and R. L. Fisher, Evolution of the east central Indian Ocean, with emphasis on the tectonic setting of the Ninetyeast Ridge, Bull. Geol. Soc. Amer., 85, 683-702, 1974.

Sclater, J. G., R. D. Jarrard, B. McGowan, and S. Gartner, Comparison of the magnetic and biostratigraphic time scales since the late Cretaceous, Ch. 13 in von der Borch, C., and Sclater, J. G., et. al, Initial Reports of the Deep Sea Drilling Project, XXII, Washington, (U. S. Gov. Printing Office) 369-376, 1974.

Sclater, J. G., B. P. Luyendyk, and L. Meinke, Magnetic lineations in the south central Indian Basin, Bull. Geol. Soc. Amer., 87, 371-378, 1976.

Smith, A. G., and A. Hallam, The fit of the southern continents, Nature, 225, 139-144, 1970.

Stephenson, P. J., Geochemistry of some Heard Island igneous rocks, in, Adie, R. J., ed., Antarctic Geology and Geophysics (Proceedings Symposium on Ant. Geol. and Geophys., Oslo, 1970), Oslo, Scand., Univ. Books, 793-802, 1972.

Thompson, G., W. Bryan, F. Frey, and C. M. Sung, Petrology and geochemistry of basalts and related rocks from sites 214, 215, 216, DSDP Leg 22, Indian Ocean, Ch. 19, in von der Borch, C., and Sclater, J. G., eds., Initial Reports of the Deep Sea Drilling Project, XXII, Washington (U. S. Gov. Printing Office) 459-468, 1974.

Veevers, J. J., J. G. Jones, and J. A. Talent, Indo-Australian stratigraphy and the configuration and dispersal of Gondwanaland, Nature, 229, 383-388, 1971.

Veevers, J. J., and J. R. Heirtzler, eds., Initial Reports of the Deep Sea Drilling Project, XXVII, Washington, (U. S. Gov. Printing Office), 1060 pp., 1974.

Vogt, P., and G. L. Johnson, Transform faults and longitudinal flow below the mid-oceanic ridge, Jour. Geophys. Res., 80, 1399-1428, 1975.

von der Borch, C., and J. G. Sclater, eds., Initial Reports of the Deep Sea Drilling Project, XXII, Washington (U. S. Gov. Printing Office), 890 pp., 1974.

Watkins, N. D., B. M. Gunn, J. Nougier, and A. K. Baksi, Kerguelen: Continental fragment or oceanic island, Bull. Geol. Soc. Amer. 85, 1974.

Weissel, J. K., and D. E. Hayes, Magnetic anomalies in the southeast Indian Ocean, in, Hayes, D. E., ed., Antarctic oceanology II, The Australian-New Zealand sector, Washington, D. C., Am. Geophys. Union, Antarctic Research Series, 19, 165-196, 1972.

CHAPTER 8. EASTERN INDIAN OCEAN DSDP SITES: CORRELATIONS BETWEEN
           PETROGRAPHY, GEOCHEMISTRY AND TECTONIC SETTING

Fred A. Frey, and John S. Dickey, Jr.

Department of Earth and Planetary Sciences
Massachusetts Institute of Technology
Cambridge, Massachusetts 02139

Geoffrey Thompson, and Wilfred B. Bryan

Woods Hole Oceanographic Institution
Woods Hole, Massachusetts 02543

Abstract. At 10 of 13 Deep Sea Drilling Project (DSDP) sites in the eastern Indian Ocean there is a correlation between inferred tectonic history and the petrographic and geochemical nature of basement basalts. Compositions of unaltered glasses and phenocrysts, in addition to abundances of Ti, Zr, and rare-earths in crystalline rocks, are used to determine the geochemical characteristics of magmas erupted at each site. At sites 212, 213, 257, 259, 260 and 261 the basalts are within the compositional range of large ion lithophile(LIL) element-depleted tholeiites dredged from spreading ridge axes, such as the Mid-Indian Ocean Ridge. These results are consistent with tectonic models indicating a seafloor spreading origin for basement basalt at these sites. In contrast, the alkali-olivine basalts at site 211 are probably related to volcanism creating the nearby Cocos, Keeling and Christmas Islands.

Rocks from the aseismic Ninetyeast Ridge (sites 214, 216 and 254) are LIL element-enriched tholeiites, ferrotholeiites and oceanic andesites. Similar suites occur on oceanic islands such as Iceland, Galapagos, Faeroes and the St. Pauls-Amsterdam complex in the Indian Ocean. The geochemical features of these Ninetyeast Ridge rocks therefore imply a petrogenesis similar to that of tholeiitic island sequences. This conclusion is consistent with tectonic models relating portions of the Ninetyeast Ridge to a hot spot trace.

Inconsistencies between tectonic models and magma geochemistry occur at sites 215, 253 and 256. The occurrence of LIL element-depleted tholeiite at site 253 on the Ninetyeast Ridge is anomalous and unexplained. Equally surprising are the high LIL element abundances in tholeiitic basalts at site 215. These basalts are similar in composition to a basalt dredged from a seamount flank in the northeast Indian Ocean, but they are distinctly different from Ninetyeast Ridge basalts. There are no anomalous bathymetric features at site 215, but the unexpectedly young, inferred basement age and the atypical composition imply that site 215 basalts were not formed at a spreading ridge axis. At site 256, LIL element-enriched ferrotholeiites are also unlike basalts formed at spreading ridge axes.

These basalts may be related to the volcanism creating the northeasterly trend of topographic highs extending from Broken Ridge through site 256.

Introduction

Although the Indian Ocean is the smallest of the three major oceanic basins, its geological history is the most complex. Major contributions (Fisher et al., 1971; McKenzie and Sclater, 1971) to understanding Indian Ocean development in terms of plate tectonics were made prior to Deep Sea Drilling Project (DSDP) Leg 22, the first DSDP Leg in the Indian Ocean. However, at that time there was very little information and understanding of the Indian Ocean east of the aseismic Ninetyeast Ridge. Yet, this oceanic region is the key to understanding the development of five lithospheric plates (Africa, Antarctica, Australia, India, and Madagascar) formed from breakup of Gondwanaland. In particular, major features of the east-central Indian Ocean which were poorly understood prior to DSDP drilling were 1) the wholly submerged Ninetyeast Ridge which extends almost north-south as a topographic high (approximately 2000 m above surrounding seafloor) from 31°S to 9°N, a distance of more than 4500 km, and 2) the origin of the Wharton Basin. Hypotheses had been presented that the Wharton Basin was Cretaceous or older (McKenzie and Sclater, 1971, p. 486) or possibly even pre-Mesozoic (Dietz and Holden, 1971). Furthermore, in the eastern Indian Ocean, near western Australia, major unanswered questions concerned the nature and direction of the rifting which initiated and continued continental separation via seafloor spreading (Veevers et al., 1971; Falvey, 1972). Also, the presence of possible continental fragments such as the Exmouth and Wallaby plateaus implied a complicated tectonic history for oceanic areas adjacent to Australia.

As indicated in this volume, the sedimentological and geophysical results obtained during the DSDP have contributed enormously to an understanding of the tectonic history of the Indian Ocean. In this paper we will focus on DSDP Indian Ocean sites east of 88°E. These sites include the Ninetyeast Ridge, Wharton Basin and oceanic basins adjacent to western and northwestern Australia. Basement rocks from these areas were obtained on DSDP Legs 22, 26 and 27. About 160 m of basalt were recovered from these sites; thus, these samples provide the first in-depth insight into the composition of the eastern Indian Ocean seafloor. Our immediate objective is to summarize petrological and geochemical characteristics of basement rocks from 13 DSDP sites (211, 212, 213, 214, 215, 216, 253, 254, 256, 257, 259, 260 and 261). We are particularly interested in evaluating the degree of correlation between petrological and geochemical features of basalts at a site, and the postulated tectonic history of that site. In addition, it is important to compare these Indian Ocean basalts with basalts obtained from similar tectonic environments in other oceans.

Our data sources are 1) the pertinent chapters in the Initial Reports of the Deep Sea Drilling Project for Legs 22, 26 and 27, and 2) the results of our detailed study on more than 50 samples from the various sites. These new data are presented in this paper.

The major goal of our research is to utilize petrological and geochemical characteristics of hard rocks drilled on DSDP Indian Ocean Legs to gain an understanding of the geological development of the Indian Ocean. In particular, we believe that the petrological and geochemical features of basalts at each DSDP site can be used to infer the tectonic environment existing during basalt petrogenesis; thus, the tectonic environment inferred from petrochemical features can be compared to that proposed on the basis of geophysical data. Studies of igneous rocks dredged and drilled from other oceans have been instrumental in establishing 1) the first order homogeneity and distinctive composition of magma upwelling along active oceanic ridges (e.g., Engel et al., 1965; Kay et al., 1970; Frey et al., 1974; Schilling, 1971, 1975b), 2) the common occurrence of fairly extensive olivine and plagioclase fractionation in oceanic ridge basalts (e.g., Miyashiro et al., 1970); however, the distinct geochemical features of ridge axis basalts are retained in highly fractionated residual liquids (Kay et al., 1970; Hart, 1971; Frey et al., 1974), 3) the common occurrence of alkaline magmas within fracture zones and intraplate eruptions (Thompson and Melson, 1972; Frey et al., 1974), and also the exposure in fracture zones of very differentiated iron-rich and cumulate gabbros (Miyashiro et al., 1970; Thompson 1973b), 4) the nature of magma compositional changes along a ridge system in the vicinity of an island (Schilling, 1973, 1975b; Dickey et al., 1974), and 5) the nature of basalts which form aseismic ridges (Hekinian, 1974a, b).

In this paper on the Eastern Indian Ocean we demonstrate that there is a high degree of correlation between petrological and geochemical features at an Indian Ocean site, and the inferred tectonic setting based on geophysical data. However, at site 256 on Leg 26 the basalt characteristics imply that previously, largely ignored, physical features are important, and at site 215 on Leg 22, the discrepancy between inferred tectonic environment based on geochemistry, and that based on geophysics, warrants caution in interpreting the site history.

Strategy

The large majority of DSDP Indian Ocean rocks are highly altered basalts with total water plus carbon dioxide content commonly exceeding 2 wt. %. In this paper we are concerned with determining magma composition, and our emphasis will be on compositional features which are usually not strongly affected by alteration processes. In particular, studies of DSDP basalts in other oceans (e.g., Frey et al., 1974) have demonstrated that the composition of palagonite-free glass and unaltered phenocryst minerals, and the abundances of various highly charged, large ionic radii trace elements (e.g., rare earth elements) are most useful in establishing the geochemical nature of oceanic basalt magmas. Normative mineral proportions determined from bulk major element composition, and abundances of singly charged alkali metals (e.g., K and Rb) are generally not as useful because they are strongly affected by alteration (e.g., Hart, 1971; Thompson, 1973a; Frey et al., 1974).

In order to evaluate the possible correlation of basaltic geochemistry with tectonic environment, it is necessary to summarize the tectonic history of the Indian Ocean east of 88°E. Following this summary we will examine the petrology and geochemistry of basalts from each major tectonic province; that is, a) Ninetyeast Ridge (sites 214, 216, 253 and 254), b) Central Indian Basin (site 215), c) Wharton Basin (sites 211, 212, 213, and 256) and d) oceanic basins adjacent to western and northwestern Australia (sites 257, 259, 260 and 261).

Tectonic History of East-Central and Eastern Indian Ocean

Sclater and Fisher (1974) demonstrated the predominantly north-south orientation of seafloor spreading immediately west and east of the aseismic Ninetyeast Ridge between magnetic anomalies 19 and 33. However, oceanic crust increases in age in opposite directions on either side of the Ninetyeast Ridge (Figure 2). This aseismic ridge appears to be tectonically associated with the Indian Plate to the west because crustal ages increase northward both on the Ninetyeast Ridge and in the Central Indian basin, whereas crustal age increases southward in the Wharton Basin. Between 93°E and the aseismic ridge, the Wharton Basin is cut by a series of long north-south deeps and shoals (Figure 1); Bowin (1973) and Sclater and Fisher (1974) have suggested that these deeps may represent an old transform fault between the Indian and Australian plates.

Several models for the Ninetyeast Ridge have been eliminated (e.g., horst-type uplift) on the basis of recent studies (Bowin, 1973; Luyendyk and Davies, 1974; Sclater and Fisher, 1974). However, consensus opinion is that different origins are required for different portions of the Ninetyeast Ridge. In particular, models combining a mantle plume or hot spot with a leaky transform fault are consistent with geophysical and age data (Luyendyk and Davies, 1974; Sclater and Fisher, 1974; Luyendyk and Rennick, 1977).

The Wharton Basin and Central Indian Basin contain major north-south oriented fracture zones. Sclater and Fisher (1974) conclude the Wharton Basin "... is made up of long thin strips, bearing easily correlatable magnetic anomalies of markedly varying time scales that are offset by major north-south fracture zones." Based on the presence of these major fracture zones, the Wharton Basin can be divided into four provinces (Figure 2) which have recorded seafloor spreading processes since the Late Cretaceous.

The topography and magnetics of the Cocos-Keeling and Christmas shoal areas in the eastern Wharton Basin (Figure 1) are not well understood. They strike northwest-southeast in contrast to the common north-south features of Wharton Basin topography. Presumably, these northwest-southeast features are of young age. On the basis of seismicity, Sykes (1970) suggested that an incipient tectonic feature (possibly a nascent island arc) was associated with a seismic zone identified between Sri Lanka and Australia. This zone includes the Cocos Islands. However, the zone is not associate with obvious topographic or magnetic features.

Preliminary evidence identified 150×10^6 yr old oceanic crust near northwest Australia (Heirtzler et al., 1973). Larson (1975) recently demonstrated that magnetic anomalies in the Argo and Gascoyne abyssal plains between the Java Trench and northwest Australia (Figure 2) can be correlated with a Late Jurassic magnetic reversal sequence. The Oxfordian age, based on nannofossils, of DSDP Site 261, is consistent with this correlation. Thus, oceanic crust sampled at site 261 is some of the oldest crust sampled by the DSDP, and it is comparable in age to the Late Jurassic crust sampled on Leg 11 in the western Atlantic.

West of southwestern Australia, Veevers and Heirtzler (1974) and Markl (1974) have utilized geological and geophysical data to infer that west of Perth, seafloor spreading was initiated in a northeast-southwest direction at least by the early Cretaceous. Veevers and Heirtzler (Figure 1, 1974) propose several narrow compartments of oceanic crust bounded by fracture zones. Markl has inferred from sparse magnetic data that an extinct Late Jurassic to Late Cretaceous spreading center was the source of oceanic crust in the southern Wharton Basin, i.e., immediately north of Broken Ridge and the Naturaliste Plateau. It is apparent that the oceanic areas adjacent to western and northwestern Australia represent the oldest known

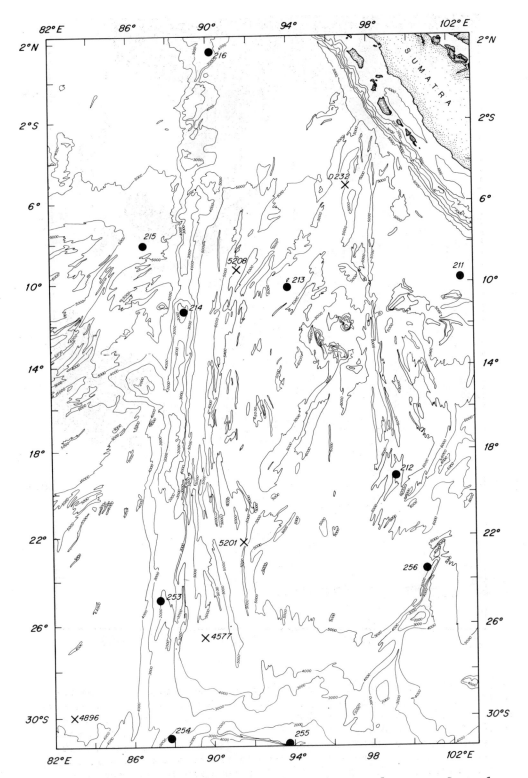

Fig. 1. Location of DSDP sites on bathymetric map of eastern Central Indian Ocean (adapted from Figure 1 of Sclater and Fisher, 1974). Dredged basalts analyzed for major elements (Tables 4a and 5a) are also indicated.

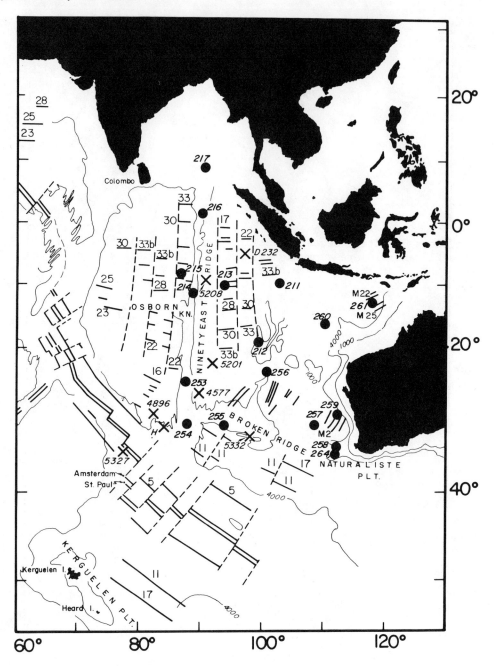

Fig. 2. DSDP site locations on a tectonic summary map of the eastern Indian Ocean (adapted from Figure 8 of Luyendyk, 1977). "X" indicates locations of dredged basalts that have been analyzed for major elements (Tables 4a and 5a). Double solid lines are ridge axes, dashed lines are fracture zones, and short solid lines (some numbered) are magnetic anomalies. The long north-south fracture zones divide the Wharton Basin into four provinces: 1) a topographically rough area between 90°E and 92°E, 2) a region between 93°E and 96°E where several magnetic anomalies are identified, 3) a region between 96°E and 98°E where magnetic anomalies are offset to the north, relative to province 2, and 4) east of 98°E.

record of seafloor spreading in the Indian Ocean, and that the ages are approximately synchronous with the Jurassic separation of North America and Africa.

Ninetyeast Ridge

Sites 214-216 (11°20.21'S, 88°43.08'E; 1°27.73'N, 90°12.48'E)

Four DSDP sites (214, 216, 253, 254) are located on the Ninetyeast Ridge (Figures 1,2). Detailed petrographic and geochemical studies of the two northernmost sites (214, ~56m basement penetration, and 216, ~27.5m basement penetration) were reported by Bougault (1974), Hekinian (1974a), and Thompson et al., (1974); the following discussion is based on their results.

As expected of a subaerial or shallow marine environment (inferred from the sedimentary sequence, Sclater et al., 1974), basalts at sites 214 and 216 are commonly vesicular and amygdaloidal. Although some textural variations are observed (Hekinian, 1974a), the typical trachytic groundmass texture of these basalts distinguishes them markedly from oceanic ridge basalts which exhibit abundant quench features (Bryan, 1972). Plagioclase, pyroxene, and magnetite are principal minerals in site 214 and 216 lavas; olivine is absent. The dominant phenocrysts and microphenocrysts are plagioclase and magnetite. The high abundance of typically euhedral magnetite (~10 vol. %) is unusual for seafloor rocks.

At site 216, only iron-rich basalt (ferrobasalt) was recovered, but at site 214 similar ferrobasalts are overlain by differentiated rocks (54-58 wt. % SiO_2) termed oceanic andesites. These andesites are <u>not</u> geochemically similar to island arc or continental margin andesites associated with subducting ocean lithosphere (Thompson et al., 1974). Despite significant compositional differences, the textures and mineral paragenesis of the ferrobasalts and oceanic andesites are similar. This similarity, in addition to close association in time and space, implies a genetic relationship between ferrobasalts and oceanic andesites at site 214.

The ferrobasalts are quite altered (average ignition loss >3 wt. %); nevertheless, this alteration does not mask the key major element abundance features of these basalts (Table 1a): high absolute iron content (FeO* (1) >12.9 wt. %) with FeO*/MgO >1.9 and TiO_2 >2 wt. %. In addition, alkali metal content is low (K_2O <1 wt. % and Na_2O <3 wt. %) for such altered and differentiated rocks. Although utmost caution is required when using major element abundances of altered rocks to infer magma composition, it is evident that rocks at these two sites follow a typical tholeiitic differentiation sequence of increasing FeO*/MgO, FeO* and TiO_2 contents at nearly constant SiO_2 abundance followed by decreasing FeO* and TiO_2 content as SiO_2 abundance increases along with FeO*/MgO (Figure 3a,b).

Further convincing evidence for the tholeiitic nature of these basalts is their relatively low (compared to alkalic basalts) large-ion-lithophile (LIL) element abundances; for example, low Ba and Zr abundances, moderate Sr abundance and moderate rare-earth element (REE) fractionation (Table 1b and Figure 4a). The low to moderate LIL element abundances in site 214 and 216 basalts are similar to those of island tholeiitic basalts. It is important to note that the

---

[1] Throughout this paper FeO* indicates total iron content in wt. % (i.e., FeO plus 0.9 Fe_2O_3 reported as FeO*.

196 FREY, DICKEY, THOMPSON, AND BRYAN

Table 1a: Major Element Composition (volatile-free) of Iron-rich Tholeiitic Basalts (wt.%)

	Ninetyeast Ridge		Wharton Basin	Galapagos Islands		Thingmuli	Faeroe Is.	Amsterdam Is.	St. Paul Is.		
	Site 214[a]	Site 216[b]	Site 256[c]	Ferrobasalt[d]	63[e]	G-146[f]	Qtz tholeiite[g]	NA-1100[h]	41[i]	AV-4[j]	AV-6[k]
SiO_2	48.1	49.5	50.4	49.0	48.6	49.9	49.5	49.73	48.39	50.09	49.94
Al_2O_3	14.9	13.5	13.3	14.3	13.8	12.9	13.5	14.4	12.6	13.6	15.6
FeO^*	14.6	13.8	13.2	13.3	12.9	15.1	14.2	11.7	17.1	13.6	11.1
MgO	6.45	6.57	6.58	4.91	6.07	5.14	5.7	6.19	5.17	3.75	4.26
CaO	9.04	8.79	10.2	9.19	10.8	9.96	10.4	11.0	8.18	9.01	11.14
Na_2O	2.75	2.57	2.76	3.59	2.80	2.90	2.5	2.45	3.02	3.87	3.53
K_2O	0.37	0.90	0.25	0.85	0.50	0.46	0.5	0.71	0.46	0.87	0.50
TiO_2	2.35	2.75	2.36	3.66	3.40	2.77	2.9	2.07	3.58	3.01	2.32
P_2O_5	0.19	0.22	0.25	0.53	0.36	0.41	-	0.22	0.82	0.41	0.28
Total	98.75	98.60	99.3	99.33	99.23	99.54	99.20	98.47	99.32	98.21	98.67
FeO^*/MgO	2.26	2.10	2.01	2.71	2.13	2.94	2.49	1.89	3.31	3.29	2.59

a) average of 6 basalts (Thompson et al., 1974).
b) average of 7 basalts (Thompson et al., 1974).
c) average of 7 basalts from Table 8a, this paper).
d) average of 12 basalts (Table 11, McBirney and Williams, 1969)
e) tholeiitic basalt containing gabbroic inclusions, Albemarle Is. (McBirney and Williams, 1969).
f) tholeiitic dike, No. 5 in table 9 (Carmichael, 1964).
g) average of 10 aphyric quartz tholeiites from lower series (Noe-Nygaard and Rasmussen, 1968).
h) Aphyric basalt with highest FeO^* and TiO_2 (Gunn et al., 1971).
i) olivine basalt (Girod et al., 1971).
j) average of 4 differentiated rocks termed "low-silica andesite" (Gunn et al., 1975).
k) average of 6 tholeiites (Gunn et al., 1975).

Table 1b: Trace Element Abundances of Iron-rich Tholeiitic Basalts (ppm)

	Ninetyeast Ridge		Wharton Basin	Galapagos	Faeroe Is.	Amsterdam Is.	St. Paul Is.
	Site 214[a]	Site 216[b]	Site 256[c]	63[d]	Lower Series[e]	NA-1100[f]	41[g]
Sc	46	42	46	45	-	-	35
V	525	445	451	310	-	-	395
Cr	38	45	108	47	-	50	55
Co	65	53	36	55	-	58	31
Ni	50	44	93	70	-	40	235
Sr	265	235	144	320	-	274	225
Ba	45	140	39	90	-	218	210
Y	26	31	45	40	-	-	67
Zr	120	159	159	255	-	-	330
La	8.4	13.2	9.8	15.8	13.1	-	19.6
Sm	5.1	5.0	4.7	5.7	6.1	-	7.26
Eu	1.7	1.45	1.5	-	2.0	-	2.30
Yb	3.0	3.6	3.3	-	2.6	-	3.6

a) average of J-42, J-44, J-46 for Sc, La, Sm Eu and Yb (Thompson et al., 1974; Frey and Sung, 1974). For all other elements avg. of 6 basalts (Table 4, Thompson et al., 1974).
b) average of J-30 and J-25 for Sc, La, Sm, Eu and Yb (Frey and Sung, 1974 and unpublished). For all other elements avg. of 7 basalts (Table 4, Thompson et al., 1974).
c) average of site 256 basalts from Table 8b (this paper).
d) from Table 7 of McBirney and Williams, 1969.
e) average of 8, Schilling and Noe-Nygaard, 1974.
f) see note (g) of Table 1a.
g) olivine basalt of Girod et al., 1971, data from this work. Also, averages for four low silica andesites (Table 1a) are Ni=20, Sr=269, and for six tholeiites are Ni=30.4, Sr=291 (Gunn et al., 1975).

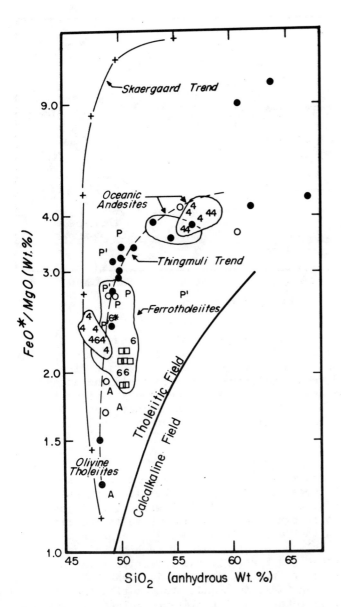

Fig. 3a. Iron enrichment trend during fractionation of tholeiitic basalts. SiO2 enrichment during late stages corresponds to fractionation of iron oxide phases. Site 214 basalts and oceanic andesites = 4, site 216 basalts = 6, site 256 basalts = □ , Amsterdam Island olivine tholeiite averages (Gunn et al., 1971, 1975) = A (although not plotted, a similar trend is followed by three Amsterdam Island basalts studied by Vinogradov et al., 1969), St. Paul Island ferrotholeiite averages (Girod et al., 1971; Gunn et al., 1975) = P' and P, respectively (although not plotted, a similar trend is followed by five St. Paul Island basalts studied by Vinogradov et al., 1969), Galapagos Island (McBirney and Williams, 1969) = O, Faeroe Island average ferrotholeiite of lower series (Noe-Nygaard and Rasmussen, 1968) = *, Skaergaard trend, solid line (Wager, 1960) = +, Thingmuli, Iceland trend, dashed line (Carmichael, 1964) = ●. Tholeiite - calcalkaline field boundary from Miyashiro, 1974.

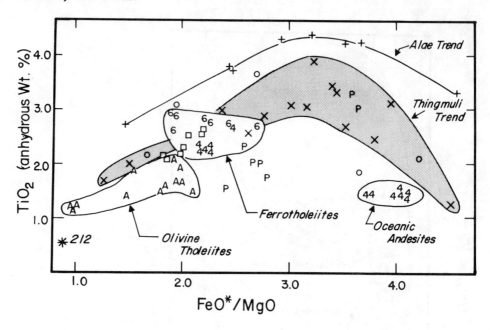

Fig. 3b. TiO$_2$ trend during fractionation of tholeiitic basalts. The initial TiO$_2$ enrichment is followed by a rapid TiO$_2$ decrease as iron-titanium oxide phases are separated from the fractionating magma. Thingmuli, Iceland trend indicated by shaded field enclosing x symbols. Trend line through + symbols indicates fractionation of Alae, Hawaii lave lake (Peck et al., 1966). * symbol indicates MgCa type of ridge basalt found at site 212. All other symbols as in Figure 3a.

relative REE distribution (Figure 4a), and other trace element abundances clearly distinguish these Ninetyeast Ridge basalts from the LIL element-depleted tholeiitic basalts typical of spreading ridges (Thompson et al., 1974). Although some ocean floor basalts, particularly those associated with the Juan de Fuca Ridge in the northeast Pacific (Kay et al., 1970; Melson, et al., 1976) have high values of FeO*/MgO (>2) and TiO$_2$ content (>2.5 wt. %), these basalts are LIL element depleted, and they are relatively depleted in light REE. The petrogenetic processes creating these iron and titanium-rich (FETI) basalts (Melson et al., 1976) retain the LIL element-depleted features characteristic of basalts formed at spreading ridge axes (e.g. Kay et al., 1970); thus site 214 and 216 basalts are unlike FETI basalts associated with spreading ridges.

Despite a 12° separation in latitude (Figures 1, 2), and possibly a 5 million year age difference (Sclater et al., 1974), basalts at sites 214 and 216 are similar in composition, and it is not difficult to develop a differentiation model relating these compositions. All of these basalts have distinctive geochemical features of high FeO*/MgO and TiO$_2$ abundance accompanied by very low Ni and Cr abundances and moderate light REE enrichment relative to a chondritic distribution (Table 1a, b and Figure 4a). Because of their higher La/Yb, TiO$_2$, Y, and Zr abundances, site 216 basalts are slightly more differentiated than site 214 basalts (Figures 3b, 4a and 5a,d). It is likely that these differences reflect magmatic differences because these geochemical features are especially resistant to change during alteration (Cann, 1970; Thompson, 1973a; Frey et al., 1974).

The evolution of the more differentiated site 216 magma can be quali-

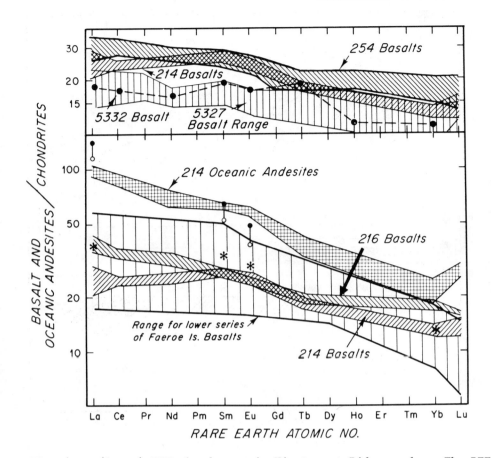

Fig. 4a. (lower) REE abundances in Ninetyeast Ridge rocks. The REE abundance range in site 214 ferrobasalts (3), site 216 ferrobasalts (2), site 214 oceanic andesites (4) and the Lower Series of Faeroe Island ferrobasalts (8) are compared to a chondrite average (Frey et al., 1968). All subsequent REE comparison diagrams are normalized to this chondrite average. Data from Frey and Sung (1974), this work from Schilling and Noe-Nygaard (1974) and Thompson et al. (1974). Asterisks indicate average for Faeroe Island lower series ferrobasalts, and vertical bar indicates values for two Galapagos icelandites (McBirney and Williams, 1969).

Fig. 4b. (upper) REE abundance range in seven site 254 basalts compared to REE range in three site 214 ferrobasalts. Also shown for comparison is the REE distribution in a Broken Ridge dredge basalt (32°26'S, 98°25'E, Figure 2) and the range of REE abundances in seven dredged basalts from the Southeast Indian Ridge axis 350 km north of St. Paul-Amsterdam Islands (34°17'S, 77°56.6'E, Figure 2). Dredged basalt data from Balashov et al. (1970), sites 5332 and 5327, respectively.

tatively modeled by considering removal of site 214 phenocryst phases from site 214 magma. Among the phenocrysts of these basalts, only clinopyroxene fractionation could cause the La/Yb increase in site 216 basalts. This observation is reinforced by lower Sc concentration in site 216 basalts (Table 1b), and the systematic decrease in Sc with increasing La/Yb (Figure 5c). The sensitivity of La/Yb and Sc abundance to clinopyroxene fractionation has been demonstrated by Frey et al.(1974).

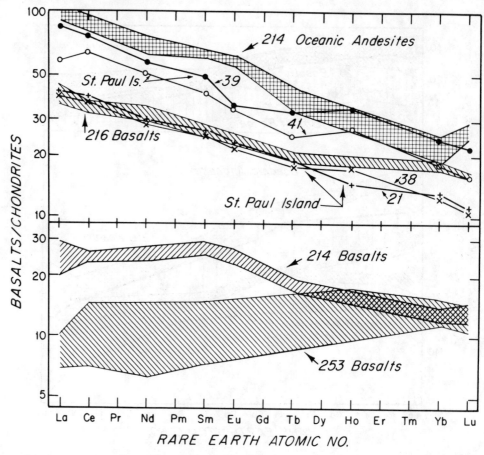

Fig. 4c. (lower) REE abundance range in site 253 picritic tholeiites (4) compared to range in site 214 ferrotholeiites (3). Data for site 253 24-1:84-85 and 58-1:2-7 are not plotted.

Fig. 4d. (upper) REE abundances in four St. Paul Island rocks (REE data from this work, samples from Girod et al., 1971) compared to range in site 216 ferrotholeiites (2) and site 214 oceanic andesites (4).

As expected from the abundance of euhedral magnetite, typically occurring as cumulates of several associated grains (Hekinian, 1974a), there is convincing geochemical evidence for the important role of magnetite fractionation; principally, the systematically decreasing V abundance with increasing LIL element abundance. Thus, V abundance decreases rapidly with increasing fractionation (Figure 5a). The strong enrichment of V in magnetite crystallizing from a silicate liquid is well established (Duncan and Taylor, 1968; Taylor et al., 1969; Ewart et al., 1973 and Miyashiro and Shido, 1975); specifically titanomagnetite/silicate liquid partition coefficients for V are >20 (Ewart et al., 1973). Apparently, these ferrobasalts have reached the fractionation stage where magnetite removal has inhibited the rapid increase in FeO*/MgO and V that is characteristic of tholeiitic basalt indifferentiation at low oxygen fugacities. The experimental studies of Presnall (1966) and Roeder and Osborn (1966) on simplified basaltic systems have demonstrated the sensitivity of compositional trends during fractionation to variations in oxygen fugacity. In particular, Figure 14 of Roeder and Osborn illustrates the oxygen fugacity changes

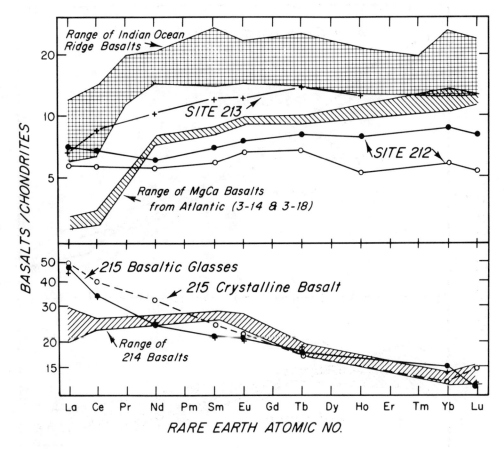

Fig. 4e. (lower) REE abundances in two site 215 basaltic glasses (●, +) and a site 215 crystalline basalt (o) compared to range in site 214 ferrotholeiites (3).

Fig. 4f. (upper) REE abundances in two site 212 metabasalts and a site 213 basalt compared to REE range of 1) five mid-Indian Ocean Ridge basalts (Schilling, 1971) and 2) MgCa ridge tholeiites from DSDP Leg 3 sites 14 and 18 (Frey et al., 1974).

required for SiO_2 enrichment in the experimental system $MgO-FeO-Fe_2O_3-CaAl_2Si_2O_8-SiO_2$. It is likely that continued magnetite removal from the ferrobasalts would result in increased SiO_2 abundances along a fractionation trend to oceanic andesites (Figure 3a).

In addition to clinopyroxene and magnetite fractionation, there is evidence for fractionation of plagioclase, the dominant phenocryst phase in these rocks. For example, on an anhydrous basis (to remove the dilution effects of high volatile contents caused by alteration) the more differentiated site 216 basalts have lower Na_2O, CaO, and Al_2O_3 (Table 1a) and higher normative Ab+Or/Ab+Or+An. However, plagioclase fractionation has not been extensive enough to cause large Eu depletions (Figure 4a). Abundance data for Sr are too variable, perhaps because of alteration effects, to be useful for evaluating the extent of plagioclase fractionation.

The oceanic andesites at site 214 are typically nonporphyritic. Some samples contain euhedral magnetite microphenocrysts (∼0.4 mm),

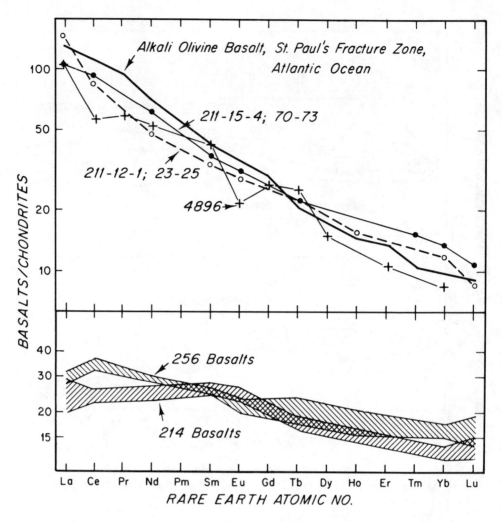

Fig. 4g. (lower) REE abundance range in site 256 ferrotholeiites (4) compared to range in site 214 ferrotholeiites (3). Except for differences in Ce (and to a lesser extent La) abundances, the REE abundance range of three site 256 basalts studied by Fleet et al. (1976) is similar to that indicated. The Ce (and La) discrepancies probably result from analytical errors. Despite these uncertainties, it is clear that all site 256 basalts are relatively enriched in light REE.

Fig. 4h. (upper) REE abundances in two site 211 alkali olivine basalts (12-3 from diabase sill, 15-4 from basal basalt), an alkali olivine basalt dredged from St. Paul's Fracture Zone in the Atlantic (Frey, 1970), and an alkali olivine basalt dredged from near the southern end of the Ninetyeast Ridge (29°56.8'S, 83°00.6'E, Figure 2), site 4896 in Balashov et al. (1970).

with plagioclase crystals >0.5 mm in length and abundant anhedral (<0.8 mm) clinopyroxene (∼30 vol. %). Although mineralogically similar to ferrobasalts, the overlying oceanic andesite lavas at site 214 differ significantly in chemical composition. Relative to the basalts these differentiated rocks are enriched in SiO_2, K_2O and Al_2O_3, but depleted in TiO_2, FeO^*, MgO and CaO (Tables 1a, 2a). In trace elements the

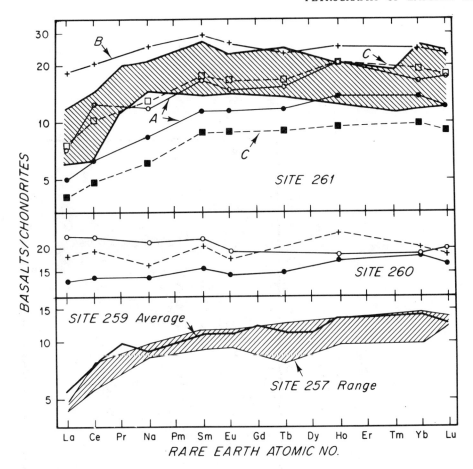

Fig. 4i. REE abundance average in site 259 basalts (Robinson and Whitford, 1974) compared to REE range in site 257 basalts (3). One basalt (not shown) studied by Fleet et al. (1976) from the top (11-2:82-84) of the site 257 core is enriched significantly in light REE. Except for anomalous Ce and Nd abundances (probably, subject to large experimental errors), the range of REE abundances in seven (Table 9b and Fleet et al., 1976) other site 257 basalts is similar to the site 257 range indicated.

Fig. 4j. REE abundances in site 260 basalts (o, +, this work; •, Robinson and Whitford, 1974).

Fig. 4k. REE abundances in site 261 basalts (unit c, ☐, ■; unit A, o, •; unit B, +) compared to range for a mid-Indian Ocean Ridge and four Carlsberg Ridge basalts (Schilling, 1971). For units A and C the lower line is from Robinson and Whitford (1974), and the upper line is from this work. Although analyses from each research group are similar in relative REE distribution, there is a constant, significant difference in absolute REE abundance.

oceanic andesites are very enriched in Sr, Ba, Zr, and light REE, and depleted in Sc, V, Cr, and Ni relative to ferrobasalts (Tables 1b, 2b and Figures 4a, 5a - 5d and 6). Also, oceanic andesites are less altered (ignition loss usually <2 wt. %) than the basalts. These magmas are differentiated to a greater degree than most ocean floor

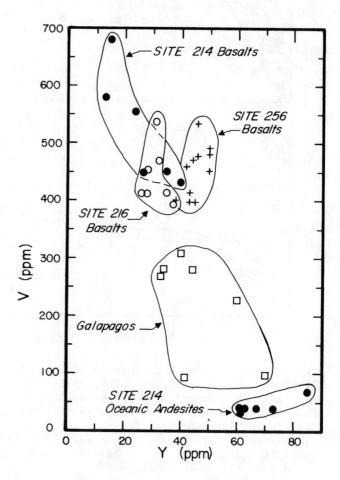

Fig. 5a. Decreasing V abundance with increasing fractionation (i.e., Y abundance). This inverse correlation from ferrotholeiites (sites 214, 216 and 256) to oceanic andesites (site 214) reflects magnetite fractionation. A similar, but less extreme range, occurs in the Galapagos Island suite (McBirney and Williams, 1969).

rocks. Moreover, these oceanic andesites are <u>not</u> compositionally similar to the rare occurrences of differentiated rocks in a spreading ridge environment (Aumento, 1969; Kay et al., 1970; Hart, 1971).

The compositional features of oceanic andesites imply that they resulted from clinopyroxene and magnetite separation from ferrobasalts. As previously discussed, the petrographic nature of ferrobasalts implies that magnetite and plagioclase are dominant liquidus phases, and thus they are likely fractionating phases. However, because the oceanic andesites have high Sr and Ba abundances and they are not strongly depleted in Eu, plagioclase does not appear to have been a major fractionating phase. Clinopyroxene fractionation is consistent

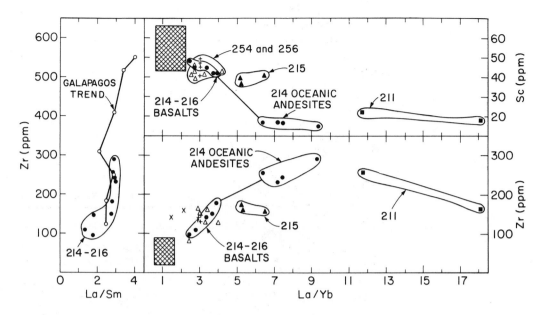

Fig. 5b. Zr abundance versus La/Sm and La/Yb. Galapagos trend from McBirney and Williams (1969). Cross-hatched rectangle encloses data for basalts from sites 212, 213, 253, 257, 259, 260 and 261 units A and C. Site 256 (+), site 254 (Δ) and two site 260 basalts (x).

Fig. 5c. Sc abundance versus La/Yb. Cross-hatched rectangle encloses data for basalts from sites 212, 213, 253, 257, 259, 260 and 261.

with the common occurrence of clinopyroxene agglomerates in some ferrobasalts. The importance of clinopyroxene fractionation in the generation of tholeiitic magmas with intermediate SiO_2 content (∼55 wt. % SiO_2) has been discussed by Wilkinson and Duggan (1973).

Site 253 (24°53'S, 87°22'E)

Site 253 (∼1 m basement penetration) is located on the western flank of the Ninetyeast Ridge about six degrees of latitude north of site 254 (Figures 1,2). A 400 m Eocene ash sequence containing two thin scoriaceous basalt flows overlies apparent basement. Kempe (1974) described the flows within the ash sequence as dark reddish-brown, partly palagonitized glass bordering glassy olivine basalt, whereas the lowermost basalt (basement?) is a porphyritic picritic olivine basalt (∼15 vol. % olivine). Frey and Sung (1974) studied one sample from a flow within the ash, and a sample of the lowermost basalt. In addition, we have studied four additional samples from the 35 cm of basalt core at the bottom of the hole.

The basalt sample from within the ash sequence has low FeO*/MgO and TiO_2 abundance (Table 3a), and it is distinct from basalts at sites 214, 216 and 254. However, in trace element composition, the high Zr (100 ppm) and Hf (>3.5 ppm) abundances, and a relative enrichment in light REE indicate island tholeiite affinities (Table 3b). However, this basalt does not have the geochemical characteristics expected of a parental composition for the high FeO*/MgO site 214 and 216 basalts. For example, despite a much lower FeO*/MgO, this site

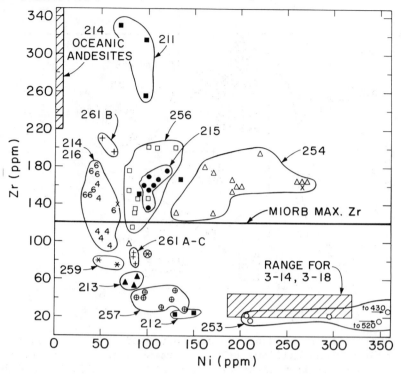

Fig. 5d. Zr versus Ni. Maximum Zr in MIORB from Fisher et al. (1968), Cann (1969), Engel and Fisher (1969), Fisher and Engel (1970) and Engel et al. (1974). ⊛ designates average MIORB (Engel et al., 1974). LIL element-depleted basalts are from sites 212, 213, 253, 257, 259 and 261 (units A and C). Range for MgCa ridge basalts from Leg 3 sites 14 and 18 from Frey et al., 1974. Moderately LIL element-enriched basalts are from sites 214, 215, 216, 254, 256, 260(x) and 261 (unit B). Highly LIL element-rich rocks are the amphibole-bearing site 211 basalts and site 214 oceanic andesites. Filled squares (2) lying outside site 211 field represent site 211 diabases.

253 basalt has Zr, Hf and La abundances similar to site 214 basalts (Tables 1a,b and 3a,b).

The lower olivine-rich flow has the geochemical and petrographic characteristics of a picritic basalt, i.e., an olivine cumulate (Frey and Sung, 1974; Kempe, 1974). Our additional analyses indicating very high MgO, Cr (>580 ppm), and Ni (200-500 ppm) abundances (Table 3a,b) corroborate this conclusion. Thus, the low Sr (<60 ppm), Ba (<30 ppm), and Zr (<30 ppm) abundances (Table 3b) are partially a result of olivine enrichment. However, these basalts are also relatively depleted in light REE like Indian Ocean ridge basalts (cf. Figures 4c,f). Since REE distributions are not affected by olivine accumulation, it is evident that these lower site 253 basalts are picritic variants of the LIL element-depleted tholeiites commonly formed at spreading ridges. Thus, site 253 basalts are very different from site 214 and 216 basalts, and it is likely that the lower basalts of site 253 were derived from a mantle of different composition than basalts from sites 214 and 216 (the problems of deriving LIL element-enriched basalts from LIL element-depleted basalts have been discussed by Schilling, 1973, and Frey et al.,1974).Yet this site is definitely associated with the topographic high that forms the Ninetyeast Ridge.

Table 2a: Major Element Composition (volatile-free) of Oceanic Andesites (wt.%)

	Ninetyeast Ridge Site 214 (a)	Galapagos Is. Basic Icelandite (b)	Galapagos Is. Icelandite (b)	Thingmuli Basaltic Andesite (c)	Faeroe Is. X-11	Faeroe Is. X-14 (d)	St. Paul Is. 39 (e)
SiO_2	56.9	55.5	60.7	54.8	55.9	55.6	55.4
Al_2O_3	15.9	14.3	14.8	13.2	14.2	11.2	12.6
FeO^*	9.84	11.5	7.90	12.5	12.5	11.9	12.9
MgO	2.48	2.72	2.15	3.33	6.0	10.6	4.61
CaO	5.79	6.66	4.99	7.16	10.3	8.45	7.53
Na_2O	3.97	4.60	4.42	3.59	2.13	1.82	3.69
K_2O	1.50	1.37	2.22	1.19	0.52	0.73	0.97
TiO_2	1.45	2.14	1.91	2.76	2.60	1.40	2.05
P_2O_5	0.64	0.68	0.50	0.83	—	—	0.21
Total	98.47	99.47	99.59	99.36	104.15	101.70	99.96
$\dfrac{FeO^*}{MgO}$	4.00	4.23	3.67	3.75	2.08	1.12	2.80

a) average of 7 from Table 7, Hekinian, 1974. Sample 41 excluded from average.
b) average of 2 from Table 12, McBirney and Williams, 1969.
c) average of 3 from Table 9, Carmichael, 1964.
d) tholeiitic andesites from section 10 of the middle series of the Faeroe lava sequence (p.87, Rasmussen and Noe Nygaard, 1969).
e) olivine-free basalt, Girod et al., 1971.

Table 2b

Trace Element Abundances of Oceanic Andesites (ppm)

	Ninetyeast Ridge Site 214 [a]	Galapagos Is. 71 [b]	Galapagos Is. 49 [c]	St. Paul Is. 39 [d]
Sc	16.5	22	16	25
V	39	98	15	250
Cr	<5	17	10	125
Co	39	24	21	38
Ni	<5	14	10	600
Sr	647	266	245	340
Ba	578	280	300	385
Y	65	70	41	95
Zr	252	570	480	355
La	32.7	47	39	28.3
Sm	11.3	13.6	9.7	8.9
Eu	3.17	3.4	2.7	2.46
Yb	4.39	-	-	4.8

a) Sc and REE are average of 4 from Table 5, Thompson et al., 1974. All others are average of 5 from Table 6, Thompson et al., 1974.
b) Duncan Is. Icelandite (55.34 Wt.% SiO_2), McBirney and Williams, 1969, Table 7.
c) Jervis Is. Icelandite (59.64 wt.% SiO_2), McBirney and Williams, 1969, Table 7.
d) St. Paul Is., Girod et al., 1971 (55.4 wt.% SiO_2).

Our conclusions about the lower basalts at site 253 differ from those of Frey and Sung (1974) who studied only a single powdered sample (253-58-1:2-7) received from D.R.C. Kempe. This sample is geochemically similar (e.g., highly altered picrite with high Cr and Ni, but low Zr and Hf abundances) to other core 58 samples except that it is relatively enriched in K_2O and light REE (Tables 3a,b). Although the degree of light REE enrichment is similar to that in the overlying volcanogenic sediments (Fleet and Kempe, 1974), we have no explanation for the light REE enrichment in the powder studied by Frey and Sung. We have verified their results on a second aliquot of this powder; however, a new powder (prepared in our laboratory) obtained from a chip from the same core interval does not have light REE enrichment. We suspect that the powder analyzed by Frey and Sung was contaminated during laboratory preparation, or that this powder contains a small amount (not detectable by x-ray diffraction) of a K and light REE-rich secondary phase.

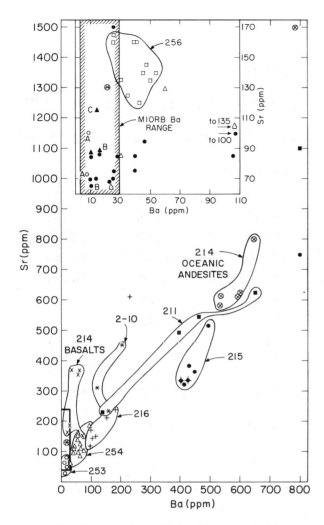

Fig. 6. Sr vs. Ba. Main diagram indicates moderately LIL element-enriched basalts from sites 214, 216 and 254, and highly LIL element-enriched oceanic andesites (214) and basalts from sites 211 and 215. Single points not within enclosed fields for 211, 215, 216 and 214 (oceanic andesites) represent very altered samples. Site 215 glasses = ✦ ; Leg 2-10 data from Frey et al., 1974; ⊗ designates MIORB average (Engel et al., 1974). Inset indicates data for LIL element-poor basalts from sites 212 (△), 213 (▲), 257 (●), 259 (o), 260 (Δ) and 261 (A,B,C), and moderately LIL element-enriched basalt from site 256. Rectangles indicate range of Sr and Ba in MIORB, Fisher et al. (1968), Cann (1969), Engel and Fisher (1969), Fisher and Engel (1970), Subbarro and Hedge (1973), and Engel et al. (1974).

Site 254 (30°58'S, 87°54'E)

Site 254 (∿45 m of basement penetration) is located on an isolated broad plateau atop the Ninetyeast Ridge near its southern terminus (Figures 1,2). It is south of the apparent intersection of Broken Ridge and the Ninetyeast Ridge. Because only two site 254 samples were analyzed during preliminary studies (Frey and Sung, 1974; Kempe, 1974), we selected 11 additional samples for analysis. On the basis of petrog-

Table 3a

Site 253: Major Element Composition (volatile-free, wt.%) [a]

	24-1:84-85	58-1:2-7	58-1 c/c(1)	58-1 c/c(3)
SiO_2	52.20	46.60	47.16	47.85
Al_2O_3	18.30	16.70	15.44	16.77
FeO*	6.92	10.50	10.49	9.42
MnO	0.05	0.12	0.17	0.19
MgO	7.87	14.80	16.02	12.44
CaO	9.67	8.38	8.68	10.48
Na_2O	3.05	1.49	1.07	1.55
K_2O	0.32	0.40	0.25	0.19
TiO_2	0.98	0.74	0.64	0.68
P_2O_5	0.35	0.09	-	-
Total	99.71	99.82	99.87	99.58
H_2O+	0.80	4.42	5.85	3.65
H_2O-	3.52	2.71	2.40	2.15
CO_2	1.62	0.38	0.46	0.16
Fe_2O_3/FeO	0.59	0.47	0.53	0.45
FeO*/MgO	0.88	0.71	0.65	0.76

a) Samples 24-1:84-85 and 58-1:2-7 from Kempe (1974); others are this work. The accuracy and precision of our data and the analytical techniques (electron microprobe, neutron activation, and emission spectrometry) have been discussed by Frey et al., (1974). Accuracy of our microprobe analyses can be estimated by comparing our results to literature values for rocks from similar positions in the DSDP cores; for example, see Table 5a. The discrepancy in Na_2O and MgO between our glass average and those of Melson et al. (Table 5a) is a systematic difference resulting from the use of different microprobe standards (see discussion in Melson et al., 1975).

Table 3b

Site 253: Trace element abundances (ppm)[a]

	24-1:84-85	58-1:2-7	58cc(1)	58cc(2)	58cc(3)	58cc(4)
B	8	2	10	7	8	8
Li	10	10	23	15	17	18
Sc	50	55±2	46	58	53	55
V	210	225	165	210	190	255
Cr	345	650	755	580	890	870
Co	42	56	55	43	48	55
Ni	110	295	520	205	210	430
Cu	90	73	30	300	205	115
Sr	77	50	30	62	40	60
Ba	14	20	8	12	6	30
Y	32	27	18	22	19	37
Zr	100	21	18	21	18	27
Hf	3.6	1.1±0.1	0.98	1.1	0.77	1.0
La	9.2	7.6±0.5	2.3	3.3	2.5	2.9
Ce	22	17.8±0.5	6.2	7.5	6.3	9.8
Nd	11	12.1±0.2	5.0	4.9	3.7	7.1
Sm	3.1	3.2±0.1	1.66	1.96	1.32	2.69
Eu	1.2	1.06±0.04	0.65	0.67	0.62	0.99
Tb	0.78	0.59±0.02	0.39	0.52	0.54	0.54
Ho	-	0.85	0.80	1.1	0.67	1.2
Yb	2.9	2.5±0.3	2.3	2.8	2.4	3.1
Lu	0.54	0.46±0.02	0.36	0.47	0.42	0.50

a) All data from this work. ± values for 58-1:2-7 indicate deviation from mean for duplicate instrumental neutron activation analyses. The accuracy and precision of emission spectrometry data can be evaluated by our data for W-1 (all in ppm, B, 11 ± 4; Ba, 180 ± 26; Co, 47 ± 3; Cr, 116 ± 8; Cu, 119 ± 4; Li, 12 ± 3; Ni, 88 ± 5; Sr, 190 ± 35; V, 279 ± 9; Y, 29 ± 2; Zr, 96 ± 10).

raphy (Kempe, 1974) there are three types of basalt: an upper coarse hyalo-ophitic basalt containing 3 mm feldspar and ophitic clusters of pyroxene and feldspar; a middle sequence of amygdaloidal lavas which contain fairly abundant altered olivine; and a lower sequence of very altered, brecciated basalts. Frey and Sung (1974) evaluated geochemical data for samples (1 each) of the upper coarse-grained basalt and the lower, very altered, brecciated basalt. They emphasized that although site 254 basalts have lower TiO_2 contents and FeO*/MgO than site 214 and 216 basalts, they are relatively enriched in light REE similar to basalts at sites 214 and 216. Thus, Frey and Sung concluded that site 254 basalts have the geochemical characteristics of island tholeiites.

This conclusion is substantiated by our additional analyses (Table 4a, b) which indicate that despite significant textural variations these basalts are similar in composition, and they are tholeiitic on a normative basis. The coarse-grained 2 m flow (core 31), about 25 m above basement, has the highest FeO*/MgO (2.09), La/Yb and lowest Ni content; however, even this basalt does not have the very low Cr and Ni abundances characteristic of site 214-216 basalts (Figure 5d). The highly altered nature of site 254 basalts (>4 wt. % total water) warrants caution when using major element features to infer magma composition. However, the island tholeiite affinity of site 254 basalts is clearly established by 1) the high abundance (compared to ridge axis tholeiites) of LIL elements such as Ba, Zr, Hf, and light REE (Table 4b), and 2) the compositional data for unaltered phenocrysts. For example, REE abundances in six site 254 basalts fall within the range of site 214 and 216 basalts (Figure 4a,b), and compared to minerals in LIL element-depleted tholeiites, site 254 clinopyroxenes have higher TiO_2 contents at a given Al_2O_3 abundance (Figure 7a) and site 254 plagioclases are significantly more K_2O-rich (Table 4c, Figure 7b).

It is unlikely that site 254, 214 and 216 basalts can be related in a simple petrogenetic model. Major element characteristics (e.g., FeO*/MgO and TiO_2 abundance) and compatible trace element abundances imply that site 254 basalt compositions are more primitive than site 214-216 basalts. However, LIL element features (e.g., La/Yb, Zr and Ba abundance) which should increase with increasing fractionation are similar in site 214-216 and 254 basalts (Figures 4a,b, 5c,d and 6). Considering the large distance and likely age difference between this southernmost Ninetyeast Ridge site, and the 214 and 216 sites, it is not surprising that these basalts cannot be related in a simple fractionation model. The more significant observation is that LIL element abundances of site 254 basalts establishes a similarity to site 214-216 basalts and island tholeiitic basalts.

Comparison of Ninetyeast Ridge Rocks to Other Areas

1. <u>Galapagos Islands, Iceland and Faeroe Islands</u>

In order to understand the petrogenesis of LIL element-enriched ferrotholeiitic basalt such as at sites 214 and 216, it is important to note that iron-rich tholeiites (with light REE enrichment relative to chondrites) and associated oceanic andesites are common in the Galapagos Islands (McBirney and Williams, 1969), at Thingmuli and Hekla, Iceland (Carmichael, 1964; Baldridge et al., 1973), and on the Iceland-Faeroe (Wyville-Thompson) Ridge; e.g., on the Faeroe Islands (Noe-Nygaard and Rasmussen, 1968; Rasmussen and Noe-Nygaard, 1969). Following, we present a comparison of basalts and associated rocks from these areas. In making these comparisons our objective is two-

Table 4a

Site 254: Major Element Composition (volatile-free) (a) and For Comparison the Composition of Basalts from Nearby Dredges

	31-1:88-90	35-1:99-101	35-3:27-30	36-1:43-46	36-3:36-39	36-3:95-97	38-1:115-117	5327	5332	5208	5201	4577	H	4896
SiO_2	48.41	46.14	47.09	47.60	48.00	47.20	49.20	50.1	49.71	50.53	49.58	51.26	49.46	48.91
Al_2O_3	17.31	14.09	14.42	16.39	16.29	14.80	14.30	18.9	14.74	16.99	14.70	15.24	15.73	13.12
FeO*	12.25	12.98	13.31	13.81	13.75	12.70	11.50	6.7	11.96	9.61	11.43	8.70	9.56	11.31
MnO	0.16	0.25	0.23	0.43	0.43	0.18	0.12	0.1	0.21	0.22	0.21	0.18	0.13	0.22
MgO	5.86	10.34	10.68	8.22	8.11	8.10	11.10	7.9	8.50	7.30	8.50	10.48	7.29	10.42
CaO	9.66	10.32	8.53	7.20	6.83	9.94	7.20	13.0	9.99	11.35	9.94	11.92	12.88	8.88
Na_2O	2.95	2.33	2.43	3.22	3.16	2.63	2.11	2.2	3.14	2.56	3.12	1.81	2.41	3.13
K_2O	0.20	0.22	0.21	0.39	0.34	0.16	0.74	0.2	0.32	0.19	0.30	0.19	0.27	1.34
TiO_2	1.80	2.27	1.98	2.29	2.14	2.62	1.72	1.0	1.33	1.52	1.54	0.35	1.06	2.61
P_2O_5	-	-	-	-	-	0.28	0.19	-	-	-	-	-	0.10	-
Total	98.59	99.04	98.90	99.61	99.08	98.61	98.18	100.10	99.90	100.27	99.32	100.13	98.89	99.94
H_2O^+	2.14	4.40	3.78	3.68	3.11	1.72	4.09	0.16	3.39	0.62	3.35	0.18	1.27	0.97
H_2O^-	2.70	2.70	1.85	4.40	2.70	2.53	3.89	-	0.73	0.35	0.54	0.07	1.70	0.16
CO_2	0.40	1.83	0.12	1.57	0.06	1.21	0.27	-	-	-	-	-	0.40	-
Fe_2O_3/FeO	1.18	0.97	0.90	2.19	1.89	0.73	1.43	0.08	0.59	0.41	0.59	0.15	1.40	0.22
FeO*/MgO	2.09	1.26	1.25	1.68	1.70	1.57	1.04	0.85	1.41	1.32	1.34	0.83	1.31	1.09

a) Samples 36-3:36-39 and 38-1:115-117 from Kempe, 1974; other DSDP samples are this work.

b) Average of 8 from Table IV, Vinogradov et al., 1969. Dredge 5327 at 34°17'S, 77°56.6'E.

c) Data from Vinogradov et al., 1969. 5332 from 32°26'S, 98°25'E; 5208 from 9°17'S, 91°22'E; 5201 from 22°21'S, 91°38'E; 4577 from 26°19'S, 89°56.5'E; 4896 from 29°56'S, 83°0.6'E.

d) H from 31°38'S, 83°48.5'E; Hekinian, 1968.

Table 4b

Site 254: Trace Element Abundances (ppm)[a] and for Comparison the Composition of Some Nearby Dredge Hauls

	Core 31-1 88-90	Core 31-1 138-141	Core 35-1 56-59	Core 35-1 99-101	Core 35-3 27-30	Core 35-3 118-120	Core 35-3 140-142	Core 36-1 43-46	Core 36-1 148-150	Core 36-3 36-39	Core 36-3 95-97	Core 38-1 115-117	Core 38-1 148-150	5327[b]	5332[c]	4896[d]
B	15	13	50	40	12	13	30	32	24	18	6	29	50			
Li	4	4	12	12	5	4	13	11	8	5	3	9	9			
Sc	43	-	-	42	39	-	-	41	-	45	42	42	-			
V	290	365	330	320	340	370	400	400	365	375	310	240	345			
Cr	495	690	595	575	560	410	530	525	525	430	270	83	410			
Co	26	45	47	50	50	44	35	42	48	50	50	55	65			
Ni	170	160	270	260	265	170	195	200	190	190	130	98	220			
Cu	200	135	38	115	150	230	165	165	130	200	120	80	80			
Sr	110	140	88	110	105	185	115	125	100	100	98	62	150			
Ba	45	55	60	65	75	95	48	65	55	40	41	29	55			
Y	43	55	50	47	49	55	55	60	49	48	39	35	60	25	28	25
Zr	130	180	165	165	165	175	160	160	155	165	130	80	195			
La	12.0	-	-	10.1	8.8	-	-	10.4	-	9.7	11.1	7.2	-	5.9	6.1	33
Ce	24	-	-	26	28	-	-	30	-	26	26	17	-	12.4	15.5	50
Nd	16	-	-	15	15	-	-	18	-	17	18	13	-	9.1	10.0	31
Sm	4.33	-	-	4.83	4.59	-	-	5.30	-	4.97	5.3	4.1	-	3.3	3.5	7.6
Eu	1.68	-	-	1.51	1.53	-	-	1.84	-	1.87	1.8	1.2	-	1.0	1.2	1.5
Tb	0.81	-	-	0.85	0.89	-	-	1.0	-	1.0	1.1	0.87	-	0.8	0.9	1.2
Ho	1.2	-	-	1.2	1.3	-	-	1.6	-	1.3	-	-	-	0.8	0.8	-
Yb	3.0	-	-	3.2	3.2	-	-	3.9	-	3.6	3.5	3.0	-	2.3	2.2	1.8
Lu	0.48	-	-	0.56	0.46	-	-	0.67	-	0.52	0.71	0.61	-	0.5	-	-

a) Sc and REE in samples 36-3:36-39 and 38-1:115-117 from Frey and Sung, 1974; others are this work.
b) Average of 7 dredged basalts from 34°17'S, 77°56'E; Balashov et al., 1970.
c) Dredged basalt from Broken Ridge at 32°26'S, 98°25'E; Balashov et al., 1970.
d) Dredged alkali olivine basalt from 29°56.8'S, 83°0.6'E; Balashov et al., 1970.

TABLE 4c
Site 254: Mineral Compositions (wt.%)

	Plagioclase				Pyroxene			
	A	B	C	D	A	B	C	D
SiO_2	55.2	54.8	54.1	50.79	50.2	49.9	49.6	49.5
TiO_2	0.11	0.12	0.10	-	1.36	1.30	1.17	1.80
Al_2O_3	27.2	28.1	28.3	29.67	2.31	3.17	3.06	3.23
FeO	0.74	0.27	0.94	-	12.1	9.62	8.98	10.76
MnO	-	-	-	-	0.28	0.20	0.22	-
MgO	0.09	0.22	0.23	-	15.1	15.2	15.2	13.74
CaO	9.42	10.6	11.1	13.62	17.9	19.6	20.2	19.86
Na_2O	6.11	5.30	5.03	3.51	0.32	0.34	0.36	0.31
K_2O	0.39	0.27	0.26	0.14	-	-	-	-
Cr_2O_3	-	-	-	-	0.08	0.36	0.38	0.28
	99.26	99.68	100.06	97.73	99.65	99.69	99.17	99.48
Cation Sum (12 O for plag, 6 O for pyx)	5.026	5.004	5.009	-	4.019	4.020	4.031	-

A. 35-1:99-101, andesine microphenocryst in groundmass
B. 35-1:99-101, small phenocryst of labradorite
C. 35-1:99-101, average for large, zoned labradorite phenocryst
D. Kempe (1974), average of 3

A,B,C, phenocrysts from 35-1:99-100
D. Kempe (1974), average of 2

fold: 1) to illustrate the striking geochemical similarities which imply similar magma petrogenesis and tectonic environments, and 2) to infer in greater detail the overall nature of the Ninetyeast Ridge; this is possible because islands allow a more thorough and representative sampling of rock types.

Site 214 and 216 basalts are similar to island ferrotholeiites in major element (e.g., FeO*/MgO, TiO_2 and K_2O), and trace element (e.g., Sr, Ba, Cr, Ni, Zr and REE) abundances (Table 1a,b). For example, the range of REE abundances in Faeroe Island plateau basalts (Schilling and Noe-Nygaard, 1974) completely overlaps the range for the site 214 and 216 basalts (Figure 4a). The island suites provide insight into the nature of the fractionation processes forming ferrotholeiites because coarse-grained cumulate textured gabbroic rocks containing euhedral calcic plagioclase, Mg-rich olivine and poikilitic clinopyroxene are occasionally found included in island ferrobasalts. Removal of such cumulates from a relatively unfractionated tholeiitic basalt creates a differentiate (ferrotholeiite) enriched in Ti, Fe, Na and K, but depleted in Al, Ca, Mg, Ni and Cr (Carmichael, 1964, McBirney and Williams, 1969).

Icelandites[1] also occur on these islands. On a FeO*/FeO*+MgO vs.

---

[1] The name icelandite was adopted by Carmichael (1964) to distinguish Icelandic lavas intermediate in composition between ferrobasalts and rhyolites that differ significantly in composition from andesites associated with subducting oceanic lithosphere. That is, icelandites have SiO_2 abundances similar to andesites, but they differ significantly in other abundances (e.g., Al_2O_3). Carmichael used icelandite to designate ∼60 wt. % SiO_2; McBirney and Williams (1969) extended this range to ∼55 wt. % SiO_2.

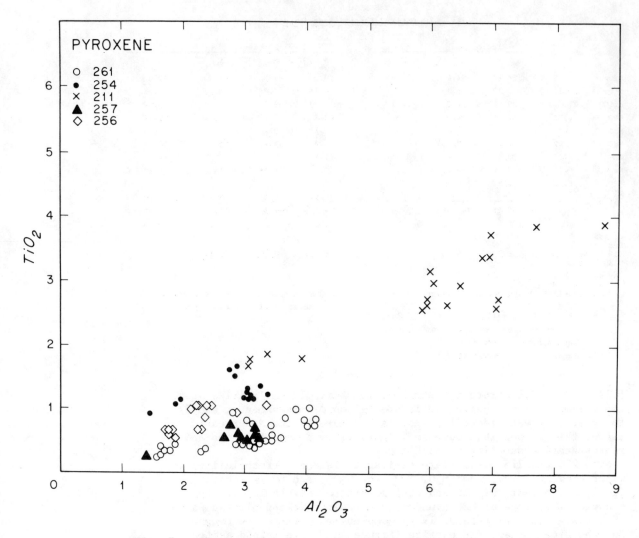

Fig. 7a. Microprobe point analyses showing overall correlation between TiO_2 and Al_2O_3. Note that LIL element-enriched sites show higher TiO_2 for a given Al_2O_3 content. The samples analyzed are listed with pyroxene analyses included in Tables 1-10.

SiO_2 diagram, these icelandites overlap with site 214 oceanic andesites (Figure 3a). On each island the icelandites differ compositionally from the associated ferrobasalts in the same way that site 214 oceanic andesites differ from site 214 and 216 ferrobasalts (cf. Tables 1a,b and 2a,b, Figure 3a,b, 4a and 5a,b). The similar geochemical trends such as $FeO*/FeO*+MgO$ vs. SiO_2 (Figure 3a), La/Sm vs. Zr (Figure 5b) and V vs. Y (Figure 5a) from ferrobasalt to icelandite (oceanic andesite) are convincing evidence of a similar petrogenetic process operating in these islands and the Ninetyeast Ridge. In each case the icelandites apparently correspond to the fractionation stage where SiO_2 and alkali metal enrichment dominate over $FeO*$ and TiO_2 enrichment. Because magnetite is a common euhedral phenocryst phase in ferrotholeiites, it is likely that the change in trend to decreasing $FeO*$ and TiO_2 reflects the onset of magnetite fractionation.

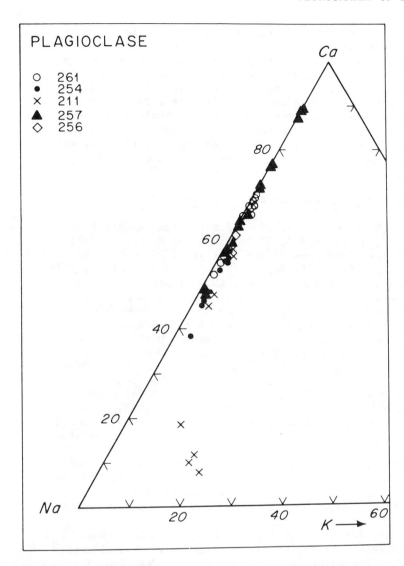

Fig. 7b. Microprobe point analyses showing overall range of variation in plagioclase. Samples analyzed are listed with plagioclase analyses included in Tables 1-10. Note the higher level of K_2O and more sodic compositions characteristic of the LIL element-enriched sites.

Lindstrom and Baitis (1974) used major and trace element data for bulk rocks and phenocryst minerals to develop fractionation paths which model closely the initial FeO^*-TiO_2 enrichment forming ferrotholeiites, and the subsequent SiO_2 enrichment forming icelandites which occur as repeated differentiation cycles on Duncan Island in the Galapagos. Extensive removal of olivine, plagioclase, and clinopyroxene are initially required to generate ferrotholeiites; eventually, opaque phases replace olivine as a significant fractionating phase, and at more than 50% crystallization, icelandite compositions are developed. The success in matching major and trace element trends during the tholeiitic differentiation sequence by crystallization and removal of phenocryst phases provides strong support for relating these rocks to an olivine tholeiite parent magma. Such differentiation is expected in shallow magma chambers.

The striking petrographical and geochemical similarities of tholeiitic rocks from sites 214 and 216 on Ninetyeast Ridge, the Galapagos Islands, Thingmuli, and the Faeroe Islands imply that in these regions 1) the mantle source rocks are of similar composition, but this composition differs markedly from the mantle source of basalt formed at spreading-ridge axes away from islands, 2) the fractional crystallization differentiation of the tholeiitic magmas proceeded similarly, and 3) the tectonic environments must have been analogous in order to have sampled similar mantle compositions, and to allow similar differentiation of the partial melt at relatively shallow levels.

2. St. Paul - Amsterdam Island Volcanic Complex

Speculative hypotheses have related aseismic ridges to hot spot traces (Carey, 1958; Wilson, 1963, 1965; Dietz and Holden, 1970); in particular, at least portions of the Ninetyeast Ridge have been proposed as a hot spot trace (Bowin, 1973; Luyendyk and Davies, 1974; Sclater et al., 1974; Sclater and Fisher, 1974; Thompson et al., 1974; Luyendyk and Rennick, 1977). In addition, the compositional similarities of some Ninetyeast Ridge rocks to rocks from "hot spots" such as Galapagos, Iceland, and Faeroes lend credence to this hypothesis. Because of the Oligocene reconstruction (Sclater and Fisher, 1974) indicating that the Ninetyeast Ridge abutted the St. Paul - Amsterdam volcanic complex, it is important to compare the rocks of this island complex with Ninetyeast Ridge rocks.

Based on recent studies of Amsterdam and St. Paul Islands (Vinogradov et al., 1969; Girod et al., 1971; Hedge et al., 1973; Gunn, et al., 1975), several geochemical similarities can be noted between these islands and Ninetyeast Ridge sites 214, 216 and 254 (Hekinian, 1974a,b; Frey and Sung, 1974).

Tholeiitic basalt is the most abundant rock type on each island. Amsterdam Island basalts have FeO*/MgO and TiO_2 contents similar to basalts from site 254, and St. Paul basalts have FeO*/MgO and TiO_2 contents similar to basalts from sites 214 and 216 (Table 1a, Figure 3a,b).

Amsterdam Island basalts are typical examples of island tholeiitic basalts which are characterized by high abundances of LIL elements relative to basalts formed at spreading ridge axes away from islands Table 1b). In addition, Sr^{87}/Sr^{86} ratios of Amsterdam Island basalts are significantly higher than mid-Indian Ocean ridge basalts (0.7040 and 0.7034, respectively, Hedge et al., 1973). Despite their lower FeO*/MgO, the Amsterdam Island basalts are similar to site 214-216 and St. Paul Island basalts in their low abundances of compatible elements such as Cr (<140 ppm) and Ni (<95 ppm) and high LIL element abundances such as Rb, Sr and Ba (Table 1b, this paper; Gunn et al., 1971, 1975; Hedge et al., 1973); however the detailed geochemical features of Ninetyeast Ridge basalts could not develop from Amsterdam Island basalts as a result of a simple fractionation model. For example, Amsterdam Island basalts have Ba abundances higher than site 214, 216 and 254 basalts, and fractionation of Amsterdam basalts would increase this discrepancy.

The analogy of the Ti-rich ferrotholeiites of St. Paul Island to ferrobasalts of sites 214 and 216 is strengthened by 1) REE abundances in St. Paul Island ferrotholeiites that fall within the range established by site 214 and 216 basalts (Figure 4a, d), and 2) the occurrence of rocks with intermediate SiO_2 basalts (∿54 wt. %) on St. Paul Island. Gunn et al. (1975) inferred that such rocks are the result of extensive alteration. While this is likely for corundum normative tuffaceous rocks with very low CaO contents (Gunn et al., 1975), we believe the high Y, Zr and REE abundance and normal CaO content of St. Paul sample 39 (Tables 2a,b) indicate a fractionated igneous rock. REE abundances in this sample are very close to the range established by site 214 oceanic andesites (Figure 4d). Also, as in Amsterdam Island and site 214, 216 and 254 basalts, St. Paul Island basalts have LIL element abundances (e.g., Sr 230-340 ppm) higher than LIL element-depleted tholeiites (Sr <150 ppm). St. Paul Island basalts, like Amsterdam Island basalts, have Sr^{87}/Sr^{86} isotopic ratios (0.7035 - 0.7065, Girod et al., 1971; Hedge et al., 1973) significantly higher than LIL element-depleted tholeiites formed at the nearby ridge axis (e.g., Mid-Indian Ocean Ridge basalts have 0.7032 - 0.7035, Subbarao and Hedge, 1973).

Apparently, the St. Paul - Amsterdam Island volcanism affects the composition of southeast Indian Ocean Ridge basalts over a wide geographic region; for example, at 34°17'S, 77°57'E (350 km. north of Amsterdam Island, Figure 2) tholeiitic basalts dredged from the axial rift zone (Udintsev and Chernysheva, 1965) are aluminous and mafic (>17.5 wt. % Al_2O_3, average FeO*/MgO = 0.85, Table 4a and Vinogradov et al., 1969), but they are relatively enriched in LREE like basalts from sites 214 and 254 (Figure 4b and Balashov et al., 1970). Similar widespread geochemical effects of island volcanism on nearby ridge segments have been documented near Iceland (Schilling, 1973), the Azores (Schilling, 1975a) and Bouvet Island (Dickey et al., 1974).

We conclude that on the basis of geochemical similarities such as tholeiitic major element compositions, similar differentiation trends to ferrotholeiites and oceanic andesites, and similar high LIL element abundances relative to tholeiitic basalts formed at spreading centers, the Ninetyeast Ridge rocks could be related to the volcanic source now forming Amsterdam and St. Paul Islands. It is important to realize that although geochemical differences require compositionally different mantle sources (or alternatively, nonequilibrium melting, O'Nions and Pankhurst, 1974) for island tholeiitic basalts (and the Ninetyeast Ridge basalts at sites 214, 216 and 254) and basalts formed at spreading ridges away from islands, these geochemical parameters do not provide direct information about mantle tectonics; in particular, the presence or absence of deep mantle plumes. However, because island tholeiites have higher LIL element abundances than tholeiites from ridge axes, a more primitive, i.e., less LIL element depleted, mantle source is implied for island basalts and the aseismic Ninetyeast Ridge.

Constraints on the nature of mantle hot spots have been developed by evaluating the evidence for relative motion among commonly proposed hot spots during the Cenozoic (Molnar and Atwater, 1973; Minister et al., 1974; Molnar and Francheteau, 1975). Plate reconstructions by Molnar and Francheteau (1975) show that a hot spot at St. Paul - Amsterdam could not have produced the Ninetyeast Ridge if this spot has remained fixed with respect to the Iceland or Tristan da Cunha - Gough hot spots. In fact, assuming fixed hot spots, they find that at 28 million years ago the Ninetyeast Ridge was 1000 km away from the Amsterdam - St. Paul hot spot. Thus, Molnar and Francheteau concluded that the Ninetyeast Ridge: 1) is related to a different hot spot, 2) is formed by mechanisms unrelated to island volcanism, or 3) the eastern Indian Ocean hot spots have moved at least 2 cm/yr with respect to Atlantic Ocean hot spots.

In general, hot spots appear to move with respect to each other at rates comparable to plate motions (i.e., few cm/yr) and/or hot spots are not localized but extend over distances larger than 600 km (Burke et al., 1973; Molnar and Atwater, 1973; Molnar and Francheteau, 1975). If the Ninetyeast Ridge is related to a hot spot at St. Paul - Amsterdam, then relative hot spot motion must be involved. These results suggest that at least some hot spots are related to convection within the upper mantle. This conclusion is not in disagreement with the geochemical data which require only the existence of two different mantle compositions in order to explain compositional differences between island and ridge axis basalts. Of course, a mechanism must exist for maintaining the identity of these two mantles.

Correlation of Geochemical Characteristics of Ninetyeast Ridge Sites with Tectonic Models for the Ninetyeast Ridge

Recent geological, geophysical and sedimentological data provide important constraints on the origin of the Ninetyeast Ridge (Bowin, 1973; Luyendyk and Davies, 1974; Sclater et al., 1974; Sclater and Fisher, 1974). Important observations are the following:

1) In cross section the ridge from 7°S to Osborn Knoll (14-16°S) has the form of an asymmetrical bell with the steepest slope to the east. However, north of about 7°S the ridge changes character to become a series of en echelon northeast-southwest trending blocks which procede northward underneath the Bengal Fan at 12°N. South of Osborn Knoll, the ridge increases significantly in width. To the east of the ridge lies a series of long, linear, north-south trending deeps and highs. All these features are evident in Figure 1.

2) Free air gravity anomalies associated with the Ninetyeast Ridge are small despite the 1500-2000 m relief; thus, the mass of the ridge must be compensated at depth.

3) Basal sediment ages at DSDP sites on the ridge become progressively older to the north (from Oligocene to Late Cretaceous). This is the same age gradient observed from magnetic anomalies in the Central Indian Basin, but it is opposite to the gradient in the Wharton Basin (Figure 2). Furthermore, much of the Ninetyeast Ridge has approximately the same age as crust immediately to the west on the Indian Plate.

4) Basal sediment ages at sites 216 and 253 are $\sim 65 \times 10^6$ yrs and $\sim 46 \times 10^6$ yrs, respectively. The separation between these sites (also oceanic crust of similar age, immediately to the west) is about 11° of latitude more than that indicated by the distance between well identified magnetic anomalies of similar age at 83-84°E (Figure 2).

5) At all DSDP sites, shallow water sediments (subaerial at site 214) immediately overlie basal volcanic units; thus, the Ninetyeast Ridge formed at, or close to, sea level. The vesicular and amygdaloidal nature of the basalts is consistent with this conclusion.

6) A systematic northward increase in water depth along the ridge is in good agreement with the empirical age-depth correlation for ocean basins.

7) Paleontologic data at each DSDP site indicates progressively warmer water species with decreasing age; that is, the sites have moved northward after formation in more southerly latitudes.

On the basis of these observations, several tectonic models have been proposed (Bowin, 1973; Luyendyk and Davies, 1974; Sclater et al., 1974; Sclater and Fisher, 1974; Luyendyk and Rennick, 1977).

These models are similar in several important respects. For example, the Ninetyeast Ridge is an extrusive feature with a low density root. It is attached to the Indian plate and was formed in shallow water very close to a spreading center crest, presumably the ancient Southeast Branch of the Central Indian Ocean Ridge. However, the topographic complexity of the ridge suggests that various parts of the ridge may have different origins.

In addition, Sclater and Fisher (1974) have reconstructed the plate tectonic history of the eastern Indian Ocean since the Early Cretaceous. Their reconstruction shows that the southeasterly edge of the Amsterdam and St. Paul Island complex abuts the southern extremity of the Ninetyeast Ridge at about 32×10^6 yrs ago. This result provides support for the proposed association of aseismic ridges, such as the Ninetyeast Ridge, with the traces of hot spots on moving oceanic plates (Carey, 1958; Wilson, 1963, 1965; Dietz and Holden, 1970).

Two features common to the tectonic models have important implications for the petrologic and geochemical nature of Ninetyeast Ridge rocks. The first is the proposal that at least part of the ridge is a manifestation of a hot spot; in this case, the composition of Ninetyeast Ridge rocks should differ markedly from basalt compositions formed at spreading ridges (e.g., Schilling, 1973, 1975a). Secondly, if various parts of the Ninetyeast Ridge have different origins, it is likely that large geochemical heterogeneities will be observed along the ridge. We have demonstrated that rocks from DSDP sites 214, 216 and 254 have geochemical affinities with tholeiitic basalts on islands (Galapagos, Iceland, Faeroes, Amsterdam, St. Paul) commonly proposed as hot spots. Furthermore, site 214 and 216 rocks appear to be genetically related, and they certainly could have been derived from the same mantle source. Only a very slightly different source composition would be required for site 254 basalts. However, the LIL element-depleted site 253 basalts require derivation from a mantle source very different than the source of site 214, 216 and 254 basalts.

It is important to ascertain if the petrological and geochemical differences of Ninetyeast Ridge basalts can be correlated with the complicated tectonic models proposed by Luyendyk and Davies (1974). Sclater et al. (1974), Sclater and Fisher (1974), and Luyendyk and Rennick, (1977). Sclater et al., proposed that north of 8°S, the Ninetyeast Ridge data are consistent with formation of the ridge "by the migration of the Indian Plate along an old active transform fault over an isolated magma chamber" (a hot spot). However, in order to account for drastically different distances between points of similar age on the Ninetyeast Ridge (and the Indian Plate immediately to the west at 86-90°E), and the Indian Plate between 83-86°E, Sclater et al. proposed that the spreading center west of the Ninetyeast Ridge jumped south 11° (see Figure 7, Sclater et al., 1974). As a result, they proposed that the nearly uniform depth portion of the Ninetyeast Ridge between 10-21°S is of equal age, and that it formed from a leaky transform fault as the ridge axis jmped south. Thus, basalt compositions from sites between 10-21°S should differ from compositions obtained at sites further to the north. According to this model, south of 21°S, the ridge should again decrease in age in the same manner as crust on the Indian Plate immediately to the west. Sclater and Fisher (1974) summarized the tectonic setting of the Ninetyeast Ridge by stating, "It is an extrusive pile attached to the Indian plate at the junction of an active spreading center and a transform fault." This model does not account for the excess elevation of the ridge or the en echelon NE-SE blocks which make up the ridge north of 7°S.

The geochemical similarity of basalts from sites 412 (11°20.21'S) and 216 (1°27.72'N) is strong evidence that the ridge basalts north of 8°S and as far south as 11°20'S formed by similar processes. Thus, it is unlikely that site 216 and 214 basalts result from different sources, i.e., from a hot spot source and from a leaky transform fault, respectively. The geochemical similarity of basalts from sites 214 and 216 to basalts from islands proposed as hot spots (e.g., Iceland) does not eliminate the Sclater and Fisher (1974) model relating the Ninetyeast Ridge to a transform fault-spreading center junction because rocks dredged from fracture zones are commonly alkalic with geochemical similarities to island rocks (Thompson and Melson, 1972). Furthermore, a tholeiitic basalt type (KP of Melson et al., 1976) with normal FeO* and TiO_2 content but with high K_2O and P_2O_5 abundances (presumably also high LIL trace element abundances) is common in dredges from fracture zones and extremely rare in normal oceanic ridge segments. Thus it is possible that rocks similar to those at sites 214, 216 and 254 occur in fracture zones; however, a similar compositional sequence has not been found in fracture zones. Ferrogabbros dredged from fracture zones demonstrate a strong Fe-Ti enrichment trend (Miyashiro et al., 1970), but they are LIL element-depleted (Thompson, 1973; Carroll, personal communication, 1975), and thus, they are not a plutonic equivalent to the LIL element-enriched site 214-216 ferrobasalts.

The anomalous nature of the basal site 253 basalts has no explanation in the models of Sclater et al. (1974) and Sclater and Fisher (1974). A possible explanation is that the volume of basalt extruded from the leaky transform fault was small at the time period corresponding to extrusion of the site 253 basal basalts, thereby enabling LIL element-depleted tholeiite magma derived from the axial rift zone to dominate at this site. However, the depth to basement at sites 253 and 254 is in accord with the expected subsidence curve based on the Ninetyeast Ridge having formed at about sea level at the crest of a spreading ridge (Table 1 of Sclater and Fisher, 1974). Thus, there is no reason to expect the extruded magma volumes to be markedly different at sites 253 and 254.

Because 1) the association of the Ninetyeast Ridge with a transform fault does not explain its excess elevation, and 2) the active Amsterdam - St. Paul Islands region connected to the southern end of the Ninetyeast Ridge during the Oligocene, Luyendyk and Davies (1974) and Luyendyk and Rennick (1977) proposed a complex tectonic model relating the ridge origin to a point volcanic source. They emphasize that because the Ninetyeast Ridge has a shallow water origin, and is in most places the same age as the Indian plate, the volcanic source and ridge crest must have been fixed relative to each other. Since the aseismic ridge exists only to the north of the Southeast Branch spreading center, the volcanic source must have been displaced slightly to the north under the Indian plate. However, northward movement of the Southeast Branch ridge crest and the postulated 11° southward jump of the ridge crest (Sclater et al., 1974) would not maintain a simple relationship between the volcanic point source and the ridge crest. As a result, Luyendyk and Davies (1974) developed a model where all plates and the point volcanic source are in relative motion. Their model (outlined in Fig. 12, Luyendyk and Davies, 1974) calls for three different origins of the ridge: (1) North of 7°S the ridge is the trace of a volcanic source. Site 216 is included in this region. (2) Between 7°S and Osborn Knoll (14-16°S) the ridge formed by leaking along a transform fault; here the ridge is presumed to be younger than the Indian plate. This region includes site 214. (3) South of Osborn Knoll the ridge was formed as a volcanic trace on the Antractic plate; however, it was later overprinted by volcanism along the transform fault formed by the southward jump of the Southeast Branch spreading center. At the southernmost tip, the ridge may again be dominantly formed from the volcanic point source.

In several respects this model is similar to that proposed by Sclater et al. (1974). Again the obvious discrepancy is that site 216 is in the region proposed as a hot spot trace, whereas site 214 is in a region formed by leaking along a transform fault; yet, basalts from these two sites are quite similar in geochemical and petrographic character. A common source and evolution are implied. The similar (or slightly older) basal sediment ages (59×10^6 yrs) of site 215 (8°07.30'S), west of the Ninetyeast Ridge, and site 214 (11°20.21'S) suggest further that the Indian plate and Ninetyeast Ridge are of similar age up to 11°S, and that the Ninetyeast Ridge may be related to a volcanic point source as far south as site 214. By including the Kerguelen hot spot in the tectonic model, Luyendyk and Rennick (1977) demonstrated that if the 11° southward ridge crest jump occurred at ~64 MYBP, the Ninetyeast Ridge would be continuously generated by a hot spot (partly by overlapping contributions from Amsterdam - St. Paul and Kerguelen) from north of site 216 to the vicinity of site 214. Our geochemical results are consistent with this version of the tectonic model.

The Luyendyk-Davies model calls for a complicated history in the region of sites 253 and 254, and the plate reconstruction of Luyendyk and Rennick (1977), suggests that basement basalts at site 254 are related to the Kerguelen hot spot rather than volcanism associated with the Amsterdam - St. Paul hot spot. None of the models account for the LIL element-depleted character of the basal site 253 basalts. However, in the reconstruction of Luyendyk and Rennick (1977) the aseismic ridge formed by the Naturaliste Plateau and Broken Ridge (Figure 2) is also related to the Kerguelen hot spot. Although DSDP site 255 on Broken Ridge did not reach basalt, a dredge haul on Broken Ridge (32°26'S, 98°25'E; Figure 2, Vinogradov et al., 1969) recovered tholeiitic basalt with a major element composition very similar (except in TiO_2) to some site 254 basalts (cf., 36-3:95-97 with 5332 in Table 4a), and a relative LREE enrichment (Balashov et al., 1970) similar to site 254 basalts (Figure 4b, the absolute REE abundance difference may result from the very different analytical techniques used). The occurrence of LIL element-enriched tholeiite on Broken Ridge supports the Luyendyk and Rennick,(1977) proposal that Broken Ridge formed as a trajectory of a hot spot. Furthermore, the geochemical similarities between site 254 basalts and the dredged Broken Ridge basalt are consistent with their model relating the southern portion of the Ninetyeast Ridge and Broken Ridge - Naturaliste Plateau to the same hot spot, i.e., Kerguelen.

Central Indian Basin

Site 215 (8°07.30'S, 84°47.50'E)

Site 215 was drilled in >5000 m of water approximately 240 km west of the Ninetyeast Ridge and 2-3° north of sites 213 and 214 (Figures 1,2). The relatively young basal sediment age at site 215 (59-60 x

10^6 yrs) compared to Central Indian Basin crust between 83 and 86°E, led Sclater et al. (1974) to place a fracture zone at 86°E (Figure 2). Apparently, between 64 MYBP and 53 MYBP (Sclater et al., 1976) the east-west ridge axis between 86°E and the Ninetyeast Ridge was offset by 11°. Sclater et al. (1974) proposed that during this southward jumping of the ridge axis, the Ninetyeast Ridge between 10° and 21°S formed from a leaky transform fault. The magnetic profiles between 86°E and the Ninetyeast Ridge are not readily interpreted, but Sclater et al. believe that there is evidence that anomalies 27-33b are north of site 215 (Figure 2), thereby supporting the existence of a major fracture zone at 86°E. Spreading rates during this time were apparently in excess of a 8 cm/yr.

At site 215 approximately 25 m of basalt were penetrated. Hekinian (1974a) described the basalt core as a succession of at least 14 pillow flows each separated by glassy surfaces (>2 cm) containing veins of calcite and palagonite. The basalts have textures and mineral paragenesis characteristic of abyssal pillow basalts; for example, plagioclase (An 68-73) is the common microphenocryst phase. However, the plagioclase K_2O content (0.15 wt. %) exceeds that of plagioclase from LIL element-depleted tholeiites (Figure 7b, Table 5c).

Major element abundances in six altered cyrstalline rocks (Hekinian, 1974a) indicate a tholeiitic major element composition similar, except for K_2O, to spreading ridge tholeiitic basalts. The unusually high and uniform K_2O (0.8 - 1.0 wt. %) abundance was attributed to alteration (total volatiles are 2.5 - 3.5 wt. %). However, Thompson et al. (1974) concluded on the basis of high Ba, Sr and Zr abundances "that the basaltic magma at site 215 had a chemical composition distinctly more 'alkaline' than basalts typical of active mid-ocean ridges." They believed that high K_2O and P_2O_5 abundances are a primary feature of these basalts.

The presence of unaltered glass in these cores enables determination of magma composition without uncertainties about chemical changes caused by alteration. Melson et al. (1976) analyzed 15 glass samples and found that high K_2O (∼1 wt. %) and P_2O_5 (0.25 wt. %) abundances are characteristic features of site 215 magma. We have studied glass from 13 areas scattered throughout the 13.3 m core, and our results (Tables 5a,b) confirm that these features are not caused by alteration or a high degree of fractionation since FeO*/MgO <1.3 (Table 5a). In addition, we have determined that site 215 glass has a larger relative enrichment in light REE than site 214-216 basalts (Figures 4e, 5b, c), and unusually high Sr and Ba abundances relative to basalts formed at spreading ridges (Table 5b, Figure 6). Because Ninetyeast Ridge basalts are highly altered, we did not use their Ba and Sr abundances to infer magma composition; however, the unusually high Ba and Sr abundance in fresh and altered site 215 basalts clearly distinguishes these basalts from all other Eastern Indian Ocean basalts, except for the alkali basalts from site 211 (Figure 6).

The magma forming the site 215 flow units was extremely homogeneous. The relative standard deviation of our 13 glass analyses (Table 5a) is less than 3.5% for all elements except MnO (7%) and TiO_2 (6%). Furthermore, alteration has not strongly changed the major element composition because the glass average is quite similar to the crystalline rock average (Table 5a). Melson et al. (1976) classify site 215 basalts as a KP type (high K_2O and P_2O_5 content), and they comment that this compositional group is extremely rare in ridge segments that are undisturbed by fracture zones or seamounts.

TABLE 5a

Site 215: Major Element Composition (Volatile-free, Wt. %)

	Crystalline Rock Avg.[a]	Glass Avg.[b]	Glass Avgs.[c] a	b	c	Dredge[d]
SiO_2	50.4±1.0	50.9±0.5	50.73	50.67	51.26	50.39
Al_2O_3	16.8±0.2	16.6±0.1	16.74	16.93	17.14	16.85
FeO^*	8.43±0.21	8.11±0.05	8.17	8.09	8.17	8.79
MnO	-- --	0.15±0.1	--	--	--	0.19
MgO	6.48±0.36	7.58±0.05	6.85	7.08	6.86	6.46
CaO	10.95±0.38	10.0±0.34	10.73	10.59	10.65	10.50
Na_2O	3.17±0.33	3.50±0.12	2.99	3.04	2.94	3.17
K_2O	0.90±0.06	0.97±0.03	0.99	0.92	1.10	0.89
TiO_2	1.71±0.02	1.56±0.09	1.75	1.67	1.74	1.70
P_2O_5	0.31±0.01	--	0.25	0.26	0.25	0.31
Total	99.15	99.37	99.20	99.25	100.11	99.25
H_2O+	2.88±0.42	--	--	--	--	1.76
H_2O-						2.24
CO_2	-- --	--	--	--	--	--
Fe_2O_3/FeO	0.60 --	--	--	--	--	2.04
FeO^*/MgO	1.30	1.07	1.19	1.14	1.19	1.36

[a] Average and standard deviation of seven analyses (Hekinian, 1974; Thompson et al., 1974).

[b] Average and standard deviation of 13 glasses from cores 17-1, 18-1, 18-2, 19-2, 20-2, 20-3 and 20-4, this work. We have not analyzed these glasses for H_2O because it is generally not possible to separate sufficient fresh glass for gas chromatography analysis. However, with a microscope it is easy to locate small fresh glass areas for electron microprobe analysis. The greater than 99% summation confirms the fresh nature of the glass analyzed by microprobe. Palagonitized glass oxide totals determined by microprobe are typically less than 93% (Table 7 of Frey et al., 1974).

[c] Three classes of Site 215 glass analyses (Melson et al., 1975); group a is avg. of 3, group b is avg. of 11, group c is represented by a single analysis.

[d] Vesicular basalt (D232) dredged from near the base of seamount in northeast Indian Ocean at about 5°S, 97°E (Engel et al., 1965).

It is puzzling that there is no apparent tectonic reason for basalts at site 215 to differ from spreading ridge basalts. The site is far to the southwest of Nikitin seamount (3'S 83°E), and the bathymetry indicates clearly that site 215 is beyond the influence of the Ninetyeast Ridge. Indeed, site 215 basalts are unlike basalts from any of the Ninetyeast Ridge sites (Tables 1a,b, 3a,b, 4a,b 5a,b and Figures 4e and 6).

Based on the geochemical nature of basalts dredged from spreading

Table 5b

Site 215: Trace Element Abundances (ppm)

	Crystalline rock Avg[a]	Altered Glass[b]	Unaltered Glass[c] 17-1: c/c	20-3: 111-112
B	4	<2	<2	<2
Li	13	8	6	5
Sc	47	–	40	37
V	245	260	310	295
Cr	250	315	325	295
Co	50	44	52	48
Ni	100	100	120	110
Cu	65	65	57	58
Sr	390	750	335	330
Ba	450	800	425	405
Y	36	42	44	43
Zr	160	135	175	165
La	16.2	–	15.5	14.7
Ce	36	–	30	30
Nd	20.1	–	14.4	14.6
Sm	4.5	–	3.91	3.94
Eu	1.5	–	1.48	1.42
Tb	0.8	–	0.81	0.89
Ho	–	–	0.84	0.78
Yb	2.5	–	3.0	2.8
Lu	0.51	–	0.44	0.44

a) average of 1 to 5 analyses from cores 18-2, 18-3, 19-1, 19-2 and 20-2 (Thompson et al., 1974; Bougault, 1974; this work).

b) 215-18-3:26-35 (Thompson et al., 1974).

c) this work.

ridge axes away from islands and from DSDP cores in oceanic crust obviously formed by seafloor spreading, it is unlikely that site 215 basalts formed at a spreading ridge axis. Melson et al. (1976) referring to von der Borch et al. (1974) describe the site 215 core as being intercalated with baked sediment, and they infer that site 215 basalt formed as a sill probably as a result of off-ridge vol-

TABLE 5c
Site 215: Plagioclase Compositions (wt.%)

	A	B	C
SiO_2	47.6	51.3	46.8
Al_2O_3	31.5	31.0	32.3
FeO	0.55	0.55	0.85
MgO	0.27	0.28	0.29
CaO	14.7	14.2	16.0
Na_2O	3.38	3.45	2.38
K_2O	0.17	0.15	0.15
	98.17	100.93	98.77
Cation Sum (12 O)	5.06	5.01	5.04

A. microphenocryst, 18-2:74-75, note because of the small grain size FeO and MgO abundances may in part reflect the adjacent glass.

B,C. microphenocrysts, 18-1:78-79

canism. If so, the surprisingly young age of the sediment-basalt contact could be explained because the basalt does not represent true basement. However, the discussion by von der Borch et al. (1974) does not mention evidence for these basalts being a sill. In fact, the pillowed nature of the basalt core, and the presence of unrecrystallized nannofossils in the contact chalk indicate that the basalt is unlikely to be a sill. Thus, we are unable to offer an explanation for the presence of LIL element-rich tholeiitic basalts at site 215.

Because the young age of presumed basal sediments at this site has played a key role in formulating tectonic models for the Ninetyeast Ridge (Sclater et al., 1974; Sclater and Fisher, 1974), we believe it is important to emphasize that it is unlikely that the presumed basement basalt was formed at a spreading center. In geochemical respects, site 215 basalts are similar to LIL element-enriched tholeiitic basalt at site 2-10 in the Atlantic (Figure 6 and Frey et al., 1974). At each site, the atypical geochemical features cannot be correlated with unusual bathymetry or tectonics.

Wharton Basin

Site 213 (10°12.71'S, 93°53.77'E)

Site 213 (Figures 1,2) in the western Wharton Basin is in the region between 93-96°E, 500 km east of the Ninetyeast Ridge, where the lineated east-west trending magnetic profiles are remarkably clear and simple (Sclater and Fisher, 1974). This site was drilled on the younger, northern side of anomaly 26, and basal sediments associated with pillow basalts gave a microfossil age of $56-58 \times 10^6$ yrs. A discrepancy of $5-6 \times 10^6$ years with the magnetic time scale of Heirtzler et al. (1968) led Sclater et al. (1974) to propose a modified time scale. The inferred spreading rate in this region is 8.1 cm/yr.

Eight meters of basalt were recovered from a penetration of about 18 m. Hekinian (1974a) recognized at least 11 successive pillow flows with upper and lower glassy chilled margins. Textural variations from chilled glassy margins to holocrystalline interiors are those expected of pillow basalts (Shido et al., 1974). Although the rocks are quite altered, agglomerations and microphenocrysts of plagiclase are abundant along with clinopyroxene and olivine pseudomorphs.

Five altered crystalline basalts (>2 wt. % volatiles) have high Al_2O_3 abundances ($\sim$18 wt. %) and Fe_2O_3/FeO, variable K_2O (0.36 to 1.09 wt. %) and low MgO (<5.6 wt. %) abundance (Table 6a). Hekinian (1974a) concluded that except for differences caused by alteration, site 213 basalts are similar in composition to Mid-Indian Ocean Ridge basalts. This conclusion based on major element abundances of altered basalts is subject to question; however, a similar conclusion was reached by Melson et al. (1976) on the basis of low K_2O (0.06 wt. %) and P_2O_5 (0.09 wt. %) in glass from this site.

Trace element abundances in altered basalts (Table 6b) confirm the similarity of site 213 basalts to those from oceanic ridge spreading centers. In particular, low Zr content and slight depletion in light REE relative to chondrites are diagnostic features of LIL element-depleted tholeiites (Figures 4f and 5b). Compared to site 215 basalts, the site 213 basalts have very low Ba and Sr abundances, lower Zr abundances at a similar Ni content, and low La/Yb (Figures 5b,d and 6). Thus, the geochemical characteristics of site 213 basalts are in accord with their formation at a spreading ridge axis as suggested by Sclater and Fisher (1974).

Site 212 (19°11.34'S, 99°17.84'E)

Site 212 was drilled in the deepest portion of the Wharton Basin (6233 m depth) at the southern end of a long linear topographic high (Investigator Fracture Zone of Sclater and Fisher, 1974) which bisects the north central Wharton Basin (Figures 1, 2). The basement contact is a 30 m section of brown clay, lacking fossils. By estimating a sedimentation rate (1 mm/yr) Sclater et al. (1974) inferred a contact age of about 100×10^6 yrs. They concluded that a) the crustal age increases to the south in the central Wharton Basin, and b) the central Wharton Basin is no older than Cretaceous. No clearly defined magnetic anomalies exist in the site vicinity, but Sclater and Fisher utilized paleontological ages at sites 211, 212, and 256 to infer that the oceanic crust at site 212 has been formed by an active ridge axis spreading south at a half-rate of 5-6 cm/yr.

Five meters of basalt were penetrated and about 4 m were recovered. Hekinian (1974a) interpreted the core to be a sequence of at least seven pillowed flows. The basalts are strongly altered, apparently by halmyrolysis and hydrothermal processes. On the basis of a chlorite-quartz-pumpellyite mineral assemblage, Hekinian classed site 212 rocks as metabasalts. Their altered nature is demonstrated by total volatiles commonly exceeding 5 wt. % and K_2O abundances ranging from 1 to >5 wt. % as a result of groundmass cryptocrystalline potassium feldspar (Table 6a). Within the basaltic core there is abundant evidence for chemical migration (analyses c

Table 6a

Sites 212 and 213: Major Element Composition (volatile-free, wt.%)

	17-3:90-99	18-1:144-150	Site 213(a) 18-2:101-103	18-2:115-117	Glass Avg.	39-1:134-136	Site 212(b) 39-3:8-14(c)	39-3:8-14(d)	39-3:145-147	Glass
SiO_2	47.85	47.72	49.60	50.13	49.66	52.93	51.91	48.51	51.56	51.33
Al_2O_3	18.65	17.97	17.86	17.89	15.92	19.52	18.55	18.15	18.79	15.60
FeO*	10.60	10.26	10.18	9.67	9.88	8.29	10.61	9.15	8.16	7.86
MnO	–	–	–	0.16	–	0.07	–	–	0.13	–
MgO	3.20	5.65	5.25	5.57	8.15	5.90	7.36	8.45	7.89	8.97
CaO	13.98	13.95	12.57	11.30	12.27	7.23	1.85	8.86	6.43	13.48
Na_2O	2.66	2.61	2.62	2.96	2.42	2.71	2.16	1.80	1.65	1.42
K_2O	1.11	0.37	0.42	0.41	0.06	2.59	3.81	2.81	3.83	0.05
TiO_2	1.09	1.06	1.05	0.87	1.07	0.59	0.69	0.62	0.57	0.61
P_2O_5	0.17	0.10	0.10	–	0.09	–	0.02	0.09	–	0.03
Total	99.31	99.69	100.55	99.01	99.52	99.95	96.96	98.44	99.01	99.35
H_2O+	2.90 (I.L)	3.50 (I.L)	2.13(I.L)	1.19	–	5.52	18.97 (I.L)	8.08 (I.L)	7.47	–
H_2O-				2.00						
CO_2	1.85	0.88	–	0.08	–	0.24	–	1.60	1.49	–
Fe_2O_3/FeO	2.39	2.52	1.23	1.36	–	–	5.17	2.52	–	–
FeO*/MgO	3.37	1.82	1.94	1.74	1.21	1.41	1.44	1.08	1.03	0.88

a) 18-2:115-117, this work; glass avg. is average of six from Melson et al., 1975; others from Hekinian, 1974.
b) 39-1:134-136 and 39-3:145-147, this work, glass from Melson et al., 1975; others from Hekinian, 1974. Additional analyses for Site 212 metabasalts are in Table 5, Hekinian, 1974.
c) pillow margin
d) pillow interior

TABLE 6b[a]
Sites 212 and 213: Trace Element Abundances (ppm)

	213					212	
	17-2:108-110	18-1:144-150	18-2:101-103	18-2:115-117	19-2:54-56	39-1:134-136	39-3:145-147
B	35	-	-	22	37	36	55
Li	17	-	-	19	18	16	50
Rb	28	-	-	23	35	60	65
Sc	-	-	-	43	-	46	46
V	210	235	235	205	200	205	225
Cr	310	280	290	325	345	1200	1200
Co	32	68	66	33	30	45	33
Ni	85	81	85	75	87	130	150
Cu	85	80	80	90	95	52	60
Sr	115	-	-	87	88	65	100
Ba	14	-	-	10	16	23	135
Y	27	-	-	22	32	12	24
Zr	53	-	-	57	62	21	23
La	-	-	-	2.2	-	1.9	2.3
Ce	-	-	-	7.6	-	5	6
Nd	-	-	-	6.2	-	3.4	3.6
Sm	-	-	-	2.18	-	1.08	1.29
Eu	-	-	-	0.85	-	0.47	0.52
Tb	-	-	-	0.64	-	0.32	0.38
Ho	-	-	-	0.86	-	0.36	0.55
Yb	-	-	-	2.55	-	1.17	1.75
Lu	-	-	-	0.41	-	0.18	0.27

a) Samples 213-18-1:144-150 and 213-18-2:101-103 from Bougault, 1974. Others are this work.

and d in Table 6a) and Hekinian concluded that it is "difficult to speculate on the origin of the metabasalts of site 212." One persistent feature of the 13 major element analyses is a uniformly low TiO_2 abundance (<0.7 wt. %); this is the only major element feature in these metabasalts which might be indicative of the magma composition (e.g., Cann, 1970). Unaltered glass is very rare at this site, but Melson et al. (1976) determined major element abundances in a single chip. Their analysis confirms the low TiO_2 content (0.61 wt. %) of these basalts and establishes other important features of the magma composition, such as very low K_2O (0.05 wt. %) and P_2O_5 (0.03 wt. %) abundances and a high 100 $(Mg/Mg+Fe^{+2})$ of 69.3 (assuming $Fe^{+3}/Fe^{+2} + Fe^{+3} = 0.1$). Because of the high MgO and CaO abundances which accompany the very low LIL element content, Melson et al. classify such basalts as MgCa type. Similar basalts occur at DSDP Leg 3 sites 14 and 18 in the south Atlantic. Because these Atlantic basalts are rich in Cr and Ni, and have high $Mg/Mg+Fe^{+2}$ Frey et al. (1974) proposed them as the best candidates for primary magma recovered from the ocean floor; i.e., their geochemical features are those expected for liquids that have undergone little fractionation since formation by partial melting of peridotite.

Our trace element analyses of site 212 basalts (Table 6b) confirm that these basalts, like those at sites 3-14 and 3-18, have high Cr and moderately high Ni abundances (Figure 5d), and very low Zr and REE abundances (Figures 4f and 5d). Some LIL elements in these very altered crystalline rocks, for example, K, Rb and Ba, have high abundances because of the extensive post-crystallization chemical changes and formation of new phases, such as potassium feldspar. We also suspect that alteration has significantly increased light REE abundances, so that site 212 metabasalts have nearly uniform REE enrichment relative to chondrites (Figure 4f). Because site 212 glass has a MgCa type composition, and the crystalline metabasalts have low REE and Zr contents but high Ni and Cr abundances, we believe site 212 magma, like that at Atlantic sites 3-14 and 3-18, experienced very little fractional crystallization after formation by partial melting. As at site 213, the geochemical nature of site 212 basalts is compatible with formation at a spreading ridge axis as suggested by Sclater and Fisher (1974).

Site 211 (09°46.53'S, 102°41.95'E)

Site 211 is bounded to the north by the Java Trench, to the west by the Nicobar Fan (an extension of the Bengal Fan), to the south by the Cocos-Keeling Plateau and to the east by Christmas Island (Figures 1,2). The site (5525 m water depth) is 2.5° south of a magnetic anomaly identified as 33 (Sclater and Fisher, 1974), and the basement-sediment contact age is about 76×10^6 yrs. Thus, the data are consistent with a continued southward spreading half rate of ∿6 cm/yr established on the basis of anomalies 29-33 to the north. A pronounced topographic hill occurs 18 km south of site 211, and other shallow areas are in the vicinity. Thus, Sclater et al. (1974) concluded the basement rocks "may be related to the northeast-trending Cocos-Keeling-Christmas Island complex, the origin of which is unknown."

An intrusive 10 m thick diabase sill (Ar^{40}/Ar^{39} age of 71 ±2 m.y., MacDougall, 1974) occurs 18 m above the presumed extrusive basement basalt. This sill consists of phaneritic holocrystalline rock with plagioclase and clinopyroxene in ophitic and subophitic texture. The diabase is nepheline normative and its alkalic nature is reflected by accessory biotite, high TiO_2 (>2 wt. %) and Na_2O + K_2O contents (typically >5 wt. %, Table 7a). However, the diabase is relatively unfractionated compared to basalts from Wharton Basin sites 213 and 256; for example, except near the lower contact, the diabase has FeO*/MgO ∿1 and Ni >125 ppm. The fine-grained basalt at the lower sediment contact was interpreted as a weathered chilled margin (Hekinian, 1974a). The much higher CaO, but lower MgO and Ni content of this basalt (Table 7a,b), may indicate that two geochemically distinct units are present in the sill.

The alkalic nature of the sill is not a result of alteration phenomena because unaltered minerals clearly indicate an alkalic composition. For example, zoned plagioclase ranges in composition from An_{55} cores to calcic anorthoclase overgrowths (Table 7c and Figure 7b) and the clinopyroxenes are calcic, aluminous titanaugites (Table 7c) which differ markedly from clinopyroxenes at sites 254, 256, 257 and 261 (Figure 7a). Also, the presence of biotite is very unusual in submarine basalts. The high Sr and Ba abundances (Figure 6, Table 7b), and a large degree of light REE enrichment (Figure 4h) further demonstrate the similarity of the diabase to alkali olivine basalts occurring on some seamounts and oceanic islands, such as nearby Christmas Island.

The 8-9 meters of presumed basement basalt at the bottom of the hole contain abundant amphibole which Hekinian (1974a) interpreted as primary amphibole, crystallized from a hydrous magma. Based on grain size variations, several basaltic flows appear to be present. All site 211 basalts are quite altered, and amphibole-rich (>50 vol. %) samples are unusually H_2O rich (>11 wt. % ignition loss for f in Table 7a). There is considerable variation in major element

TABLE 7a

Site 211: Major Element Composition (volatile-free, wt.%)

	Diabase Sill					Amphibole-Bearing Basalts					Christmas Island
	a	b	c	d	e	f	g	h	i	j	k
SiO_2	48.66	48.55	46.36	44.98	45.73	42.39	45.55	45.55	51.18	48.86	51.97
Al_2O_3	15.61	15.95	15.30	15.75	15.59	20.09	18.40	18.03	18.30	17.23	16.52
FeO*	10.43	11.05	10.95	9.27	10.26	11.03	9.67	9.79	9.74	8.73	9.97
MnO	0.16	-	-	-	-	-	-	-	-	0.13	0.08
MgO	11.46	10.19	10.10	4.40	5.08	7.18	4.68	5.99	2.45	7.48	4.79
CaO	6.62	6.17	6.11	12.54	11.20	9.70	9.73	8.42	7.08	7.66	7.43
Na_2O	3.02	3.39	3.84	3.75	3.59	3.17	3.60	4.28	3.90	3.64	4.20
K_2O	1.41	2.18	2.00	3.10	2.58	1.21	2.24	1.91	3.66	3.04	2.65
TiO_2	2.06	2.42	2.37	2.39	2.39	3.18	2.79	2.71	2.84	2.45	2.13
P_2O_5	-	0.64	0.60	0.61	0.62	0.84	0.77	0.74	0.80	-	-
Total	99.44	100.54	97.63	96.79	97.04	98.79	97.43	97.42	99.95	99.22	99.74
H_2O^+	4.17	5.54	6.72	4.46	3.60	11.43	7.72	6.13	5.96	3.62	1.16
H_2O^-	1.50	(I.L.)	(I.L.)	(I.L.)	(I.L.)	(I.L.)	(I.L.)	(I.L.)	(I.L.)	1.35	1.30
CO_2	0.14	-	-	4.90	2.55	0.97	2.38	0.44	1.76	1.95	-
Fe_2O_3/FeO	0.53	0.65	0.44	2.91	2.29	0.38	0.82	0.35	0.23	7.11	0.68
FeO*/MgO	0.91	1.08	1.08	2.10	2.02	1.54	2.01	1.63	3.98	1.17	2.08

a) 211-12-1:23-25, coarse diabase (this work)
b) 211-12-1:64-66, coarse diabase (Hekinian, 1974)
c) 211-12-1:143-145, coarse diabase (Hekinian, 1974)
d) 211-12-2:106-109, fine-grained basalt (chill zone?) (Hekinian, 1974)
e) 211-12-2:114, fine-grained basalt (chill zone?) (Hekinian, 1974)
f) 211-15-1:130-134, >70% amphibole (Hekinian, 1974)
g) 211-15-2:0-5 (Hekinian, 1974)
h) 211-15-2:76-82 (Hekinian, 1974)
i) 211-15-2:90-97 (Hekinian, 1974)
j) 211-15-4:70-73 (this work)
k) Christmas Island trachybasalt, Table 3, Smith and Mountain, 1925.

TABLE 7b

Site 211: Trace Element Abundances (ppm)[a]

	Diabase			Amphibole-Bearing Basalt			
	12-1:23-25	12-1:143-145	12-2:100-102[b]	15-2:0.5	15-2:14-16	15-3:40-45	15-4:70-73
B	46	–	27	–	74	75	30
Li	22	–	53	–	110	44	95
Rb	38	–	55	–	60	80	85
Sc	21	–	–	–	–	–	18
V	175	140	205	115	200	155	185
Cr	335	275	270	70	73	40	70
Co	30	69	24	36	15	16	21
Ni	135	127	90	75	95	67	95
Cu	35	31	47	21	29	29	22
Sr	540	–	490	–	230	1100	625
Ba	460	–	395	–	140	800	650
Y	26	–	27	–	38	39	34
Zr	165	–	150	–	315	330	255
La	47.5	–	–	–	–	–	32.8
Ce	74	–	–	–	–	–	82
Nd	28	–	–	–	–	–	37
Sm	6.12	–	–	–	–	–	6.76
Eu	1.95	–	–	–	–	–	2.19
Tb	1.0	–	–	–	–	–	1.0
Ho	1.1	–	–	–	–	–	0.85
Yb	2.6	–	–	–	–	–	2.8
Lu	0.29	–	–	–	–	–	0.37

a) samples 12-1:143-145 and 15-2:0.5 from Bougault, 1974; all others are this work.
b) sample from fine-grained basalt near lower contact of sill.

TABLE 7c

Site 211 (core 12-1:23-25cm): Mineral Compositions (wt.%)

	Plagioclase			Pyroxene			
	A	B	C	A	B	C	D
SiO_2	65.4	61.7	55.8	45.8	48.3	47.6	46.4
TiO_2	0.04	0.07	0.07	3.17	2.77	2.61	2.73
Al_2O_3	20.9	23.0	28.0	6.54	5.93	4.83	6.49
FeO	0.24	0.32	0.35	8.34	8.27	8.13	7.75
MnO	-	-	-	0.15	0.13	0.16	0.11
MgO	0.03	0.18	0.05	12.4	12.8	13.3	13.1
CaO	1.76	4.02	9.71	22.5	22.2	22.4	22.5
Na_2O	8.21	8.05	5.74	0.59	0.51	0.52	0.51
K_2O	3.46	1.63	0.46	0.01	-	0.01	-
Cr_2O_3	-	-	-	0.01	0.02	-	0.08
	100.04	98.97	100.18	99.51	100.93	99.56	99.67

Cation Sum
(12 O for plag, 4.990   5.002   5.003   4.041   4.011   4.036   4.041
6 O for pyx)

A. Calcic anorthoclase overgrowth on zoned phenocryst
B. Potassic oligoclase inner zone of phenocryst
C. Calcic andesine core of phenocryst

A,B. Titanaugite phenocrysts
C.   Fractured titanaugite grain
C.   Large Titanaugite xenocryst

composition (f-j of Table 7a), but it is not possible to ascertain whether this variation reflects alteration effects or the presence of geochemically distinct flow units. Several samples are nepheline normative, and they all have the high TiO_2 (>2.1 wt. %) and total alkalies (>4.3 wt. %) expected of alkalic basalts. Compared to the overlying diabase sill, these basalts have higher FeO^*/MgO and Zr abundances (Figure 5d), but lower Sc, Cr, Co and Ni contents (Table 7b). As in the diabase, Sr, Ba, Zr and light REE in these basal basalts are markedly enriched relative to LIL element-depleted tholeiites (Figures 4h, 5b,d and 6). Among the Eastern Indian Ocean basalts we have studied, site 211 basalts have the highest Zr content and La/Yb.

The basalts of site 211 are akin to the alkali olivine basalt suite characteristic of many oceanic islands. In particular, Christmas Island, 3° to the east, is composed of an alkali-olivine basalt series characterized by high abundances of Na_2O, K_2O and TiO_2, and the common presence of "pinkish-buff" augite (titan-augite?). As an example, a Christmas Island trachy-basalt (Smith and Mountain, 1925) has a similar major element composition to the lower, MgO-poor basalts at site 211 (Table 7a).

Thus, neither the sill or underlying basalts at site 211 have chemical compositions expected in a spreading ridge environment. As implied by Hekinian (1974a) and Sclater et al. (1974), the basalts at this site are almost certainly related to Cocos-Keeling and Christmas shoal areas which extend from 95°E to the Java Trench in a trend slightly north of due east. The Christmas Island rocks are interbedded with limestones, and Smith and Mountain (1925) concluded that the Upper Volcanic series is Miocene age or younger, and the Lower Volcanic series is Eocene or older. Although the volcanic activity certainly postdates that of seafloor spreading, the tectonic origin of this complex is poorly understood.

Site 256 (23°27.35'S, 100°46.46'E)

Site 256 is located in an area of abyssal hills (5361 m water depth) in the southern Wharton Basin, west of the Wallaby Plateau which extend to about 103°E (Figures 1,2). Just south of the site there is a 6000 m deep trough oriented in a northeast-southwest direction. On the basis of nannofossils, the minimum basement age is 102×10^6 yrs. Luyendyk and Davies (Fig. 18, 1974) have shown that the postulated basement ages of site 212 (95×10^6 yrs) and site 256 (102×10^6 yrs) are in accord with northern Wharton Basin magnetic anomalies (east of the Investigator Fracture Zone) which increase in age to the south (Figure 2). Thus, sites 212 and 256 are consistent with this portion of the southern Wharton Basin forming by southerly spreading at a half rate of about 6 cm/yr.

Site 256 basalt recovery was 12 m from a penetration of 19 m. The basalt core is a sequence of flows with each flow grading downwards from fine to coarser-grained basalt. Plagioclase laths and clinopyroxene clusters are abundant along with glass and microphenocrysts of magnetite. Olivine is rare (Kempe, 1974). The basalts have textural and mineralogical features typical of submarine pillow lavas. For example, common features are conspicuous strain shadows in pyroxene, tending toward an hourglass form, and traces of sector-zoning in plagioclase. Plagioclase is zoned (Table 8c) from calcic cores (An_{67-70}) to more sodic rims (An_{55-60}). The relatively calcic plagioclase distinguishes site 256 basalts from more alkalic basalts such as at sites 211, 215 and 254 (Figure 7b).

Eleven site 256 basalts are remarkably similar in composition (Tables 8a,b) except for one sample highly enriched in Cu. Because of the homogeneous composition, systematic differences between our analyses and those of Kempe (1974) and Fleet et al. (1976) are apparent (Tables 8a,b). For example, our analyses are higher in MgO and Ce but lower in TiO_2, Ni, Sr, La, Y and Zr. However, these discrepancies do not affect the major observations that the FeO* and TiO_2-rich site 256 basalts (Figures 3a,b) are examples of FETI basalts (Melson et al., 1975), and that relative to spreading ridge basalts, site 256 basalts are anomalously rich in Sr, Ba, Y, Zr and light REE (Table 8b and Figures 4g, 5a,b,d, and 6). In fact, almost all key major and LIL trace element characteristics of site 256 basalts are within the range observed for site 214 and 216 basalts (cf. Tables 1a,b, and 8a,b, and Figures 3a,b, 4g, 5a,b, c, and 6). However, some transition metals (e.g., Cr, Ni and Cu) are more abundant in site 256 basalts (Figure 5d). This is consistent with the rare presence of olivine in site 256 basalts, whereas olivine is absent in site 214 and 216 basalts which are apparently slightly more fractionated. As at sites 214 and 216, site 256 basalts are akin to the ferrotholeiites of oceanic island tholeiitic basalt sequences. The similar light REE enrichment (relative to chondrites) of site 214, 216, 256 basalts and oceanic island tholeiites (Figure 4a, g) distinguishes these basalts from the Fe-Ti-rich basalts found at some spreading ridges such as the Juan de Fuca Ridge (Kay et al., 1970).

The compositional similarity of site 214, 216 and 256 basalts strongly suggests that a similar mantle composition was parental to these basalts, and further, that similar processes of fractionation led to the ferrotholeiite composition. Because basalts dredged from axial valleys of spreading ridges have distinctly different trace element abundances, it seems unlikely that site 256 basalts formed at a spreading ridge axis. Since it is reasonable to expect some correlation of basaltic composition with tectonic environment, such as on the Ninetyeast Ridge, it is expected that site 256 should be topographically anomalous. Interestingly, in the site 256 description in the DSDP preliminary report (p. 306, Davies et al., 1974), the site location was described as "on the flank of a basement high which forms the northern lip of a deep trough, probably a fracture zone." This unique bathymetric location is particularly evident in the bathymetric profile along the track of Leg 26 (Davies et al., 1974). Furthermore, charts of Eastern Indian Ocean bathymetry (Markl, 1974; Sclater and Fisher, 1974) clearly show a north by northeast trend of topographic highs (<1500 m at 98°20'E, 29°S and <2000 m at 100°30'E and 25°40'S) extending from the vicinity of Broken Ridge through the site 256 location (Figures 1, 2). We suggest that site 256 basalts are related to the volcanism causing this linear feature, rather than to formation at a spreading axis in the northern Wharton Basin.

Southeast Wharton Basin

Two sites (257 and 259) are located near the southwest coast of Australia (Figure 2). Because it has been suggested (Markl, 1974) that basalts at these sites formed from a northeast-southwest oriented spreading ridge in the southern Wharton Basin, we consider these sites together.

Site 257: 39°59'S, 108°21'E
Site 259: 29°37'S, 112°42'E

TABLE 8a

Site 256: Major Element Composition (volatile-free, wt.%) (a)

	9-3:45-47	9-3:52-54	10-1:141-143	10-2:87-90	10-4:114-116	11-1:27-30	11-3:110-113
SiO_2	50.57	50.27	50.32	50.39	50.16	50.28	50.51
Al_2O_3	13.57	13.32	13.13	13.36	13.35	13.01	13.64
FeO*	13.22	13.18	13.54	13.46	12.88	13.23	13.11
MnO	0.23	0.25	0.21	0.18	0.24	0.21	0.18
MgO	6.18	7.07	6.70	6.22	6.86	6.64	6.38
CaO	10.64	10.10	9.74	10.04	10.47	10.38	9.86
Na_2O	2.71	2.66	2.95	2.76	2.82	2.65	2.75
K_2O	0.25	0.23	0.29	0.25	0.20	0.26	0.26
TiO_2	2.55	2.18	2.28	2.64	2.11	2.20	2.54
P_2O_5	0.25	-	-	0.23	-	-	0.27
Total	100.17	99.27	99.15	99.53	99.12	98.88	99.50
H_2O+	0.60	1.56	0.69	0.90	0.66	0.93	0.65
H_2O-	1.05	1.20	1.20	1.37	0.85	1.05	1.43
CO_2	0.28	0.37	0.10	0.97	0.22	0.24	0.39
Fe_2O_3/FeO	0.30	0.54	0.48	0.36	0.44	0.48	0.37
FeO*/MgO	2.14	1.86	2.02	2.16	1.88	1.99	2.05

a) 9-3:45-47, 10-2:87-90, 11-3:101-113 from Kempe, 1974; others are this work.

TABLE 8b

Site 256: Trace Element Abundances (ppm) [a]

	9-3: 15-17	9-3: 45-47	9-3: 52-54	9-3: 138-140	10-1: 141-143	10-2: 87-90	10-4: 62-64	10-4: 114-116	11-1: 27-30	11-3: 110-113	11-3: 148-150
B	<5	--	6	<5	<5	--	9	<5	17	--	10
Li	12	7	6	6	7	8	8	5	7	7	7
Sc	--	--	41	--	46	--	--	49	48	--	--
V	535	450	475	460	470	490	400	395	400	470	415
Cr	105	105	110	120	100	110	95	130	120	110	85
Co	32	--	37	44	35	--	38	32	33	--	40
Ni	80	130	85	83	85	110	87	87	100	100	78
Cu	>2000	17	200	240	225	12	220	210	215	12	215
Sr	140	165	125	140	135	160	160	135	145	160	120
Ba	53	25	35	45	50	40	42	30	47	25	42
Y	46	50	46	42	44	50	38	45	43	50	43
Zr	175	200	130	115	135	200	145	145	145	200	155
La	--	12.6	9.0	--	10.1	12.2	--	10.4	9.8	12.7	--
Ce	--	23.7	33	--	29	15.1	--	30	30	24.2	--
Nd	--	17.4	17	--	18	15.1	--	17	18	15.8	--
Sm	--	5.0	4.31	--	4.86	4.0	--	4.73	4.85	5.0	--
Eu	--	1.8	1.37	--	1.55	1.2	--	1.59	1.57	1.8	--
Tb	--	0.8	1.1	--	0.83	0.9	--	1.1	1.1	0.8	--
Ho	--	--	1.1	--	1.4	--	--	1.4	1.4	--	--
Yb	--	3.1	3.0	--	3.4	2.4	--	3.4	3.3	3.2	--
Lu	--	0.4	0.47	--	0.65	0.4	--	0.54	0.53	0.5	--

[a] Samples 9-3:45-47, 10-2:87-80 and 11-3:110-113 from Kempe, 1974 and Fleet et al., 1976; others are this work.

TABLE 8c

Site 256: Mineral Compositions (wt.%)

	Plagioclase			Pyroxene				
	A	B	C	A	B	C	D	E
SiO_2	53.4	51.2	53.09	51.3	51.4	52.4	52.3	50.97
TiO_2	0.11	0.06	–	0.93	0.72	0.74	0.75	1.03
Al_2O_3	29.1	30.1	28.24	2.16	2.18	2.45	2.16	2.93
FeO	0.88	0.68	–	12.7	10.4	7.95	8.92	10.51
MnO	0.02	0.04	–	0.32	0.26	0.18	0.20	–
MgO	0.18	0.22	–	16.4	15.9	16.7	17.2	16.74
CaO	12.7	13.9	12.27	15.6	18.4	19.5	18.8	16.76
Na_2O	4.54	3.69	4.50	0.27	0.31	0.27	0.22	0.31
K_2O	0.07	0.06	0.10	–	–	–	–	–
Cr_2O_3	–	–	–	0.02	0.12	0.38	0.21	0.33
	101.00	99.95	98.20	99.70	99.69	100.57	100.76	99.54
Cation Sum (12 O for plag, 6 O for pyx)	5.010	5.002	–	4.006	4.010	4.004	4.010	–

A. 11-1:27-30, labradorite, average of large-zoned phenocryst

B. 11-1:27-30, labradorite, unzoned phenocryst

C. average of 3, Kempe (1974).

A,B,C,D. augite from 11-1:27-30, A in groundmass, B,C are microphenocrysts, D is a phenocryst.

E. average of 3, Kempe (1974).

Site 257 is located in an abyssal hill area in the southeastern Wharton Basin, northeast of the Naturaliste Plateau (Figure 2). It was hoped that the basement age difference between sites 256 and 257 would provide evidence for the age gradient and spreading direction during the initial formation of the Indian Ocean. The inferred basement ages at Wharton Basin sites 212, 256 and 259 are consistent with an extrapolation from the southward increasing age progression established by the northern Wharton Basin east-west magnetic anomalies (Figure 2). At site 257, mid-Albian ($\sim 102 \times 10^6$ yrs) nannofossils and forminifera occur 13 m above the sharp apparently conformable sediment-basalt contact (Luyendyk and Davies, 1974). This minimum basement age (102×10^6 yrs) for site 257 falls below the extrapolated age trend (Figure 18, Luyendyk and Davies, 1974); however, Luyendyk and Davies (1974) cite evidence that a basement age of 130×10^6 yrs is more reasonable, and it is broadly consistent with the range of basalt K-Ar ages (Rundle et al., 1974). Thus, Luyendyk and Davies conclude that Wharton Basin ocean floor age increases systematically southward which is consistent with Wharton Basin formation by north-south spreading as suggested by Sclater and Fisher (1974).

Markl (1974) emphasized that although site 257 is 1100 km southeast of site 256, the oldest dateable sediments at each site have a similar age of about 100×10^6 yrs. If basement at site 257 is of comparable age to that at site 256, then basalt at these sites has not been formed by southerly spreading of the east-west ridge axis which formed the northern Wharton Basin. Markl (1974) identified early Cretaceous magnetic anomalies striking approximately northeast between the northern edge of the Naturaliste Plateau and the Australian continent. Magnetic age increases towards the continent where the oldest identified anomaly is about 127×10^6 yrs. Site 259 in the Perth Abyssal Plain lies within these anomalies (Figure 2) and its inferred basement age ($>110 \times 10^6$ yrs) is consistent with a half spreading rate of ~ 2.3 cm/yr. If it is postulated that basalt at site 257 formed from the same spreading ridge axis as site 259, a spreading rate of 2.1 cm/yr is required. Although the data density is poor, Markl (Figure 3, 1974) suggests, on the basis of these Mesozoic magnetic anomalies, that most of the southern Wharton Basin between the Ninetyeast Ridge and the Australian continent was formed by a northeast-southwest trending, late Jurassic-late Cretaceous spreading ridge. The extinct ridge axis is postulated to be midway between the similar age sites 256 and 257.

Veevers and Heirtzler (1974a) proposed a similar preliminary tectonic model for site 259, and by inference, site 257. They propose that $\sim 120 \times 10^6$ yrs. ago rifting began off southwest Australia. They deduce, like Markl (1974), that spreading initiated in a northeast-southwest direction (Figure 1e, Veevers and Heirtzler, 1974a). Evidence cited for this spreading direction is a) the northeast trending magnetic anomalies west of Perth, b) the northwest trending ridges south of the Wallaby Plateau which trend at right angles to these magnetic anomalies, c) the orientation of the Wallaby-Perth scarp, and d) the southeast and northeast borders of the Cuvier Abyssal Plain.

Although magnetic anomalies in the southern Wharton Basin are not sufficiently well understood to construct a detailed history of seafloor spreading in this region, there is a consensus that site 257 and 259 basalts formed at a spreading ridge axis. According to Markl (1974) and Veevers and Heirtzler (1974a), basalt at sites 257 and 259 formed on the southeast side of a ridge in the southern Wharton Basin, whereas Luyendyk and Davies (1974) and Sclater and

Fisher (1974) inferred that site 257 basalts formed, like site 212 basalts, from an east-west oriented ridge in the northern Wharton Basin.

At site 257 basalt recovery was 32 m from a penetration of 64.5 m. The basalt sequence alternates between medium and fine-grained vesicular basalt, and at least 7 or 8 flows are present (Kempe, 1974). Especially in the upper portions, the basaltic core is highly fractured. Kempe (1974) has interpreted the sequence to be a series of flows with the upper portion of each flow consisting of fine-grained, highly vesicular glassy or subvariolitic basalt, whereas the more abundant lower portions are coarser and often porphyritic.

The phenocrysts are zoned plagioclase (An 48-89), clinopyroxene and olivine (as pseudomorphs). The low orthoclase component in the plagioclase (Table 9c, Figure 7b), and the occurrence of calcic plagioclase xenocrysts are typical of LIL element-depleted tholeiitic basalts. Although there is a wide range in compositions from groundmass to phenocryst pyroxene (Table 9c), all site 257 pyroxenes have low TiO_2 and Al_2O_3 contents in the range established by pyroxenes from LIL element-depleted tholeiites (Figure 7a).

Kempe (1974) found considerable compositional variation in site 257 basalts, and he concluded that Ca/Ca+Na and Fe/Fe+Mg variations correlated with depth in the core. Our analyses, coupled with those of Kempe (1974) and Fleet et al. (1976) (Tables 9a,b), indicate the presence of three compositional groups (A,B,C), but we find no systematic trend with depth. These groups are recognized on the basis of FeO*/MgO and abundances of CaO, TiO_2, Cr, Ni, Zr and REE (Tables 9a,b). (Note, that in comparing Zr abundances of site 257 basalts, we utilize only our abundance data because we believe that the common value of 100 ppm reported by Kempe (1974) is not accurate.) Group A (core 11-2 through 14-2) is the least fractionated group in terms of major elements (e.g., FeO*/MgO <0.93), and it is clearly distinguished by its relatively high Cr and Ni abundance coupled with low Zr and heavy REE abundance (Figure 5d, Table 9b). Fleet et al. (1976) found relative light REE enrichment in a sample from core 11-2. Possibly, the relatively higher abundances of Ti, Sr, Ba, La and Ce in this sample indicate that Group A should be subdivided. The occurrence of basalt with light REE enrichment in the upper part of this core has important implications and more detailed work on this core is in progress. Group B (core 14-4 and 14-5) is the most differentiated group in terms of major (Table 9a) and trace elements (Figure 5d), and Group C (cores 15-1 through 17-5) is intermediate between Groups A and B (Tables 9a,b; Figure 5d).

Between Groups B and C at the top of core 15 (37 m into the basalt) there is a clay layer (60 cm) with consolidated, graded sand and subrounded basalt pebbles. Kempe (1974) and Luyendyk and Davies (1974) concluded that a significant intrabasalt hiatus occurred. According to K-Ar dating (Rundle et al., 1974), the upper basalts at site 257 are $\sim 100 \times 10^6$ yrs. old (in agreement with paleontologic estimates), and the lower basalts beneath the interbasaltic sediment layer are older than 150×10^6 yrs. Of course, the accuracy of K-Ar dates in these moderately altered basalts is questionable.

The important feature of all site 257 basalts (except 11-2: 82-84) is that they have the geochemical characteristics of LIL element-depleted tholeiites. This is evident from their low Ba, Zr and TiO_2 abundances (Tables 9a,b; Figures 5d, 6) and their relative depletion in light REE (Figures 4i, 5b). Furthermore, if only the least altered basalts (H_2O + <1 wt. % and Fe_2O_3/FeO <0.38) are considered, the K_2O content is ≤ 0.12 wt. %.

At site 259, 20 m of basalt were recovered from a penetration

TABLE 9a

Site 257: Major Element Composition (volatile-free, wt.%) [a]

	Group A					Group B					Group C				
	11-2: 82-84	11-3: 25-27	12-2: 79-81	13-3: 43-44	14-2: 78-80	14-4: 124-126	14-5: 25-27	15-1B: 126-128	15-2: 37-39	15-2: 122-124	16-2: 94-96	16-3: 31-33	17-1: 91-93	17-3: 26-28	17-5: 65-67
SiO_2	51.36	50.95	51.28	50.72	50.99	51.20	50.34	51.15	50.46	50.52	50.61	51.04	50.24	50.92	51.42
Al_2O_3	16.97	16.07	15.85	14.98	15.92	14.95	15.25	16.36	15.44	16.45	17.72	15.82	15.00	15.22	14.71
FeO*	6.78	8.36	7.70	8.32	7.42	10.24	11.24	8.17	9.00	9.34	8.00	8.26	8.89	8.54	9.61
MnO	0.29	0.35	0.21	0.25	0.17	0.19	0.25	0.17	0.26	0.23	0.16	0.19	0.16	0.18	0.17
MgO	8.24	10.50	9.58	8.96	9.55	7.55	7.79	8.01	9.01	8.24	7.30	8.94	8.46	8.81	8.20
CaO	12.15	9.81	12.18	12.95	12.66	11.88	10.76	12.87	12.49	11.36	13.25	12.27	13.00	13.08	12.46
Na_2O	2.47	2.38	2.34	2.12	2.35	2.09	2.42	2.12	1.91	2.23	2.04	2.33	2.14	2.11	2.06
K_2O	0.12	1.09	0.25	0.05	0.43	0.06	0.49	0.04	0.11	0.76	0.04	0.07	0.47	0.11	0.08
TiO_2	1.23	0.73	0.90	0.85	0.91	1.01	0.96	0.90	0.90	0.93	0.85	0.79	0.92	0.97	0.94
P_2O_5	0.21	—	0.10	0.10	0.10	0.10	—	0.07	0.08	0.09	0.08	—	0.12	0.10	0.09
Total	99.82	100.28	100.39	99.30	100.50	99.27	99.52	99.86	99.66	100.15	100.05	99.71	99.40	100.04	99.74
H_2O+	0.89	2.77	1.07	0.74	1.08	0.48	1.07	0.40	0.56	1.43	0.65	0.90	0.88	0.84	0.63
H_2O-	1.99	1.95	2.74	1.59	2.94	0.73	1.22	1.26	1.55	2.53	0.90	1.30	2.35	1.68	0.84
CO_2	0.36	0.57	0.41	1.05	1.26	0.32	0.10	0.55	0.42	0.42	0.23	0.19	1.29	0.77	0.24
Fe_2O_3/FeO	0.34	1.78	0.66	0.36	0.66	0.21	1.03	0.34	0.36	1.16	0.38	0.99	0.72	0.27	0.21
FeO*/MgO	0.82	0.80	0.80	0.93	0.78	1.36	1.44	1.02	1.00	1.13	1.10	0.92	1.05	0.97	1.17

a) samples 11-3:25-27, 14-5:25-27 and 16.3:31-33 from this work; others from Kempe, 1974.

of 38.5 m. The upper basalt core is altered breccia which grades downward into dark, fine-grained, sparsely porphyritic basalt (Robinson and Whitford, 1974). Plagioclase (An 55-60) is the dominant phenocryst, but clinopyroxene occurs as rare microphenocrysts. Altered olivine is rare. This basalt sequence is believed to represent basement of Early Cretaceous age ($\sim$112 x 10^6 yrs. based on age of overlying sediments).

Although site 259 basalts are moderately altered (1.6 - 3 wt. % loss on ignition), the major and trace element abundances (Tables 10a,b) clearly indicate that these basalts are within the compositional range of basalts from mid-oceanic ridges. Their LIL element-depleted nature is demonstrated by low Sr, Ba and Zr contents (Figures 5b,d, 6), and a relative depletion in light REE (Figure 4i).

Because basalts from sites 257 and 259 may have originated from the same ridge axis, it is of interest to compare basalts from these sites. Although there are some compositional differences (CaO and TiO_2), site 259 basalts are similar in FeO*/MgO, Al_2O_3 and trace element abundances to the most differentiated site 257 basalts, that is, Group B (core 14-4 and 14-5). This similarity is particularly evident in their REE distributions (Figure 4i). The compositional differences between site 257 and 259 basalts are smaller than differences commonly found in basalts from a single dredge haul (e.g., Rhodes and Rogers, 1972); therefore, the geochemical data are consistent with site 257 and 259 basalts originating at a common spreading axis as inferred by Markl (1974).

Northeast Wharton Basin

Two sites (260 and 261) are located near the northwest coast of Australia (Figure 2). Because it has been suggested (Veevers and Heirtzler, 1974a) that basement basalts at these sites formed from a northeast-southwest spreading ridge off northwestern Australia, we

TABLE 9b

Site 257: Trace Element Abundances (ppm) [a]

	A					B						C								
	11-2: 82-84	11-3: 25-27	12-2: 79-81	13-3: 43-44	14-2: 78-80	14-4: 124-126	14-5: 25-27	15-1B: 126-128	15-2: 37-39	15-2: 122-124	16-1: 65-67	16-2 94-96	16-3: 31-33	17-1: 70-72	17-1: 91-93	17-1: 97-99	17-3 26-28	17-5: 65-67	17-5: 67-79
B	--	25	--	--	--	--	30	--	--	--	13	--	<5	75	--	<5	--	--	<5
Li	18	43	16	10	12	7	8	7	7	8	23	7	7	11	13	10	13	9	6
Sc	--	47	--	--	--	--	44	--	--	--	--	--	47	--	--	--	--	--	--
V	300	280	310	310	330	330	280	310	310	310	275	290	240	245	290	255	290	330	235
Cr	390	670	200	300	335	165	160	290	325	360	460	350	435	195	220	225	220	200	200
Co	--	43	--	--	--	--	43	--	--	--	53	--	40	27	--	41	--	--	34
Ni	120	140	180	140	150	115	97	130	130	130	130	115	115	88	150	125	160	100	95
Cu	17	40	75	38	50	15	140	15	25	95	42	17	135	88	88	50	75	12	120
Sr	170	85	85	75	100	70	70	65	70	70	95	75	87	85	85	85	85	75	68
Ba	25	105	40	40	150	25	13	10	≤10	≤10	46	40	16	28	≤10	29	≤10	25	23
Y	40	24	50	50	25	50	27	50	40	25	23	50	26	23	25	27	25	50	25
Zr	150	28	100	100	100	100	45	100	100	50	47	100	30	40	75	38	100	100	40
La	6.9	1.44	1.2	2.0	--	2.7	1.58	1.9	--	2.3	--	1.2	1.41	--	--	--	--	2.6	--
Ce	13.8	6.8	2.7	5.1	--	2.7	6.0	2.3	--	2.3	--	4.3	4.9	--	--	--	--	4.9	--
Nd	11.9	4.5	4.8	7.2	--	8.7	5.8	5.6	--	11.8	--	4.9	4.9	--	--	--	--	4.0	--
Sm	3.4	1.68	1.4	1.8	--	1.9	2.09	1.7	--	1.8	--	1.9	1.66	--	--	--	--	1.8	--
Eu	1.2	0.64	0.5	0.7	--	0.7	0.79	0.5	--	0.7	--	0.6	0.71	--	--	--	--	0.7	--
Tb	0.6	0.46	0.4	0.4	--	0.5	0.60	0.4	--	0.2	--	0.4	0.36	--	--	--	--	0.4	--
Ho	--	0.68	--	--	--	--	0.93	--	--	--	--	--	0.79	--	--	--	--	--	--
Yb	2.1	2.0	1.6	1.9	--	2.2	2.9	2.2	--	2.0	--	2.0	2.5	--	--	--	--	2.0	--
Lu	0.2	0.43	0.2	0.4	--	0.3	0.48	0.3	--	0.3	--	0.3	0.42	--	--	--	--	0.3	--

[a] Samples 11-3:25-27, 14-5:25-27, 16-1:65-67, 16-3:31-33, 17-1:70-72, 17-1:97-99, 17-5:67-69 from this work; others from Kempe (1974), and Fleet et al. (1976).

consider these sites together.
<u>Site 260</u>: 16°9'S, 110°18'E
<u>Site 261</u>: 12°57'S, 117°54'E

Site 260 is in the Gascoyne Abyssal Plain and site 261 is on the north side of the Argo Abyssal Plain (Figure 2). In a preliminary discussion of the age of eastern Indian Ocean DSDP drilling sites, Heirtzler et al. (1973) noted the importance of sites 260 and 261 to an understanding of Mesozoic seafloor spreading which caused the breakup of Gondwanaland. In the northeast Wharton Basin, Falvey (1972) had identified nearly northeast trending (∿60°) magnetic anomalies. Utilizing this trend and other data, Veevers and Heirtzler (1974a) proposed that seafloor spreading near northwestern Australia began ∿160 x 10^6 years ago by the initiation of a spreading ridge near the continental margin along the southern edge of the Tethys Ocean. According to their model, basalt from DSDP sites 260 and 261 originated from this northeast-southwest oriented ridge axis (Figures 1a,b, Veevers, and Heirtzler, 1974a).

Larson (1975) has correlated the magnetic anomaly lineations associated with this spreading ridge axis in the Argo Abyssal Plain with Late Jurassic magnetic reversals giving rise to anomalies M-22 to M-25 (Figure 2). This correlation is in good agreement with the paleontological inferred basement age (152 x 10^6 yrs.) of site 261. Thus, there is strong evidence for suggesting that site 261 basalt formed at this Late Jurassic spreading center.

Magnetic lineations just south of site 260 may be related to the northeast trending anomalies in the Argo and Gascoyne Abyssal Plains

TABLE 9c
Site 257: Mineral Compositions (wt.%)

	Plagioclase*				Pyroxene								
	A	B	C	D	A	B	C	D	E	F	G	H	I
SiO_2	54.1	51.7	49.1	45.7	50.8	51.5	51.25	53.10	52.25	51.98	50.67	52.92	53.06
TiO_2	0.07	0.07	0.02	–	0.59	0.51	0.64	0.17	0.32	0.51	0.68	0.65	0.93
Al_2O_3	28.1	30.0	32.3	33.9	3.05	2.64	3.34	2.09	2.74	3.43	4.24	2.25	3.14
FeO	1.22	0.91	0.53	0.39	9.77	8.67	5.30	4.96	7.19	6.30	7.06	13.73	6.46
MnO	–	0.01	0.01	0.02	0.21	0.22	–	–	–	–	–	–	–
MgO	0.22	0.36	0.34	0.25	15.8	18.0	17.30	18.62	17.55	18.04	15.94	16.22	17.46
CaO	11.6	13.7	15.7	18.0	18.8	17.3	20.01	20.09	18.52	18.17	20.23	15.08	19.92
Na_2O	4.79	3.70	2.57	1.34	0.18	0.18	0.18	–	–	0.18	0.22	0.13	0.18
K_2O	0.05	0.02	–	–	–	–	–	–	–	–	–	–	–
Cr_2O_3	–	–	–	–	–	0.07	0.75	0.53	–	0.36	0.33	–	0.19
	100.15	100.47	100.57	99.60	99.2	99.09	98.77	99.56	98.57	98.97	99.37	100.98	101.34
Cation Sum (12 O for plag, 6 O for pyx)	4.999	5.002	5.004	5.010	4.014	4.010	–	–	–	–	–	–	–

* Partial analyses of plagioclase groundmass, phenocryst cores and rims (An 63-91) from 5 other site 257 basalts are in Table 12 of Kempe, 1974

A. 16-3:31-33, labradorite groundmass microlite
B. 16-3:31-33, labradorite microphenocryst
C. 16-3:31-33, bytownite rim on large xenocryst
D. 16-3:31-33, calcic bytownite, core of xenocryst

A,B. 16-3:31-33, groundmass augite
C. 11-2, avg. of 2, Kempe, 1974
D. 14-4, phenocryst, Kempe, 1974
E. 14-4, groundmass, Kempe, 1974
F. 15-1B, avg. of 2, Kempe, 1974
G. 16-2, phenocryst, avg. of 2, Kempe, 1974
H. 16-2, groundmass, Kempe, 1974
I. 17-5, avg. of 3, Kempe, 1974

(p. 341, Veevers and Heirtzler, 1974b). The magnetic anomalies suggest that oceanic basement at site 260 is slightly older than at site 261, and thus, Veevers and Heirtzler (1974a) proposed an age of 155×10^6 yrs. for basement basalt at site 260. Therefore, the basaltic sill at site 260 which is overlain by mid-Albian ($10^5 \times 10^6$ yrs.) sediments must be underlain by considerable amounts of sediment.

At site 260, ∼0.5 m of core were recovered from a penetration of 9 m. The basalt is believed to be a sill because of 1) the relatively young age and red discoloration of the overlying greenish Albian sediments, and 2) the fresh, medium-grained, glass-free, unbrecciated nature of the basalt core (Robinson and Whitford, 1974). The basalts are sparsely porphyritic. Zoned laths and microphenocrysts of plagioclase are abundant along with anhedral and rare microphenocrysts of clinopyroxene and groups of small magnetite octahedra. Olivine, or its pseudomorphs, have not been positively identified (Robinson and Whitford, 1974).

Site 260 basalts are quite variable in composition (Tables 10a, b). Basalt from core 18-2 has the highest compatible trace element (Cr, Co, Ni) abundances, and the lowest FeO*/MgO, but it also has the highest TiO_2 and LIL element abundances (Tables 10a,b). More detailed study of site 260 basalts may reveal the presence of more than one sill. Important geochemical characteristics of each site 260 basalt are relatively high Ba and Zr contents (compared to LIL element-depleted tholeiites, Figures 5d, 6), and a nearly chondritic relative REE distribution (Figure 4j). In petrographical and geochemical respects, these basalts are intermediate between LIL element-depleted tholeiites and island tholeiites.

At site 261, 21 m of basalt were recovered from 47 m of penetration. Robinson and Whitford (1974) divided the basalt core into three units: Unit C, a basal pillow basalt breccia complex, is interpreted as basement; Unit B, the middle sequence is a 3 m section of fine-grained, highly altered and fractionated basalt; Unit A, the upper unit, is a 10 m-thick, coarse-grained sill with a steep intrusive contact with baked and discolored sediments. Upper Jurassic (Oxfordian, $\sim 152 \times 10^6$ yrs.) calcareous nannoplankton in these sediments establish that these oceanic basement rocks are the oldest recovered from the Indian Ocean.

In addition to basalt, Unit C contains abundant veins and irregular masses of calcite along with bright green, waxy smectite and irregular masses of hematite (Frontispiece, Veevers et al., 1974c). The basalts are fine-grained and nonporphyritic olivine tholeiites. Plagioclase and clinopyroxene are the dominant minerals. Basalt from core 37-1 is nepheline normative (Robinson and Whitford, 1974), but this is almost certainly a result of alteration (Table 10a). The low LIL element abundances (Table 10b, Figures 5b,d, 6) and a relative light REE depletion (Figure 4k), indicate that this basement basalt is akin to LIL element-depleted tholeiites.

The altered basalt of Unit B is cut by numerous calcite veins. It is nonporphyritic and consists of a fine-grained mixture of plagioclase, clinopyroxene, ilmenite and olivine (Robinson and Whitford, 1974). Although on a normative basis Unit B basalts are also olivine tholeiites, these basalts differ markedly in composition from Unit C basalts. Notable, are the very high TiO_2 content (>3.3 wt. %) and the high abundances of FeO^*, P_2O_5 and total alkalies in Unit B basalts (Table 10a). High abundances of LIL trace elements (Table 10b, Figures 5b,d, 6) also indicate that this unit is compositionally distinct from Units A and C. The presence of olivine and only a moderately high FeO^*/MgO (Table 10a) indicate this magma was not highly fractionated with respect to olivine; however, the very low Cr abundance (Table 10b) and only a slight light REE depletion (Figure 4k) are consistent with a LIL element-depleted tholeiitic magma which has undergone extensive clinopyroxene ± spinel fractionation.

The relatively fresh basalt of Unit A varies considerably in grain size and texture. "It is typically slightly glomeroporphyritic with clusters of plagioclase, augite and olivine crystals in a fine-grained variolitic groundmass" (Robinson and Whitford, 1974). In other areas, patches of fine plagioclase-augite intergrowths alternate with much coarser patches (~ 3 mm) of subophitic plagioclase-augite intergrowths. On the average, clinopyroxene composes about 1/3 and plagioclase 2/3 of the basalt. Plagioclase ranges from An_{52} to An_{70}, and it contains very little K_2O (Table 10c, Figure 7b). Similar to clinopyroxenes from LIL element-depleted tholeiites (Frey et al., 1974), Unit A pyroxenes are relatively low in TiO_2 at moderate Al_2O_3 levels (Table 10c, Figure 7a). Mineralogically and texturally, this unit resembles DSDP basalts from Leg 3, site 15 and Leg 2, site 11A in the Atlantic. Such augite-rich basalts are more common in eastern Indian Ocean DSDP cores than in the Atlantic Ocean DSDP cores and dredge hauls. None of the site 261 basalts are mineralogically similar to Atlantic Ocean basalt of presumed similar Jurassic age at site 105 of Leg 11.

Except for one sample with a very low K_2O content, only small major element differences occur among Unit A basalts (Table 10a). TiO_2 content is nearly constant. Basalt from core 34-1 of Unit A is nearly identical (within experimental error) in all major and

TABLE 10a

Sites 259-260-261: Major Element Composition (volatile-free, wt.%)[a]

	259				260			261 (A)			261 (B)		261 (C)		
	37-2: 32-38	38-2: 116-129	39-1: 85-92	40-1: 140-145	18-2: 140-142	20 pc. 5	20-1: 16-18	33 cc pc. 2	34-1: 75-77	34-2: 142-150	35-3: 13-20	35-4: 87-97	37-1: 126-133	38-5: 110-120	39-1: 11-13
SiO_2	50.69	51.65	51.62	51.23	50.73	51.86	51.69	50.22	51.48	49.84	48.95	49.69	48.49	49.94	51.53
Al_2O_3	16.67	15.61	16.49	16.58	17.90	13.84	15.45	14.82	15.01	14.40	15.75	15.60	15.96	15.28	14.71
FeO*	10.48	10.03	8.90	9.38	10.04	12.25	11.63	10.58	9.09	10.52	11.47	10.49	8.93	9.45	9.28
MnO	0.13	0.19	0.16	0.18	0.29	0.20	0.27	0.21	0.18	0.22	0.43	0.15	0.22	0.25	0.18
MgO	7.70	7.19	7.91	8.38	7.36	5.84	7.55	7.15	8.52	7.41	7.74	6.91	7.08	7.60	7.87
CaO	7.82	10.04	9.52	8.42	6.52	10.25	7.28	11.33	10.80	12.03	6.34	7.09	13.09	11.84	11.18
Na_2O	1.89	2.34	3.12	2.70	3.98	2.72	3.38	2.16	2.85	2.16	3.32	3.64	3.16	2.19	2.81
K_2O	0.65	0.31	0.13	0.27	0.38	0.14	0.20	0.64	0.54	0.01	0.92	0.83	0.67	0.61	0.37
TiO_2	1.61	1.27	1.27	1.37	1.97	1.75	1.64	1.39	1.30	1.32	3.46	3.34	0.94	1.00	1.32
P_2O_5	0.22	0.13	0.09	0.15	–	0.13	–	0.07	–	0.19	0.33	0.29	0.17	0.08	–
Total	97.86	98.76	99.21	98.66	99.17	98.98	99.09	98.57	99.77	98.10	98.71	98.03	98.71	98.24	99.26
H_2O^+	2.65	1.59	2.41	2.91	1.80	1.15	1.53	1.70	1.08	0.27	3.10	2.25	3.05	1.68	1.01
H_2O^-	–	–	–	–	3.85	–	2.70	–	1.60	–	–	–	–	–	1.95
CO_2	–	–	–	–	0.39	–	0.18	–	0.28	–	–	–	–	–	0.45
Fe_2O_3/FeO	–	–	–	–	1.27	–	0.96	–	0.75	–	–	–	–	–	0.71
FeO*/MgO	1.36	1.39	1.13	1.12	1.36	2.10	1.54	1.48	1.07	1.42	1.48	1.52	1.26	1.24	1.18

a) samples 260-18-2:140-142, 260-20-1:16-18, 261-34-1:75-77, 261-39-1:11-13 from this work; others from Robinson and Whitford (1974).

247

TABLE 10b

Sites 259-260-261: Trace Element Abundances (ppm)[a]

	259			260			261[A]			261[B]			261[C]	
	38-1: 65-67	41-1: 101-104	Avg.	18-2: 140-142	20-1: 16-18	20 pc. 5	33-1: 101-105	34-1: 75-77	Avg.	35-2: 120-123	36-1: 60-63	Avg.	39-1: 11-13	Avg.
B	25	30		9	18		34	17		28	70		20	
Li	12	7		15	10		12	11		17	16		8	
Sc	-	-		66	57		-	52		-	-		52	
V	265	265		430	340		305	330		445	480		335	
Cr	125	170		90	72		145	170		10	7		150	
Co	52	15		110	36		22	50		13	22		41	
Ni	68	48		265	67		88	85		63	50		83	
C	65	40		75	60		95	65		55	70		45	
Sr	73	100		130	85		72	97		90	65		115	
Ba	7	8	6.1-9.3	60	30	45	<2	7	3.4	20	15	35	8	9.2-35
Y	34	46	22	46	46	-	41	46	-	70	60	-	48	
Zr	75	80	58	160	140	85	78	88	62	195	210	150	85	
La	-	-	1.8	7.7	6.0	4.3		2.34	1.6	-	-	6.0	2.45	1.3
Ce	-	-	6.7	20	17	12		11	5.5	-	-	18	9	4.2
Nd	-	-	5.3	13	9.7	8.3		7.1	5.0	-	-	15	7.8	3.6
Sm	-	-	2.0	4.03	3.75	2.8		3.11	2.1	-	-	5.3	3.16	1.6
Eu	-	-	0.76	1.32	1.21	0.98		1.03	0.79	-	-	1.8	1.17	0.61
Tb	-	-	0.53	-	-	0.70		0.73	0.55	-	-	1.1	0.79	0.42
Ho	-	-	0.95	1.3	1.7	1.2		1.5	0.99	-	-	1.8	1.5	0.67
Yb	-	-	2.8	3.7	4.1	3.7		3.4	2.8	-	-	5.1	3.9	2.0
Lu	-	-	0.43	0.68	0.65	0.57		0.62	0.43	-	-	0.79	0.60	0.31

a) Data for 259 Avg., 260 -20 pc.5, 261 Avg's from Robinson and Whitford, 1974. Others are this work.

TABLE 10c

Site 261: Mineral Compositions (wt.%)

	Plagioclase					Pyroxene			
	A	B	C	D	E	A	B	C	D
SiO_2	54.6	52.2	51.6	54.0	51.3	50.6	53.1	51.5	52.3
TiO_2	0.10	0.05	0.04	0.08	0.05	0.84	0.34	0.76	0.43
Al_2O_3	27.8	29.3	30.5	28.5	30.4	3.10	2.04	3.92	3.07
FeO	1.10	0.67	0.69	0.98	0.64	11.1	6.63	7.35	5.50
MnO	0.01	0.01	-	0.10	0.03	0.27	0.18	0.21	0.16
MgO	0.16	0.31	0.31	0.21	0.31	15.7	18.6	17.4	17.8
CaO	11.4	13.3	14.4	11.8	14.1	17.6	18.3	18.4	20.0
Na_2O	5.37	4.08	3.29	4.88	3.47	0.20	0.16	0.34	0.36
K_2O	0.04	-	0.05	0.07	0.01	-	-	-	-
Cr_2O_3	-	-	-	-	-	-	.27	.22	0.56
	100.58	99.62	100.88	100.62	99.92	99.41	99.62	100.10	100.18
Cation Sum (12 O for plag, 6 O for pyx)	5.019	5.013	4.988	5.007	4.996	4.011	3.998	4.005	4.009

A-B-C. 34-1:75-77, labradorite groundmass microlite, phenocryst, and average of zoned phenocryst

D-E. 34-2:15 labradorite microphenocrysts

A-B. 34-2:15, augite in groundmass and microphenocryst

C-D. 34-1:75-77, augite microphenocrysts

trace element abundances to basalt from core 39-1 in the lower section of Unit C (Tables 10a,b, Figure 4k). Thus, we agree with the conclusion of Robinson and Whitford (1974) that basalts from Units A and C of site 261 lie well within the compositional range of basalts dredged from active axes of modern oceanic ridges. Since a considerable time gap must have occurred between emplacement of the basement (Unit C), and the overlying sill (Unit A), the similarity in composition between these units indicates that magmas with a composition identical to axial basalts can be intruded away from an active ridge axis (e.g., compare 34-1 and 39-1 in Tables 10a,b). This compositional similarity has important implications for oceanic ridge basalt petrogenesis.

Because basement basalts from sites 260 and 261 are inferred to have originated from the same spreading ridge axis, it is important to compare the site 260 basaltic sill composition to site 261 compositions. Site 260 basalt is most similar to Unit B of site 261. Particularly striking are their high, nearly chondritic, REE distributions (Figures 4i,k), and high LIL element abundances (Figures 5d, 6). However, there are significant differences in TiO_2 and Cr content. Because of the large compositional differences between Unit B and other site 261 units, it is evident that site 260 and 261 magmas could have originated from, or in, the vicinity of the same spreading ridge axis. Thus, the petrographic and geochemical results are consistent with the inference of Veevers and Heirtzler (1974a,b) and Larson (1975) that basalts from these sites are related to volcanism from a spreading ridge axis off the northwest coast of Australia.

Summary

On the basis of petrography and geochemistry, igneous rocks from 13 eastern Indian Ocean DSDP sites can be classified into seven groups:

<u>Groups 1 and 2</u>. Basalts from Ninetyeast Ridge sites 214 and 216, and Wharton Basin site 256 are titaniferous, iron-rich (TiO_2 >2.3 wt. %, FeO*/MgO > 2) tholeiites with moderately high LIL element abundances, and a marked relative enrichment in light REE (Table 11). Identical compositional features characterize ferrotholeiites from islands such as the Galapagos, Iceland, Faeroes and the Indian Ocean, St. Pauls - Amsterdam Island Complex (Figures 3a,b, 4a,d,g, 5a,b). In addition, rocks with intermediate SiO_2 abundances, oceanic andesites or icelandites, such as those overlying basalt at site 214, are commonly associated on oceanic islands with ferrotholeiites. These similarities imply that the petrogenesis of aseismic Ninetyeast Ridge volcanics is similar to that of some oceanic islands. Most tectonic models for the Ninetyeast Ridge (Luyendyk and Davies, 1974; Sclater et al., 1974; Sclater and Fisher, 1974) utilize a combination of hot spot and a leaky transform fault as magma sources. In particular, sites 214 and 216 have been proposed as originating from different sources; however, the petrographic and geochemical similarity of site 214 and 216 basalts indicates nearly identical petrogenesis. This conclusion is consistent with a tectonic model involving two hot spots (Amsterdam - St. Paul and Kerguelen; Luyendyk and Rennick, 1977).
Site 256, more than 10°E of the Ninetyeast Ridge (Figure 1,2), is in an area of the Wharton Basin believed to have been formed by seafloor spreading (Markl, 1974; Sclater and Fisher, 1974). However, the geochemical characeristics of site 256 basalts are not consistent with an origin at a spreading ridge axis (Figures 5b,c,d, 6). We

suggest that these ferrotholeiites resulted from non-ridge volcanism which is represented by a north by northeast trend of topographic highs extending from the vicinity of Broken Ridge through the site 256 location (Figures 1,2).

Group 3. Tholeiitic basalts with moderate FeO*/MgO occur at sites 215 and 254 (Table 11). These basalts have high LIL element abundances, particularly at site 215 (Figures 5d, 6), and they are relatively enriched in light REE (Figures 4b,e, 5b,c). It is unlikely that these basalts formed at spreading ridge axes away from islands. In the case of site 254, on the Ninetyeast Ridge, this result is consistent with the occurrence of LIL element-enriched ferrotholeiites at sites 214 and 216; that is, site 254 basalts may represent a less fractionated product of the same volcanic processes that created the more northern portion of the Ninetyeast Ridge, or alternatively, the geochemical similarities between site 254 basalts and basalts dredged from Broken Ridge are consistent with the proposal of Luyendyk and Rennick (1977) that the southern portion of the Ninetyeast Ridge, Broken Ridge and the Naturaliste Plateau were formed by volcanism from the Kerguelen hot spot.

Site 215 is bathymetrically distinct from the Ninetyeast Ridge (Figure 1), and the very high K_2O, Sr and Ba abundances (Table 5a, b, Figure 6) and high La/Yb (Figures 4e, 5b,c) distinguish fresh site 215 basaltic glass from Ninetyeast Ridge basalts. Such LIL element-rich tholeiites (KP group of Melson et al., 1976) are usually associated with fracture zones and major seamounts. For example, a dredged basalt from the flanks of a seamount in the northeast Indian Ocean ($\sim$5°S, 97'E) has a major element composition nearly identical to site 215 glass (Table 5a). We have no explanation for the presence of this basalt composition at the topographically featureless area of site 215.

Group 4. Site 211 contains alkali-olivine basalts with characteristically high LIL element abundances (Table 11, Figures 5d, 6), and highly fractionated REE distributions (Figures 4h, 5b,c). Similar basalts occur on Christmas Island, $\sim$3° to the east. Basalts at this site are almost certainly related to volcanism creating the Cocos-Keeling and Christmas shoal areas which extend from 95°E to the Java Trench in a trend slightly north of due east.

Groups 5 and 6. Basalts from sites 213, 257, 259 and 261 (Units A and C) have major and compatible trace element abundances similar to tholeiitic basalts dredged from spreading ridge axes (Table 11); in addition, they have the diagnostic feature of low LIL element abundances (Table 11, Figures 5b,d, 6). Geochemically less differentiated (e.g., FeO*/MgO <0.9, TiO_2 <0.7 wt. %, Ni >140 ppm) basalts occur at sites 212 and 253. Basaltic glass at site 212 is similar in composition to the MgCa group of ridge basalts (Frey et al., 1974; Melson et al., 1976), whereas site 254 picritic basalts have an unfractionated nature because of a high abundance of cumulate olivine.

On the basis of Sr isotope ratios, Subbarao and Hedge (1973) proposed "that the source of Mid-Indian Ocean Ridge basalts was depleted in alkalies more recently and/or to a lesser degree than ridge basalts from other oceans." Except for the neutron activation data of Schilling (1971), REE analyses of dredged Indian Ocean basalts (Balashov et al., 1970; Nichols and Islam, 1971) indicate that a surprisingly large proportion of these sites contain basalts with relative light REE enrichment compared to a chondritic average (it should be noted that the analytical techniques used by Balashov et al. and Nichols and Islam yield lower quality data than the neutron activation tech-

TABLE 11

Geochemical Features (averages) of Eastern Indian Ocean Basalts[1]

BASALT TYPE		SITE	FeO*/MgO	Ni	TiO$_2$	Zr	Ba	La/Yb
Picrite	⎫ Group 6	253	0.71	341	0.69	21	14	1.04
Mg Ca	⎬	212	0.88	140	0.61	22	23	1.44
	⎭	3-14	0.84	320	0.79	35	<5	0.38
		4577 (dredge)	0.83	-	0.35	-	-	-
LIL-Element	⎫	213	1.21	83	1.07	57	13	0.86
Depleted	⎪ Group 5	257	1.02	127	0.92	38	34	0.60
Tholeiites	⎬	259	1.25	58	1.38	71	7	0.64
	⎪	261 (A,C)	1.28	85	1.21	71	5	0.64
	⎭	MIORB	1.07	109	1.23	71	17	0.95
Transitional Tholeiites	⎫							
and Highly Fractionated	⎬ Group 7	260	1.67	67-265	1.79	128	45	1.57
Group 5 Basalts	⎭	261 (B)	1.50	56	3.40	185	23	1.18
Island	⎫ Group 3	215	1.07	115	1.56	170	415	5.21
Tholeiites	⎭	254	1.51	194	2.12	156	57	2.96
	St. Paul Is. (38)	2.43	265	1.50	180	205	5.26
		5327 (dredge)	0.85	-	1.00	-	-	2.57
		5332 (dredge)	1.41	-	1.33	-	-	2.77
Ferrotholeiites	⎫ Groups	214	2.26	50	2.35	120	45	2.80
	⎬ 1 & 2	216	2.10	44	2.75	159	140	3.67
	⎭	256	2.01	93	2.36	159	39	2.97
	Galapagos (63)	2.13	70	3.40	255	90	-
Oceanic		214	4.00	<5	1.45	252	578	7.45
Andesites	Galapagos (47 and 91)	3.67	12	1.91	525	290	-
Alkali Olivine	⎫ Group 4	211	2.07	83	2.79	300	530	11.7
Basalts	⎬	St. Paul F.Z. (Atlantic)	1.14	270	2.70	200	300	21.4
	⎭	4896 (dredge)	1.09	-	2.61	-	-	18.3

(1) 3-14 data from Frey et al., 1974; MIORB data Engel and Fisher, 1975 and Schilling, 1971; St. Paul Is. data from Girod et al., 1971 and this work; Galapagos data from McBirney and Williams, 1969; St. Paul Fracture Zone data from Frey, 1970 and Melson et al., 1972; dredge data from Vinogradov et al., 1969, and Balashov et al., 1970; others from this work.

nique). An abundance of tholeiitic basalts with relative light REE enrichment implies that the mantle source of Indian Ocean Ridge basalts is less depleted in LIL elements than the mantle sources in the Atlantic and Pacific. However, we believe there is not yet sufficient high quality data to fully evaluate this important implication, and that our DSDP data demonstrate that much of the eastern Indian Ocean floor is composed of basalt with LIL element depletion equivalent to that in Atlantic and Pacific ridge basalts. Furthermore, the extensive dredge and DSDP sampling of the western Indian Ocean have demonstrated that western Indian Ocean ridge basalts (except near the triple junction) have LIL element abundances remarkably similar to ocean floor basalts from the Atlantic and Pacific (Table 9 of Engel and Fisher, 1975). Thus, we believe it is premature to conclude that a large portion of Indian Ocean Ridge basalts have originated from a mantle source less depleted in LIL elements than the source of Atlantic and Pacific seafloor basalts.

The petrographic and geochemical based inference for a spreading ridge origin of basalts at sites 212, 213, 257, 259 and 261 is consistent with tectonic histories proposed for these sites (Luyendyk and Davies, 1974; Markl, 1974; Sclater et al., 1974; Sclater and Fisher, 1974; Veevers and Heirtzler, 1974a,b; Larson, 1975). However, the presence of olivine-rich, LIL element-depleted tholeiitic basalt at site 253 is not consistent with basalt compositions at other Ninetyeast Ridge DSDP sites (Table 11). Interestingly, dredged basalts from the vicinity of the southern Ninetyeast Ridge range widely in major element composition. Tholeiitic basalts occur northeast of site 253 (dredges 5201 and 5208 in Figures 1, 2 and Table 4a), and southeast of site 253 at 26°18'S, 90°00'E (4577 in Figures 1 and 2) the basalts appear to be MgCa tholeiites (Tables 4a and 11). Southwest of site 253 tholeiitic basalt was recovered at 31°38'S, 83°48.5'E (H in Figures 1, 2 and Table 4a), but alkali olivine basalt was dredged at 3805 m from 29°57'S, 83°01'E (4896 in Figures 1, 2 and Tables 4a,b). The alkalic nature of the latter basalt is well illustrated by its large relative enrichment in light REE (Figure 4h, Table 4b). It is evident that additional ocean floor sampling coupled with trace element analyses and more geophysical data are needed in order to understand the tectonic development of the region between site 253 and the Amsterdam - St. Paul Island volcanic complex.

Group 7. Site 260 and 261 (Unit B) basalts are tholeiites with LIL element abundances and REE distributions intermediate between island tholeiites (such as at sites 254 and 215), and LIL element-depleted tholeiites (such as at sites 213, 257, 259 and 261 A, C). Because of their generally higher FeO*/MgO and low compatible trace element abundance (Table 11), these basalts may be highly fractionated variants of LIL element-depleted tholeiites. Thus, as suggested by tectonic models for sites 260 and 261, these basalts could have formed at a spreading ridge axis.

The petrographic and geochemical results for 10 of these 13 DSDP sites are consistent with proposed tectonic models. Igneous rocks similar to those from oceanic islands occur at 3 of 4 Ninetyeast Ridge sites. Basalts similar to those at modern spreading ridge axes occur at 6 sites believed to have been formed by seafloor spreading. Unlike Kempe (1975), we find, on the basis of alteration-resistant trace element abundances, no consistent differences between dredged and drilled basalts presumed to have been formed by seafloor spreading. Significant inconsistencies between proposed tectonic models and basalt characteristics occur at sites 256, 253 and 215. At site 256, a re-examination of bathymetric features suggests a non-seafloor spreading history for this site. The presence of LIL element-depleted tholeiites at site 253 on the Ninetyeast Ridge, and of LIL element-enriched tholeiitic basalt at site 215 in the Central Indian Basin are inconsistent features without adequate explanation.

Acknowledgements. The DSDP samples were supplied through the assistance of the National Science Foundation, and this research was supported by NSF Grants DES-74-00147 and 00268. Nuclear reactor irradiations were made at the National Bureau of Standards (Gaithersburg, Md.) and Georgia Tech. (Atlanta, Ga.) reactors). We thank D. Bankston, J. Guertler, R. Houghton, S. Roy, M. Sulanowski and C. Sung for assistance in obtaining experimental data, P. Young and P. Thompson for manuscript preparation, and D.R.C. Kempe and H. Dick for reviews of an early draft. Woods Hole Oceanographic Institution Contribution Number 3782.

References

Aumento, F., Diorites from the Mid-Atlantic Ridge at 45°N, Science, 165, 1112-1113, 1969.

Balashov, Y. A., L. V. Dmitriev, and A. Y. Sharas'kin, Distribution of the rare earths and yttrium in the bedrock of the ocean floor, Geochem. Int., 7, 456-468 (Translated from Geokhimiya, 6, 647-660, 1970), 1970.

Baldridge, W. S., T. R. McGetchin, and F. A. Frey, Magmatic evolution of Hekla, Iceland, Contr. Min. and Petrol., 42, 245-258, 1973.

Bezrukov, O. L., A. Y. Krylov, and V. I. Chernysheva, Petrography and absolute age of basalts from the floor of the Indian Ocean, Oceanology, 6, 210-214, 1966.

Bougault, H., Distribution of first series transition metals in rocks recovered during DSDP Leg 22 in the northeastern Indian Ocean, in von der Borch, C. C., Sclater, J. G., et al., Initial Reports of the Deep Sea Drilling Project, Washington (U. S. Government Printing Office) 22, 449-457, 1974.

Bowin, C., Origin of the Ninety East Ridge from studies near the Equator, Jour. Geophys. Res., 78, 6029-6043, 1973.

Bryan, W. B., Morphology of quench crystals in submarine basalts, Jour. Geophys. Res., 77, 5812-5819, 1972.

Burke, K., W.S.F. Kidd, and J. T. Wilson, Relative and latitudinal motion of Atlantic hotspots, Nature, 245, 133-137, 1973.

Cann, J. R., Spilites from the Carlsberg Ridge, Indian Ocean, Jour. Petrol., 10, 1-19, 1969.

Cann, J. R., Rb, Sr, Y, Zr and Nb in some ocean floor basaltic rocks, Earth Planet. Sci. Lett. 10, 7-11, 1970.

Carey, S. W., The tectonic approach to continental drift, in S. W. Carey, convenor, Continental Drift - A Symposium, Hobart, University of Tasmania, 177-355, 1958.

Carmichael, E.S.E., The petrology of Thingmuli, a Tertiary volcano in eastern Iceland, Jour. Petrol., 5, 435-460, 1964.

Davies, T. A., B. P. Luyendyk, et al., Initial Reports of the Deep Sea Drilling Project, Washington (U. S. Government Printing Office) 26, 295-325, 1974.

Dickey, J. S., F. A. Frey, and E. B. Watson, Basalts from the South Atlantic triple junction, Abstracts with Programs, Geol. Soc. Amer. Ann. Mtg. 6 (7), 1974.

Dietz, R. S., and J. C. Holden, Reconstruction of Pangea: Breakup and dispersion of continents, Permian to Present, Jour. Geophys. Res., 75, 4934-4956, 1970.

Dietz, R. S., and J. C. Holden, Pre-Mesozoic oceanic crust in the eastern Indian Ocean (Wharton Basin)?, Nature, 229, 309-312, 1971.

Duncan, A. R., and S. R. Taylor, Trace element analyses of magnetites from andesitic and dacitic lavas from Bay of Plenty, New Zealand, Contr. Min. and Petrol., 20, 30-33, 1968.

Engel, A. E., C. G. Engel, and R. G. Havens, Chemical characteristics of oceanic basalts and the upper mantle, Bull. Geol. Soc. Amer., 76, 719-734, 1965.

Engel, C. G., and R. L. Fisher, Lherzolite, anorthosite, gabbro, and basalt dredged from the Mid-Indian Ocean Ridge, Science, 166, 1136-1141, 1969.

Engel, C. G., R. L. Fischer, and A. E. Engel, Igneous rocks of the Indian Ocean floor, Science, 150, 605-609, 1965.

Engel, C. G., and R. L. Fisher, Granitic to ultramafic complexes of the Indian Ocean ridge system, Western Indian Ocean, Bull. Geol. Soc. Amer., 18, 1553-1578, 1975.

Ewart, A., W. B. Bryan, and J. B. Gill, Mineralogy and geochemistry of the younger volcanic islands of Tonga, S. W. Pacific, J. Petrol., 14, 429-465, 1973.

Falvey, D. A., Sea-floor spreading in the Wharton Basin (northeast Indian Ocean) and the breakup of eastern Gondwanaland, J. Australian Petrol. Explor. Assoc., 12, 86-88, 1972.

Fisher, R. L., and C. G. Engel, Lherzolite, anorthosite, gabbro and basalt dredged from the cross-fractures and rifted zone of the Mid-Indian Ocean Ridge (in Russian), Geokhiimia, 6, 661-677, 1970.

Fisher, R. L., C.G. Engel, and T.W.C. Hilde, Basalts dredged from the Amirante Ridge, western Indian Ocean, Deep Sea Res., 15, 521-534, 1968.

Fisher, R. L., J. G. Sclater, and D. P. McKenzie, Evolution of the Central Indian Ridge, western Indian Ocean, Bull. Geol. Soc. Amer., 82, 553-562, 1971.

Fleet, A. J., and D.R.C. Kempe, Preliminary geochemical studies of the sediments from DSDP Leg 26, southern Indian Ocean, in Davies, T. A. Luyendyk, B. P., et al., Initial Reports of the Deep Sea Drilling Project, Washington (U. S. Government Printing Office) 26, 541-551, 1974.

Fleet, A. J., P. Henderson, and D.R.C. Kempe, Rare earth element and related chemistry of some drilled southern Indian Ocean basalts and volcanogenic sediments, Jour. Geophys. Res., 81, 4257-4268, 1976.

Frey, F. A., Rare earth and potassium abundances in St. Paul's rocks, Earth Planet. Sci. Lett., 7, 351-360, 1970.

Frey, F. A., W. B. Bryan, and G. Thompson, Atlantic Ocean floor: Geochemistry of basalts from Legs 2 and 3 of the Deep Sea Drilling Project, Jour. Geophys. Res., 79, 5507-5527, 1974.

Frey, F. A., M. Haskin, J. Poetz, and L. Haskin, Rare earth abundances in some basic rocks, Jour. Geophys. Res., 70, 6085-6097, 1968.

Frey, F. A., and C. M. Sung, Geochemical results for basalts from sites 253 and 254, in Davies, T. A., Luyendyk, B. P. et al., Initial Reports of the Deep Sea Drilling Project, Washington, (U. S. Government Printing Office) 26, 567-572, 1974.

Girod, M., G. Camus, and Y. Vialette, Discovery of tholeiitic rocks at Saint-Paul Island (Indian Ocean) (in French), Contr. Mineral. and Petrol., 33, 108-117, 1971.

Gunn, B. M., N. D. Watkins, W. E. Trzcienski, and J. Nougier, The Amsterdam-St. Paul volcanic province and the formation of low Al tholeiitic andesites, Lithos, 8, 137-149, 1975.

Hart, S. R., K, Rb, Cs, Sr, and Ba contents, and isotope ratios of ocean floor basalts, Phil. Trans. Roy. Soc., London, A, 268, 573, 1971.

Hedge, C. E., N. D. Watkins, R. A. Hildreth, and W. P. Doering, Sr^{87}/Sr^{86} ratios in basalts from islands in the Indian Ocean, Earth Planet. Sci. Lett., 21, 29-34, 1973.

Heirtzler, J. R., G. O. Dickson, E. M. Herron, W. C. Pitman, and S. LePichon, Marine magnetic anomalies, geomagnetic reversals, and motions of the ocean floor, Jour. Geophys. Res., 73, 2119-2136, 1968.

Heirtzler, J. R., J. J. Veevers et al., Age of the floor of the Eastern Indian Ocean, Science, 180, 952-954, 1973.

Hekinian, R., Rocks from the Mid-Oceanic Ridge in the Indian Ocean, Deep Sea Res., 15, 195-213, 1968.

Hekinian, R., Petrology of igneous rocks from Leg 22 in the northeastern Indian Ocean, in von der Borch, C. C., Sclater, J. G., et al., Initial Reports of the Deep Sea Drilling Project, Washington (U. S. Government Printing Office) 22, 413-447, 1974a.

Hekinian, R., Petrology of the Ninety East Ridge (Indian Ocean) compared to other aseismic ridges, Contr. Mineral. and Petrol., 43, 125-147, 1974b.

Kay, R., N. Hubbard, and P. Gast, Chemical characteristics and origin of oceanic ridge volcanic rocks, Jour. Geophys. Res., 75, 1585-1613, 1970.

Kempe, D.R.C., The petrology of the basalts, Leg 26, in Davies, T.A., Luyendyk, B. P., et al., Initial Reports of the Deep Sea Drilling Project, Washington (U. S. Government Printing Office) 26, 465-503, 1974.

Kempe, D.R.C., Normative mineralogy and differentiation patterns of some drilled and dredged oceanic basalts, Contr. Mineral. Petrol., 50, 305-320, 1975.

Larson, R. L., Late Jurassic sea-floor spreading in the Eastern Indian Ocean, Geology, 3, 69-71, 1975.

Lindstrom, M. M., and H. Baitis, Fractional crystallization of tholeiitic magma (Abstract), EOS, 55, 460, 1974.

Luyendyk, B. P., and T. A. Davies, Results of DSDP Leg 26 and the geologic history of the southern Indian Ocean, in Davies, T. A., Luyendyk, B. P., et al., Initial Reports of the Deep Sea Drilling Project, Washington (U. S. Government Printing Office) 26, 909-943, 1974.

Luyendyk, B. P., Deep Sea Drilling on the Ninetyeast Ridge: synthesis and a tectonic model, in Indian Ocean Geology and Biostratigraphy, AGU, Washington, this volume, 1977.

Luyendyk, B. P., and R. Rennick, Tectonic origin of aseismic ridges in the eastern Indian Ocean, Bull. Geol. Soc. Amer., 1977 (in press).

MacDougall, I., Potassium-argon ages on basaltic rocks recovered from DSDP Leg 22, Indian Ocean, in von der Borch, C. C., Sclater, J. G., et al., Initial Reports of the Deep Sea Drilling Project, Washington (U. S. Government Printing Office) 22, 377-379, 1974.

McBirney, A. A., and H. Williams, Geology and petrology of the Galapagos Islands, Geol. Soc. Amer. Mem., 118, 1969.

McKenzie, D., and J. G. Sclater, The evolution of the Late Cretaceous, Geophys. J. R. Astr. Soc., 25, 437-528, 1971.

Markl, R. G., Evidence for the breakup of eastern Gondwanaland by the early Cretaceous, Nature, 251, 196-200, 1974.

Markl, R. G., Bathymetric map of the eastern Indian Ocean, in Davies, T. A., Luyendyk, B. P., et al., Initial Reports of the Deep Sea Drilling Project, Washington (U. S. Government Printing Office) 26, 967-968, 1974.

Melson, W. G., S. R. Hart, and G. Thompson, St. Paul's rocks, Equatorial Atlantic: petrogenesis, radiometric ages and implications on sea-floor spreading, Geol. Soc. Amer. Mem., 132, 241-272, 1972.

Melson, W. G., T. L. Vallier, T. L. Wright, G. Byerly, and J. Nelen, Chemical diversity of abyssal volcanic glass erupted along Pacific, Atlantic and Indian Ocean seafloor spreading centers in the Geophysics of the Pacific Ocean basin and its Margin, Amer. Geophys. Union Monograph Series, 19, 351-368, 1975.

Minister, J. B., T. H. Jordan, P. Molnar, and E. Haines, Numerical modeling of instantaneous plate tectonics, Geophys. J. Roy. Astr. Soc., 36, 541-576, 1974.

Miyashiro, A., Volcanic rock series in island arcs and active continental margins, Amer. Jour. Sci., 274, 321-355, 1974.

Miyashiro, A., and F. Shido, Tholeiitic and calc-alkalic series in relation to the behaviors of titanium, vanadium, chromium, and nickel, Amer. Jour. Sci., 275, 265-277, 1975.

Miyashiro, A., F. Shido, and M. Ewing, Crystallization and differentiation in abyssal tholeiites and gabbros from mid-oceanic ridges, Earth Planet. Sci. Lett. 7, 361-365, 1970.

Molnar, P., and T. Atwater, Relative motion of hot spots in the mantle, Nature, 246, 288-291, 1973.

Molnar, P., and J. Francheteau, The relative motion of "hot spots" in the Atlantic and Indian Oceans during the Cenozoic, Geophysical Jour., 43, 763-774, 1975.

Nicholls, G. D., and M. R. Islam, Geochemical investigations of basalts and associated rocks from the ocean floor and their implications, Phil. Trans. Roy. Soc. Lond. A., 268, 469-486, 1971.

Noe-Nygaard, A., and J. Rasmussen, Petrology of a 3000 metre sequence of basaltic lavas in the Faeroe Islands, Lithos, 1, 286-304, 1968.

O'Nions, R. K., and R. J. Pankhurst, Petrogenetic significance of isotope and trace element variations in volcanic rocks from the Mid-Atlantic, Jour. Petrol., 15, 603-634, 1974.

Peck, D. L., T. L. Wright, and J. G. Moore, Crystallization of tholeiitic basalt in Alae Lava Lake, Hawaii, Bull. Volcanol., 29, 629-655, 1966.

Presnall, D. C., The join forsterite-diopside-iron oxide and its bearing on the crystallization of basaltic and ultramafic magmas, Amer. J. Sci., 264, 753-809, 1966.

Rasmussen, J., and A. Noe-Nygaard, Beskrivelse til geologisk kort over Faeroerne (in Danish), Geol. Survey Denmark I, Series 24, 370 pp., 1969.

Rhodes, J. M., and K. U. Rodgers, Chemical diversity of Mid-Ocean ridge basalts from a single dredge haul (Abstract), EOS, 53, 1133, 1972.

Robinson, P. T., and D. J. Whitford, Basalts from the eastern Indian Ocean, DSDP Leg 27, in Veevers, J. J., Heirtzler, J. R., et al., Initial Reports of the Deep Sea Drilling Project, Washington (U. S. Government Printing Office) 27, 551-559, 1974.

Roeder, P. L., and E. F. Osborn, Experimental data for the system $MgO-FeO-Fe_2O_3-CaAl_2Si_2O_8-SiO_2$ and their petrologic implications, Amer. J. Sci., 264, 428-480, 1966.

Rundle, C. C., M. Brook, N. J. Snelling, P. H. Reynolds, and S. M. Barr, Radiometric age determinations, in Davies, T. A., Luyendyk, B. P. et al., Initial Reports of the Deep Sea Drilling Project, Washington (U. S. Government Printing Office) 26, 513-516, 1974.

Schilling, J.-G., Sea-floor evolution: rare-earth evidence, Phil. Trans. Roy. Soc. Lond. A, 268, 663-703, 1971.

Schilling, J.-G., Iceland mantle plume: Geochemical evidence along Reykjanes Ridge, Nature, 242, 565-571, 1973.

Schilling, J.-G., Azores mantle blob: rare-earth evidence, Earth Planet. Sci. Lett., 25, 103-115, 1975a.

Schilling, J.-G., Rare-earth variations across 'normal segments' of the Reykjanes Ridge 60°-53°N, Mid-Atlantic Ridge, 29°S, and East Pacific Rise, 2°-19°S and evidence on the composition of the underlying low-velocity layer, Jour. Geophys. Res., 80, 1459-1473, 1975b.

Schilling, J.-G., and A. Noe-Nygaard, Faeroe-Iceland plume: rare earth evidence, Earth Planet. Sci. Lett., 24, 1-14, 1974.

Sclater, J.G., C.C. von der Bosch, et al., Regional synthesis of the deep sea drilling results from Leg 22 in the eastern Indian Ocean, in von der Borch, C. C., Sclater, J. G., et al., Initial Reports of the Deep Sea Drilling Project, Washington (U. S. Government Printing Office) 22, 815-831, 1974.

Sclater, J. G., and R. L. Fisher, Evolution of the east central Indian Ocean with emphasis on the tectonic setting of the Ninetyeast Ridge, Bull. Geol. Soc. Amer., 85, 683-702, 1974.

Sclater, J. G., B. P. Luyendyk, and L. Meinke, Magnetic lineations in the southern part of the Central Indian Basin, Bull. Geol. Soc. Amer., 87, 371-378, 1976.

Shido, F., A. Miyashiro, and M. Ewing, Compositional variation in pillow lavas from the Mid-Atlantic Ridge, Marine Geol., 16, 177-190, 1974.

Smith, N. C., and E. D. Mountain, The volcanic rocks of Christmas Island (Indian Ocean), Geol. Soc. Quart. J., 82, 44-66, 1925.

Subbarao, K. V., and C. E. Hedge, K, Rb, Sr, and $^{87}Sr/^{86}Sr$ rocks from the Mid-Indian Oceanic Ridge, Earth Planet. Sci. Lett., 18, 223-228, 1973.

Sykes, L. R., Seismicity of the Indian Ocean and a possible nascent island arc between Ceylon and Australia, Jour. Geophys. Res., 75, 5041-5055, 1970.

Taylor, S. R., M. Kaye, A.J.R. White, A. R. Duncan, and A. Ewart, Genetic significance of Co, Cr, Ni, Sc and V content of andesites, Geochim. Cosmochim. Acta, 33, 275-286, 1969.

Thompson, G., A geochemical study of the low-temperature interaction of sea-water and oceanic igneous rocks, EOS, 54, 1015-1018, 1973a.

Thompson, G., Trace element distributions in fractionated oceanic rocks - II. Gabbros and related rocks, Chem. Geol. 12, 99-111, 1973b.

Thompson, G., W. B. Bryan, F. A. Frey, and C. M. Sung, Petrology and geochemistry of basalts and related rocks from sites 214, 215, 216, DSDP Leg 22, Indian Ocean, in von der Borch, C. C., Sclater, J. G., et al., Initial Reports of the Deep Sea Drilling Project, Washington (U. S. Government Printing Office) 22, 459-468, 1974.

Thompson, G., and W. G. Melson, The petrology of oceanic crust across fracture zones in the Atlantic Ocean: Evidence of a new kind of sea-floor spreading, Jour. Geol., 80, 526-538, 1972.

Udintsev, G. B., and V. I. Chernysheva, Rock samples from the upper mantle of the Indian Ocean rift zone, Doklady, Earth Sci. Sections, 85-88. AGI translation from Akad. Nauk SSSR, Doklady, 165, 1147-1150, 1965.

Veevers, J. J., and J. R. Heirtzler, Tectonic and paleogeographic synthesis of Leg 27, in Veevers, J. J., Heirtzler, J. R., et al., Initial Reports of the Deep Sea Drilling Project, Washington, (U. S. Government Printing Office) 27, 1049-1054, 1974a.

Veevers, J. J., and J. R. Heirtzler, Bathymetry, seismic profiles, and magnetic-anomaly profiles, in Veevers, J. J., Heirtzler, J. R., et al., Initial Reports of the Deep Sea Drilling Project, Washington (U. S. Government Printing Office) 27, 339-381, 1974b.

Veevers, J. J., J. R. Heirtzler., et al., Initial Reports of the Deep Sea Drilling Project, Washington, (U. S. Government Printing Office) 27, Frontispiece, 1974c.

Veevers, J. J., J. G. Jones, and J. A. Talent, Indo-Australian stratigraphy and the configuration and dispersal of Gondwanaland, Nature, 229, 383-388, 1971.

Vinogradov, A. P., G. B. Udintsev, L. V. Dmitriev, V. F. Kanaev, Y. P. Neprochnov, G. N. Petrova, and L. N. Rikunov, The structure of the mid-oceanic rift zone of the Indian Ocean and its place in the world rift system, Tectonophysics, 8, 377-401, 1969.

von der Borch, C. C., J. G. Sclater, et al., Initial Reports of the Deep Sea Drilling Project, Washington (U. S. Government Printing Office) 22, 193-212, 1974.

Wager, L. R., The major element variation of the layered series of the Skaergaard intrusion and a re-estimation of the average composition of the hidden layered series and of the successive residual magmas, Jour. Petrol., 1, 364-398, 1960.

Wilkinson, J.F.G., and N. T. Duggan, Some tholeiites from the Inverell area, New South Wales, and their bearing on low pressure tholeiite fractionation, Jour. Petrol., 14, 339-348, 1973.

Wilson, J. T., A possible origin of the Hawaiian Islands, Can. J. Phys., 41, 863-870, 1963.

Wilson, J. T., Evidence from oceanic islands suggesting movement in the earth, Phil. Trans. Roy. Soc. Lond. Ser. A., 258, 145-167, 1965.

CHAPTER 9. LARGE ION LITHOPHILE ELEMENTS AND Sr AND Pb ISOTOPIC
VARIATION IN VOLCANIC ROCKS FROM THE INDIAN OCEAN

K. V. Subbarao

Indian Institute of Technology
Powai, Bombay, India

R. Hekinian

Centre Oceanologique de Bretagne
Brest, France

D. Chandrasekharam

Indian Institute of Technology
Powai, Bombay, India

Abstract. The normative mineralogy of relatively unaltered major element analyses of Mid-Indian Oceanic Ridge (MIOR) - including Carlsberg Ridge (CR) display limited variation, and concentrated towards the diopside apex of the Di-Hy-Ol-Az-Ne tetrahedron, while the Mid-Atlantic Ridge (MAR) and East Pacific Rises (EPR) rocks display greater variation from nepheline normative tholeiites to quartz normative alkali basalts. The MIOR basalts are enriched in Large Ion Lithophile (LIL) elements (K, Rb, Sr, Cs and U) and Sr and Pb isotopes suggesting that the source of MIOR was depleted at a much later time and also to a lesser degree than the sources of MAR and EPR. Chemical and Sr isotopic differences exist between MIOR and CR. The volcanic rocks from aseismic 90°E Ridge include pyroxene basalts, picritic basalts and oceanic andesites (variation from olivine normative to quartz normative) suggesting wide range of differentiation which appears to be not duplicated in other oceanic ridges. The major element analyses, higher concentrations of LIL elements, and Sr^{87}/Sr^{86} ratios fairly match with the geochemistry of alkali island basalts. Geochemical differences between various tectonically different regimes including seismic and aseismic ridges and islands in the Indian Ocean may probably suggest the presence of systematic heterogeneities with alkali-poor and alkali-rich zones in the mantle giving rise to a chemically zoned structure.

Introduction

Earlier petrological investigations on the dredged rocks from the Indian Ocean were primarily limited to a small portion of the Mid-Indian Oceanic Ridge (MIOR) (Balashov, Dmitriev and Sharaskin, 1970; Bezrukov, Krylov, and Chernysheva, 1966; Dmitriev, 1974; Engel, Fisher and Engel, 1965; Engel and Fisher, 1975; Hekinian, 1968). Very little is known about the trace element distribution and isotopic composition of the Indian oceanic crust (Subbarao and Hedge, 1973; Thompson, Bryan, Frey and

Sung, 1973). During legs 22 and 26 of the GLOMAR CHALLENGER in the Indian Ocean additional information on basaltic rocks from various structural provinces were gathered (Hekinian, 1974; Kempe, 1974). Further, significant contributions have been made by Frey, Dickey, Thompson and Bryan (1977) on the petrography and geochemistry of basalts from eastern Indian Ocean DSDP sites.

The purpose of the present study is to present new data on some LIL elements (Rb, Sr, Cs and U) and Sr and Pb isotopes from both dredge hauls and DSDP samples from the Indian Ocean (Fig. 1). Existing differences between volcanics found on the seismically active Mid-Indian Oceanic Ridges and the aseimic Ninetyeast Ridge are noted. Both isotopic and dispersed element studies are used to understand the differentiation processes and mantle source chemistry operating within the lithosphere of the Indian Ocean. It is also the aim of the present work to make a comparative study with other oceanic ridges such as the East Pacific Rise (EPR) and the Mid-Atlantic Ridge (MAR) which could give insights into compositional heterogeneities of the source material giving rise to the various ridge systems.

Methods of Study and Presentation of Data

The major element analyses were carried out using XRF and atomic absorption techniques (Hekinian, 1974). Rb and Cs concentrations were measured using neutron activation, while U concentrations were determined by both delayed fission neutron counting and fission track techniques. Sr contents were made by X-ray spectrography. The analytical uncertainties are as follows: K ± 3%, Rb and Cs ± 2%, Sr ± 5% and U ± 10%. The two new strontium isotopic analyses were measured by J. Matsuda at the Geophysical Institute, University of Tokyo, Tokyo, Japan, on a 25 cm radius, single focussing mass spectrometer using single tantalum filament mode of ionisation, with a digital output (analytical precision ± 0.0002 (2 σ). The Sr^{87}/Sr^{86} ratio of E & A Sr CO_3 is 0.7080. The two new lead isotopic analyses were made by R. E. Zartman, U. S. Geological Survey, Denver, USA, using the facilities of the Charles Arms Laboratory, California Institute of Technology. The analytical procedures used were similar to those developed for lunar samples and have been described by Tera and Wasserburg (1972). The lead was analysed in a mass spectrometer using surface ionization from a single tantalum filament with $H_3 PO_4$ - silica gel emitter. The Pb isotopic data reported in Table 2 have been normalized to absolute values using conversion factors obtained from the analysis of Caltec shelf lead and is considered to be accurate to 0.2 M.

We have extensively used published chemical analyses but selected only those samples having less than 1% H_2O in our discussions, as these samples would possibly represent nearly fresh and original material. Further, chemical analyses with more than 1.5% Fe_2O_3 are reduced to 1.5% and recalculated the C.I.P.W. norms to minimise oxidation and alteration effects. These two precautions were taken to avoid altered samples in our discussions. Therefore the number of published analyses used in various diagrams in the present study has been considerably reduced due to these precautions, although reasonable number of analyses are available in the literature.

While compiling the chemical data we encountered some difficulties particularly with 90° E rocks. We would be forced to exclude most of the limited number of available analyses from the 90° E Ridge we have included these data also in our compilation diagrams as an exception. This should not be construed as a reflection on our conclusions due to the relatively

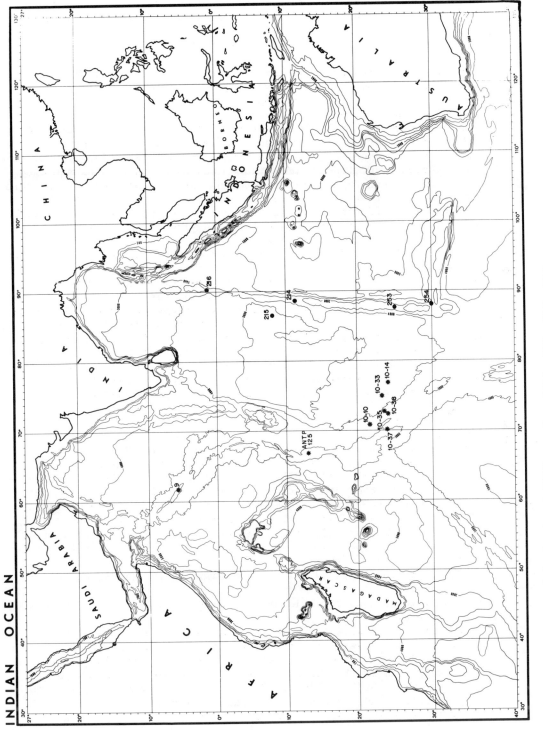

Fig. 1. Location of dredged and DSDP samples (214, 215, 216, 253, 254) from the Indian Ocean (Belousov et al., 1962). Sample 9 was reported by Cann (1969). Samples 10-10, 10-14, 10-33, 10-35, 10-37, and 10-38 are from Engel et al., 1965b; Hekinian, 1968; Subbarao and Hedge, 1973. Sample ANTP-125 from Engel and Fisher (1975).

altered nature of the samples. In fact the interpretations based on our unpublished REE data (which are considered to be unaffected by alteration) support our interpretations and conclusions reached in this paper on 90° E Ridge rocks. Therefore this suggests that the samples under study may not be too altered to be excluded from scientific study.

Major Elements Chemistry

In order to bring out possible differences and/or similarities between the three major oceanic ridge systems, we have included published major element analyses on rocks from MAR, EPR, MIOR (including CR). The normative mineralogy of the basalts are presented on the Di-ol-Hy-Ne-Qz tetrahedron of Yoder and Tilley (1962). The normative mineralogy of MAR and EPR rocks show a continuous variation from quarts normative tholeiites to nepheline normative alkali types (Figure 2a), whereas the MIOR (including CR) rocks concentrate towards diopside apex and also show variation (Figure 2b). Therefore the normative mineralogy of the three ridges suggest that MAR and EPR are relatively more basic than MIOR. Perhaps this difference may also suggest different degrees of partial melting at different depths as well as varying differentiation processes within the sources of these three ridges.

Hekinian (1974) and Kempe (1974) reported major elements analyses for volcanic rocks from aseismic 90° E Ridge obtained through DSDP legs 22 and 26 (sites 214, 216, 253, 254). The volcanic rocks include pyrozene-basalts, amygdular, vesicular and homogeneous basalts, picrite basalts and oceanic andesites. The normative mineralogy of these rocks is well depicted in figure 2c, with a complete variation from olivine normative to quartz normative alkali-rich basalts and andesites. This suggests that the 90° E Ridge is characterized by a wide range of differentiation as revealed by progressive chemical and normative variation which does not appear to be duplicated in other mid-oceanic ridges. Furthermore, the chemical characteristics of the aseismic 90° E Ridge are similar to island volcanic series (Hekinian, 1974b).

We also report new major element analyses for six basalts (Table 1) from site 215), which is located in the central Indian Ocean and very close to the western flank of the 90° E Ridge. The location of this site is important in the sense that this site lies between a major fracture zone at 89°E and 90°E Ridge. Further, the relatively young age (59-60 M.Y.) of the basal sediment at this site is also significant and crucial to the tectonics and genesis of the Central Indian Ocean basin (Sclater and Fisher, 1974). The samples are moderately fresh to weathered and buried under 150.8 meters of sediment. The rocks from this site form a succession of pillow lavas, showing texturally different zones:

1) Upper and lower glassy surface, sometimes with olivine and plagioclase phenocrysts.
2) Transitional zone which is hypocrystalline, aphanitic and moderately weathered.
3) Innermost crystalline and less weathered zone of pillow lavas consists of holocrystalline rock with intergranular and subophitic texture. Prismatic and elongated laths of plagioclase (An_{45}-An_{62}) occurs closely with fresh anhedral plates and rods of clinopyroxene (2V = 49° to 55°). Olivine (2 V = 82° - 84°) occurs marginal to plagioclase and pyroxene, surrounded by a rim of serpentine.

Few small vesicles are scattered throughout the thin section of samples from site 215. The normative mineralogy of crystalline (Table 1) and glassy basalts (Frey et al., 1977) are plotted in Figure 2d.

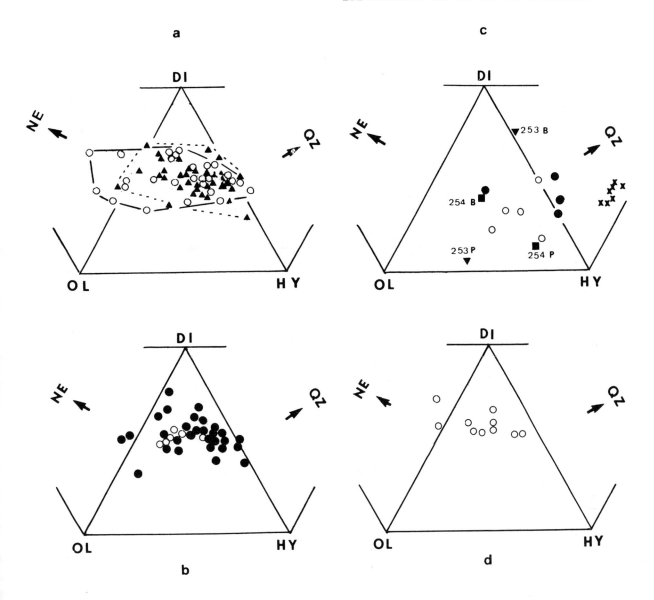

Fig. 2. Normative plots of oceanic ridge rocks in the Di-Hy, ol-Qz-Ne diagrams. Norms are from analyses recalculated with a fixed Fe_2O_3 1.5%. Figure 2a: Plots of rocks from Mid-Atlantic Ridge (▲ MAR) and East Pacific, Juan de Fuca and Gorda Ridges (O EPR). References: MAR: Aumento (1969); Bougault and Hekinian (1974); Kay et al. (1970); Miyashiro et al. (1969); Muir and Tilley (1964) and Shido et al. (1971). EPR: Engel and Engel (1964); Engel et al. (1965); Hekinian (1971); Kay et al. (1970) and Moore (1970). Figure 2b: Plots of rocks from Mid-Indian Oceanic Ridge (● MIOR) and Carlsberg Ridge (O, CR). MIOR: Engel et al.(1965b); Hekinian (1968); (Balashov et al. (1970); Engel and Fisher, 1975) CR: Cann (1969). Figure 2c: Plots of 90°E Ridge rocks recovered during leg 22 (o Basalt and * andesite of site 214; ● basalts site 216) and leg 26 (B - basalt and P - picrite) (Kempe, 1974) of Deep Sea Drilling Project. Figure 2d: Plots of basalts from leg 22, site 215 (this study, Hekinian, 1974, Melson et al., 1975).

TABLE 1. Chemical analyses and normative composition of basalts from Site 215, DSDP, Indian Ocean

Oxides	215-18-2 (47-53)	215-18-3 (110-112)	215-19-1 (23-32)	215-19-1 (87-93)	215-19-2 (145-150)	215-20-2 (20-23)
SiO_2	47.45	49.68	50.23	48.92	49.12	48.29
Al_2O_3	16.29	16.51	16.58	16.53	16.06	16.21
Fe_2O_3	4.84	2.59	2.72	2.46	3.90	2.54
FeO	3.98	5.63	5.69	5.92	5.00	5.77
MgO	5.93	6.82	6.47	6.15	5.93	6.48
CaO	10.81	10.52	10.39	10.55	11.28	10.24
Na_2O	3.70	2.87	3.00	3.07	2.86	2.95
K_2O	0.96	0.83	0.84	0.80	0.93	0.88
TiO_2	1.67	1.63	1.66	1.66	1.67	1.67
P_2O_5	0.30	0.28	0.29	0.31	0.30	0.30
CO_2	0.17	0.00	0.00	0.00	0.00	0.00
H_2O	3.18	2.59	2.74	2.55	3.59	2.63
Total	99.27	99.94	100.60	98.91	100.63	97.95

Norm

Qz		0.00	0.00	0.00	0.00	0.00	0.00
Ne		3.86	0.00	0.00	0.00	0.00	0.00
Or		5.67	4.90	4.96	4.73	5.50	5.20
Ab		24.19	24.29	25.39	25.98	24.20	24.96
An		25.01	29.17	29.29	29.10	28.24	28.39
Di	Wo	10.68	8.62	8.50	8.86	10.76	8.54
Di	En	6.35	5.45	5.24	5.39	6.32	5.30
Di	Fs	3.79	2.63	2.76	2.98	3.91	2.73
Hy	En	0.00	8.15	8.50	4.82	4.27	5.16
Hy	Fs	0.00	3.94	4.48	2.66	2.64	2.66
Fo		5.90	2.38	1.66	3.58	2.93	3.98
Fa		3.88	1.27	0.96	2.18	2.00	2.66
Mt		2.17	2.17	2.17	2.17	2.17	2.17
Ilm		3.17	3.10	3.15	3.15	3.17	3.17
Ap		0.70	0.65	0.67	0.72	0.70	0.70

Eight of these basalts are diopside-rich whereas two samples fall in the nepheline normative field. In general normative composition of these rocks are akin to oceanic ridge basalts.

Large Ion Lithophile Elements Chemistry

The variation in LIL elements in rocks from oceanic ridges is shown in Figure 3. We have included our new elemental and isotopic analyses (Table 2) and available published analyses from the literature in this diagram.

It is implied that in a compilation of this type it is not possible to use the very same samples to bring out the range of values for all the elements and isotopes under consideration, as we have more data on some elements and scanty data on the rest. For example, the number of samples involved in bringing out the range of K for MAR rocks is not the same for Rb from the same ridge.

We have chosen some of the LIL elements (K, Rb, Sr, Cs, U) in oceanic basalts from seismic and aseismic ridges to obtain information on the

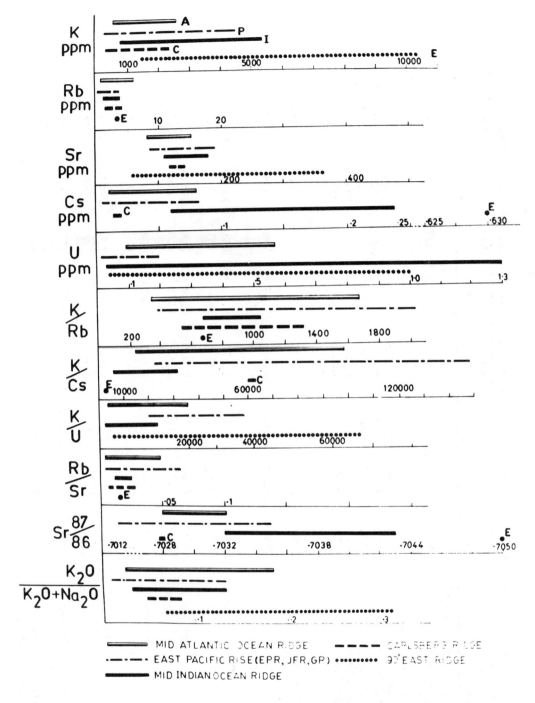

Fig. 3. Variation of LIL elements, their corresponding ratios and Sr^{87}/Sr^{86} for oceanic ridges. References: Refer to Table 2.

differentiation history and mantle source chemistry (including mantle depletion processes in the geological part). The variation of K, Rb, Sr, Cs and U in the major oceanic ridges is clear from Figure 3. The MIOR rocks are enriched in K, Rb, Cs and U relative to MAR and EPR rocks.

TABLE 2. Analytical data on dredged and drilled (DSDP Leg 22) rocks from the Indian Ocean

Sample No.	K ppm	Rb ppm	Sr ppm	Cs ppm	U ppm	K/Rb	K/Cs	Rb/Sr
*Mid-Indian Oceanic Ridge:								
IO-10 Porphyritic Alkali basalt	1820	2.71	144	0.082	1.3	672	22200	0.019
IO-14 Glassy olivine tholeiite	1910	2.80	179	0.236	–	682	8093	0.016
IO-33 Tholeiite (Glass)	1410	1.33	110	0.057	–	1060	24700	0.012
IO-35 Tholeiite	2570	3.48	136	0.238	0.2	720	10800	0.026
IO-38 Tholeiite	2320	2.56	137	0.070	0.4	910	33140	0.019
IO-37 Tholeiite	2320	2.37	142	0.088	0.15	979	26360	0.017
DSDP Leg 22:								
214-53-1 (30-35) Pyroxene basalt	2400	3.55	220	0.630	0.6	670	3809	0.016
215-18-1 (68-73) Basalt (Glass)	6900	26.80	390	0.416	0.29	256	16586	0.067
MIOR: Average	1700 (19)	2.54 (6)	141 (6)	0.128 (6)	0.51 (4)	831 (6)	20882 (6)	0.018 (6)
MAR: Average	1276 (73)	2.28 (18)	117 (16)	0.036 (10)	0.28 (17)	869 (18)	44843 (10)	0.016 (16)
EPR: Average	1470 (61)	1.4 (26)	120 (22)	0.023 (15)	0.075 (8)	1062 (27)	73193 (15)	0.015 (16)
CR: Average	1730 (12)	2.1 (5)	142 (4)	0.018 (1)	–	938 (5)	62000 (1)	0.017 (4)
Indian Ocean Average alk. island basalts	7500 (81)	22.98 (42)	389.5 (42)	–	0.75 (14)	357 (20)	–	0.057 (42)
Alk. basalts Average from seamounts of four	14276	22.48	605	–	–	947	–	0.032

TABLE 2 (continued)

Sample No.	K/U	$\frac{K_2O}{K_2O+Na_2O}$	$\frac{(La)}{(Sm)}_{e.f.}$	$\frac{Sr^{86}/Sr^{87}}{}$ (+)	$\frac{Pb206}{Pb204}$	$\frac{Pb207}{Pb204}$	$\frac{Pb208}{Pb204}$	References
*Mid-Indian Oceanic Ridge:								
IO-10 Porphyritic Alkali basalt	1400	0.064	–	0.7033	–	–	–	
0-14 Glassy olivine tholeiite	–	0.059	–	0.7043	–	–	–	
IO-33 Tholeiite (Glass)	–	0.056	–	0.7032	17.55	15.47	37.42	
IO-35 Tholeiite	12850	0.095	–	0.7033	18.21	15.57	38.22	
IO-38 Tholeiite	5800	0.082	–	0.7035	–	–	–	
IO-37 Tholeiite	15466	0.081	1.03	0.7032	–	–	–	
DSDP Leg 22:								
214-53-1 (30-35) Pyroxene basalt	4000	0.135	–	0.7050	–	–	–	1
215-18-1 (68-73) Basalt (Glass)	23793	0.225	–	0.7047	–	–	–	1
MIOR: Average	8879 (4)	0.069 (19)	1.03 (1)	0.7034 (6)	17.88 (2)	15.52 (2)	37.82 (2)	1,2,3,4,11
MAR: Average	9119 (17)	0.071 (64)	–	0.7028 (6)	18.25 (3)	15.43 (3)	37.56 (3)	5,6,7,8,9,10,12, 13,17,18,21,24,27
EPR: Average	20664 (8)	0.057 (42)	0.588 (4)	0.7026 (23)	18.19 (3)	15.40 (3)	37.60 (3)	5,6,8,11,14,15,16, 17,18,19,20,25,26
CR: Average	–	0.066 (8)	0.52 (4)	#0.7028 (1)	–	–	–	3,5,11,22,23,24
Indian Ocean alk. island basalts Average	–	0.305 (50)	–	0.7043 (38)	–	–	–	28,29,30,31,32,33
Alk. basalts from seamounts Average of four	–	0.257	–	0.7031	–	–	–	34

TABLE 2. (continued)

*Mid-Indian Oceanic Ridge: References: Rb - Average of this study and Subbarao and Hedge (1973); Cs and U and Pb isotope ratios - This study: K, Sr, Sr, Sr^{87}/Sr^{86} - Subbarao and Hedge (1973); (La/Sm)e.f. - Schilling (1971).

References:

1) This study; 2) Subbarao and Hedge (1973); 3) Hekinian (1968); 4) Engel et al. (1965b); 5) Hart (1971); 6) Kay et al. (1970); 7) Cann (1970); 8) Engel et al. (1965); 9) Miyashiro et. al. (1969); 10) Nichols et al. (1964);11) Schilling (1971); 12) Aumento (1971); 13) Bougault and Hekinian (1974); 14) Subbarao (1972); 15) Engel and Engel (1964); 16) Hekinian (1971); 17) Hart (1969); 18) Tatsuroto et al. (1965); 19) Fisher (1972); 20) Funkhouser et al. (1968); 21) Aumento (1968); 22) Cann (1969); 23) Cann and Vine (1966); 24) Cann (1970); 25) Moore (1970); 26) Hedge and Peterman (1970); 27) Tatsumoto (1966); 28) Hedge et al. (1973); 29) Gunn et al. (1970); 30) McDougall and Compston (1965); 31) Upton and Wadsworth (1972); 32) Oversby (1972); 33) Zielinski (1975); 34) Subbarao et al. (1973); # S. R. Hart (Personal conversation).

(+) The Sr^{87}/Sr^{86} ratio of E & A $SrCO_3$ is 0.7080.

( ) Number of analyses representing the average values.

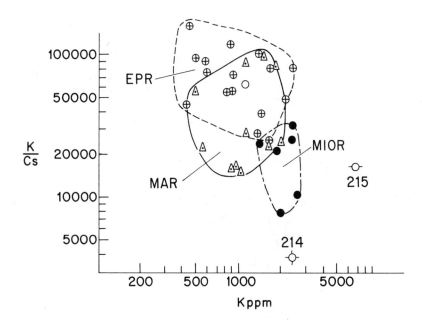

Fig. 4. Variation of K/Cs vs K for rocks from Mid-Indian Oceanic Ridge, Mid-Atlantic Ridge and Pacific Ocean Ridges and Deep Sea Drilling Project (leg 22, sites 214 and 215). References: MAR: Hart (1971) and Kay et al. (1970). EPR: Hart (1971) and Kay et al. (1970). MIOR: This study. Sites 214 and 215: This study. The symbols are the same as in Figure 2.

Similarly the K/Cs - K and K/U - K indeed, form distinctly different fields for these ridge rocks (Figures 4 and 5).

The K/U-K diagram displays a positive trend with separate fields for different ridge basalts (Figure 5). Although there is some overlap between MAR and EPR fields, the MIOR field is clearly separated as a narrow band running parallel to the MAR basalt trend. The MIOR rocks indeed show higher concentration of K (2 570 ppm) and lower K/U ratio (1 400). The K/Cs data for different ridge basalts appears to show a negative correlation with K (Figure 4). Although we have drawn fields for MAR, EPR and MIOR basalts, there is considerable overlap particularly between MAR and EPR fields. However, the lower K/Cs ratio (8090) with higher K value (2570 ppm) and a distinct chemical field for MIOR basalts is significant. We interpret these results including negative correlation in terms of a depletion model for the mantle to bring out chemical differences that are resulted during the early differentiation history of the earth. Perhaps the higher K/Cs and lower K samples (such as EPR) appear to have been highly depleted in dispersed elements including K, Rb, Sr, Cs, whereas the lower K/Cs and higher K may perhaps be least depleted. Therefore, we possibly have different degrees of depletion (at different times) below the three major oceans. The overlap between the three fields (MAR, EPR and MIOR) may perhaps be due to continuous depletion processes in the mantle. In other words this implies that the source of MIOR is relatively less depleted than EPR and MAR.

Sr and Pb ISOTOPIC Ratios

The variation in Sr^{87}/Sr^{86} ratio between oceanic ridges is shown in Table 2 and Figure 3. The range of MAR and EPR is nearly identical (Table 2) but MIOR rocks display consistently higher ratios (.7031 -

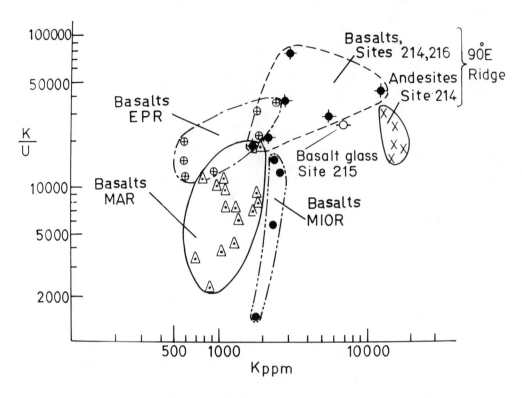

Fig. 5. Variation of K/U vs K for volcanic rocks from oceanic ridges. References: MAR: Aumento (1971); Frey et al. (1974); Tatsumoto et al. (1965). MIOR: Funkhouser et al. (1968); Tatsumoto et al.)1965). MIOR: This study. Deep Sea Drilling Project leg 22, sites 214 and 216: This study and Thomspon et al. (1973).

.7043 rocks close to the triple junction - Subbarao and Hedge, 1973; rocks further north of triple junction - Engel and Fisher , 1975), Peterman and Hedge (1971), Subbarao and Hedge (1973) and Subbarao, Clark and Forbes (1973), have shown positive correlation between Sr^{87}/Sr^{86} and K_2O/K_2O+Na_2O for rocks from different tectonic environments from the ocean suggesting different degrees of depletion of dispersed elements in the mantle. While the MAR and EPR rocks follow this general trend, the MIOR rocks differ in having higher Sr^{87}/Sr^{86} ratios similar to some of the island basalts but having K_2O/K_2O+Na_2O ratios in the range of other oceanic ridge basalts (Figure 6). Various degrees of depletion of LIL elements would be reflected by variation in Rb/Sr, K_2O/K_2O+Na_2O whereas different times of depletion would be reflected only in the Sr^{87}/Sr^{86} ratios. Therefore this anomaly in the MIOR appears to be primarily due to later time of depletion of K, Rb and other LIL elements in the mantle source area. In other words, the depletion of the source of MIOR basalts may have occurred several hundred million years later and also to a lesser degree than the depletion of the source of the basalts from MAR and EPR (Subbarao and Hedge, 1973). A similar interpretation was offered by Peterman, Coleman and Hildretti (1971) for Troodos Massif Rocks which is considered to be Mesozoic ocean-plate remnant, even with higher Sr^{87}/Sr^{86} ratios (Figure 6). Our new Pb isotope analysis (Table 2) on two of the MIOR basalts are strikingly higher (Pb^{206}/Pb^{204} = 17.55 - 18.21; Pb^{207}/Pb^{204} = 15.47 - 15.57; Pb^{208}/Pb^{204} = 37.42 - 38.22) than the

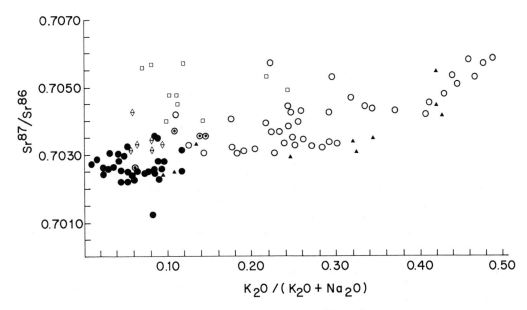

Fig. 6. Variation of K_2O/K_2O+Na_2O versus Sr^{87}/Sr^{86} ratios for oceanic rocks from different tectonic environments (●) indicates Mid-Atlantic Ridge and East Pacific Rise tholeiites (Peterman and Hedge, 1971; Subbarao, 1972; Subbarao and Clark, unpublished data): (◊) indicates the Mid-Indian Oceanic Ridge tholeiites (Subbarao and Hedge, 1973): (o) indicates oceanic island basalts (Peterman and Hedge, 1972; Hedge, Watkins, Hildrath and Doering, 1973): (△) indicates seamount basalts (Subbarao et al., 1973; Subbarao and Hekinian, 1975, under review); (▫) indicates the Troodo's Massif mafic rocks (Peterman et al., 1971). (⊙) Abyssal hill basalt (Subbarao, 1972); -o- Site 215 basalt, -⊹- Site 214 basalt (this study).

ranges thus far reported for ocean ridge basalts (Pb^{206}/Pb^{204} = 17.70 - 18.70; Pb^{207}/Pb^{204} = 15.39 - 15.53; Pb^{208}/Pb^{204} = 37.03 - 38.15; Tatsumoto, 1966) but match with our earlier higher Sr^{87}/Sr^{86} ratios and support our earlier interpretation on the Sr isotopes also. Thus these Sr and Pb isotopic data clearly suggest that MIOR rocks are distinctly different both isotopically and chemically from MAR and EPR rocks.

90°East Ridge

We present concentrations of K, Rb, Sr, Cs and U and Sr^{87}/Sr^{86} ratio from site 214 (Table 2). This sample from site 214 has higher K, Rb, Cs and U but fall within the ranges of samples studied from islands (Table 2). The Rb/Sr value of 0.016 is compatible with higher Rb concentration (3.55 ppm) and higher Sr^{87}/Sr^{86} ratio (.7050). The higher concentrations of K and Cs are also consistent with the higher Rb/Sr and Sr^{87}/Sr^{86} ratios. The basalts and andesites from 90° E Ridge are distinctly different with higher concentrations of K (1494-10407 ppm) and K/U ratio (14940-62250) (Figures 3 and 5; Table 2), from ocean ridge basalts but similar to alkali island basalts. It appears that differentiation has progressed further than in the seismic mid-ocean ridges, as is revealed by the presence of complete sequence of rocks from picrite basalts to andesites. K/U - K relationship for 90° E Ridge rocks, in fact lends support to the above interpretation.

Similarly K concentrations correlate well with K/Cs ratios (Figure 4).

The lower K/Rb (670) and K/Cs (3809) match fairly well with ocean island basalts (Table 2). Further the higher Sr^{87}/Sr^{86} (.7050) and K_2O/K_2O+Na_2O (.135) ratios of basalts are strikingly similar to alkali island basalts and appears to be least depleted in LIL elements (Figure 6).

We also report LIL and Sr^{87}/Sr^{86} data for a glassy basalt from site 215 (Table 1) which lies close to the western flank of 90°E Ridge, particularly to site 214.

Site 215 glassy basalt has higher alkali element concentrations and Sr^{87}/Sr^{86} ratio and corresponding ratios such as K/Cs, K/Rb, Rb/Sr and K_2O/K_2O+Na_2O and lie within the seamount and/or island basalt range (Table 2, Figure 6). Frey et al. (1977) reported relatively higher REE concentrations for 215 basalts than basalts from sites 214 and 216 from 90°E Ridge and suggested "that site 215 is beyond the influence of Ninetyeast Ridge." Indeed site 215 basalts are unpublished REE range for 215 basalts (2 samples) overlap the 90°E Ridge basalt range (site 214 : 3 samples, site 216 : 1 sample). For example, La varies between 11 to 31 ppm in 214 and 216 basalts; and 17.6 to 17.8 ppm in 215 basalts. Similarly our unpublished Sr^{87}/Sr^{86} ratios for 90°E Ridge rocks match fairly well with higher Sr^{87}/Sr^{86} ratio of 0.7047 for 215 glassy basalt. Thus these geochemical similarities between sites 215 and 214 and 216 basalts may possibly suggest (1) some sort of genetic relationship between 215 and 90°E Ridge rocks, or (2) 215 rocks probably resulted from off ridge volcanism which is nearly identical to 90°E Ridge volcanism. In fact, Bonatti and Fisher (1971) have shown that East Pacific Rise and non-ridge basalts erupted at various distances from the axis of the ridge do not differ systematically, suggesting that both types of basalts are generated in the upper mantle within the same range of depths (pressures) and conditions. However, Subbarao (1972) presented contrary geochemical evidences from the abyssal hill rocks from the Pacific Ocean to that of Bonatti and Fisher (1971). Therefore, we perhaps have a situation similar to that shown by Bonatti and Fisher (1971), in the Indian Ocean also particularly between site 215 and 90°E Ridge. However, with this meagre data it would be premature to conclude definitely on the relationship between site 215 and 90°E Ridge rocks. A detailed geochemical account on the 90°E Ridge rocks forms another paper which is now under preparation.

Where or not site 215 basalts form part of 90°E Ridge, the geochemical data for both these suites strongly resemble alkali basalts from islands and seamounts, but significantly different from seismically active mid-oceanic ridge rocks. Thus we regard this as an additional evidence to bring out tholeiitic and alkali types of volcanism on the seismic Mid-Indian Oceanic Ridge and aseismic 90°E Ridge respectively.

Geochemical Differences between Mid-Indian Oceanic Ridge and Carlsberg Ridge

We made an attempt to see whether any difference exists between MIOR and CR as well as other portions of the ridge. Although the normative composition of Di-ol-Hy-Ne-Qz (Figure 2b) does not hold much promise due to clustering of MIOR and CR plots in a narrow area, the other elemental parameters such as $MgO-TiO_2$ variations appear to indicate some chemical differences. Earlier Hekinian (1968) showed a crude separation between MIOR and CR rocks using $MgO-TiO_2$ diagram. We have now included some additional data for CR (Cann and Vine, 1966) and MIOR (Balashov et al., 1970; Engel and Fisher, 1975) in the diagram and notice two prominent positive and negative trends for MIOR and CR rocks respectively (Figure 7).

Based on major element chemistry, Hekinian (1968) suggested a pyroxene

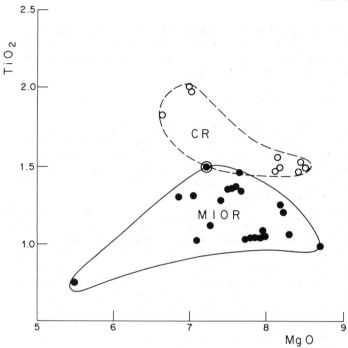

Fig. 7. Variation of TiO_2 vs MgO for basalts from MIOR and CR.
References: MIOR: Engel et al. (1965b), Hekinian (1968);
Balashov et al. (1970); Engel and Fisher (1975); CR: Cann and Vine
(1966); Cann (1970) and Hekinian (1968).

olivine-rich source depleted in plagioclase and a plagioclase enriched source for CR and MIOR rocks respectively. We tentatively interpret the two trends exhibited by CR and MIOR rocks in TiO_2-MgO diagram as a reflection of low pressure crystallization-differentiation possibly with a shift or migration of the source from the northern end (CR) towards central portions of MIOR. The mafic components appear to have concentrated in the CR suite while residual silicic phases are enriched in MIOR.

The characteristic concentration of K, Rb, Cs, Sr and Sr^{87}/Sr^{86} and $(La/Sm)_{e.f.}$ ratios are presented for a limited number of rocks from CR (Cann, 1970; Hart, 1971 and personal comm; Schilling, 1971). The basalt from CR shows lower concentrations of Cs (.018 ppm) and higher K/Cs ratio (62000) and lower $(La/Sm)_{e.f.}$ (.52) and Sr^{87}/Sr^{86} (.7028). Schilling (1971, p. 668) reported $(La/Sm)_{e.f.}$ values of 1.03 (1 sample) and 0.52 (average of 4 samples) for basalts from MIOR and CR rocks respectively. Schilling (1975) subdivided the mid-oceanic ridges into three different genetic kinds based on the $(La/Sm)_{e.f.}$ ratio, i.e., plume ridge segments $(La/Sm)_{e.f.} <1$) normal ridge segments $(La/Sm)_{e.f.} <1$) and transitional ridge segments $(La/Sm)_{e.f.} \sim 1$). Whether or not we accept this suggestion in total, we regard the $(La/Sm)_{e.f.}$ difference between the MIOR and CR rocks is significant. Perhaps this variation in $La/Sm)_{e.f.}$ may reflect the chemical difference in the source area(s) and also confirm the Sr isotopic difference between these two ridges.

The limited elemental and $(La/Sm)_{e.f.}$ and Sr^{87}/Sr^{86} data suggest a relatively undepleted or less depleted source for MIOR rocks while the CR rocks might be tapping a relatively depleted source similar to MAR and EPR. It is rather difficult to infer whether these differences are due to totally different sources or chemical variations within a single source due to crystallization-differentiation possibly during migration of the source from north to south. Hence these interpretations cannot be accepted without reservation until more analytical and isotopic data are available from different parts of the MIOR.

Summary and Conclusions

1) Relatively unaltered major element analyses of oceanic basalts from MAR, EPR and MIOR are presented on the normative Di-Hy-Ol-Ne-Qz tetrahedron of Yoder and Tilley (1962) using a fixed 1.5% Fe_2O_3. MAR and EPR rocks display variation from nepheline normative tholeiites to quartz normative alkali types, whereas MIOR volcanics are concentrated in the upper half of the diagram more towards the diopside apex. This suggests that MIOR are less basic than MAR and EPR. These differences may perhaps be due to varying partial melting processes possibly at different depths of the mantle and also at different times in the geological past.

2) The volcanic rocks from the aseismic 90° E Ridge (DSDP legs 22 and 26) include pyroxene basalts, amygdular-vesicular basalts, picritic basalts and intermediate differentiated rocks oceanic andesites), which display similar variation in the normative tetrahedron from olivine normative picrite types to quartz normative alkali basalts. This variation is suggestive of wide range of differentiation which is not duplicated in other oceanic ridges. The basalts have an alkali affinity similar to seamount and island basalts; similarly the oceanic andesites also follow the comparable fractionation trends to the St-Paul and New Amsterdam volcanic suites. The basalts from the aseismic 90°E Ridge differ from the seismic mid-oceanic ridges in having higher Fe_2O_3+FeO (> 10%), TiO_2 (> 2%) and K_2O (0.2 to 1.5%).

Six new major element analyses from site 215 which is very close to the western flank of the 90° E Ridge (close to site 214 on the 90° E Ridge) are also reported. These samples are diopside-rich (normative) except for one nepheline normative-rich basalt. Fresh basalt glass is also encountered from this site. Site 215 crystalline and glassy basalts are enriched in LIL, REE, Sr^{87}/Sr^{86} similar to 90°E Ridge basalts (sites 214 and 216 basalts) suggesting possible genetic relationship between site 215 and sites 214 and 216 basalts or the presence of nearly similar volcanism for 90° E Ridge rocks (sites 214 and 216) and site 215. In general the rocks from site 215 and 90°E Ridge (sites 214 and 216) are transitional tholeiite alkali basalts and fairly matches with that of island and/or seamount basalts.

3) The variation of some of the LIL elements (K, Rb, Sr, Cs and U) and respective elemental ratios of oceanic rocks suggest that the sources of MAR and EPR are highly depleted in the LIL elements while the source of MIOR is least or less depleted. Ninetyeast Ridge is in fact enriched in these elements similar to alkali island basalts. Similarly K/Cs versus K, and K/U versus K for all these rocks show negative and positive trends respectively. While the MAR and EPR rocks display some overlap, the MIOR rocks show well defined fields in these chemical diagrams. This variation is interpreted in terms of continuous depletion in the mantle. Interestingly, the plots of basalts from sites 214 show relatively undepleted nature of their source areas in these chemical diagrams.

The Se and Pb isotopes of MIOR are higher than those cited for MAR and EPR suggesting that the source of MIOR was depleted in LIL elements more recently and/or to a less degree than the basalts from the other oceans.

4) The higher LIL elemental concentrations, K_2O/K_2O+Na_2O (0.135-0.225), Sr^{87}/Sr^{86} (0.7047-0.7050) ratios and lower K/U (4000) and

K/Cs (3809-16586) ratios of 90°E Ridge rocks fairly match with the
geochemistry of alkali island basalts. We interpret these geochemical differences between seismic and aseismic ridges and similarities
between aseismic ridges and islands in terms of a "zoning-depletion"
model, with the presence of alkali poor and alkali enriched zones in
the mantle which probably were produced during early differentiation
history of the earth's mantle and continued to retain their heterogeneity due to continuous depletion and volcanic processes. This
interpretation is purely tentative and needs to be tested as more
chemical and isotopic analyses are available on the Indian Ocean
rocks in the future. However, our unpublished rare earth element
and Sr^{87}/Sr^{86} data on 90°E Ridge rocks also support our observations
and tentative interpretations reached in this paper. Subbarao
(1972), and Subbarao and Hekinian (1976), under review, demonstrated
similar zonal structure for the mantle beneath the Pacific Ocean.

5) Geochemical differences between MIOR and CR rocks are brought
out using $MgO-TiO_2$ diagram. Negative and positive trends are
exhibited by these two suites of rocks possibly due to low pressure
crystallization-differentiation in the magma(s).

The lower LIL elemental concentrations and $(La/Sm)_{e.f.}$ (0.52) and
Sr^{87}/Sr^{86} (0.7028) ratios of CR rocks contrast with higher LIL
elemental concentrations and $(La/Sm)_{e.f.}$ (1.03) and Sr^{87}/Sr^{86}
(0.7031-0.7043) ratios of MIOR rocks. These data probably indicate
the presence of relatively depleted and undepleted mantle zones as
the sources of CR and MIOR rocks. However, with this meager data,
particularly on CR rocks, it is rather difficult to conclude whether
these differences are entirely due to different sources or merely
a reflection of chemical variation within a single source. Perhaps
the large isotopic difference between portions of the ridge (i.e.,
CR and MIOR) need to be evaluated carefully. Whether or not we
resolve this problem of source material, the presence of pronounced
geochemical and isotopic variations along the Indian Ocean Ridge
is significant.

Acknowledgments. We are extremely grateful to M. Sankar Das and G.R.
Reddy of the Bhabha Atomic Research Center, Bombay, for providing neutron
activation laboratory facilities and assistance and instruction during the
course of this work. One of us (K.V.S.) is indebted to R. L. Fleischer,
General Electric Company, for introducing him to the fission track analysis
and also providing G.E. standard glass.

We are grateful to the DSDP organizers for providing leg 22 basalt
samples. Thanks are also due to J. Matsuda, University of Tokyo and
R.E. Zartman, United States Geological Survey, for analyzing two samples
for Sr and Pb isotopes respectively, and S. R. Hart, Department of
Terrestrial Magnetism, Carnegie Institution, Washington, for making
available Sr^{87}/Sr^{86} ratio for Carlsberg Ridge basalt. We also thank
S. Ramanan, Indian Instiute of Technology for his assistance in computer
calculations.

References

Aumento, F., 1968, The Mid-Atlantic Ridge near 45°N. II. Basalts from
the area of Confederation Peak, Can. Jour. Earth Sci., 5, 1-20, 1968.
Aumento, F., Uranium content of mid-oceanic basalts, Earth and Planet. Sci.
Lett., 11, 90-94, 1971.
Balashov, Yu. A., L. V. Dmitriev, and A. Ya., Sharaskin, Distribution of
the rare earths and Yttrium in the bedrock of the ocean floor,
Geokhimiya, 6, 647-660 (In Russian), translated in Geochem. Internat.
7, no. 1, 456-468, 1970.

Belousov, I. M., L. K. Zatonskii, V. F. Kanaev, and H. A. Marova, Bathymetric map of the Indian Ocean, publ. by the Inst. of Oceanology, U.S.S.R., 1962.

Bezrukov, P. L., A. Ya., Krylov, and V. I. Chernysheva, Petrography and absolute age of the Indian Ocean floor basalts, Okeanologiia Aka. Nauk, USSR, 2, 261-266, 1966.

Bonatti, E., and D. E. Fisher, Oceanic basalts: Chemistry versus distance from oceanic rises, Earth Planet. Sci. Lett., 11, 307-311, 1971.

Bougault, H., and R. Hekinian, Rift Valley in the Atlantic Ocean near 36°50'N: Petrology and geochemistry of basaltic rocks, Earth Planet. Sci. Lett., 24, 249-261, 1974.

Cann, J. R., and F. J. Vine, An area on the crest of the Carlsberg Ridge: Petrology and magnetic survey, Phil. Trans. Roy. Soc. London, A259, 198-217, 1966.

Cann, J. R., Spilites from the Carlsberg Ridge, Indian Ocean, Jour. Petr., 10, 1-19, 1969.

Cann, J. R., Rb, Sr, Y, Zr and Nb in some ocean floor rocks, Earth Planet. Sci. Lett., 10, 7-11, 1970.

Dmitriev, L. V., Petrochemical study of the basaltic basement of the Mid-Indian Ridge: Leg 24, Djibouti to Mauritius, Initial Reports of the Deep Sea Drilling Project, 24, 767-799, 1974.

Engel, A.E.J., and C. E. Engel, Igneous rocks of the East Pacific Rise, Science, 146, 477-485, 1964.

Engel, A.E.J., C. E. Engel, and R.G. Haven, Chemical characteristics of oceanic basalts and the upper mantle, Bull. Geol. Soc. Amer. 76, 719-734, 1965b.

Engel, C. E., R. L. Fisher, and A.E.J. Engel, Igneous rocks of the Indian Ocean floor, Science, 150, 605-610, 1965.

Engel, C. G., and R. L. Fisher, Granitic to ultramafic rock complexes of the Indian Ocean ridge system, western Indian Ocean, Bull. Geol. Soc. Amer., 86, 1553-1578, 1975.

Fisher, D. E., U/He ages as indicators of excess Argon in deep sea basalts, Earth Planet. Sci. Lett., 14, 255-258, 1972.

Frey, F., J. R. Dickey, G. Thompson, and W. B. Bryan, Eastern Indian Ocean DSDP sites: correlations between petrography, geochemistry and tectonic setting, in Indian Ocean Geology and Biostratigraphy, AGU, Washington, this volume, 1977.

Funkhouser, J. G., D. E. Fisher, and E. Bonatti, Excess Argon in deep sea rocks, Earth Planet. Sci. Lett., 5, 95-100, 1968.

Gunn, B. M., E. E. Abranson, J. Nougier, R. Coy-Yll, and N. D. Watkins, Geochemistry of an oceanite-Ankaramite basalt suite from East Island-Crozet Archipelago, Contr. Mineral. Petrol., 28, 319-339, 1970.

Hart, S. R., K, Rb, Cs contents and K/Rb, K/Cs ratios of fresh and altered submarine basalts, Earth Planet. Sci. Lett., 6, 295-303, 1969.

Hart, S. R., K, Rb, Cs, Sr and Ba contents and Sr isotope ratios of ocean floor basalts, Phil. Trans. Roy. Soc. London, A268, 573-582, 1971.

Hedge, C. E. and Z. E. Peterman, The strontium isotopic composition of basalts from the Gorda and Juan de Fuca Rises, northeastern Pacific Ocean, Contr. Mineral. Petrol., 27, 114-120, 1970.

Hedge, C. E., N. D. Watkins, R. A. Hildreth, and W. P. Doering, Sr^{87}/Sr^{86} ratios in basalts from islands in the Indian Ocean, Earth Planet. Sci. Lett., 21, 29-34, 1973.

Hekinian, R., Rocks from the Mid-Oceanic Ridge in the Indian Ocean, Deep Sea Res., 15, 195-213, 1968.

Hekinian, R., Chemical and mineralogical differences between abyssal hill basalts and ridge tholeiites in the eastern Pacific Ocean, Mar. Geol., 11, 77-91, 1971.

Hekinian, R., Petrology of igneous rocks from leg 22 in the northeastern Indian Ocean, Initial Reports of the Deep Sea Drilling Project, 22, Washington (U. S. Government Printing Office), 413-447, 1974a.

Hekinian, R., Petrology of the Ninetyeast Ridge (Indian Ocean) compared to other aseismic ridges, Contr. Mineral. Petrol., 43, 125-147, 1974b.

Kay, R., N. J. Hubbard, and P. W. Gast, Chemical characteristics and origin of oceanic ridge volcanic rocks, J. Geophys. Res., 75, 1585-1613, 1970.

Kempe, D.R.C., The petrology of the basalts, leg 26, Initial Reports of the Deep Sea Drilling Project, 26, Washington (U. S. Government Printing Office), 465-503, 1974.

McDougal, T., and W. Compston, Strontium isotope composition and potassium-rubidium ratios in some rocks from Reunion and Rodriguez Islands, Indian Ocean, Nature, 207, 252-253, 1965.

Miyashiro, A., F. Shido, and M. Ewing, Diversity and origin of abyssal tholeiite from the Mid-Atlantic Ridge near 24° and 30°N latitude, Contr. Mineral. Petrol., 23, 38-52, 1969.

Melson, W. G., T. L. Vallier, T. L. Wright, G. Byerly, and J. Nelen, Chemical diversity of abyssal volcanic glass erupted along Pacific, Atlantic, and Indian Ocean sea floor spreading centers in Geophysics of the Pacific Ocean Basin, Jour. Geophys. Res. Monograph Ser. (1977).

Moore, J. G., Water content of basalt erupted on the ocean floor, Contr. Mineral. Petrol., 28, 272-279, 1970.

Muir, I. D., and C. E. Tilley, Basalts from the northern part of rift zone of the Mid-Atlantic Ridge, Jour. Petr., 5, 409-434, 1964.

Nicholls, G. D., A. J. Nalwalk, and E. E. Hays, The nature and composition of rock samples dredged from the Mid-Atlantic Ridge between 22°N and 52°N, Mar. Geol. 1, 333-343, 1964.

Oversby, V., Genetic relations among the volcanic rocks of Reunion: chemical and Pb isotopic evidence, Geochim. Cosmochim. Acta, 36, 1167-1179, 1972.

Peterman, Z. E., and C. E. Hedge, Related Sr isotopic and chemical variations in oceanic basalts, Bull. Geol. Soc. Amer., 82, 493-500, 1971.

Peterman, Z. E., R. G. Coleman, and R. Hildreth, Sr^{87}/Sr^{86} in mafic rocks of the Troodos Massif, Cyprus, U. S. Geol. Surv. Prof. Paper 750D, D157-D161, 1971.

Schilling, J. G., Sea-floor evolution: rare earth evidence, Phil. Trans. Roy. Soc. London, A268, 663-706, 1971.

Schilling, J. G., Rare earth variations across normal segments of the Reykjanes Ridge, 60°-53°N, Mid-Atlantic Ridge, 29°S and East Pacific Rise, 2°-19°S and evidence on the composition of the underlying low-velocity layer, J. Geophys. Res., 80, 1459-1473, 1975.

Sclater, J. G., and R. L. Fisher, Evolution of the east Central Indian Ocean, with emphasis on the tectonic setting of the Ninetyeast Ridge, Bull. Geol. Soc. Amer., 85, 683-702, 1974.

Shido, F., and A. Miyashiro, Crystallization of abyssal tholeiites, Contr. Mineral. Petrol, 31, 251-266, 1971.

Strong, D. F., Petrology of the island of Moheli, Western Indian Ocean, Bull. Geol. Soc. Amer., 83, 389-406, 1972.

Subbarao, K. V., The strontium isotopic composition of basalts from the East Pacific and Chile Rises and abyssal hills in the East Pacific Ocean, Contr. Mineral. Petrol., 37, 111-120, 1972.

Subbarao, K. V., and C. E. Hedge, K, Rb, Sr, and Sr^{87}/Sr^{86} in rocks from the Mid-Indian Oceanic Ridge, Earth Planet. Sci. Lett., 18, 223-228, 1973.

Subbarao, K. V., G. S. Clark, and R. B. Forbes, Strontium isotopes in some seamount basalts from the northeastern Pacific Ocean, Jour. Earth Sci., 10, 1479-1484, 1973.

Subbarao, K. V., and R. Hekinian, Rocks from the Mathematician seamount, Northeast Pacific Ocean (under review).

Tatsumoto, M., C. E. Hedge, and A.E.J., Engel, Potassium, rubidium, strontium, thorium, uranium and the ratio of strontium-87 to strontium-86 in oceanic tholeiitic basalts, Science, 150, 886-888, 1965.

Tatsumoto, M., Genetic relations of oceanic basalts as indicated by lead isotopes, Science, 153, 1094-1101, 1966.

Tera, F., and G. J. Wasserburg, U-Th-Pb analyses of soil from the Sea of Fertility, Earth Planet. Sci. Lett., 13, 457-466, 1972.

Thompson, G., W. B. Bryan, F. A. Frey, and C. M. Sung, Petrology and geochemistry of basalts and related rocks from sites 214, 215, 216, DSDP leg 22, Indian Ocean, Initial Reports of the Deep Sea Drilling Project, 22, Washington (U. S. Government Printing Office), 456-468, 1973.

Upton, B.G.J., W. J. Wadsworth, Aspects of magmatic evolution on Reunion Island, Phil. Trans. Roy. Soc. London, 271, 105-130, 1972.

Yoder, H. S., C. E. Tilley, Origin of basaltic magmas: an experimental study of natural and synthetic rock systems, Jour. Petr., 3, 342-532, 1962.

Zielinski, R. A., Trace element evaluation of a suite of rocks from Reunion Island, Indian Ocean, Geochim. Cosmochim. Acta, 39, 713-734, 1975.

CHAPTER 10. SEISMIC VELOCITIES AND ELASTIC MODULI OF IGNEOUS AND METAMORPHIC ROCKS FROM THE INDIAN OCEAN

Nikolas I. Christensen

Department of Geological Sciences and Graduate Program in Geophysics
University of Washington
Seattle, Washington

Introduction

Within the past decade, over 60 deep crustal refraction profiles have been completed in the Indian Ocean. The interpretation of data from these profiles requires detailed information on seismic velocities of Indian Ocean rocks at elevated pressures. This paper is intended to summarize experimental studies of seismic velocities in rocks from the Indian Ocean and compare the laboratory measurements with seismic refraction velocities. Emphasis is placed on the nature of layer 2. Several new measurements of velocities at elevated pressures are given for Indian Ocean rocks obtained from the Deep Sea Drilling Project and by dredging.

In addition to compressional (V_p) and shear (V_s) velocities, several parameters useful in describing elastic properties and interpreting composition from seismic refraction studies are presented for Indian Ocean rocks. These include the shear modulus or rigidity (μ), Lamé's constant (λ), the bulk modulus (K), Young's modulus (E), compressibility (β), Poisson's ratio (σ) and the seismic parameter (ϕ) defined as K/ρ where ρ is density. Although there are several techniques for obtaining these constants, they all can be conveniently calculated at various pressures from V_p, V_s and ρ. A detailed tabulation of the relationships between the various elastic moduli and velocities and density is given by Birch (1961).

Velocity and Density Data

Variations of V_p and V_s with pressure for a basalt are shown in Figure 1. The rapid increase in velocity at low pressures is a consequence of the closure of grain boundary cracks and the accompanying reduction of porosity. Above a few kilobars the pore spaces are reduced in volume and the changes in velocity at higher pressures are often primarily related to the effect of pressure on the individual mineral components.

The importance of using velocities from water saturated samples in the interpretation of relatively shallow seismic data has been discussed in several studies (e.g., Dortman and Magid, 1968; Nur and Simmons, 1969; Christensen and Salisbury, 1972) and the influence of water saturation on velocities is illustrated in Figure 1. Here "air dry" refers to a sample with a moisture content related to the humidity of the laboratory prior to the measurements. During the runs pore pressures were kept minimal; water from the saturated samples was allowed to drain

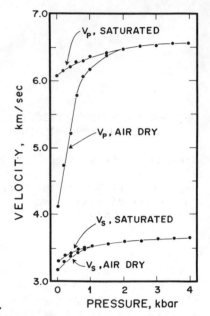

Fig. 1. The effect of water saturation on velocities for a typical oceanic basalt.

from the rock pore spaces into 100 mesh screens placed between the rock cores and copper jackets.

Of the 53 sites drilled in Legs 22 through 27, basalt was recovered at 30 sites. Compressional wave velocities have been reported at elevated pressures for basalts from 17 of these sites and shear wave velocities have been measured for samples from 7 sites. These measurements and bulk densities of the basalts are summarized in Table 1.

Sixty new measurements of velocities in Indian Ocean DSDP rocks to 6 kbar are presented in Table 2. Included are measurements of V_p and V_s for basalt from 8 sites for which no velocities have been previously reported for igneous rocks and additional measurements of V_s for 10 sites for which only V_p have been previously reported. Additional data are also given for holes in which only air dry samples were studied. All samples were water saturated prior to the measurements. Because of earlier findings that DSDP basalts are nearly isotropic (e.g., Christensen and Salisbury, 1972) velocities were measured in only one direction for each interval sampled.

With the exceptions of andesites from Site 214, a limestone from Site 213 and a volcanic sediment from Site 253, the rocks of Table 2 are basalts. Bulk densities of the samples are included in Table 2. Petrographic descriptions can be found in the initial reports and chemical analyses have been reported for several rocks at intervals close to where the samples for velocity measurements were selected.

The velocities were obtained by measuring the transit times of compressional and shear waves through cylindrical rock cores using a similar technique to that described by Birch (1960). Barium titanate and AC-cut quartz transducers with natural resonance frequencies of 1 to 2 MHz were used to generate the compressional and shear waves, respectively. Temperatures of all runs were between 20° and 30°C.

The elastic constants (K, E, μ, λ, β and σ), V_p/V_s, and ϕ calculated from the velocities and densities at selected pressures are given in Table 3. The velocities and densities used in the calculations were corrected for length changes at elevated pressures using an iterative routine and the dynamically determined compressibilities.

TABLE 1. Summary of Indian Ocean Basalt
Velocity Measurements at 0.5 kbar

Sample	Density (g/cm³)	V_p (km/sec)	V_s (km/sec)	Reference
24-231-64-1	2.74	4.74*	-	Schreiber et al. (1974)[1]
24-231-64-3	2.75	5.11	-	"
24-233-9-1	2.67	4.42	-	"
24-233-13-1	2.73	4.96*	-	"
24-235-20-2	2.71	5.26*	-	"
24-235-20-4	2.75	5.40*	-	"
24-235-20-5	2.67	5.31*	-	"
24-236-33-3	2.35	3.68	-	"
24-236-34-2	2.68	4.80*	-	"
24-236-36-1	2.73	4.93*	-	"
24-236-37-1	2.76	5.53	-	"
24-238-55-2	2.81	5.32	-	"
24-238-57-3	2.82	5.33*	-	"
24-238-59-2	2.90	5.96	-	"
24-238-61-4	2.82	5.54	-	"
24-238-64-1	2.71	5.57	-	"
25-239-20-1 (125-128)	2.72	5.43	2.97	Christensen et al. (1974a)[2]
25-239-21-1 (46-49)	2.76	5.47	2.99	"
25-240-7-1 (120-123)	2.80	5.87	3.26	"
25-245-19-1 (37-40)	2.89	6.11	3.30	"
25-248-15-1 (35-38)	2.68	5.39	2.91	"
25-248-17-2 (122-125)	2.76	5.88	3.23	"
25-249-33-2 (27-30)	2.32	2.61	2.15	"
25-249-33-2 (126-129)	2.19	3.84	1.97	"
25-249-33-3 (128-131)	2.39	4.23	2.23	"
26-250A-26-2 (140)	2.85	6.22	-	Hyndman (1974)[2]
26-250A-26-6 (58)	2.82	6.17	-	"
26-251A-31-2 (84)	2.82	5.61	-	"
26-251A-31-3 (50)	2.86	6.49	-	"
26-251A-31-4 (48)	2.93	6.13	-	"
26-251A-31-5 (105)	2.94	6.06	-	"
26-254-31-1 (111)	2.74	5.52	-	"
26-254-35-1 (107)	2.75	4.97	-	"
26-254-36-3 (105)	2.82	5.36	-	"
26-256-10-2 (68)	2.96	5.88	-	"
26-256-10-3 (85)	2.96	6.47	-	"
26-257-11-2 (74)	2.74	5.21	-	"
26-257-12-1 (130)	2.73	6.11	-	"
26-257-12-3 (35)	2.73	6.13	-	Hyndman (1974)[2]
26-257-13-3 (15)	2.82	5.60	-	"
26-257-14-2 (95)	2.75	5.27	-	"
26-257-15-1 (133)	2.89	6.13	-	"
27-259-35-1 (128-131)	2.26	3.78	1.92	Christensen et al. (1974b)[2]
27-259-36-2 (66-70)	2.08	3.47	1.78	"

TABLE 1 (continued)

Sample	Density (g/cm^3)	V_p (km/sec)	V_s (km/sec)	Reference
27-259-37-2 (105-109)	2.43	4.27	2.27	"
27-259-39-1 (135-138)	2.55	4.66	2.51	"
27-259-41-3 (53-56)	2.66	4.95	2.78	"
27-261-33-1 (55-59)	2.59	4.97	2.63	"
27-261-33-1 (131-134)	2.79	5.55	2.86	"
27-261-34-3 (69-71)	3.00	6.50	3.60	"
27-261-35-3 (84-87)	2.60	4.47	2.30	"
27-261-37-2 (92-95)	2.72	5.35	2.93	"
27-261-38-2 (23-26)	2.75	5.50	2.81	"
27-261-38-5 (94-97)	2.80	5.77	3.18	"

* Mean of measurements in more than one direction.
1 Air dry.
2 Water saturated.

Some samples included in the present study were obtained from intervals close to those in which measurements have been reported from other laboratories. Comparisons of these data are presented in Table 4. Considering that different cores were studied, as is illustrated by density comparisons, the agreement of the velocity data of Table 1 with that of Hyndman (1974) is excellent. This agreement is probably a consequence of similarities in the initial preparation of the samples (i.e., water saturation) prior to the measurements. The compressional wave velocities of Schreiber et al. (1974) for air dry samples, on the other hand, are significantly lower than the velocities measured for saturated specimens.

In addition, velocities and densities have been measured in basalt, serpentinite and metagabbro dredged from the Central Indian Ridge. Velocities as a function of pressure, densities and calculated elastic constants are given in Tables 5 and 6. For each sample velocities were measured to 10 kbar in three cores cut in mutually perpendicular directions.

The basalt, recovered at depths of 3540 to 3880 meters at 12°25'S and 65°56'E, is very fresh and contains subradiating laths of plagioclase with minor pyroxene and olivine in a glassy matrix. The serpentinite and metagabbro were dredged at depths of 4240 to 4880 meters at 17°04'S and 66°50'E. The serpentinite contains lizardite with minor chrysotile and opaques. Relic olivine and enstatite constitute less than 10% of the rock. The metagabbro shows abundant evidence of recrystallization. Hornblende and clusters of fibrous actinolite after clinopyroxene and enstatite form approximately 60% of the rock. Plagioclase is variable in composition ranging from An_{30} to An_{50}. Less than 5% relic enstatite is also present. Preferred mineral orientation is lacking though evidence of deformation is indicated by bent plagioclase lamellae. Brief descriptions of the dredge hauls from which the samples of Tables 5 and 6 were obtained are given by Engel and Fisher (1969).

TABLE 2. Compressional (P) and Shear (S) Wave Velocities

Sample	Bulk Density g/cc	Mode	Velocity (km/sec) at Varying Pressures							
			0.2 kb	0.4 kb	0.6 kb	0.8 kb	1.0 kb	2.0 kb	4.0 kb	6.0 kb
211-12-1, 70-73 cm	2.670	P S	5.18 2.57	5.21 2.58	5.24 2.59	5.25 2.60	5.27 2.62	5.33 2.66	5.41 2.72	5.48 2.75
211-15-2, 69-71 cm	2.430	P S	4.52 2.19	4.55 2.22	4.58 2.25	4.61 2.28	4.63 2.31	4.70 2.44	4.84 2.54	4.98 2.58
211-15-4, 67-70 cm	2.480	P S	4.91 2.30	4.93 2.34	4.95 2.37	4.96 2.40	4.97 2.43	5.01 2.53	5.08 2.63	5.15 2.67
212-39-1, 58-61 cm	2.290	P S	3.99 1.96	4.02 2.00	4.04 2.02	4.06 2.04	4.08 2.06	4.15 2.09	4.29 2.14	4.43 2.19
213-17-2, 43-49 cm (limestone)	2.642	P S	6.34 3.24	6.35 3.25	6.36 3.26	6.37 3.27	6.37 3.27	6.39 3.29	6.43 3.30	6.47 3.31
214-48-1, 87-90 cm (andesite)	2.329	P S	4.43 1.80	4.44 1.86	4.46 1.93	4.48 1.98	4.50 2.03	4.57 2.25	4.71 2.51	4.84 2.59
214-48-2, 7-10 cm (andesite)	2.677	P S	5.74 3.09	5.75 3.11	5.76 3.12	5.76 3.13	5.77 3.14	5.79 3.17	5.84 3.19	5.88 3.22
214-49-1, 143-146 cm (andesite)	2.673	P	5.95	5.97	5.98	5.99	6.00	6.05	6.09	6.12
214-53-1, 50-53 cm	2.738	P S	5.00 2.62	5.02 2.65	5.04 2.67	5.06 2.69	5.06 2.71	5.13 2.75	5.23 2.79	5.34 2.89
216-37-2, 77-80 cm	2.602	P S	4.66 2.37	4.69 2.41	4.72 2.43	4.74 2.46	4.76 2.48	4.82 2.54	4.90 2.63	4.98 2.69

TABLE 2 (continued)

Sample	Bulk Density g/cc	Mode	Velocity (km/sec) at Varying Pressures							
			0.2 kb	0.4 kb	0.6 kb	0.8 kb	1.0 kb	2.0 kb	4.0 kb	6.0 kb
220-18-4, 75-79 cm	2.605	P	4.48	4.55	4.60	4.64	4.67	4.79	4.89	4.98
221-19-2, 18-21 cm	2.765	P S	5.50 3.02	5.53 3.03	5.56 3.05	5.58 3.07	5.61 3.09	5.69 3.15	5.80 3.22	5.92 3.24
223-40-2, 46-50 cm	2.084	P S	2.89 1.44	3.02 1.46	3.11 1.49	3.17 1.52	3.24 1.54	3.51 1.66	3.88 1.83	4.21 1.97
231-64-1, 60-64 cm	2.781	P S	5.54 2.94	5.57 2.95	5.59 2.98	5.61 2.99	5.62 3.00	5.65 3.01	5.70 3.01	5.76 3.02
233A-8-2, 63-68 cm	2.756	P S	5.21 2.71	5.24 2.73	5.26 2.75	5.28 2.76	5.30 2.77	5.37 2.80	5.46 2.84	5.55 2.89
235-19-1, 124-128 cm	2.681	P S	5.30 2.75	5.33 2.76	5.35 2.77	5.36 2.79	5.37 2.80	5.42 2.85	5.49 2.90	5.57 2.93
236-34-1, 147-150 cm	2.672	P S	5.06 2.70	5.10 2.70	5.14 2.71	5.17 2.71	5.19 2.72	5.25 2.75	5.36 2.80	5.47 2.85
238-55-2, 10-13 cm	2.860	P S	5.90 3.17	5.93 3.19	5.95 3.21	5.96 3.22	5.98 3.23	6.05 3.27	6.16 3.32	6.29 3.35
238-59-2, 26-29 cm	2.920	P S	6.22 3.37	6.24 3.39	6.25 3.39	6.26 3.41	6.27 3.42	6.29 3.43	6.34 3.44	6.38 3.45
238-62-2, 75-79 cm	2.859	P S	5.93 3.30	5.95 3.31	5.97 3.31	5.98 3.31	6.00 3.32	6.04 3.33	6.10 3.34	6.15 3.36

TABLE 2 (continued)

Sample	Bulk Density g/cc	Mode	Velocity (km/sec) at Varying Pressures								
			0.2 kb	0.4 kb	0.6 kb	0.8 kb	1.0 kb	2.0 kb	4.0 kb	6.0 kb	
238-64-1, 29-131 cm	2.864	P S	5.99 3.25	6.02 3.27	6.04 3.29	6.06 3.30	6.08 3.31	6.14 3.33	6.18 3.34	6.23 3.35	
250-A-25-3, 97-101 cm	2.821	P S	5.83 2.10	5.86 3.00	5.88 3.01	5.89 3.01	5.90 3.02	5.94 3.13	6.01 3.15	6.09 3.17	
251A-31-2, 82-87 cm	2.792	P S	5.40 2.81	5.47 2.83	5.54 2.86	5.56 2.88	5.59 2.90	5.69 2.98	5.79 3.04	5.90 3.10	
253-24-1, (volcanic arenite) 133-136cm	2.221	P S	4.30 1.87	4.36 2.01	4.40 2.10	4.43 2.17	4.45 2.22	4.51 2.33	4.62 2.43	4.72 2.46	
254-35-2 47-52 cm	2.741	P S	4.87 2.47	4.97 2.49	5.03 2.51	5.08 2.53	5.12 2.54	5.30 2.62	5.35 2.70	5.46 2.76	
254-36-3, 110-114 cm	2.785	P S	5.36 2.69	5.39 2.71	5.42 2.73	5.44 2.74	5.46 2.76	5.52 2.81	5.60 2.89	5.68 2.93	
256-9-2, 50-53 cm	2.652	P S	4.68 2.32	4.75 2.34	4.80 2.37	4.83 2.39	4.86 2.42	4.95 2.51	5.12 2.61	5.25 2.68	
257-11-2, 57-61 cm	2.665	P S	4.96 2.73	5.03 2.77	5.08 2.79	5.13 2.81	5.16 2.83	5.27 2.86	5.38 2.91	5.50 2.95	
257-15-2, 1-4 cm	2.811	P S	5.94 2.95	5.97 2.97	5.98 2.99	6.00 3.00	6.01 3.02	6.05 3.06	6.13 3.11	6.21 3.14	
257-17-5, 126-129 cm	2.933	P S	6.29 3.49	6.30 3.49	6.32 3.50	6.33 3.51	6.34 3.51	6.39 3.53	6.45 3.56	6.51 3.60	
260-18 cc	2.307	P S	3.48 1.91	3.54 1.94	3.59 1.96	3.64 1.97	3.69 1.99	3.83 2.02	4.04 2.08	4.24 2.15	

TABLE 3. Elastic Constants

Sample	Pressure (kb)	V_p/V_s	σ	ϕ (km/sec)^2	K (Mb)	β (Mb^{-1})	μ (Mb)	E (Mb)	λ (Mb)
211-12-1, 70-73 cm	0.4	2.02	0.34	18.3	0.49	2.05	0.18	0.48	0.37
	1.0	2.02	0.34	18.6	0.50	2.01	0.18	0.49	0.38
	2.0	2.00	0.33	18.9	0.51	1.97	0.19	0.51	0.38
	6.0	1.99	0.33	19.8	0.54	1.87	0.20	0.54	0.40
211-15-2, 69-71 cm	0.4	2.06	0.35	14.2	0.35	2.90	0.12	0.32	0.27
	1.0	2.00	0.33	14.3	0.35	2.88	0.13	0.35	0.26
	2.0	1.93	0.32	14.1	0.35	2.91	0.14	0.38	0.25
	6.0	1.93	0.32	15.7	0.39	2.57	0.16	0.43	0.28
211-15-4, 67-70 cm	0.4	2.10	0.35	17.0	0.42	2.37	0.14	0.37	0.33
	1.0	2.05	0.34	16.8	0.42	2.39	0.15	0.39	0.32
	2.0	1.98	0.33	16.5	0.41	2.44	0.16	0.42	0.30
	6.0	1.93	0.32	16.9	0.42	2.36	0.18	0.47	0.31
212-39-1, 58-61 cm	0.4	2.01	0.34	10.8	0.25	4.03	0.09	0.24	0.19
	1.0	1.98	0.33	10.9	0.25	3.98	0.10	0.26	0.19
	2.0	1.98	0.33	11.3	0.26	3.83	0.10	0.27	0.19
	6.0	2.02	0.34	13.1	0.31	3.28	0.11	0.30	0.23
213-17-2, 43-49 cm	0.4	1.95	0.32	26.3	0.69	1.44	0.28	0.74	0.51
	1.0	1.95	0.32	26.3	0.70	1.44	0.28	0.75	0.51
	2.0	1.94	0.32	26.4	0.70	1.43	0.29	0.76	0.51
	6.0	1.96	0.32	27.1	0.72	1.39	0.29	0.77	0.53
214-48-1,	0.4	2.39	0.39	15.1	0.35	2.84	0.08	0.23	0.30
	1.0	2.21	0.37	14.7	0.34	2.92	0.10	0.26	0.28
	2.0	2.03	0.34	14.1	0.33	3.03	0.11	0.32	0.25
	6.0	1.87	0.30	14.3	0.34	2.95	0.16	0.41	0.23
214-48-2, 7-10 cm	0.4	1.85	0.29	20.2	0.54	1.86	0.26	0.67	0.37
	1.0	1.84	0.29	20.1	0.54	1.85	0.26	0.68	0.36
	2.0	1.83	0.29	20.1	0.54	1.85	0.27	0.69	0.36
	6.0	1.83	0.29	20.7	0.56	1.79	0.28	0.72	0.37
214-53-1, 50-53 cm	0.4	1.89	0.31	15.8	0.43	2.31	0.19	0.50	0.31
	1.0	1.87	0.30	15.8	0.43	2.30	0.20	0.52	0.30
	2.0	1.86	0.30	16.1	0.44	2.25	0.21	0.54	0.31
	6.0	1.85	0.29	17.2	0.48	2.09	0.23	0.59	0.32
216-37-2, 77-80 cm	0.4	1.95	0.32	14.3	0.37	2.69	0.15	0.40	0.27
	1.0	1.92	0.31	14.4	0.38	2.66	0.16	0.42	0.27
	2.0	1.90	0.31	14.6	0.38	2.63	0.17	0.44	0.27
	6.0	1.85	0.29	15.0	0.40	2.53	0.19	0.49	0.27
221-19-2, 18-21 cm	0.4	1.82	0.28	18.3	0.51	1.97	0.25	0.65	0.34
	1.0	1.81	0.28	18.7	0.52	1.93	0.26	0.68	0.34
	2.0	1.80	0.28	19.1	0.53	1.89	0.28	0.70	0.35
	6.0	1.82	0.28	20.8	0.58	1.72	0.29	0.75	0.39
223-40-2 46-50 cm	0.4	2.06	0.35	6.2	0.13	7.68	0.04	0.12	0.10
	1.0	2.10	0.35	7.3	0.15	6.56	0.05	0.13	0.12
	2.0	2.11	0.36	8.6	0.18	5.52	0.06	0.16	0.14

TABLE 3. (continued)

Sample	Pressure (kb)	V_p/V_s	σ	ϕ (km/sec)^2	K (Mb)	β (Mb^{-1})	μ (Mb)	E (Mb)	λ (Mb)
	6.0	2.13	0.36	12.3	0.26	3.81	0.08	0.22	0.21
231-64-1, 60-64 cm	0.4	1.89	0.30	19.4	0.54	1.85	0.24	0.63	0.38
	1.0	1.87	0.30	19.5	0.54	1.84	0.25	0.65	0.38
	2.0	1.88	0.30	19.8	0.55	1.81	0.25	0.66	0.38
	6.0	1.91	0.31	70.9	0.59	1.71	0.25	0.66	0.42
233A-8-2, 62-68 cm	0.4	1.92	0.31	17.5	0.48	2.07	0.21	0.54	0.35
	1.0	1.91	0.31	17.8	0.49	2.03	0.21	0.56	0.35
	2.0	1.92	0.31	18.4	0.51	1.97	0.22	0.57	0.36
	6.0	1.92	0.31	19.5	0.54	1.84	0.23	0.61	0.39
235-19-1, 124-128 cm	0.4	1.94	0.32	18.2	0.49	2.05	0.20	0.54	0.35
	1.0	1.92	0.31	18.4	0.49	2.03	0.21	0.55	0.35
	2.0	1.90	0.31	18.4	0.50	2.01	0.22	0.57	0.35
	6.0	1.90	0.31	19.4	0.53	1.90	0.23	0.61	0.37
236-34-1, 147-150 cm	0.4	1.89	0.31	16.3	0.44	2.30	0.19	0.51	0.31
	1.0	1.91	0.31	17.0	0.46	2.19	0.20	0.52	0.32
	2.0	1.91	0.31	17.5	0.47	2.13	0.20	0.53	0.33
	6.0	1.92	0.31	18.9	0.51	1.96	0.22	0.57	0.37
238-55-2, 10-13 cm	0.4	1.86	0.30	21.5	0.62	1.62	0.29	0.75	0.42
	1.0	1.86	0.30	21.8	0.62	1.60	0.30	0.77	0.43
	2.0	1.85	0.29	22.2	0.64	1.57	0.31	0.79	0.43
	6.0	1.88	0.30	24.4	0.70	1.42	0.32	0.84	0.49
238-59-2, 26-29 cm	0.4	1.84	0.29	23.6	0.69	1.45	0.34	0.87	0.46
	1.0	1.83	0.29	23.7	0.69	1.44	0.34	0.88	0.46
	2.0	1.83	0.29	23.9	0.70	1.43	0.34	0.89	0.47
	6.0	1.85	0.29	24.7	0.73	1.37	0.35	0.90	0.50
238-62-2, 75-79 cm	0.4	1.80	0.28	20.8	0.60	1.68	0.31	0.80	0.39
	1.0	1.81	0.28	21.3	0.61	1.64	0.31	0.80	0.40
	2.0	1.82	0.28	21.7	0.62	1.61	0.32	0.81	0.41
	6.0	1.83	0.29	22.6	0.65	1.53	0.32	0.83	0.44
238-64-1, 129-131 cm	0.4	1.84	0.29	21.9	0.63	1.59	0.31	0.79	0.42
	1.0	1.84	0.29	22.3	0.64	1.57	0.31	0.81	0.43
	2.0	1.85	0.29	22.9	0.66	1.52	0.32	0.82	0.45
	6.0	1.86	0.30	23.7	0.68	1.46	0.32	0.84	0.47
250A-25-3, 97-101 cm	0.4	1.95	0.32	22.3	0.63	1.59	0.25	0.67	0.46
	1.0	1.96	0.32	22.7	0.64	1.56	0.26	0.68	0.47
	2.0	1.90	0.31	22.2	0.63	1.59	0.29	0.72	0.44
	6.0	1.92	0.31	23.5	0.67	1.49	0.28	0.75	0.48
253-24-1, 133-136 cm	0.4	2.17	0.37	13.6	0.30	3.30	0.09	0.25	0.24
	1.0	2.00	0.33	13.2	0.29	3.41	0.11	0.29	0.22
	2.0	1.94	0.32	13.1	0.29	3.42	0.12	0.32	0.21
	6.0	1.92	0.31	14.0	0.32	3.15	0.14	0.36	0.23
254-35-2, 47-52 cm	0.4	2.00	0.33	16.4	0.45	2.22	0.17	0.45	0.34
	1.0	2.01	0.34	17.6	0.48	2.07	0.18	0.47	0.36

TABLE 3. (continued)

Sample	Pressure (kb)	V_p/V_s	σ	ϕ (km/sec)^2	K (Mb)	β (Mb^{-1})	μ (Mb)	E (Mb)	λ (Mb)
	2.0	2.02	0.34	18.9	0.52	1.93	0.19	0.50	0.39
	6.0	1.98	0.33	19.5	0.54	1.85	0.21	0.56	0.40
254-36-3, 110-114 cm	0.4	1.99	0.33	19.3	0.54	1.86	0.20	0.54	0.40
	1.0	1.98	0.33	19.7	0.55	1.82	0.21	0.56	0.41
	2.0	1.96	0.32	19.9	0.56	1.80	0.22	0.58	0.41
	6.0	1.94	0.32	20.7	0.58	1.72	0.24	0.63	0.42
256-9-2, 50-53 cm	0.4	2.03	0.34	15.3	0.41	2.47	0.15	0.39	0.31
	1.0	2.01	0.34	15.8	0.42	2.38	0.15	0.41	0.32
	2.0	1.97	0.33	16.1	0.43	2.33	0.17	0.44	0.32
	6.0	1.96	0.32	17.9	0.48	2.09	0.19	0.41	0.35
257-11-2, 57-61 cm	0.4	1.82	0.28	15.1	0.40	2.48	0.20	0.52	0.27
	1.0	1.83	0.29	16.0	0.43	2.34	0.21	0.55	0.28
	2.0	1.84	0.29	16.7	0.45	2.23	0.22	0.56	0.30
	6.0	1.86	0.30	18.5	0.50	2.01	0.23	0.61	0.34
257-15-2, 1-4 cm	0.4	2.01	0.34	23.8	0.67	1.49	0.25	0.66	0.51
	1.0	1.99	0.33	24.0	0.67	1.48	0.26	0.68	0.50
	2.0	1.98	0.33	24.1	0.68	1.47	0.26	0.70	0.50
	6.0	1.98	0.33	25.3	0.72	1.39	0.28	0.74	0.53
257-17-5, 126-129 cm	0.4	1.80	0.28	23.4	0.69	1.46	0.36	0.92	0.45
	1.0	1.81	0.28	23.8	0.70	1.43	0.36	0.93	0.46
	2.0	1.81	0.28	24.2	0.71	1.41	0.37	0.94	0.47
	6.0	1.81	0.28	25.0	0.74	1.35	0.38	0.97	0.49
260-18 cc	0.4	1.82	0.29	7.5	0.17	5.77	0.09	0.22	0.12
	1.0	1.86	0.30	8.3	0.19	5.19	0.09	0.24	0.13
	2.0	1.90	0.31	9.2	0.21	4.67	0.09	0.25	0.15
	6.0	1.97	0.33	11.7	0.27	3.64	0.11	0.28	0.20

Velocity-Density Relationships

Because of the limited time for sea floor weathering to effect rock properties, dredge basalts from or near ridge crests are relatively high in velocity and density (e.g., Christensen, 1972a; Barrett and Aumento, 1970). Older basalts obtained from deep sea drilling, however, show wide ranges in density and velocities which correlate with age (Christensen and Salisbury, 1972, 1973). Combined Pacific and Atlantic data give rates of decreasing velocity of 1.89 and 1:35 km/sec per 100 my for V_p and V_s, respectively.

The vertical extent of weathering in oceanic crust is still an important unknown, which is likely controlled by the depth and continuity of fractures and the downward diffusion of sea water through these fractures and grain boundary cracks. For Sites 259 and 261 of Leg 27 it appears that weathering extends downward from the sediment-basalt interface to at least 50 meters (Christensen et al., 1974b). Data of Hyndman (1974) for Site 257 of Leg 26 suggest a similar depth. Older sites most likely are weathered to greater depths.

As expected from the wide range in age of basalt samples recovered

TABLE 4. Comparisons of Compressional Wave Velocities of Indian
Ocean Basalts with Previously Reported Velocities

Sample	Density (g/cm^3)	Velocities (km/sec) at Varying Pressure (kbar)			Reference
		0.5	1.0	2.0	
251-A-31-2 (85 cm)	2.79	5.51	5.59	5.69	Table 2
251-A-31-2 (84 cm)	2.84	5.61	5.68	5.76	Hyndman (1974)
254-36-3 (112 cm)	2.79	5.41	5.46	5.52	Table 2
254-36-3 (105 cm)	2.82	5.36	5.48	5.55	Hyndman (1974)
257-11-2 (59 cm)	2.67	5.06	5.16	5.27	Table 2
257-11-2 (74 cm)	2.74	5.21	5.31	5.38	Hyndman (1974)
257-15-2 (2 cm)	2.81	5.97	6.01	6.05	Table 2
257-15-1 (133 cm)	2.89	6.13	6.22	6.31	Hyndman (1974)
231-64-1 (62 cm)	2.78	5.58	5.62	5.65	Table 2
231-64-1A	2.76	4.76	5.04	5.17	Schreiber et al. (1974)
231-64-1B	2.72	4.53	4.66	4.84	Schreiber et al. (1974)
238-55-2 (12 cm)	2.86	5.94	5.98	6.05	Table 2
238-55-2A	2.81	5.32	5.41	5.54	Schreiber et al. (1974)
238-59-2 (28 cm)	2.92	6.24	6.27	6.29	Table 2
238-59-2A	2.90	5.76	6.00	6.04	Schreiber et al. (1974)

from the Indian Ocean, densities and velocities vary significantly
(Figure 2). For comparisons with the Indian Ocean DSDP velocities,
mean velocities of the dredge samples of Table 5 and Atlantic and Pacific
Ocean DSDP data are included in Figure 2. Only velocities from water
saturated samples have been included. The linear and nonlinear solutions are least-squares regression lines for 77 water saturated DSDP
basalts with the following parameters (Christensen and Salisbury,
1975):

$$V_p = 3.56\rho - 4.26$$
$$V_p = 2.33 + 0.081\rho^{3.63}$$
$$V_s = 2.17\rho - 3.07$$
$$V_s = 1.33 + 0.011\rho^{4.85}$$

Figure 2 illustrates several important differences in velocities for
the various rock types. In particular, andesite, serpentinite and
metagabbro have higher compressional wave velocities than ocean floor
basalts having similar densities. Serpentinite and andesite have lower
mean atomic weights and thus the observed higher velocities are in
agreement with predictions based on chemistry (Birch, 1961). The metagabbro, on the other hand, is probably similar in chemistry to the
fresh basalts and its relatively high velocity is related to its coarser
grain size and lower porosity. The differences in shear velocities for
the DSDP basalts and the andesites, serpentinite and metagabbro are less
pronounced than those observed for compressional wave velocities.

TABLE 5. Compressional (P) and Shear (S) Wave Velocities in Dredged Rocks

Sample	Bulk Density (g/cm³)	Mode	Velocity (km/sec) at Varying Pressures (kbar)								
			0.2	0.4	0.6	0.8	1.0	2.0	4.0	6.0	8.0	10.0
Basalt	2.781	P	5.77	5.95	6.09	6.20	6.22	6.37	6.51	6.56	6.58	6.60
	2.818	P	5.86	6.04	6.16	6.24	6.29	6.45	6.57	6.62	6.64	6.65
	2.813	P	5.86	6.04	6.15	6.22	6.27	6.42	6.55	6.60	6.61	6.62
Mean	2.804	P	5.83	6.01	6.13	6.22	6.26	6.41	6.54	6.59	6.61	6.62
Basalt	2.781	S	2.96	3.09	3.21	3.28	3.32	3.45	3.52	3.61	3.62	3.62
	2.818	S	2.97	3.10	3.26	3.32	3.37	3.51	3.62	3.65	3.66	3.66
	2.813	S	2.97	3.10	3.26	3.32	3.36	3.50	3.59	3.63	3.64	3.65
Mean	2.804	S	2.97	3.10	3.24	3.31	3.35	3.49	3.58	3.63	3.64	3.65
Serpentinite	2.686	P	5.64	5.68	5.69	5.71	5.72	5.78	5.89	5.95	6.03	6.08
	2.698	P	5.81	5.86	5.87	5.89	5.90	5.95	6.05	6.12	6.18	6.21
	2.715	P	6.03	6.07	6.09	6.11	6.13	6.19	6.28	6.35	6.41	6.45
Mean	2.700	P	5.83	5.87	5.88	5.90	5.91	5.97	6.07	6.15	6.21	6.25
Serpentinite	2.686	S	2.57	2.58	2.59	2.60	2.60	2.64	2.68	2.69	2.70	2.71
	2.698	S	2.64	2.65	2.66	2.67	2.67	2.71	2.75	2.78	2.80	2.80
	2.715	S	2.81	2.83	2.84	2.85	2.86	2.90	2.94	2.95	2.96	2.96
Mean	2.700	S	2.67	2.69	2.70	2.71	2.71	2.75	2.79	2.81	2.82	2.83
Metagabbro	2.926	P	6.77	6.80	6.81	6.82	6.83	6.87	6.91	6.95	6.97	6.99
	2.921	P	6.74	6.76	6.78	6.80	6.82	6.86	6.92	6.98	7.02	7.04
	2.882	P	6.54	6.59	6.62	6.64	6.66	6.73	6.82	6.90	6.95	6.98
Mean	2.910	P	6.68	6.72	6.74	6.75	6.77	6.82	6.88	6.94	6.98	7.00
Metagabbro	2.926	S	3.59	3.61	3.62	3.64	3.64	3.69	3.75	3.78	3.80	3.81
	2.921	S	3.64	3.66	3.67	3.69	3.69	3.74	3.79	3.81	3.83	3.84
	2.882	S	3.57	3.60	3.63	3.65	3.67	3.73	3.78	3.80	3.82	3.83
Mean	2.910	S	3.60	3.62	3.64	3.66	3.67	3.72	3.77	3.80	3.82	3.82

TABLE 6. Elastic Constants

Sample	Pressure (kb)	V_p/V_s	σ	ϕ (km/sec)^2	K (Mb)	β (Mb)	μ (Mb)	E (Mb)	λ (Mb)
Basalt	0.4	1.94	0.32	23.3	0.65	1.52	0.27	0.71	0.47
	1.0	1.87	0.30	24.2	0.68	1.47	0.32	0.82	0.47
	2.0	1.84	0.29	24.9	0.70	1.43	0.34	0.88	0.47
	6.0	1.82	0.28	25.7	0.73	1.37	0.37	0.95	0.48
	10.0	1.82	0.28	25.9	0.74	1.36	0.37	0.96	0.49
Serpentinite	0.4	2.18	0.37	24.8	0.67	1.48	0.20	0.54	0.54
	1.0	2.18	0.37	25.1	0.68	1.46	0.20	0.55	0.55
	2.0	2.17	0.37	25.6	0.70	1.44	0.21	0.56	0.56
	6.0	2.19	0.37	27.2	0.74	1.35	0.21	0.59	0.60
	10.0	2.22	0.37	28.6	0.79	1.27	0.22	0.60	0.64
Metagabbro	0.4	1.86	0.30	27.7	0.81	1.24	0.38	0.99	0.55
	1.0	1.84	0.29	27.8	0.81	1.23	0.39	1.01	0.55
	2.0	1.83	0.29	28.0	0.81	1.22	0.40	1.04	0.55
	6.0	1.83	0.29	28.8	0.84	1.18	0.42	1.08	0.56
	10.0	1.83	0.29	29.3	0.86	1.16	0.43	1.10	0.58

Since V_s is inherently more difficult to measure than V_p, any relationship between the two velocities is useful in estimating V_s once V_p has been measured. In Figure 3, V_p is plotted against V_s for the Indian Ocean Rocks at 0.4 kbar. A linear relationship between V_p and V_s for the basalts is apparent; the least squares solution being $V_s = 0.59\ V_p - 0.32$. For this solution the correlation coefficient is 0.98 and the standard error of estimate of V_s from V_p is 0.10 km/sec. Also of importance is the anomalous nature of serpentinite, whereas the andesites and metagabbro velocities fall close to the least squares solution. The high ratios of V_p to V_s for serpentinite also have been observed for serpentinites dredged from the Mid-Atlantic Ridge and have been used as an argument against models with abundant serpentinite in the lower oceanic crust (Christensen, 1972b).

Since the velocities are related to density, many elastic constants should also vary systematically with density, the trends being related to the degree of sea floor weathering. This is shown in Figure 4 where elastic constants at 0.4 kbar decrease markedly with decreasing density. The elastic constants of the dredged rocks and andesites are also shown in Figure 4 for comparison with the DSDP basalts. Of significance are the observations that (1) μ and E of the andesites and dredged rocks agree well with DSDP basalts of similar density and (2) λ and K of andesite and serpentinite are much higher than basalts of equivalent density.

Layer 2 Seismic Refraction Velocities

Locations of Indian Ocean refraction survey sites and DSDP sites are shown in Figure 5. Drilling sites for which basalt velocities have been measured (Tables 1 and 2) are indicated by solid circles. Seismic refraction locations are labeled S-1 through S-64. Water depths, sediment thicknesses, basement velocities, and basement thicknesses are summarized in Table 7. Lower crustal velocities and thicknesses and

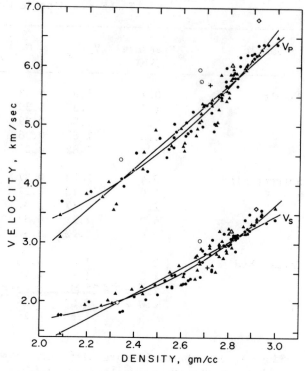

Fig. 2. Velocity-density relations at 0.5 kbar for Indian Ocean DSDP basalts (solid triangles), other DSDP basalts (dots), Indian Ocean andesite (circles), serpentinite (plus), dredge basalt (open triangle) and metagabbro (diamond).

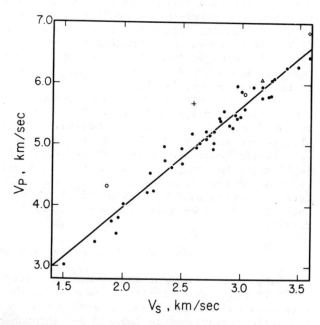

Fig. 3. The relationship between V_p and V_s for Indian Ocean DSDP basalts (dots), andesites (circles), serpentinite (plus), dredge basalt (triangle) and metagabbro (diamond).

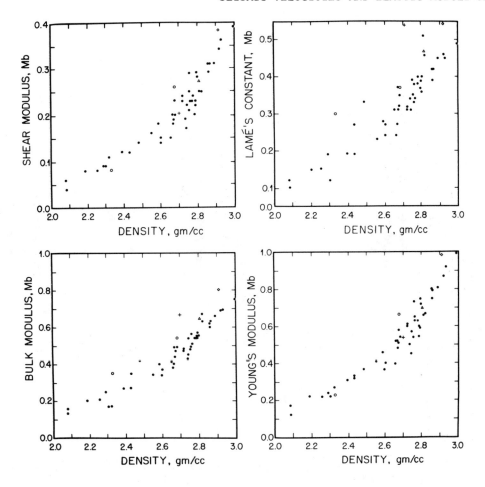

Fig. 4. Elastic constants at 0.5 kbar versus density for Indian Ocean DSDP basalts (dots), andesites (circles), serpentinite (plus), dredge basalt (triangle) and metagabbro (diamond).

upper mantle velocities, when reported, can be found in the references of Table 7.

The proximities of several seismic refraction surveys to sites in which velocities have been measured for basement rocks allow comparisons of laboratory velocities with refraction velocities (e.g., Schreiber et al., 1974; Christensen et al., 1972b). Such comparisons, however, are complicated by many factors. Of critical importance is the influence of fractures which, if abundant in the upper oceanic crust, are expected to reduce velocities. Laboratory measurements are usually from specimens which have been selected to be as free of fractures as possible. Also significant is the depth of weathering, as has been illustrated by the studies of velocities from Site 257 (Hyndman, 1974) and Sites 259 and 261 (Christensen et al., (1974b). For these sites weathering decreases with depth producing velocity gradients within the upper portion of basalt.

Weathering has been found to lower velocities and densities of layer 2 basalts. Of particular significance, it has been demonstrated that weathering increases with age such that compressional wave velocities of basalts from the uppermost portion of layer 2 decrease approximately

TABLE 7. Indian Ocean Seismic Refraction Measurements

Station Number	Water Depth (km)	Sediment Thickness (km)	Basement Velocity (km/sec)	Basement Thickness (km)	Reference
S-1*	1.18*	0.22*	4.28*	---	Tramontini & Davies (1969)
S-2	0.71	0.47	3.98	2.03	Laughton & Tramontini (1969)
S-3	1.39	1.16	4.33	2.88	"
S-4	1.10	0.42	3.99	1.71	"
S-5	1.98	0.53	3.94	1.83	"
S-6	1.4	1.23	4.25	1.71	"
S-7	2.17	0.41	5.22	2.81	"
S-8	2.4	0.34	4.07	2.00	"
S-9	2.24	1.14	4.60	1.59	"
S-10	1.61	1.52	5.30	1.51	"
S-11	2.22	0.65	4.81	1.82	"
S-12	2.33	0.39	4.11	1.11	"
S-13	1.98	1.87	5.0	2.9	Francis & Shor (1966)
S-14	2.84	1.04	6.13	4.4	"
S-15	0.06	1.13	6.22	?	Shor & Pollard (1963)
S-16	3.23	0.67	4.48	1.4	Francis & Shor (1966)
S-17	3.97	0.45	5.41	1.9	"
S-18	4.04	0.66	5.79	2.2	"
S-19	4.27	0.15	5.02	2.4	"
S-20	4.31	0.78	5.0(?)	?	"
S-21	3.86	0.65	6.31	3.6	"
S-22	0.91	1.42	4.76	4.2	"
S-23	4.84	0.23	5.0	0.11	"
S-24	3.95	0.60	5.37	2.10	"
S-25	3.15	0.35	5.48	2.8	"
S-26	2.66	0.51	5.5	2.5	"
S-27	0.12	1.31	4.55	?	Shor & Pollard (1963)
S-28	0.12	1.25	4.36	?	"
S-29	2.71	0.71	4.67	2.2	Francis & Raitt (1967)
S-30	5.73	0.15	4.80	0.6	"
S-31	5.00	0.21	5.38	1.4	Francis & Raitt (1967)
S-32	5.22	0.08	5.09	1.2	"
S-33	5.95	0.13	5.37	1.7	"

TABLE 7. (continued)

Station Number	Water Depth (km)	Sediment Thickness (km)	Basement Velocity (km/sec)	Basement Thickness (km)	Reference
S-34	5.59	0.27	5.38	1.3	"
S-35	5.68	0.49	5.18	2.4	"
S-36	5.67	0.56	4.89	0.8	"
S-37	5.74	0.14	4.78	1.4	"
S-38	5.10	0.07	4.86	2.4	"
S-39	5.35	0.35	4.98	1.3	"
S-40	5.35	0.60	4.66	1.1	"
S-41	5.57	0.47	5.98	1.2	"
S-42	5.00	0.40	5.0	0.9	"
S-43	2.21	0.60	4.70	2.0	Francis & Rairr (1967)
S-44	1.97	0.68	4.80	2.1	"
S-45	4.83	0.55	6.57	1.6	"
S-46	3.75	0.32	4.57	0.5	"
S-47	1.90	0.37	4.52	1.5	"
S-48	3.20	0.33	4.43	1.2	"
S-49	4.04	0.26	5.03	0.9	"
S-50	4.99	0.12	5.46	1.7	"
S-51	4.82	0.18	5.62	1.4	"
S-52	4.79	0.22	5.78	1.2	"
S-53	4.33	0.31	6.24	1.8	"
S-54	5.49	0.21	5.48	1.2	"
S-55	4.91	0.11	5.00	0.7	"
S-56	5.03	0.48	5.02	2.11	Ludwig et al. (1968)
S-57	3.60	3.5	5.28	3.1	Francis et al. (1966)
S-58	4.17	3.7	6.56	4.6	"
S-59	4.81	1.7	5.28	2.5	"
S-60	5.04	0.7	4.20	3.0	"
S-61	5.06	0.3	6.24	3.1	"
S-62	4.38	0.3	4.86	1.8	"
S-63	0.05	0.3	5.72	3.3	"
S-64	3.91	0.81	5.4	0.65	Francis & Shor (1966)

* Average of 10 closely spaced stations.

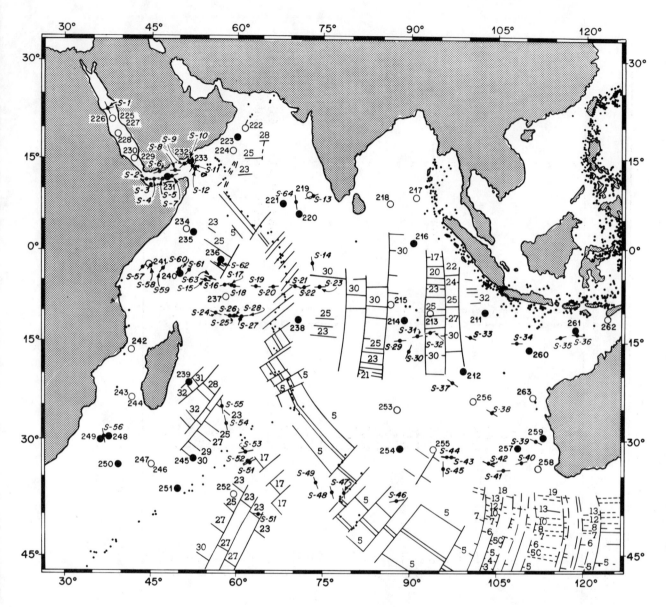

Fig. 5. Locations of seismic refraction surveys (S-1 through S-64) and DSDP sites. Lines through the refraction sites indicate azimuth of profile. Magnetic anomalies are after the tabulation of Pitman et al. (1974).

1.9 km/sec per 100 my (Christensen and Salisbury, 1972, 1973). Fractures, on the other hand, are likely to be abundant in relatively young crust. Examination of DSDP cores shows that fractures in older basalts are often filled with secondary minerals such as carbonates. At greater depths within layer 2 metamorphism is likely to play an important role in healing fractures created at ridge crests. Thus the influence of fractures on layer 2 velocities will also be age dependent, but velocities will increase with age because of fracture closure.

A difficulty encountered in comparing refraction and laboratory velocities of layer 2 is related to the techniques commonly used in refraction

SEISMIC VELOCITIES AND ELASTIC MODULI OF ROCKS   297

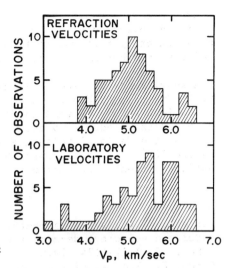

Fig. 6. Histograms of measured seismic velocities at 0.4 kbar for Indian Ocean basalts and Indian Ocean layer 2 seismic refraction velocities.

profiling. Since most seismic experiments within the Indian Ocean have been designed to investigate lower crustal and upper mantle structure as well as layer 2 velocities and thicknesses, a limited number of arrivals are usually recorded from layer 2. Thus the reported seismic velocities are low in accuracy (Raitt, 1963) and fine structure such as subdivisions of layer 2 which have been found in the Atlantic Ocean from sonobuoy studies (Talwani et al.,1971), often can be missed.

The new data in Table 2 allow several additional comparisons of refraction and laboratory velocities. However, as discussed above, these comparisons must be viewed with caution. The velocities of basalts from Sites 231, 236, 240 and 248 are higher than basement velocities reported in the vicinities of these sites (see Figure 5 and Table 7). It is concluded that cracks and/or sediments interlayered within the basalts significantly lower basement velocities in these regions. The laboratory velocities for Site 260, on the other hand, are much slower than the 5.38 km/sec basement refraction velocity reported at S-34. It appears that the velocity structure at this site is similar to Sites 259 and 261, in which thin layers of highly weathered basalt form the top of layer 2 (Christensen et al.,1974b).

Good agreement exists between the basement refraction velocity at S-33 (5.37 km/sec) and the laboratory velocities for basalt from Site 211 (5.21 km/sec at 0.4 kbar). Also the basement velocity of 5.4 km/sec at S-64 is similar to the 5.53 km/sec velocity for the basalt from Site 221.

The variability in layer 2 has been emphasized in summaries of oceanic crustal refraction data (Raitt, 1963; Shor et al., 1971; Christensen and Salisbury, 1975) and is illustrated for the Indian Ocean in Figure 6, where the 64 layer 2 seismic refraction velocities for sites located in Figure 5 and tabulated in Table 7 are shown in histogram form. Also illustrated in Figure 6 are 67 laboratory measured velocities of water saturated Indian Ocean basalt at 0.4 kbar. Note that the maximum basalt velocities agree well with maximum layer 2 velocities. Layer 2 seismic refraction velocities lower than those included in the histogram of Figure 6 may very well be present in the Indian Ocean. These could either be confused with sediment velocities or be masked by overlying higher velocity sediments (Christensen et al., 1973). Thus the ranges of refraction and laboratory velocities could be in even closer agreement than that illustrated in Figure 6. The histogram of refraction velocities is fairly symmetric, whereas the laboratory velocities are skewed toward the higher velocities. The latter may simply result from a biased selection of young relatively unweathered samples for the measurements, or it may be that the symmetrical shape of the refraction velocity histogram is influenced by fracturing in layer 2.

Acknowledgments. The assistance of M. Brown, R. Carlson, D. Fountain, R. L. Gresens, K. Hubert, R. McConaghy, M. Salisbury and R. J. Stewart in various phases of this study is greatly appreciated. R. L. Fisher provided the dredge samples. Samples were supplied through the assistance of the National Science Foundation. This research was supported by the Office of Naval Research Contract N-00014-67-A-0103-0014 and National Science Foundation grant GA-36138.

References

Barrett, D. L., and F. Aumento, The Mid-Atlantic Ridge near 45°N. XI. Seismic velocity, density and layering of the crust, Can. J. Earth Sci., 7, 1117-1124, 1970.

Birch, F., The velocity of compressional waves in rocks to 10 kb, 1, J. Geophys. Res., 65, 1083-1102, 1960.

Birch, F., The velocity of compressional waves in rocks to 10 kb, 2, J. Geophys. Res., 66, 2199-2224, 1961.

Christensen, N. I., Compressional and shear wave velocities at pressures to 10 kilobars for basalts from the East Pacific Rise, Geophys. J. Roy. Astron. Soc., 28, 425-429, 1972a.

Christensen, N. I., The abundance of serpentinites in the oceanic crust, J. Geol., 80, 709-719, 1972b.

Christensen, N. I. and M. H. Salisbury, Sea floor spreading, progressive alteration of Layer 2 basalts, and associated changes in seismic velocities, Earth Planet. Sci. Lett., 15, 367-375, 1972.

Christensen, N. I., Velocities, elastic moduli and weathering-age relations for Pacific Layer 2 basalts, Earth Planet. Sci. Lett., 19, 461-470, 1973.

Christensen, N. I., Structure and constitution of the lower oceanic crust, Rev. Geophys. Space Phys., 13, 57-86, 1975.

Christensen, N. I., D. M. Fountain, and R. J. Stewart, Oceanic crustal basement: a comparison of seismic properties of DSDP basalts and consolidated sediments, Mar. Geol., 15, 215-226, 1973.

Christensen, N. I., D. M. Fountain, R. L. Carlson, and M. H. Salisbury, Velocities and elastic moduli of volcanic and sedimentary rocks recovered on DSDP Leg 25. in Simpson, E.S.W., Schlich, R., et al., Initial Reports of the Deep Sea Drilling Project, XXV, Washington (U. S. Government Printing Office), 357-360, 1974a.

Christensen, N. I., M. H. Salisbury, D. M. Fountain, and R. L. Carlson, Velocities of compressional and shear waves in DSDP Leg 27 basalts. in Veevers, J. J., Heirtzler, J. R., et al., Initial Reports of the Deep Sea Drilling Project, XXVII, Washington (U. S. Govern. Printing Office), 445-449, 1974b.

Dortman, N. B., and H. Sh., Magid, Velocity of elastic waves in crystalline rocks and its dependence on moisture content, Dokl. Akad. Nauk SSSR, Earth Sci. Sec., English Transl., 179, 1-8, 1968.

Engel, C. G., and R. L. Fisher, Lherzolite, anorthosite, gabbro, and basalt dredged from the mid-Indian Ocean ridge, Science, 166, 1136-1141, 1969.

Francis, T.J.G., R. W. Raitt, Seismic refraction measurements in the southern Indian Ocean, J. Geophys. Res., 72, 3015-3041, 1967.

Francis, T.J.G., and G. G. Short, Seismic refraction measurements in the northwest Indian Ocean, J. Geophys. Res., 71, 427-449, 1966.

Francis, T.J.G., D. Davies, and M. N. Hill, Crustal structure between Kenya and the Seychelles, Roy. Soc. London Phil. Trans., Series A, 259, 240-261, 1966.

Hyndman, R. D., Seismic velocities of basalts from DSDP Leg 26, in Luyendyk, B. P. and Davies, T.A., et al., Initial Reports of the Deep Sea Drilling Project, XXVI, Washington (U. S. Govern. Printing Office), 509-512, 1974.

Laughton, A. S., C. Tramontini, Recent studies of the crustal structure in the Gulf of Aden, Tectonophysics, 8, 359-375, 1969.

Ludwig, W. J., J. E. Nafe, E.S.W. Simpson, and S. Sacks, Seismic-refraction measurements on the southeast African continental margin, J. Geophys. Res., 73, 3707-3719, 1968.

Nur, A., and G. Simmons, The effect of saturation on velocity in low porosity rocks, Earth Planet. Sci. Lett., 7, 183-193, 1969.

Pitman, W. C., R. L. Larson, and E. M. Herron, Age of the ocean basins determined from magnetic anomaly lineations. Printed by the Geol. Soc. of Amer., Inc., Boulder, Colorado, 1974.

Raitt, R. W., The crustal rocks, in Hill, M. N. (ed.), The Sea: v. 3, New York, (Interscience), 85-102, 1963.

Schreiber, E., M. R. Perfit, and P. J. Cernock, Compressional wave velocities in samples of basalt recovered by DSDP Leg 24, in Fisher, R. L., Bunce, E.T., et al., Initial Reports of the Deep Sea Drilling Project, XXIV, Washington (U. S. Govern. Printing Office), 787-790, 1974.

Shor, G. G., Jr., and D. D. Pollard, Seismic investigations of Seychelles and Saya de Malha Banks, Northwest Indian Ocean, Science, 48-49, 1963.

Shor, G. G., Jr., H. W. Menard, and R. W.Raitt, Structure of the Pacific Basin, in Maxwell, A.E. (ed.) The Sea, 4, New York (Wiley and Sons), 3-27, 1971.

Talwani, M., C. C. Windisch, and M. G. Langseth, Jr., Reykjanes ridge crest: a detailed geophysical study, J. Geophys. Res., 76, 473-517, 1971.

Tramontini, C. and D. Davies, A seismic refraction survey in the Red Sea, Geophys. J. R. Astron. Soc., 17, 225-241, 1969.

CHAPTER 11. THE MAGNETIC PROPERTIES OF INDIAN OCEAN BASALTS

M. W. McElhinny

Research School of Earth Sciences
Australian National University
Canberra, A.C.T., Australia

Abstract. The magnetic properties of DSDP basalts have been investigated at 24 sites in the Indian Ocean. The data represent about one-half of those acquired from all the oceans. Summary histograms are compiled for the natural remanent magnetization intensity, susceptibility, Königsberger ratio, median destructive field and initial Curie temperature. They confirm the overall patterns of these parameters derived for the Atlantic and Pacific Oceans. The basalts generally have stable magnetizations and only in isolated cases is instability and the acquisition of viscous remanent magnetization likely to be significant. Königsberger ratios confirm the validity of the Vine and Matthews model. Initial Curie temperatures indicate that oceanic basalts are extensively maghematized, probably due to sea floor weathering. The process accounts for the rapid drop in average intensity of magnetization within a short distance from the ridge axis. A world-wide synthesis of all the magnetic parameters shows that there are significant differences between drilled and dredged oceanic basalts, undoubtedly reflecting sampling distribution. The average intensity of magnetization of drilled samples of 2.4×10^{-3}G is rather less than that generally used in modelling marine magnetic anomalies. This suggests that a uniformly magnetized Layer 2 is a more realistic model than one in which the upper 0.5 km is very strongly magnetized.

Introduction

The interpretation of marine magnetic anomalies in terms of the Vine and Matthews (1963) hypothesis requires that the basalts comprising the oceanic crust have certain magnetic characteristics. Essentially these are that the remanence should dominate the induced magnetization and largely be responsible for the anomaly. An appropriate value must then be chosen for the average magnetization and the thickness of the magnetized layer. The earliest models assumed an average intensity of magnetization of 5×10^{-3} Gauss for a uniformly magnetized Layer 2 of thickness 2 km (Vine, 1966; Pitman and Heirtzler, 1966). The models also required that the source of the central anomaly have a magnetization about twice that for regions further from the ridge axis. Later Talwani et al. (1971) suggested that the magnetic layer responsible for the anomalies may be as thin as 0.4 km and constitute only the upper part of Layer 2, referred to as Layer 2A with average magnetization of 1.2×10^{-2} Gauss on the flanks and 3×10^{-2} Gauss in the axial zone. A similar thin magnetic layer was used by Herron (1972). The decrease in magnetization away from the axial zone has been shown by Irving et al. (1970) to be due

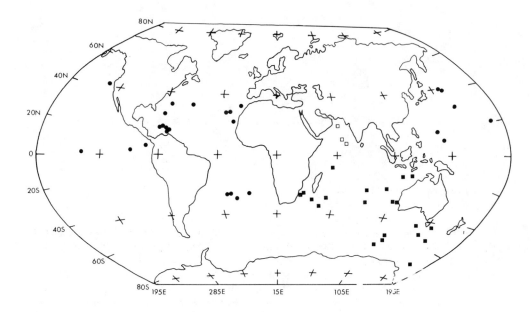

Fig. 1. Locations of Deep Sea Drilling Sites from which the magnetic properties of oceanic basalts have been investigated. The squares are the Indian Ocean sites reviewed in this paper. The sites marked with open squares have only intensity of magnetization data.

to the oxidation of titanomagnetites due to sea floor weathering. The underlying Layer 3 must be more weakly magnetized whatever the model chosen for Layer 2. From a detailed magnetic study of dredged samples Fox and Opdyke (1973) showed that serpentinized peridotite cannot be an important component of this layer. The magnetic properties of gabbro are much more compatible with this representing the main constituent of Layer 3. These effectively exclude the model of Hess (1962) in which Layer 3 is composed of serpentinized peridotite.

The analysis of the magnetic properties of basaltic samples drilled in the Deep Sea Drilling Project has provided further insight into the problem of modelling magnetic anomalies at sea. Lowrie (1974) has shown that the average magnetic properties of samples drilled largely from the Atlantic and Pacific Oceans are significantly different from those of dredged samples. In particular intensities of magnetization are much lower for the drilled samples so that models based upon the properties of dredged samples could be significantly in error. The magnetic properties of DSDP basalts from the Indian Ocean have been more extensively studied than those from the other oceans. The data acquired on legs 22 to 29 represent nearly 50 percent of results derived from the world's oceans (Figure 1).

The limited amount of material available for study from the basalts from each hole has made it difficult to analyse magnetic inclination. None of the published data have been analysed by cooling unit and consequently none of the paleolatitudes that have been deduced can be considered reliable. In any case the time span covered by the thickness of basalt in each hole could well be too short to average out the effects of secular variation. Furthermore there appear to be significant differences between the paleomagnetic properties of submarine basalts and those of subaerial basalts (Ade-Hall and Kitazawa, 1973). The data for submarine basalts are much more scattered. The spread of submarine NRM inclinations about predicted paleoinclinations is about twice that ob-

TABLE 1. Summary of the magnetic properties of Indian Ocean basalts. Geometric mean values are given for the Natural Remanent Magnetization (NRM), susceptibility (k), Königsberger ratio (Q_n) and median destructive field (MDF). Arithmetic means are given for the Curie temperature (T_c). N is the number of samples.

LEG (Reference	SITE	N	NRM (10^{-3}G)	N	k (10^{-3}G.Oe^{-1})	Q_n	N	MDF (Oe)	N	T_c (°C)
23	220	17	2.32	–	–	–	1	83	–	–
(Whitmarsh	221	6	1.07	–	–	–	1	124	–	–
et al., 1974)	223	7	3.83	–	–	–	1	90	–	–
25	239	2	1.42	2	0.691	5.79	2	168	–	–
(Wolejszo	245	2	4.75	1	2.83	5.75	1	35	–	–
et al., 1974)	248	3	5.09	2	1.05	11.70	2	60	–	–
	249	2	4.20	2	1.23	10.66	2	227	–	–
26	250	6	10.39	6	1.08	30.17	1	40	2	231
(Peirce et al.,	251	4	29.88	4	0.986	8.90	4	74	4	420
1974; Ade-	253	1	8.8	1	3.11	0.57	1	145	1	536
Hall, 1974b)	254	9	1.18	9	0.552	4.11	3	382	3	390
	256	17	5.84	16	3.06	3.53	4	70	6	251
	257	30	2.94	29	1.775	3.14	10	105	11	347
27*	259	4	2.86	4	0.447	11.03	1	120	–	–
(McElhinny,	260	1	11.94	1	0.930	26.20	1	68	–	–
(1974)	261	4	2.21	4	0.707	6.37	1	51	–	–
28	265	2	2.61	2	0.174	23.26	2	395	3	271
(Lowrie and	266	2	0.34	2	0.239	2.14	2	323	3	362
Hayes, 1975)	267	1	0.54	1	0.218	3.86	1	325	1	352
	274	4	6.08	4	1.412	6.48	4	80	3	320
29	279A	4	2.66	4	1.480	2.79	4	84	3	250
(Lowrie and	280A	7	0.012	7	0.089	0.20	11	257	2	412
Israfil, 1975)	282	10	1.16	10	0.166	11.09	13	263	2	309
	283	2	0.69	2	2.13	0.52	2	50	1	392

* Susceptibility measurements previously unpublished.

served for subaerial basalts. In this paper therefore the overall synthesis of the paleomagnetic measurements made on basalts from legs 22 to 29 will be restricted to an analysis of magnetic properties only.

Summary of Magnetic Properties

The magnetic properties of 147 basalt samples drilled from 24 sites from Indian Ocean legs 22 to 29 have been investigated. No determinations were made from legs 22 and 24 and on leg 23 only intensities of magnetization are available. From the remaining 21 sites the intensities and susceptibilities of 113 samples have been measured. Studies of magnetic viscosity, Curie temperatures and opaque petrological characteristics have also been made at a few sites.

The distribution of the 24 sites in the Indian Ocean is shown in Figure 1. Overall geometric mean values of natural remanent magnetization (NRM), susceptibility (k) and Königsberger ratio (Q_n) at each site are summarized in Table 1. Irving et al. (1966) have shown that intensities and susceptibilities of paleomagnetic collections exhibit a log normal distribution. Königsberger ratios were calculated assuming the total field intensity at each site is that given by the 1965 IGRF.

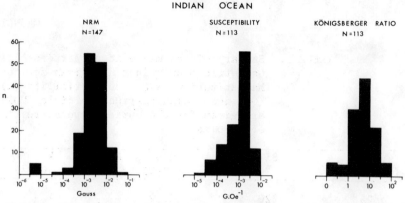

Fig. 2. Histograms of natural remanent magnetization intensity, susceptibility and Königsberger ratio for DSDP Indian Ocean basalt samples.

Figure 2 shows histograms for the three parameters for all the Indian Ocean samples and a summary by sites and by samples of overall geometric mean values is given in Table 2. The average intensities of magnetization are very close to those for the Pacific and Atlantic Oceans (Lowrie, 1974) but susceptibilities appear to be slightly higher and consequently the Königsberger ratios slightly lower. In spite of this remanence still dominates the induced magnetization as required by the Vine and Matthews hypothesis.

Alternating field demagnetization studies carried out on most samples showed that the remanence was generally stable. However, the demagnetization curves almost all exhibit a rapid fall off of intensity with initial demagnetizing field. Median destructive fields (the peak alternating field at which the NRM intensity is reduced to one-half its value) average only about 120 oersteds (Tables 1 and 2). The majority of values are less than 150 oersteds (Figure 3) contrasting with that found for the Atlantic and Pacific Oceans (Lowrie, 1974) where values in the range 200 to 400 oersteds were representative of the largest number of sites. Samples with MDFs lower than about 100 oersteds tend to be very susceptible to the acquisition of viscous remanent magnetization (VRM). A number of VRM acquisition tests were undertaken by Peirce et al. (1974), Lowrie and Hayes (1975) and Lowrie and Israfil (1975).

Lowrie (1973) has shown that samples with MDFs of less than 50 oersteds have high enough viscosity coefficients that their magnetizations could

TABLE 2. The magnetic properties of Indian Ocean DSDP basalts. Geometric mean values by sites and by samples of NRM, k, Q_n and MDF and arithmetic means for T_c. Symbols as in table 1.

	N	NRM(10^{-3}G)	k(10^{-3}G.Oe^{-1})	Q_n	MDF(Oe)	T_c(°C)
Site Means	24	2.29	–	–	116	–
	21	2.31	0.772	4.81	–	–
	14	–	–	–	–	346
Sample Means	147	2.14	–	–	–	–
	113	2.22	0.885	4.19	–	–
	75	–	–	–	128	–
	45	–	–	–	–	133

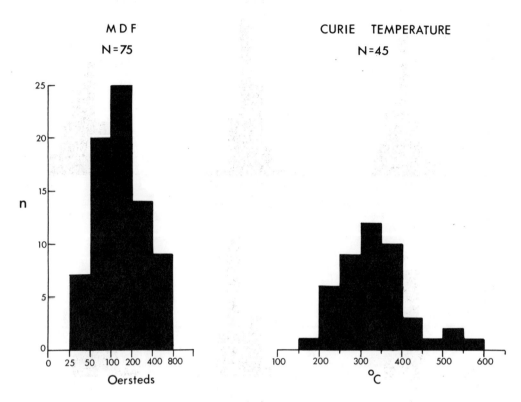

Fig. 3. Histograms of median destructive field and initial Curie temperature for DSDP Indian Ocean basalt samples.

be entirely viscous, acquired since the beginning of the Brunhes epoch. He suggested this could be an explanation of the magnetic quiet zones or regions of uncorrelatable magnetic anomalies (Heirtzler and Hayes, 1967). An opportunity to test this hypothesis occurred on leg 20 where site 282 was located in one of these quiet zones. MDFs were especially high averaging 263 oersteds (Table 1) and VRM acquisition test on such material confirmed that this is unlikely as an explanation of this quiet zone (Lowrie and Israfil, 1975). In the Indian Ocean less than 10 percent of MDFs lie in the range 25-50 oersteds (Figure 3) and they occur in isolated samples at a number of sites. Although VRM may be significant in some individual samples, the instances where this represents an overall property of all the basalts from a single site are rare. Particular examples are site 57 in the equatorial western Pacific (Lowrie, 1973) and site 319 in the equatorial eastern Pacific (Lowrie and Kent, 1977). In the southern Indian Ocean site 283 may also be in this category (Lowrie and Israfil, 1975). In the two Pacific Ocean sites the VRM properties appear to be associated with titanomagnetites with Curie temperatures around 150°C, considerably lower than is generally observed in oceanic basalts.

Most submarine basalts described in the literature contain fine (less than 10μ) titanomagnetite grains of skeletal form when viewed in reflected light (Irving et al., (1970); Lowrie, 1974). Electron microscope studies show large populations of discrete titanomagnetite grains much finer

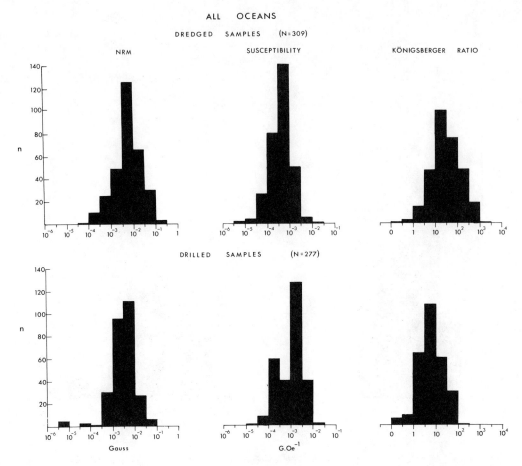

Fig. 4. Histograms of natural remanent magnetization intensity, susceptibility and Königsberger for world-wide collections of dredged and drilled oceanic basalt samples.

than 0.1μ in diameter (Evans and Wayman, 1972). These features suggest rapid quenching of the basalt. The titanomagnetite grains do not generally show the features of high temperature oxidation commonly seen in subaerial basalts. Most samples studied to date contain Class 1 (least oxidized) titanomagnetites according to the classification of Wilson et al. (1968). This overall picture could well be biased because sampling of the ocean floor has been restricted essentially to the surface region of Layer 2. Ade-Hall (1974a) has investigated some of the leg 26 basalts and for the first time has identified magnetites of much large grain size. Furthermore about one-third of the samples studied contained magnetite in deuteric oxidation Class 2, 3 and 4. As has also been noted in previous studies extensive maghematization (low temperature oxidation) is present in all samples.

Thermomagnetic analysis provides a quick and convenient method of distinguishing titanomagnetite from titanomaghemite. The technique that has been used in investigating submarine basalts involves rapidly heating and cooling specimens in air to temperatures in excess of 600°C in the presence of a strong magnetic field of 3,000 oersteds or more. An initial Curie temperature typically in the range 200 to 400°C is observed on heating but on cooling a second Curie temperature above 500°C is observed replacing the initial ones and the room temperature magnetization in-

TABLE 3. The magnetic properties of world wide oceanic basalts. A comparison between drilled samples from the DSDP and dredged samples. Geometric site mean values are given with symbols as in table 2.

Site Mean	N	NRM(10^{-3}G)	k(10^{-3}G.Oe^{-1})	Q_n
Drilled samples (DSDP)	47	2.37	0.729	6.28
Dredged samples	110	5.76	0.318	39.9

creases by a factor of two or three. The non-reversible nature of such curves is largely due to maghematization of the original titanomagnetite.

Initial Curie temperatures have been measured at a number of sites (Table 1), the majority of values falling in the range 250°-400°C (Figure 3). The average value is around 340°C (Table 2). This confirms the presence of extensive maghematization. Samples with Curie temperatures below about 250°C and showing reversible thermomagnetic curves contain single phase little altered titanomagnetite of Class 1 oxidation state. Those with Curie temperatures above about 459°C contain low titanium titanomagnetite produced by the subsolidus exsolution of ilmenite following high temperature oxidation during initial cooling. The intermediate Curie temperatures arise largely from low temperature oxidization causing elevation of the Curie temperature from an initially low value below 250°C and reduction of the spontaneous magnetization (Readman and O'Reilly, 1970; Ade-Hall et al., 1971; Marshall and Cox, 1972; Lowrie et al., 1973a).

World-Wide Summary of Magnetic Properties of Oceanic Basalts

Lowrie (1973) has recently compiled overall statistics of the magnetic properties of DSDP basalts from 26 sites mainly in the Pacific (Lowrie et al., 1973a) and Atlantic (Lowrie et al., 1973b) Oceans, but including 4 sites from the Indian Ocean (Lowrie and Hayes, 1975) also summarized in this paper. Since the magnetic data from the Indian Ocean constitute about half the total available from all oceans, it is convenient to provide an overall world-wide synthesis of the magnetic properties of oceanic basalts. In addition to the data compiled by Lowrie (1974) and in this paper the results obtained by Larson and Lowrie (1975) from leg 32 in the Pacific have also been included. The distribution of the DSDP sites for which extensive magnetic data are available are shown in Figure 1. The world-wide coverage is now fairly extensive. Figure 4 compares the distribution of NRM, susceptibility and Königsberger ratio of 309 dredged samples with that of 277 drilled samples from the world's ocean. This is an updated version of the compilation of Lowrie (1974). The mean NRM of dredged samples is more than twice that of drilled samples and the mean susceptibility less than one-half (Table 3). This almost certainly reflects the sampling distribution in which dredged samples are preferentially collected from near ridge crests where topography is higher and sampling easier. Irving et al. (1970) have shown that the NRM of samples collected from the ridge axis is exceedingly strong at about 1×10^{-1}G. This value decreases by a factor of ten or more within the first 10 km from the axis and thereafter levels off to values of around 5×10^{-3}G.

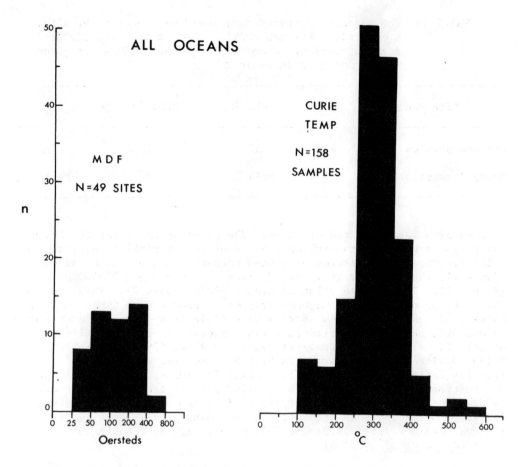

Fig. 5. Histograms of median destructive field and initial Curie temperature for world-wide DSDP oceanic basalt samples.

The comparison between the drilled and dredged samples suggests that the use of magnetic properties derived from dredged samples in modelling the magnetic characteristics of the oceanic crust could lead to errors. The average magnetization of 2.37×10^{-3}G for drilled samples is much less than that commonly used for modelling marine magnetic anomalies. This suggests that the model of a uniformly magnetized Layer 2 of thickness about 2 km is much more reasonable than one in which only the upper 0.5 km is strongly magnetized. Königsberger ratios of the dredged samples average more than six times those of drilled samples. However, the value of around 6 is still sufficient to justify remanence being the dominant magnetization component.

The distribution of median destructive fields at 49 DSDP sites is fairly evenly spread between 50 and 400 oersteds (Figure 5) showing that stability characteristics cover a wide range. Curie temperatures however have a sharp peak around 300°C (Figure 5) with by far the majority of values lying between 250 and 350°C. This suggests that the state of low temperature oxidation over the world's oceans is fairly uniform consistent with the view that this process takes place largely within a short distance from the ridge crest (Irving et al., 1970).

References

Ade-Hall, J. M., The opaque mineralogy of basalts from DSDP leg 26, in Davies, T.A., Luyendyk, B. P., et al., Initial Reports of the Deep Sea Drilling Project, 26, Washington (U. S. Government Printing Office) 533-539, 1974a.

Ade-Hall, J. M., The strong field magnetic properties of basalts from the DSDP leg 26, in Davies, T.A., Luyendyk, B. P., et al., Initial Reports of the Deep Sea Drilling Project, 26, Washington (U. S. Government Printing Office), 529-532, 1974b.

Ade-Hall, J. M., and K. Kitazawa, A quantitative assessment of supposed differences between the paleomagnetic properties of oceanic submarine and subaerial basalts, Trans. Amer. Geophys. Un., 54, 1022-1024, 1973.

Ade-Hall, J. M., H. C. Palmer, and T. P. Hubbard, The magnetic and opaque petrological response of basalts to regional hydrothermal alteration, Geophys. J., 24, 137-174, 1971.

Evans, M. E., and M. L. Wayman, The mid-Atlantic ridge near 45°N, XIX. An electron microscope investigation of the magnetic minerals in basalt samples, Can. J. Earth Sci., 9, 671-678, 1972.

Fox, P. J., and N. D. Opdyke, Geology of the oceanic crust: Magnetic properties of oceanic rocks, J. Geophys. Res., 78, 5139-5154, 1973.

Heirtzler, J. R., and D. E. Hayes, Magnetic boundaries in the North Atlantic Ocean, Science, 157, 185-187, 1967.

Herron, E. M., Sea-floor spreading and the Cenozoic history of the East Central Pacific, Bull. Geol. Soc. Amer., 83, 1671-1692, 1972.

Hess, H. H., History of ocean basins, in Petrologic Studies: a volume to honor A. F. Buddington, edited by A.E.J. Engel, H. L. James, and B.F. Leonard, Bull. Geol. Soc. Amer., 599-620, 1962.

Irving, E., L. Molyneux, and S. K. Runcorn, The analysis of remanent intensities and susceptibilities of rocks, J. Geophys. Res., 10, 451-464, 1966.

Irving, E., J. K. Park, S. E. Haggerty, F. Aumento, and B. Loncarevic, Magnetism and opaque mineralogy of basalts from the mid-Atlantic ridge at 45°N, Nature, 228, 974-976, 1970.

Larson, R. L., and W. Lowrie, Paleomagnetic evidence for motion of the Pacific plate from leg 32 basalts and magnetic anomalies, in Larson, R. L., Initial Reports of the Deep Sea Drilling Project, 32, Washington (U. S. Government Printing Office), 751-758, 1975.

Lowrie, W., Viscous remanent magnetization in oceanic basalts, Nature, 243, 27-29, 1973.

Lowrie, W., Oceanic basalt magnetic properties and the Vine and Matthews hypothesis, J. Geophys. Res., 40, 513-536, 1974.

Lowrie, W., and D. E. Hayes, Magnetic properties of oceanic basalt samples, in Hayes, D.E., Frakes, L.A., et al., Initial Reports of the Deep Sea Drilling Project, 28, Washington (U. S. Government Printing Office), 869-878, 1975.

Lowrie, W., and M. N. Israfil, Paleomagnetism of basalt samples from leg 29, in Kennett, J. P., Houtz, R. E., et al., Initial Reports of the Deep Sea Drilling Project, 29, Washington (U. S. Government Printing Office), 1109-1115, 1975.

Lowrie, W., and D. E. Kent, Viscous remanent magnetization in basalt samples, in Initial Reports of the Deep Sea Drilling Project, 34, Washington (U. S. Government Printing Office), 1977.

Lowrie, W., R. Løvlie, R., and N. D. Opdyke, Magnetic properties of Deep Sea Drilling Project basalts from the North Pacific Ocean, J. Geophys. Res., 78, 7647-7660, 1973a.

Lowrie, W., R. Løvlie, and N. D. Opdyke, The magnetic properties of Deep Sea Drilling Project basalts from the Atlantic Ocean, Earth Planet. Sci. Lett., 17, 338-349, 1973b.

Marshall, M., and A. Cox, Magnetic changes in pillow basalt due to sea-floor weathering, J. Geophys. Res., 77, 6459-6469, 1972.

McElhinny, M. W., Paleomagnetism of basalt samples, leg 27, in Veevers, J. J., Heirtzler, J. R., et al., Initial Reports of the Deep Sea Drilling Project, 27, Washington (U. S. Government Printing Office), 403-404, 1974.

Peirce, J. W., C. R. Denham, and B. P. Luyendyk, Paleomagnetic results of basalt samples from DSDP Leg 26, Southern Indian Ocean, in Davies, T.A., Luyendyk, B. P., et al., Initial Reports of the Deep Sea Drilling Project, 26, Washington (U. S. Government Printing Office), 517-527, 1974.

Pitman III, W. C., and J. R. Heirtzler, Magnetic anomalies over the Pacific-Antarctic Ridge, Science, 154, 1164-1171, 1966.

Readman, P. W., and W. O'Reilly, The synthesis and inversion of non-stochiometric titanomagnetites, Phy. Earth Planet. Int., 4, 121-128, 1970.

Talwani, M., C. Windisch, and M. G. Langseth, Jr., Reykjanes ridge crest: A detailed geophysical study, J. Geophys. Res., 76, 473-517, 1971.

Vine, F. J., Spreading of the ocean floor: new evidence, Science, 154, 1405-1415, 1966.

Vine, F. J., and D. H. Matthews, Magnetic anomalies over oceanic ridges, Nature, 199, 947-949, 1963.

Whitmarsh, R. B., O. E. Weser, D. A. Ross, et al., in Initial Reports of the Deep Dea Drilling Project, 23, Washington (U. S. Government Printing Office), 129-130, 179, 313, 1974.

Wilson, R. L., S. E. Haggerty, and N. D. Watkins, Variation of paleomagnetic stability and other parameters in a vertical traverse of a single Icelandic lava, J. Geophys. Res., 16, 79-96, 1968.

Wolejszo, J., R. Schlich, and J. Segoufin, Paleomatnetic studies of basalt samples, Deep Sea Drilling Project, leg a5, in Simpson, E.S.W., Schlich, R., et al., Initial Reports of the Deep Sea Drilling Project, 25, Washington (U. S. Government Printing Office), 55-572, 1974.

CHAPTER 12.   INTRODUCTION TO STRATIGRAPHY AND PALEONTOLOGY

Hans M. Bolli

Geology Department
Swiss Federal Institute of Technology
Zurich, Switzerland

John B. Saunders
Museum of Natural History
Basel, Switzerland

Ten papers of this synthesis deal with the paleontology-biostratigraphy of the Indian Ocean drilling results. Three of them concern the Mesozoic, Paleocene to Eocene and Oligocene to Holocene calcareous nannoplankton. The Cretaceous, Maastrichtian to Eocene, Oligocene and Neogene planktonic foraminifera are the subject of four contributions while two give an account of the Cretaceous and Neogene benthonic foraminifera. Finally, one paper summarized investigations on 18 other fossil groups, including the radiolaria, which were published in the Initial Reports.

Figure 1 gives the ages of sediments recovered from each Leg drilled in the Indian Ocean. That figure also shows where basaltic basement was reached and the completeness of the geologic record obtained.

The principal fossil groups which were studied for biostratigraphic and ecologic interpretations and published for each Leg are calcareous nannoplankton, planktonic foraminifera and radiolaria. Regrettably, the present synthesis volume does not contain a contribution on the Indian Ocean radiolaria recovered by DSDP. However, Figure 2 in Bolli (1977) shows the radiolaria reported in the Initial Reports by Leg, Site and Age. Further, references are given for the radiolaria publications in the Initial Report volumes.

In higher latitudes (Legs 28, 29) diatoms and also silicoflagellates proved useful for biostratigraphic dating of Neogene and Paleogene sediments. Of the siliceous microfossils, diatoms were also studied in some sites of the lower latitude Legs 22, 24, 25 and Site 262 of Leg 27.

Pollen and spores were investigated in certain stratigraphic intervals of some sites of Legs 22 and 26-29, dinoflagellates/acritarchs in Legs 27-29.

Of the other microfossil groups, isolated Calcisphaerulidae were for the first time reported in detail from the Jurassic to Upper Cretaceous in Leg 27, and also briefly mentioned from Site 257 of Leg 26. Contributions on bryozoans and sponge spicules appeared in the Leg 29 Initial Reports volume, on ostracods in the Legs 24 and 27, on bivalves and cephalopods in Leg 27, on fish denticles in Legs 26 and 27, on otoliths in Leg 25, on fecal pellets in Leg 28 and on organisms incertae sedis (small calcareous spheres) in the Leg 27 Initial Reports volume.

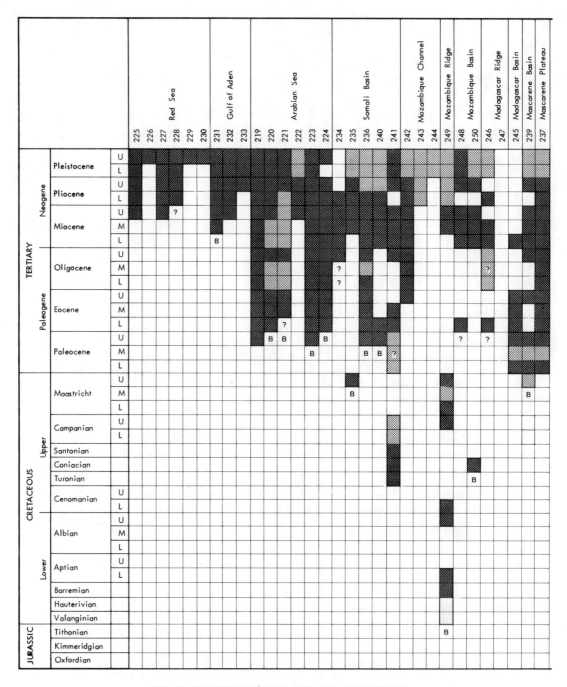

AGES OF SEDIMENTS RECOVERED FROM THE INDIAN OCEAN
ON DSDP LEGS 22-29 SITES 211-269 AND 280-282

Stratigraphy

Basement was reached in 33 of 62 sites drilled in the Indian Ocean (Figure 2). This is a very high percentage considering that for numerous sites such as 218 (Bengal Fan), 225-230 (Red Sea), 262 (Timor Trough) and others, basement had previously been thought out of reach. Where basement was reached, one knew that the sedimentary

INTRODUCTION TO STRATIGRAPHY AND PALEONTOLOGY   313

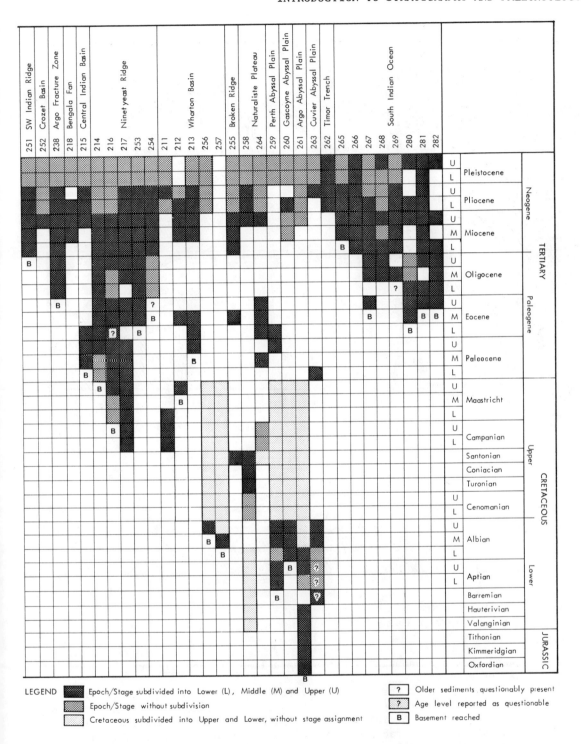

and therefore the biostratigraphic section is as complete as can be expected for that site.

The oldest Indian Ocean sediments penetrated by DSDP are of <u>Upper Jurassic</u> (Tithonian to Oxfordian) age cored in the Argo Abyssal Plain Site 361. They are dated primarily by calcareous nannoplankton, dinoflagellates and benthonic foraminifera. Present in these sedi-

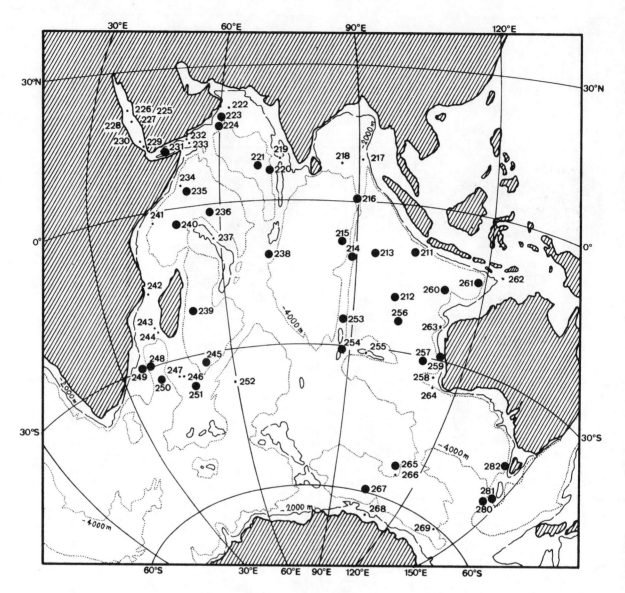

Indian Ocean Sites drilled by DSDP Legs 22-29
● indicates Basement reached and recovered

ments are also abundant Calcisphaerulidae, Inoceramus prisms and a few ostracods.

Of the eight sites which recovered Lower Cretaceous (Figure 3), sediments of all but one are situated west of the Australian continental margin. On the African side only Site 249 on the Mozambique Ridge entered Lower Cretaceous. Six of the eight sites reached basement and penetrated all Lower Cretaceous sediments present. Biostratigraphic resolution based on calcareous nannoplankton, foraminifera, palynomorphs and in a few instances molluscs is in most sites reasonably good at this level and subdivisions of stages could be distinguished. The oldest sediments in the Wharton Basin and in the Gascoyne Abyssal Plain are Albian, in the Perth Abyssal Plain, Aptian. No definite age for the oldest stage present is available

INTRODUCTION TO STRATIGRAPHY AND PALEONTOLOGY   315

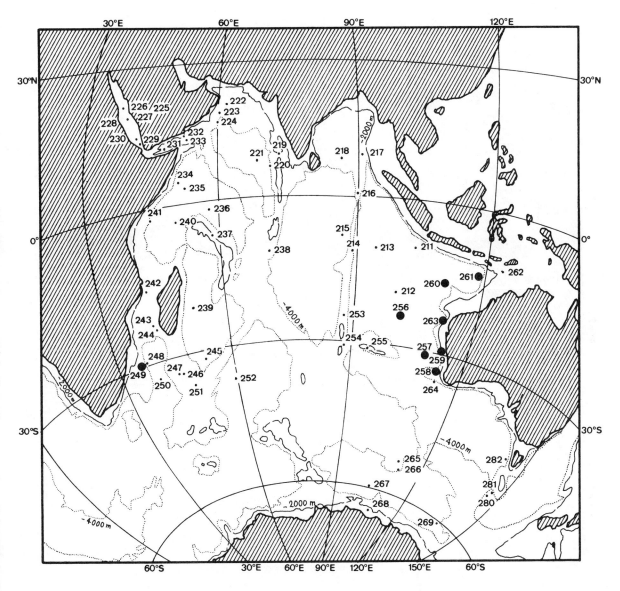

Indian Ocean Sites drilled by DSDP Legs 22-29
● indicates Lower Cretaceous sediments recovered

from the Mozambique Ridge and the Naturaliste Plateau. A virtually complete Lower Cretaceous sequence which continues into the Upper Jurassic was penetrated in the Argo Abyssal Plain Site 361.

Upper Cretaceous sediments were entered or penetrated in 17 sites (Figure 4) slightly more than double those reaching the Lower Cretaceous. Twelve sites are situated in the eastern and five in the western Indian Ocean, all towards the continental margins. From the information obtained through these sites it must be assumed that the Upper Cretaceous Cenomanian to Maastrichtian sedimentary sequence is incomplete over wide areas of the Indian Ocean. Due to lack of diagnostic faunas and floras, assignment of the sediments to biozones or even stages is often difficult or impossible. This applies in particular to the Upper Cretaceous obtained from the sites

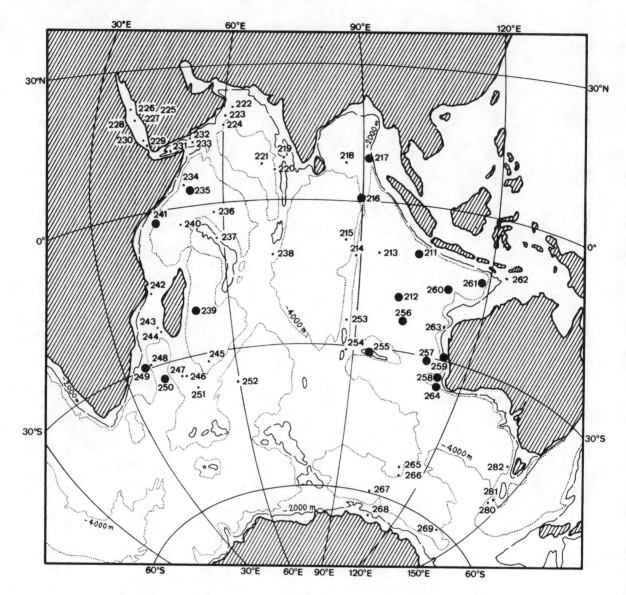

Indian Ocean Sites drilled by DSDP Legs 22-29
● indicates Upper Cretaceous sediments recovered

located between the Australian continental margin and the Ninetyeast Ridge (Figures 1, 4). Basement underlies Upper Maastrichtian in the Somali Basin Site 235, Maastrichtian s.l. in the Ninetyeast Ridge Site 216, and Coniacian in the Mozambique Basin Site 250. Before entering Basement the Wharton Basin Site 212 penetrated Upper Cretaceous, but faunal evidence is too poor to indicate which part of the Upper Cretaceous is represented. The only definitely established Cenomanian is from the Mozambique Ridge Site 249. Except for Sites 255 and 258 the Upper Cretaceous of all Sites west of the Australian continental shelf could only be determined in a tentative way without subdivision into stages or zones. Upper Cretaceous sediments of these sites appear much reduced in thickness and stratigraphically, and may even be totally absent as is the case in the

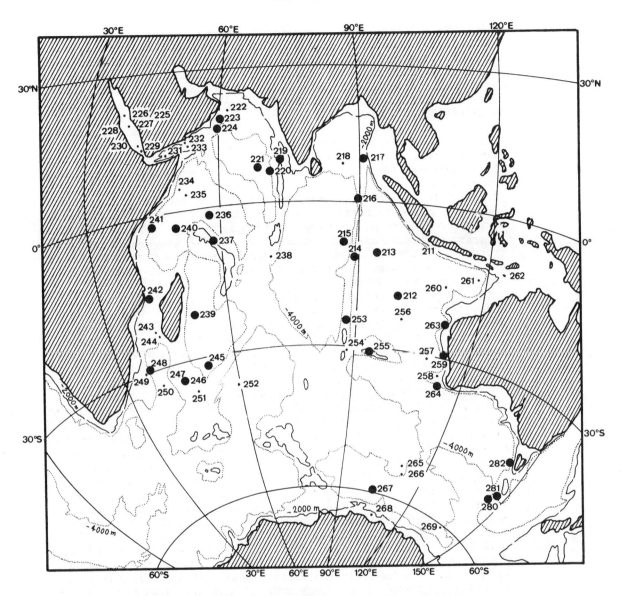

Indian Ocean Sites drilled by DSDP Legs 22-29
● indicates Eocene and/or Paleocene sediments recovered

Cuvier Abyssal Plain Site 261 and possibly also in Site 259 in the Perth Abyssal Plain. One Site (249, Mozambique Ridge) in the western Indian Ocean and seven sites between the Ninetyeast Ridge and the Australian continental margin continue from Upper Cretaceous into Lower Cretaceous sediments.

Paleocene-Eocene sediments were recovered from 29 sites (Figure 5), in 13 they directly overlie basement. In the four South Indian Ocean sites between Antarctica and Australia which reached basement, the oldest sediments are Upper-Middle Eocene. The four sites that entered basement in the Arabian Sea had Lower Eocene or Upper Paleocene as oldest sediments. The most complete Paleocene-Eocene sequence was found in the Ninetyeast Ridge Sites 214, 216 and 217, and in the Madagascar Basin (245)-Mascarene Basin (239)-Mascarene Plateau (237) area. The Paleocene-Eocene section is fragmentary or totally absent particularly in the eastern Indian Ocean sites East of the Ninetyeast Ridge and in sites in the Mozambique Ridge and Basin. This may well be due in part to the selection of sites at points of reduced sediment thickness and therefore only a local feature.

It is of significance that the <u>Oligocene</u> has been recovered from fewer sites (24) than the Eocene which is recognized in 29 sites. This "deficit" is accounted for by the widespread absence or non-recognition of Oligocene sediments in the Eastern Indian Ocean between the Ninetyeast Ridge and Indonesia/Australia and also in the Mozambique Ridge/Basin area, as has been noted for the Paleocene-Eocene also.

A determining factor for this poor Oligocene record throughout the Indian Ocean may be ascribed to the selection of many drill sites in deep basins and in other areas of strongly reduced sediment thickness in order to reach basement. Conclusions on the development and distribution of the Oligocene in the Indian Ocean based on these spotty observations in mostly unfavorably located sites must therefore be drawn with great caution.

The <u>Miocene</u> was recovered complete or in part in 46 of the 62 Indian Ocean sections. As is true for the Oligocene, the poorest representation of Miocene sediments, with this interval completely or partially missing, is in the sites situated between the Ninetyeast Ridge and Indonesia/Australia. In most of the other sites which penetrated into older sediments the Miocene is reasonably complete and continuous on paleontological evidence.

<u>Pliocene</u> and <u>Pleistocene</u> were recovered in 53 and 57 sites respectively, that is in almost all sites. In a few (226, 229, 230 of the Red Sea) drilling already terminated within the Upper Pleistocene. In some others (212, 257, Wharton Basin, 234, Somali Basin, and 245, Madagascar Basin) the Pleistocene is either totally missing or of such a reduced thickness that it was not recognized in the top core.

The <u>Cretaceous-Tertiary</u> boundary was recovered in cores from a number of sites. In the northeastern Indian Ocean sites 216 and 217 on the Ninetyeast Ridge, sedimentation was found to be nearly continuous with apparently only a minor hiatus in that the lowermost Paleocene could not be recognized.

With the exception of Site 239 in the Mascarene Basin where Lower Paleocene is underlain by a few centimeters of Maastrichtian above basalt, all other western and eastern South Indian Ocean sites entering Cretaceous show a very considerable hiatus with much of the Tertiary and also the Upper Cretaceous missing. In the southwestern Indian Ocean Site 249, Middle Miocene rests on Upper Maastrichtian, and in 250 Lower Miocene on Coniacian. Of the southeastern Indian Ocean sites the Cretaceous-Tertiary boundary was recovered in Site 255 with Eocene resting on Santonian and in Site 258 with Miocene overlying Santonian.

Though not continuously cored and the boundaries not actually recovered in cores, one must conclude from the small intervals between cored Tertiary and Cretaceous that very considerable parts of the sedimentary sequence between the recovered deepest Tertiary and the highest Cretaceous are missing in the following eastern Indian Ocean sites: Sites 256 and 257 (Pliocene/Cretaceous s.l.), Site 259 (Upper Paleocene/Albian), Site 260 (Miocene/Upper Cretaceous), Site 261 (Upper Miocene/Upper Cretaceous), Site 263 (Paleocene/Lower Albian), Site 264 (Paleocene/Campanian), Site 211 (Pliocene/Lower Maastrichtian, Site 212 (Middle Eocene/Upper Maastrichtian).

<u>Reduced Sections, Hiatuses</u>: When discussing and interpreting con-

tinuities of Tertiary and Cretaceous sedimentary sequences drilled in the Indian Ocean, one has to keep in mind that most sites were located in areas of reduced sediment thickness in order to reach basement with the penetration capabilities of the GLOMAR CHALLENGER. Such reduced sediment thickness and stratigraphic non-continuity were necessarily caused by non-deposition, condensed deposition or erosional effects. How far reduced and incomplete sedimentary sequences recorded in the drilled Indian Ocean sites extend remains unknown until thicker and consequently stratigraphically more complete sections have been drilled.

Virtually all sections with Tertiary and Mesozoic sediments are reduced in thickness and stratigraphically incomplete. This is apparent in Figure 1, "Ages of sediments recovered from the Indian Ocean". What is also evident here is that Tertiary sections directly overlying basement, as in the Arabian Sea to the North or in the South Indian Ocean sites are stratigraphically continuous with no or only minor hiatuses. It is further of interest to note that while Miocene was recovered from 46 sites, Oligocene is known only from 24, while Eocene-Paleocene is again known from 29. In none of the above-mentioned western and eastern South Indian Ocean sites which extended into or through the Cretaceous were Oligocene sediments reported, with the only exception of Site 241 where some Upper Oligocene is found.

In particular the minor stratigraphic gaps appearing on Figure 1 may in fact indicate existing hiatuses, but they may also be caused by insufficient coring density or an inadequate fossil record.

Paleontology

Planktonic foraminifera are present in most Indian Ocean sections. Because of their value for the biostratigraphic subdivision of Cretaceous and Tertiary sediments they have been the subject of detailed investigations through all sedimentary sections where they occurred. The investigations on this faunal group in the Indian Ocean, ranging from low to high latitudes and shallow to deep water sediments brought to light the same limitations planktonic foraminiferal biostratigraphy is subject to in the other oceans: 1) the high resolution planktonic foraminiferal zonal schemes, largely established on tropical index species are difficult to apply in higher latitudes, and 2) $CaCO_3$ dissolution, which with increasing depth progressively destroys the tests of the planktonic foraminifera, makes age dating of the sediments at great depth less accurate or even impossible. The biostratigraphic and ecologic significance of the planktonic foraminifera are reported in this volume in four separate contributions covering the Cretaceous (Herb and Scheibnerova, 1977; Scheibnerova, 1977), Maastrichtian to Eocene (McGowran, 1977), Oligocene (Fleisher, 1977) and the Neogene (Boltovskoy, 1977).

Comparison and correlation of the DSDP results with marine sections from land areas adjacent to the Indian Ocean have, though in a rather limited way, been attempted by some contributing authors. Cretaceous planktonic and benthonic foraminifera and their distribution are particularly well known from Australia, including the basins immediately adjacent to the Indian Ocean. More recently, published information on the Cretaceous and Tertiary of the western margin of the Indian Ocean, such as SE Africa and Madagascar have also become available. The Oliogocene Lindi section in coastal Tanzania serves as an example of how stratigraphically well dated land sections can now be tied in with the results obtained from the Indian Ocean drilling. Published modern studies on the Indonesian Neogene planktonic foraminifera still remain almost nonexistent, with the exception perhaps of the Bodjonegoro well 1 section in Java, with which the tropical Indian Ocean faunas are readily correlatable.

The Indian Ocean faunal and biostratigraphic results should also be better tied in with those known from the Indian subcontinent and other land areas bordering the northern Indian Ocean.

Because of the widespread absence of Neogene sediments and the scarce information available from areas where they do occur in the coastal areas surrounding the Indian Ocean, there is at present little to compare the extensive Indian Ocean Neogene data with the bordering land sequences. Exceptions are the deep water Neogene sections of Bodjonegoro well mentioned above and the recent publications on the faunas from the Andaman-Nicobar Islands in the Bay of Bengal. Some data on the younger Neogene planktonic foraminifera of the Indian Ocean, in particular on those of Pliocene to Recent age, was already available prior to the DSDP drilling. It was obtained through numerous plankton tows, dredgings and piston cores for which Vincent (1977) gives a condensed account in her introduction.

The study of the Indian Ocean Cretaceous planktonic foraminifera has been adversely affected by stratigraphically incomplete sections and by the presence of cold water, high latitudinal faunas in most sections. The Albian faunas which occur as the oldest in several sites, all situated between the Ninetyeast Ridge and Australia, are of the temperate to cold water type with Hedbergellids strongly dominating the assemblages. Where sites entered or penetrated Upper Cretaceous sediments their record is particularly incomplete. Cenomanian faunas were recovered from only two sites. In the few Coniacian to Santonian intervals recovered, planktonic foraminiferal assemblages were still found to be predominantly of the globigerinid and heterohelicid type, characteristic for the non-tropical Austral type province, and with tropical index forms largely absent. Only during the terminal Cretaceous penetrated on Ninetyeast Ridge and on the Mozambique Ridge do tropical tethyan type assemblages begin to appear, an indication for the rapid northward shift of the Indian plate at that time. Faunas indicative of higher latitudes with their low species diversity made precise age determinations difficult in particular in the Albian-Cenomanian but also in the higher Upper Cretaceous except for the Maastrichtian.

The investigations on the Paleocene-Eocene planktonic foraminifera in this volume by McGowran (1977) are restricted to selected sites in the northern and eastern Indian Ocean. Excluded are thus the faunas recovered from the southwestern Indian Ocean around and South of Madagascar, and the southern Indian Ocean sites of Leg 28 where the only recorded Eocene planktonic foraminifera occur in a core of Site 367B.

The most complete Eocene and Paleocene sequences were recovered from the Madagascar Basin, Mascarene Basin and Plateau, and from several sites from the Ninetyeast Ridge area, while good Eocene and some Upper Paleocene is also present in the Arabian Sea and Somali Basin sites. In contrast, the Paleocene-Eocene is poorest and very incomplete in the sites drilled East of the Ninetyeast Ridge.

The Paleocene-Eocene planktonic foraminifera of the tropical Indian Ocean region compare well with those reported earlier from the Atlantic and Mediterranean sections on which the zonal schemes were originally established, and which are readily applicable to the Indian Ocean sections. McGowran also extended these zonal subdivisions to land sections of the same age in basins in southern and western Australia, Indonesia and India.

Oligocene planktonic foraminifera from Indian Ocean sites have with few exceptions been reported only rather tentatively. The reasons for this are a total or partial absence of Oligocene sediments in many sites. Where present, planktonic foraminiferal tests often proved to be strongly affected by dissolution, as is also true in the Lower Miocene, making accurate identification impossible. There are, however, exceptions to this, where the established Oligocene zonal scheme could be applied, as for instance in sites on Ninetyeast Ridge and the Arabian Sea, and also from the land based Lindi section in the coastal area of Tanzania.

Neogene sediments and with them planktonic foraminifera have been recovered from nearly all Indian Ocean sites. It was realized that their biostratigraphic application is, however, often influenced by a number of adverse factors, principally because of the latitudinal restrictions in species distribution including many index forms. This becomes progressively more pronounced towards the younger Neogene. The standard planktonic foraminiferal zonal schemes, largely based on tropical Atlantic taxa were therefore not always fully applicable in the Indian Ocean, because of the scarceness or even absence of index species. $CaCO_3$ dissolution in particular in the sections situated in deeper basins and pronounced also in the Lower Miocene further adversely influenced the application of planktonic foraminifera to Indian Ocean Neogene biostratigraphy.

Despite these drawbacks, however, this faunal group still proved essential in the dating and ecologic interpretation of the Indian Ocean Neogene sediments. This is well documented by Vincent (1977) in this volume for the Neogene planktonic foraminifera. She discusses plankton zonations, planktonic foraminiferal biostratigraphy and their application to the Indian Ocean. Discussed are paleoclimatic changes, stage boundaries, foraminiferal events calibrated to the paleomagnetic time scale. An account is further given on the Neogene planktonic foraminiferal taxa described as new from the Indian Ocean, sediment accumulation rates and on the paleogeography during the Neogene.

Benthonic Foraminifera. Scheibnerova (1977) in her contribution to the present volume, discusses what is known of the Late Jurassic and Cretaceous from the DSDP sites. She also makes some comparisons with coeval faunas known from the surrounding land areas, particularly from Australia.

In the Initial Reports, benthonic foraminifera have received closest attention in Leg 27. Site 261 in the Argo Abyssal Plain south of the eastern end of the Java Trench, is particularly well covered for this group of organisms having four papers (Leg 27 Initial Volume; papers by Bartenstein, Krasheninnikov, Kuznetsova, Scheibnerova) devoted to it.

The Late Jurassic - Early Cretaceous section is dominated by agglutinated forms, often to the exclusion of calcareous ones. The complete fauna is of considerable diversity with over 90 species in 44 genera according to Kuznetsova (1974). This author, together with Bartenstein (1974) who also studied the arenaceous forms, considers the faunas to represent deep water deposition below the lysocline. With this interpretation Scheibnerova (this volume) disagrees. She considers them to be of shallow water origin in cool temperatures and with restricted connections to the open ocean. She maintains that coeval faunas known from Australia support this view. At the present time this question is unresolved.

In Sites 260 (Gascoyne Abyssal Plain) and 261 (Argo Abyssal Plain), Upper Cretaceous zeolitic-clays contain a fauna of agglutinated benthonic forms described by Krasheninnikov (1974). The fauna is presumed to be autochthonous and to have been deposited below the lysocline. Of the 44 species, 28 are described as new either here or from Leg 20 in the western Pacific (Krasheninnikov, 1973). This is an indication that much work remains to be done on benthonic foraminiferal assemblages from the deep oceans, the results of which can be expected to be of value, not only for dating, but also for their palaeobathymetric implications.

Important genera and species for the Late Jurassic and Cretaceous are listed by Scheibnerova in the present volume.

The Paleocene and Eocene benthonic foraminiferal faunas of Sites 236 (NE of the Seychelles Bank) and 237 (on the Mascarene Plateau) have been studied by Vincent, Gibson and Brun (1974). Though benthonic fossils are very rare compared with the planktonic component, they were able to identify 134 species and to erect four depth assemblages. The palaeobathymetry suggested is lower bathyal in Site 236 and upper bathyal progressing to lower bathyal in Site 237. This evidence is useful in dating movements on the Mascarene Plateau.

McGowran (1974 and this volume) lists assemblages indicating shelf depths for Site 214 (Ninetyeast Ridge) in the Paleocene and deeper water assemblages from Sites 213 and 215. Again, the paleobathymetric implications are important. In this volume McGowran also discusses other benthonic foraminiferal occurrences that have a bearing on the Paleogene history of the region.

Boltovskoy, in the present volume, has studied the Neogene sections of 15 sites from which he reports 306 species. These occur as low percentages of predominantly planktonic assemblages.

It is of interest to note that in 10 of the 15 sites shallow water benthonics such as Amphistegina, Peneroplis, species of Elphidium, etc. were found as displaced faunas. Though normally of small size and present in very small numbers, sometimes relatively large specimens are found. The ten sites are amongst those drilled in relative proximity to shallow water areas bordering the continents. Boltovskoy rightly points out the importance of a study of the benthonics for the recognition of displaced faunas as these would probably be difficult to recognize using planktonic species. The correctness of the age interpretations might suffer as a consequence.

The use of benthonic foraminifera as age markers in the Neogene is less certain.

Calcareous Nannofossils. Prior to the GLOMAR CHALLENGER cruises in the Indian Ocean, only an Upper Cretaceous locality in South Africa and one in southeast Australia had been investigated for calcareous nannofossils in this part of the world. Paleocene and/or Eocene occurrences had been described from India, Pakistan, Egypt and Magagascar and only a few Oliogcene and Neogene occurrences of calcareous nannofossils were known from the Indian Ocean. Thus the contributions dealing with Mesozoic, Paleocene to Eocene and Oligocene through Holocene calcareous nannofossils by Thierstein (1977). Proto Decima (1977) and Müller (1977) respectively, are mainly based on the published results from the 62 sites drilled in the Indian Ocean during Legs 22 through 29 of the Deep Sea Drilling Project.

All Oxfordian to Maastrichtian sediments containing more than a few percent carbon carbonate could be dated by their coccolith content. Changes in species composition and relative abundances were found to reflect the higher paleolatitudinal positions of the localities investigated. The only Jurassic recovered (Site 261) includes a quite similar assemblage to those of southeastern France or the northwest Atlantic. The species diversity in the Lower Cretaceous DSDP samples is generally low and characteristic tethyan taxa such as Nannoconus are absent at the 5 relevant sites. In the Upper Cretaceous, the assemblages show distinct differences indicating higher or lower paleolatitudinal positions of the sites.

No continuous sequence over the Cretaceous/Tertiary boundary was recovered at Sites 216, 217 and 239, the 3 sites where Upper Maastrichtian and Lower Paleocene are represented. The low diversity of the nannoflora of the higher latitudes together with the absence of some marker species reduces the high resolution biostratigraphy applicable in the lower latitudes of the Indian Ocean, especially in the Upper Eocene. As in other oceans, Isthmolithus recurvus and Chiasmolithus oamaruensis were found mainly in the high, Discoaster barbadiensis and D. saipanensis mainly in the lower latitudes.

In the Oligocene and the Neogene, the calcareous nannofossil assemblages of low- and mid-latitudinal areas of the Indian Ocean are similar. In the mid-latitudes, a decrease in frequency of some groups (discoasters, ceratoliths, sphenoliths) can be observed, but only in high latitudes and in areas influenced by the Circumpolar Current was it necessary to apply a modified zonation. During the Oligocene, species diversity was generally low and this is also so in the Indian Ocean. It remained low through the early Miocene, increased in the Middle Miocene and was relatively high through the Early Pliocene. Since then it has decreased again.

Silicoflagellates, Archaeomonads and Ebridians. The investigation of Silicoflagellates, Archaeomonads and Ebridians is not routinely performed on board GLOMAR CHALLENGER, but may be undertaken by specialists of calcareous nannofossils on legs to relative high latitudes where coccolith diversity and abundance decrease or where calcareous nannofossils are lost completely. As a result of the Deep Sea Drilling Project, these siliceous micro- and nannofossils became known from the Indian Ocean, from where they had not been previously reported. The more common and conspicuous Silicoflagellates were reported from tropical as well as from high latitude sites from the Maastrichtian through the Pleistocene of Legs 22, 25, 28 and 29. From the last site the Archaeomonads, Ebridians and some endoskeletal dinoflagellates were also investigated.

References

Bartenstein, H., Upper Jurassic-Lower Cretaceous primative arenaceous Foraminifera from DSDP sites 259 and 261 Eastern Indian Ocean in Veevers, J. J., Heirtzler, J. R., et al., Initial Reports of the Deep Sea Drilling Project, Washington (U. S. Government Printing Office), 27, 683-696, 1974.

Bolli, Hans M., Paleontological-biostratigraphical investigations, Indian Ocean Sites 211-269 and 280-282, DSDP Legs 22-29 in Indian Ocean Geology and Biostratigraphy, AGU, Washington, this volume, 1977.

Boltovskoy, E., Neogene deep water benthonic Foraminifera of the Indian Ocean, in Indian Ocean Geology and Biostratigraphy, AGU Washington, this volume, 1977.

Fleisher, R. L., Oligocene planktonic foraminiferal assemblages from Deep Sea Drilling Project sites in Indian Ocean Geology and Biostratigraphy, AGU, Washington, this volume, 1977.

Herb, R., and V. Scheibnerova, Synopsis of Cretaceous planktonic

Foraminifera from the Indian Ocean in Indian Ocean Geology and Biostratigraphy, AGU, Washington, this volume, 1977.

Krasheninnikov, V., Cretaceous benthonic Foraminifera, Leg 20, Deep Sea Drilling Project, in Heezen, B. C., MacGregor, I. D., et al., Initial Reports of the Deep Sea Drilling Project, Washington (U. S. Government Printing Office), 20, 205-220, 1973.

Krasheninnikov, V. A., Cretaceous and Paleogene planktonic Foraminifera, Leg 27 of the Deep Sea Drilling Project in Veevers, J. J., Heirtzler, J. R. et al., Initial Reports of the Deep Sea Drilling Project, Washington (U. S. Government Printing Office), 27, 663-672, 1974.

Kuznetsova, K. I., Distribution of benthonic Foraminifera in Upper Jurassic and Lower Cretaceous deposits at Site 261, DSDP Leg 27, in the Eastern Indian Ocean in Veevers, J. J., Heirtzler, J. R., et al., Initial Reports of the Deep Sea Drilling Project, Washington (U. S. Government Printing Office), 27, 673-682, 1974.

McGowran, B., Foraminifera in von der Borch, C. C., Sclater, J. G., et al., Initial Reports of the Deep Sea Drilling Project, Washington (U. S. Government Printing Office), 22, 609-627, 1974.

McGowran, B., Maastrichtian to Eocene foraminiferal assemblages in the northern and eastern Indian Ocean region: Correlations and historical patterns in Indian Ocean Geology and Biostratigraphy, AGU, Washington, this volume, 1977.

Müller, C., Distribution of calcareous nannoplankton in Oligocene to Holocene sediments of the Red Sea and the Indian Ocean reflecting paleoenvironment, in Indian Ocean Geology and Biostratigraphy, AGU, Washington, this volume, 1977.

Proto Decima, F., Paleocene to Eocene calcareous nannoplankton of the Indian Ocean in Indian Ocean Geology and Biostratigraphy, AGU, Washington, this volume, 1977.

Scheibnerova, V., Synthesis of the Cretaceous benthonic Foraminifera recovered by the Deep Sea Drilling Project in the Indian Ocean, in Indian Ocean Geology and Biostratigraphy, AGU, Washington, this volume, 1977.

Thierstein, H. R., Mesozoic calcareous nannofossils from the Indian Ocean, DSDP Legs 22 to 27, in Indian Ocean Geology and Biostratigraphy, AGU, Washington, this volume, 1977.

Vincent, E., J. M. Gibson, and L. Brun, Paleocene and Early Eocene microfacies, benthonic Foraminifera, and paleobathymetry of the Deep Sea Drilling Project sites 236 and 237 western Indian Ocean, in Fisher, R. L., Bunce, E. T., et al., Initial Reports of the Deep Sea Drilling Project, Washington (U. S. Government Printing Office), 24, 859-885, 1974.

Vincent, E., Indian Ocean Neogene planktonic foraminiferal biostratigraphy of the Indian Ocean and paleooceanographic implications in Indian Ocean Geology and Biostratigraphy, AGU, Washington, this volume, 1977.

CHAPTER 13. PALEONTOLOGICAL-BIOSTRATIGRAPHICAL INVESTIGATIONS, INDIAN
OCEAN SITES 211-269 AND 280-282, DSDP LEGS 22-29

Hans M. Bolli

Geology Department, Swiss Federal Institute of Technology
Zurich, Switzerland

Introduction

Seventy-three papers published in the Initail Reports series of the Deep Sea Drilling Project deal with the paleontology and biostratigraphy of the 62 sites drilled during the six Indian Ocean Legs 22-27 and Legs 28 and 29 which drilled in the Southeastern Indian Ocean (Sites 264-269 and 280-282). Forty-seven of the contributions deal with the three major groups foraminifera (25), calcareous nannoplankton (14) and radiolaria (8). The remaining 26 papers describe or contain information on:

>Diatoms
>Silicoflagellates
>Archaeomonads
>Ebridians
>Calcisphaerulids
>Pollen, Spores
>Dinoflagellates
>Acritarchs
>Bryozoans
>Sponge spicules
>Ostracods
>Bivalves
>Cephalopods
>Fish denticles
>Otoliths
>Fecal pellets
>Organisms incertae sedis

In addition to the 73 contributions, paleontological/biostratigraphic information on each Site is summarized in the individual Site chapters. It is restricted there largely to foraminifera, calcareous nannoplankton and radiolaria. Furthermore, biostratigraphic and paleontologic results are synthesized in separate chapters for Legs 22, 23 and 27.

Indian Ocean foraminifera, in particular planktonics, and calcareous nannoplankton are dealt with in other contributions of this volume. This paper, therefore, only covers the above listed minor fossil groups. Because the third major group of fossils, the radiolaria, are not treated separately in this volume, a short resume on their Site occurrence and age is also included (Figure 2).

This contribution is set up to be a guide to the individual fossil groups other than foraminifera and calcareous nannoplankton described

FOSSIL GROUP	LEG 22	LEG 23	LEG 24	LEG 25	LEG 26	LEG 27	LEG 28	LEG 29
Diatoms	211, 213, 215, 216 Quat.-Mioc., Maastr. Schrader, Bukry		236, 238 Quat. - Miocene Schrader	240, 241 Quaternary Bukry		262 Quaternary Jousé & Kazarina	264 - 269 Quat.-Mioc., U. Eoc. Mc. Collum	280 - 282 Quat.- Oligocene Site chapters
Silicoflagellates (S) Archaeomonads (A) Ebridians (E)	211, 213 - 217 (S) Quat. - Cretaceous Bukry			240, 241 (S) Quaternary Bukry			264, 266, 267, 269(S) Quat.- U. Eocene Ciesielski, Bukry	280, 281 (A, S, E) Plioc.- U. Eocene Perch-Nielsen, Bukry
Calcisphaerulids					257 Lower Cretaceous Herb (p. 747)	259 - 261, 263 U. Cret.-U. Jurassic Bolli		
Pollen, Spores	214 Paleocene Harris				254 L. Miocene or older Kemp	259, 261, 263 Lower Cretaceous Wiseman & Williams	268, 269 Tertiary, rew. Perm. Kemp	280 - 282 Miocene - Eocene Haskell & Wilson
Dinoflagellates Acritarchs						259, 261, 263 Lower Cretaceous Wiseman & Williams	264, 266, 268, 269 Plioc., Mioc., Eoc. Kemp	280 - 282 Miocene - Eocene Haskell & Wilson
Bryozoa								282 Quat., Miocene Wass & Yoo
Sponge spicules								280 - 282 Quaternary - Eocene Site chapters
Ostracods			237, 238 Quat.-Lower Eocene Benson			259 - 251, 263 L. Cret.- U. Jurassic Oertli		
Bivalves						259, 260, 263 Lower Cretaceous Speden		
Cephalopods						263 Lower Cretaceous Stevens		
Fish denticles					250, 251, 256 Neogene-Paleogene Doyle et al	259 - 261, 263 Quat. - L. Cretaceous Bolli (Leg Synthesis)		
Otoliths (O) Fecal pellets (F)				242, 246 (O) Quaternary Martini			266 (F) Pliocene Kaneps	
Organisms incertae sedis						260, 261, 263 Lower Cretaceous Bolli		

Fig. 1. Fossil groups (other than Foraminifera, calcareous Nannoplankton, Radiolaria) described or listed in the Indian Ocean DSDP Initial Reports, Volumes 22 - 29. Shown are Sites, Ages and Authors.

and listed in the Indian Ocean DSDP Initial Reports, rather than a synthesis of them. The information given is restricted therefore mainly to occurrence, age, author and short notes on each group.

Compared with the foraminifera, calcareous nannoplankton and radiolaria which were reported on in the Initial Reports rather extensively from all Indian Ocean Sites, the other faunas and floras were treated in a rather inconsistent way (see Figure 1). Of these, the diatoms had the best coverage in that descriptions or mention of them exist from six out of eight legs. In particular silicoflagellates, palynomorphs and ostracods which, like foraminifera, calcareous nannoplankton and radiolaria, are present in most Sites, should still be studied more completely for each Leg. It is in particular within these groups also that many as yet unknown taxa were discovered in the few studied Sites. From this it is apparent that at least certain provinces of the Indian Ocean Cretaceous and Tertiary contain faunas and floras not known or nonexistent in age equivalent beds of surrounding land masses.

Diatoms

Diatoms have been reported from Leg 22 (Schrader, 1974, Bukry 1974; both in Initial Report v. 24), Leg 24 (Schrader, 1974), Leg 25 (Bukry, 1974a), Leg 27 (Jouse & Kazarina, 1974), Leg 28 (McCollum, 1975) and Leg 29 (in Site chapters).

A Tropical Indian Ocean Diatom zonal system (TID), subdividing the Upper Miocene to Quaternary into 21 zones is proposed by Schrader (1974, Figure 4). The Leg 22 and 24 Sites are zoned and correlated based on them (1974, Figure 12). Leg 22 Site 213, E of the Ninetyeast Ridge at 10°S, and Site 215, W of the Ninetyeast Ridge at 8°S contain excellently preserved diatom assemblages. The Holocene to Upper Miocene TID zones 1-14 were recognized in Cores 1-8 of Site 213, and in Cores 1-6 of Site 215. In addition to their more comprehensive treatment in Sites 213 and 215, diatom species are also mentioned in Bukry's report from these sites, and further from Site 211 (Pliocene-Quaternary) and 216 (Upper Maastrichtian).

Leg 24 Cenozoic diatoms were studied in Site 236, N of the Seychelles at 1°S, where they are rare to absent in most samples, and in Site 238 at 11° at the NE end of the Argo Fracture Zone, where they are common to abundant in the upper 215 meters (Quaternary to Upper Miocene), to decrease in number and preservation from 215 - 293m (Upper-Middle Miocene), and to be absent from 293-568.5 m (Middle Miocene-Oligocene). The Holocene to Upper Miocene zones 1-21 are distinguished in Site 238 between top and 272m (Cores 1-29).

Reporting on diatoms from Leg 25 is restricted to the listing of an abundant population in the Upper Quaternary Core 1 of Site 240, drilled at 3°S in the Abyssal Plain of the Somali Basin, 800 km E of Africa. An Upper Quaternary diatom assemblage, similar to that of Site 240 also occurs in Cores 1-4 (0-64m) of Site 241, drilled at 2°S on the Continental Rise of East Africa, at the shoreward margin of the Somali Basin, about 270 km from the coast.

Site 262 of Leg 27, situated in the Timor Trough between Australia and the island of Timor provides an extended, continuously cored Pleistocene section of some 347m thickness, with abundant siliceous and calcareous planktonic microfossils. The siliceous forms, radiolaria and diatoms are restricted to the upper approximately 260m, i.e. the Upper and Middle Pleistocene. Five diatom zones are distinguished in this interval of which the most extended are the Upper Pleistocene Zone I with a diatom association comparable to the recent Indian Ocean warm water flora, and a Middle Pleistocene Zone III with a flora indicating somewhat colder water conditions.

Leg 22	Leg 23	Leg 24	Leg 25	Leg 26	Leg 27	Leg 28	Leg 29
211 Q P	219 Q P M O E	231 Q P	239 M	250 Q P M C	259 E Pa C	264 E	280 P M O E
212 P M	220 O E	232 Q P	240 Q	251 Q P M	260 Q P O C	265 Q P M	281 P M E
213 Q P M	221 E	233 Q P	241 P M	252 P M	261 Q C	266 Q P M	282 O E
214 Q P M O E	222 Q P	234 P	242	253 Q E	262 Q	267 Q P M O	
215 Q P M	223 Q P M E	235 Q	243/244	254 Q	263 Q C	268 Q P M	
216 Q P M O E	224 Q M	236 Q P M O E Pa	245	255 C		269 Q P M	
217 Q P M O E	225–230 Red Sea	237 Q P M E	246/247	256 Q C			
218 Q P M	225, 227 P	238 Q P M	248 E	257 O E C			
	226, 228–230 no Radiolaria		249 C	258 Q P M C			

Fig. 2. Age of Indian Ocean Radiolaria described or listed in DSDP Initial Reports, Volumes 22–29, Sites 211–269, and 280–282. Q = Quaternary, P = Pliocene, M = Miocene, O = Oligocene, E = Eocene, Pa = Paleocene, C = Cretaceous.

A number of planktonic foraminifera (Globorotalia cf tosaensis, G. tosaensis tosaensis, G. tosaensis tenuitheca, G. crassaformis hessi) disappear within the planktonic foraminiferal Globorotalia crassaformis hessi Zone, or at about the time when siliceous microfossils begin to appear. These changes at about the same level within the Site 262 Pleistocene apparently signal a fairly radical ecological change, probably from more open marine circulation to the rather restricted conditions that exist in the area today.

In the Indian Ocean part of Leg 28, diatoms occur in Sites 265-269.

Except for an extremely poor and badly preserved Upper Eocene flora in Core 2cc, all investigated core catcher samples in Site 264 proved to be barren of diatoms. Pleistocene to Middle Miocene diatoms were determined in Cores 1-7 of Site 265, and Pleistocene to Lower Miocene forms in Cores 1-24 of Site 266.

Diatoms in Site 267 are generally poor, they occur in Cores 1 and 2 and in Core 1 of 267A where they range from Miocene to Pliocene. Site 268 diatoms are rare, poorly preserved and restricted to the Pleistocene to Miocene Cores 1-12. Abundance and presence of diatoms in the Pleistocene to Miocene Cores 1-9 in Site 269 is variable.

A scheme of 15 diatom range zones, defined as local range zones, are proposed by McCollum (1975).

The following diatom information is taken from the site chapters of the Leg 29 Sites 280-282: Site 280: Upper Miocene to lower Pliocene, and a Pleistocene assemblage are present in Core 1. In 280A, Cores 1 to 7 contain abundant and well preserved diatoms and Core 8 includes a poorer assemblage, all of Upper Eocene to Lower Oligocene age.

Site 281: Core 1 contains poorly preserved diatoms of Upper Pleistocene age, Core 3 a Middle Pliocene assemblage and Core 6 a late Middle Miocene assemblage. Rich Upper Eocene diatom assemblages occur in Cores 14 to 16. In Site 282 fragments only of diatoms were found in Cores 5, 6, 7 and 15.

Silicoflagellates, Archaeomonads, Ebridians

Indian Ocean fossil silicoflagellates are reported in separate papers within the Initial Reports series only from Leg 28 (Ciesielsky, 1975; Bukry, 1975) and Leg 29 (Perch-Nielsen, 1975; Bukry, 1975). However, their occurrence in several Sites of Leg 22 (Bukry, 1974) and Leg 25 (Bukry, 1974b) are briefly dealt with together with calcareous nannoplankton and diatoms. Archaeomonads and Ebridians are reported from the Leg 29 Sites 280A and 281 (Perch-Nielsen, 1975a).

Silicoflagellates

In the tropical Eastern Indian Ocean Leg 22, silicoflagellate species are listed from the following Sites:

Latitude	Site	Cores	Age
10°S	211	1-6	Quaternary-Pliocene
10°S	213	1-7	Quaternary-Upper Miocene
11°S	214	38	Lower Paleocene
8°S	215	1-7	Quaternary-Upper-Middle, Upper Miocene
1°N	216	30-32	Upper Maastrichtian (exceptionally abundant, shown with table)
9°N	217	4	Miocene

A remarkably similar association of silicoflagellate species to that present in Site 216 (1°N) also occurs at the Campbell Plateau Site 275 of Leg 29, drilled at 50°S.

Silicoflagellates were recorded from the Quaternary of Site 240, Core 1, and 241, Cores 1-4 of Leg 25, drilled at 3°S in the Somali Basin, and at 2°S at the base of the Continental Rise of East Africa, at the shoreward margin of the Somali Basin.

Silicoflagellates were described from the Indian Ocean Sites 264, 266, 267 and 269 of Leg 28.

While the Naturaliste Plateau Site 264 yielded silicoflagellates only in the Upper Eocene Core 2 (core catcher), they were found to be abundant in the Pliocene-Pleistocene of Sites 266, 267 and 269, all located in the highly productive waters considerably further South than Site 264. Upper to Middle Miocene sections of these sites are generally barren of silicoflagellates or contain only low diversity, coldwater assemblages. Lower Middle to Lower Miocene silicoflagellates are again more diverse, including frequent warm water cosmopolitan species. The silicoflagellate association in Core 5 of Site 367 indicates Upper Oligocene.

Very close coring and good recovery in Site 266 provided a nearly complete stratigraphic sequence of Pleistocene to Lower Miocene silicoflagellate assemblages which were subdivided into 13 zones. At this site many of the Middle and Lower Miocene silicoflagellate zones could be related to previously described calcareous nannoplankton zones. This section thus serves as reference for a correlation of these two zonal schemes.

In Leg 29 Upper Eocene and Oligocene silicoflagellates were found in Cores 1 to 7 of Site 280A. In 281, Core 1 contains Pleistocene, Cores 6-8(?) Upper Miocene to Lower Pliocene and Cores 14-16 Upper Eocene silicoflagellate assemblages. No silicoflagellates were reported from Site 282.

Archaeomonads, Ebridians

Site 280A furnished rare endoskeletal dinoflagellates, rare to few ebridians and rare to few archaeomonads of Oligocene (Cores 1 to 6) and Upper Eocene (Core 7) age. At Site 281, the above mentioned groups were found in the Upper Eocene (Cores 14-16). Miocene ebridians and endoskeletal dinoflagellates also occur at this site, the latter were found also in the Pleistocene.

Calcisphaerulids

Calcisphaerulids are small, calcareous, single chambered fossils placed in the incertae sedis family Calcisphaerulidae Bonet 1956. Their tests are spherical to ovoid to distinctly elongate with rounded ends. Their average diameters is 40-70 micron and their maximum length up to 120 micron. The walls are formed by calcite crystals of varying size, shape and arrangement. They may consist of one, two and occasionally also three layers, whose thickness may vary from 1-12 micron. A more or less circular aperture of variable size is usually present, but speciments may also be without. It is thought that calcisphaerulids may be a cyst form of a yet unknown organism, possibly a calcareous alga. Their known stratigraphic range is Jurassic, Lower and Upper Cretaceous. More recently, calcisphaerulids have also been reported from the Paleocene collected on Leg 35 (Antarctic) and Leg 40 (South Atlantic).

Calcisphaerulids have long been known from Jurassic/Cretaceous limestones where they may contribute to a considerable part of their volume.

So far they have been described and figured almost exclusively from thin sections. The occurrence of such forms in the soft Upper Jurassic and Cretaceous sediments of Leg 27 for the first time made it possible to isolate them in large numbers and good preservation, and to study them under the SEM. This resulted in the description of 12 new species and 4 forms in open nomenclature, all assigned to the genus Pithonella.

The stratigraphic distribution of individual species is restricted in the Leg 27 sections in that Upper Jurassic/Lowermost Cretaceous, Aptian/Albian, and Upper Cretaceous taxa can be clearly distinguished. Calcisphaerulids were found present in all Leg 27 Cretaceous sections drilled in the basins W and NW of the Australian continental margin. In Site 259 in the Albian, in 260 in the Albian and Upper Cretaceous, and in 261 restricted to the Upper Jurassic-Lower Cretaceous. In Site 263 they were noted in only three samples, probably of Albian age.

Numerous calcisphaerulid species were described and figures previously under several generic names from thin sections of Jurassic-Cretaceous outcrops on the East Indian Archipelago. They were published by a number of authors, in particular by Wanner (1940) from several of the islands, including Timor, Ceram, Misol, and by Vogler (1941) from Misol. It was of particular interest therefore to find calcisphaerulids in three dimensional form in deep sea sites close to those described in thin sections from the East Indian Archipelago. A direct comparison of the isolated Leg 27 specimens investigated by SEM, with the thin sectioned forms from the East Indian Archipelago land sections has not yet been attempted, but it is expected that the age equivalent forms will be closely comparable and thus contribute to stratigraphic correlations within the general area.

Following the investigations of Leg 27, Albian samples from Leg 26, Site 257, were also checked for calcisphaerulids. Some species were found present there that are also recorded from the Leg 27, Site 259 (I.R. Leg 26, p. 747, 756-757, and I.R. Leg 27, p. 852.

Pollen and Spores

Indian Ocean pollen and spores were reported from Leg 22 (Harris, 1974), Leg 26 (Kemp, 1974), Leg 27 (Wiseman and Williams, 1974), Leg 28 (Kemp, 1975) and Leg 29 (Haskell and Wilson, 1975).

Palynological data are recorded from the Paleocene Cores 36-53 (337-448m subbottom) of Leg 22 Site 214, situated on the Ninetyeast Ridge at about 11°S. The upper part of the sequence, to about 390m, is of predominantly marine aspect. Its spores, pollen and dinoflagellate cysts correlate with the Australasian region rather than tropical Africa, South America, India or Southeast Asia. Below 390m, from Core 42, the sequence is entirely non-marine, paludal to lacustrine. The floral association suggests that at Paleocene time the northern part of the Ninetyeast Ridge was in a temperate region of high rainfall, separated from the Australian and Antarctic plates (many characteristic genera and species of the Australasian early Tertiary are absent), and also not connected with either India or Southeast Asia.

Leg 26 Site 254 was drilled on the South end of the Ninetyeast Ridge, at 31°S. Spores and pollen were investigated in five samples of Cores 28-33 and given a minimum tentative age of Lower Miocene. A more detailed analysis was carried out on Core 29 where the floral preservation was found to be outstanding. Few of the 25 angiosperm species could be identified and tied in with described Tertiary species. Of paleogeographic interest is a group of pollen species that is common in the early Tertiary of Australasia, some species of which are thought to have been derived from a distinctive southern flora which in the early

Tertiary was distributed through Australia, Antarctica, New Zealand and southern South America. The well mixed pollen assemblage suggests deposition at some distance from the plant communities, which are regarded as possibly having been lagoonal. The high diversity of angiosperms, pteridophyte spores and epiphytic fungal remains suggests a warm and extremely humid climate.

The Site 254 flora was compared with the modern one of the Indian Ocean islands Amsterdam and St. Paul at 37°S, and found to be much richer, in that the at least 25 flowering plants of 254 are matched by only 17 on these islands. This reduction was probably caused by the cooling of the southern Indian Ocean during Neogene time. Almost no species are common to the Site 254 microflora, and those known from the Tertiary of southern India, Assam and the Andaman islands.

Lower Cretaceous pollen, spores and dinoflagellates of the Leg 27 Sites 259, 261 and 263, drilled in the Perth, Cuvier and Gascoyne Abyssal Plains resp., situated W and NW of the western Australian continental shelf, were investigated and their distribution given on range charts. Intervals and ages of the studied Sites are:

Site	Core	Age
259	Core 18-33	Aptian, probably Lower to Upper
261	16, 24-26	Lower Aptian to Upper Aptian or Lower Albian
263	5-29	Barremian to Upper Aptian or Lower Albian

Spores and pollen comprise only a minor part compared with the marine palynomorphs. The assemblages contain mainly pteridophytic, bryophytic and gymnospermous elements previously reported especially from Australia, India and South America. A partially successful attempt was made to tie the floral distribution into the zonal scheme proposed by Dettmann and Playford. The ages indicated by the pollen and spores compare with those obtained from dinoflagellates present in the same samples.

In Leg 28 Sites 268 and 269 Tertiary pollen are rare to very rare, but reworked Permian spores such as Verrucosisporites pseudoreticulatus occur in unusual abundance.

In Leg 29 pollen and spores were recorded in the Indian Ocean Sites as follows: Cores 10 to 22 of Site 280A contain generally rich Eocene floras with fairly low diversity. Cores 14 to 17 were found to contain Upper Eocene floras. In the catcher samples of Site 282, Cores 4 to 17 are present fairly poor floras of Lower Miocene, Oligocene and Upper Eocene age.

Dinoflagellates and Acritarchs

Sinoflagellates and acritarchs were reported from Leg 27 (Wiseman and Williams, 1974), Leg 28 (Kemp, 1975) and Leg 29 (Haskell and Wilson, 1975).

The Leg 27 dinoflagellates were studied from the same sites and the same samples as the spores and the pollen. The ranges and ages of 86 known species encountered in Sites 259, 261 and 263 are plotted on a chart which provides valuable information for the distribution of Upper Jurassic to Upper Cretaceous dinoflagellates. A number of more recently published papers served as the main references for this chart. The ranges of some species however, known to be rather more restricted in western Australia than shown on world wide charts, were treated accordingly. As for the pollen and spores, the dinoflagellate distribution in each site is shown on individual charts.

Bryozoans

Of all Indian Ocean sites, bryozoans have only been reported from the Leg 29 Site 282 W of Tasmania (Wass and Yoo, 1975).

More than 15,000 bryozoan fragments have been identified from 17 samples of Upper Miocene and Upper Pleistocene age from Core 1, Sections 1 to 6 of this site. The assemblages are closely related to those found in the Tertiary of southern Australia and the Recent of the southern Australian continental shelf.

Sponge Spicules

The only siliceous sponge remains reported from the Indian Ocean sites are from Leg 29, Site 282 W of Tasmania, where they are common in the Oligocene to Upper Eocene Cores 5-17. They are mentioned in the site report (I.R. v. 29, p. 327) and figures on plates 1-3, p. 323-327. In general, sponge spicules were found to be abundant and large in Oligocene and Eocene sediments also in other Leg 29 sites, while in the Pleistocene, Pliocene and Miocene they are poor both in diversity and abundance.

Ostracods

Ostracods were described and figured from the Indian Ocean Legs 24 (Benson, 1974) and 27 (Oertli, 1974). Except for one Site in Leg 27 where the Upper Jurassic ostracods may be of shallow water nature, both reports deal with deep sea forms which are still little known. Most of the taxa, therefore, were identified on a generic level only. The ostracods of Leg 24 are from Site 237 where they range in age from the middle Lower Eocene to Quaternary, and from Site 238 where they are reported from Middle Oligocene to Upper Pliocene. Site 237 lies on the Mascarena Plateau SE of the Seychelles, Site 238 is situated near the northeastern end of the Argo Fracture Zone on the Central Indian Ridge.

The reported Leg 27 ostracods are of Upper Oxfordian to Albian age and come from Sites 259-261 and 263 drilled in the basins W and NW of the Australian continental margin. The Cretaceous associations are dominated by the genus Arculicythere. Forms assigned to the Bairdiids are also frequent. Like in the deep sea Cenozoic forms of Leg 24, almost no specific determinations were possible in these poorly known Mesozoic deep water faunas. Therefore, no comparisons with published shallow water faunas from Australia, India and South Africa were possible.

Bivalves

Bivalves, mainly in fragmentary form, were reported from the Lower Cretaceous of the Leg 27 Sites 259, 260 and 263, drilled in the Perth-, Cuvier, and Gascoyne Abyssal Plains, all off the western Australian continental shelf (Speden, 1974).

They occur as large fragments in Site 263, Core 17cc, as small ones (recovered from foraminiferal washings) in Site 259, Core 14cc, and Site 260, Core 9cc and 10/2/34-36 cm.

The large specimens were determined as ?Aucellina sp. indet.. If this generic placement is correct the age would be Neocomian to Turonian. The small fragments Aucellina sp. A are tentatively placed in the Aptian-Albian. Its specific affinity to A. cf. gryphaeoides would suggest an Albian age.

A species identity of Aucellina sp. A with the Australian A. cf. gryphaeoides is considered possible because of the close geographic relationship of Sites 259 and 260 with the Great Artesian Basin, where A. cf. gryphaeoides is common. Paleographic considerations suggest that the cool to cold Austral bioprovince or a transitional, more temperate zone extended to include Sites 259 and 260, or that these sites were under the influence of a northward extending colder water mass.

Cephalopods

A guard of a belemnite determined as of Parahibolites sp. indet. was recovered from Leg 27, Site 263, Core 26/2/108-110 cm (Stevens, 1974). It is dated as probable Upper Albian, an age in agreement with the calcareous nannoplankton. The presence of this genus in the Cuvier Abyssal Plain is interpreted as possibly related to an appearance of Tethyan migrants which in Albian time entered the northwestern and western regions of Australia. On the other hand the occurrence of Austral elements in nearby areas indicates that Parahibolites was living in a warm-temperate rather than in a tropical habitat.

A test fragment and internal mold of an ammonite determined as ?cf. Prohysteroceras (Goodhallites) richardsi Whitehouse occurred in Site 263, Core 18/5/105 cm (Stevens, 1974). The genus is known from such widespread places as Queensland, southern India, southern Africa, Madagascar, England and Texas. From this it may be assumed that the form arrived, like Parahibolites, in Upper Albian time as a Tethyan element, migrating along the Tethyan seaway into the Indo-Pacific region.

Fish Denticles

A fundamental paper entitled "Stratignathy" (= study of the time relationships of ichthyoliths sometimes referred to as fish skeletal debris) by Doyle, Kennedy and Riedel (1974) is published in the Leg 26 Initial Report. It deals with microscopic fish denticles present in pelagic sediments and their use for biostratigraphic correlation. This paper is a follow-up of a first study of this kind by Helms and Riedel (1971) in Leg 7 Initial Report. Both investigations found their impetus because, due to solution effects, deep sea samples often contain no microfossils other than microscopic skeletal debris and teeth of fishes. In such sediments they offer therefore the only possible basis for biostratigraphic interpretation.

The paper contains chapters on "Taxonomic approach", a clearly illustrated "System description", "Samples investigated and techniques employed", "Fish skeletal debris at some Leg 26 Sites", "Stratigraphic results", and a "Descriptive section". It is illustrated by 26 plates showing Eocene to Quaternary fish denticles.

The Linnean taxonomic system could not be applied because of the impossibility of determining at present the nature of the fishes. Instead a scheme was devised for describing and naming the two-dimensional shapes of the various kinds of skeletal debris as they appear in transmitted light. These names and descriptions do not imply biological relationships or functions. The scheme was designed to be computer-compatible.

Samples investigated are from the tropical Atlantic, Pacific and Indian Ocean, ranging in age from Lower Eocene to Quaternary. The Indian Ocean samples come from a series of DODO stations between 6°N and 23°S and from Leg 26 Sites 250, 251 and 256. The distribution of the ichthyoliths is given on a series of tables where they are generally correlated with the stratigraphic framework based on calcareous nannoplankton (Bukry reports). Figure 2, p. 841 in I.R. v. 26 shows the distribution of stratigraphically restricted forms based on which 10

subdivisions are distinguished from the Lower Eocene to the Quaternary.
The microscopic fish remains, in particular teeth, offer a welcome way of dating such sediments, because they remain preserved as the only fossils in sediments where calcareous and siliceous microfaunas and floras became dissolved.

To serve as a guide for future stratignathic studies on Leg 27 sites, occurrence and frequency of fish debris in each of the predominantly Cretaceous cores of Sites 259-261 and 263 are listed on Figures 1-3 and 5 of the "Synthesis of the Leg 27 Biostratigraphy and Paleontology" (Bolli, 1974a).

Fecal Pellets

Fecal pellets occur in the Pliocene Cores 5-8 of the Leg 28 Site 266 (Kaneps, 1975). They are uniform sized, mostly bilobate pellets of .26 to .27 mm length, consisting of two appressed spheres. They are massive and made up of diatom fragments and calcite crystals.

Organisms Incertae Sedis

Small, well preserved spheres measuring between 50 and 80 micron, with comparatively large, regularly distributed circular pores occur in the Eastern Indian Ocean Sites 260, 261 and 263 of Leg 27 (Bolli, 1974). They are restricted there to the Lower Cretaceous, Albian/Aptian. The thin walls of the scarce specimens consist of calcite. Strangely, they were recorded only from non-calcareous sediments, together with radiolaria, arenaceous foraminifera, dinoflagellates and fish debris. Despite the fact that they strongly resemble spumelline radiolaria in shape, they could not be placed there nor into any other known fossil group.

References

Diatoms

Bukry, D., Coccolith and Silicoflagellate stratigraphy, Eastern Indian Ocean, Deep Sea Drilling Project, Leg 22, Initial Reports of the Deep Sea Drilling Project, Washington, D. C. (U. S. Government Printing Office) 22, 601-607, 1974.

Bukry, D., Phytoplankton stratigraphy, offshore East Africa Deep Sea Drilling Project, Leg 25, Initial Reports of the Deep Sea Drilling Project, Washington, D. C. (U. S. Government Printing Office) 25, 635-646, 1974a.

Jousé, A. P., and G. H. Kazarina, Pleistocene Diatoms from Site 262, Leg 27, Deep Sea Drilling Project, Initial Reports of the Deep Sea Drilling Project, Washington, D. C. (U. S. Government Printing Office) 27, 925-945, 1974.

McCollum, D. W., Diatom stratigraphy of the Southern Ocean, Leg 28, Initial Reports of the Deep Sea Drilling Project, Washington, D. C. (U. S. Government Printing Office) 28, 515-571, 1975.

Schrader, H. J., Cenozoic marine planktonic diatom stratigraphy of the tropical Indian Ocean, Initial Reports of the Deep Sea Drilling Project, Washington, D. C. (U. S. Government Printing Office) 24, 887-967, 1974.

Silicoflagellates, Archaeomonads, Ebridians

Bukry, D., Coccolith and Silicoflagellate stratigraphy, Eastern Indian Ocean, Deep Sea Drilling Project, Leg 22, Initial Reports of the Deep

Sea Drilling Project, Washington, D. C. (U. S. Government Printing Office) 22, 601-607, 1974.

Bukry, D., Phytoplankton stratigraphy, offshore East Africa, Deep Sea Drilling Project, Leg 25, Initial Reports of the Deep Sea Drilling Project, Washington, D. C. (U. S. Government Printing Office) 25, 635-646, 1974b.

Bukry, D., Coccolith and Silicoflagellate stratigraphy near Antarctica, Deep Sea Drilling Project, Leg 28, Initial Reports of the Deep Sea Drilling Project, Washington, D. C. (U. S. Government Printing Office) 28, 709-723, 1975.

Bukry, D., Silicoflagellate and Coccolith stratigraphy, Deep Sea Drilling Project, Leg 29, Initial Reports of the Deep Sea Drilling Project, Washington, D. C. (U. S. Government Printing Office) 28, 845-872, 1975.

Ciesielski, P. F., Biostratigraphy and paleoecology of Neogene and Oligocene Silicoflagellates from cores recovered during Antarctic Leg 28, Deep Sea Drilling Project, Initial Reports of the Deep Sea Drilling Project, Washington, D. C. (U. S. Government Printing Office 28, 625-691, 1975.

Perch-Nielsen, K., Late Cretaceous to Pleistocene Silicoflagellates from the southern Southwest Pacific, Deep Sea Drilling Project Leg 29, Initial Reports of the Deep Sea Drilling Project, Washington, D. C. (U. S. Government Printing Office) 29, 677-721, 1975.

Perch-Nielsen, K., Late Cretaceous Archaeomonadaceae, Diatomaceae, and Silicoflagellata from the South Pacific, Deep Sea Drilling Project, Leg 29, Initial Reports of the Deep Sea Drilling Project, Washington D. C. (U. S. Government Printing Office) 28, 873-907, 1975a.

Calcisphaerulids

Bolli, H. M., Jurassic and Cretaceous Calcisphaerulidae from Deep Sea Drilling Project, Leg 27, Eastern Indian Ocean, Initial Reports of the Deep Sea Drilling Project, Washington, D. C. (U. S. Government Printing Office) 27, 843-907, 1974.

Herb, R., Cretaceous planktonic foraminifera from the Eastern Indian Ocean, Initial Reports of the Deep Sea Drilling Project, Washington, D. C. (U. S. Government Printing Office) 27, 745-769, 1974.

Vogler, J., Ober-Jura und Kreide von Misol (Niederländisch-Ostinden), Palaeontographica, 4, 245-293, 1941.

Wanner, J., Gesteinsbildende Foraminifera aus Malm und Unterkreide des östlichen Ostindischen Archipels, Pal. Z., 22, 75-99, 1940.

Pollen, Spores, Dinoflagellates, Acritarchs

Harris, W. K., Palynology of Paleocene sediments at Site 214, Ninetyeast Ridge, Initial Reports of the Deep Sea Drilling Project, Washington, D. C. (U. S. Government Printing Office) 22, 503-508, 1974.

Kemp, Elizabeth, M., Preliminary palynology of samples from Site 254, Ninetyeast Ridge, Initial Reports of the Deep Sea Drilling Project, Washington, D. C. (U. S. Government Printing Office) 26, 815-823, 1974.

Wiseman, Julie F., and A. J. Williams, Palynological investigation of samples from Sites 259, 261, and 263, Leg 27, Initial Reports of the Deep Sea Drilling Project, Washington, D. C. (U. S. Government Printing Office) 27, 915-924, 1974.

Kemp, Elizabeth M., Palynology of Leg 28 drill sites, Deep Sea Drilling Project, Initial Reports of the Deep Sea Drilling Project, Washington, D. C. (U. S. Government Printing Office) 28, 599-623, 1975.

Haskell, T. R., and G. J. Wilson, Palynology of sites 280-284, Deep Sea Drilling Project Leg 29, off southeastern Australia and western New Zealand, Initial Reports of the Deep Sea Drilling Project, Washington, D. C., (U. S. Government Printing Office) 29, 723-741, 1975.

Bryozoans

Wass, R. E., and J. J. Yoo, Bryozoa from Site 282 West of Tasmania, Initial Reports of the Deep Sea Drilling Project, Washington, D. C., (U. S. Government Printing Office) 29, 809-831, 1975.

Ostracods

Benson, R. H., Preliminary report on the Ostracods of Leg 24, Initial Reports of the Deep Sea Drilling Project, Washington, D. C. (U. S. Government Printing Office) 24, 1037-1043, 1974.

Oertli, H. J., Lower Cretaceous and Jurassic Ostracods from the Deep Sea Drilling Project, Leg 27 - A preliminary account, Initial Reports of the Deep Sea Drilling Project, Washington, D. C. (U. S. Government Office) 27, 947-965, 1974.

Bivialves, Cephalopods

Speden, I. G., Cretaceous Bivalvia from cores, Leg 27, Initial Reports of the Deep Sea Drilling Project, Washington, D. C. (U. S. Government Printing Office) 27, 977-981, 1974.

Stevens, G. R., Leg 27 Cephalopoda, Initial Reports of the Deep Sea Drilling Project, Washington, D. C. (U. S. Government Printing Office) 27, 983-989, 1974.

Fish denticles

Helms, Phyllis B., and W. R. Riedel, Skeletal debris of fishes, Initial Reports of the Deep Sea Drilling Project, Washington, D. C. (U. S. Government Printing Office) 7, 1709-1720, 1971.

Otoliths

Martini, E., Quaternary fish otoliths from Sites 242 and 246, Leg 25, Initial Reports of the Deep Sea Drilling Project, Washington, D. C. (U. S. Government Printing Office) 25, 647-655, 1974.

Fecal Pellets

Kaneps, A. G., Fecal pellets in Pliocene Antarctic Deep Sea sediments, Leg 28, Deep Sea Drilling Project, Initial Reports of the Deep Sea Drilling Project, Washington, D. C. (U. S. Government Printing Office) 28, 585-587, 1975.

Organisms Incertae Sedis

Bolli, H. M., Calcareous organisms incertae sedis from the Lower Cretaceous of Deep Sea Drilling Project Leg 27, Initial Reports of the Deep Sea Drilling Project, Washington, D. C. (U. S. Government Printing Office) 27, 909-913, 1974.

Radiolaria

Chen, Pei-Hsin, Antarctic Radiolaria, Initial Reports of the Deep Sea Drilling Project, Washington, D. C. (U. S. Government Printing Office) 28, 437-513, 1975.

Johnson, D. A., Radiolaria from the Eastern Indian Ocean, Deep Sea Drilling Project Leg 22, Initial Reports of the Deep Sea Drilling Project, Washington, D. C. (U. S. Government Printing Office) 22, 521-575, 1974.

Nigrini, Catherina, Cenozoic Radiolaria from the Arabian Sea, Deep Sea Drilling Project Leg 23, Initial Reports of the Deep Sea Drilling Project, Washington, D. C. (U. S. Government Printing Office) 26, 1051-1121, 1974.

Petrushevskaya, M. G., Cenozoic Radiolarias of the Antractic Leg 29, Deep Sea Drilling Project, Initial Reports of the Deep Sea Drilling Project, Washington, D. C. (U. S. Government Printing Office) 29, 541-675, 1975.

Renz, G. W., Radiolaria from Leg 27 of the Deep Sea Drilling Project, Initial Reports of the Deep Sea Drilling Project, Washington, D. C. (U. S. Government Printing Office) 27, 769-841, 1974.

Riedel, W. R., and Annika Sanfilippo, Radiolaria from the Southern Indian Ocean, Deep Sea Drilling Project Leg 26, Initial Reports of the Deep Sea Drilling Project, Washington, D. C. (U. S. Government Printing Office) 26, 771-813, 1974.

Sanfilippo, Annika and W. R. Riedel, Radiolaria from the West-Central Indian Ocean and Arabian Sea, Deep Sea Drilling Project Leg 24, Initial Reports of the Deep Sea Drilling Project, Washington, D. C. (U. S. Government Printing Office) 25, 657-661, 1974.

CHAPTER 14. MESOZOIC CALCAREOUS NANNOFOSSILS FROM THE INDIAN OCEAN, DSDP LEGS 22 to 27

Hans R. Thierstein

Lamont-Doherty Geological Observatory of Columbia University
Palisades, New York

Abstract. All Mesozoic (Oxfordian - latest Maeshrichtian) sediment with a calcium carbonate content exceeding a few percent have been dated by their nannofossil assemblages. Age correlation remains tentative in the early Cretaceous intervals at sites 249, 260, and 263, mainly because of dilution of the nannoplankton by detrital clay. Changes in species compositions and relative abundances reflect the paleolatitudinal positions of the DSDP sites and of some additional sample localities (piston cores and land sections on the surrounding continents). Species diversities, however, are correlated most closely with the degree of dissolution of the assemblages.

Introduction

Except for two localities, one in South Africa (Pienaar, 1968, 1969) and one in southwestern Australia (Deflandre, 1959) Mesozoic calcareous nannofossils were virtually unknown from the Indian Ocean and the remnants of ancient Gondwanaland prior to the GLOMAR CHALLENGER cruises. During Legs 22 through 27 of the Deep Sea Drilling Project, approximately 2200 m of Mesozoic sediment containing calcareous nannofossils were drilled in 16 holes from which about 630 m were recovered (Figure 1). The stratigraphic range of the encountered assemblages is Oxfordian through latest Maestrichtian. This synthesis is based on the contributions of coccolith specialists in the published Initial Reports of the Deep Sea Drilling Project, Vol. 22 through 27 and on a light-microscopic examination of 32 fossiliferous DSDP samples (Leg 22: 8 samples, Leg 25: 13 samples, Leg 27: 11 samples) not previously investigated by the author. For comparative purposes a number of additional assemblages were examined.

Review of Initial Reports Information

Leg 22: Sites 211, 212, 216, 217

A brief definition of the zonal scheme and the zonal assignment of the recovered cores was given by Gartner (1974). He included distribution charts of the more important species showing their presence or absence. Similarities in species composition between the late Maestrichtian assemblages on the Ninetyeast Ridge and those known from northern Europe and the North Atlantic were pointed out by Gartner (1974) and Bukry (1974a).

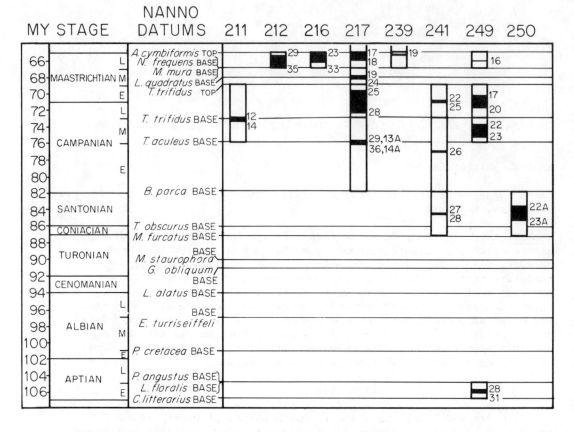

Fig. 1a,b. Mesozoic cores recovered from the DSDP sites in the Indian Ocean in relation to the calcareous nannoplankton datum sequence proposed by Thierstein (1976). The total biostratigraphic range which the cores may contain is shown, with the black intervals indicating the relative position of the estimated recovery.

Leg 25: Sites 239, 241, 249

A very short account of Mesozoic assemblages from the western Indian Ocean was given by Müller (1974). A reexamination of samples from Core 19, Site 239 confirmed a late Maestrichtian age for that interval, as suggested by Bukry (1974b).

Leg 26: Sites 250, 255, 256, 257, 258

An outline of the Mesozoic zonation, distribution charts showing relative abundances of most of the species present, as well as relative abundance and preservation of the encountered assemblages, were given by Thierstein (1974). The influence of differential solution and diagenesis on the morphology of some Cretaceous species was discussed and illustrated. Bukry (1974d) listed zonal and age assignments of the studied samples, including some species lists of selected assemblages.

Leg 27: Site 259, 260, 261, 263

Presence and absence criteria for nannofossils were given in range charts by Proto Decima (1974). The stratigraphic correlation of some

MESOZOIC CALCAREOUS NANNOFOSSILS 341

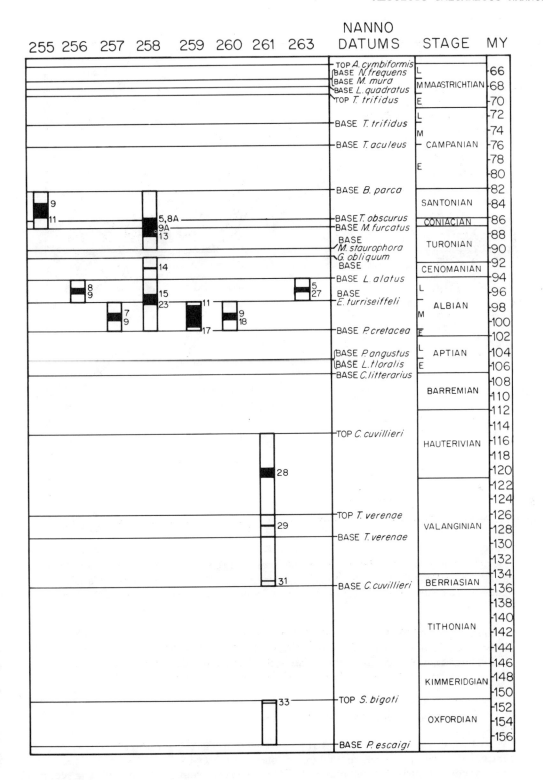

intervals with scarce atypical assemblages was discussed. Discrepancies in the age assignments of the nannofossil assemblages from sites 260 and 263 between Proto Decima (1974) and Bukry (1974c) are caused by scarce occurrence of only a few marker species, which are either con-

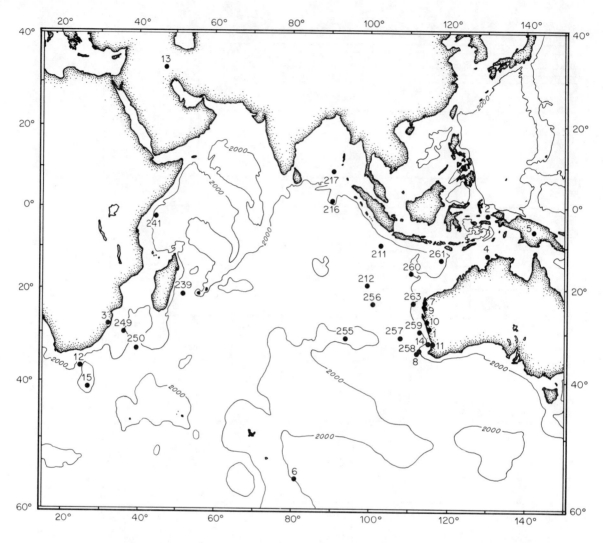

Fig. 2. Geographic location of Mesozoic calcareous nannofossil assemlages recovered from DSDP Sites 211 through 263. Additional samples studied for comparison (localities 1 through 15) are listed in Table 2.

taminated or reworked. A reexamination of a few samples from these intervals by the author only confirmed the difficulties of stratigraphic interpretation of these poor assemblages. A discussion of the problems in correlating these Mesozoic intervals may also be found in Bolli (1974).

Nannofossil-Bearing Sediment from the Indian Ocean

The biostratigraphic correlation of the nannofossiliferous DSDP cores recovered from the Indian Ocean is shown in Figure 2. The relative abundances of calcareous nannofossils, their preservation and the general lithology of the main stratigraphic intervals are given in Table 1. The sedimentation rates are calculated assuming a continuous sedimentation between the highest and the lowest cores of a determined biostratigraphic interval and are based on the absolute time scale proposed by

Thierstein (1976). All of the reasonably well preserved nannofossil assemblages from Campanian and Maestrichtian sediment recovered on GLOMAR CHALLENGER's Indian Ocean Legs 22 and 25 are similar to assemblages known elsewhere (Bukry and Bramlette, 1970; Cita and Gartner, 1971; Manivit, 1971; Roth, 1973). Cenomanian through Santonian assemblages, however, could only be correlated with those in the type sections (Manivit, 1971; Thierstein, 1976) after the effect of diagenetic alteration on the morphology of some marker fossils had been recognized and taken into account (Thierstein, 1974, 1976). The Albian nannofossil communities from sites 256, 257, 258, and 259 are very similar to those known from Great Britain (Thierstein, 1973). The scarce occurrence and unusual species composition of the nannofossil assemblages recovered at sites 260 (Cores 9-18) and 263 (Cores 5-27) together with either downhole contamination or reworking of some specimens, still leaves dating of these intervals tentative. Only careful, closely spaced resampling and examination of these cores may improve the reliability of their correlation. For this study the hypothesis of reworking has been assumed to be the more likely and the intervals in question are considered to be of Albian age. The earliest Cretaceous and late Jurassic assemblages in Cores 28 through 33 at site 261 show limited species diversities as well, but nevertheless led the various investigators (Proto Decima, 1974; Bukry, 1974c; Thierstein, 1976) to similar stratigraphic conclusions.

With a few exceptions, the calculated rates and corresponding lithologies are in agreement with those from other oceans, taking into account the possible bias of the above mentioned assumptions. Exceptionally rapid accumulation is suggested for the late Cretaceous claystones at site 241 (Cores 22-25), for the early Cretaceous clays at site 258 (Cores 15-23) and especially at site 263 (Cores 5-27). Both dating uncertainties and rapid terrigeneous influx (Veevers, Heirtzler et al., 1974, p. 293)) may be responsible for the enormous accumulation rate of almost 200 meters per million years at site 263.

Paleobiogeographic and Paleoecologic Interpretation of Mesozoic Calcareous Nannofossils from the Indian Ocean

In Mesozoic times, most of the sites were located at considerably higher latitudes than today (Sclater, von der Borch et al., 1974). The following interpretations are based on qualitative observations only. The recovered nannofossil assemblages will prove invaluable for future quantitative interpretations of paleodistribution patterns, since only very few high-latitude assemblages from the Southern Hemisphere have become available so far.

Additional samples from the Indian Ocean and the surrounding continents were investigated for comparative purposes, but only a small number of them contained calcareous nannofossils. The stratigraphic and geographic provenance, abundance and preservation of these assemblages are given in Table 2 and their geographic locations are shown in Figure 2.

Jurassic

The composition of the nannofossil assemblages recovered at site 261 was described by Proto Decima (1974). They are the first pre-Cretaceous nannofossils known from the Southern Hemisphere. Although their preservation and abundance is only moderate, they are quite similar in species composition to those from southeastern France (Castillon, Orpierre) near the Oxfordian/Kimmeridgian boundary, as well as to those from DSDP site

TABLE 1. Mesozoic lithologic and stratigraphic units recovered in the Indian Ocean during Legs 22 through 27 of the Deep Sea Drilling Project. Preservation key: P, poor; M, moderate; G, good; O, secondary overgrowth; E, etching.

Site	Core	Recovery m	Recovery %	Thickness m	General Lithology	Calcareous Nannofossils Abund	Calcareous Nannofossils Pres	Age	Sed. Rate m/m.y.
211	12	0.4	5	-	Clay	F	PE	L. Campanian -E. Maestrichtian	-
211	13-14	1.8	10	19	Clay	C	PE	M. Campanian	6
212	29-35	42.0	65	67	Chalk	A	MO	L. Maestrichtian	33
216	23-33	51.5	50	105	Chalk + volc. clay	A	MO	L. Maestrichtian	52
217	17-18	9.5	50	19	Chalk	A	NEO	L. Maestrichtian	10
217	19-24	12.6	20	57	Clay-rich chalk	A	GEO	M. Maestrichtian	29
217	25-28	25.0	65	38	Clay-rich chalk	A	MEO	L. Campanian -E. Maestrichtian	10
217	29-36	21.8	20	100	Clay micarb chalk + silica clay micrite	C	MEO	E.-M. Campanian	11
217A	13-14	0.6	5	-	Brown clay	C	PE	L. Maestrichtian	-
239	19								
241	22-25	19.0	10	218	Claystone	A	ME	L. Campanian -E. Maestrichtian	55
241	26	4.5	5	-	Claystone	C	GE	E. Campanian ?	-
241	27-28	10.3	10	95	Claystone	C	ME	Coniacian -Santonian	19
249	16	0.4	5	-	Clay-rich chalk	A	G	L.Maestrichtian	-
249	17-20	33.0	50	68	Clay-rich chalk	A	G	L. Campanian -E. Maestrichtian	17
249	21	9.5	100	10	Clay rich chalk	A	G	M. Campanian	-
249	22-23	16.0	90	18	Claystone	A	ME	E. Campanian	-
249	28-31	16.0	25	66	Claystone + siltstone	R	ME	E. Aptian ?	-
250A	22-23	8.0	40	19	Clay	C	PE	Coniacian	4
255	9-11	2.6	10	29	Limestone	F	PO	Coniacian -Santonian	6
256	8-9	9.3	50	19	Clay	A	ME	L. Albian	3
257	7-9	14.0	25	57	Clay	A	ME	M. Albian	14

Site	Cores	Recovery m	Recovery %	Thickness m	General Lithology	Calcareous Nannofossils Abund	Calcareous Nannofossils Pres	Age	Sed. Rate m/m.y.
258	5-11	30.0	25	121	Chalk + limestone	A	G	Coniacian-Santonian	24
258A	8-9	14.5	50	29	Chalk	A	MEO	Turonian	5
258	12-13	1.5	15	10	Chalk	A	GE	Cenomanian	5
258	14	28.6	30	86	Clay	A	GE	L.Albian	32
258	15-19	13.6	10	105	Clay	C	GE	M. Albian	26
258	20-23	49.0	75	67	Clay + clay rich ooze	A	GE	M. Albian	17
259	11-17	22.3	25	90	Clay + clay rich ooze	A	GEO	M. Albian	22
260	9-18	22.3	50	48	Claystone	A	MEO	M. Valanginian	16
261	28-30	10.8	30	29	Claystone	C	MEO	Kimmeridgian-Valanginian	-
261	31-32	0.2	-	-	Claystone	A	MO	Oxfordian	-
261	33	125.0	20	589	Clay	C	MO	L. Albian	196
263	5-27								

100, Core 2 in the Northwest Atlantic (Thierstein, 1976). Their species diversity is lower than at DSDP site 105 in the Northwest Atlantic. The differences, however, might be due to the preservation of these assemblages.

Berriasian - Barremian

Nannofossils from this interval were recovered at site 261 in the Argo Abyssal Plain and have been described by Proto Decima (1974) and Bukry (1974c). The assemblages are dominated by Watznaueria barnesae and contain only a few other species that are commonly encountered in assemblages from southern Europe, the northwestern Atlantic (DSDP Sites 4, 99, 100, 101, and 105 and the western Pacific (DSDP Sites 49, 50). The coccolith communities from the Argo Abyssal Plain are also less diverse than those from the Island of Misool (north of Ceram), which show similar preservation. The samples from Misool contain common Watznaueria barnesae, few Nannoconus colomii and Parhabdolithus embergeri, and rare Parhabdolithus asper, Lithraphidites carniolensis, Zygodiscus diplogrammus, Reinhardtites fenestratus, Cyclagelosphaera margerelii, Conusphaera mexicana, Micrantholithus obtusus, and Cretarhabdus surirellus. Some of these species (N. colomii, C. mexicana, M. obtusus) are found, so far, only in tropical marginal sea paleoenvironments (Thierstein, 1976). These differences suggest a considerably greater latitudinal separation between the Argo Abyssal Plain and the Island of Misool during the early Cretaceous than exists today.

Aptian

The Aptian nannofossil assemblages from the Mozambique Plateau (DSDP Site 249, Cores 28-31) have been listed by Bukry (1974b). They show extremely low species diversities (Bukry, 1974b) and differ in species composition from contemporaneous assemblages in the Central Atlantic at DSDP sites 135, 136 (Roth and Thierstein, 1972) and southern Europe (Thierstein, 1973) by the absence of characteristic Tethyan taxa such as Nannoconus, Rucinolithus irregularis and Parhabdolithus infinitus. Common specimens of the nearshore or shallow sea indicator Micrantholithus obtusus were detected by Bukry (1974b).

Albian

Sediments with Albian nannofossils were recovered in the Wharton Basin (Sites 256, 257, 259, 260, 263) and on Naturaliste Plateau (Site 258). Species lists may be found in Thierstein (1974). Only one fossiliferous sample from South Africa from an Albian outcrop along the Umzinene River, Zululand (Lambert and Scheibnerova, 1974) was available for comparison. The scarce but well-preserved nannofossils in this sample include the following species: Parhabdolithus angustus, Parhabdolithus asper, Watznaueria barnesae, Braarudosphaera bigelowii, Lithraphidites carniolensis, Cruciellipsis chiastia, Vagalapilla compacta, Biscutum constans, Prediscosphaera cretacea, Octopodorhabdus decussatus, Zygodiscus diplogrammus, Zygodiscus elegans, Tranolithus gabalus, Chiastozygus litterarius, Tranolithus orionatus, Manivitella pemmatoidea, Lithastrinus septentrionalis, Parhabdolithus splendens, Vagalapilla stradneri and Cretarhabdus surirellus. This assemblage lacks Tethyan species and is considered to have been deposited in shallow water (Lambert and Scheibnerova, 1974). It is similar to the earliest Middle Albian assemblages at site 258 (Cores 22 and 23) except for the absence of the mid- to high-latitude species Cribrosphaerella primitiva.

TABLE II. Nannofossil assemblages studied for comparison. Locality numbers refer to Figure 2. See Table I for preservation key.

	Sample Locality	No. of Samples	Lat.	Long.	Calcareous Nannofossils Abund	Calcareous Nannofossils Pres	Age	References
1	Newmarracarra limestone W-Australia	1	29°S	115°E	R	M	M. Bajocian	Arkell and Playford, 1954
2	Fatjet limestone, Misool (near Vogler's sample #42)	2	2°S	130°E	C	MO	Berriasian	Vogler, 1941
3	Umzinene River, Zululand, S-Africa	1	28°S	32°E	R	GE	Albian	Lambert and Scheibnerova, 1974
4	Bathurst Island, #2 Bore, 250'-950', N-Australia	4	12°S	130°E	F	ME	Cenomanian	
5	South Strickland River, Papua	2	6°S	142°E	F	ME	Turonian	
6	Eltanin Core 54-7, 112-204 cm, Kerguelen Plateau	4	56°S	81°E	A	MEO	Cenomanian	Quilty, 1973
6	Eltanin Core 54-7, 38-92 cm, Kerguelen Plateau	5	56°S	81°E	A	MEO	Turonian	Quilty, 1973
7	Gearle Siltstone, Rough Range South #1 Bore, Core 68	1	23°S	114°E	C	ME	Turonian	Belford and Scheibnerova, 1971
8	RC-56, 200 cm, L-DGO core, Naturaliste Plateau	1	34°S	112°E	A	MO	Turonian	Burckle, Saito and Ewing, 1967
9	Toolonga lutite, Grierson #1 Bore, Core 6, W-Australia	1	24°S	114°E	A	GEO	Santonian	
10	Toolonga lutite, Murchinson River, W-Australia	3	28°S	114°E	A	GEO	Santonian	Johnstone et al., 1958
11	Gingin chalk, Gingin, W-Australia	4	32°S	116°E	A	GEO	Santonian	Belford, 1960
12	V24-213, 310 cm, L-DGO core, Agulhas Plateau	1	37°S	25°E	A	MO	E. Campanian	
7	Miria marl, C-Y Creek, W-Australia	1	23°S	114°E	A	GEO	M. Maestrichtian	Condon et al., 1956
13	Garau, West Central Iran	1	33°N	47°E	C	MO	M. Maestrichtian	Haq, 1971
14	Warnbro #1 Bore, 2194', W-Australia	1	32°S	115°E	A	GE	L. Maestrichtian	Johnstone et al., 1973
15	V16-56, 220-230 cm, L-DGO core, Agulhas Plateau	2	41°S	27°E	C	MEO	L. Maestrichtian	Herman, 1963

Cenomanian - Turonian

Moderately well preserved nannofossils from this interval were recovered in Cores 12, 13, and 14 at site 258 on the Naturaliste Plateau (Thierstein, 1974). For comparative purposes Turonian assemblages from piston core RC 8-56 (Burckle, Saito and Ewing, 1967) taken on Naturaliste Plateau and Cenomanian and Turonian assemblages in piston core E 54-7 (Quilty, 1973) taken on Kerguelen Plateau were examined. Additional assemblages from the Gearle Siltstone (Turonian) in West Australia (Belford and Scheibnerova, 1971), Bathurst Island No. 2 Bore (Cenomanian) and South Strickland River, Papua (Turonian) were kindly made available for study by the Bureau of Mineral Resources in Canberra.

Although preservation and abundance of the Cenomainian nannofossils from the Bathurst Island and the Kerguelen Plateau are only moderate, they are similar in species composition to the well-preserved assemblages on Naturaliste Plateau. Only in the latter, however, the fragile Lithraphidites alatus could be identified. Mid- to high-latitude indicating Cribrosphaerella primitiva are present in all three localities. Whether some differences in relative abundances between the Turonian assemblages from Kerguelen Plateau, Naturaliste Plateau, West Australia and Papua may be explained by etching or are due to changing paleoecologic conditions must await quantitative evaluation and comparison with well-preserved assemblages from other geographic areas.

Coniacian - Santonian

A comparison of the assemblages from sites 250 (Mozambique Basin) and 255 (Broken Ridge) with those from site 258 (Naturaliste Plateau) revealed major differences in species diversities (Thierstein, 1974) which are most probably caused by differential preservation. The Mozambique Basin assemblages are intensely etched, those from Broken Ridge strongly overgrown, whereas those from Naturaliste Plateau are well preserved. Santonian assemblages from the Gingin Chalk and Toolonga Calcilutite (West Australia) contain more abundant Lucianorhabdus cayeuxii and Tetralithus obscurus, and less abundant Eiffellithus trabeculatus (= Chiastozygus cuneatus) than those on nearby Naturaliste Plateau. Only quantitative analyses can eventually yield clues whether these changes are due to paleoecologic or minor stratigraphic differences. A number of species considered to be characteristic of higher latitudes, such as Gartnerago obliquum, Kamptnerius magnificus, Lucianorhabdus cayeuxii, Tetralithus obscurus and Vagalapilla octoradiata are far more abundant in the southern Indian Ocean and in Australia than in contemporaneous assemblages from the Carribean Sea (DSDP Sites 150, 151, 153) and the Central Pacific (DSDP Sites 167, 170, 171).

Campanian - Middle Maestrichtian

Nannofossiliferous sediments from this interval were recovered from the Wharton Basin (Site 211), the Ninetyeast Ridge (Site 217), the Somali Basin (Site 241) and the Mozambique Plateau (Site 249). A comparison with assemblages of the same age from the Agulhas Plateau (L-DGO Core V24-213), West Australia (Miria Marl), Iran (Garau Valley), and the Central Pacific (DSDP Sites 165, 167, 170 and 171) revealed the following observations: with increasing latitude the frequencies of Watznaueria barnesae, Micula staurophora and Cretarhabdus surirellus all decrease, whereas Tetralithus aculeus, Lucianorhabdus cayeuxii, and Broinsonia parca become more abundant. The assemblages from Sites 211 and 217 also correspond to this general trend, when traced back to their paleopositions (Sclater et al., 1974, p. 830). Species diversities seem to depend mainly on the degree of etching of the assemblages.

Late Maestrichtian

Nannoplankton assemblages of this age were recovered at Site 212 (Wharton Basin), Sites 216, 217 (Ninetyeast Ridge) and Site 239 (Mascarene Basin). Two samples from L-DGO Core V 16-56, taken on the Agulhas Plateau, were studied for comparison. The late Maestrichtian latitudinal sequence of these assemblages from North to South was as follows: Site 239, Site 217, Site 216, Site 212, V 15-56 (Sclater et al., 1974; McKenzie and Sclater, 1971). At the first two sites Micula mura has been found (Bukry, 1974b; Gartner, 1974) a species that is characteristic for low latitudes (Worsley and Martini, 1970). The high latitude Nephrolithus frequens is present at Site 212 (Gartner, 1974) and in core V 16-56 (personal observation). Latest Maestrichtian species diversities in this area seem to correspond to the preservational patterns.

Acknowledgments. Samples for comparative purposes were kindly provided by Drs. D. J. Belford (Bureau of Mineral Resources, Canberra), R. M. Goll (Lamont-Doherty Geological Observatory, Palisades), B. U. Haq (Woods Hole Oceanographic Institution), P. G. Quilty (West Australian Petroleum Pty., Ltd., Perth), V. Scheibner (Geological Survey, N.S.W., Sydney), S. W. Wise (Florida State University, Tallahassee), and by the curator of the Lamont-Doherty Geological Observatory core collection. The manuscript was reviewed by Peter R. Thompson, K. Perch-Nielsen and D. Bukry.

The author was supported by a post-doctoral fellowship of the Swiss National Science Foundation.

References

Arkell, W. J., and P. E. Playford, The Bajocian ammonites of Western Australia, Philos. Transact. Roy. Soc. London, Ser. B, 237, 547-604, 1954.

Belford, D. J., Upper Cretaceous Foraminifera from the Toolonga Calcilutite and Gingin Chalk, Western Australia, Bur. Min. Res., Bull., 57, 1960.

Belford, D. J., and V. Scheibnerova, Turonian foraminifera from the Carnarvon Basin, Western Australia, and their paleogeographical significance, Micropaleontology, 17, 331-344, 1971.

Bolli, H. M., Synthesis of the Leg 27 biostratigraphy and paleontology, in Veevers, J. J., Heirtzler, J. R., et al., Initial Reports of the Deep Sea Drilling Project, 27, Washington (U. S. Govt. Printing Office, 993-999, 1974.

Bukry, D., Coccolith and silicoflagellate stratigraphy, Eastern Indian Ocean, Deep Sea Drilling Project, Leg 22, in Von der Borch, C.C., Sclater, J. G. et al., Initial Reports of the Deep Sea Drilling Project, 22, Washington, (U. S. Govt. Printing Office) 601-608, 1974a.

Bukry, D., Phytoplankton stratigraphy, offshore East Africa, Deep Sea Drilling Project Leg 25, in Simpson, E.S.W., Schlich, R., et al., Initial Reports of the Deep Sea Drilling Project, 25, Washington, (U. S. Govt. Printing Office) 635-646, 1974b.

Bukry, D., Coccolith stratigraphy, offshore western Australia, Deep Sea Drilling Project Leg 27, in Veevers, J. J., Heirtzler, J. R., et al., Initial Reports of the Deep Sea Drilling Project, 27, Washington, (U. S. Govt. Printing Office) 623-630, 1974c.

Bukry, D., Cretaceous and Paleogene coccolith stratigraphy, Deep Sea Drilling Project, Leg 26, in Davies, T. A., Luyendyk, B. P., et al., Initial Reports of the Deep Sea Drilling Project, 26, Washington, (U. S. Govt. Printing Office) 669-673, 1974d.

Bukry, D., and M. N. Bramlette, Coccolith age determinations, Leg 3 Deep Sea Drilling Project, in Maxwell, A. E., et al., Initial Reports of the Deep Sea Drilling Project, 3, Washington (U. S. Govt. Printing Office) 589-611, 1970.

Burckle, L. H., T. Saito, and M. Ewing, A Cretaceous (Turonian) core from the Naturaliste Plateau southeast Indian Ocean, Deep-Sea Research, 14, 421-426, 1967.

Cita, M. B., and S. Gartner, Deep Sea Upper Cretaceous from the western North Atlantic, in Farinacci, A., Proc. II. Plankt. Conf., Roma 1970, 1, 287-319, 1971.

Condon, M. A., D. Johnstone, C. E. Prichard, and M. H. Johnstone, The Giralia and Marilla Anticlines, Northwest Division, W. A., Bull. Miner. Resour. Austr. Bull., 25, 1956.

Deflandre, G., Sur les nannofossiles calcaires et leur systematique, Rev. Micropaleont., 2, 127-152, 1959.

Gartner, S., Nannofossil biostratigraphy, Leg 22, Deep Sea Drilling Project, in Von der Borch, C. C., Sclater, J. G., et al., Initial Reports of the Deep Sea Drilling Project, 22, Washington (U. S. Govt. Printing Office) 577-600, 1974.

Haq, Bilal Ul, Paleogene calcareous nannoflora, Part I: The Paleocene of West-Central Persia and the upper Paleocene of West Central Pakistan, Stockholm Contrib. Geology, 25, no. 1, 56 p., 1971.

Herman, Y., Cretaceous, Paleocene, and Pleistocene sediments from the Indian Ocean, Science, 140, 1316-1317, 1963.

Johnstone, D., M. A. Condon, and P. E. Playford, The stratigraphy of the lower Murchinson River area and Yaringa North Station, Western Australia, Jour. Roy. Soc. W. Austr., 41, 13-16, 1958.

Johnstone, M. H., D. C. Lowry, and P. G. Quilty, The geology of the southwestern Australia - a review, Jour. Roy. Soc. W. Austr., 56, no. 1 + 2, 5-15, 1973.

Lambert G., and V. Scheibnerova, Some Albian foraminifera from Zululand (South Africa) and Great Artesian Basin (Australia), Micropaleontology, 20, 76-96, 1974.

Manivit, H., Les Nannofossiles calcaires du Cretace francais (de l' Aptien au Danien). Essai de biozonation appuyee sur les stratotypes, These. Fac. Sci., Univ. de Paris, 1971.

McKenzie, D., and J. G. Sclater, The evolution of the Indian Ocean since the Late Cretaceous, Geophys. Journ. Roy. Astron. Soc., 25, 437-528, 1971.

Müller, C., Calcareous nannoplankton, Leg 25 (Western Indian Ocean) in Simpson, E.S.W., Schlich, R. et al., Initial Reports of the Deep Sea Drilling Project, 25, Washington, (U. S. Govt. Printing Office) 579-634, 1974.

Pienaar, R. N., Upper Cretaceous coccolithophorids from Zululand, South Africa, Paleontol., 11, no. 3, 361-367, 1968.

Pienaar, R. N., Upper Cretaceous calcareous nannoplankton from Zululand, South Africa, Paleontol. Africana, 12, 75-149, 1969.

Proto Decima, F., Leg 27 Calcareous Nannoplankton, in Veevers, J. J., Heirtzler, J. R., et al., Initial Reports of the Deep Sea Drilling Project, 27, Washington (U. S. Govt. Printing Office) 589-622, 1974.

Quilty, P. G., Cenomanian-Turonian and Neogene sediments from Northeast of Kerguelen Ridge, Indian Ocean, Jr. Geol. Soc. Australia, 20, no. 3, 361-368, 1973.

Roth, P. H., Calcareous Nannofossils - Leg 17, Deep Sea Drilling Project, in Winterer, E. L., Ewing, J. I. et al., Initial Reports of the Deep Sea Drilling Project, 17, Washington (U. S. Govt. Printing Office) 695-793, 1973.

Roth, P. H., and H. R. Thierstein, Calcareous Nannoplankton: Leg 14, in Hayes, D. E., A. C. Pimm, et al., Initial Reports of the Deep Sea Drilling Project, 14, Washington (U. S. Govt. Printing Office) 421-485, 1972.

Sclater, J. G. et al., Regional synthesis of the Deep Sea Drilling Results from Leg 22 in the Eastern Indian Ocean, in Von der Borch, C. C., Sclater, J. G., et al., Initial Reports of the Deep Sea Drilling Project, 22, Washington (U. S. Govt. Printing Office) 815-832, 1974.

Thierstein, H. R., Lower Cretaceous calcareous nannoplankton biostratigraphy, Abh. Geol. B. A., Wien, 29, 1973.

Thierstein, H. R., Calcareous nannoplankton - Leg 26, Deep Sea Drilling Project, in Davies, T. A., Luyendyk, B. P., et al., Initial Reports of the Deep Sea Drilling Project, 26, Washington, (U. S. Govt. Printing Office) 619-667, 1974.

Thierstein, H. R., Mesozoic calcareous nannoplankton biostratigraphy of marine sediments. Mar. Micropaleontol., 1, 325-362, 1976.

Veevers, J. J., J. R. Heirtzler et al., Initial Reports of the Deep Sea Drilling Project, 27, Washington (U. S. Govt. Printing Office) 1060 p., 1974.

Vogler, J., Ober-Jura und Kreide von Misol (Niederländisch-Ostindien), Paleontolographica, supp., 4, no. 4, 243-293, 1941.

Worsley, T., and E. Martini, Late Maastrichtian nannoplankton provinces, Nature, 225, no. 5239 (March 28, 1970), 1242-1243, 1970.

CHAPTER 15. PALEOCENE TO EOCENE CALCAREOUS NANNOPLANKTON OF THE
             INDIAN OCEAN

Franca Proto Decima

Instituto di Geologia dell'Universita di Padova (Italia)

Abstract. The information on the Paleocene to Eocene calcareous nannoplankton of the Indian Ocean Sites of the Deep Sea Drilling Project is summarized. With the exception of the lower Paleocene Markalius inversus, Cruciplacolithus tenuis s.s. and Ellipsolithus macellus zones of Martini, all tropical/subtropical Paleocene and Eocene nannoplankton zones of Martini (1971) and of Bukry (1973) could be recognized. Different biostratigraphic units had to be applied for the Southern Indian Ocean where the low diversification of the higher latitude nannofloras, and the absence of some markers prevented the high biostratigraphic resolution applicable in low latitudes. The possibility of a gap at the Cretaceous/Tertiary boundary is discussed. The paleodistribution patterns of some climatically restricted species are reported.

Introduction

Prior to the GLOMAR CHALLENGER cruises the knowledge on Paleocene-Eocene calcareous nannoplankton of the Indian Ocean and surrounding regions was very scarce, being limited to some localities of India (Narasimhan, 1963; Pant and Mamgain, 1969), Pakistan (Haq, 1967, 1971), the Red Sea coast of Egypt (Shafik and Stradner, 1971; El-Dawoody and Barakat, 1973) and Madagascar (Perch-Nielsen and Pomerol, 1973). The sixty-two sites drilled by the GLOMAR CHALLENGER in the Indian Ocean during the Leg 22 to 29 of the Deep Sea Drilling Project (see Figure 1) therefore furnish most of the available information about the nannoplankton of this region.

The authors involved in the nannoplankton studies of the Indian Ocean are, from Leg 22 to 29, Gartner, Boudreaux, Roth, Müller, Thierstein, Proto Decima, Burns and Perch-Nielsen. Bukry, in addition has contributed to the nannoplankton biostratigraphy for all the Legs and Edwards is co-author with Perch-Nielsen in the high latitude nannoplankton stratigraphy of the South Indian Ocean.

The present article is based on the results published by the abovementioned authors in the Initial Reports of the Deep Sea Drilling Project, Volumes 22 to 29.

The occurrence and distribution of Paleocene and Eocene sediments in the sites drilled are shown in Figure 1. Because some different zonal schemes are used by different authors the sites are reported following the biostratigraphic schemes used in Figures 2-5, where the recognized nannoplankton zones are also indicated. The sites are thus reported twice when they were object of two different nannoplankton zonal schemes. If no special comments are made in the following chapters, the differences

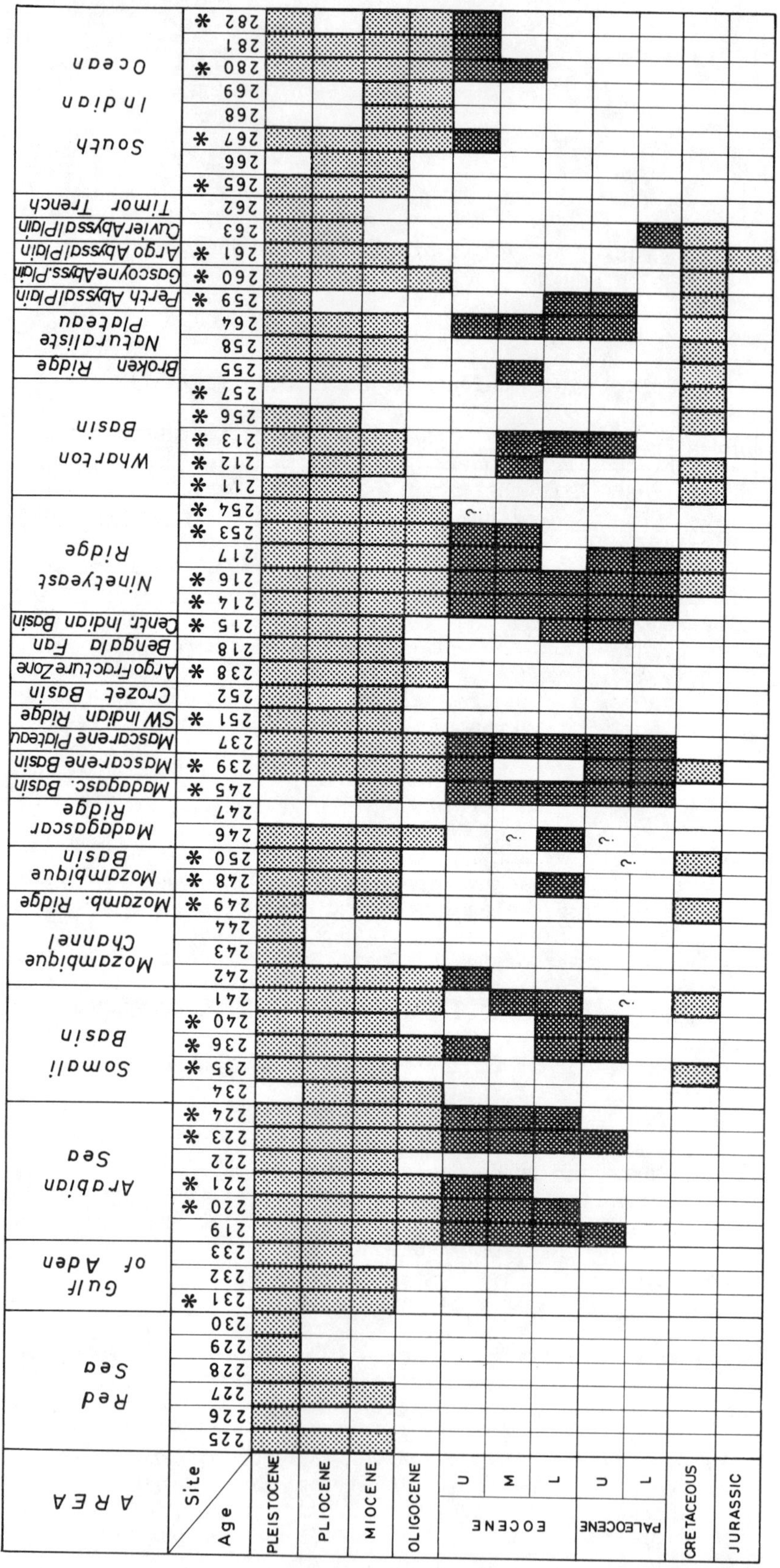

Figure 1. Age of sediments in Indian Ocean sites. * Indicates basement reached.

AGE		Recognized zones		Somali Basin 240	Somali Basin 241	Mozamb. Chann. 242	Mozambique Basin 248	Madagasc. Basin 245	Mascarene Basin 239	Ninetyeast Ridge 253	Broken Ridge 255	Naturaliste Plateau 264	Perth Abissal Plain 259
EOCENE	UPPER	NP 20	*Sphenolithus pseudoradians*			■			■	■			
EOCENE	UPPER	NP 19	*Isthmolithus recurvus*			■		■					
EOCENE	UPPER	NP 18	*Chiasmolithus oamaruensis*							■			
EOCENE	MIDDLE	NP 17	*Discoaster saipanensis*					■		■		■	
EOCENE	MIDDLE	NP 16	*Discoaster tani nodifer*							■	■	■	
EOCENE	MIDDLE	NP 15	*Chiphragmalithus alatus*					■				■	
EOCENE	MIDDLE	NP 14	*Discoaster sublodoensis*		■			■				■	
EOCENE	LOWER	NP 13	*Discoaster lodoensis*		■			■				■	■
EOCENE	LOWER	NP 12	*Marthasterites tribrachiatus*	■			■					■	■
EOCENE	LOWER	NP 11	*Discoaster binodosus*					■					■
EOCENE	LOWER	NP 10	*Marthasterites contortus*					■					■
PALEOCENE		NP 9	*Discoaster multiradiatus*	■				■					■
PALEOCENE		NP 8	*Heliolithus riedeli*					■					
PALEOCENE		NP 7	*Discoaster gemmeus*					■					
PALEOCENE		NP 6	*Heliolithus kleinpelli*					■				■	
PALEOCENE		NP 5	*Fasciculithus tympaniformis*					■	■			■	
PALEOCENE		NP 4	*Ellipsolithus macellus*										
PALEOCENE		NP 3	*Chiasmolithus danicus*					■	■				
PALEOCENE		NP 2	*Cruciplacolithus tenuis*						■				
PALEOCENE		NP 1	*Markalius inversus*										

Figure 2. Sites where Martini's biozones are recognized.

in number and/or name of the recognized zones for the same site, as they appear from Figures 2 to 4, are to be considered the result of a partial section examination and/or of a different species nomenclature and zonal definition, without real biostratigraphic discrepancies.

Paleocene to Eocene Calcareous Nannoplankton Distribution in the Geographical Units of the Indian Ocean

Arabian Sea

Five sites drilled in the Arabian Sea (219, 220, 221, 223 and 224) contain Eocene or Eocene and Paleocene sediments. Calcareous nannofossils are generally common to abundant in Eocene chalk and ooze. The

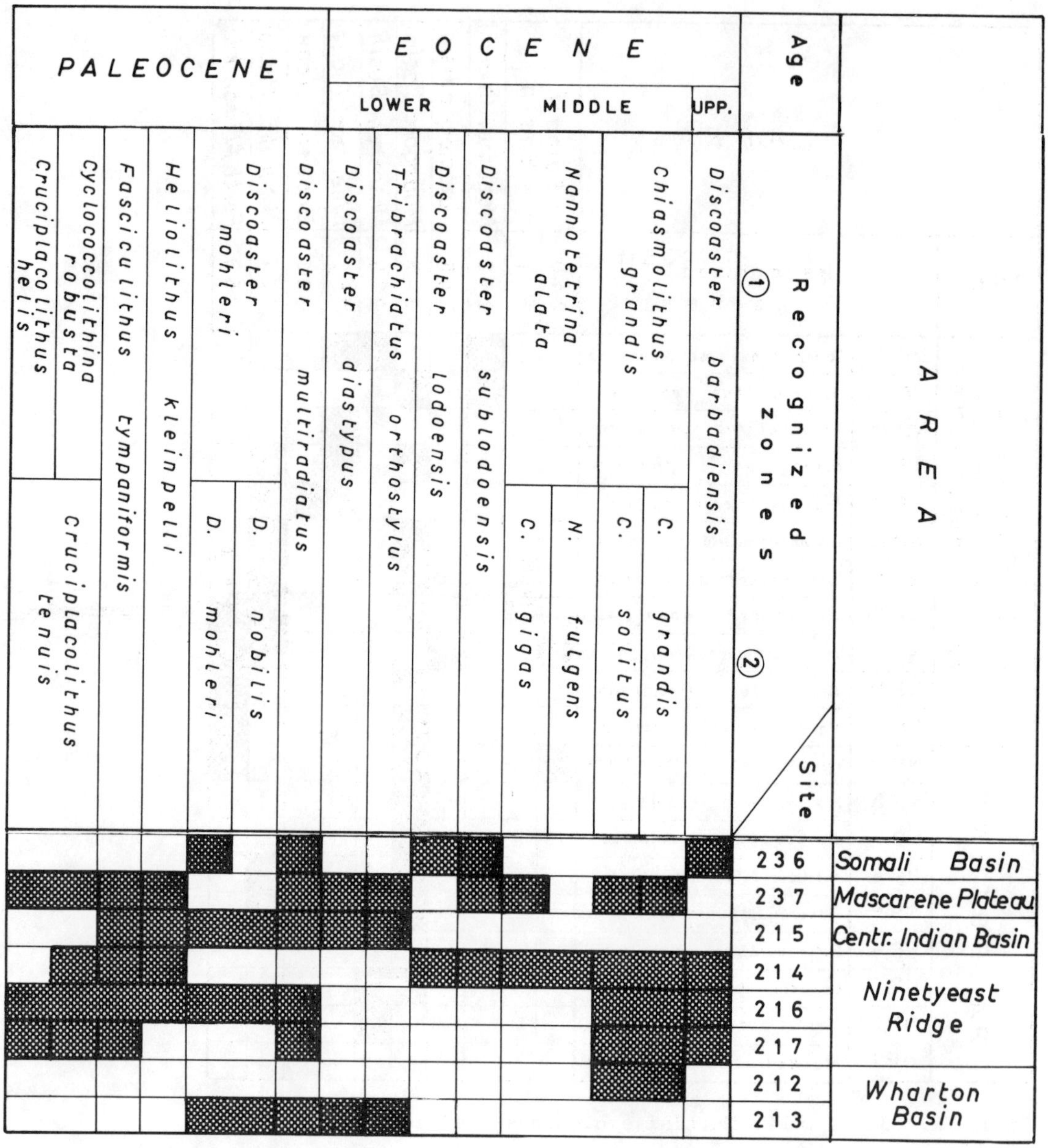

Figure 3. Sites where Gartner's 1974 (1) and Roth's 1974 (2) biozones are recognized.

upper Eocene assemblages are typical for low latitudes with abundant Discoaster saipanensis, D. barbadiensis, D. tani, Reticulofenestra umbilica, Sphenolithus pseudoradians. Absent are Isthmolithus recurvus and Chiasmolithus oamaruensis. Chiasmolithus grandis, C. gigas, Sphenolithus spiniger and S. obtusus are present in the middle Eocene and Discoaster sublodoensis in the lowermost middle Eocene. The lower Eocene is represented by the Discoaster lodoensis and Tribrachiatus orthostylus zones. Nannoplankton becomes rare to absent in the Paleocene of site 219, which is characterized by marine continental shelf-type deposition. At site 223 the common occurrence of Discoaster multiradiatus and associated species characterizes the upper Paleocene. The underlying

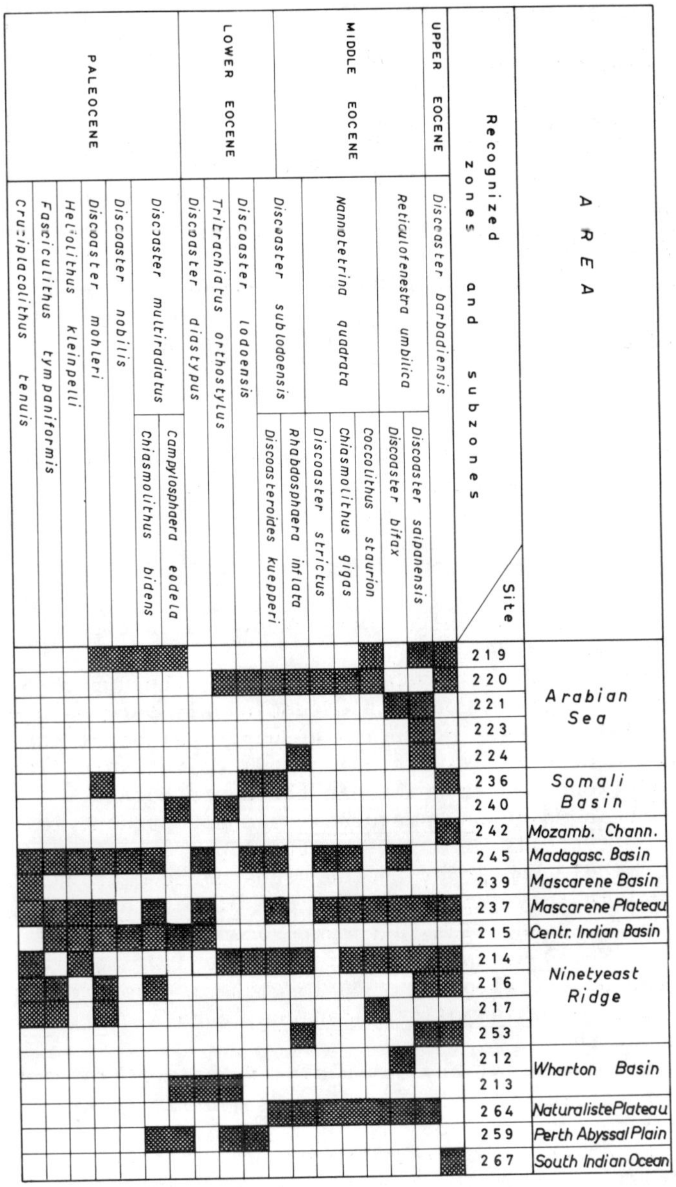

PALEOCENE TO EOCENE CALCAREOUS NANNOPLANKTON 357

Figure 4. Sites where Bukry's 1973 biozones are recognized.

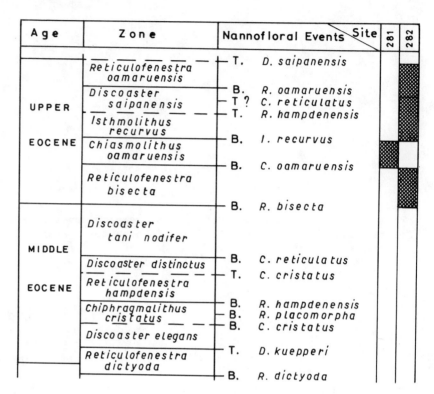

Figure 5. Sites where Edwards and Perch Nielsen's 1975 biozones are recognized.

portion of the sedimentary section consists of brown clay that contains only few Paleocene nannofossils; here the absence of D. multiradiatus suggests an age older than uppermost Paleocene.

Somali Basin

Paleocene-Eocene nannoplankton assembalges were recovered in the Somali Basin at sites 236, 240 and 241.

At site 236, 270 km northeast of the Seychelles Islands, the upper Paleocene-Eocene sedimentary sequence consists of dominantly pure nannochalks or oozes. Substantial hiatuses occur in the upper Paleocene and Eocene. Nannofossils show considerable alteration and are recrystallized. The Discoaster barbadiensis zone is recognized in the upper Eocene. Isthmolithus recurvus and Chiasmolithus oamaruensis are not present in these assemblages which are typical for a low-latitude open oceanic environment. Most of the middle and lower Eocene seems to be missing. This stratigraphic interval is only represented by the Discoaster sublodoensis and D. lodoensis zones. The Paleocene-Eocene boundary is marked by an unconformity. The Discoaster multiradiatus and D. mohleri zones are recognized in the upper Paleocene. Of interest is the presence in the Paleocene of this site of Chiasmolithus danicus, a basal Tertiary species typical for high latitudes.

At site 240, situated in the central abyssal plain of the western part of the Somali Basin, upper Miocene to Recent sediments overlie lower Eocene sediments, with a large gap of more than 40 m.y. Nannoplankton of the Marthasterites tribrachiatus Zone (=Tribrachiatus orthostylus Zone) was found in the lower Eocene. The assemblages are composed of only a few solution-resistant species such as T. orthostylus, Discoaster lodoensis, D. barbadiensis, D. diastypus, Discoasteroides kuepperi, Sphenolithus radians and Coccolithus pelagicus. Richer and more diversified associations of the D. multiradiatus Zone are present in the upper Paleocene.

At Site 241, on the East African continental rise of the southern Somali Basin, the Paleocene is almost devoid of fossils. Poor assemblages composed of discoasters and a few other resistant species are present in the middle and lower Eocene Discoaster sublodoensis and D. lodoensis zones.

Mozambique Channel

Three holes were drilled in the Mozambique channel, but only at Site 242 was a complete sequence from Pleistocene to upper Eocene penetrated. The generally well preserved calcareous nannofossils are the dominant component in these sediments. The very rare occurrence of Discoaster saipanensis in the uppermost Eocene renders it difficult to identify in this section the Eocene/Oligocene boundary based on the extinction of this species. The Sphenolithus pseudoradians and Istmolithus recurvus zones were recognized in the upper Eocene. Specimens of I. recurvus are present but Chiamolithus oamaruensis is not recorded. The relative abundance of Discoaster versus Chiasmolithus (499 to 1) suggests a warm-water deposition for the upper Eocene of this Site, at a latitude similar to the actual one.

Mozambique Basin

An incomplete sequence of Pleistocene to lower Eocene or Paleocene was drilled in Site 248 in the north-western Mozambique Basin. The Paleogene is represented by pelagic sediments deposited below the regional CCD. The only observed calcareous biogenic level contains a nannoplankton assemblage of Marthasterites tribrachiatus (NP 12) Zone, which indicate a lower Eocene age.

Madagascar Ridge

The Paleogene sediments of Site 246, near the crest of the Madagascar Ridge consist of shelly carbonate sand and glauconitic sand accumulated in an agitated and relatively shallow-water environment. Calcareous nannofossils are absent or only very rare. The few and poorly preserved species indicate Eocene but do not allow a more detailed biostratigraphic assignment.

Madagascar Basin

An almost complete sequence of Paleocene and Eocene nannoplankton zones was recovered at Site 245 in the southern Madagascar Basin. The preservation of middle Eocene to lower Oligocene nannofossil assemblages varies from moderate to strongly etched. In the Paleocene to lower Eocene coccoliths are overgrown and fragmented by diagenetic processes. This change in preservation is closely related to one in lithology, i.e. from a brown silty clay above to a pale orange and pink chalk below. The discrimination between upper Eocene and Oligocene assemblages is rendered difficult due to the presence of reworked Eocene species such as Chiasmolithus grandis, Discoaster barbadiensis and D. lodoensis. More reliable coccolith assemblages are present from middle Eocene downwards, with common and diversified populations of Nannotetrina in the middle Eocene. The occurrence of Tribrachiatus orthostylus in the middle Eocene Nannotetrina quadrata Zone has been reported by Bukry (1974), who also recorded it at the same level from several localities on the west coast of North America. The lower Eocene sediments contain diagnostic coccolith assemblages. They suggest warm-temperature, open-ocean conditions by the absence of near-coast and shallow-water indicators and by a moderate to high Discoaster/Chiasmolithus ratio. One or two of the species Discoaster multiradiatus, D. mohleri, Cyclolithella robusta, Fasciculithus involvutus, Heliolithus kleinpelli dominate in turn the assemblages of the Paleocene where the lower zones are characterized by Chiasmolithus danicus, Cruciplacolithus tenuis and Markalius inversus.

Mascarene Basin

At Site 239, situated in the abyssal plain of the southern Mascarene Basin, nannoplankton assemblages are in general not very rich in the Paleocene-Eocene brown and silty clays. The probable presence of the Sphenolithus pseudoradians Zone was determined in the upper Eocene. A well-established break of about 20 m.y. duration seems to occur between the upper Eocene and the middle Paleocene Fasciculithus tympaniformis Zone. This Site includes the Cretaceous/Tertiary boundary. An asssociation with Cruciplacolithus tenuis, Zygodiscus sigmoides, Markalius inversus, rare Chiasmolithus danicus and some Micula mura indicating the Cruciplacolithus tenuis Zone (NP2)/Chiasmolithus danicus Zone (NP3) is present in the lowermost Tertiary. An upper Campanian age, based on the frequence of Tetralithus aculeus is suggested for the underlying Cretaceous.

Mascarene Plateau

Site 237 is located on the Mascarene Plateau, between the Seychelles and the Saya de Malha Bank. The sedimentary sequence ranges from lower Paleocene to Pleistocene, and consists mainly of nanno ooze and chalk rich in calcareous plankton. The Paleocene through middle Eocene section measures some 500 m. The nannofossils show slight etching and moderate overgrowth which becomes still more pronounced in the lower Paleocene. An almost complete sequence of Paleocene and Eocene nannoplankton zones was recognized at this Site.

Central Indian Basin

Calcareous nannofossil ooze ranging in age from lower Eocene to middle Paleocene was recovered at Site 215 in the Central Indian Basin. The lower Eocene Tribrachiatus orthostylus and Discoaster diastypus zones and the Paleocene D. multiradiatus, D. mohleri, Heliolithus kleinpelli and Fasciculithus tympaniformis zones are represented. For the most part the nannofossil assemblages are well preserved and of normal diversification. Large size specimens are present in the Chiasmolithus bidens Subzone of the Discoaster multiradiatus Zone. Heliolithus kleinpelli is very frequent in the nominal zone and a great abundance of Cyclolithella robusta was observed in the Fasciculithus tympaniformis Zone. Poor preservation and low diversity characterize the nannofossil assemblages at the top and bottom of the calcareous sequence.

Ninetyeast Ridge

Four (214, 216, 217 and 253) of the five sites drilled along the crest of the Ninetyeast Ridge contain Eocene or Eocene and Paleocene calcareous nannofossils.

At the northernmost Site 217 calcareous nannofossils were recovered throughout the cored interval ranging in age from Pleistocene to Upper Cretaceous. The preservation is poor in the Paleocene-Eocene and there seem to be several gaps in the sedimentary record. Mixed assemblages consisting of Upper Cretaceous and lower Paleocene species, interpreted as the result of bioturbation, are present at the Cretaceous/Tertiary boundary (see Table 2).

The stratigraphic column at Site 216 ranges from Upper Maastrichtian to upper Pleistocene. Calcareous nannofossils were recovered throughout this interval. The preservation of nannofossils is generally poor and species diversity low. The assemblages are unusual in their great dominance of some species, and the absence of others. In the upper Eocene, for instance, Reticuofenestra reticulata is very abundant and R. umbilica is lacking; Fasciculithus tympaniformis is the predominant species in the upper Paleocene Discoaster mohleri Zone. The low diversity recorded could be attributable to shallow-water conditions or high latitude. Relatively undisturbed sediments were recovered at the Cretaceous/Tertiary boundary. Predominantly Danian assemblages with some Maastrichtian contaminants were recovered in the vicinity of the boundary (see Table 1). Cretaceous assemblages contain Nephrolithus frequens and this indicates youngest Maastrichtian age. Cruciplaceolithus tenuis, Markalius inversus and Chiasmolithus danicus are present together in the basal Tertiary sediments, but this Danian nannoplankton assemblage is not the oldest known for the Tertiary. Thus a hiatus seems to be present at the Cretaceous/Tertiary boundary. According to Gartner (1974) the sedimentary structure indicates that the Maastrichtian sediments were relatively firm when calcareous sediments again began to accumulate in early Danian time. Thus the evidence for a hiatus, probably related to a time of non deposition, is also indicated by the lithologic features.

Site 214 contains a more or less continuous record from Holocene to the early Eocene Discoaster lodoensis Zone. Much of the lower Eocene and upper Paleocene may, however, not be present. Poor assemblages of Paleocene and the presence of pentaliths indicate a restricted oceanic influence that would suggest lagoonal or shallow shelf environments.

At Site 253 the stratigraphic column consists on the evidence of foraminifera/nanno ooze of Pleistocene to upper Eocene age; the latter overlies a volcanic ash sequence containing scarce nannofossils of middle Eocene age. All nannofossil assemblages show strong overgrowth, and those in the volcanic ash were also showing partial dissolution. Although the contemporary appearance in the sequence of Isthmolithus recruvus and Sphenolithus pseudoradians prevents recognition of the Isthmolithus recurvus Zone as defined by Martini (1971), the upper Eocene sequence seems to be a complete one. The poor middle Eocene assemblages suggest correlation with the Discoaster tani nodifer Zone of Martini, or with the D. saipanensis Subzone of Bukry. Bukry (1974) also recognises the lower middle Eocene Rhabdosphaera inflata Subzone. The presence in the middle Eocene of Zygolithus dubius, Micrantholithus sp., Braarudosphaera bigelowi, B. discula, Pemma rotundum and Zygrablithus bijugatus and the scarcity of discoasters indicates a near-shore or shallow-water environment and the disappearance of the pentaliths in the upper Eocene points to a continuous deepening at this site. Transitional to subtropical paleoconditions for the Paleogene, and subtropical to tropical ones for the Neogene are suggested by the nannoplankton assemblages at this Site.

All the sites located on the Ninetyeast Ridge penetrated a thick sequence of mostly Tertiary calcareous oozes and furnish nannofossil evidence for a transition of older shallow-water into younger deep water sediments, with volcanogenic material present in Sites 214, 216 and 253. The presence of the high latitude species Nephrolithus frequens in the upper Maastrichtian and Chiasmolithus danicus in the lower Paleocene and the low diversity of some assemblage could be related to the Cretaceous-Lower Tertiary high latitude position of the Indian plate.

Wharton Basin

Sediments containing Paleocene to Eocene calcareous nannoplankton are scattered in the Wharton Basin. The Upper Cretaceous through Lower Tertiary sequences are thin, implying low sedimentation rates, erosion or hiatuses during this period. Moreover, the sections are often barren of calcareous fossils because the sedimentation took place below the calcium carbonate compensation depth. Nannoplankton of Paleocene-Eocene age was recovered only at Sites 212 and 213. At Site 212, in the deepest part of the Wharton Basin, a chalk unit about 80 meters thick is present between two units of brown clay barren of calcareous nannofossils. This chalk was referred by Gartner to the Chiasmolithus grandis Zone and by Bukry to the Discoaster bifax Subzone both of late middle Eocene age. It falls in the time interval from the first occurrence of Reticulofenestra umbilica about 47 m.y., to the last occurrence of Chiasmolithus grandis, about 43 m.y. The high sedimentation rate and the selective solution of the nannofossils suggest that these sediments probably were carried in and redeposited by bottom currents below the regional calcium carbonate compensation depth.

Nannofossil ooze of lower Eocene to upper Paleocene age was recovered at Site 213. The sedimentary history of this Site is very similar to that of 215 on the opposite side of the Ninetyeast Ridge. The nanno ooze underlies a non-fossiliferous zeolitic brown clay and overlies a basal iron manganese oxide sediment which in turn rests on weathered pillow basalt. The youngest nannofossils assemblages are referred to the early Eocene Discoaster lodoensis and Tribrachiatus orthostylus zones. Downwards the same assemblage becomes progressively richer in species. The Discoaster diastypus Zone and the upper Paleocene D. multiradiatus and D. mohleri zones were also recognized. The nannofossil assemblages have a normal oceanic aspect, except for the top and bottom of the calcareous interval where they are composed of only the most resistant species which show strong evidence of solution.

Broken Ridge

The calcareous sediments at Site 255 on Broken Ridge contain common to few strongly overgrown nannofossils of middle Eocene age. The correlation with the Discoaster tani nodifer Zone is based on the presence of Reticulofenestra umbilica and the absence of younger markers. The lack of discoasters in this assemblage could indicate a nearshore depositional environment.

Naturaliste Plateau

Paleocene and Eocene sediments were recovered at Site 264, near the southern edge of the Naturaliste Plateau. The Eocene section ranging in age from lower Upper Eocene is extremely thick; it follows a middle Paleocene-lower Eocene unconformity. The nannofossil assemblages are quite diversified and show a transitional character. They are composed of Discoaster, Chiasmolithus, Sphenolithus, Reticulofenestra, Cyclococcolithus, Heliconthosphaera, Nannotetrina. Some near-shore indicators like Lanternithus minutus and Zygrablithus bijugatus are also present. Evidence for cool-temperate conditions is furnished by the abundance of placoliths, the low Discoaster/Chiasmolithus ratio, the low proportion of rosette- to free-rayed discoaster specimens, and the scarceness of Sphenolithus and Cyclococcolithus formosus. The middle Palecene is represented by the Heliolithus kleinpelli and Fasciculithus tympaniformis zones.

Perth Abyssal Plain

A complete sequence of lower Eocene to upper Paleocene nannoplankton zones was cored at Site 259, in the Perth Abyssal Plain. The assemblages appear to be residual, almost wholly composed of the solution resistant taxa Discoaster, Fasciculithus, and some Coccolithus. The lower Eocene sequence of 9.5 meters is strongly condensed. The upper Paleocene, considerably thicker than the lower Eocene above, overlies zeolitic clay, barren of calcareous nannofossils.

Southern Indian Ocean

Eocene calcareous nannoplankton bearing sediments were recovered in the Southern Indian Ocean at four Sites (267, 280, 281 and 282).

At Site 280, 100 km south of the South Tasmanian Rise, the Paleogene sedimentary sequence consists of Oligocene and upper Eocene silty diatom ooze. It is underlain by lower to middle Eocene glauconitic clayey silts with chert and highly organic silty claystone with almost no biogenic components. The calcareous nannofossils in these sediments are rare, small, poorly preserved and not diversified. The suggested ages are based on only very few specimens per assemblage. The lower part of the section is referred to the lower to middle Eocene based on Chiasmolithus solitus. An upper Eocene to lower Oligocene age is suggested for the upper part of the sequence that contains Reticulofenestra bisecta, Chiasmolithus oamaruensis and R. placomorpha. The low diversity of the nannoflora reflects adverse depositional and diagenetic conditions.

At Site 281, on the South Tasmanian Rise, Eocene sediments are represented by 28.5 meters of rich biogenic, glauconitic silty sands above and 19 meters of glauconitic sandstone and probable micaschist breccias below. Nannofloras are common, moderately well preserved and diversified in the upper unit, but scarce and poorly preserved with a low diversity in the lower one. The assemblages are in good agreement with those of the early upper Eocene Chiasmolithus oamaruensis Zone. Discolithina pulchra and D. pulcheroides in the upper lithologic unit suggest a shallow-water environment. The paucity of the nannoflora in the lower units indicates restricted environmental conditions that are in agreement with upper Eocene paleolatitude of this Site, near Antarctica.

At Site 282, west of Tasmania, the Eocene sediments consist of 103 meters of dark brown silty clay to clayey silts with sponge spicules, nannofossils and carbon. These overlie a fine grained, extrusive pillow basalt. The lower part of the sedimentary unit above the basalt contains poorly preserved nannofloras with Reticulofenestra placomorfa, Cyclicargolithus reticulatus and Discoaster tani nodifer. They are referred to the upper Eocene but do not allow for a more detailed stratigraphic assignment. Discoaster saipanensis and C. reticulatus are present in the upper part of the unit assigned to the upper Eocene. The Eocene/Oligocene boundary is placed above the last occurrence of D. saipanensis.

The sedimentary section at Site 267, in the deep basin south of the Southeast Indian Ridge, consists of an upper siliceous and a lower calcareous unit. Abundant, moderately preserved nannofossils are contained in the calcareous section. They represent the southernmost Paleogene assemblage presently known in the Indian Ocean. Their accurate age is difficult to assess because most of the previously described zonal index fossils are absent from these high latitude sediments. An upper Eocene or lower Oligocene age is inferred for a part of the sequence containing Chiasmolithus expansus, Dictyococcites bisectus and Reticulofenestra hillae. The lowermost part of the sequence is considered upper Eocene on the basis of the common occurrence of Isthmolithus recurvus, Chiasmolithus altus, C. expansus and C. oamaruensis. The great abundance of Chiasmolithus specimens also indicates a cool-water environment.

Paleobiogeographic and Paleoecologic Remarks

The paleodistribution patterns of some climatically restricted species are indicated on Figure 6. The high latitude species Chiasmolithus danicus is reported from all the sites where lower Paleocene nannoplankton have been recovered. This could be an indication for higher latitudes of these sites during the early Tertiary. At Site 216, on the Ninetyeast Ridge, moreover, C. danicus bearing sediments overlie upper Maastrichtian with Nephrolithus frequens also considered a high latitude species. However, at Site 217, about seven degrees further North, the underlying upper Maastrichtian contains the tropical species Micula mura. Thus evidence for the climatic significance of C. danicus is rather questionable in the Indian Ocean region.

The cold-water indicators Isthmolithus recurvus and Chiasmolithus oamaruensis were not found in the upper Eocene of the Somali Basin, Arabian Sea, Mascarene Plateau, Mascarene Basin and the northern part of the Ninetyeast Ridge. This indicates a tropical position for these regions since upper Eocene. Of these two species, I. recurvus shows a greater ecologic tolerance. It is recorded from Sites, 242 and 245, where C. oamaruensis is still missing.

Finally, as expected, the warm-water indicators Discoaster barbadiensis and D. saipanensis are absent from the late Eocene of the South Indian Ocean.

Cretaceous/Tertiary Boundary

Sediments with nannoplankton associations of the Cretaceous/Tertiary boundary were recovered at Sites 216 and 217 on the Ninetyeast Ridge and 239 in the Mascarene Basin. The distribution of some marker species in the vicinity of the boundary at Site 216 and 217 is shown in Table 1 and 2. A similar situation also characterized Site 239.

Such a distribution if compared with Martini's standard zonation (1971) makes evident a gap in the nannoplankton zonal sequence. The Markalius inversus Zone cannot be recognized; the Cruciplacolithus tenuis Zone is missing at Site 216 and only very questionable at Sites 217 and 239. In addition to a zonal gap a break in sedimentation is also suggested by the lithologic features at Site 216.

In contrast, a Markalius inversus (M. astroporus Auct.) Zone of 10 meters and a Cruciplacolithus tenuis Zone of about 30 meters were reported by El-Dawoody and Barakat (1973) from the shelf-type basal Paleocene of the Red Sea coast of Egypt. The Cretaceous/Tertiary boundary is here also marked by a disconformity between the Maastrichtian Arkhangelskiella cymbiformis Zone and the overlying Markalius inversus Zone which contains in its basal part reworked late Cretaceous nannoplankton.

For further remarks on the Cretaceous/Tertiary boundary see also the following chapter.

PALEOCENE TO EOCENE CALCAREOUS NANNOPLANKTON   365

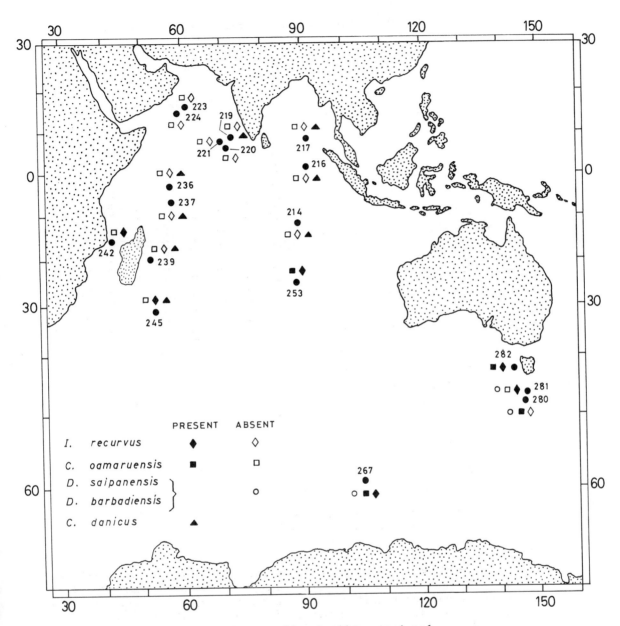

Figure 6. Paleodistribution of some climatically restricted nannoplankton species.

Biostratigraphic Remarks

The Paleocene-Eocene calcareous nannoplankton associations of the low and mid latitude open sea environment are essentially the same the world over. They furnish one of the most impressive and reliable means for a detailed inter-regional, inter-continental and inter-oceanic correlation of the marine sediments of this age. Such correlations are based on a number of different zonal schemes which during the past 15 years have been proposed by a number of authors. Except for some zones and subzones which are based on fossils subject to a

TABLE 1. Distribution of marker species at the Cretaceous/Tertiary boundary.

Ninetyeast Ridge Deep Sea Drilling Project Site 216.

Core / Section / cm	Arkhangelskiella cymbiformis	Cretarhabdus conicus	Cylindralithus gallicus	Nephrolithus frequens	Markalius inversus	Zygodiscus sigmoides	Cruciplacolithus tenuis	Chiasmolithus danicus	Fasciculithus tympaniformis
22 / CC						+	+		+
23 / 2 / 108					+	+	+	+	
23 / 2 / 109						+	+	+	
23 / 2 / 110	x		x						
23 / 2 / 117	x				+	+	+	+	
23 / 2 / 120			x	x	+			cf.	
23 / 2 / 139	x		x	x					
23 / 3 / 1	x	x	x	x					

Note: + = Tertiary species
x = Cretaceous species

narrower climatic or geographic control, the different zonal schemes are well correlatable, are often complementary in the same stratigraphic sequence and can readily be substituted one for the other. The zonations of Martini (1971), Bukry (1973), Gartner (1974), Roth (1974) and Edwards and Perch-Nielsen (1975) were used for the subdivision of the Indian Ocean Paleocene and Eocene sediments. In many cases two different zonal scheme could be applied to the same sequence (see Figures 2 to 5).

Compared with others, the Martini zones allow the most detailed subdivision of the lower Paleocene. In the examined Indian Ocean section, however, Markalius inversus, Cruciplacolithus tenuis and Chiasmolithus danicus were already found present together in the oldest recovered Paleocene. Thus, the two lowermost Paleocene (Markalius inversus and Cruciplacolithus tenuis s.s. zones of Martini could not be recognized. If these two zones are to be considered as distinct biostratigraphic units one must therefore assume that in the examined Indian Ocean sections a hiatus exists between the Upper Cretaceous and the Lower Tertiary. Further investigations will be necessary to ascertain whether

Table 2. Distribution of marker species at the Cretaceous/Tertiary boundary.

Ninetyeast Ridge Deep Sea Drilling Project - Site 217.

Core / Section / cm	Arkhangelskiella cymbiformis	Cretarhabdus conicus	Cylindralithus gallicus	Micula mura	Markalius inversus	Biantholithus sparsus	Zygodiscus sigmoides	Cruciplacolithus tenuis	Chiasmolithus danicus	Ellipsolithus macellus	Fasciculithus tympaniformis	Prinsius bisulcus
14 / CC							+	+			+	+
15 / 1 / 121							+	+	cf.	+		
15 / 5 / 20					+		+	+	+			
16 / 6 / 39					+		+		+			
16 / CC	x		x		+	+	+	+				
17 / CC	x		x									
20 / CC	x	x	x	x								

the Markalius inversus and the Cruciplacolithus tenuis s.s. zones, in fact, represent effective steps in the Tertiary nannoflora evolution or are really the result of impoverished and selected assemblages. The Markalius inversus and the Chiasmolithus tenuis zones were recognized in the oceanic sediments of some Caribbean Sea DSDP sites, but the planktonic foraminifera there do not furnish any valid support for the solution of the problem. The Ellipsolithus macellus Zone characterizing the uppermost lower Paleocene was also not identified in the Inidan Ocean sediments. Because the nominal species is present in younger levels, its first occurrence could not be utilized here to define the bottom of this zone.

Some problem also remains concerning a detailed biostratigraphy of the upper Eocene. Chiasmolithus oamaruensis and Isthmolithus recurvus are absent from the typical low-latitude upper Eocene sediments of the Arabian Sea and Somali Basin. The simultaneous first occurrence elsewhere, i.e. at Site 253, of Sphenolithus pseudoradians and Isthmolithus recurvus moreover, prevents the identification of the Isthmolithus recurvus Zone s.s., which, based on planktonic foraminiferal evidence, might however be present. Numerous nannofloral events are indicated by Edwards and Perch-Nielsen (1975) for the high-latitude upper Eocene in the southern Southwest Pacific. However, the poor nannofloras of low diversification reported from the southern Indian Ocean Sites do not allow such high biostratigraphic resolution.

References

Boudreaux, J. E., Calcareous nannoplankton ranges, Deep Sea Drilling Project Leg 23, in Whitmarsh, R. B., Weser, O. E., Ross, D. A., et al., Initial Reports of the Deep Sea Drilling Project, Washington (U. S. Government Printing Office) 23, 1073-1090, 1974.

Bukry, D., Low-latitude coccolith biostratigraphic zonation, in Edgar, N. T., Saunders, J. B., et al., Initial Reports of the Deep Sea Drilling Project, Washington (U. S. Government Printing Office) 15, 685-703, 1973.

Bukry, D., Coccolith and silicoflagellate stratigraphy, Eastern Indian Ocean, Deep Sea Drilling Project Leg 22, in von der Borch, C. C., Sclater, J. G., et al., Initial Reports of the Deep Sea Drilling Project, Washington (U. S. Government Printing Office) 22, 601-608, 1974a.

Bukry, D., Coccolith stratigraphy, Arabian and Red Seas, Deep Sea Drilling Project Leg 23, in Whitmarsh, R. B., Weser, O. E., Ross, D. A., et al., in Initial Reports of the Deep Sea Drilling Project, Washington (U. S. Government Printing Office) 23, 1091-1093, 1974b.

Bukry, D., Coccolith zonation of cores from the Western Indian Ocean and the Gulf of Aden, Deep Sea Drilling Project Leg 24, in Fisher, R. L., Bunce, E. T., et al., Initial Reports of the Deep Sea Drilling Project, Washington (U. S. Government Printing Office) 24, 995-996, 1974c.

Bukry, D., Phytoplankton stratigraphy, offshore East Africa, Deep Sea Drilling Project Leg 25, in Simpson, E.S.W., Schlich, R., et al., Initial Reports of the Deep Sea Drilling Project, Washington (U. S. Government Printing Office) 25, 635-646, 1974d.

Bukry, D., Coccolith stratigraphy, offshore western Australia, Deep Sea Drilling Project Leg 27, in Veevers, J. J., Heirtzler, J. R., et al., Initial Reports of the Deep Sea Drilling Project, Washington (U. S. Government Printing Office) 27, 623-630, 1974e.

Bukry, D., Cretaceous and Paleogene coccolith stratigraphy, Deep Sea Drilling Project Leg 26, in Davies, T. A., Luyendyk, B. P., et al., Initial Reports of the Deep Sea Drilling Project, Washington (U. S. Government Printing Office) 26, 669-673, 1974f.

Bukry, D., Silicoflagellate and coccolith stratigraphy, Deep Dea Drilling Project, Leg 29, in Kennett, J. P., Houtz, R. E., et al., Initial Reports of the Deep Sea Drilling Project, Washington (U. S. Government Printing Office) 29, 845-858, 1975a.

Bukry, D., Coccolith and silicoflagellate stratigraphy near Antarctica, Deep Sea Drilling Project, Leg 28, in Hayes, D. E., Frakes, L. A., et al., Initial Reports of the Deep Sea Drilling Project, Washington (U. S. Government Printing Office) 28, 709-723, 1975b.

Burns, D. A., Nannofossil biostratigraphy for Antarctic sediments, Leg 28, Deep Sea Drilling Project, in Hayes, D. E., Frakes, L. A., et al., Initial Reports of the Deep Sea Drilling Project, Washington (U. S. Government Printing Office) 28, 589-598, 1975.

Davies, T. A., B. P. Luyendyk, et al., Initial Reports of the Deep Sea Drilling Project, Washington (U. S. Government Printing Office) 26, 1129 p., 1974.

Edwards, A. R., A calcareous nannoplankton zonation of the New Zealand Paleogene, Proc. II Plankt. Conf., Roma 1970, 1, 381-419, 1971.

Edwards, A. R., and K. Perch-Nielsen, Calcareous nannofossils from the southern Southwest Pacific, Deep-Sea Drilling Project, Leg 29, in Kennett, J. P., Houtz, R. E., et al., Initial Reports of the Deep Sea Drilling Project, Washington (U. S. Government Printing Office) 29, 469-539, 1975.

El-Dawoody, A. S. and M. G. Barakat, Nannobiostratigraphy of the Upper

Cretaceous-Paleocene contact in Duui range, Quseir District, Egypt, Riv. Ital. Paleont., 79, 103-124, 1973.

Fisher, R. L., E. T. Bunce et al., Initial Reports of the Deep Sea Drilling Project, Washington (U. S. Government Printing Office) 24, 1183 p., 1974.

Gartner, S., Nannofossil biostratigraphy, Leg 22, Deep Sea Drilling Project, in von der Borch, C. C., Sclater, J. G., et al., Initial Reports of the Deep Sea Drilling Project, Washington (U. S. Government Printing Office) 22, 577-599, 1974.

Haq, U.Z.B., Calcareous nannoplankton from the Lower Eocene of the Zinda Pir, District Dera Ghazi Khan, West Pakistan, Geol. Bull. Panjab Univ., 6, 55-83, 1967.

Haq, U.Z.B., Paleogene calcareous nannoflora Part I: The Paleocene of West-Central Persia and the Upper Paleocene-Eocene of West Pakistan, Stockholm Contrib. Geol., 25, 1-56, 1971.

Kennett, J. P., R. E. Houtz, et al., Initial Reports of the Deep Sea Drilling Project, Washington (U. S. Government Printing Office) 29, 1197 p., 1975.

Martini, E., Standard Tertiary and Quaternary nannoplankton zonation, Proc. II plankt. Conf., Roma 1970, 2, 739-785, 1971.

Müller, C., Calcareous nannoplankton, Leg 25 (Western Indian Ocean) in Simpson, E.W.W., Schlich, R., et al., Initial Reports of the Deep Sea Drilling Project, Washington (U. S. Government Printing Office) 25, 579-633, 1974.

Narashimhan, T., Coccolithophorids and related nannoplanktons from the Cretaceous-Tertiary sequence of Khasi Hills, Assam, J. Geol. Soc. India, 4, 109-11, 1963.

Pant, S. G., and V. D. Mamgain, Fossil nannoplanktons from the Indian sub-continent, Rec. Geol. Survey India, 97, 108-128, 1969.

Perch-Nielsen, K., and Ch, Pomerol, Nannoplankton calcaire a la limite Crétacé-Tertiaire dans le Bassin de Majunga (Madagascar), C. R. Acad. Science Paris, 276, serie D, 2435-2438, 1973.

Proto Decima, F., Leg 27 calcareous nannoplankton in Veevers, J. J., Heirtzler, J. R., et al., Initial Reports of the Deep Sea Drilling Project, Washington (U. S. Government Printing Office) 27, 589-621, 1974.

Roth, P. H., Calcareous nannofossils from the north-western Indian Ocean, Leg 24, Deep Sea Drilling Project, in Fisher, R. L., Bunce, E. T., et al., Initial Reports of the Deep Sea Drilling Project, Washington (U. S. Government Printing Office) 24, 969-994, 1974.

Shafik, S., and H. Stradner, Nannofossils from the Eastern Desert, Egypt with reference to Maastrichtian Nannofossils from the USSR, Jb. Geol. B. A., Sonderband, 17, 69-104, 1971.

Simpson, E.S.W., R. Schlich, et al., Initial Reports of the Deep Sea Drilling Project, Washington (U. S. Government Printing Office) 25, 884 p., 1974.

Thierstein, R., Calcareous nannoplankton - Leg 26, Deep Sea Drilling Project, in Davies, T. A., Luyendyk, B. P., et al., Initial Reports of the Deep Sea Drilling Project, Washington (U. S. Government Printing Office) 26, 619-668, 1974.

Veevers, J. J., Heirtzler, J. R. et al., Initial Reports of the Deep Sea Drilling Project, Washington (U. S. Government Printing Office) 27, 106 p., 1974.

von der Borch, C. C., Sclater, J. G. et al., Initial Reports of the Deep Sea Drilling Project, Washington (U. S. Government Printing Office) 22, 890 p., 1974.

Whitmarsh, R. B., Weser, O. E., Ross, D. A. et al., Initial Reports of the Deep Sea Drilling Project, Washington (U. S. Government Printing Office) 23, 1180 p., 1974.

CHAPTER 16. DISTRIBUTION OF CALCAREOUS NANNOPLANKTON IN OLIGOCENE
TO HOLOCENE SEDIMENTS OF THE RED SEA AND THE INDIAN
OCEAN REFLECTING PALEOENVIRONMENT

Carla Müller

Geolog.-Paläontolog. Institut, Johann-Wolfgang Goethe-Universität,
Senckenberg-Anlage 32-34, 6000 Frankfurt a/M, West Germany

Abstract. Results of the nannoplankton investigations of Oligocene
to Holocene sediments from 52 sites, drilled in the Indian Ocean
(Fig.1, Fig.2) during the Deep Sea Drilling Project (Leg 22 to Leg 29)
are presented.
Nannofossils are the main constituent of the pelagic sediments,
especially the Cenozoic chalks and oozes, covering the ridges and
deep basins of the Indian Ocean. Their distribution and preservation
in the sediments recovered from the different sites is closely related
to depositional conditions reflecting the paleoenvironment which in
turn is strongly connected with the tectonic evolution of the Indian
Ocean.
Nannoplankton assemblages of low- and mid-latitude regions of the
Indian Ocean are similar. A decrease in frequency of some groups
(discoasters, ceratoliths, sphenoliths) can be observed in mid-latitudes. However, most index fossils occur also there. This makes it
possible to use the same zonation in nearly all parts of the Indian
Ocean. Assemblages of high-latitudes or areas influenced by the
Circumpolar Current (Crozet Basin) are distinguished by low species
diversity, thus a modified zonation must be used.

I. Environment and Nannoplankton Distribution in the Sediments

Bottom water circulation and the productivity of the surface water
masses are the main factors controlling the nature and rate of accumulation of sediments and the distribution of microfossils. The distribution of nannoplankton is primarily determined by temperature,
salinity, nutrients and currents. Secondary factors affecting the
nannoplankton distribution in the sediments are:
 1) dissolution
 2) dilution, either due to a high amount of terrigenous material
(turbidites, glacial sediments) or a high increase of siliceous
microfossils (in zones of high productivity due to upwelling,
regions influenced by the Circumpolar Current
 3) winnowing, caused by bottom current activities after deposition,
or by suspension transport by currents before sedimentation
 4) reworking and redeposition of nannofossils which were originally
deposited in another environment due to turbidity currents, slumping.

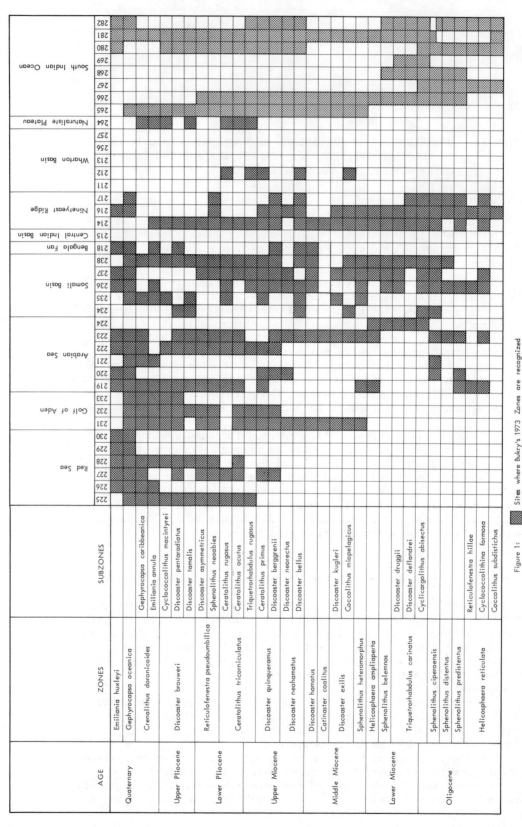

Figure 1: Sites where Bukry's 1973 Zones are recognized
South Indian Ocean Sites where Bukry's zonal scheme is not clearly applicable. Here only larger stratigraphic intervals can be recognized

OLIGOCENE TO HOLOCENE CALCAREOUS NANNOPLANKTON 373

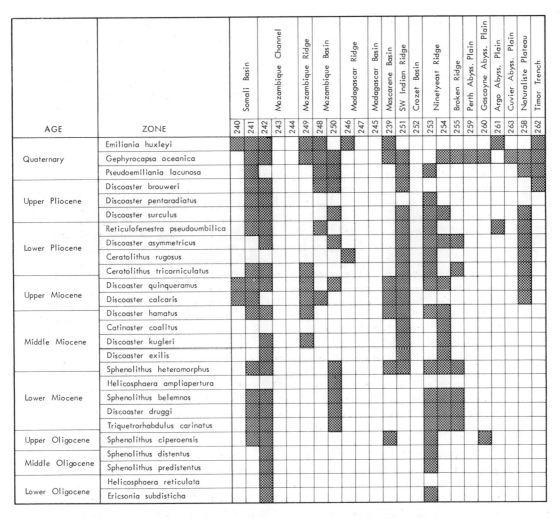

Figure 2: Sites where the Zones of the standard zonation by Martini, 1971, are recognized

Preservation

Dissolution may be the most important factor controlling the distribution of nannoplankton in the sediments. It changes the amount of nannoplankton in the sediments as well as the assemblages. Selective dissolution of coccoliths can change the temperature aspect of an assemblage (McIntyre and McIntyre, 1971, Berger (1973), Schneidermann, (1973) and results in an enrichment of dissolution-resistant species (discoasters, certain coccoliths), for further discussions see Bukry (1973), Roth (1973), Schneidermann (1973).

Sediments deposited below the calcium carbonate compensation depth (CCD) are barren of nannoplankton. Sometimes it is possible to find some strongly etched species in micro-nodules in which they have been protected (Müller, 1974). Dissolution rate of the nannoplankton depends on surface productivity, water depth, temperature, water chemistry and Ca CO_3 content of the sediments. Nannofossils in coarse-grained sediments in general are more affected by dissolution and are often destroyed mechanically compared to fine-grained sediments. This may be due to the lower content of nannofossils and the better circulation of aggressive water. On the other hand, nannoplankton will show an overgrowth by secondary calcite which affects discoasters and ceratoliths more than coccoliths. The degree of overgrowth depends upon the lithology, and is independent from the age of the sediments. Nannofossils are considerably more overgrown in pure oozes, chalks and limestones than in marls and clayey nannofossil oozes.

II. Nannoplankton Distribution

Red Sea

During Leg 23 six holes were drilled in the Red Sea (Sites 225 to 230). The oldest fossiliferous sediments recovered are of late Miocene age (Discoaster quinqueramus Zone, NN 11) intercalated or overlying the evaporitic sequence. This sequence was encountered at Sites 225, 227, 228, and is suggested at Site 230 based on the high salinity of the pore water. The overlying sediments consist of an almost complete late Miocene to Holocene section rich in calcareous nannofossils.

Dark colored beds with a high amount of organic carbon are described from the Upper Pliocene to Lower Pleistocene at Sites 225, 227 and 228 (Whitmarsh, Weser, Ross et al., 1974). Such horizons were also observed in the Upper Pleistocene from the middle and northern Red Sea (Müller, 1977). The excellently preserved nannoplankton assemblages of these layers are marked by a high species diversity. The dark layers are underlain by horizons distinguished by a low diversified assemblage with heavily overgrown nannofossils. It is presumed thatt his impoverishment is caused by a restricted influx of water from the Indian Ocean which caused an increase in salinity of the Red Sea. These horizons are well developed in the middle and northern part of the Red Sea at the glacial/postclacial boundary, but are missing in the southernmost part of the Red Sea and the Gulf of Aden.

The preservation of the Quaternary nannofossils is excellent in the southern Red Sea, while they are more or less overgrown in the other parts of the Red Sea (Müller, 1977). The assemblages consist of the typical species described by McIntyre and Bé (1967) for the tropical zone with Geophyrocapsa oceanica, Umbilicosphaera mirabilis. Cyclococcolithus leptoporus, Umbellosphaera irregularis, U. tenuis, Syracosphaera pulchra, Helicosphaera carteri, Thoracosphaera heimi, T. albatrosiana, Oolithotus fragilis and Rhabdosphaera clavigera. Holococcoliths are abundant in the Upper Pleistocene sediments of the Red Sea (Müller, 1977).

The assemblages of the southernmost part of the Red Sea differ from those above and are comparable with the assemblages of the Gulf of Aden, consisting of species typical for the subtropical and transitional zones (McIntyre and Be, 1967).

The sulfide and heavy-metal-rich mud of the Quaternary recovered at Site 226 from the Atlantic "Hot-Brine" area contain only a few strongly dissolved nannofossils. Generally nannoplankton is missing within these sediments. The nannoplankton assemblages of the Pliocene and Upper Miocene are dominated by discoasters, Sphenolithus abies and Reticulofenestra pseudoumbilica. It is possible to determine all stratigraphic zones established for this interval (Boudreaux, 1974).

Western Indian Ocean

Gulf of Aden. Three holes were drilled in the Gulf of Aden during Leg 24 (Sites 231, 232, and 233).

Site 231 is located at the south coast of the Gulf of Aden. A sequence of 566.5 meters of hemipelagic mud with some intercalated sand layers was recovered, and ranges in age from upper middle Miocene to Holocene. It is underlain by basalt including some thin layers of middle Miocene (Sphenolithus heteromorphus Zone, NN 5). The nannofossils are slightly etched in samples, but they are heavily overgrown in the sediments which are included in the basalt.

Site 232 is situated near the western edge of the Alula-Fartak Trench, and Site 233 is on the back side of the eastern flank of this trench. At Site 232, a 434 meter section of Pliocene and upper Miocene was recovered. The nannoplankton assemblages are rich and diversified. Nannofossils of the Upper Miocene are slightly etched and reworked species were found throughout the larger part of the section indiccating displaced material.

At Site 233 a lower Pliocene to Quaternary sequence was recovered. Reworked species are less frequent compared to Site 232.

Sediments from these sites consist mainly of nannofossil chalk with several layers of sand or broken detrital material and lower Pliocene volcanic ash layers. The sediments are abundant in well-preserved nannofossils which are slightly etched especially in the sand layers. Discoasters are common in the Neogene section. Ceratoliths are poorly represented. Coccolithus pelagicus increasing in the upper Pliocene, disappears abruptly at the Plio-Pleistocene boundary (Roth, 1974). Discoaster intercalcaris, a cool-water species was only observed in the Gulf of Aden (Roth, 1974). Investigations of Upper Quaternary sediments from the Gulf of Aden (Müller, 1977) have shown that the nannoplankton assemblage belongs to the subtropical and transitional zones (McIntyre and Be, 1967). The species diversity is lower than in the Red Sea except for the southernmost part of the Red Sea. Furthermore, the forms are generally smaller and more fragile than in the Red Sea, indicating lower water temperatures. Some species, which are abundant in the Quaternary of the Red Sea and the tropical Indian Ocean, are missing in the Gulf of Aden. Only a few specimens of Syracosphaera pulchra were found. These "cooler" nannoplankton assemblages may in part be the result of selective dissolution of the warm-water species.

Arabian Sea. Six holes were drilled during Leg 23 in the Arabian Sea. Sites 222, 223, and 224 are located in the area of the Owen Fracture Zone and Owen Ridge. Site 222 was located on the western margin of the Indus Cone just east of the Owen Fracture Zone in 3546 m of water. The sediments include a 1300 meter thick complete sequence of green and gray clay ranging in age from late Miocene (Discoaster quinqueramus Zone, NN 11) to Quaternary. Autochthonous nannoplankton is rare and restricted mainly to some intercalated green detrital clay layers. The amount of nannofossils is strongly diluted by the high amount of terrigenous sediments. Reworked Cretaceous species are present within the upper Pliocene and were found throughout most of the underlying sediments. The Pleistocene sediments are abundant in well-preserved nannoplankton. Although the Pliocene and Upper Miocene silty clays contain only few nannoplankton due to dilution, all typical index fossils were found. The nannoplankton have only moderate preservation in this section.

Site 223 is located near the edge of an abyssal plain which laps onto the west slope of the Owen Ridge. The influence of the Indus Cone is absent at this site due to the protection afforded by the Owen Ridge (no reworked Cretaceous species were found at Site 223). The sedimentary sequence includes the upper Paleocene to Quaternary, with unconformities within the Eocene and lower Miocene. The sediments are rich in nannoplankton. Helicosphaera ampliaperta was observed only at Site 223 and at the other sites of the Arabian Sea, but it seems to be missing in all other regions of the Indian Ocean, except south of Australia, Bukry (1974) described Discoaster formosus from the Sphenolithus heteromorphus Zone at Site 223, which also seems to have a restricted distribution.

Site 224 is located near the crest of the Owen Ridge in water 2500 m deep. A sedimentary sequence of middle lower Eocene to Pleistocene was recovered with nannoplankton-rich sediments from lower Miocene to Pleistocene. Few nannofossils are present in the Oligocene due to dissolution. This indicates a rise of the lysocline during the Oligocene. Sites 223 and 224 are located in a region where upwelling occurs during the present time. The presence of siliceous microfossils since middle-upper Miocene indicates that upwelling was initiated at this time.

Site 220 was drilled in the Arabian Sea on a broad platform just west of the Chagos-Laccadive Ridge. Site 219 was placed closest to the ridge crest. At both sites a complete section was recovered ranging in age from early Eocene to Pleistocene. The sediments are abundant in well-preserved nannofossils typical for the tropical and subtropical zones.

Site 221 is located at the southern end of the Arabian Abyssal Plain in 4650 m of water. 128 meters of Pleistocene to upper Pliocene sediments were penetrated consisting of clayey silts with graded silts, sand beds and local nannofossil ooze. This material seems to come from the Indian shelf (nannoplankton ooze). It is rich in nannoplankton with some reworked species. These turbidites are underlain by 42 meters of brown clay deposited below CCD, including the stratigraphic interval of Miocene/Pliocene. These sediments are underlain by nannoplankton ooze of late Oligocene to middle Eocene age.

Somali Basin. Seven holes were drilled to investigate the history of the Somali Basin. Site 234 was located near the western margin of the basin in 4738 m of water, Site 241 was placed at the East African continental rise in 4054 m of water.

The upper Pliocene and Pleistocene at Site 234 is missing. The presence of manganese-nodules at the top of the section indicates slow or no deposition in an oxidizing environment. The few nannofossils observed at the top of the sequence give an early Pliocene age (Reticulofenestra pseudoumbilica Zone). Nannoplankton are strongly etched and only dissolution-resistant species are common in some horizons. This site was close to the CCD throughout the time of deposition.

The sediments recovered at Site 241 consist of 470 meters of late Oligocene to Pleistocene nannofossil oozes with minor foraminiferal, diatom and radiolarian ooze, silty clay and silts. The Quaternary sediments are rich in well-preserved nannofossils. Some reworked species of Pliocene and Miocene age were found. Alternation of sediments without nannoplankton or few corroded species (lower and middle Miocene, lower Pliocene) connected with an increase of terrigenous sediments, and those containing more and better preserved nannofossils indicates that these fluctuations are caused by dilution of the nannoplankton. In the upper Oligocene at Site 241 a variety of Sphenolithus ciperoensis is present with very large spines (Müller, 1974). This form was also found in a piston core taken northeast of Mauritius (OSIRIS cruise of the "Marion Dufresne" in 1973, unpublished), containing uppermost Oligocene to lowermost Miocene sediments.

Olausson et al. (1971) described four piston cores from the Somali Basin, and illustrated a close connection between the Somali Current movements and the foraminifera concentration in late Pleistocene sediments. They conclude that the foraminiferal ooze is related to the presence of high nutrient levels in the overlying water masses. However, high foraminifera contents may indicate either the presence of high productivity zones or the winnowing of nannoplankton by currents. The effect of current activities can be recognized by the high amount of reworked nannofossils (10% of the assemblages) in these sediments. The relationship between an increase of Coccolithus pelagicus and the increase of reworked species in the same level indicates that the presence of C. pelagicus seems not to be related to a cold period, but only to an increase of reworking by which also more specimens of C. pelagicus are reworked. Similar observations of the relationship between an increase of C. pelagicus and reworked species was made in the Mediterranean Sea (Müller, unpublished) and in the Gulf of Aden (Müller, 1977).

Site 235 was located in the northwest Somali Basin at the westernmost edge of the abyssal plain that onlaps the eastern flank of the Chain Ridge. Basement was encountered at 651.5 meters, being overlain by about 52 meters of brown clay with only traces of nannoplankton. Dissolution-resistant species indicate a late Cretaceous age for the sediment inclusions in the basalt. The overlying 503 meters of sediment of middle Miocene to Pleistocene age are common to abundant in fairly well-preserved nannoplankton. Some reworked Eocene/Oligocene species were found in the sand layers. An influx of displaced shallower sediments can be assumed because the preservation of the calcareous material is better than would normally be expected in a water depth greater than 5000 meters.

At Site 240, located in the central abyssal plain of the West Somali Basin, 157 meters of terrigenous/pelagic silty nannofossil or radiolarian ooze and clay and silty sands were recovered, ranging in age from upper Miocene to Quaternary. They are underlain by lower Eocene nannoplankton ooze and basement. The Quaternary sediments contain common well-preserved nannoplankton with some Plio/Miocene reworked species. The Pliocene and Miocene section contain very few poorly preserved nannofossils. The assemblages consist mainly of discoasters, while the coccoliths are dissolved.

Site 236 is located 270 km northeast of the Seychelles Bank in the outermost foothills southwest of the Carlsberg Ridge. Basement, encountered at 306 m, was overlain by a sequence of upper Paleocene to Pleistocene. The lower Oligocene to Quaternary sediments (253 m) consist mainly of nannoplankton chalk and ooze. They were abundant in slightly etched nannofossils with a high species diversity. The Quaternary and Pliocene sediments are marked by a high amount of reworked species of Eocene, Oligocene and Miocene age. The proportion of nannoplankton decreases in the middle Miocene and upper lower Miocene. The Oligocene sediments are again abundant in nannofossils. Site 237, is located in 1630 meters of water on the saddle joining the Seychelles Bank to the Saya de Malha Bank. Sediments of Oligocene to Quaternary age are present by an uninterrupted sequence of nannofossils chalk and ooze with a rich, diversified, well-preserved nannoplankton assemblage. The sediments were deposited above CCD. Productivity increases during the upper Miocene as indicated by the abundance of foraminifera and siliceous microfossils, causing the high sedimentation rate of 11.3 m/m.y., as compared to about 2.7 m/m.y. from Oligocene to middle Miocene.

Site 238, at the extreme northeast end of the Argo Fracture Zone, has a complete section from middle Oligocene (Sphenolithus predistentus Zone, NP 23) to Holocene. The section consists of nannofossil ooze with a well-preserved highly diversified assemblage. In the Quaternary a fairly high amount of reworked Oligocene and Pliocene species was observed. In the upper Miocene, lower Miocene, and Oligocene reworked species were found. The high sedimentation rate since upper Miocene points to a considerable organic productivity (average sedimentation rate 25 m/m.y.).

From Sites 236, 237 and 238, Roth (1974) described the more frequent occurrence of Braarudosphaera bigelowi (a near-shore species) in the lower and middle Oligocene. This species was also frequently found in the same stratigraphic interval in the Atlantic. There is still no adequate explanation for the high production of braarudospaerids. Special oceanographical conditions seem to be most likely.

Mascarene Basin. Site 239 is located in the western Mascarene Basin. The upper section of detrital silt and clay and silty sands contains a mixed Pleistocene/Pliocene nannoplankton assemblage. The sediments are abundant in well-preserved nannofossils. Specimens of the genus Scyphosphaera are common in the Upper Miocene. Dissolution decreases the amount of nannofossils in the Upper Oligocene to middle Miocene. Most of the coccoliths are dissolved in the middle Miocene, while the discoasters are enriched in the brown clay and silty clay. Only few specimens were found in the Upper Oligocene.

Mozambique Channel. Site 242 is located at the eastern flank of the Davie Ridge. The complete section, ranging in age from late Eocene to Holocene, consists of pelagic sediments of nannoplankton ooze and nannoplankton chalk. The sediments are abundant in well-preserved or slightly overgrown nannoplankton. Species of the genus Scyphosphaera are common in the upper Miocene and lower Pliocene. In the lowermost Miocene (Triquetrorhabdulus carinatus Zone, NN 1) a large number of sphenoliths are found. Sphenolithus capricornutus and Sphenolithus delphix seem to be restricted to this zone, and probably to the uppermost Oligocene.

Sites 243 and 244 were located about 5 km apart in the Zambesi Canyon in the southeastern Mozambique Channel. Only few centimeters of coarse sands and gravels were recovered. Nannoplankton is missing due to the absence of the fine fraction, with only a few Pleistocene species found in some clay pebbles. It can be assumed that the fine fraction was not deposited at all in this region due to the high current velocity, but was transported in suspension further south to the Mozambique Basin.

Mozambique Basin. Two holes were drilled in the Mozambique Basin. Site 248 and Site 250. The lithologic sequences at both sites are strongly influenced by the Zambesi Canyon.

Site 248 is located in the northwestern part of this basin, east of a very steep slope of the Mozambique Ridge in a water depth of 4994 m. Nannofossils, common in the Quaternary, are less frequent in the Pliocene and Miocene due to dissolution. In the sandy silts and coarse quartz sands of the lower-middle Miocene nannofossils are missing. Dilution of the nannoplankton, as well as the deposition below CCD, may be the reason for the absence of nannofossils in this part of the section.

Site 250 was located in the southeastern Mozambique Basin in a water depth of 5119 meters. An extremely thick (400 meters) Pleistocene-Pliocene section of detrital silty clay was drilled. The Pleistocene sediments contain common to abundant, slightly to strongly etched nannofossils, decreasing within the Pliocene. This indicates the deposition close to the CCD. Nannofossils are almost missing in deposits older than early Pliocene.

Mozambique Ridge. Site 249 was located close to the crest of the Mozambique Ridge in a water depth of 2088 m. The section includes only two meters of Pleistocene, underlain by 176 meters of upper and middle Miocene foraminifera-rich nannoplankton ooze. The upper Miocene (Discoaster quinqueramus Zone, NN 11) becomes extremely thick (about 67 meters). This may indicate either a high productivity of the overlying water masses or a redeposition of nannoplankton transported in suspension from other areas. The sediments are abundant in well-preserved nannoplankton with a high species diversity. The high foraminifera content in the middle Miocene sediments may indicate high productivity as well as a concentration of foraminifera by winnowing out of the nannoplankton. The condensed Pliocene-Pleistocene section at Site 249 and the mixed foraminifera assemblages described by Saito and Fray (1964) from two piston cores from the Mozambique Ridge make winnowing of the fine fraction more likely caused by the initiation of the Agulhas current at late Middle Miocene time (Leclaire, 1974). The middle Miocene is underlain by Cretaceous.

Madagascar Ridge. Sites 246 and 247 were drilled on the Madagascar Ridge in a water depth of 1030 m. The section includes 125 meters of unconsolidated foraminiferal ooze ranging in age from early Miocene to Holocene. This sequence overlies a calcareous sand (lower Eocene). Nannofossils are common in the Quaternary and Pliocene foraminiferal ooze, but are rare in the Miocene sediments. The decreases may be due to winnowing by bottom currents, but more likely it is an effect due to washing during drilling.

Madagascar Basin and Crozet Basin. The Madagascar Basin and the Crozet Basin are separated by the southwest branch of the mid-Indian Ocean Ridge. Site 245 was located in the southern Madagascar Basin about 200 miles northwest of the Southwest Indian Ridge axis in 4857 m of water. The uppermost 63 meters of the section consist of brown clay barren of nannoplankton deposited below the CCD. Only Core 1, CC contains a nannoplankton assemblage of middle Miocene (Discoaster hamatus Zone, NN 9), however these sediments were probably brought in by slumping. Strongly etched nannofossils were found in micronodules at 54 m to 81 m at Site 245A, indicating a late Eocene age.

Significant differences exist between the western and eastern Crozet Basin concerning the nannoplankton assemblages and their distribution due to different influence of circumpolar water masses.

Site 252 is located in the Crozet Basin southeast of the Southwest Indian Ridge in a water depth of 5032 m. The section consists of radiolarian clay and diatom-bearing radiolarian clay of middle-late Miocene to Quaternary. This sequence shows the effects of the Circumpolar Current. Nannoplankton is missing due to dissolution, only few nannofossils were found in some marly patches on manganese nodules (Pleistocene).

Investigations of piston cores taken during the cruise of the "Marion Dufresne" (OSIRIS cruise in 1973) in the southern Crozet Basin (42°15'S, 51°20'E, 43°03'S, 51°19'E, 43°49'S, 51°19'E, 44°59'S, 53°17'E, 46°01'S, 55°01'E, 49°26'S, 61°45'E, 45°17'S, 72°50'E, 43°28'S, 73°07'E) have shown a nannoplankton assemblage mostly restricted to some layers throughout the Quaternary, which sometimes consist only of nannoplankton and foraminifera, with Emiliania huxleyi, Coccolithus pelagicus, Gephyrocapsa sp. (small), and a few specimens of Cyclococcolithus leptoporus, Ceratolithus cristatus, Syracosphaera pulchra and Helicosphaera carteri. C. pelagicus, a cold water species becomes frequent in some horizons. Generally, there are only few nannoplankton probably diluted by the high onset of siliceous microfossils as well as affected by dissolution. The species are smaller and more fragile than in the tropical and subtropical regions of the Indian Ocean.

In the Pliocene section (determined by Radiolaria), nannofossils are present only in some layers of nannofossil ooze indicating an influx of warmer water. The assemblages consist of Reticulofenestra gelida (Geitzenauer and Huddlestun, 1973) nov. comb. (investigations by the author with the SEM have shown that the central area is covered by a grill).Coccolithus pelagicus and a few specimens of Cyclococcolithus leptoporus. Discoasters are missing or occur only sporadically. The same assemblage of the Pliocene-Pleistocene is described by Geitzenauer and Huddlestun (1973) from the subantarctic Pacific Ocean. The scarcity and low diversity of nannofossils, mainly in the Pliocene, are due to the low water temperatures which restrict the presence of most of the nannoplankton species, adapted to warmer water. Sediments of the eastern Crozet Basin in general are rich in nannoplankton (piston cores taken during the cruise of the "Marion Dufresne", OSIRIS cruise in 1973, 42°01'S, 67°36'E, 34°33'S, 63°32'E, 31°22'S, 61°52'E). The nannoplankton assemblages are of a lower diversity which at least partially is caused by selective dissolution of more fragile species. Cyclococcolithus leptoporus, Coccolithus pelagicus, Gephyrocapsa caribbeanica, Helicosphaera carteri and Pseudoemiliania lacunosa are the most important species of the Pleistocene. A relative enrichment of Cyclococcolithus leptoporus in the Quaternary or of Cyclococcolithus macintyrei in the Pliocene sediments can be observed very often caused by selective dissolution.

Discoasters are frequent in the Neogene sediments of the eastern Crozet Basin. Depending from water depth the coccoliths are more or less etched while the discoasters are well preserved. Ceratoliths are rare, species of the genera Scyphosphaera and Thoracosphaera are missing.

Southwest Indian Ocean Ridge. Site 251 was located about 180 km north of the crest of the Southwest Indian Ocean Ridge in water depth of 3489 m. The complete 489 meter sequence consists of upper lower Miocene (Sphenolithus heteromorphus Zone, NN 5) to Holocene nannoplankton ooze and chalk underlain by basalt. The section is abundant in well-preserved or slightly etched nannofossils. They show signs of overgrowth in the lowermost 50 meters of the profile. The assemblages can be compared with those described from the subtropical zone. However, the scarcity of discoasters and sphenoliths in the Neogene section documents transitional paleotemperatures.

The Eastern Indian Ocean

Central Bengal Fan. Site 218 was located in the Central Bengal Fan in 3749 m of water depth. The turbidite sequence ranges in age from Pleistocene to Middle Miocene (Catinaster coalitus Zone, NN 8). Content of nannoplankton is diluted by the high amount of terrigenous material. The preservation is poor in the coarser grained sediments, the nannofossils are often broken. The preservation is better in the intercalated pelagic oozes and fine detrital sediments. Reworked species of Cretaceous and Tertiary were observed throughout the section except in the nannofossil oozes.

Central Indian Basin and Wharton Basin. To compare the biostratigraphic sequences of the Central Indian Ocean and the Wharton Basin, two holes were drilled on either side of the Ninetyeast Ridge.

Site 215 was located in the Central Indian Basin west of the Ninetyeast Ridge in 5309 m of water; Site 213 was drilled east of the ridge in water depth of 5601 m. The Quaternary and Neogene sediments at both sites consist of radiolarian-diatom ooze barren of nannoplankton, being deposited below the CCD. An additional eight holes were drilled in the Wharton Basin of the eastern Indian Ocean.

Site 211 lies in the northeastern part of the Indian Ocean approximately 130 miles south of the axis of the Java Trench in 5525 meters of water. The stratigraphic section of 435 meters is comprised of a siliceous ooze, ranging in age from late Pliocene to Quaternary, deposited below the CCD. It is underlain by a thick lower Miocene to Pliocene unit of turbidite silts, sands and radiolarian clay. These sediments are barren of nannofossils, as well as the underlying brown clay of unknown age.

The Tertiary sediments encountered at Sites 212, 259, 260, 261 and 263 consist mainly of well-stratified calcareous turbidites unconformably overlying the Cretaceous. The calcareous oozes were originally deposited on broad marginal plateaus and were brought into the basins by slumping and/or turbidite activities indicated by in general well-preserved mixed assemblages. This can be observed at Site 259 located near the eastern side of the Perth Abyssal Plain at the foot of the continental slope, at Site 263, near the eastern edge of the Cuvier Abyssal Plain on the foot of the continental rise, at Site 260, in the Gascoyne Abyssal Plain at the foot of the Exmouth Plateau, and at Site 261, in the northeast Argo Abyssal Plain at the foot of the Scott Plateau.

Site 212 is located at the deepest part of the Wharton Basin. Upper Cretaceous to Pliocene sediments are 521 meters thick. The sequence consists of 4 units of calcareous ooze and chalk separated by brown clay without nannofossils. The brown clay is regarded as normal deep-basin sediment, whereas the calcareous material is postulated to be exotic and to have been transported to the basin by slumping or bottom current activities from areas above the CCD. The calcareous units (Pliocene, lower-middle Miocene) contain mixed nannoplankton assemblages.

The detrital brown clays encountered at Site 256 (southern Wharton Basin, water depth 5361 m) and Site 257 (southeast Wharton Basin, water depth 5278 m) are barren of nannoplankton.

Timor Trench. Hole 262 was drilled near the axis of the western part of the Timor Trough. The sediments consist of 414 meters of pelagic ooze and shallow-water sediments (upper Pliocene, _Discoaster brouweri_ Zone, NN 18 and Quaternary). They are rich in well-preserved nannoplankton with reworked Cretaceous, Paleocene and Neogene species. The determination of the Plio/Pleistocene boundary is difficult because _Discoaster brouweri_ is present rarely in all samples due to reworking. _Cyclococcolithus macintyrei_ and _Ceratolithus rugosus_, which can be used for the determination of the boundary are missing.

Ninetyeast Ridge. Five sites were located on the Ninetyeast Ridge from North to South: Site 217 on the eastern flank, Site 216 and Site 214 on the crest, Site 253 on the western flank and Site 254 at the southern end. Comparison of the stratigraphic sequence for all sites shows that the age of the basal sediments overlying the basalt becomes progressively older northwards, from Miocene at Site 254 to older than Campanian at Site 217. A sinking of the Ridge is indicated at all sites by a lower unit deposited in a shallow water environment (indicated by shallow water nannoplankton species) which is overlain by a unit of pelagic sediments. The change from the shallow water environment to an oceanic environment is diachronous. It is younger in the south and becomes older northwards. The preservation of the nannofossils is good in the Quaternary down to the Upper Miocene. Generally, below this stratigraphic level, the nannoplankton is heavily overgrown. At Sites 253 and 254, only Quaternary nannofossils are well-preserved.

The lower Oligocene shallow-water deposits at Site 253 are distinguished by the presence of _Zygrhablithus bijugatus_. The presence of _Chiasmolithus oamaruensis_ in the lower and middle Oligocene at this Site indicates either lower water temperatures as compared with the Miocene which contains a tropical-subtropical nannoplankton assemblage with discoasters, ceratoliths, sphenoliths and scyphospheres, or a northward drift of this site by plate motions into regions of lower latitudes. Sediments at Site 254 are abundant in nannoplankton. Reworked Pliocene species are present in the Quaternary. The portion of reworked species increases in the foraminiferal-nannofossil oozes of the Pliocene and Miocene, probably due to current activities. The presence of Braarudosphaera bigelowi in the early Miocene indicates nearshore or shallow-water environment.

Broken Plateau. Site 255 is on top of the Broken Plateau. The Quaternary and Neogene sequence includes 33 meters of a deep-water foraminiferal-nannofossil ooze. Transport and redeposition of nannofossils is indicated by the mixed assemblages. Nannofossils are abundant and overgrown. Pelagic sedimentation began in the lower Miocene in conjunction with the subsiding Broken Plateau.

Naturaliste Plateau. Two sites were located on the Naturaliste Plateau. Site 258 at the northern slope in a water depth of 2787 m, and Site 264 at the southern slope in a water depth of 2873 m. At Site 258 an upper Miocene to Holocene section (114 m) was recovered with a rich and well-preserved nannoplankton assemblage, overlying Cretaceous deposits. Transport by currents is indicated by reworked species.

At Site 264 the thin Neogene section (31 m) of upper Miocene to Holocene foraminiferal-nannofossil ooze with a rich and well-preserved assemblage overlies a Paleogene section. The nannoplankton assemblages of the Neogene from Sites 258 and 264 show a cool subtropical environment. Discoasters and ceratoliths are rare.

The South Indian Ocean. Sediments recovered in the South Indian Ocean are distinguished by the presence of only few nannofossils in assemblages of extremely low diversity. Three holes were drilled (Sites 265, 266 and 267) at different distances south from the Southeast Indian Ridge. The sedimentary sequences encountered at these sites are similar. The upper part consists of predominantly diatom ooze passing gradually downwards into nannofossil ooze and chalk. The diatom ooze is devoid of nannoplankton except for a few scattered horizons with poor assemblages consisting mainly of long ranging species which are not very useful for the zonation. The proportion of nannofossils is diluted by the high onset of siliceous microfossils which are responsible for the high sedimentation rates in the Quaternary (107 m/m.y. at Site 265, 40 m/m.y. at Site 266). On the other hand living conditions are unfavourable for nannoplankton (cold water). The forms are in general small and fragile. Nannofossils become more frequent in the lower Pliocene to middle Miocene sediments with Reticulofenestra pseudoumbilica, Coccolithus pelagicus, and probably Reticulofenestra gelida, Discoasters are missing. These assemblages correspond to those described from the Neogene of the southern Crozet Basin. The lower Miocene assemblages observed at Site 266 are distinguished by the presence of discoasters, mainly of the Discoaster deflandrei-group. The lithologic change is diachronous (lower Miocene at Site 267, and upper middle Miocene at Sites 265 and 266) indicating a cooling extending progressively further north during the Miocene.

The sedimentary sequences recovered at Site 268 on the lower continental rise of the Knox Coast of Antarctica and at Site 269 on the southeast edge of the South Indian Abyssal Plain consist mainly of silts, clays and diatom oozes. The oldest sediments recovered at Site 268 are probably of late Oligocene age and at Site 269 of middle Oligocene age.

Nannofossils are present sporadically in some isolated horizons in the upper Oligocene lower Miocene. The assemblages are poorly preserved and of a low species diversity. They consist of Chiasmolithus altus, Discoaster defandrei, Spenolithus moriformis, Dictyococcites dictyodus and Zygrhablithus bijugatus.

Site 282 was located west of Tasmania, Sites 281 and 280 were located south of Tasmania. At Site 282 one meter of a Pleistocene siliceous foraminiferal-nannofossil ooze was recovered. Nannofossils are common, moderately to poorly preserved. This sequence is underlain by siliceous nannofossil ooze and detrital clay of early Pliocene or late Miocene age. Nannoplankton are abundant but poorly preserved. The assemblage is of low species diversity. The underlying Oligocene consists of silty diatom ooze, with rare, poorly preserved nannofossils.

At Site 281 a sequence of 112 meters of early Miocene to Holocene nannofossil-foraminifera ooze was encountered overlying a section of early Miocene glauconitic sands. A major unconformity eliminates almost the Oligocene. Nannofossils are abundant throughout the sequence, being moderately preserved. Nannoplankton assemblages observed at Sites 280 to 282 are given in Table 5.

At Site 280 a thin Pleistocene section of nannofossil and foraminifera oozes was recovered underlain disconformably by 7 meters of late Miocene nannofossil ooze. This sequence is in turn disconformably underlain by lower Miocene, followed by the Oligocene section. Nannoplankton is common to abundant throughout the section being moderately to poorly preserved. The assemblages are comparable with those described from Site 281. Additional Coronocyclus nitescens, Helicosphaera ampliaperta, H. euphratis, H. obliqua, H. recta and Triquetrorhabdulus carinatus were found in the upper Oligocene and lower Miocene. Recognized nannoplankton zones at all Sites of Leg 22 to Leg 29 are summarized in Tables 1 and 2.

III. Discussion of Nannoplankton Biostratigraphy

Low- and Mid-Latitude Areas

Results of the Indian Ocean Legs have shown that the low- and mid-latitude nannoplankton assemblages in the Indian Ocean and in the Red Sea consist largely of the same species. A decrease in number of discoasters, spenoliths, ceratoliths and scyphospheres can be observed in the mid-latitude regions, as compared to those of tropical areas, as for example, the Gulf of Aden (Roth, 1974), Naturaliste Plateau (Sites 258 and 264), Broken Ridge (Site 255). However, there is a great difference if compared with southern high-latitude assemblages. Nannoplankton species of low- and mid-latitudes and high latitudes of the Indian Ocean are listed below.

The presence of the same index fossils in low- and mid-latitudes makes it possible to use the same biostratigraphic zonation in nearly all regions of the Indian Ocean. However, in the high-latitudes, the standard zonation (Martini, 1971) cannot be used due to the absence of most of the index fossils (discoasters, sphenoliths and ceratoliths).

Based on the zonation of Bramlette and Wilcoxon (1967) the boundaries of the Oligocene to lowermost middle Miocene are mainly determined by the first or last occurrence of certain species of the genus Sphenolithus. These are generally common in the low- and mid-latitude areas of the Indian Ocean, while they are missing in high-latitudes. They are relatively resistent against dissolution, thus being found even in sediments in which nannoplankton is affected by dissolution.

Sphenolithus predistentus is present in the lower Oligocene (NP 21 and

Table 1. Occurrence of calcareous nannofossil species in the Red Sea and Indian Ocean as reported from Legs 22 to 29

OCCURRENCE OF CALCAREOUS NANNOFOSSIL SPECIES IN THE READ SEA AND INDIAN OCEAN AS REPORTED FROM LEG 22 TO 29		
* reported from Red Sea only ** reported from Gulf of Aden and southern Red Sea only *** reported from Arabian Sea; Ninetyeast Ridge **** reported from Arabian Sea; south of Australia + additional species, not reported in Leg volumes	Low and mid- Latitude	High Latitude
Braarudosphaera bigelowii (Gran & Braarud) Deflandre 1947	x	
Catinaster calyculus Martini & Bramlette 1963	x	
Catinaster coalitus Martini & Bramlette 1963	x	
Catinaster mexicanus Bukry 1971	x	
Ceratolithus armatus Müller 1974	x	
Ceratolithus cristatus Kamptner 1954	x	x
Ceratolithus primus Bukry & Percival 1971	x	
Ceratolithus rugosus Bukry & Bramlette 1968	x	
Ceratolithus tricorniculatus Gartner 1967	x	
Chiasmolithus altus Bukry & Percival 1971		x
Coccolithus? abisectus Müller 1970	x	x
Coccolithus doronicoides Black & Barnes 1963	x	x
Coccolithus pelagicus (Wallich) Schiller 1930	x	x
Corisphaera? obscura Müller 1975*+	x	
Coronocyclus nitescens (Kamptner) Bramlette & Wilcoxon 1967	x	x
Craspedolithus declivus Kamptner 1963+	x	
Cyclococcolithus floridanus (Roth & Hay) Hay 1970	x	x
Cyclococcolithus leptoporus (Murray & Blackman) Kamptner 1954	x	x
Cyclococcolithus macintyrei Bukry & Bramlette 1969	x	x
Cyclococcolithus rotula Kamptner 1956	x	
Cyclococcolithus subtilis Müller 1975**+		
Dictyococcites dictyodus (Deflandre & Fert) Martini 1969	x	x
Discoaster altus Müller 1974	x	
Discoaster asymmetricus Gartner 1969	x	
Discoaster berggrenii Bukry 1971	x	
Discoaster bollii Martini & Bramlette 1963	x	
Discoaster brouweri Tan Sin Hok 1927	x	
Discoaster calcaris Gartner 1967	x	
Discoaster challengeri Bramlette & Riedel 1954	x	
Discoaster deflandrei Bramlette & Riedel 1954	x	x
Discoaster druggi Bramlette & Wilcoxon 1967	x	x
Discoaster exilis Martini & Bramlette 1963	x	
Discoaster formosus Martini & Worsley 1971***	x	
Discoaster hamatus Martini & Bramlette 1963	x	
Discoaster kugleri Martini & Bramlette 1963	x	
Discoaster neohamatus Bukry 1969	x	
Discoaster nephados Hay 1967	x	x
Discoaster pentaradiatus Tan Sin Hok 1927	x	
Discoaster perplexus Bramlette & Riedel 1954	x	
Discoaster pseudovariabilis Martini & Worsley 1971	x	
Discoaster quinqueramus Gartner 1969	x	
Discoaster saundersii Hay 1967	x	x
Discoaster surculus Martini & Bramlette 1963	x	
Discoaster tamalis Kamptner 1967	x	
Discoaster tanii nodifer Bramlette & Riedel 1954	x	
Discoaster trinidadensis Hay 1967	x	x
Discoaster variabilis Martini & Bramlette 1963	x	
Discolithina japonica Takayama 1967	x	
Discosphaera tubifera (Murray & Blackman) Kamptner 1944	x	
Emiliania huxleyi (Lohmann) Hay & Mohler 1967	x	x

Table 2. Same as Table 1

	Low and mid-Latitude	High Latitude
Florisphaera profunda Okada & Honjo 1973+	x	
Gephyrocapsa aperta Kamptner 1963+	x	
Gephyrocapsa caribbeanica Boudreaux & Hay 1967	x	
Gephyrocapsa oceanica Kamptner 1943	x	x
Gephyrocapsa protohuxleyi McIntyre 1970	x	x
Helicosphaera ampliaperta Bramlette & Wilcoxon 1967****	x	x
Helicosphaera carteri (Wallich) Kamptner 1954	x	x
Helicosphaera compacta Bramlette & Wilcoxon 1967	x	
Helicosphaera euphratis Haq 1966	x	x
Helicosphaera obliquipons Bramlette & Wilcoxon 1967	x	
Helicosphaera perch-nielseniae (Haq) Jafar & Martini 1975	x	
Helicosphaera recta Haq 1966	x	x
Helicosphaera sellii (Bukry & Bramlette) Jafar & Martini 1975	x	x
Heliolithus crassus Müller 1975****+	x	
Isthmolithus recurvus Deflandre 1954	x	x
Colithotus fragilis (Lohmann) Martini & Müller 1972	x	
Orthorhabdus serratus Bramlette & Wilcoxon 1967	x	
Pontosphaera sracusana Lohmann 1902	x	
Pseudoemiliania lacunosa (Kamptner) Gartner 1969	x	x
Reticulofenestra gelida (Geitzenauer & Huddlestun) nov. comb.+		x
Reticulofenestra pseudoumbilica (Gartner) Gartner 1969	x	x
Reticulofenestras umbilica (Levin) Martini & Ritzkowski 1968	x	x
Rhabdosphaera clavigera Murray & Blackman 1898	x	
Rhabdosphaera procera Martini 1969	x	
Rhabdosphaera stylifera Lohmann 1902	x	
Scapholithus fossilis Deflandre 1954	x	
Scyphosphaera amphora Deflandre 1942	x	
Scyphosphaera apsteini Lohmann 1902	x	
Scyphosphaera campanula Deflandre 1942	x	
Scyphosphaera conica Kamptner 1955	x	
Scyphosphaera cylindrica Kamptner 1955	x	
Scyphosphaera deflandrei Müller 1974	x	
Scyphosphaera globulata Bukry & Percival 1971	x	
Scyphosphaera intermedia Deflandre 1942	x	
Scyphosphaera kamptneri Müller 1974	x	
Scyphosphaera pulcherima Deflandre 1942	x	
Scyphosphaera recurvata Deflandre 1942	x	
Sphenolithus abies Deflandre 1954	x	
Sphenolithus belemnos Bramlette & Wilcoxon 1967	x	
Sphenolithus capricornutus Bukry & Percival 1971	x	
Sphenolithus ciperoensis Bramlette & Wilcoxon 1967	x	
Spenolithus delphix Bukry 1973	x	
Sphenolithus distentus (Martini) Bramlette & Wilcoxon 1967	x	
Sphenolithus heteromorphus Deflandre 1953	x	
Sphenolithus moriformis (Brönnimann & Stradner) Bramlette & Wilcoxon 1967	x	x
Sphenolithus predistentus Bramlette & Wilcoxon 1967	x	
Syracosphaera pulchra Lohmann 1902	x	x
Umbellosphaera irregularis Paasche 1955	x	
Umbellosphaera tenuis (Kamptner) Markali & Paasche 1955	x	
Umbilicosphaera jafarii Müller 1974+	x	
Umbilicosphaera mirabilis Lohmann 1902	x	x
Thoracosphaera albatrosiana Kamptner 1963	x	
Thoracosphaera heimii (Lohmann Kamptner 1941	x	
Thorosphaera flabellata Halldal & Markali 1955++	x	
Triquetrorhabdulus carinatus Martini 1965	x	
Triquetrorhabdulus rugosus Bramlette & Wilcoxon 1967	x	
Triquetrorhabdulus striatus Müller 1974	x	
Zygrhablithus bijugatus (Deflandre) Deflandre 1959	x	

NP 22), last specimens were found in the upper Oligocene (NP 25). Sphenolithus distentus has its first occurrence in the middle Oligocene at the base of the Sphenolithus predistentus Zone (NP 23) and disappears at the top of the Sphenolithus distentus Zone (NP 24), lower part of the upper Oligocene. Sphenolithus ciperoensis ranges from the base of the Sphenolithus distentus Zone (NP 24) probably to the lowermost part of the Triquetrorhabdulus carinatus Zone (NN 1) of the upper Oligocene. Some specimens with very thick spines were observed at Site 241 in the Somali Basin (Müller, 1974) and in a piston core from the Madagascar Ridge (OSIRIS cruise of the "Marion Dufresne" in 1973, unpublished). In the lowermost Miocene (NN 1) additional Sphenolithus capricornutus and S. delphix are present. S. capricornutus was described by Bukry and Percival (1971) from the upper Oligocene to lower Miocene of the Atlantic. S. delphix is described by Bukry (1973) as upper Oligocene and lowermost Miocene from the Atlantic and Pacific. Both species were found in the lowermost Miocene at Site 242 (Davie Ridge) and in a piston core taken south of the Rodriquez Fracture Zone (OSIRIS cruise of the "Marion Dufresne" in 1973). The are also observed by Martini (1976) in the Triquetrorhabdulus carinatus Zone (NN 1) at Site 317B in the Pacific and seem to be of stratigraphic value. Sphenolithus capricornutus is reported by Gartner (1974) from the uppermost Oligocene and lowermost Miocene of Site 216 (Ninetyeast Ridge) and by Roth (1974) from the uppermost Oligocene of Site 238 (NE end of the Argo Fracture Zone). The first specimens of Sphenolithus belemnos are observed in the uppermost Triquetrorhabdulus carinatus Zone (NN 1) to the top of the Sphenolithus belemnos Zone (NN 3) where it becomes frequent. Sphenolithus heteromorphus is typical for the uppermost part of the Sphenolithus belemnos Zone (NN 3) to the top of the Sphenolithus heteromorphus Zone (NN 5). Sphenolithus moriformis is present throughout the Oligocene and lower Miocene. It is also present in high-latitudes. The first occurrence of Sphenolithus abies is not well known. This species becomes abundant in the middle Miocene and throughout the lower Pliocene.

The nannoplankton zones of the middle Miocene and Pliocene are generally determined by discoasters. The extinction of Discoaster brouweri marks the Pliocene/Pleistocene boundary in tropical and subtropical areas. This species is missing or is present only sporadically in high-latitudes, where it disappears in the lower Pliocene. This indicates that the extinction of this species is climatically controlled. Discoasters are abundant in the Neogene of low and mid-latitude areas of the Indian Ocean. They become less frequent in regions of temperate environment (Gulf of Aden, Broken Plateau and Naturaliste Plateau), and are missing almost completely in the sediments recovered from the high-latitudes. They are the most dissolution-resistant group of nannoplankton, and therefore often enriched in the sediments due to selective dissolution of the more fragile coccoliths.

Discoasters are rare in the Oligocene and of a low species diversity caused by cold water temperatures. In the lower Oligocene only Discoaster tani nodifer was observed. It has its last occurrence in the Ericsonia subdisticha Zone (NP 21). The nannoplankton assemblage of the upper Oligocene and lower Miocene are characterized by discoasters of the Discoaster deflandrei-group. They were also found in cool subtropical assemblages of this stratigraphic interval in the southern Indian Ocean (Leg 28 and 29). Discoaster druggii and Discoaster hamatus are generally rare in the Indian Ocean. Discoaster formosus, described by Martini and Worsley (1971) from the Sphenolithus heteromorphus Zone (NN 5) from the Pacific, was found only in the Arabian Sea (Bukry, 1974) and in a few specimens by Thierstein (1974) at Site 254 (Ninetyeast Ridge) and by Edwards and Perch-Nielsen (1975) at Site 282 (west of

Tasmania). Discoaster bollii seems to be restricted to some parts of
the Indian Ocean. This species is only reported by Müller (1974) from
Sites 241, 242 (Somali Basin) and Site 249 (Mozambique Ridge) and by
Roth (1974) from Site 231 (Gulf of Aden). Discoaster altus described by
Müller (1974) from Site 242 (Davie Ridge) seems to be restricted to the
lower Pliocene (upper part of NN 13 to lower part of NN 15). This
species was also found in the Discoaster asymmetricus Zone (NN 14) of
three piston cores taken during OSIRIS cruise of the "Marion Dufresne"
in 1973 (Müller, unpublished) east of the Seychelles Islands, northeast
of Mauritius and southeast of Madagascar. It was also observed by
Hekel (1973), Leg 20 and by Martini (in press, Leg 33, Site 317B) in
the same stratigraphic interval from the tropical zone of the Pacific.
Probably this species is restricted to tropical areas.

Ceratoliths are first abundant in the upper Miocene (Discoaster quinqueramus Zone, NN 11) in low- and mid-latitude areas. They are less
frequent in temperate environment and are missing in high-latitudes.
This group shows great variations and can be heavily overgrown by
secondary calcite thus making identification very difficult. The first
occurrence and the extinction of Ceratolithus tricorniculatus Gartner
and Ceratolithus primus Bukry and Percival at the same stratigraphic
level makes it likely that they belong to the same species. Ceratolithus primus is regarded only as a variety of Ceratolithus tricorniculatus. Ceratolithus rugosus occurs in the lower Pliocene (base of
NN 13) to lowermost Pleistocene (NN 19). Ceratolithus cristatus ranges
from the upper Pliocene (probably upper NN 16) to Recent. Species of
the genus Scyphosphaera are abundant and of high diversity in the upper
Miocene and lower Pliocene (Site 242 Davie Ridge, Site 239 Mascarene
Basin, Site 249 Mozambique Ridge, 250 Mozambique Basin, Site 253
Ninetyeast Ridge). They disappear almost completely in the late Pliocene. Most of them have a long range and are not very useful for age
determination. Species of the genus Helicosphaera are of stratigraphic
value. Helicosphaera reticulata is typical for the middle Eocene to
lower Oligocene and has its last occurrence in the Helicosphaera reticulata Zone (NP 22). Helicosphaera recta has its first occurrence in the
middle Oligocene (base of the Sphenolithus distentus Zone, NP 24). Its
extinction determined the upper boundary of the Sphenolithus ciperoensis Zone (NP 25). Helicosphaera ampliaperta, characteristic for part
of lower Miocene, has been found only in the Arabian Sea (Boudreaux,
1974) and at Site 282 south of Tasmania (Edwards and Perch-Nielsen,
1975). It seems to be missing in all other areas of the Indian Ocean.
Therefore, the lower boundary of the Sphenolithus heteromorphus Zone
(NN 5) is determined by the first occurrence of Discoaster exilis
(Martini and Worsley, 1971). H. ampliaperta seems to prefer cooler
water temperatures. It is also missing in the tropical region of the
Pacific Ocean (Martini and Worsley, 1971). Conversely Helicosphaera
recta is present in the tropical area of the Indian Ocean, while it was
not observed in the tropical region of the Pacific Ocean (Martini and
Worsley, 1971). Helicosphaera carteri has its first occurrence in the
lower Miocene (upper part of the Triquetrorhabdulus carinatus Zone, NN 1).
It is common throughout Neogene sediments.

Species of the genera Coccolithus and Cyclococcolithus are abundant in
the Oligocene to Holocene. However, often their amount in a nannoplankton assemblage is diminished by selective dissolution. Coccolithus
pelagicus is present as two varieties. The smaller one without the central bridge dominates in the Oligocene through lower Pliocene, while the
large variety with the central bridge becomes frequent in the uppermost
Pliocene and lower Pleistocene (Müller,1974 and results of the OSIRIS
cruise of the "Marion Dufresne" in 1973, unpublished). Roth (1974)

described an increase of Coccolithus pelagicus in the upper Pliocene, and an abrupt disappearance of this species at the Pliocene/Pleistocene boundary in the Gulf of Aden. Coccolithus pelagicus is missing in the Quaternary sediments of the tropical zone of the Indian Ocean. This species is present south of 32°S throughout the Quaternary (investigation results of piston cores taken during the OSIRIS cruise in 1973, unpublished).

Coccolithus abisectus appears in the middle Oligocene at the boundary of the Sphenolithus predistentus/Sphenolithus distentus Zone (NP 23/NP 24), and ranges into the lowermost middle Miocene (NN 5/NN 6). The extinction of this species was used by Roth (1974) to define the Oligocene/Miocene boundary (top of the Reticulofenestra abisecta Zone). However, this is not possible since a higher stratigraphic range for this species is known.

Cyclococcolithus rotula is one of the dominate species of the middle to upper Miocene nannoplankton assemblages. The extinction of Cyclococcolith macintyrei lies at the Pliocene/Pleistocene boundary, and can be used to determine this boundary in areas where discoasters are missing or are present only sporadically.

Geophyrocapsa oceanica and Gephyrocapsa aperta are abundant in Quaternary sediments. These species are most frequent in the tropical and subtropical areas. Probably a small variety of Geophyrocapsa oceanica is present in sediments from high-latitudes. Thierstein (1974) mentions the first occurrence of Geophyrocapsa sp. below the extinction of Discoaster brouweri. Reticulofenestra pseudoumbilica ranges at least from the Sphenolithus heteromorphus Zone, (NN 5) to the top of the lower Pliocene (Reticulofenestra pseudoumbilica Zone, NN 15). Species of the genus Reticulofenestra, which are abundant in many Oligocene land sections of Europe and New Zealand, are not present in the deep sea sediments of the Indian Ocean. They seem to prefer shallow-water or nearshore environment, as well as a cooler environment. Nannoplankton species indicating shallow water such as Lanthernithus minutus, Zygrhablithus bijugatus, Braarudosphaera bigelowi and species of the genus Discolithina, are largely missing in the deep-sea sediments of the Indian Ocean. Braarudosphaera bigelowi was reported by Roth (1974) at Site 237 and at Site 238 from the Mascarene Plateau and the northeast end of the Argo Fracture Zone in the lower-middle Oligocene, and by Thierstein (1974) from Sites 253 and 254 from the Ninetyeast Ridge. Zygrhablithus bijugatus was observed by Gartner (1974) in the Oligocene at Site 214 from the Ninetyeast Ridge, and by Edwards and Perch-Nielsen (1975) at Site 282 south of Australia. Lanthernithus minutus is reported by Thierstein (1974) from the lower Oligocene at Site 253 (Ninetyeast Ridge).

In Table 3 stratigraphic events are summarized which help distinguish the nannoplankton zones of the standard zonation (Martini, 1971) in the Indian Ocean.

Nannoplankton Zonations

Zonations used by the different authors for age determinations of Oligocene to Holocene sediments of the Indian Ocean are those of Martini and Worsley (1970), Martini (1970, 1971) Bukry (1973) and Roth (1974). Their correlation is given in Table 4. Some notable discrepancies are discussed below. The Eocene/Oligocene boundary is determined by the extinction of Discoaster saipanensis.

The Helicopontosphaera reticulata Zone of Bukry (1973) includes the Ericsonia? subdisticha Zone (NP 21) and the Helicopontosphaera reticulata Zone (NP 22) used by Martini (1971) and Roth (1974). The top of

Table 3. Quaternary to Lower Oligocene calcareous nannoplankton zones (from Martini, 1971) and biostratigraphic events

AGE			NANNOPLANKTON ZONES		FIRST OCCURRENCE AT BASE (B) OR WITHIN ZONE	LAST OCCURRENCE AT TOP (T) OR WITHIN ZONE
QUATERNARY			NN 21	Emiliania huxleyi	Emiliania huxleyi B	
			NN 20	Gephyrocapsa oceanica		
			NN 19	Pseudoemiliania lacunosa	Gephyrocapsa oceanica	Pseudoemiliania lacunosa (T) Ceratolithus rugosus
PLIOCENE	UPPER		NN 18	Discoaster brouweri		Cyclococcolithus macintyrei Discoaster brouweri (T)
			NN 17	Discoaster pentaradiatus		Discoaster pentaradiatus (T)
			NN 16	Discoaster surculus	Ceratolithus cristatus Pseudoemiliania lacunosa	Discoaster surculus (T) Discoaster asymmetricus Discoaster variabilis Discoaster tamalis
	LOWER		NN 15	Reticulofenestra pseudoumbilica		Reticulofenestra pseudoumbilica (T) Discoaster altus Sphenolithus abies
			NN 14	Discoaster asymmetricus	Discoaster tamalis Discoaster asymmetricus (B)	Ceratolithus tricorniculatus (T)
			NN 13	Ceratolithus rugosus	Discoaster altus Ceratolithus rugosus (B)	
			NN 12	Ceratolithus tricorniculatus		Discoaster challengeri Triquetrorhabdulus rugosus
MIOCENE	UPPER		NN 11	Discoaster quinqueramus	Ceratolithus primus Ceratolithus tricorniculatus Discoaster surculus Discoaster quinqueramus (B)	Discoaster quinqueramus (T) Discoaster calcaris Discoaster neohamatus Discoaster pseudovariabilis
			NN 10	Discoaster calcaris		Catinaster calyculus Discoaster bollii
			NN 9	Discoaster hamatus	Discoaster bollii Discoaster neohamatus Discoaster pentaradiatus Discoaster hamatus (B)	Discoaster hamatus (T) Discoaster exilis Catinaster coalitus
	MIDDLE		NN 8	Catinaster coalitus	Discoaster calcaris Catinaster calyculus Catinaster coalitus (B)	Discoaster kugleri
			NN 7	Discoaster kugleri	Discoaster kugleri (B)	Coronocyclus nitescens
			NN 6	Discoaster exilis	Cyclococcolithus macintyrei Discoaster brouweri Discoaster challengeri Discoaster pseudovariabilis Triquetrorhabdulus rugosus	Cyclococcolithus floridanus Coccolithus? abisectus
			NN 5	Sphenolithus heteromorphus	Sphenolithus abies Reticulofenestra pseudoumbilica Discoaster formosus Discoaster exilis	Discoaster formosus Sphenolithus heteromorphus (T) Discoaster druggi
	UPPER		NN 4	Helicopontosphaera ampliaperta	Discoaster variabilis	Helicosphaera ampliaperta (T)
			NN 3	Sphenolithus belemnos	Sphenolithus heteromorphus	Sphenolithus belemnos (T)
			NN 2	Discoaster druggi	Discoaster druggi Cyclococcolithus leptoporus Sphenolithus belemnos Discoaster druggii (B)	Triquetrorhabdulus carinatus (T)
			NN 1	Triquetrorhabdulus carinatus	Discoaster trinidadensis Sphenolithus cf. belemnos Helicosphaera carteri	Dictyococcites dictyodus Sphenolithus delphix Sphenolithus capricornutus
OLIGOCENE	UPPER		NP 25	Sphenolithus ciperoensis	Sphenolithus delphix Sphenolithus capricornutus	Helicosphaera recta (T) Helicosphaera compacta Zygrhablithus bijugatus
	MIDDLE		NP 24	Sphenolithus distentus	Helicosphaera recta (B) Sphenolithus ciperoensis Coccolithus abisectus	Sphenolithus distentus (T)
			NP 23	Sphenolithus predistentus	Sphenolithus distentus	Sphenolithus pseudoradians
	LOWER		NP 22	Helicosphaera reticulata		Reticulofenestra umbilica (T) Helicosphaera reticulata Istmolithus rectus
			NP 21	Ericsonia? subdisticha		Cyclococcolithus formosus (T) Discoaster tani nodifer

Table 4. Correlations of calcareous nannoplankton zones used by different authors

Series or Subseries	MARTINI 1971		Boundary species	BUKRY 1973 zones	subzones	Boundary species	ROTH 1974	Boundary species
Quaternary	NN 21	Emiliania huxleyi	▼ E. huxleyi*	Emiliania huxleyi		▼ E. huxleyi*	Emiliania huxleyi	▼ E. huxleyi*
Quaternary	NN 20	Gephyrocapsa oceanica	▼ P. lacunosa*	Gephyrocapsa oceanica		▼ G. oceanica*	Gephyrocapsa oceanica	▼ G. oceanica*
Quaternary	NN 19	Pseudoemiliania lacunosa		Gephyrocapsa doronicoides	Gephyrocapsa caribbeanica	▼ G. caribbeanica*	Pseudoemiliania lacunosa	▼ G. caribbeanica*
Upper Pliocene	NN 18	Discoaster brouweri	▼ D. brouweri*		Emiliania annula	▼ G. brouweri*	Cyclococcolithus macintyrei	▼ D. brouweri*
Upper Pliocene	NN 17	Discoaster pentaradiatus	▼ D. pentaradiatus*	Discoaster brouweri	Cyclococcolithus macintyrei	▼ D. pentaradiatus*, D. surculus*	Discoaster pentaradiatus	▼ D. pentaradiatus*
Upper Pliocene	NN 16	Discoaster surculus	▼ D. surculus*		Discoaster pentaradiatus	▼ D. tamalis*	Discoaster tamalis	▼ D. tamalis*
Lower Pliocene	NN 15	Reticulofenestra pseudoumbilica	▼ R. pseudoumbilica*	Reticulofenestra pseudoumbilica	Discoaster tamalis	▼ R. pseudoumbilica	Reticulofenestra pseudoumbilica	▼ R. pseudoumbilica
Lower Pliocene	NN 14	Discoaster asymmetricus	▼ C. tricorniculatus*		Discoaster asymmetricus	▼ D. asymmetricus A*		▼ C. tricorniculatus*
Lower Pliocene	NN 13	Ceratolithus rugosus	▼ D. asymmetricus*	Ceratolithus tricorniculatus	Sphenolithus neoabies	▼ C. primus*, C. tricorniculatus*	Ceratolithus rugosus	▼ C. primus*
Lower Pliocene	NN 12	Ceratolithus tricorniculatus	▼ C. rugosus*		Ceratolithus rugosus	▼ C. rugosus*, C. acutus*	Ceratolithus acutus	▼ C. rugosus*, C. acutus*
Upper Miocene	NN 11	Discoaster quinqueramus	▼ D. quinqueramus*	Discoaster quinqueramus	Ceratolithus acutus	▼ C. acutus*, T. rugosus*	Ceratolithus tricorniculatus	▼ D. quinqueramus*
Upper Miocene	NN 11				Triquetrorhabdulus rugosus	▼ C. quinqueramus*	Ceratolithus primus	▼ C. primus*
Upper Miocene	NN 10	Discoaster calcaris	▼ D. quinqueramus*		Ceratolithus primus	▼ C. primus*	Discoaster berggrenii	▼ D. berggrenii*
Upper Miocene	NN 10			-Discoaster neohamatus	Discoaster berggrenii	▼ D. berggrenii*, D. neorectus*	Discoaster neohamatus	▼ D. neorectus*
Upper Miocene	NN 9	Discoaster hamatus	▼ D. hamatus*		Discoaster neorectus	▼ D. neorectus*	Discoaster bellus	▼ D. hamatus*
Upper Miocene	NN 9			Discoaster hamatus	Discoaster bellus	▼ D. hamatus*	Discoaster hamatus	▼ D. hamatus*
Middle Miocene	NN 8	Catinaster coalitus	▼ D. hamatus*	Catinaster coalitus			Catinaster coalitus	▼ C. coalitus*
Middle Miocene	NN 7	Discoaster kugleri	▼ C. coalitus*	Discoaster exilis	Discoaster kugleri	▼ C. coalitus*, D. kugleri*	Discoaster kugleri	▼ D. kugleri*
Middle Miocene	NN 6	Discoaster exilis	▼ D. kugleri*		Coccolithus miopelagicus	▼ D. kugleri*	Discoaster exilis	▼ S. heteromorphus*
Middle Miocene	NN 5	Sphenolithus heteromorphus	▼ S. heteromorphus*	Sphenolithus heteromorphus		▼ S. heteromorphus*	Sphenolithus heteromorphus	
Lower Miocene	NN 4	Helicopontosphaera ampliaperta	▼ H. ampliaperta*	Helicopontosphaera ampliaperta		▼ D. deflandrei A*, H. ampliaperta*	Helicopontosphaera ampliaperta	▼ H. ampliaperta
Lower Miocene	NN 3	Sphenolithus belemnos	▼ S. belemnos*	Sphenolithus belemnos		▼ S. heteromorphus*, S. belemnos*	Sphenolithus belemnos	▼ S. heteromorphus*
Lower Miocene	NN 2	Discoaster druggii	▼ T. carinatus*		Discoaster druggii	▼ S. belemnos*	Discoaster druggii	▼ T. carinatus*
Lower Miocene	NN 1	Triquetrorhabdulus carinatus	▼ D. druggi*	Triquetrorhabdulus carinatus	Discoaster deflandrei	▼ D. druggii, O. serratus*	Triquetrorhabdulus carinatus	▼ D. druggii
Upper Oligocene	NP 25	Sphenolithus ciperoensis	▼ H. recta*	Sphenolithus ciperoensis	Cyclicargolithus abisectus	▼ C. abisectus A*	Reticulofenestra abisectus	▼ R. abisectus
Upper Oligocene	NP 25					▼ S. ciperoensis*, D. bisectus*	Sphenolithus ciperoensis	▼ S. ciperoensis
Middle Oligocene	NP 24	Sphenolithus distentus	▼ S. distentus*	Sphenolithus distentus		▼ S. ciperoensis*	Sphenolithus distentus	▼ S. distentus
Middle Oligocene	NP 23	Sphenolithus predistentus	▼ S. ciperoensis*	Sphenolithus predistentus		▼ S. distentus*	Sphenolithus predistentus	▼ S. ciperoensis*
Lower Oligocene	NP 22	Helicopontosphaera reticulata	▼ R. umbilica*	Helicopontosphaera reticulata	Reticulofenestra hillae	▼ R. hildae, R. umbilica*	Helicopontosphaera reticulata	▼ R. umbilica*
Lower Oligocene	NP 22		▼ C. formosus*		Cyclococcolithus formosa	▼ C. formosa*		▼ C. formosa*
Lower Oligocene	NP 21	Ericsonia ? subdisticha	▼ D. saipanensis*		Coccolithus subdistichus	▼ C. subdistichus A*	Ericsonia subdisticha	▼ D. saipanensis*
Lower Oligocene	NP 21					▼ D. saipanensis*		

this zone is defined by the extinction of Reticulofenestra umbilica.

The Sphenolithus predistentus to Sphenolithus ciperoensis Zone of Bukry (1973) cannot be correlated to the corresponding zones of this stratigraphic interval described in the standard zonation. Bukry (1973) defines the top of the Sphenolithus predistentus Zone by the first occurrence of Sphenolithus distentus and the top of the Sphenolithus distentus Zone by the first occurrence of Sphenolithus ciperoensis. However, Sphenolithus distentus is already present near the base of the Sphenolithus predistentus Zone (NP 23, Bramlette and Wilcoxon, 1967, Martini, 1971), and the first occurrence of Sphenolithus ciperoensis determines the base of the Sphenolithus distentus Zone (NP 24). According to this, the Sphenolithus distentus Zone and the Sphenolithus ciperoensis Zone of Bukry (1973) correspond to the stratigraphic interval of the Sphenolithus predistentus/Sphenolithus distentus Zone (NP 23/NP 24) of the standard zonation.

Neither the last occurrence of Coccolithus? abisectus nor the end of the abundance peak of this species can be used for the definition of the Reticulofenestra abisecta Zone of Roth (1974) or the Cyclicargolithus abisectus Subzone of Bukry (1973) of the uppermost part of the Oligocene. Coccolithus? abisectus was described by Müller (1974) from the Sphenolithus belemnos Zone (NN 3) and by Thierstein (1974) from the Discoaster exilis Zone (NN 6). The Sphenolithus belemnos Zone of Bukry (1973) is defined as interval from the first occurrence of Sphenolithus belemnos to the first occurrence of Sphenolithus heteromorphus. Thus it does not correspond to the Spenolithus belemnos Zone (NN 3), but it includes the Discoaster druggi Zone (NN 2), as well as the Sphenolithus belemnos Zone (NN 3). The Discoaster exilis Zone of Bukry (1973) includes the Discoaster exilis Zone (NN 6) and the Discoaster kugleri Zone (NN 7) of standard zonation. The Discoaster neohamatus Zone of Bukry (1973) can be correlated to the Discoaster calcaris Zone (NN 10). Discoaster neohamatus is more characteristic for this zone, while Discoaster calcaris occurs less frequent in the Indian Ocean. The top of this zone is marked by the first occurrence of Discoaster berggrenii, which is regarded only as a variety of Discoaster quinqueramus.

The Ceratolithus tricorniculatus Zone of Bukry (1973) corresponds to the Ceratolithus tricorniculatus Zone (NN 12), Ceratolithus rugosus Zone (NN 13) and Discoaster asymmetricus Zone (NN 14). The top of this zone is defined by the last occurrence of Ceratolithus primus and Ceratolithus tricorniculatus. The same event is used to determine the top of the Discoaster asymmetricus Zone (NN 14).

The Discoaster brouweri Zone of Bukry (1973) includes the Discoaster surculus Zone (NN 16), Discoaster pentaradiatus Zone (NN 17) and the Discoaster brouweri Zone (NN 18) of the standard zonation. The top of the Discoaster pentaradiatus Subzone of Bukry (1973) is marked by the extinction of Discoaster pentaradiatus and Discoaster surculus. In fact, it is very difficult to recognize the Discoaster pentaradiatus Zone (NN 17) because it is extremely thin and both species become extinct nearly at the same time. The Geophyrocapsa oceanica Zone of Bukry (1973) and Roth (1974) is defined by the first occurrence of Geophyrocapsa oceanica. This zone overlaps with the Pseudoemiliania lacunosa Zone (NN 19) of the Pleistocene. Gephyrocapsa oceanica is present in the lower part of the Pseudoemiliania lacunosa Zone (NN 19).

High-Latitude Area

A zonation for high-latitudes is given by Edwards and Perch-Nielsen (1975) Table 5.

The nannoplankton assemblages of the southern high-latitudes are dis-

Table 5. High latitude zonation (Edwards and Perch-Nielsen, 1975) and typical species of zones.

ADOPTED AGE	NANNOFOSSIL ZONES	IMPORTANT NANNOFOSSILS
Upper Pleistocene	Coccolithus pelagicus	Coccolithus pelagicus Cyclococcolithus leptoporus Emiliania huxleyi Helicosphaera carteri Gephyrocapsa oceanica Gephyrocapsa sp. (small)
Lower Pleistocene to Upper Pliocene	Pseudoemiliania lacunosa	Coccolithus pelagicus Cyclococcolithus leptoporus Cyclococcolithus macintyrei Helicosphaera carteri Helicosphaera sellii Gephyrocapsa oceanica (uppermost part) Gephyrocapsa sp. (small, uppermost part) Pseudoemiliania lacunosa
Lower Pliocene to late Middle Miocene	Reticulofenestra pseudoumbilica	Coccolithus pelagicus Cyclococcolithus leptoporus Cyclococcolithus macintyrei Helicosphaera carteri Helicosphaera sellii (uppermost part) Pseudoemiliania lacunosa Reticulofenestra pseudoumbilica Discoaster variabilis Sphenolithus moriformis (lower part) Sphenolithus neoabies (lower part) Coccolithus? abisectus (lower part)
early Middle Miocene	Cyclicargolithus neogammation	Coccolithus pelagicus Cyclococcolithus leptoporus Cyclococcolithus macintyrei Helicosphaera carteri Cyclococcolithus floridanus Coccolithus? abisectus Reticulofenestra pseudoumbilica Sphenolithus moriformis Sphenolithus neoabies
Lower Miocene	Discoaster deflandrei	Coccolithus pelagicus Cyclococcolithus macintyrei Cyclococcolithus floridanus Helicosphaera carteri Reticulofenestra pseudoumbilica Sphenolithus moriformis Sphenolithus neoabies Discoaster deflandrei Discoaster saundersi Helicosphaera ampliaperta Helicosphaera euphratis
Middle to Upper Oligocene	Reticulofenestra bisecta	Coccolithus pelagicus Coccolithus? abisectus Chiasmolithus altus Cyclococcolithus floridanus Discoaster deflandrei Discoaster saundersi Reticulofenestra bisecta Reticulofenestra placomorpha Sphenolithus moriformis Zygrhablithus bijugatus Helicosphaera euphratis Helicosphaera recta Chiasmolithus oamaruensis
Lower Oligocene	Reticulofenestra placomorpha	Coccolithus pelagicus Cyclococcolithus floridanus Chiasmolithus altus Reticulofenestra placomorpha Sphenolithus moriformis
Lower Oligocene	Blackites rectus	Coccolithus pelagicus Chiasmolithus altus Chiasmolithus oamaruensis Isthmolithus recurvus Reticulofenestra placomorpha Reticulofenestra bisecta Cyclococcolithus floridanus

tinguished by low species diversity of most long ranging species, which are less useful for a zonation. The assemblage of the Quaternary is given in Table 5. The lower Pliocene to middle Miocene assemblages include Reticulofenestra pseudoumbilica, R. gelida, Coccolithus pelagicus, and few specimens of Cyclococcolithus leptoporus and some discoasters. In the lower Miocene and upper Oligocene Coccolithus? abisectus, Chiasmolithus altus and a few discoasters of the Discoaster deflandrei-group were observed. The extinction of the discoasters is climatically controlled. Last discoasters were found in the lower Pliocene.

The abundance of various species of the genus Chiasmolithus in the Oligocene sediments indicates cold water temperatures during this time. Chiasmolithus altus occurs in the middle and late Oligocene sequences. This species seems to be restricted to cool water. It was also found in the North Atlantic (Perch-Nielsen, 1972) and in the Norwegian-Greenland Sea (Leg 38). Chiasmolithus oamaruensis normally disappears in the upper Eocene but ranges into Oligocene in certain areas (Martini, 1971). Thierstein (1974) reported Chiasmolithus oamaruensis from the Sphenolithus distentus Zone (NP 24) of the middle Oligocene at Site 253 located on the Ninetyeast Ridge. The presence of this species in the middle Oligocene is due to lower water temperatures and is also described by Edwards (1971) from New Zealand, and by Edwards and Perch-Nielsen (1975) from the late-middle Oligocene south of Australia. This species is present only sporadically in the tropical areas of the Indian Ocean.

Isthmolithus recurvus, present in the lower Oligocene, is a cold water form missing in the tropical region of the Indian Ocean. It is present south of 15°S (Sites 242 and 212) and becomes frequent further south. The extinction of this species lies at the top of the Helicopontosphaera reticulata Zone (NP 22). However, in northern regions (Germany, Netherlands) a few specimens were observed in the lowermost middle Oligocene, Sphenolithus predistentus Zone (NP 23) Müller (unpublished). Its sporadic occurrence in the middle Oligocene is also reported by Edwards and Perch-Nielsen (1975) from the southern Indian Ocean, indicating that it has a longer range in areas of cold water temperatures.

Conclusions

Nannoplankton assemblages as well as their distribution, proportion and preservation in the sediments of the Red Sea and the Indian Ocean reflect the paleoenvironment and paleocirculation patterns, which in turn are closely related to the evolution of the Indian Ocean since Oligocene time. Changes of paleocirculations to the present current system are the most important factors.

A close correlation probably exists between lower water temperatures during the Oligocene low production of calcareous microfossils, lower species diversity, and the presence of condensed series due to dissolution facies near or below CCD in the deep sea basins of the western (Sites 224, 234, 235, 239, 248, and 250) as well as of the eastern Indian Ocean (Sites 215, 213, 212, 211, 256, 257) during this time. Kennett et al. (1972) suggested that these conditions are caused by the influx of cold water masses into low latitude areas of the Indian Ocean due to the northward deflection of "circumpolar water" by Australia. The Oligocene unconformities, observed in the western Indian Ocean in different basins such as the Somali Basin (Sites 240, 241), Mascarene Basin (Site 239) and Mozambique Basin (Sites 250 and 248?) are related to strong bottom currents initiated during Oligocene time (Leclaire, 1974). Unconformities were brought to a close at the uppermost Oligocene or lowermost Miocene, probably due to the initiation of the present Circumpolar Current at the end of the Oligocene by the final separation of Australia from Antarctica south of the South Tasman Rise, or as assumed by Leclaire (1974) by the uplift of the southwest branch of the mid-Indian Ocean Ridge effecting as a barrier. This Oligocene unconformity does not exist in the Arabian Sea (Sites 219, 220, 221, 223 and 224) in the eastern Somali Basin (Sites 236, 237 and 238) and on the Ninetyeast Ridge (Sites 217, 216, 214 and 253) where complete sequences from Oligocene to Holocene sediments were encountered. However, in the eastern Indian Ocean this unconformity cannot be recognized with certainty due to the formation of condensed series in the deep sea basins. They may or may not include unconformities.

The early Miocene nannoplankton assemblages are still of lower diversity as compared to middle and late Miocene assemblages. Also the productivity is lower reflected by lower sedimentation rates.

Since middle Miocene an increase in tempo of diversification process of nannofossils, mainly discoasters, takes place, which comes to an end in the early Pliocene. This may be connected with an increase of temperatures. The highest species diversity occurs in the upper Miocene and in the lower part of the lower Pliocene with a high diversity of discoasters, ceratoliths and scyphospheres. The nannoplankton zones of the upper Miocene (Discoaster calcaris Zone, (NN 10) and Discoaster quinqueramus Zone (NN 11) are thick at all sites of the Indian Ocean. This indicates the high productivity during this time. The nannoplankton assemblages of this stratigraphic interval reflect tropical to subtropical environment, which seems to be constant during the upper Miocene and lower Pliocene. This high productivity of nannoplankton as well as the rich and high diversified assemblages since late-middle Miocene-upper Miocene are caused by a significant change of the paleoenvironment due to the initiation of the modern Indian Ocean Equatorial Current system when India moved north to the Equator (Leclaire, 1974). This is connected with the onset of upwelling induced by divergences which besides other changes produced high productivity resulting in sedimentation rates about three times as high as those during the preceding Oligocene and lower Miocene. This trend is observed at all sites of the tropical Indian Ocean: Gulf of Aden, (Sites 231, 232 and 233), Somali Basin (Sites 240, 241, 234, 235, 236, 237 and 238), Arabian Sea (Sites 219, 223 and 224).

Signs of upwelling since late Miocene time can be observed in the northern Wharton Basin (Sites 211 and 213), in the central Indian Ocean (Site 215), and on the Ninetyeast Ridge (Site 214). This observation probably indicates that the sites located in the eastern Indian Ocean moved northward into the zone of high productivity during middle Miocene due to plate motion, while no significant motion and change of site-positions in regard to paleolatitudes can be observed in the western Indian Ocean.

The high production of calcareous nannoplankton caused a lowering of the lysocline which results in a higher proportion of nannofossils in the sediments and a better preservation. This becomes more evident in the sediments deposited in the deep sea basins in which calcareous nannofossils are mostly affected by dissolution (Somali Basin Sites 241 and 235, Mascarene Basin Site 239).

An almost complete disappearance of the scyphospheres in the latest early Pliocene, the extinction of the genus Sphenolithus, as well as the extinction of at least four species of the genus Discoaster, may indicate a renewed decrease of water temperatures. This assumption is confirmed by the presence of the large variety of Coccolithus pelagicus in the late Pliocene and lowermost Pleistocene sediments (Roth, 1974,

Müller, 1974 and results of the OSIRIS cruise of the "Marion Dufresne" in 1973, unpublished).

The modern nannoplankton assemblages in the Red Sea and the Indian Ocean, determined by water temperatures, salinity, nutrients and currents, correspond to those of the floristic zones described by McIntyre and Be (1967). These zones are less distinguished by the restricted occurrence of certain species than by the different abundance of certain species, Coccolithus pelagicus is present south of 32°S in the surface sediments, indicating lower water temperatures in this region.

The assemblages of the high-latitude areas or of parts of the Crozet Basin which are influenced by the Circumpolar Current, are of low diversity. The forms are smaller and more fragile as compared to forms of tropical or subtropical regions.

The dilution of nannofossils due to the high amount of terrigeneous material, in general connected with poor preservation and an increase of reworked species in sediments since early middle Miocene, recovered from the Indus Cone (Site 222) and the Bengal Fan (Site 218) as well as from the basins east of Africa reflect the uplift of the Himalayas (Laughton, McKenzie and Sclater, 1972) respectively of Africa and Madagascar (Simpson, Schlich et al., 1974).

Acknowledgements. I wish to thank Dr. L. Leclaire and Dr. S. White for critically reading the manuscript and editorial help.

References

Adelseck, C. G., G. W. Geehan, and P. H. Roth, Experimental evidence for selective dissolution and overgrowth of calcareous nannofossils during diagenesis, Bull. Geol. Soc. Amer., 84, 2755-2762, 1973.

Berger, W. H., Deep-sea carbonates: evidence for a coccolith lysocline, Deep-Sea Res., 20, 917-921, 1973.

Boudreaux, J. E., Calcareous nannoplankton ranges, Deep Sea Drilling Project Leg 23, in Whitmarsh, R. B., Weser, O. E., Ross, D. A. et al., Initial Reports of the Deep Sea Drilling Project, 23, Washington (U. S. Government Printing Office), 1073-1093, 1974.

Bramlette, M. N., and J. A. Wilcoxon, Middle Tertiary calcareous nannoplankton of the Cipero Section, Trinidad, W. I., Tulane Stud. Geol., 5, 93-132, 1967.

Bukry, D., Low-latitude coccolith biostratigraphic zonation, in Edgar, N. T., Saunders, J. B., et al., Initial Reports of the Deep Sea Drilling Project, 15, Washington (U. S. Government Printing Office), 685-703, 1973.

Bukry, D., Coccolith stratigraphy, Arabian and Red Seas, Deep Sea Drilling Project Leg 23, in Whitmarsh, R. B., Weser, O. E., Ross, D. A., et al., Initial Reports of the Deep Sea Drilling Project, 23, Washington (U. S. Government Printing Office) 1091-1093, 1974.

Bukry, D., and S. F. Percival, New Tertiary calcareous nannofossils, Tulane Stud. Geol. Paleont.,8, 123-146, 1971.

Davies, T. A., B. P. Luyendyk, et al., Initial Reports of the Deep Sea Drilling Project, 26, Washington (U. S. Government Printing Office) XX + 1129, 1974.

Edwards, A. R., A calcareous nannoplankton zonation of the New Zealand Paleogene, Proc. II. plank. Conf., Roma 1970, 1, 381-419, 1971.

Edwards, A. R., and K. Perch-Nielsen, Calcareous nannofossils from the southern Southwest Pacific, Deep Sea Drilling Project, Leg 29, in Kennett, J. P., Houtz, R. E., et al., Initial Reports of the Deep

Sea Drilling Project, 29, Washington (U. S. Government Printing Office) 469-539, 1975.

Ewing, M., S. Eittreim, M. Truchan, and J. I. Ewing, Sediment distribution in the Indian Ocean, Deep-Sea Res., 16, 231-248, 1969.

Fisher, R. L., E. T. Bunce, et al., Initial Reports of the Deep Sea Drilling Project, 24, Washington (U. S. Government Printing Office) XX + 1183, 1974.

Frakes, L. A., and E. M. Kemp, Influence of continental positions on the early Tertiary climates, Nature, 240, 97-100, 1972.

Gartner, S., Nannofossil biostratigraphy, Leg 22, Deep Sea Drilling Project, in von der Borch, C. C., Sclater, J. G., et al., Initial Reports of the Deep Sea Drilling Project, 22, Washington (U. S. Government Printing Office), 577-599, 1974.

Geitzenauer, K. R., and P. Huddlestun, An upper Pliocene-Pleistocene calcareous nannoplankton flora from a subantarctic Pacific deep-sea core, Micropaleontology, 18, 405-408, 1973.

Harris, T.F.W., Sources of the Agulhas current in the springs of 1964, Deep-Sea Res., 19, 633-650, 1972.

Heezen, B. C., and C. D. Hollister, Deep sea current evidence from abyssal sediments, Mar. Geol., 1, 141-174, 1964.

Hekel, H., Nannofossil biostratigraphy, Leg 20, Deep Sea Drilling Project, in Heezen, B. C., MacGregor, I. D., et al., Initial Reports of the Deep Sea Drilling Project, 20, Washington (U. S. Government Printing Office), 221-247, 1973.

Kennett, J. P., R. E. Burns, et al., Australian-Antarctic continental drift, paleocirculation changes and Oligocene deep sea erosion, Nature, Phys. Sci., 239, 51-55, 1972.

Kennett, J. P., R. E. Houtz, et al., Initial Reports of the Deep-Sea Drilling Project, 29, Washington (U. S. Government Printing Office) 1975.

Laughton, A. S., D. P. McKenzie, and J. G. Sclater, The structure and evolution of the Indian Ocean, in Int. Geol. Congr. 24th, Montreal, Section 8, 65-73, 1972.

Leclaire, L., Late Cretaceous and Cenozoic pelagic deposits-paleoenvironment and paleooceanography of the central western Indian Ocean, in Simpson, E.S.W., Schlich, R., et al., Initial Reports of the Deep Sea Drilling Project, 25, Washington (U. S. Government Printing Office), 481-513, 1974.

Le Pichon, X., The deep sea circulation in the southwest Indian Ocean, J. Geophys. Res., 65, 4061-4074, 1960.

Martini, E., Standard Paleogene calcareous nannoplankton zonation, Nature, 226, 560-561, 1970.

Martini, E., Standard Tertiary and Quaternary nannoplankton zonation, Proc. II. plank. Conf., Roma 2, 739-785, 1971.

Martini, E., and T. Worsley, Standard Neogene calcareous nannoplankton zonation, Nature, 225, 289-290, 1970.

Martini, E., and T. Worsley, Tertiary calcareous nannoplankton from the western equatorial Pacific, in Winterer, E. L., et al., Initial Reports of the Deep Sea Drilling Project, 7, Washington (U. S. Government Printing Office), 1471-1507, 1971.

McIntyre, A., and A.W.H. Be, Modern Coccolithophoridae of the Atlantic Ocean - I. Placoliths and cyrtoliths, Deep-Sea Res., 14, 561-597, 1967.

McIntyre, A., and R. McIntyre, Coccolith concentrations and differential solution in oceanic sediments, in Funnell, B. M. and Riedel, W. R. (eds.) The micropaleontology of oceans, Cambridge (Cambridge Univ. Press), 253-261, 1971.

McKenzie, D., and J. G. Sclater, The evolution of the Indian Ocean since the late Cretaceous, Geophys. J. Roy. Astro. Soc., 25, 437-528, 1971.

Müller, C., Calcareous nannoplankton, Leg 25 (western Indian Ocean), in Simpson, E.S.W., Schlich, R., et al., <u>Initial Reports of the Deep Sea Drilling Project</u>, <u>25</u>, Washington (U. S. Government Printing Office) 579-633, 1974.

Müller, Nannoplankton-Gemeinschaftern aus dem Jung-is Quartär des Golfs von Aden und dem Roten Meer, <u>Geol. Jb.</u>, 1977.

Olausson, E., B. U. Haq, G. B. Karlson, and I. U. Olausson, Evidence in Indian Ocean cores of late Pleistocene changes in oceanic and atmospheric circulation, <u>Geol. Fören. Stockholm Forbanlingar</u>, <u>93</u>, 51-84, 1971.

Perch-Nielsen, K., Remarks on late Cretaceous to Pleistocene coccoliths from the North Atlantic, in Laughton, A. S., Berggren, W. A., et al., <u>Initial Reports of the Deep Sea Drilling Project</u>, <u>12</u>, Washington (U. S. Government Printing Office) 1003-1069, 1972.

Pimm, A. C., R. H. Burroughs, and E. T. Bunce, Oligocene sediments near Chain Ridge, northwest Indian Ocean: structural implication, <u>Mar. Geol.</u>, <u>13</u>, M14-M18, 1972.

Proto Decima, F., Leg 27 calcareous nannoplankton, in Veevers, J. J., Heirtzler, J. R., et al., <u>Initial Reports of the Deep Sea Drilling Project</u>, <u>27</u>, Washington (U. S. Government Printing Office) 589-621, 1974.

Roth, P. H., Calcareous nannofossils Leg 17, Deep Sea Drilling Project, in Winterer, E. L., Ewing, J. J., et al., <u>Initial Reports of the Deep Sea Drilling Project</u>, <u>17</u>, Washington (U. S. Government Printing Office) 675-796, 1973.

Roth, P. H., Calcareous nannofossils from the northwestern Indian Ocean, Leg 24, Deep Sea Drilling Project, in Fisher, R. L., Bunce, E. T., et al., <u>Initial Reports of the Deep Sea Drilling Project</u>, <u>24</u>, Washington (U. S. Government Printing Office) 969-994, 1974.

Saito, T., and C. Fray, Cretaceous and Tertiary sediments from the southwestern Indian Ocean, <u>Bull. Geol. Soc. Amer. (Special Papers)</u>, <u>82</u>, 171-172, 1964.

Schneidermann, N., Deposition of coccoliths in the compensation zone of the Atlantic Ocean, <u>Proc. Symp. calc. nannofossils</u>, 140-151, 1973.

Sclater, J. G., and C.G.A., Harrison, Elevation of the Mid-Ocean Ridges and the evolution of the Southwest Indian Ridge, <u>Nature</u>, <u>230</u>, 175-177, 1971.

Simpson, E.S.W., R. Schlich, et al., <u>Initial Reports of the Deep Sea Drilling Project</u>, <u>25</u>, Washington (U. S. Government Printing Office XX + 884 p., 1974.

Thierstein, R., Calcareous nannoplankton - Leg 26, Deep Sea Drilling Project, in Davies, T. A., Luyendyk, B. P. et al., <u>Initial Reports of the Deep Sea Drilling Project</u>, <u>26</u>, Washington (U. S. Government Printing Office) 619-668, 1974.

Veevers, J. J., J. R. Heirtzler, et al., <u>Initial Reports of the Deep Sea Drilling Project</u>, <u>27</u>, Washington (U. S. Government Printing Office) XXI + 1060 p., 1974.

von der Borch, C. C., J. G. Sclater, et al., <u>Initial Reports of the Deep Sea Drilling Project</u>, <u>22</u>, Washington (U. S. Government Printing Office) XX + 890 p., 1974.

Warren, B. A., Evidence for a deep western boundary current in the south Indian Ocean, <u>Nature Phys. Sci.</u>, <u>229</u>, 18-19, 1971.

Whitmarsh, R. B., O. E. Weser, D. A. Ross, et al., <u>Initial Reports of the Deep Sea Drilling Project</u>, <u>23</u>, Washington (U. S. Government Printing Office) XX + 1180 p., 1974.

CHAPTER 17. SYNOPSIS OF CRETACEOUS PLANKTONIC FORAMINIFERA FROM THE INDIAN OCEAN

René Herb

Geological Institute, University of Berne
Berne, Switzerland

Viera Scheibnerova

Geological and Mining Museum
Sydney, Australia

Abstract. The oldest planktonic foraminifera in cores from the Indian Ocean are of Barremian or questionable Early Aptian age and occur at DSDP site 249 on the Mozambique Ridge. Typical Aptian planktonic foraminifera, however, have not been found in the Indian Ocean. Albian assemblages, recovered at five sites in the eastern Indian Ocean, contain a few species with the genus Hedbergella being predominant, and Ticinella, Praeglobotruncana and Rotalipora being very rare. Cenomanian planktonic foraminifera have so far been recovered only from the Naturaliste Plateau and the Kerguelen Plateau. Assemblages of this age contain a few species with slightly more species being present in samples from the Kerguelen Plateau than in the former area. The species diversity of the Indian Ocean assemblages increased during later periods of the Late Cretaceous, but important tropical index forms are still missing; notably in the Coniacian to Santonian interval of the Naturaliste Plateau. Globigerine-shaped and heterohelicid taxa dominate over double-keeled Globotruncana. These features are typical for the non-tropical, cool Austral biogeoprovince of the Cretaceous. Terminal Cretaceous sediments were penetrated on the northern Ninetyeast Ridge and on the Mozambique Ridge. Despite their content in good index species, they are still of a similar Austral or transitional aspect, however with greater Tethyan affinities than earlier in the Cretaceous, especially in the Late Maastrichtian of the Ninetyeast Ridge, where a tropical assemblage occurs for the first time as effect of the rapid northward shift of the Indian plate.
  Based on the distributional patterns of benthonic and planktonic foraminifera an evaluation of depositional water depth for different periods of the Cretaceous in the Eastern Indian Ocean is attempted. Deep open basin facies can be assumed for the Albian of the Eastern Indian Ocean at some DSDP-sites (256, 257 and upper part of 260), while others (sites 258, lower part of 260, and 263) show shallow water assemblages. An increase of water depth and deposition below the $CaCO_3$ compensation depth prevailed in the basins of the eastern Indian Ocean during the Late Cretaceous, whereas on the Naturaliste Plateau and on the Broken Ridge shallow open marine conditions are indicated by planktonic and benthonic foraminifera during earlier

parts of the Late Cretaceous, similar to what can be observed in the Perth basin of Western Australia.

The biostratigraphic significance of the Austral planktonic assemblages is discussed. The low species diversity, combined with the lack of many index species known from tropical areas, is a major difficulty for precise age determinations in many cases, particularly with respect to the Albian-Cenomanian boundary and a subdivision of the Albian and Cenomanian. In the Late Cretaceous the lack or rare occurrence of index forms, such as Globotruncana concavata or species of the G. fornicata-group makes precise age determinations equally difficult.

Introduction

During the Deep Sea Drilling Project operations in the Indian Ocean, carried out by the GLOMAR CHALLENGER in 1972 (Legs 22-27 and site 264 of Leg 28) Cretaceous sediments were penetrated in three main areas: 1) the western Indian Ocean adjacent to the coast of the African continent and Madagascar, 2) the basins of the eastern Indian Ocean east of approximately 98°E, and 3) on parts of the aseismic structures: the northern Ninetyeast Ridge, the Broken Ridge and the Naturaliste Plateau. As shown in Figure 1, many of these drill holes furnished Cretaceous planktonic foraminifera while others did not. Of the five drill sites at which Cretaceous sediments were cored in the western Indian Ocean only site 249 has shown good planktonic assemblages at different stratigraphic intervals of the Cretaceous. At the other locations the dating of Cretaceous sediments is mainly based on the occurrence of nannoplankton or benthonic foraminifera.

In the eastern Indian Ocean there is a widespread occurrence of Albian planktonic foraminifera in the deep-sea basins, the Wharton Basin (site 256), the Perth Abyssal Plain (sites 257 and 259), the Gascoyne Abyssal Plain (site 260) and on the Naturaliste Plateau (site 258). Late Cretaceous planktonic foraminifera, on the other hand, are limited in their occurrence or very rare in the deep sea basins (sites 212 and 257). However, they are abundant in the cores taken on the aseismic structures such as the Campanian-Maastrichtian on the northern Ninetyeast Ridge (sites 216 and 217), the Santonian on the Broken Ridge (site 255) and the Cenomanian-Santonian interval on the Naturaliste Plateau (sites 258 and partly 264). The occurrence, abundance and preservation at these sites is shown in the stratigraphic sections of Figures 2-5.

Western Indian Ocean (Site 249)

Cretaceous planktonic foraminifera were recovered during Deep Sea Drilling operations at only one location in the western Indian Ocean, site 249. The site is located on the Mozambique Ridge, in a present day latitude of approximately 30° south and in a water depth of 2088 m. The microfossils have been discussed and figured by Sigal (1974, pp. 689-693, Pl. 1-2, in Simpson, Schlich et al., 1974, pp. 297-300) and are briefly summarized here.

According to Sigal assemblages of three main time intervals can be distinguished: Barremian or Early Aptian, Late Albian or Early Cenomanian and Late Campanian to Late Maastrichtian. The occurrences in the stratigraphic section of this site are shown in Figure 2.

The Barremian or Early Aptian assemblages are sparse, moderately well preserved and characterized by a variety of tiny, apparently highly variable species. Among them Sigal identified Globigerina cf. tardita

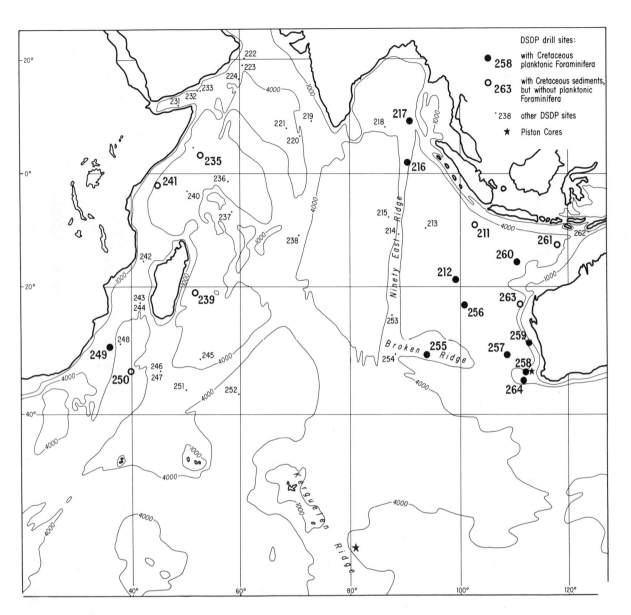

Fig. 1. Location map of DSDP drilling sites and piston cores showing occurrences of Cretaceous sediments and planktonic foraminifera.

or cf. quadricamerata Antonova (= G. 19963 Sigal of the Barremian stratotype) and G. cf. tuschepensis Antonova (= G. 19968 Sigal of the Barremian stratotype). A Barremian age of the respective level in site 249 is therefore highly probable, but Early Aptian cannot be excluded.

Late Albian or Early Cenomanian assemblages with moderately to well preserved planktonic foraminifera occur higher in the section (core 23, sections 3 and 4, see Simpson, Schlich et al., 1974, p. 298), after an interval with abundant radiolarians.

They are dominated by globigerine-shaped forms, and only from the top of core 23, section 3, Sigal noted a more diverse fauna with

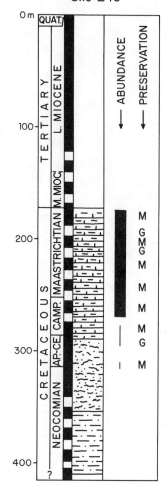

Fig. 2. Stratigraphic section of site 249 (Mozambique Ridge), showing abundance and preservation of Cretaceous planktonic foraminifera. For explanations see Figure 3.

Rotalipora balernaensis Gandolfi, Ticinella sp. (?T. roberti or Rotalipora praebalernaensis?), which indicates a Late Albian or possibly Early Cenomanian age.

From the lower part of the Late Cretaceous, Sigal mentions only sparse occurrences of Globotruncana. The upper part (Campanian/Maastrichtian), however, is characterized by abundant and well preserved planktonic foraminifera with a high species diversity. From the biostratigraphical point of view, the occurrences of Globotruncana calcarata Cushman in the lower part, and Globotruncana contusa (Cushman) var. galeoides Herm and Abathomphalus mayaroensis (Bolli) in the upper part are of particular interest and determine a time interval from Late Campanian to Late Maastrichtian.

On the other hand a number of species typical for this time interval are absent (Globotruncana gansseri Bolli) or rare (Globotruncana stuarti De Lapparent, Globotruncana falsostuarti Sigal, Globotruncana caliciformis (De Lapparent), Globotruncana contusa (Cushman).

Albian of the Eastern Indian Ocean

Drilling in the deep-sea basins of the eastern Indian Ocean has penetrated sediments of early Cretaceous age at six sites (256, 257, 259, 260, 261 and 263). At four of them (256 in the Wharton Basin, 257 and 259 in the Perth Abyssal Plain, and 260 in the Gascoyne Abyssal Plain) as well as in the section of the Naturaliste Plateau

CRETACEOUS PLANKTONIC FORAMINIFERA 403

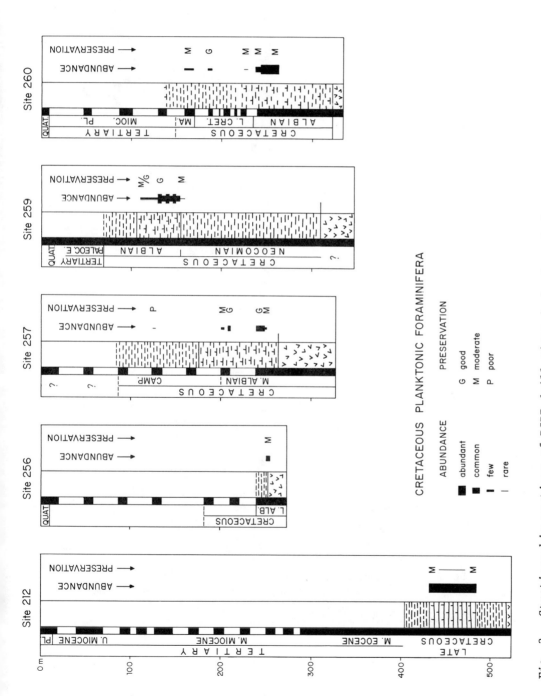

Fig. 3. Stratigraphic sections of DSDP drill sites in basins of the Eastern Indian Ocean, showing abundance and preservation of Cretaceous planktonic foraminifera.

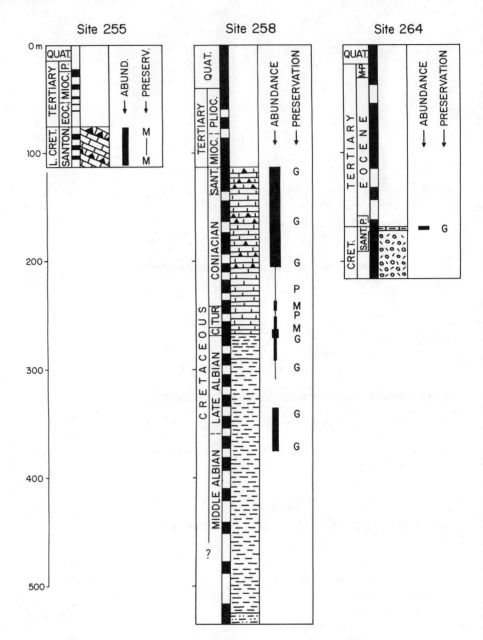

Fig. 4. Stratigraphic sections of DSDP drill sites on Broken Ridge and Naturaliste Plateau, showing abundance and preservation of Cretaceous planktonic foraminifera. For explanations see Figure 3.

(site 258) good assemblages of Albian planktonic foraminifera were found. However, all are characterized by a low species diversity. Typical representatives of the genus Ticinella, and early forms of Rotalipora and Praeglobotruncana are missing, with the exception of primitive forms of Ticinella found at site 257 (Herb 1974, Pessagno and Michael 1974).

Detailed examination of these assemblages reveals some significant differences between them. Site 256 in the Wharton Basin has fur-

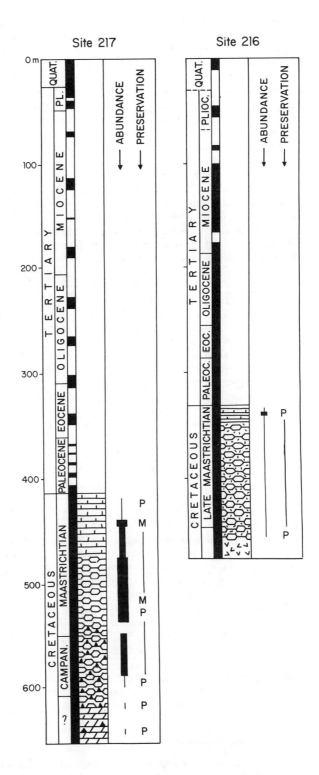

Fig. 5. Stratigraphic sections of DSDP drill sites on the northern Ninetyeast Ridge, showing abundance and preservation of Cretaceous planktonic foraminifera. For explanations see Figure 3.

nished a virtually monospecific assemblage of small Hedbergella planispira (Tappan) with transitional forms to H. infracretacea (Glaessner), and rare Schackoina cf. cenomana (Herb 1974, Plate 1). A greater diversity is reached at site 257 with Hedbergella planispira, H. aff. infracretacea and Hedbergella aff. sigali Moullade being the most frequent forms, with less frequent occurrence of Globigerinelloides caseyi (Bolli, Loeblich and Tappan), Ticinella cf. roberti (Gandolfi) and a probable new species, Ticinella sp. A (Herb 1974, Plates 1 and 2). The occurrence of these primitive Ticinella indicates a Middle Albian age, as determined by the nannoplankton (Thierstein 1974).

The Albian fauna of site 258 is similar to 257, except for the complete lack of Ticinella and the occurrence of Hedbergella amabilis Loeblich and Tappan[1]. The most frequent species in the Albian of site 258 is Hedbergella planispira.

In comparing planktonic foraminifera of sites 259 and 260, Krasheninnikov (1974b) showed that the assemblages of site 259 had considerably fewer species than those of site 260. Both assemblages, however, contain predominantly Hedbergella with such genera as Ticinella, Planomalina, Clavihedbergella, Schackoina and early representatives of Rotalipora and Praeglobotruncana missing.

The assemblages at site 259 are characterized by the continuous presence of Hedbergella aff. infracretacea, H. planispira, rare H. amabilis and H. sigali (core catcher 14). This is basically the same species composition as at site 257.

Site 260 shows a more diverse assemblage with Globigerinelloides eaglefordensis (Moreman), G. bentonensis (Morrow), G. ultramicra (Subbotina), G. aff. maridalensis (Bolli), sporadic G. gyroidinaeformis Moullade, Hedbergella brittonensis, H. infracretacea, H. globigerinelloides (Subbotina), H. aff. delrioensis (Carsey) and H. trocoidea (Gandolfi), of which H. infracretacea and H. delrioensis (Carsey) and H. trocoidea (Gandolfi), of which H. infracretacea and H. delrioensis are predominant. A characteristic feature is the absence of Hedbergella planispira, a major constituent of the other four Albian assemblages.

Based mainly on the presence of Globigerinelloides eaglefordensis, G. bentonensis, G. ultramicra and Hedbergella brittonensis Krasheninnikov (1974b) assumed a Late Albian to Early Cenomanian age with preference for Late Albian. Species, such as Globigerinelloides aff. maridalensis, G. gyroidinaeformis and H. trocoidea would however indicate an Early or Middle Albian age. G. gyroidinaeformis has been found only rarely and in the lower part of the section. A reexamination of Hedbergella trocoidea has shown that it is not identical with the typical form from the Early and Middle Albian; it has fewer chambers in the final whorl. Risch (1971, p. 47) mentioned such forms from the Late Albian and Early Cenomanian. The chronostratigraphic significance of the foraminiferal assemblages remains therefore ambiguous. Similar problems were encountered by the investigation of the nannoplankton. Proto Decima (1974) and Theirstein (this volume) placed this interval in the Middle Albian, but Proto Decima pointed out the presence of

---

[1] Following Caron (1966), Herb (1974) considered H. amabilis to be a junior synonym of H. simplicissima. Dr. J. Sigal (oral communication), however, recognizes differences in the form of the chambers. The problem must remain open until a new documentation of the type of H. simplicissima becomes available.

reworked Aptian forms and Thierstein the lack of important index forms.

Krasheninnikov (1974b) interpreted the richer and more diverse foraminiferal assemblage from site 260 to have been deposited in a different, probably warmer environment than the fauna of site 259. Based on the combined study of the Leg 26 and 27 sites this interpretation is possible. The southern sites (257, 258 and 259) form a group with similar specific composition and may represent a relatively cool paleoenvironment. The significance of the site 256 fauna is not definite, but it shows more affinities to the southern sites than to site 260.

The benthonic assemblages in the upper parts of sites 259 and 260 contain identical species (Scheibnerova, this volume). This indicates that at site 260 only the uppermost layers of sea water masses were affected by their latitudinal position closer to, but still not yet within the influence of the Tethyan tropical biogeoprovince.

Late Cretaceous of the Eastern Indian Ocean

In the cores drilled in the deep sea basins of the eastern Indian Ocean planktonic foraminiferal assemblages of Late Cretaceous age were only recovered at site 212 in the Wharton Basin. Here the lowest carbonate unit (cores 29 to 35) contained abundant planktonic and some calcareous benthonic foraminifera, uniformly minute in size (McGowran 1974, p. 625). The planktonic component is, according to this author, a Globigerinelloides-"Hedbergella"-Heterohelix assemblage with no juvenile Globotruncana, and has been size-sorted. Specimens of Globotruncanella citae (Bolli) indicate a Late Campanian to Maastrichtian age.

In cores of the other sites in the Wharton Basin, the Perth Abyssal Plain and the Gascoyne Abyssal Plain no planktonic foraminifera of Late Cretaceous age occurred, except for one single specimen of Globotruncana cf. elevata (Brotzen) found in a siliceous clay of site 257, core 5 (Herb 1974, p. 747). The area has either been situated below the lysocline at that time, or postdepositional dissolution has removed the tests of calcareous organisms.

Late Cretaceous planktonic foraminifera were repeatedly recovered, however, in the sites located on the aseismic ridges and plateaus of the eastern Indian Ocean, the Ninetyeast Ridge, the Broken Ridge, the Kerguelen Plateau and the Naturaliste Plateau.

Cenomanian-Turonian assemblages were recovered at site 258 (Leg 26) on the Naturaliste Plateau and in a piston core taken by the Eltanin on the eastern side of the southern Kerguelen Plateau (Quilty 1973). Earlier, Burckle, Saito and Ewing (1967) recorded Turonian planktonics in a piston core from the Naturaliste Plateau. Comparing the Cenomanian of the Naturaliste Plateau with the assemblages described by Quilty from the Kerguelen Plateau, the slightly greater diversity of the Kerguelen fauna becomes evident. Rotalipora reicheli (Mornod), of which only a fragment of a single specimen was found at site 258, occurred in a good population on the Kerguelen Plateau. Hedbergella delrioensis and H. amabilis were listed from both areas. Schackoina cenomana, Sch. bicornis Reichel and a greater variety of Globigerinelloides is typical for the Kerguelen Plateau, whereas the generally most frequent species of the Naturaliste Plateau, H. planispira was not found by Quilty.

The Turonian assemblages of site 258 are poorly preserved and some of the smaller species are probably removed by diagenetic dissolution processes. Praeglobotruncana algeriana Caron, P. hagni Scheibnerova, possible P. stephani (Gandolfi), P. cf. imbricata (Mornod), Hedbergella (?) paradubia (Sigal) and Praeglobotruncana helvetica (Bolli) were identified. The latter, also found in the CONRAD piston core in the same area by Burckle, Saito and Ewing (1967), indicates a Middle Turonian age.

The composition of the Kerguelen Turonian assemblage is slightly different. This might be due to its greater age which, based on the occurrence of Rotalipora greenhornensis (Morrow), cannot be younger than Early Turonian. As a whole, however, the stratigraphic sequences as well as the Cenomanian-Turonian planktonic foraminifera are similar on the Naturaliste and Kerguelen Plateaus.

On the Broken Ridge, at site 255, indurated and partly silicified limestones of Santonian age with Inoceramus were reached. At some levels these limestones contain abundant planktonic foraminifera, predominantly globigerine-shaped forms and very rarely Globotruncana. They are transgressively overlain by Eocene shallow-water deposits, which supports the idea of a different geologic history for Kerguelen Plateau and Broken Ridge (Quilty 1973).

Abundant Coniacian-Santonian planktonic foraminifera occurred in the uppermost parts of the Cretaceous sequence in site 258 (Leg 26). The most striking feature of these assemblages is the proliferation of mostly small globigerine-shaped and heterohelicid taxa. Double-keeled globotruncanid species were relatively minor constituents.

The most abundant species in all samples is Globigerinelloides ehrenbergi followed mostly by Heterohelix. Among the globigerine-shaped forms the following species were distinguished by Herb (1974): Archaeoglobigerina bosquensis, Whiteinella baltica and Hedbergella crassa. H. aff. simplicissima, H. planispira, Schackoina cenomana and S. mulitspinata occurred in some samples. Globotruncana tricarinata was the most frequent of the double-keeled globotruncanids. Some assemblages contained only G. lapparenti and G. coronata. G. marginata, G. angusticarinata, G. aff. ventricosa, G. arca and G. fornicata manaurensis occurred discontinuously. G. concavata and the fornicata group were extremely rare or missing. This made precise stratigraphic determinations difficult.

Pessagno and Michael (1974b) have applied a different taxonomic nomenclature to species from the same stratigraphic intervals in some cases, but usually reached the same stratigraphic conclusions. Their Albian/Cenomanian boundary, however, is placed considerably lower than in the report of Herb (1974). It has to be pointed out that, based on the present planktonic foraminifera alone, this boundary cannot be fixed precisely. The age determinations made on the basis of the nannoplankton seem to be more reliable in this case and have therefore been adopted here.

The second site (264), which was drilled by the GLOMAR CHALLENGER on the Naturaliste Plateau, was selected at a location where a particularly thin sedimentary cover could be expected. In fact, only 1,15 m of Cretaceous clay-rich nannochalk was recovered above a volcanoclastic conglomerate and below a Paleocene-Eocene sequence. The foraminiferal assemblage in this sediment is dominated by well preserved tests of Heterohelix reussi, Globigerinelloides asperus, Hedbergella planispira, H. delrioensis and Schackoina multispinata. Globotruncana coronata and Globotruncana linneiana were found in very small numbers. This mixture of species indicates mixed assemblage with a total range from Albian/Cenomanian through Santonian/Campanian. However, the main feature of the site 258 assemblage, i.e. the domination by heterohelicid and globigerinid taxa, can also be observed here.

Comparing the sections of the Naturaliste Plateau, the Broken Ridge and the Kerguelen Plateau with each other we can recognize that sediments of the same or similar ages were deposited on these structures and that major unconformities occur at comparable time intervals:

Late Cretaceous is present in all three areas, Eocene is found on the
Naturaliste Plateau and the Broken Ridge, and Late Neogene again on all
three plateaus. This signifies at least some structural relations
between the three areas. On the other hand, the Cretaceous sections
show a number of significant differences as outlined above. Even the
relatively closely spaced sites 258 and 264 on the Naturaliste Plateau
differ from each other: site 258 shows a particularly thick sequence
of Albian, a thin and possible incomplete interval of Cenomanian and
Turonian, as well as good sections of Coniacian and Santonian; site 264
thin Late Cretaceous sediments (probably lower Senonian) with reworked
Cenomanian above a volcanoclastic conglomerate of probably Albian age.
This indicates at least some tectonic activity on these structures in
earlier parts of the Late Cretaceous and possibly the beginning of the
separation of these plateaus, as proposed by Quilty (1973).

Campanian - Maastrichtian of the Northern Ninetyeast Ridge

In the northern part of the Ninetyeast Ridge, at site 217, a sequence
of Campanian to Maastrichtian sediments was drilled. Planktonic
foraminifera were listed and discussed by McGowran (1974), and Pessagno
and Michael (1974a). McGowran recognized the following three intervals
from top to bottom:

1) 57 m with a diverse planktonic foraminiferal assemblage of the
Abathomphalus mayaroensis-zone. The diversity is increasing upwards and
indicates tropical conditions at the top.

2) 75 m with abundant planktonic foraminifera, dominated by Rugoglobigerina rugosa, double-keeled Globotruncana (especially G. arca),
Gublerina, Heterohelix and Globigerinelloides. Single-keeled Globotruncana are virtually absent. Abathomphalus mayaroensis, however, occurs
in the uppermost part of this zone and indicates still a Late Maastrichtian age for at least this interval. McGowran (1974) places the whole
interval 2 into the Maastrichtian and considers it to be an extratropical
assemblage.

3) 57 m with an association of Archaeoglobigerina, Globotruncana
linneana s.l., Globigerinelloides and Heterohelix, with Globotruncana
becoming prominent near the top. McGowran assigned this assemblage to
the Campanian. Benthonic foraminifera are more common than in units 1
and 2, and indicate shallow water conditions with a possible deepening
trend in the upper part.

The superposition of these three zones indicates, according to Mc-
Gowran (1974), Gartner, Johnson and McGowran (1974) and McGowran (this
volume) a transition from a shallow water environment at the bottom to
open marine conditions at the top, as well as a rapid shifting of this
site from an extratropical to a tropical bioprovince in the latest
Cretaceous, corresponding to the extensive northward shift of the Indian
plate.

At site 216, located 8 degrees south of site 217, only Maastrichtian
was recovered below the Tertiary. Shallow water sediments with rare
planktonic foraminifera in the lower part are again followed by assemblages with an increasing amount of planktonics, including Abathomphalus
mayaroensis, thus indicating Late Maastrichtian.

Paleoecology and Paleobiogeography

Cretaceous planktonic assemblages recovered by the DSDP in the Indian
Ocean generally show the following features, significant for paleoecologic and paleobioprovincial interpretations;

1) Low specific diversity in the Albian and in the lower and middle part of the Late Cretaceous. The genera Ticinella and Rotalipora are extremely rare in the Cenomanian. Equally restricted is the Globotruncana group, especially the G. fornicata and G. concavata groups which are characteristic of the tropical zone of that time.

2) Non-keeled and mostly small species of Hedbergella, Globigerinelloides, Whiteinella, Archaeglobigerina and Heterohelix predeominated over the keeled globotruncanid species.

Such assemblages can generally be interpreted as either coming from a restricted environment (shallow water, closed basin) or from a relatively cool-water area (Bandy, 1967). They are known to occur in the Boreal and northern Transitional Biogeoprovinces of the Northern Hemisphere and in the Austral and Southern Transitional Biogeoprovinces in the Southern Hemisphere. In the Late Cretaceous of North America Douglas (1972) distinguished a Pacific province north of the Tethyan province, characterized by low species diversity, by the presence of hedbergellid and double-keeled species and by the high frequency of cosmopolitan species. Such assemblages are also typical in New Zealand, South Africa (Lambert 1972), South America (Herm 1966) and especially Western Australia, where a gradual increase of tropical elements in the Late Cretaceous globotruncanid assemblages can be observed in more northerly situated Cretaceous deposits (for more detail see Belford 1958 and 1960, Belford and Scheibnerova 1971, Edgell 1957, 1961). In the Carnarvon Basin for instance the Tethyan influence was quite strong considering the paleolatitudinal position, and a warm current was postulated to account for this (Scheibnerova 1971b, Belford and Scheibnerova 1971).

Non-tropical planktonic assemblages in Cretaceous deposits are thus characterized by impoverished faunas of low specific diversity, composted of predominating Hedbergella species before the Turonian, and Hedbergella, Globigerinelloides, Heterohelix as well as a number of mainly double-keeled species of Globotruncana as well as Rugoglobigerina in the later parts of the Late Cretaceous. In the Coniacian to Santonian of the Naturaliste Plateau (site 258), the genus Globotruncana is essentially restricted to the G. lapparenti-group and to G. coronata, G. marginata, G. cretacea (Coniacian) and G. sp. aff. ventricosa (Santonian), whereas the G. fornicata-group is rare and G. concavata missing. We can see definite changes in the specific diversity of non-tropical planktonic assemblages during the Late Cretaceous. The increase in the specific diversity of planktonic assemblages can be related to the world-wide temperature fluctuations, and in some parts of the Indian Ocean to the relative northward movement of the Indian plate. This is especially significant in the Maastrichtian of the northern Ninetyeast Ridge where the effect of rapid northward movement of the Indian plate seems to be superimposed upon a general warming trend, as outlined above. Also in the southwestern Indian Ocean, at site 249, the available data (Sigal 1974; Sigal in Simpson, Schlich et al. 1974) suggest a much greater diversity of the planktonic foraminifera in the Maastrichtian than in the Albian to Cenomanian. However, a number of important index forms for defining the tropical Maastrichtian assemblages were not found here (Globotruncana gansseri) or are rare (G. stuarti-group, G. contusa). This indicates a slightly extratropical position, corresponding to the present-day latitude of the site. Latitudinal temperature control is therefore also quite distinct. The greater diversity of the Albian assemblage at site 260 compared with those from more southern locations in the Perth Abyssal Plain and on the Naturaliste Plateau may serve as an example for this.

The distributional pattern of Rotaliporids in the Cenomanian and early Turonian seems to be more complicated, but as a whole it is probably also controlled by temperature. Rotalipora reicheli was extremely rare in the Cenomanian of site 258 on the Naturaliste Plateau (Herb 1974). It occurred in greater numbers in sediments of the same age on the Kerguelen Plateau (Quilty 1973), but it is the only species of Rotalipora to be found in the Cenomanian of the southeastern Indian Ocean. A greater species diversity of Rotalipora and Praeglobotruncana is reached in the Turonian of the Naturaliste Plateau (Burckle, Saito and Ewing 1967; Herb 1974). It is however, the only place in the Indian Ocean where Turonian planktonic foraminifera were recovered so far.

It thus seems possible to define the Austral biogeoprovince in the sense of Scheibnerova (1970, 1971a, 1974) stage by stage by the specific composition of planktonic foraminiferal assemblages. However, due to the limited number of drillholes and the great distances between them, the available data are so far still very incomplete, making such a definition premature in this synthesis; further data as well as detailed comparison with land sections will be needed.

As far as the Cretaceous bathymetry is concerned, benthonic assemblages are of particular value (see Scheibnerova 1976a). In many deep sea samples planktonic foraminifera were either substantially restricted in their numbers or missing. Signs of partial dissolution of their calcitic tests were frequently observed. These observations were in most cases attributed to the effects of dissolution in depositional depths near or below the lysocline.

However, in most cases where calcareous benthonic foraminifera occur without planktonic assemblages being present at all, they can well be correlated, on the specific level, with those known to occur in coeval sediments on adjacent land sections where we must assume a shallow water environment. It would be difficult to imply completely different depth conditions for the same benthonic species occurring on land, eg. in South Peninsular India, Western Australia, and the Great Australian Basin (Narayanan and Scheibnerova 1975, Scheibnerova 1972-1976) and on the present bottom of the Indian Ocean (Scheibnerova 1974 and 1977 in press) when even the sediments are strikingly similar. Bathymetrically these sediments fall within the inner shelf, the outer shelf and the upper (?middle) slope (Sliter 1972, Scheibnerova 1976). In the case of the Lower Cretaceous of Leg 27 this question was discussed in detail by Scheibnerova (1974).

Recently, Scheibnerova (1976) proposed the following classification of paleoenvironments of the Great Australian Basin, based on the distribution pattern of benthonic foraminifera:

1) Shallow basin margin facies
2) Deep basin margin facies
3) Basinal depression facies
4) Intrabasinal ridge facies
5) Shallow open basin facies
6) Deep open basin facies

The first four facies would fall within the inner and outer shelf of Sliter and Baker (1972), and facies 5 and 6 within the upper (?middle) slope.

Based on the composition of benthonic foraminiferal assemblages, particularly the presence or absence of Lingulogavelinella frankei as a shallow-water indicator on one side, and the presence or absence of planktonic foraminifera on the other side, the following correlation of the eastern Indian Ocean sites with this classification can be attempted:

Sites 256 and 257. Deep open basin facies (presence of planktonic foraminifera and absence of L. frankei in the Albian, dissolution facies in the Late Cretaceous).

Site 258. Shallow open basin facies (presence of planktonic foraminifera and L. frankei in the Albian and Cenomanian, planktonic foraminifera and shallow-water benthonic foraminifera in the Turonian to Santonian).

Site 259. Deep open basin facies in the Albian (presence of planktonic foraminifera and absence of typical L. frankei).

Site 260. Shallow basin margin facies (?) in the lower part of the section (absence of planktonic foraminifera, shallow-water benthonic foraminifera similar to those of site 263) deep open basin facies in the upper part of the section (presence of planktonic foraminifera and absence of L. frankei).

Site 263. Shallow basin margin facies (?) (absence of planktonic foraminifera and of L. frankei).

In the Albian sediments of site 263 partial or complete dissolution of calcitic tests was attributed to dissolution in the sediment probably shortly after the deposition under reducing conditions, and may explain the lack of Lingulogavelinella frankei in this shallow water assemblage. This would be equivalent to the dissolution observed in similar deposits on adjacent land, in the Great Australian Basin in particular. Here also, a shallow water depth can be expected based on the benthonic assemblage. However, it cannot be decided whether planktonic foraminifera were primarily missing or have been diagenetically dissolved in the Albian of site 263.

According to these criteria, the Leg 26 occurrences of Albian and Cenomanian planktonic foraminifera in the Eastern Indian Ocean, with the exception of site 258 on the Naturaliste Plateau, belong to the Deep open basin facies, which corresponds to the upper (?) middle slope.

In the Late Cretaceous of the Eastern Indian Ocean basins the assemblages of characteristic agglutinated benthonic foraminifera indicate a definite increase of the water depth, as is particularly well documented by Krasheninnikov (1974a) from sites 260 and 261. These forms are so far unknown from coeval sediments on land. A purely agglutinated assemblage with Labrospira pacifica Krasheninnikov was also found at site 256 (core 8, section 6, 115-119 cm). A deep-water origin is also supported by the lithology of these sediments - brown zeolitic clays. However, considering the outlined faunal evidences as well as sedimentological criteria, it seems probable that in the Indian Ocean the lysocline occurred in a considerably higher position during the Cretaceous than today.

The complete absence of planktonic foraminifera in the Eastern Indian Ocean prior to Albian time is a striking feature of these distributional patterns. It could well be the effect of still very limited marine connections in the newly opened Indian Ocean. In the Albian, however, open marine conditions already prevailed over large areas between Australia and India, although the water depth did not, as outlined above, exceed depths corresponding to the upper or possibly middle slope, and greater depths were only reached by Late Cretaceous time.

Sigal (1974) mentioned "an unusual microfaunal assemblage that is quite different (by the benthonic species present and by the absence of planktonic species) from the one that is known either in the Majunga

Basin or in the Diego region in Madagascar". This microfauna may probably be similar or identical with the one from Leg 27 as described by Krasheninnikov (1974a). Here, the observed complete lack of planktonic foraminifera can be attributed to dissolution below the lysocline.

Biostratigraphic Value of Cretaceous Planktonic Foraminifera
Recovered by the DSDP in the Indian Ocean

Except for the Campanian and Maastrichtian, planktonic foraminifera recovered by the DSDP in the Indian Ocean are represented by impoverished assemblages with few species. As outlined in the previous paragraph such assemblages are typical of non-tropical Cretaceous seas; in our case they were classified as characteristic for the Austral biogeoprovince (Scheibnerova 1970, 1971a, b). Secondary factors such as selective dissolution may be of influence in some cases (site 256?), but generally they are not of major importance.

The detailed biostratigraphic zonation of Late Cretaceous pelagic sediments is based on tropical assemblages of planktonic foraminifera with a high diversity. However, major difficulties arise when this tropical biostratigraphic zonation is to be used for dating Cretaceous sediments of non-tropical origin. Recently, the problem became especially obvious with recognition and study of the Austral Biogeoprovince (Scheibnerova 1970, 1971a, b).

Generally, the southernmost sites contain the least diversified assemblages and there is a slight increase in the number of planktonic species in more northerly sites. However, as a rule the planktonic foraminiferal assemblages in Cretaceous sediments recovered by the DSDP in the Indian Ocean are typical non-tropical assemblages. They have been used for biostratigraphic dating by most authors participating in their study. However, many problems have been encountered and have hampered exact dating. In many cases the lack of the tropical index species makes the task of a micropaleontological stratigrapher extremely difficult. Albian deposits for instance mostly lack representatives of the genus _Ticinella_. Rare specimens of _Schackoina_ found in the beds transitional between the Albian and Cenomanian were of a relatively greater importance than the same specimens in the tropical zone. In the author's opinion at least 2 species of _Schackoina_ occurred in these deposits. One species, occurring in the late Albian, with noticeably stouter chambers and another one, typical for Cenomanian and younger deposits with more slender chambers.

Most of the Albian assemblages of the eastern Indian Ocean could be interpreted as being Cenomanian as well (see e.g. Pessagno and Michael (1974). Their Albian age is indicated by the nannoplankton assemblages of the _Eiffelithus turriseiffeli_-Zone, which allows a worldwide correllation for the Late Albian.

With respect to the incomplete stratigraphic record of planktonic foraminifera from the Indian Ocean area it would be inappropriate to establish a planktonic foraminiferal zonation for the Austral biogeoprovince at the present time.

References

Bandy, O. L., Cretaceous planktonic foramniferal zonation, _Micropaleontology_, 13, 1-13, 1967.
Belford, D. J., Stratigraphy and micropaleontology of the Upper Cretaceous of Western Australia, _Geol. Rundsch._, 47, 629-647, 1958.
Belford, D. J., Upper Cretaceous foraminifera from the Toolonga Calcilu-

tite and Gingin Chalk, Western Australia, Australia Bur. Min. Res., Geol. Geophys., Bull., 57, 1-198, 1960.

Belford, D. J., and V. Scheibnerova, Turonian foraminifera from the Carnavron Basin, Western Australia, and their paleogeographic significance, Micropaleontology, 17, no. 3, 331-344, 1971.

Bolli, H. M., The genera Praeglobotruncana, Rotalipora, Globotruncana, and Abathomphalus in the Upper Cretaceous of Trinidad, B.W.I., U. S. Nat. Mus. Bull., 215, 51-60, 1957.

Bolli, H. M., Planktonic foraminifera from the Cretaceous of Trinidad, B.W.I., Bull. Amer. Pal., 39, no. 179, 257-277, 1959.

Burckle, L. H., T. Saito, and M. Ewing, A Cretaceous (Turonian) core from the Naturaliste Plateau, southeast Indian Ocean, Deep-Sea Res., 14, 421-426, 1967.

Caron, M., Globotruncanidae du Cretace superieur du synclinal de la Gruyere (Prealpes Medianes, Suisse), Rev. Micropal., 9, no. 2, 68-93, 1966.

Douglas, R. G., Paleozoogeography of Late Cretaceous planktonic foraminifera in North America, J. Foram. Res., 2, no. 1, 14-34, 1972.

Edgell, H. S., The genus Globotruncana in northwest Australia, Micropaleontology, 3, no. 2, 101-126, 1957.

Gartner, S., Jr., D. A. Johnson, and B. McGowran, Paleontology synthesis of deep sea drilling results from Leg 22 in the northeast Indian Ocean in von der Borch, C., Sclater, J., et al., Initial Reports of the Deep Sea Drilling Project, Washington (U. S. Government Printing Office), 22, 805-814, 1974.

Herb, R., Cretaceous planktonic foraminifera from the Eastern Indian Ocean, in Davies, T. A., Luyendyk, B. P. et al., Initial Reports of the Deep Sea Drilling Project, Washington (U. S. Government Printing Office) 26, 745-769, 1974.

Herm, D., Micropaleontological aspects of the Magellanese geosyncline Proc. 2nd African Micropaleont. Coll., Ibadan, 72-86, 1966.

Krasheninnikov, V. A., Upper Cretaceous benthonic agglutinated foraminifera, Leg 27 of the Deep Sea Drilling Project, in Veevers, J. J., Heirtzler, J. R. et al., Initial Reports of the Deep Sea Drilling Project, Washington (U. S. Government Printing Office), 27, 631-661, 1974a.

Krasheninnikov, V. A., Cretaceous and Paleogene planktonic foraminifera, Leg 27 of the Deep Sea Drilling Project, in Veevers, J. J., Heirtzler, J. R. et al., Initial Reports of the Deep Sea Drilling Project, Washington (U. S. Government Printing Office), 27, 663-671, 1974b.

Lambert, G., A study of the Cretaceous foraminifera from northern Zululand, M. Sc. thesis, Univ. of Natal, Durban, 375 pp., 1972.

McGowran, B., Foramnifera, in von der Borch, Ch. C., Sclater, J. G., et al., Initial Reports of the Deep Sea Drilling Project, Washington (U. S. Government Printing Office) 22, 609-627, 1974.

Narayanan, V., and V. Scheibnerova, Lingulogavelinella and Orithostella (Foraminifera) from the Utatur group of the Trichinopoly Cretaceous, South (Peninsular) India. Rev. esp. Micropaleont., 7, no. 1, 25-36, 1975.

Pessagno, E. A., and F. Y. Michael, Mesozoic foraminifera, Leg 22, Site 217, in von der Borch, Ch. C., Sclater, J. G. et al., Initial Reports of the Deep Sea Drilling Project, Washington (U. S. Government Printing Office) 22, 629-634, 1974a.

Pessagno, E. A., and F. Y. Michael, Results of shore laboratory studies on mesozoic planktonic foraminifera from Leg 26, sites 255, 257 and 258, in Davies, T. A., Luyendyk, B. P. et al., Initial Reports of the Deep Sea Drilling Project, Washington (U. S. Government Printing Office) 26, 969-972, 1974b.

Proto Decima, F., Leg 27 calcareous nannoplankton, in Veevers, J. J., Heirtzler, J. R. et al., Initial Reports of the Deep Sea Drilling Project, Washington (U. S. Government Printing Office) 27, 589-621, 1974.

Quilty, P. G., Cenomanian-Turonian and Neogene sediments from Northeast of Kerguelen Ridge, Indian Ocean, J. Geol. Soc. Australia, 20, no. 3, 361-368, 1973.

Risch, H., Stratigraphie der höheren Unterkreide der bayerischen Kalkalpen mit Hilfe von Mikrofossilien, Palaeontographica, 138A, 1-80, 1971.

Scheibnerova, V., Some notes on paleoecology and paleogeography of the Great Artesian Basin, Australia, during the Cretaceous, Search, 1, no. 3, 125-126, 1970.

Scheibnerova, V., Paleoecology and Paleogeography of Cretaceous deposits of the Great Artesian Basin, Australia, Records Geol. Surv. New South Wales, 1, 13, 5-48, 1971a.

Scheibnerova, V., The great Artesian Basin, Australia, a type area of the Austral biogeoprovince of the Southern hemisphere, equivalent to the Boreal biogeoprovince of the Northern hemisphere. Proc. II Plankt. Conf., Roma 1970, 1129-1138, 1971b.

Scheibnerova, V., Implications of Deep Sea Drilling in the Atlantic for studies in Australia and New Zealand - Some new views on Cretaceous and Cainozoic paleogeography and biostratigraphy, Search, 2, no. 7, 251-254, 1971c.

Scheibnerova, V., Aptian-Albian benthonic foraminifera from DSDP leg 27, sites 259, 260 and 263, eastern Indian Ocean, in Veevers, J. J., Heirtzler, J. R., et al., Initial Reports of the Deep Sea Drilling Project, Washington, D. C. (U. S. Government Printing Office) 27, 1974.

Scheibnerova, V., Cretaceous foraminifera of the Great Australian Basin, Mem. Geol. Surv. New South Wales, Paleont., no. 17, 277 pp., 1976a.

Scheibnerova, V., Synthesis of Cretaceous benthonic foraminifera recovered by DSDP in the Indian Ocean in Indian Ocean Geology and Biostratigraphy, AGU, Washington (this volume), 1977.

Scheibnerova, V., Some Cretaceous benthonic foraminifera from the DSDP Leg 26 in the Eastern Indian Ocean, Volume in honour of Dr. Irene Crespin, 1977.

Sigal, J., Comments on Leg 25 sites in relation to the Cretaceous and Paleogene stratigraphy in the Eastern and southeastern Africa coast and Madagascar regional setting, in Simpson, E.S.W., Schlich, R. et al., Initial Reports of the Deep Sea Drilling Project, Washington (U. S. Government Printing Office) 25, 1974.

Sliter, W. V., Cretaceous foraminifera - depth habitats and their origin, Nature, 239, no. 5374, 514-515, 1972.

Sliter, W. V., and R. A. Baker, Cretaceous bathymetric distribution of benthic foraminifera, J. Foram. Res., 2, no. 4, 167-193, 1972.

Thierstein, H. R., Calcareous nannoplankton - Leg 26, Deep Sea Drilling Project in Davies, T. A., Luyendyk, B. P., et al., Initial Reports of the Deep Sea Drilling Project, Washington (U. S. Government Printing Office) 26, 619-667, 1974.

CHAPTER 18. MAASTRICHTIAN TO EOCENE FORAMINIFERAL ASSEMBLAGES IN THE NORTHERN AND EASTERN INDIAN OCEAN REGION: CORRELATIONS AND HISTORICAL PATTERNS

Brian McGowran

Department of Geology, University of Adelaide
Adelaide 5001, Australia

Abstract. Foraminiferal data are summarized from on and adjacent to the Mascarene Plateau, Chagos-Laccadive Ridge, Ninetyeast Ridge, Naturaliste Plateau. Most of the records are Paleocene and Eocene although Maastrichtian strata were encountered on the Ninetyeast Ridge. The review is extended to the margins of India-Pakistan and Australia where three stratigraphic sequences (sensu Sloss, 1963, on a finer scale) are recognized: Campanian-Maastrichtian, Paleocene - Early Eocene, and Middle-Late Eocene. Correlations are based on the recognition of important (i.e. widespread and consistent) biostratigraphic events plotted against Berggren's (1972) time scale.

The distribution of Maastrichtian foraminifera supports the notion of rapid transform movement along the Ninetyeast Ridge at that time, whilst the Ridge was sinking rapidly. Foraminiferal biofacies reflect rapid and concerted sinking interrupted by an isochronous hiatus across the Paleocene/Eocene boundary on the Mascarene Plateau and Chagos-Laccadive and Ninetyeast Ridges; the same hiatus occurs on the Naturaliste Plateau. The resumption of sedimentary accumulation on the topographic highs is coeval with a change from calcareous ooze to brown clay in the Central Indian and Wharton Basins. The timing of the interruption matches suggestively Australia/Antarctica separation and other major geotectonic events, and forshadows the demise of the Ninetyeast Ridge as a transform fault.

The earliest Tertiary sequence on the continental margins ends within the Early Eocene with regression, hiatus, or nonidentification, seemingly regardless of whether in detrital (terrigenous) or carbonate facies, or of whether the margin was a leading edge or a trailing edge. Thus, rapid seafloor spreading and sinking in the ocean (Maastrichtian-Paleocene) was followed after an interruption by rapid sinking (Early Eocene) and this caused marginal regression.

Early in the Middle Eocene there is a terminal oceanic event marked by the cherty "Horizon A" which is isochronous; marginal strata were regressive or absent at that time. Later in the Middle Eocene, when the oceanic record is noteworthy mainly for hiatus, there is a synchronous marginal transgression marking the onset of extensive shelf carbonate accumulation. Whereas a tectonic explanation may come to suffice for the Paleocene-Early Eocene, paleo-oceanographic changes become discernible in the Middle Eocene: Australia/Antarctica separation allowed deflection of southern waters to the south, expansion of the tropical belt on the evidence of planktonic and benthonic foraminifera, and a shelf/basin carbonate fractionation with reduced

Table 1. Data on DSDP Sites.

	21-209	21-210	22-213	22-214	22-215	22-216	22-217	23-219	23-220	23-223	24-236	24-237	26-253	28-264
Location	Queensland Plateau	Coral Sea Basin	Wharton Basin	Ninetyeast Ridge	Central Indian Basin	Ninetyeast Ridge	Ninetyeast Ridge	Chagos-Laccadive Ridge	Arabian Basin	Owen Ridge	north of Mascarene Plateau	Mascarene Plateau	Ninetyeast Ridge	Naturaliste Plateau
latitude longitude	15°56.2'S 152°11.3'E	13°46.0'S 152°53.8'E	10°12.7'S 93°53.8'E	11°20.2'S 88°43.1'E	08°07.3'S 86°47.5'E	01°27.7'N 90°12.5'E	08°55.6'N 90°32.3'E	09°01.8'N 72°52.7'E	06°31.0'N 70°59.0'E	18°45.0'N 60°07.8'E	01°40.6'S 57°38.9'E	07°05.0'S 58°07.5'E	24°52.6'S 87°22.0'E	34°58.1'S 112°02.7'E
water depth	1428	4643	5611	1665	5319	2247	3020	1764	4036	3633	4504	1640	1962	2873
Oligocene/Eocene	—140/150—	—542—		—257—		—258—	—312—	—173—	—168/198—	—548—		—215—	—116—	—35—
	sand limestone chert	nanno ooze		nanno ooze		nanno ooze	chalk	ooze chalk	chalk	chalk		nanno ooze	nanno ooze	
Late/Middle Eocene	—259/272—	—554—		—267—		—263—	—316/344—	—186—		—566—		—216—	—142—	
	sand limestone	nanno ooze clay		nanno ooze		chalk chert	chalk ?chert	ooze chalk chert	chalk chert	chalk		nanno ooze	chalk volcanic ash	nanno ooze chalk chert
Middle/Early Eocene		—605—		—305—		—292—	—354/374—	—228/237—	—263—	—599/609—				—154—
		nanno ooze clay	brown clay nanno ooze	nanno ooze glauconitic	brown clay nanno ooze			chalk	chalk chert	chalk	chalk chert	chalk chert		chalk chert
Eocene/Paleocene			—142—	—336—	—104—	—292—	—354/374—	—264—		—637/656—	—295—	—348—		—163—
			nanno ooze Fe oxide	glauconitic lignitic- volcanogenic	nanno ooze chalk	ooze chalk chert	chalk chert	limestone			chalk	chalk chert		chalk
Tertiary/ Cretaceous						—332—	—421—							—169—
						chalk glauc.-volc. claystone	chalk chert dolomite							chalk basalt- conglomerate
Total depth (sediment)	344	711	152 basalt	478 basalt	156 basalt	457 basalt	664	411	331 basalt	717 basalt	306 basalt	694	555 basalt	216
Total depth	344	711	173	500	175	478	664	411	346	740	328	694	559	216

oceanic accumulation in response to the accumulation of the marginal sequence.

Introduction

The planning for Leg 22, the first stage of the Deep Sea Drilling Project's program in the Indian Ocean, was influenced rather strongly by marine geomagnetic data (later published by Sclater and Fisher, 1974) which suggested major tectonic changes during the early Tertiary. The results of drilling have thoroughly confirmed that prediction, and have added to the already substantial data indicating that the Paleocene-Eocene was a critical time in the history of the Indo-Australian Plate. To give only three examples: Powell and Conaghan (1973) suggested a continent-continent collision between India and a proto-Tibetan landmass before the Middle Eocene; Davies and Smith (1971) proposed, admittedly on slender geochronological evidence, the emplacement of the Papuan Ultramafic Belt at about the same time; Weissel and Hayes (1972) and McGowran (1973) drew attention to geomagnetic and stratigraphic evidence respectively for the final separation of Australia from Antarctica at about the same time.

This review of (mostly planktonic) foraminiferal studies attempts to correlate data, in the time interval Maastrichtian-Eocene, in an arc from the Mascarene Plateau to the Naturaliste Plateau with observations to the north and east: India - Pakistan and the Australian region (Figure 1). Taking Berggren's (1972) scale of biostratigraphy and geochronology as the standard, I have placed foraminiferal-biostratigraphic events and occurrences against time, rather than against sediment thickness as in the usual lithostratigraphic columns; what is lost in facies changes and accumulation rates is more than gained in the view of synchronism provided at the refinement available. Thus, the aim is to correlate, and to point out some of the ways in which the stratigraphic and biofacies patterns bear on tectonic and oceanic geohistory in the Indian Ocean region. The review does not attempt to summarize all the biostratigraphic information published in the DSDP volumes, but there are some biostratigraphic items, one being an attempt to relate southern extratropical biostratigraphy in the Eocene to the tropical "standard" zones - a problem that has been exercising southern extratropical biostratigraphers for years. In the interest of communication it was deemed better to use a more or less stable biostratigraphy for the early Tertiary (Berggren, 1972) (there is not comparable stability in the Maastrichtian) rather than use local zones where the defining events of the P-Zones were not observed.

Scope of Review: Sources

Foraminiferal records from the Coral Sea Basin (Sites 209, 210) were taken from the Site Reports (J. P. Kennett, in Burns, Andrews et al., 1973), and Olsson (1974).

For the Wharton Basin (213), northern part of the Ninetyeast Ridge (214, 216, 217) and Central Indian Basin (215): McGowran (1974); Pessagno and Michael (1974); Berggren, Lohmann and Poore (1974); Gartner, Johnson and McGowran (1974); personal observations.

For the Chagos-Laccadive Ridge (219, 220) and Own Ridge (223): Site Reports (R. L. Fleisher, and V. D. Mamgain et al., both in Whitmarsh, Weser et al., 1974); Fleisher (1974); Akers (1974b).

For the Mascarene Plateau and vicinity (236, 237): Site Reports (E. Vincent, in Fisher, Bunce et al., 1974); Heiman, Frerichs and Vincent 1974; Vincent (1974); Vincent, Gibson and Brun (1974).

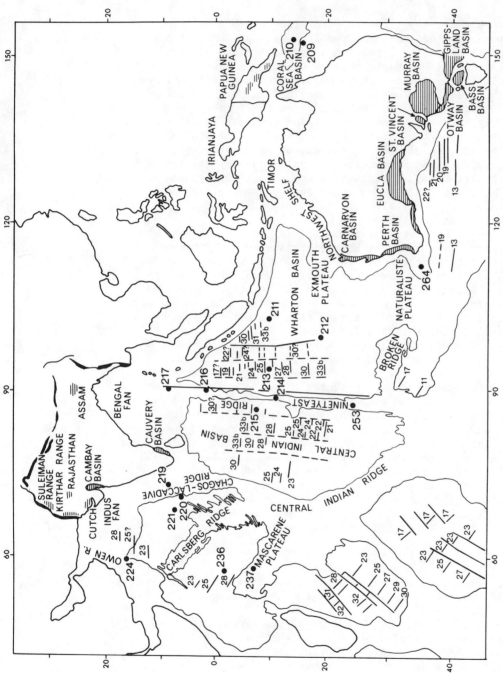

Figure 1. Locality map. Oceanic topography defined by 4000 meter isobath. Geomagnetic anomalies from Weissel and Hayes (1972), McKenzie and Sclater (1972), Sclater and Fisher (1974), and compilation by Pitman et al. (1974); younger anomalies not shown; for ages of Anomalies 29 to 16, see text figure 9. DSDP sites shown are those discussed or correlated, plus Sites 211 and 212. Shaded areas on continents are basins, localities or sections discussed. The Indian subcontinent is outlined by ophiolite zones and other tectonic lineaments (Gansser, 1966) merely to show that the sections discussed are stratotectonically on the Indian side of the presumed continent/continent suture.

For the southern end of the Ninetyeast Ridge (253): Site Report (R. C. Herb, in Davies, Luyendyk et al., 1974); Akers (1974a).

For the Naturaliste Plateau: Site Report (P. N. Webb, in Hayes, Frakes et al., 1975); personal observations.

Records from the southwest Indian Ocean (Sigal, 1974) and from the Wharton Basin other than Site 213 (McGowran, 1974; Krasheninnikov, 1975) are not included here.

Discussion of foraminiferal assemblages and successions in Pakistan, India and the Australian region is in several respects an updating and extension of a previous review (McGowran, 1968).

Maastrichtian Biostratigraphy

Biostratigraphic Standard

The Maastrichtian is taken as the interval between the extinction of Globotruncana calcarata and the extinction of the Cretaceous microfauna. Defining events (Figure 2) allow the recognition of four zones (Premoli Silva and Bolli, 1973). The subsidiary events shown are mostly clustered together within the zones as a compromise between the spacing asserted by Premoli Silva and Bolli and the assemblage zone approach of Pessagno (1967; Smith and Pessagno, 1973). The succession of events appears to be fairly consistent and confirmed; perhaps the main anomaly is that van Hinte (1972) shows Globotruncana stuarti appearing before G. contusa. This is unlikely. In Figure 2 Globotruncana falsocalcarata is added tentatively to the list of important events.

Ninetyeast Ridge

At Site 217, 66.5 m of chalk are assigned to the Globotruncanella mayaroensis Zone, and G. intermedia extends for another 9.5 m (McGowran, 1974). No other significant events of correlative significance could be recognized even though a further 113 m of sediment, often rich in planktonic foraminifera, was drilled. Thus, whereas Globotruncanella mayaroensis is the last of the commonly known and important elements of the late Maastrichtian fauna to appear (Figure 2), at Site 217 it is the first. Below about 478m (and therefore from within the G. mayaroensis Zone downwards) the assemblages have the aspect of the Campanian-Early Maastrichtian of the Carnarvon Basin: Inoceramus prisms are common to abundant and the planktonic microfauna is of low diversity but high numbers, with Rugoglobigerina and Globotruncana arca and G. lapparenti/linneiana prominent. It was concluded that there is no significant unconformity because the lower and "older-looking" assemblages was explained by the rapid migration of Site 217 from an extratropical position, when it was initially very shallow, to a tropical position when it was much deeper (see below).

In contrast, Pessagno and Michael (1974) analyse the Maastrichtian section at Site 217 in quite a different way. In applying the biostratigraphy of the Gulf Coast region (Pessagno, 1967; Smith and Pessagno, 1973) they have ascribed to upward reworking all specimens of Globotruncana lapparenti, G. linneiana and ventricosa which are in assemblages younger than expected. (This reworking was recognized almost up to the top of the Maastrichtian where G. calcarata was identified, but the specimens are probably juvenile G. falsocalcarata.) As a result, Pessagno's Globotruncana gansseri Subzone (equivalent to the gansseri plus contusa zones in Figure 2) has to occupy - if it is present at all - a 150 cm interval at about 488 m.

On geological grounds, it is difficult to see from where all the re-

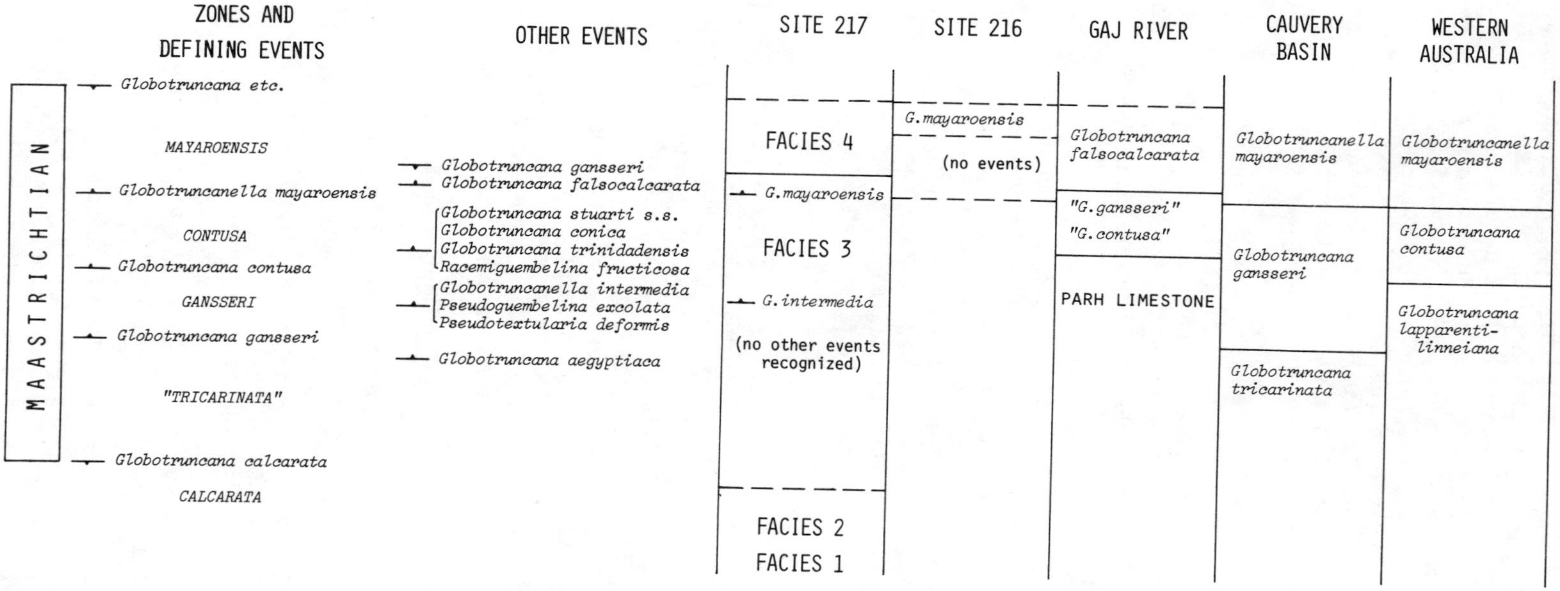

Figure 2. Maastrichtian biostratigraphic events, zones and correlations, as discussed in text. In this and subsequent text figures – first occurrence historically, i.e. up-section – last occurrence. Most events are believed to be evolutionary (phyletic emergence, or extinction) although not all have been demonstrated thoroughly.

worked material could come. On biostratigraphic grounds, it is much easier to postulate that a lower Maastrichtian microfauna lingered outside the tropics beyond the time when it was superseded in the tropical zone by the more diverse, younger microfauna. This is supported by the inverse stratigraphic relationship of Globotruncanella mayaroensis to coeval species.

At Site 216, 8 degrees south from Site 217, there are some 16 m of chalk with Globotruncanella mayaroensis. McGowran (1974) has contrasted the assemblage with the coeval assemblage at Site 217 as being of lower diversity, with G. mayaroensis occurring in unusual numbers. Below 348 m Globotruncana arca and Globotruncanella citae occur, but almost all of the sediment down to the basalt at 457 m contains only a sporadic association of Heterohelix, Guembelitria, and small "globigerinid" tests. However, Gartner (1974) recorded Nephrolithus frequens throughout the section, which means that it correlates approximately with the Globotruncanella mayaroensis Zone.

Pakistan and India

The western Gaj River section includes limestones (Parh Limestone) overlain by calcareous shales with interbedded limestones, and then shales (Jamburo Group); the limestone/shale transition was dated as within the middle part of the Maastrichtian (McGowran, 1968). The section was not analysed in detail. In studying the same section, Dorreen (1974) places this transition at the base of the Maastrichtian and divides the Maastrichtian into three parts: in ascending order, the Globotruncana contusa, G. gansseri and G. falsocalcarata zones. Both authors note the absence of Globotruncanella mayaroensis.

A sample from immediately above the Parh Limestone contains Globotruncana gansseri, G. conica, G. stuarti sensu stricto, Pseudotextularia deformis, and Pseudoguembelina excolata; Globotruncana contusa s.s. is present in the next sample; Racemiguembelina fructicosa then appears but is always sporadic. This assemblage is no older than the Globotruncana contusa Zone. However, Dorreen's distinction of a Globotruncana falsocalcarata Zone is important: G. falsocalcarata overlaps G. gansseri slightly but becomes common, even abundant and dominant, higher, as the assemblage becomes sparser towards the top of the Maastrichtian. An assemblage with prominent Globotruncana falsocalcarata and no Globotruncanella mayaroensis seems to be most unusual.

Bhatia (1974) summarizes important new observations on sections on the northern and eastern margins of India. Late Cretaceous assemblages (but without definite Maastrichtian elements) are found in the Indus Flysch. From Upper Assam, a curious list includes Globotruncanella mayaroensis but no Globotruncana. In Central Assam, late Maastrichtian assemblages include Globotruncana stuarti, Racemiguembelina fructicosa, Pseudoguembelina excolata. The "large" benthonics Siderolites and Orbitoides also occur (Nagappa, 1959). In the Baratang Formation in the Andaman Islands, the Globotruncana duwi Zone includes Pseudoguembelina excolata, Globotruncana stuarti, G. gansseri, and Globotruncanella mayaroensis.

Govindan (1972) has zoned what appears now to be a reasonably complete Maastrichtian sequence in the Cauvery Basin (Figure 2). Globotruncana (= G. aegyptiaca) and G. gansseri are succeeded by G. contusa, and then by G. stuarti and Globotruncanella mayaroensis. Govindan's paper is important in extending the biostratigraphy above the Globotruncana gansseri Zone for the first time.

Papua New Guinea

Owen (1973) has described material from the Lagaip Beds in the Western Highlands of New Guinea. Samples with Globotruncana ventricosa and Rugoglobigerina scotti, and lacking Globotruncana calcarata and G. gansseri, were placed in Pessagno's (1967) zonule equivalent to the G. "tricarinata" Zone (Figure 2). An assemblage with Pseudotextularia elegans, Racemiguembelina fructicosa, Globotruncana contusa and G. stuarti, but neither G. gansseri nor Globotruncanella mayaroensis, was assigned to the equivalent of the Globotruncana contusa Zone. The Maastrichtian assemblage listed earlier (McGowran, 1968) has now been observed personally; it is very rich, including all expected elements of the Globotruncanella mayaroensis Zone including the zone fossil itself.

Similarly, an assemblage from southeastern Papua (McGowran, 1968) contains in addition G. mayaroensis. Owen (1973) has described this assemblage from another sample.

Western Australian Margin

Shelf carbonates in the Carnarvon Basin include the Korojon Calcarenite and Miria Marl (McGowran, 1968, 1969; Gartner et al., 1974). The planktonic assemblage in the Korojon Calcarenite is dominated by Rugoglobigerina pennyi/rugosa, Globotruncana lapparenti/linneiana/arca, Heterohelix and Gublerina; Globotruncana elevata, Globotrancanella citae and others are very rare. Racemiguembelina aff. fructicosa at the top of the unit indicates that it extends into the mid-Maastrichtian. The contact with the Miria Marl is visibly disconformable in the field, and the Miria Marl has an assemblage of Globotruncana arca, G. elevata, G. contusa, G. cf. falsostuarti, Globotruncanella citae, G. mayaroensis, Rugoglobigerina rugosa/pennyi, Pseudotextularia elegans/deformis, Gublerina cuvillieri, Heterohelix spp. Both assemblages are noteworthy in lacking numerous Maastrichtian taxa: a critical observation in Maastrichtian biogeography (below). Further north, on the Northwest Shelf, an assemblage is recognized with Globotruncana contusa, G. aegyptiaca and the G. stuarti group but not G. gansseri or Globotruncanella mayaroensis. The diversity is higher than in the Carnarvon Basin.

Paleocene Biostratigraphy

Biostratigraphic Standard

There has been little change in the zone scheme for the Paleocene and in its application and recognition in the Indian Ocean region (McGowran, 1968, 1974). The Cretaceous/Tertiary boundary, still to be explained adequately as a geohistorical and biohistorical event, is stabilized at the level marked by the disappearance of the Late Cretaceous planktonic microfauna. The top of the Paleocene is recognized here at the last occurrence of the species group of Morozovella velascoensis, although this group includes the new "Globorotalia" edgari which extends a short way above Morozovella velascoensis in the Caribbean region (Premoli Silva and Bolli, 1973). McGowran (1971) argued that Pseudohastigerina - with attendant taxonomic problems - appears in the later part of the Paleocene, and now Premoli Silva and Bolli (1973) record forms intermediate between Planorotalites compressa and Pseudohastigerina wilcoxensis from as low as the Morozovella angulata Zone (Zone P.3). In any case, Pseudohastigerina tends to be elusive in the critical interval in oceanic sediments.

The zonation (Figure 3) is from Berggren (1972); Premoli Silva and Bolli (1973, Figure 4) demonstrate that there is widespread consensus in Paleocene foraminiferal biostratigraphy, variations in nomenclature and the choice of defining events notwithstanding. In Figure 3 I add other events, from various sources, which now seem to be pinned down with reasonable confidence as occurring consistently in the right chronological order.

Ninetyeast Ridge

The Cretaceous/Tertiary boundary was drilled at Sites 216 and 217. At Site 216 the critical interval is in Core 23, Section 2 (photographed in Gartner, 1974, Figure 10). There is a lithological change, in that the lower sediment contains well defined burrows, some glauconite, and is mottled, whereas the overlying sediment is more uniform, but clearly bioturbated. The "contact" is at about 105 cm. At 110-111 cm the assemblage includes Globotruncana, Rugoglobigerina and Pseudotextularia, but no Danian elements were found. At 99-100 cm the assemblage contained rare Globotruncana and Rugoglobigerina along with Planorotalites compressa, "Subbotina" inconstans and others, thus Zones P.1d, with Zones P.1a to P.1c not distinguished. Nannofossil evidence tends to support this, although the more detailed sampling indicates Danian down to 117 cm (below Maastrichtian at 110 cm) (Gartner, 1974). All evidence, taken together, indicates an hiatus with the development of a deep sea hardground. At Site 217 the record is noteworthy (McGowran, 1974) for (1) an overlap of Cretaceous taxa and Tertiary taxa closely similar, biostratigraphically, to the extent of the hiatus at the shallower Site 216; and (2) the successive disappearance downhole of Subbotina inconstans, Planorotalites compressa, Subbotina pseudobulloides, and Subbotina triloculinoides, leaving Eoglobigerina eobulloides, Globoconusa daubjergensis and Chiloguembelina. There is a striking concentration of cf. Hedbergella monmouthensis at this level, which is at the base of Core 16; Globotruncanella mayaroensis and diverse contemporaries disappear at the top of Core 17.

The Paleocene biostratigraphic record on the Ninetyeast Ridge is pieced together from the crest Sites 214, 216 and 217, and the flanking basin Sites 213 and 215. The events listed in Figure 3 include almost all the defining events and several others whose importance is confirmed (Planorotalites chapmani and Morozovella acutispira in Zone P.4). Unlike the Maastrichtian and the Eocene, the Paleocene here has few problems of age determination arising out of biogeography.

The Paleocene changes in lithofacies and biofacies at Site 214 have been discussed by Pimm, McGowran and Gartner (1974; see Figure 10). The lowest significant foraminiferal assemblages were recovered from about 390 m (Core 41). The planktonics at this level are not directly comparable with assemblages in good biostratigraphic reference sections: their aspect is influenced by the shallowness and/or extratropical position of the site. However, cf. Globoconusa of the daubjergensis/kozlowskii lineage, primitive acarinids, and Morozovella aff. angulata suggested a correlation with (rather than identification of) Zone P.3. But the absence of Zone P.4 fossils does not necessarily imply an older age. Higher (335-361 m), P.4 was identified on Planorotalites pseudomenardii at the top and a few poor specimens of Planorotalites chapmani lower in that interval. At about 335 m the Zone P.4 assemblage also includes Planorotalites imitata, Morozovella laevigata/convexa and Acarinina mckannai. This is in Core 36, Section 3, whereas Section 2 contains Morozovella aragonensis, M. lensiformis, M. marginodentata/subbotinae, and Acarinina soldadoensis. This assemblage is no lower than Zone P.7, and thus the hiatus within

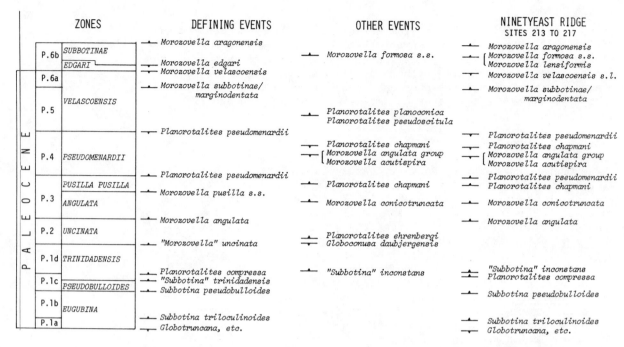

Figure 3. Paleocene biostratigraphic events, zones and correlations, as discussed in text. In this and all subsequent text figures, P-Zones are plotted in relative time value after Berggren (1972) (see Figure 9). Other zones from Bolli (1966) and Premoli Silva and Bolli (1973).

glauconitic sediments at about 335 m is represented biostratigraphically by Zones P.5, P.6, and perhaps part of P.7.

Further north (Sites 216, 217), the sequence of events in Zone P.4 is consistent (as it is at Site 215), indicating that the zone could be subdivided on the disappearance of Morozovella acutispira and Planorotalites chapmani (Figure 4). Below Zone P.4 the record is poor. It is likely that there is a break between Zones P.1 and P.3 or P.4 at both crest sites; recovery was not quite complete enough to confirm this. Two basalts at Site 215 sandwich a thin, highly indurated chalk with a poor Zone P.3 assemblage. A basalt at Site 213 underlies sediment of Zone P.5 or P.4 age.

Above Zone P.4 and across the Paleocene/Eocene boundary there is parallelism of events in time (Figures 4,9). In the deep basin Sites 215 and 213, a facies change from ooze to clay occurs close to the Zone P.6/P.7 boundary, and close to the time of resumption of sedimentation on the Ridge crest at Site 214. The accumulation of calcareous ooze on the Ridge crest is interrupted by a hiatus: Zones P.5 to P.10 are missing at Site 217 (where coring was not quite continuous), and Zones P.6b to P.10 are missing at Site 216 (where coring was continuous).

Other Records

Paleocene was identified on the Chagos-Laccadive Ridge (Fleisher, 1974) and on the Mascarene Plateau (Vincent, Gibson and Brun, 1975; Heiman, Frerichs and Vincent, 1974); it is discussed below because the biofacies changes are more significant than the biostratigraphy

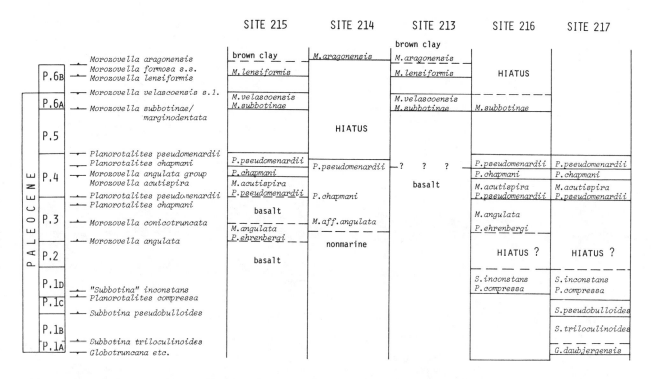

Figure 4. Paleocene biostratigraphic sections, DSDP Sites on and flanking Ninetyeast Ridge. The intervals labelled with species names are not zones, but records of species which identify - or can refine- the P-Zones. Except for the obvious case at Site 215, the age of basalt is not implied.

(Table 5, 6). Paleocene on the Naturaliste Plateau (Table 3) and Late Paleocene-Early Eocene on the Perth Abyssal Plain to the north (Krasheninnikov, 1974) are not considered further here.

Eocene Biostratigraphy

Biostratigraphic Standard

Figure 5 tabulates the defining events (Bolli, 1966; Blow, 1969); the P-Zones are drawn in relative thickness after Berggren (1972).

Ninetyeast Ridge

Although the succession is by no means uniform (Table 2), Site 214 provides the most complete oceanic section through the Eocene in the Indian Ocean. The following summary of recognized biostratigraphic events considered to be important has to incorporate biogeographic and "climatic" evidence too, because it has considerable bearing not only on the succession itself, but also on relating extratropical successions to the tropical standard.

In the Early Eocene, the disappearance of the keeled morozovellids (Morozovella gracilis, M. marginodentata, etc., closely parallel to M. acutispira in the Paleocene and M. spinulosa etc. in the Middle Eocene) is the most visible event between the incoming of M. aragonensis (at a disconformity) and the incoming of M. caucasica and Acarinina densa.

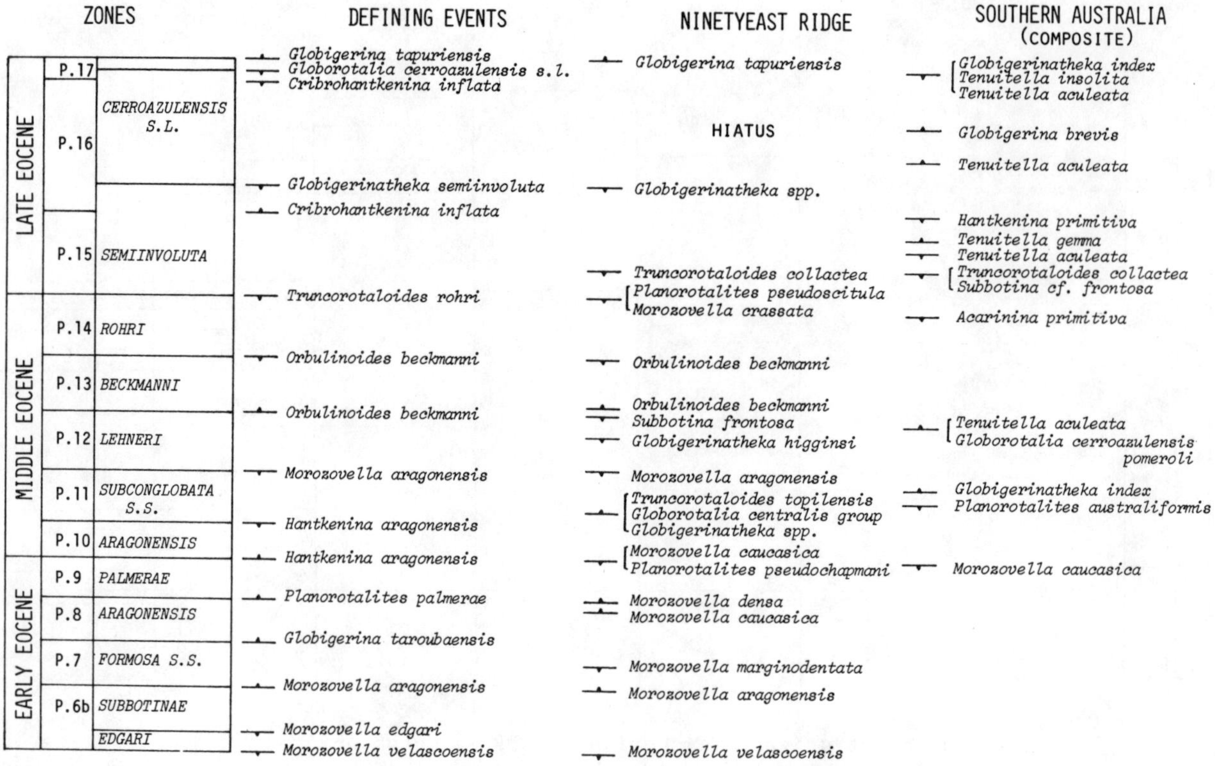

Figure 5. Eocene biostratigraphic events, zones and correlations, as discussed in text. The succession for the Ninetyeast Ridge, upwards from base Morozovella aragonensis, is based on Site 214 (Table 2). The succession for southern Australia is pieced together from the southern marginal basins, plus Site 264 (Figures 6, 7; Table 3; also McGowran, in press, b). P-Zones from Berggren (1972); other zones from Bolli (1966, 1972) and Premoli Silva and Bolli (1973).

The disappearance of M. caucasica is considered generally to be close to the Early/Middle Eocene boundary, placed at the Zone P.9/P.10 boundary. A poorly known species, identified tentatively as Planorotalites pseudochapmani, disappears at the same level. The disappearance of the Lower Eocene acarininids of the A. soldadoensis - A. angulosa group in the vicinity supports this recognition of the P.9/P.10 boundary; Planorotalites palmerae remains confined to a few records (Schmidt and Raju, 1973) and Hantkenina is poor value in most oceanic sections.

There are problems in the P.10-P.11 interval in the lowest Middle Eocene, between top Morozovella caucasica and top M. aragonensis (McGowran, 1974). Hantkenina does not occur below Truncorotaloides topilensis or "Globorotalia" cerroazulensis possagnoensis. It is concluded, first, that Zone P.10 cannot be rigorously identified at Site 214, and, second, that the primitive members of Hantkenina (aragonensis, mexicana, dumblei) need further ontogenetic and stratigraphic study before they can be used as defining events with any confidence. There seem to be "climatic" reasons for the difficulty in recognizing the P.10-P.11 interval. Thus the range of keeled Morozovella is disjunct, the species M. aragonensis, M. crassata and M. coronata disappearing temporarily; one residue is extremely sparse with only the robust

MAASTRICHTIAN TO EOCENE FORAMINIFERA

Table 2. Summary of planktonic foraminiferal succession in the Eocene, Site 214, Ninetyeast Ridge.

Core-section	Depth	Assemblage	Zone	Comments
27-6	255	*Globigerina tapuriensis*, *Subbotina angiporoides*, *Tenuitella gemma*, *Pseudohastigerina micra*, *P. barbadoensis*, etc.	P.18	Preservation very poor. Hiatus above Zone P.15
27-cc 28-cc	257 266.5	Large and variable *Globigerinatheka* including *semiinvoluta*, *subconglobatus/luterbacheri*, *index*. *Globorotalia cerroazulensis* and *cocoaensis*; *Hantkenina primitiva* and *brevispina*; *Chiloguembelina cubensis*, large globigerinids (*pseudovenezuelana*, *tripartita*, *galavisi*, *gortanii*) and *Globigerinita*; *Pseudohastigerina micra*. *Tenuitella aculeata*; *Truncorotaloides collactea* at base.	P.15	Preservation very poor. Assemblage tropical, but strongly modified by dissolution. *T. collactea*, the last acarininid, extends distinctly above top range of *Morozovella* and *Planorotalites*: important for cross-correlating tropical and southern extratropical biostratigraphic events (Figures 5,6).
29-2 29-5	268 273	*Morozovella crassata*, but morozovellids poor and sporadic. *Planorotalites pseudoscitula*, *Globorotalia cerroazulensis* (not *cocoaensis*). Large globigerinids; *Globigerinatheka* smaller and less variable than in Zones P.15 or P.13. No *Hantkenina*.	P.14	Preservation poor. An interregnum between Zones P.13 and P.15. Disjunct distributions of large *Globigerinatheka*, *G. cerroazulensis cocoaensis* and *Hantkenina* reflects watermass fluctuations (in the case of *Hantkenina*, perhaps via increased dissolution).
29-6 29-cc	275 276	*Orbulinoides beckmanni*, *Globigerinatheka* spp., "*Globigerinita*" *echinatus*, large globigerinids, *Subbotina linaperta/angiporoides*. Flourishing morozovellid complex including *M. coronata*, *M. crassata*, *M.* aff. *lehneri*. *Truncorotaloides topilensis*, *M. densa*, *Hantkenina primitiva/australis*, *Planorotalites pseudoscitula*, *Globorotalia pomeroli/cerroazulensis/cocoaensis*.	P.13	Preservation relatively good. Presence of *Orbulinoides*, large diverse *Globigerinatheka*, large diverse *Morozovella* and *G. cerroazulensis cocoaensis* (oldest occurrence recorded) indicate expansion southward of tropical zone. Thus, correlates well with (a) extensive transgression on continents, (b) southward migration of *Discocyclina* (Australia, Ninetyeast Ridge).
30-1 31-4	276.5 290	Lowest *Globorotalia cerroazulensis pomeroli*; highest *Subbotina frontosa* and *Globigerinatheka higginsi*. Disjunct range of *Hantkenina*: *primitiva* group above and *dumblei* group below. Morozovellids (*lehneri*, *coronata*, *crassata*) occur sporadically. Lowest *Tenuitella aculeata* (rare but convincing) and *T. gemma*. *Truncorotaloides topilensis* and diverse acarininids (*densa*, *spinuloinflata*, *rotundimarginata*, *collactea*). *Globigerinatheka index/tropicalis* and *curryi/euganea*. *Planorotalites pseudoscitula*, *Pseudohastigerina*, *Chiloguembelina*. *Guembelitria* prominent in one sample.	P.12	Preservation poorer than in Zones P.11 or P.13. *Globigerinatheka*, subbotinids, globigerinids and acarininids occur more consistently than do *Hantkenina* (gap between older and younger complexes) and *Morozovella*. Presence of *Tenuitella aculeata* and *Globorotalia cerroazulensis pomeroli* near top of zone used to fix events in extratropical succession (Figure 5). *Guembelitria* horizon estimated to be slightly younger than *Guembelitria* interval on Naturaliste Plateau (Figure 6; Table 3).
31-5 31-cc	291.5 295	*Morozovella aragonensis*, *M. crassata*, *M. coronata*, *M. broedermanni*, *Truncorotaloides topilensis*, diverse acarininids as above. *Hantkenina dumblei* group, *Planorotalites pseudoscitula*, *Pseudohastigerina wilcoxensis/micra*, *Chiloguembelina* spp., *Globorotalia cerroazulensis possagnoensis*. Lowest large and well developed *Globigerinatheka index/tropicalis* and *curryi/euganea*.	P.11	Preservation relatively good, deteriorating upwards. Base of interval noteworthy for *Globigerinatheka* specimens: taken as reference for base *G. index* in southern Australia.
32-1 32-2	295 298	Lowest *Globigerinatheka* (except *senni*), lowest *G. cerroazulensis possagnoensis*, lowest *Hantkenina* (*aragonensis/mexicana/dumblei*), lowest *Truncorotaloides topilensis*. Morozovellids sporadic; acarininids diverse and include *A. primitiva*. *Planorotalites australiformis*, *P. pseudoscitula*, and lowest (common, variable) *Chiloguembelina*.	P.11	Interval 245-300m believed to record sharp fluctuations in oceanic conditions. One sample at 298m has very sparse residue with only robust *Globigerinatheka senni*, *Morozovella densa* and subbotinids/globigerinids. Crusts on tests, especially acarininids, particularly heavy. Keeled *Morozovella* has a sporadic distribution. Extratropical species *Planorotalites australiformis* and *Acarinina primitiva* occur as only a fraction of their known ranges (*P. australiformis* near its extinction). All evidence indicates a sharp cooling in P.10 and early P.11 followed by a warming with numerous first appearances (including - presumably delayed - primitive *Hantkenina*) in a rebound effect.
32-3 32-cc	298 304.5	Lowest common *Globigerinatheka senni* and bullate globigerinids. *Morozovella aragonensis*, *M. coronata*, *M. crassata*. Diverse acarininids including *A. primitiva* and "pre-*topilensis*". Turborotaliids resemble *pseudomayeri* and *wilsoni*. *Planorotalites pseudoscitula* (not *australiformis*), *Pseudohastigerina wilcoxensis*.	P.11 or P.10	Top range *P. australiformis* just below base range *G. index*: consistent with extratropical succession (Figure 5). Zone P.10 not identified clearly, since *Hantkenina* (including *aragonensis*) believed to appear first in Zone P.11.
33-1 34-5	305 320	*Morozovella aragonensis*, *M. caucasica*, *M. crassata*, *M. densa*. Acarininids of Lower Eocene aspect: *soldadoensis*, *angulosa*, *pentacamerata*, *triplex*, *broedermanni*. *Planorotalites pseudoscitula*, *P. pseudochapmani*, *Pseudohastigerina wilcoxensis* (very variable). Subbotinids include *S. frontosa*.	P.9	Preservation relatively good. Top *Morozovella caucasica* taken as Lower/Middle Eocene boundary, as on Naturaliste Plateau (Table 3). Base *Morozovella densa* taken as Zone P.8/P.9 boundary.
34-6 34-cc	321.5 323.5	*Morozovella aragonensis*, *M. formosa* s.s., *M. lensiformis*. Acarininids as above; other taxa mostly as above.	P.8 to P.7	
35-1 36-2	324 334.5	*Morozovella subbotinae/marginodentata* complex, *M. aragonensis*, *M. lensiformis*. Other taxa as above. Decreasing up-section: neritic benthonics (*Cibicides*, *Anomalinoides*), molluscan fragments, echinoid spines, fecal pellets, glauconite.	P.7	Evidence indicates facies change from neritic to oceanic. Eocene section ends in hiatus: Zone P.7 at 334.5m over Zone P.4 at 336m.

Globigerinatheka senni (dominant), Morozovella densa and Subbotina; the biostratigraphically important extratropical forms Acarinina primitiva and Planorotalites australiformis are distinct, with greatly abbreviated ranges, only in this part of the section; acarininids are especially common and calcite crusts on the tests are thicker here. All of these observations can be explained by watermass changes toward cooling, the changes controlling both the distribution of populations and the preservation of the skeletons.

In contrast, the narrow interval identified as Zone P.13 on the presence of Orbulinoides beckmanni seems to record an episode of warming. O. beckmanni is restricted to the base of one core (two samples) and this record almost certainly represents an excursion southwards by the species and a part-range, not the total range. Specimens of Morozovella and Globigerinatheka are more robust and flourishing than they are in Zones P.12 or P.14, and preservation is better than below or above. In the initial report on Site 214 (McGowran, 1974) the association of Globorotalia cerroazulensis cocoaensis with Orbulinoides beckmanni was ignored as a probable contaminant, because G. cerroazulensis cocoaensis is well established in a phylogenetic context as an excellent marker in the Late Eocene (Tourmarkine and Bolli, 1970). However, subsequent careful study convinces me that the association is real. The premature occurrence of G. cerroazulensis cocoaensis would, then, be an instance of allochronous parallelism, perhaps stimulated by climatic change but never becoming more than a peripheral isolate in the lineage, unlike the main occurrence of the morphotype in Zone P.15-P.16. The occurrence is being described elsewhere.

The Late/Middle Eocene boundary is marked clearly by the disappearance of Morozovella, Acarinina and Planorotalites. Assemblages above this level contain large, variable Globigerinatheka; although the morphotypes recognized nomenclaturally by Bolli (in 1972) have not been distinguished consistently on the Ninetyeast Ridge, the presence of a Globigerinatheka complex, including G. semiinvoluta and G. subconglobatus luterbacheri, is a good indication of Zone P.15. I now recognize an hiatus between top Globigerinatheka and base Globigerina tapuriensis.

The Eocene record elsewhere on the Ninetyeast Ridge is poor (Figure 6). At Site 253 the upper part of the Middle Eocene is recognizable on the presence of acarinids and Globigerinatheka higginsi. Zone P.15 was identified on the presence of Globigerinatheka semiinvoluta and sampling is good enough to identify an hiatus above this level. At Site 216 an assemblage with Globorotalia cerroazulensis pomeroli and G. cerroazulensis s.s., Morozovella crassata and others is placed in Zone P.14; immediately below, an association of M. aragonensis and Truncorotaloides topilensis identified Zone P.11. At Site 217 the latter association is also present. An assemblage with G. cerroazulensis cunialensis and Hantkenina may be the only record of (in situ) uppermost Upper Eocene in the oceanic sediments under review.

Chagos-Laccadive Ridge, Owen Ridge, Mascarene Plateau

Site 219 has a Zone P.14-P.15 interval (Fleisher, 1974) similar to that on the Ninetyeast Ridge. This interval is separated from an earlier interval with Morozovella aragonensis; an hiatus is clearly defined by the absence of Zone P.12 and P.13. At Site 220 only the assemblage with M. aragonensis and Globigerinatheka was encountered. At both sites the Early/Middle Eocene boundary is not clearly marked although Hantkenina aragonensis was found: Morozovella caucasica does not have a range top comparable with Site 214. Both Eocene sedimentary sections

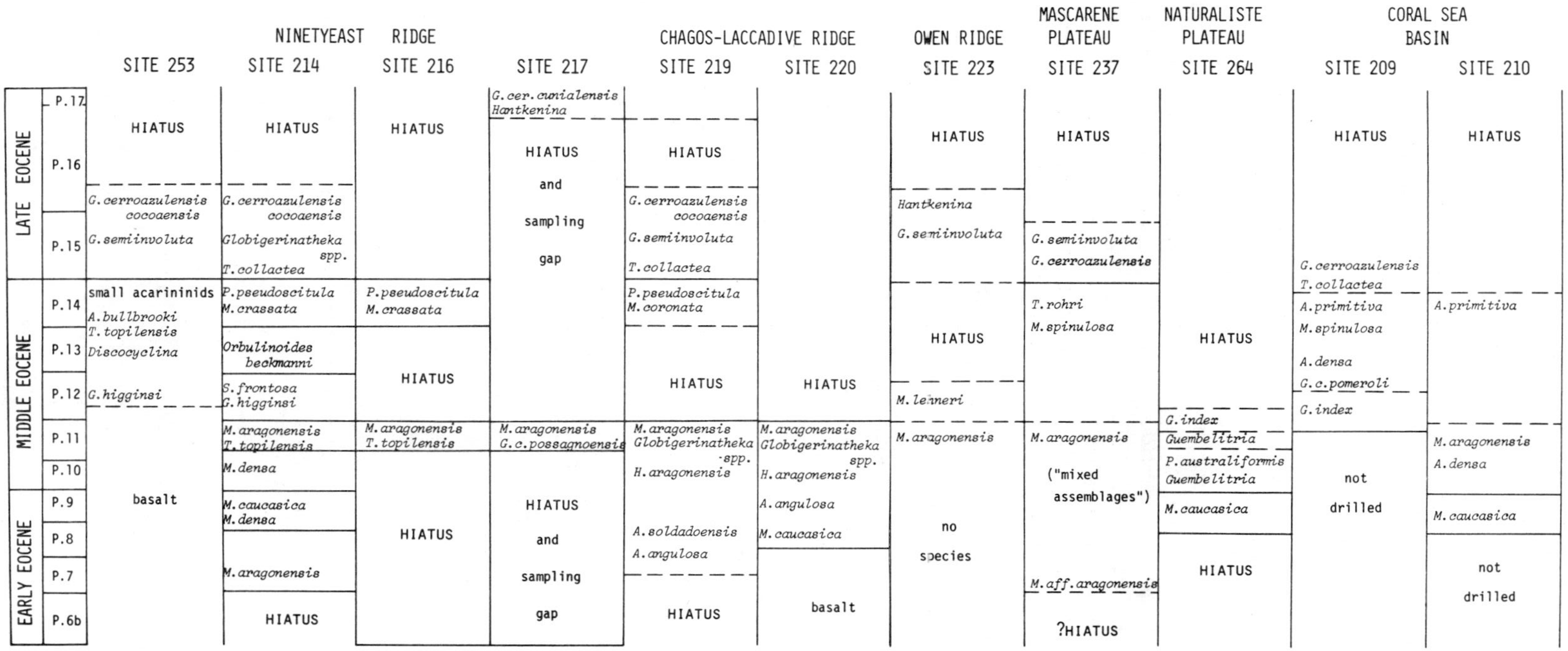

Figure 6. Eocene biostratigraphic sections, DSDP Sites, northeastern Indian Ocean and Coral Sea. The species listed in each section are selected as characteristic of the interval and as providing stratigraphic markers, with reference to Table 2, Figure 5 and discussion text.

begin in the vicinity of Zones P.7-P.8, Site 219 with an hiatus over Zone P.4, and Site 220 encountering basalt.

On the Owen Ridge and Mascarene Plateau the Eocene record is rather poor, but the general pattern is discernible (Figure 6): early Middle Eocene with Morozovella aragonensis, late Middle Eocene with Truncorotaloides rohri and M. spinulosa, early Late Eocene with Globigerinatheka mexicana/semiinvoluta, and hiatuses separating these intervals.

Naturaliste Plateau

A summary of brief personal observations and conclusions on Site 264 (two holes) is given here in Table 3. The planktonic foraminiferal assemblages are very different from those at Site 214, the nearest significant, coeval oceanic section in the Indian Ocean. The Lower Eocene is recognized on the presence of Morozovella causasica. There is a sharp change, in that this assemblage is succeeded by one dominated by Guembelitria, within which Planorotalites australiformis disappears. The disappearance of Guembelitria and initial appearance of Globigerinatheka index are the next significant events, in that order. At the top of the Eocene section one specimen of Globorotalia cerroazulensis pomeroli was found in association with Acarinina primitiva and Morozovella densa. The absence of convincing Tenuitella aculeata is considered good evidence that the Neogene/Eocene hiatus at Site 264 extends below the foot of the Eocene transgression in southern Australia (high in Zone P.12 or P.13: see below). P. N. Webb (in Hayes, Frakes et al., 1975: Site Report for Site 264) identified Late Eocene immediately below the Miocene/Eocene unconformity on the absence of Acarinina primitiva, but this species is present in samples contaminated by Miocene material. Site 264 provides the first significant section below Zone P.13 in or near southern Australia. It is typical of the southern extratropical regime in: (1) the presence of Morozovella caucasica in the Lower Eocene; (2) the poor record, otherwise, of morozovellids; (3) the dominance of Acarinina primitiva (especially) and A. collactea among acarininids; (4) the presence of Planorotalites australiformis. Radiolarians occur in some abundance at the top of the section where the preservation of foraminifera is poor; this horizon is of about the same age as the initial appearance of radiolarians on the Ninetyeast Ridge at Sites 216 and 217, but not 214, where they were not found below the Neogene (Johnson, 1974).

Early Tertiary Foraminiferal Succession,
Indian and Australian Continental Margins

Pakistan and India

A previous review of the foraminiferal succession from the Cretaceous/Tertiary boundary into the Early Eocene (McGowran, 1968) revealed (among other things) two points, important at the time. First, there was no convincing succession across the Erathem boundary: there are demonstrable hiatuses, or regressive or shallow water facies (e.g. Pab Sandstone), or trap (volcanics). With few exceptions, this remains the case. However, Bhatia (1974) records an identification of the "Globigerina" eugubina Zone in Assam, and the "Globigerina" edita Zone in the Andaman Islands. A second main conclusion was that the Laki Series, characterized by mollusca and larger foraminifera and dated traditionally as Early Eocene, was in large part of Paleocene age and extended even as low as Zone P.4; thus, the preceding Ranikot Series would have to be squeezed into a rather narrow interval in the mid-

Table 3. Summary of planktonic foraminiferal succession, Site 264, Naturaliste Plateau. Paleocene: from P. N. Webb, in Site Report (Hayes, Frakes, et al., 1975, Chapter 2). Eocene: personal observations.

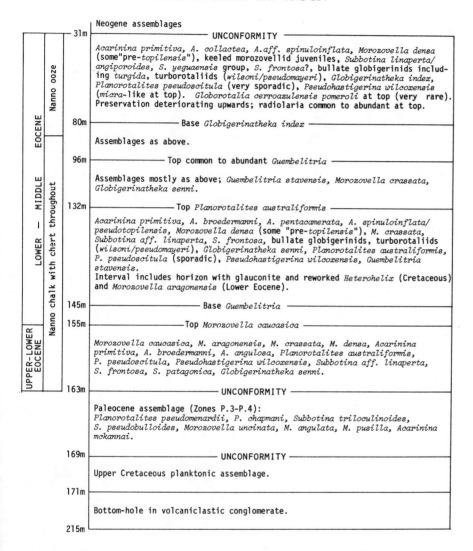

Paleocene, although it was evident too that the Ranikot and Laki were not adequate as chronostratigraphic units. These points have been substantiated by subsequent work (e.g. Sigal, Singh and Lys, 1971; Datta, Banerji and Soodan, 1970).

On the western margin, the Gaj River section (Kirthar Range), and Rakhi Nala section (Suleiman Range) are marginal to the continental platform. The sequence in the Gaj River has been divided recently into three Paleocene, two Early Eocene and two Middle Eocene zones (Dorreen, 1974) whereas a previous, very brief summary (McGowran, 1968) and further observations on the Early-Middle Paleocene and Early Eocene intervals indicated that the standard zones could be recognized (Figure 7). Assemblages with <u>Globotruncana falsocalcarata</u> are succeeded by an

434 MCGOWRAN

Fig. 7a.

			TIMOR	IRIANJAYA	NEW GUINEA HIGHLANDS	PORT MORESBY	CARNARVON BASIN	EUCLA BASIN	OTWAY BASIN
EOCENE	LATE	P.17		G.cerroazulensis Pellatispira			G.cerroazulensis	G.increbescens T.gemma	G.index — G.increbescens T.gemma Haytkenina primitiva T.aculeata T.collactea
		P.16	limestone with larger forams, T.topilensis G.cerroazulensis		limestones with larger forams and T.topilensis	limestones with larger forams			
		P.15							
	MIDDLE	P.14		?			A.primitiva G.index S.frontosa	A.primitiva G.index S.frontosa	HIATUS A.primitiva
		P.13							
		P.12							
		P.11	?		not known	not known	HIATUS	HIATUS	HIATUS P.australiformis
		P.10							
		P.9							
	EARLY	P.8	both deep and shallow water foram facies; former with extensive reworking up to Zone P.7				M.aragonensis M.subbotinae M.aequa		M.aequa M.acuta
		P.7		not known					
		P.6b					M.velascoensis P.pseudoscitula		
		P.6a							
PALEOCENE	LATE	P.5		M.velascoensis P.pseudomenardii		limestones	P.pseudomenardii M.acutiespira		
		P.4			P.pseudomenardii	P.pseudomenardii			
		P.3		?		M.conicotruncata	M.conicotruncata		P.ehrenbergi
	EARLY	P.2			limestones	G.daubjergensis P.compressa	G.daubjergensis P.compressa		
		P.1					HIATUS		HIATUS

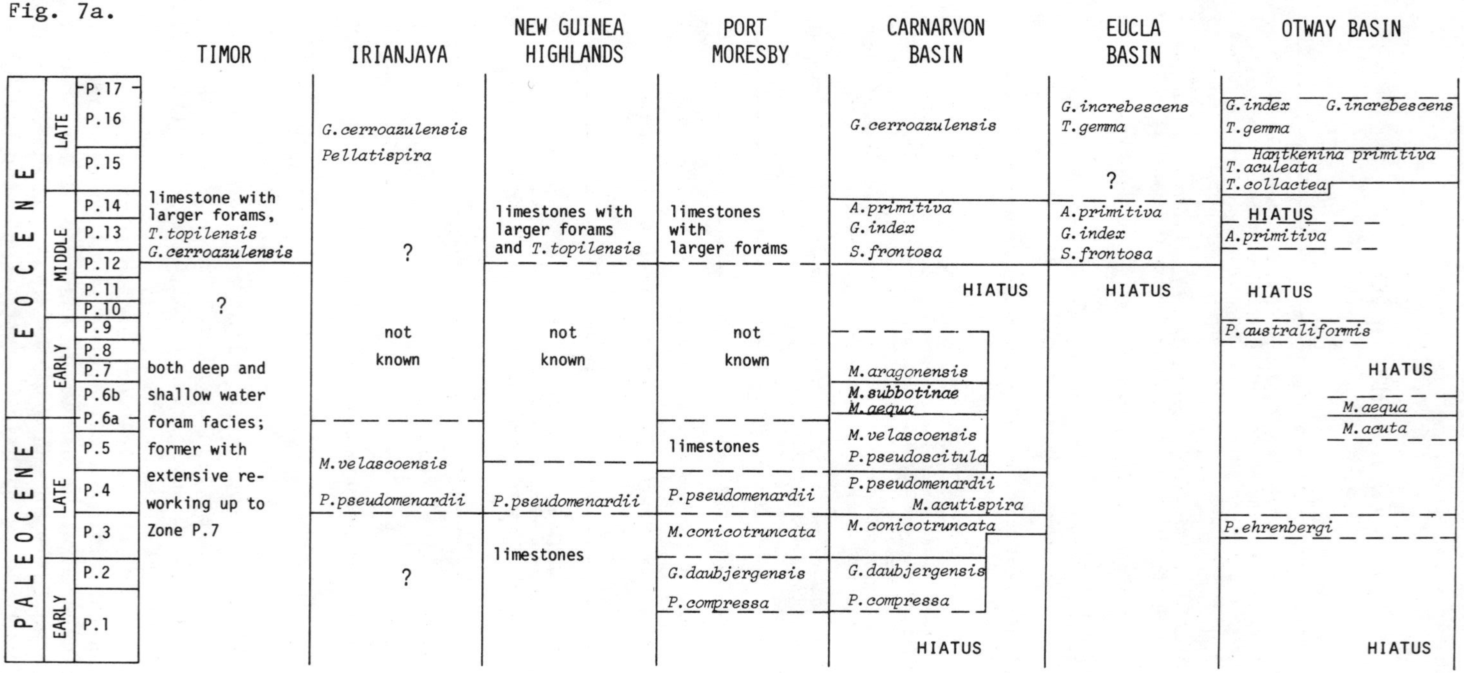

Figure 7. Early Tertiary biostratigraphic records and correlations, northern and eastern margins of the Indian Ocean. With reference to Figure 1, the Gaj River section is in the Kirthar Range and the Rakhi Nala is in the Suleiman Range. Carnarvon Basin: left, offshore (west); right, onshore (east). Otway Basin, ingressions pre-Late Eocene: left, Gambier Embayment (west) to right, Port Campbell Embayment (east). General comments as for Figure 6.

MAASTRICHTIAN TO EOCENE FORAMINIFERA 435

Age	Zone	CUTCH	CAMBAY BASIN	GAJ RIVER	RAJASTHAN	RAKHI NALA	ASSAM	CAUVERY BASIN
EOCENE LATE	P.17	HIATUS	G. cerroazulensis Crib. inflata	G. cerroazulensis Cribrohantkenina	HIATUS	impoverished assemblage	G. cerroazulensis Crib. inflata	G. cerroazulensis Crib. inflata
	P.16							G. semiinvoluta
	P.15			hard limestones			Globigerinatheka spp.	
EOCENE MIDDLE	P.14	T. rohri	T. rohri	M. spinulosa	T. rohri	T. rohri	T. rohri	T. rohri
	P.13	O. beckmanni		M. lehneri	M. lehneri	M. lehneri	M. lehneri	O. beckmanni
	P.12	G. cerroazulensis	?				T. topilensis	T. topilensis
	P.11	HIATUS LATERITE	agglutinated assemblage	?HIATUS agglutinated assemblage	HIATUS LATERITE agglutinated assemblage "Ypresian" larger foraminifera	no planktonics	NONMARINE	G. kugleri
EOCENE EARLY	P.10							H. aragonensis (partly regressive)
	P.9					Acarininids		M. aragonensis
	P.8			M. aragonensis		M. subbotinae		M. subbotinae
	P.7	Nummulites	Nummulites Assilina	M. velascoensis s.l. limestone with Miscellanea Alveolina Discocyclina		M. formosa	limestones with larger foraminifera	M. aequa
	P.6b							
	P.6a				M. velascoensis	M. velascoensis		M. velascoensis
PALEOCENE LATE	P.5			P. ehrenbergi	P. pseudomenardii	P. pseudomenardii		P. pseudomenardii
	P.4			M. angulata	(M. pusilla Zone)	M. conicotruncata	perhaps all planktonic zones present	M. angulata
	P.3			M. uncinata	regressive lower Ranikot			M. uncinata
	P.2			P. compressa		no planktonics		S. trinidadensis
PALEOCENE EARLY	P.1	TRAP	TRAP	HIATUS				?

Fig. 7b.

association in calcareous shales of Planorotalites compressa, P. imitata, Subbotina pseudobulloides, S. triloculinoides, S. trinidadensis, S. inconstans, Globoconusa daubjergensis: this is Zone P.1d. Morozovella uncinata appears subsequently (Zone P.2) and then M. angulata M. conicotruncata (Zone P.3). Hard limestones, assigned by Dorreen to his Globorotalia velascoensis Zone, are here placed in Zones P.4-P.5 because the shales immediately below have Planorotalites ehrenbergi but not P. pseudomenardii. Shales above the limestones have rare Morozovella velascoensis s.l. (probably occlusa) at the base (P.6a) but are an Early Eocene assemblage with diverse acarininids (A. pseudotopilensis group, A. soldadoensis/angulosa group), Morozovella marginodentata group, M. aequa, Pseudohastigerina wilcoxensis (Zone P.6b). Morozovella aragonensis appears near the top of the planktonic sequence, which is replaced by an agglutinated benthonic assemblage of shallow water aspect with some reworked planktonics. Interestingly, Dorreen records the benthonic species Nuttallides truempyi associated with M. aragonensis; N. truempyi is considered by some to indicate bathyal depths (Table 6). Dorreen does not recognize a regressive interval between his Globorotalia aragonensis Zone (Early Eocene) and Hantkenina dumblei Zone (Middle Eocene). However, he does not record Morozovella aragonensis in the latter zone; this contains M. lehneri, Globorotalia centralis (= G. cerroazulensis pomeroli) and Globigerinatheka and can be correlated with Zone P.12, in which the species group of Hantkenina including H. dumblei also occurs. Orbulinoides, thus Zone P.13, was not recognized, but the top of the Middle Eocene is seen on the disappearance of morozovellids and acarininids. The Late Eocene is sparse because of hard limestones.

The Gaj River succession is paralleled very closely in the Rakhi Nala section, discussed most recently by Samanta (1972, 1973a). The Rakhi Nala is different at the base, where thick, coarse clastics have no determinable planktonic foraminifera. The biostratigraphic sequence begins within Zone P.3 (Morozovella conicotruncata), and Planorotalites pseudomenardii and M. velascoensis occur. The Early Eocene is very similar to the Gaj River: M. subbotinae group, M. aequa, Pseudohastigerina. M. aragonensis is restricted to a very short interval, disappearing as the assemblage becomes sharply restricted in diversity and dominated by acarininids without keeled morozovellids. Middle Eocene assemblages are separated from this assemblage by strata without planktonics, except where reworked. The Middle Eocene has Morozovella lehneri, Globorotalia centralis (= G. cerroazulensis pomeroli), Globigerinatheka, primitive Hantkenina and Truncorotaloides topilensis but no Morozovella aragonensis: this is Zone P.12. Thus, the timing of two intervals with rich assemblages and separated by a regressive interval is virtually identical in the Gaj River and the Rakhi Nala.

This stratigraphic parallelism extends over the foreland to the east. In Cutch (Mohan and Soodan, 1970; see also Samanta, 1970b), the Middle Eocene is clearly transgressive, lying over laterite and beginning with gypsiferous shales in which the planktonic assemblages are low in diversity. The association of Globorotalia "centralis", Truncorotaloides topilensis, Morozovella lehneri, Subbotina frontosa and others in the absence, again, of M. aragonensis indicates Zone P.12; Orbulinoides identifies Zone P.13 and Zone P.14 is present, but the Late Eocene is missing. In Rajasthan (Khosla, 1972a,b; Singh, 1969; Sigal et al., 1971) the higher Early Eocene includes marginal marine, agglutinated foraminiferal assemblages, and a laterite; the Middle Eocene transgression begins with a ferruginous sandstone succeeded by larger foraminiferal assemblages (Discocyclina Zone and Flosculina Zone) from the lower of which are recorded Morozovella spinulosa and M. lehneri but not M. aragonensis. Again, the Late Eocene is missing. In the

Himalayan Foothills on the northern continental margin (Datta et al., 1970; Raju and Guha, 1970) there is perhaps a rather more complete larger foraminiferal succession, but Raju and Guha record a barren zone between the Lower Eocene Dictyoconoides and Assilina assemblages and the Middle Eocene Nummulites assemblages.

In Assam in the northeast, the Paleocene planktonic succession mentioned already is overlain by larger foraminiferal assemblages in the Late Paleocene to Early Eocene. Samanta (1968, 1969, 1970a, 1971, 1973c) has shown that the Middle Eocene marine section, with abundant planktonics and larger benthonics, overlies non-marine strata. Samanta correlates the lowest assemblage with the lowest Middle Eocene assemblage in the Rakhi Nala and with Zones P.12 and P.13. The evidence is good: the presence of Globorotalia "centralis", Truncorotaloides topilensis, Morozovella lehneri, and the absence of M. aragonensis.

In the Cauvery Basin in South India the planktonic succession through the Paleocene-Eocene seems to be rather more complete than it is anywhere around the leading margin of the subcontinent (Raju, 1968; Raju and Guha, 1970; Banerji, 1973). Samanta (1970c, 1973b) has restudied the Paleocene near Pondicherry, about which there had been some confusion (Rajagopalan, 1968; McGowran, 1968) but he compounds this confusion by refusing to accept Planorotalites pseudomenardii as a total range marker for Zone P.4, which includes algal and discocyclinid limestones and marlstones. The Eocene is relatively complete, and includes records, perhaps unique on the subcontinent, of Morozovella aragonensis extending into the Middle Eocene. Even so, there are some indications of regression in the vicinity of the Early/Middle Eocene boundary. Schmidt and Raju (1973) show convincingly that "Globorotalia" palmerae is an end member of a Planorotalites lineage.

There is, then, considerable new and more precise evidence to support the old notion that there is widespread hiatus, or regressive facies, between the "Laki" (= "Ypresian", Early Eocene) and the "Khirthar" (= "Lutetian", Middle Eocene)(see Nagappa 1959) for an outstanding review:" ....Khirthar time witnessed perhaps the most important and widespread of all the transgressions in the history of the Early Tertiary of the region"). This transgression was virtually isochronous within Zone P.12. Indeed, there is little evidence of larger foraminiferal assemblages in the interval spanning late Early Eocene to early Middle Eocene (Figure 8). In terms of the East Indian letter classification of larger foraminiferal zones (Adams, 1970), the Ta_2/Ta_3 (= Early/Middle Eocene) boundary records a most abrupt faunal change which reflects stratigraphic changes rather than inadequate knowledge (Adams is rather more cautious but does not relate the succession to a planktonic foraminiferal zonation, as is attempted in the present Figure 8). Detailed lithostratigraphy in Pakistan, where the only described sections across the Ta_2/Ta_3 boundary are found (Adams, 1970), consistently yields evidence of regression and a break in the faunal succession e.g. Fatmi,1974). The youngest "Laki" (Ta_2) assemblages extend as high as Zone P.7, with Morozovella aragonensis (e.g. Dorreen, 1974); the oldest "Khirthar" assemblages are within Zone P.12. The gap is bigger in the Cauvery Basin where, as noted, the planktonic succession is rather more complete: one larger foraminiferal assemblage is centred on the Zone P.4-P.5 interval, and the next begins in Zone P.12-P.13.

Australian Region

The marine Early Tertiary of the Australian region can be treated conveniently in three parts: the northern margin (Timor, Irianjaya, Papua New Guinea), the northwest-southwest continental margin, and the southern margin (Figures 1,7). The foraminiferal biofacies vary

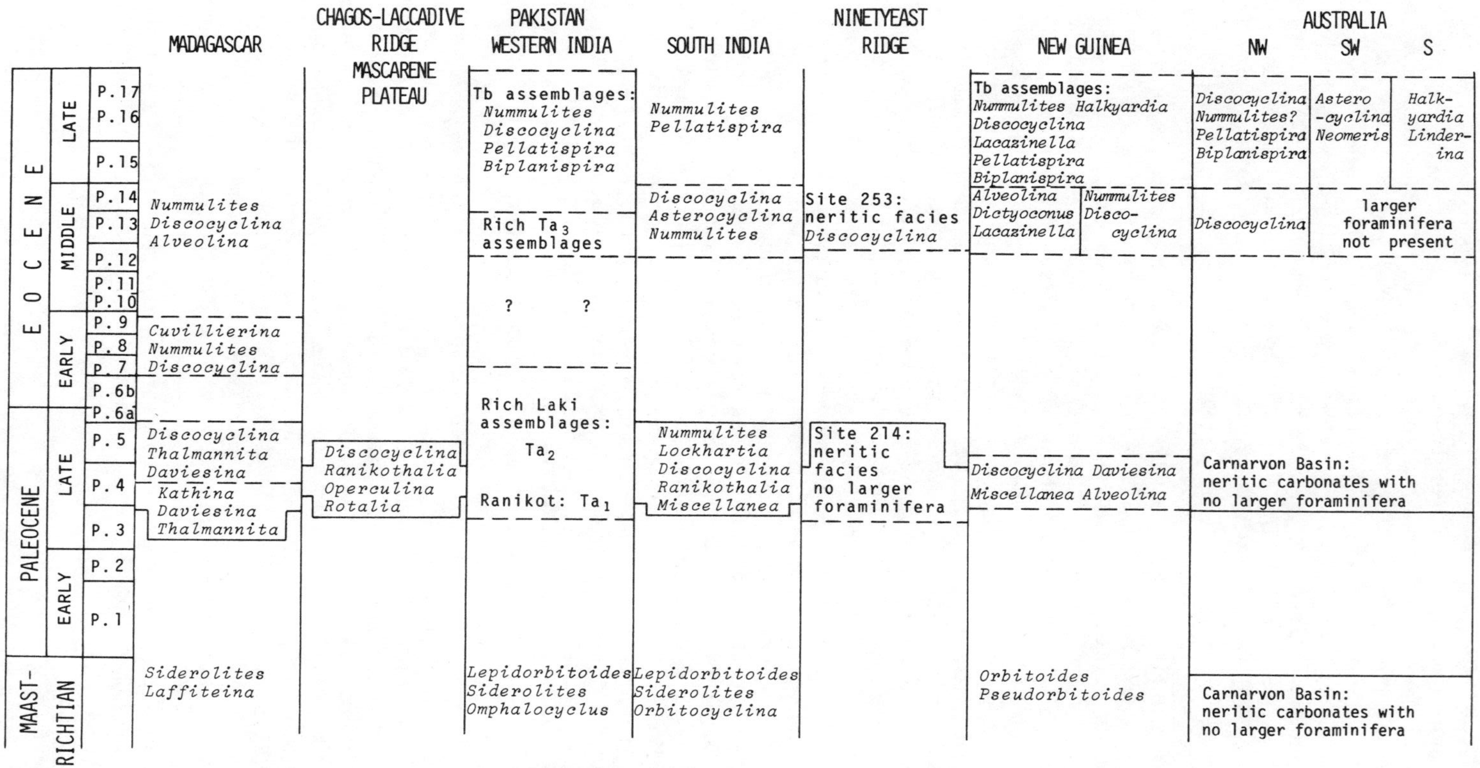

Figure 8. Maastrichtian to Eocene assemblages of larger foraminifera, Indian Ocean region. Data from Nagappa (1959), Lys (1960), Adams (1969) and references and observations used to compile Text figures 6 and 7 and Tables 5 and 6. Indo-West Pacific Letter Classification ("Stages"; actually assemblage zones): Ta1, traditionally Late Paleocene; Ta2, Early Eocene, Ta3, Middle Eocene; Tb, Late Eocene.

considerably, partly for stratotectonic reasons and partly because some 40° latitude are covered, but parallelisms and similarities are emerging, especially in the timing of geohistorical events.

The situation on Timor is complex and variable (Audley-Charles and Carter, 1972). Paleocene is very poorly known although most of the zonal indices - and several from the Maastrichtian - are found reworked into the Early Eocene (Kolbano facies) especially up to Zone P.7 (Morozovella aragonensis). However, younger assemblages have been reported but not detailed. The facies is deep-water: virtually all the indigenous and reworked specimens are planktonic. An Early Eocene, shallow-water ("back-reef") larger foraminiferal assemblage (Mosu facies) would seem to be a very rare occurrence (Adams, 1970). The younger Wiluba-Dartollu facies is much more familiar: a Discocyclina-Nummulites assemblage is associated with Truncorotaloides topilensis and Globorotalia centralis and is no older than Zone P.12; the facies either extends into or is repeated in the Late Eocene (with Pellatispira).

On the island of New Guinea, it may be that most of the Early Eocene and the lower part of the Middle Eocene is entirely absent. Thus, in Irianjaya, Visser and Hermes (1962) do not differentiate in their reference biostratigraphy any planktonic foraminiferal zones between a Morozovella velascoensis Zone (= Zones P.4-P.5; Planorotalites pseudomenardii is present) and a Globorotalia centralis Zone which is Late Eocene because Pellatispira and Biplanispira are shown as ranging below it. In the vicinity of Port Moresby, and in the central Highlands region of Papua New Guinea, deep water planktonic assemblages are now known to be quite widespread around the edge of the Australian Platform (McGowran, 1968, for older records; Belford, 1967; numerous unpublished personal observations). Most of the records range from Zone P.3, with Morozovella conicotruncata, through Zone P.4 especially (Planorotalites pseudomenardii) and perhaps into Zone P.5, but Palmieri (1971) has described a Zone P.1d assemblage with Globoconusa daubjergensis and Planorotalites compressa. Early Eocene is, in contrast to the Late Paleocene, almost entirely unknown and may not extend above Zone P.6. The Late Eocene, also in contrast, is widespread, as it is throughout the Indo-Pacific region (Adams, 1970), but it is becoming clearer that these shelf carbonate sequences begin in the Middle Eocene. At least two distinct Middle Eocene biofacies are known: Alveolina (Fasciolites)-Dictyoconus (Bain and Binnekamp, 1973) and a more "open shelf" association of Nummulites, Discocyclina and rotaliids which can be mixed with acarininid and morozovellid planktonics including Truncorotaloides topilensis. This association occurs in the Central Highlands and at the head of the Gulf of Papua (personal observations).

On the continental margin of Western Australia (Quilty, 1974b), an Early Tertiary section crops out in the Carnarvon Basin (Condon et al., 1956). Paraconformable shelf carbonates range in age from Senonian to Late Eocene. The base of the Paleocene lies diachronously (Zones P.3 into P.4) over Maastrichtian with Globotruncanella mayaroensis, and this sequence is restricted to Zone P.4 (McGowran, 1964, 1968, 1969). The next packet of carbonates is Middle Eocene, on the evidence of the morozovellid Globorotalia spinulosa (faunal list in Condon et al., 1956) to Late Eocene, with Pellatispira. In the subsurface to the west and on the northwest shelf, the Paleocene is of deeper water facies and also more complete, beginning with a Zone P.1d association of Globoconusa daubjergensis and Planorotalites compressa, and extending above Zone P.4 and into the Early Eocene, where the highest recognized biostratigraphic horizon is base *Morozovella aragonensis*. The base of the Middle-Late Eocene carbonates can be dated as Zone P.12, on the presence of *Globigerinatheka* and *Subbotina frontosa* and the absence of *Morozovella aragonensis*. Thus, the "major unconformity" recognized by petroleum geologists (Thomas and Smith, 1974) straddles the Early/Middle Eocene boundary. As in the Gaj River and the Rakhi Nala, in Timor, and on the Naturaliste Plateau, there is an horizon of reworked foraminifera at this level, but not below or above. However, the critical interval usually is obscured by recrystallization of carbonates (Wright, 1973, plus unpublished range chart). *Discocyclina* is associated with *Acarinina* and *Morozovella* in about Zone P.13. Biogeographically, the Carnarvon Basin succession reflects the plate-wide changes in Eocene stratigraphy, in that larger foraminifera have not been found in the Paleocene-Early Eocene sequence but are present from the Middle Eocene onwards, although reduced in diversity compared to the "pool" to the north (Figure 8).

In the Perth Basin to the south (Quilty, 1974a) a stratigraphically isolated episode of terrigenous sedimentation in a shallow to marginal marine environment was dated as Zone P.4 on the presence of *Planorotalites chapmani* and *P. pseudomenardii* (McGowran 1968, 1969). Subsequent work (Quilty, 1974a) has extended the range of the episode to Zone P.6 or low P.7, with *Acarinina primitiva*, *Pseudohastigerina* sp., and *Morozovella dolabrata* (? = *M. lensiformis*). That is, there is a close similarity in time, albeit in a quite different environment, to the Paleocene-Early Eocene carbonate sequence to the north.

On the southern Australian margin the Paleocene to Early Eocene is recorded in a rapidly deposited sequence of terrigenous clastics of approximately the same range in age as the sequence in the Perth Basin. The environment is mostly marginal marine, punctuated by marine ingressions with shelly fossils including foraminifera with a variable planktonic component. Four ingressions in the Otway Basin with characteristic planktonics are so labelled in Figure 7; ingressions also occur in a similar lithological sequence on the continental margin south of the Eucla Basin. Correlation is a continual problem (McGowran, 1965, 1968, 1970, 1977, b) because we are dealing with extratropical assemblages in a marginal marine environment. The assemblage with *Planorotalites australiformis* occurs in an episode of detrital sedimentation distinctly younger than the lower sedimentary episode; palynobiostratigraphic studies (Harris, 1971) have identified a major terrestrial microfloral change between the ingressions here labelled *Morozovella aequa* and *P. australiformis*.

The Middle Eocene transgression marks the onset of the extratropical, bryozoal, carbonate sedimentational regime which characterizes the southern Australian continental margin facing the ocean today (McGowran, 1977, a). Correlation of the Middle to Late Eocene calcareous sections across southern Australia has been hampered by access, in that much of the critical material is in the subsurface on this tectonically relatively stable margin, and by biogeography, in that the microfaunas were not only extratropical, but changed rapidly and reversibly as a response to climatic and oceanographic changes foreshadowing the major climatic deterioration across the Eocene/Oligocene boundary. These phenomena and problems are now well known in local micropaleontology although published documentation is virtually nonexistent. The biostratigraphic events listed in Figure 5 are a composite sequence derived from: Site 264 on the Naturaliste Plateau (across Early/Middle Eocene boundary; Table 3); the Eucla

Basin (intra-Middle Eocene; McGowran and Lindsay, 1969); and the Otway Basin (Upper Eocene; McGowran et al., 1971; McGowran, 1977, b). In most respects there is good agreement with the succession of events to the southeast, including New Zealand (Jenkins, 1971).

The oldest Eocene assemblages in the Eucla Basin include Globigerinatheka index, Tenuitella aculeata, Globorotalia cerroazulensis pomeroli, Subbotina frontosa, and acarininids some of which tend toward Truncorotaloides topilensis. Thus, the foot of the transgression is dated as within Zone P.12; it could be a little higher since the absence of Orbulinoides (also not known from the Carnarvon Basin) is a biogeographic, not temporal, effect. T. aculeata occurs, very rarely, down to Zone P.12 at Site 214 on the Ninetyeast Ridge. The Middle Eocene limestones with cherts which suggest the presence in the Eucla Basin of the oceanic "Horizon A" (McGowran, 1973) are both thicker, relative to the Late Eocene, and more widespread than Quilty (1974b) indicates (Lindsay and Harris, 1975), but foraminiferal-biostratigraphic changes are obscured by lithification.

In the Otway Basin an ingression preceding the main transgression (McGowran, 1973) brought in Acarinina primitiva, which is known (in the Carnarvon Basin) to range as high as Zone P.14. Acarininids (Truncorotaloides collactea) disappear just above the Middle/Late Eocene boundary, as they do in oceanic sections in the Indian Ocean (Figure 6; Table 2). Tenuitella aculeata has a disjunct range, as shown in Figure 5; indeed, in the Late Eocene in the Otway Basin several species (e.g. Tenuitella gemma, Pseudohastigerina micra) fluctuate strongly in their abundance through the section. Hantkenina primitiva occurs sporadically in a narrow interval estimated to correlate with the highest part of Zone P.15. The top of the Eocene is an unconformity in most sections, and it is, really, a hopeful guess to place the upper limit of Globigerinatheka index against Zone P.17 (Figure 5). In more complete sections in southern Australia, Lindsay (1969 and references therein; also unpublished studies) places the Eocene/Oligocene boundary slightly above the disappearance of G. index and T. aculeata, at the point where the morphotype Subbotina linaperta disappears from the S. linaperta-angiporoides lineage (see also McGowran, Lindsay and Harris, 1971).

Biofacies in the Indian Ocean: Changes with Time

Maastrichtian, Ninetyeast Ridge

As mentioned in a biostratigraphic context, the Maastrichtian succession at Site 217 is most unusual. In the usual situation in the pantropical belt, Globotruncanella mayaroensis is the last of the important species to appear (Figure 2). At Site 217, this event is the first: below it, single-keeled Globotruncana, double-keeled Globotruncana, and the various genera of the Heterohelicidae are less diverse than might be expected, or represented by long- and wide-ranging forms, or absent. McGowran (in Gartner et al., 1974) proposed that this inverted succession records a rapid migration by the Site from an extratropical into a tropical situation, thus supporting in a qualitative way the geophysical evidence for rapid transform movement during the Maastrichtian (Sclater, von der Borch et al., 1974; Sclater and Fisher, 1974). The argument developed in Gartner and others to explain the succession outlined in Table 4 was summarized as follows:

1) The assemblages in Facies 2 and Facies 3 at Site 217 parallel the succession in shelf carbonates in the Carnarvon Basin, which was

considerably further south at that time, as well as being on another plate.

2) On the Indian Plate during the late Maastrichtian, Site 216 was discernibly extratropical whereas Site 217 (Facies 4) was closer, in complexion of assemblages, to India-Pakistan. On the Australian Plate, assemblages in the Carnarvon Basin (the Miria Marl) were extratropical in contrast to New Guinea. (Subsequently, Ms M. Apthorpe (personal communication) has confirmed that there is a gradient on the Northwest Shelf on the evidence of Globotrunacana aegyptiaca and G. gansseri; the latter is restricted to the far north.)

3) With the passage of (Maastrichtian) time, the assemblages at Site 217 changed from resembling those in the Carnarvon Basin to resembling those in New Guinea. The critical observation was that Globotruncanella intermedia and G. mayaroensis appeared, up-section, before their diverse contemporaries, not after, as would be expected.

4) Factors including microfaunal evolution, height of water column, deep-sea solution, and paleoclimatic change were acknowledged, but considered to be insufficient to explain the observed similarities and contrasts. (Isotopic data subsequently published by Saito and van Donk (1974) reinforce that conclusion, which was based partly on an abstract of their work.)

5) Accordingly, the change in foraminiferal assemblages through the Maastrichtian at Site 217 was explained by a substantial movement of the Indian Plate relative to the Australian Plate during that time, the Ninetyeast Ridge being on the Indian Plate.

Biofacies Changes and Sinking History: Ninetyeast Ridge, Chagos-Laccadive Ridge, Mascarene Plateau

The recognition of successional changes in foraminiferal assemblages has thrown considerable light on the tectonic history of these morphological elements (McGowran, 1974; Gartner, Johnson and McGowran, 1974; Pimm, McGowran and Gartner, 1974; Fleisher, 1974; Site Report for Site 219, in Whitmarsh, Weser et al., 1974; Vincent, Gibson and Brun, 1974). The results are summarized in time-facies profiles (Tables 5,6).

At Site 214 the sedimentary succession begins with nonmarine facies including terrestrial palynomorphs. The lowest marine sediments have diverse benthonic foraminiferal assemblages. Prominent elements include Karreria pseudoconvexa, Cibicides umbonifer and related forms, a complex of Gyroidinoides including G. octocamerata, Cibicidina ekblomi, Alabamina westraliensis, Epistominella cf vitrea, Baggatella aff. coloradoensis, Ceratobulimina sp. The assemblage is very similar to various extratropical, shallow-water, Paleocene assemblages (more so than nomenclature implies: see McGowran, 1965) but is especially close in its complexion to the one found near the base of the Paleocene sedimentary episode in the Otway Basin (McGowran, 1965) (Planorotalites ehrenbergi horizon in Figure 7). As noted above, horizons with shelly fossils, including calcareous-perforate foraminifera, occur in southern Australia as ingressions in a terrigenous clastic succession in which agglutinated foraminifera and organic-walled phytoplankton are the only marine indicators. There can be no doubt that the marine environment was very shallow. Similar assemblages occur in Zone P.4 in the Perth Basin (McGowran, 1964, 1965). In the next biofacies unit distinguished (Table 5), the planktonic component increases upwards toward the hiatus; poorer preservation in lithified carbonates obscures comparison of the benthonic elements with the lower unit. How-

Table 4. Biofacies changes, Late Cretaceous, Site 217, Ninetyeast Ridge.

AGE	DEPTH	SUMMARY OF SUCCESSION	INTERPRETATION AND COMMENT
LATE MAASTRICHTIAN	420m Core 17 — Core 22 480m	FACIES 4 Nannofossil chalk. Planktonic foraminifera of oceanic aspect; close to CCD towards top. Diverse assemblages: *Globotruncanella (mayaroensis, intermedia)*; single-keeled *Globotruncana (stuarti, stuartiformis, elevata, †falsocalcarata, conica)*; double-keeled *Globotruncana (arca, contusa, aegyptiaca, trinidadensis, etc.). Globotruncana gansseri; Rugoglobigerina (rugosa/pennyi);* Heterohelicidae (*Pseudotextularia, Racemiguembelina, Planoglobulina,* Pseudoguembelina, Gublerina, Heterohelix). Reduction downsection in overall diversity: disappearance of *G. aegyptiaca*; reduction in numbers and morphotypic variation in single-keeled *Globotruncana*; same in Heterohelicidae. *Globotruncana lapparenti/linneiana* present.	Oceanic, deepening up-section to below lysocline. Assemblages becoming increasingly tropical up-section. Similar in general aspects to Late Maastrichtian in India-Pakistan and New Guinea, more tropical than Carnarvon Basin, or Site 216 (where diversity is less, but *Globotruncanella mayaroensis* is prominent).
CAMPANIAN TO LATE MAASTRICHTIAN	Core 23 — Core 30 550m	FACIES 3 Micarb chalk, partly shelly: robust *Inoceramus* abundant in lower samples, decreasing upwards. Planktonic foraminifera abundant, increasing up-section; corrosion increases up-section. Assemblages dominated by *Rugoglobigerina rugosa/pennyi;* Heterohelicidae reduced to, mostly, *Gublerina* and *Heterohelix;* double-keeled *Globotruncana (arca, lapparenti/linneiana,* some *ventricosa);* single keeled *Globotruncana* rare to absent. Benthonics include *Osangularia, Valvalabamina, Cibicides voltziana* etc., *Bolivina incrassata gigantea,* rare *Bolivinoides* including *B. miliaris*.	Outer neritic, deepening up-section to oceanic carbonate facies. Assemblages extratropical and similar to Campanian-Maastrichtian in Carnarvon Basin (abundance of *Rugoglobigerina, Globotruncana* dominated by *arca/lapparenti/linneiana;* absence of numerous taxa although planktonic numbers high).
CAMPANIAN	Core 31 — Core 36 600m	FACIES 2 Micarb chalk; silicified or cherty. Planktonic foraminifera rare at base, increasing up-section. An association of *Archaeoglobigerina bicui, Globotruncana lapparenti/linneiana* (with *arca* prominent toward top), Globigerinelloides and Heterohelix. Benthonics more common than in higher units: *Gosella, Gaudryina,* lentic- ulinids, nodosariids, *Ellipsoidella, Praebulimina, Allomorphina, Valvalabamina, Gyroidinoides, Karreria ribbingae/excavata, Angulogave-linella rakauroana, Nuttallinella coronula;* miliolids rare. *Inoceramus* present but less abundant than in Facies 3; echinoid remains, sponge spicules.	Inner neritic carbonate shelf facies, deepening up-section. Assemblages extratropical and similar to Carnarvon Basin.
	Core 37 Core 12A Core 17A 664m	FACIES 1 Dolarenite, cherty, shelly in part. Foraminifera absent to very rare.	Inner neritic, marginal marine carbonate shelf facies.

Table 5. Facies and biofacies changes reflecting concerted early sinking histories, Chagos-Laccadive Ridge and Ninety-east Ridge.

SITE 219: CHAGOS-LACCADIVE RIDGE
9°01.75'N, 72°52.67'E   Water depth 1764 meters.

EARLY EOCENE	Zone P.8 or P.7	Cores 25 to 26	Chalk, rich in foraminifera; locally with phosphatic concretions. Planktonics dominant, although low in diversity. Benthonics typical of deep water assemblages.	Oceanic
			HIATUS: ABSENCE OF ZONES P.5 AND P.6	
LATE PALEOCENE	Zone P.4	Core 27 82m	Limestone, often glauconite-rich, underlain by calcareous, glauconitic sandstone. Planktonic foraminifera relatively rare. Benthonic foraminifera dominantly calcareous-perforate, including *Rotalia*, *Operculina*, *Discocyclina*; agglutinated and porcellanous forms not common. Closely similar to inner neritic carbonate biofacies in Cauvery Basin.	Tropical inner neritic, salinity normal
		Core 8A-1	Common bivalves, bryozoans, echinoids.	
	Zone P.4	Core 8A-2 75m	Sediment mostly terrigenous; clay component dominated by montmorillonite. Planktonic foraminifera more common than above. Benthonic foraminifera less common; *Discocyclina* and *Operculina* occur sporadically. Some layers rich in shell fragments especially bivalves; also bryozoans, red algae.	Tropical mid-neritic, salinity normal
		Core 14A	Basement not reached.	

SITE 214: NINETYEAST RIDGE
11°20.2'S, 88°43.1'E   Water depth 1665 meters

EARLY EOCENE	Zones P.8 to P.7	Core 34	Calcareous ooze. Abundant planktonic foraminifera. Rare benthonic foraminifera.	Oceanic
	Zone P.7	Core 35 to Core 36-2	Sediment is glauconitic, with echinoid spines, fecal pellets, shelly material. Planktonic foraminifera abundant at top, decreasing downwards; *Planorotalites* common to abundant. Benthonic foraminifera include *Cibicides* attached to plant stems.	Outer neritic deepening upwards, warmer than below
			HIATUS: ABSENCE OF ZONES P.5 AND P.6	
LATE PALEOCENE	Zone P.4	Core 36-3 36m	Glauconitic, calcareous silt and sand, with shells. Planktonic foraminifera decrease downwards; *Morozovella* and *Planorotalites* subordinate to absent. Benthonic foraminifera increase downwards in numbers but variable in numbers and dominances; preservation poor. Main types: *Cibicides*, *Karreria*, *Alabamina*, *Gyroidinoides*, *Cibicidina*, uvigerinids. Agglutinated and porcellanous forms rare to absent.	Inner neritic deepening upwards "extratropical"
	Zone P.4 or P.3	Core 39	Sponge spicules, echinoidal and molluscan fragments, ostracods.	
		Core 40 19m	Glauconitic shelly calcareous silt and limestone with volcanic components. Planktonic foraminifera present but subordinate to benthonics. Benthonics better preserved than above; assemblages dominated by *Karreria*, *Alabamina*, *Gyroidinoides*, *Cibicides*, *Baggatella*, *Epistominella*, *Cancrisulimina*, uvigerinids, lagenids; no porcellanous or agglutinated forms. Sponge spicules, echinoidal and molluscan fragments, ostracods.	Inner neritic, very shallow, salinity close to normal "extratropical"
		Core 41		
PALEOCENE		Core 42 100m Core 53	Lignite, volcanic clay, tuff, and lapilli tuff interlayered with differentiated flow. Terrestrial microflora.	Nonmarine
			Basalt	

ever, it is clear that the site is deepening: a lagoon or platform is becoming exposed increasingly to oceanic influence. Above the hiatus, in the Early Eocene, the change from glauconitic, shelly carbonate to calcareous ooze is matched by the disappearance of Cibicides whose morphology indicates attachment to plant stems, and echinoid spines, fecal pellets, etc.

The Shipboard Scientific Party for Leg 23 (Whitmarsh, Weser et al., 1974) compared Site 219 on the Chagos-Laccadive Ridge with Site 217 on the Ninetyeast Ridge, but the startling comparison is with Site 214. Drilling of Site 219 ceased in the vicinity of the Zone P.3/P.4 boundary, on the evidence of the lowest dateable assemblage with Planorotalites pseudomenardii (Fleisher, 1974). This species extends up to a sharp, between-core faunal change, being succeeded by an Acarinina soldadoensis/angulosa assemblage of Zone P.7 or P.8. That is, the hiatuses at Sites 214 and 219 are virtually identical in duration. The Paleocene foraminiferal assemblages suggested to Fleisher that the site was shoaling from mid-neritic to inner neritic: forms similar to Hanzawaia and Cancris occur in the lowest part of the section, where large rotaliids were not found but planktonics were more common. Higher, planktonics are rare, and an association of Discocyclina, Operculina and Rotalia occurs. This record means that Site 219 is similar in foraminiferal biofacies to the South Indian continental margin (Cauvery Basin) whereas Site 214 is similar to the western and southern Australian margin in lacking larger benthonics (Figure 8). V. D. Mamgain et al. (in Site Report for Site 219) record Discocyclina and Operculina as ranging from the base of the drilled section (Early Paleocene) to the top of the Middle Eocene with one record of Discocyclina in the Early Oligocene. The ages given for the section differ from Fleisher's, perhaps because of the confusion over correlations at Pondicherry; Fleisher's analysis is followed here. Even so, there would appear to be some downslope, up-section reworking of Discocyclina into the sediment of deep water aspect above the hiatus. Apart from the presence of larger foraminifera, the section at Site 219 contrasts with that at Site 214 in being much thicker in Zone P.4. This can be explained by both more rapid carbonate accumulation at a lower latitude and by the substantial terrigenous component in the sediment: hence the shoaling.

Vincent, Gibson and Brun (1974) and Heiman, Frerichs and Vincent (1974) recognized a between-core hiatus straddling the Paleocene/Eocene boundary at Site 236, north of the Mascarene Plateau: chalk with Planorotalites pseudomenardii and Morozovella velascoensis (Zone P.4), over basalt, is succeeded by chalk with M. subbotinae and M. aragonensis (Zone P.7). On the Mascarene Plateau itself (Site 237) mixing was evident in the interval (chalk) between cores 37 and 41, which is also the interval between the topmost occurrence of Planorotalites pseudomenardii (Zone P.4) and "the base of the common and persistent occurrence of Morozovella aragonensis" (Zone P.7). The evidence for mixing consists of records of morozovellids (M. aequa, M. velascoensis, M. formosa, M. aragonensis) which do not fit the established succession. It is suggested here that the evidence might be explained by an hiatus involving Zones P.5 and P.6, with downslope (up-section) displacement of material into Zone P.6b or P.7. The drilled section remained in fine-grained, silicified carbonate, passing through Zone P.4 (base of Planorotalites pseudomenardii), Zone P.3 (Morozovella angulata, M. abundocamerata = conicotruncata), into Zones P.2 and P.1d (M. uncinata, Planorotalites ehrenbergi, P. compressa, Subbotina pseudobulloides).

Table 6. Facies and biofacies changes reflecting early sinking history, Mascarene Plateau; note close parallelism with Table 5.

SITE 237: MASCARENE PLATEAU
7°04.99'S, 58°07.48'E   Water depth 1640 meters.

EARLY EOCENE	Zones P.9 to P.7	Cores 32 to 38 Cores 39 40: no recovery	Calcareous ooze, underlain by partly lithified and silicified chalk, chert. Benthonic Assemblage 4. Dominated by *Nuttallides truempyi*; *Anomalina dorri* replaces Paleocene morphotypes. Presence of *Gyroidina planata*, *Stilostomella kressenbergensis*, etc.	Lower bathyal
	Zone P.7?	Core 41	Benthonic Assemblage 3. *Nuttallides truempyi*, *Pseudovalvulineria beccariiformis*.	Lower bathyal
			POSSIBLE HIATUS : POSSIBLE ABSENCE OF ZONES P.5 AND P.6	
LATE PALEOCENE	Zone P.4	Core 42	Chalk, partly silicified; some chert; glauconitic. Benthonic Assemblage 3. Dominated by *Nuttallides truempyi*; also *Pseudovalvulineria beccariiformis*, *Gavelinella* aff. *danica*, etc.	Lower bathyal
		182m	Benthonic Assemblage 2. "Midway-type fauna," as below.	Lower neritic to upper bathyal with slumping from upper neritic
			Limestone (0.2m) with Benthonic Assemblage 1. Larger foraminifera *Discocyclina*, *Ranikothalia*, *Miscellanea*, various rotaliids, and calcareous alga *Archaeolithothamnium*.	
			Partly silicified chalk as above. Benthonic Assemblage 2. "Midway-type fauna" includes *Anomalina midwayensis*, *Cibicides alleni*, *Gavelinella danica*, *Gyroidina girardana*, *G. globosa*, *G. nitida*, *Loxostomum applinae*, *Marssonella oxycona*, *Pleurostomella paleocenica*, *Pullenia coryelli*, *Stilostomella* spp., *Vaginulina longiforma*, and many others.	
	Zone P.3	Core 61	Three horizons with Benthonic Assemblage 1 (0.2m each) but lacking larger foraminifera.	
EARLY PALEOCENE	Zones P.2 to P.1	Core 62 122m Core 67	Partly silicified chalk as above. "Midway-type fauna", Benthonic Assemblage 2, as above. [Planktonics predominant (90-99%) at all Paleocene-Eocene levels except where Benthonic Assemblage 1 is identified (less than 30%).]	Lower neritic to upper bathyal

Basement not reached.

Vincent et al. (1974) recognized three foraminiferal assemblages in the Paleocene at Sites 236 and 237, each being attributed to a range of depths: upper neritic (<50 m), lower neritic to upper bathyal (50-600 m) and lower bathyal (600-2500 m); a fourth assemblage in the Early Eocene is also lower bathyal but the Paleocene species have been replaced by others. The Benthonic Assemblages 1 to 4 are listed in highly summarized form on Table 6. Benthonic Assemblage 1, with tropical larger foraminifera (Discocyclina, Miscellanea, Ranikothalia) and the calcareous alga Archaeolithothamnium, occurs as a thin layer which slumped from an environment within the photic zone; three other slumped horizons contain calcareous algae and molluscan and echinoderm debris but not larger foraminifera. The inner neritic, extratropical benthonic assemblage found on the Ninetyeast Ridge was not found here. Benthonic Assemblage 2 was compared with the "Midway-type fauna" on the Gulf Coast (Plummer, 1927; Cushman, 1951). At Site 237 this assemblage changes to B.A.4 (incoming of Nuttallides truempyi) within Zone P.4; the change is transitional. As they point out themselves, the depth ranges attributed by Vincent et al. to the assemblages on actualistic grounds are arguable because of ecological displacement into deeper water since the early Tertiary (thus, I agree with Berggren (1974a,b) in restricting the Midway fauna to the neritic environment); however, the conclusion that there is a deepening with time at this deepwater site is convincing, as is the consistency of biofacies changes with an expected sinking of the plate. At Site 236, the water was too deep to show similar biofacies changes. Vincent et al., also conclude that sedimentation rates at Site 237 change rather abruptly near the top of Zone P.4: as the water deepened, the rate of accumulation dropped from 68 m/m.y. in Zone P.4 to 7.5 m/m.y. This contrast is heightened if the hiatus suggested here exists between the Paleocene and the Eocene, and the similarity to Site 219 is increased.

There is, then, a pronounced parallelism in age/biofacies relationships, in environments ranging from inner neritic to bathyal, between sites on the Ninetyeast Ridge, Chagos-Laccadive Ridge, and Mascarene Plateau. Each structure has a sedimentary cover clearly recording rapid sinking during Late Paleocene to Early Eocene time, with a stratigraphic break in the same place and occupying virtually the same span of time. A time/facies profile across the Paleocene/Eocene boundary on the Ninetyeast Ridge (Figure 9) heightens the sense of critical change in the Indian Ocean at that time. At Site 215, a long calcareous section through Zone P.4 has poor planktonic faunas and benthonics including Nuttallides truempyi and Angulogavelinella beccariiformis, which would indicate the presence of "lower bathyal" Benthonic Assemblage 3 (Vincent et al., 1974). Nuttallides is still present in the Lower Eocene, but Zone P.6 is represented by substantially better planktonic assemblages: specimens are more abundant and preservation is better. Then, brown clay replaces calcareous ooze within Zone P.7. The same facies change takes place at about the same time at Site 213 on the other side of the Ridge, and both events are virtually synchronous with the resumption of sediment accumulation at the crestal Site 214. Finally, and extending the theme still further, there is an hiatus on the Naturaliste Plateau between Zone P.4 (Planorotalites pseudomenardii) and approximately Zone P.8 (Morozovella caucasica) (Table 3; Figure 6).

Diachronous or allochronous, south-to-north facies changes implying sinking with spreading along the Ninetyeast Ridge through the Maastrichtian-Eocene (Pimm et al., 1974) can be extended to Site 253 in the south (Figure 10).

Early Tertiary Stratigraphic Patterns and Events

Cretaceous/Tertiary Boundary

In the ocean, the two relatively complete sections, both on the Ninetyeast Ridge, contain evidence of disconformity, one (the shallower) with hiatus, the other with evidence for microfaunal mixing. In deeper water continental sections (Pakistan, Assam, Andaman Islands, Papua New Guinea, possibly Timor; all on the leading edge of the plate, thus in mobile belts) the Cretaceous/Tertiary contact has in common with these oceanic sections latest Maastrichtian (Globotruncanella mayaroensis Zone) overlain by Danian sediments to a varying degree of completeness. Where the "Globigerina" eugubina Zone or Zone P.1a have been identified, details are not available (Assam, Andaman Islands), but it would seem that the deep water situation with an abrupt microfaunal) change documented by Luterbacher and Premoli Silva (1964) is being repeated. On more stable margins, there is a disconformity, or volcanics, or regressive sequence. Thus, the situation elsewhere is recognized now in the Indian Ocean region.

Paleocene

Two striking aspects of a pattern emerge. The first is that sediment has accumulated, in an extremely wide range of facies, in successions that begin locally in the Zone P.1-P.3 interval and end in the vicinity of Zone P.7 with regression or hiatus (continental), or in a change from calcareous ooze to brown clay (deep oceanic: Central Indian and

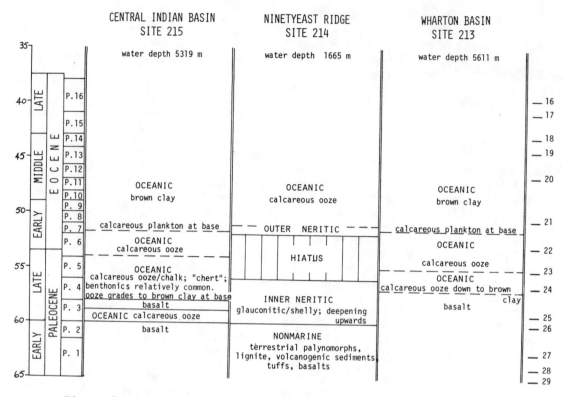

Figure 9. Time-facies profile across the Ninetyeast Ridge. Left, geochronological/chronostratigraphic/biostratigraphic interrelationships, from Berggren (1972). Right, geomagnetic anomalies, from Heirtzler et al. (1968) with revised ages from Sclater et al. (1974). No ages implied for basalts (except in Zone P.3, Site 215).

Wharton Basins). This succession seems to reach a maximum in Zone P.4, in that Planorotalites pseudomenardii, the total range indicator for the zone, is the most widespread and commonly recognized index fossil in the region (Figure 7). Similarly, a plot of larger foraminiferal assemblages against planktonic zones (Figure 8) suggests a tropical biofacies event in neritic shelf carbonates centred on Zones P.4- P.5. In southern Australia, the same interval is recorded in shelf carbonates without those fossils (Carnarvon Basin) or in marginal marine detrital facies (Perth and Otway Basins).

The second aspect of the pattern is that the strata on three oceanic topographic units, or structures, record rapid sinking in concert with a virtually isochronous unconformity across the Paleocene/Eocene boundary interrupting that sinking. Zones P.5 and P.6 are lacking (Tables 5, 6). The unconformity is present on the Naturaliste Plateau, too (Table 3). On present evidence, this unconformity is not clearly matched on the continents, where assemblages with Morozovella velascoensis and M. subbotinae/marginodentata or their correlatives are rather widespread (Figure 7). However, that point deserves further scrutiny.

Early Eocene

Sedimentary accumulation resumes above these oceanic disconformities in Zone P.7 - at about the same time as the facies changes from ooze

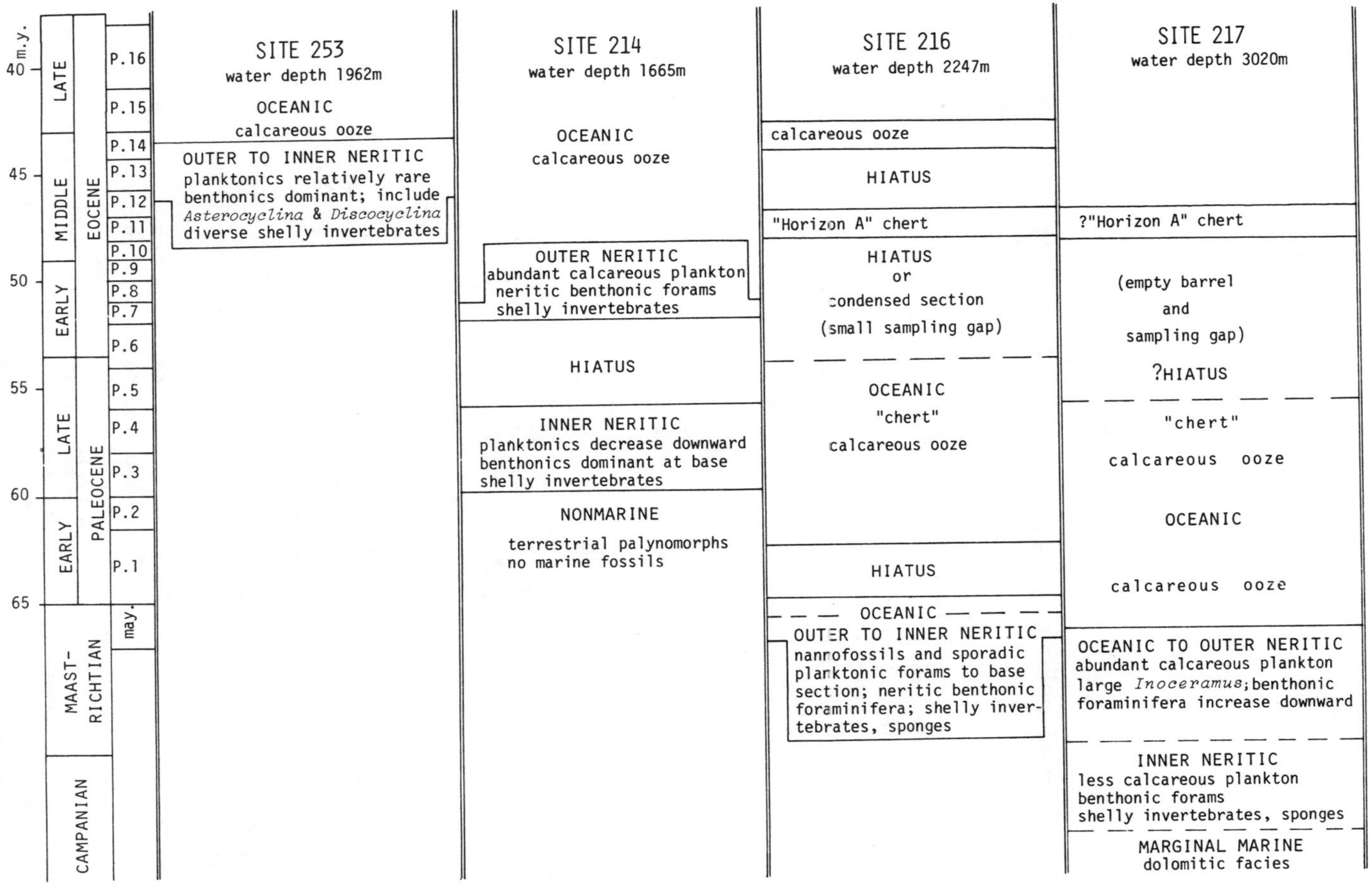

Figure 10. Time-facies profile along the Ninetyeast Ridge, indicating allochronously homotaxial successions becoming younger to the south.

to clay at two deep basin sites (Figure 9), and at about the same time as the earliest Tertiary stratigraphic sequence ends on the continental margins. By this latter generalization is meant: hiatus, demonstrated biostratigraphically, sometimes with reworked, older fossils; regression, where planktonic assemblages are succeeded by inshore benthonic, agglutinated assemblages or nonmarine sediments, evaporites or laterites; or simply by the fact that planktonic foraminiferal identification of Late Paleocene and Middle Eocene have accumulated during the past decade whilst records of Early Eocene have not (Figure 7). Again, it must be stressed that this "gap" in the Early Eocene and into the Middle Eocene occurs regardless of facies, of latitude, or of stratotectonic position, being found in mobile belts and on platforms.

Middle Eocene, Lower Part

The regression/hiatus on the continental margins continues. In contrast, oceanic sections reveal a widespread sedimentary event terminating at about the top of Zone P.11 (disappearance of Morozovella aragonensis) or sometimes within Zone P.12. It is within this accumulation of calcareous ooze that the foraminiferal evidence for a "cool interval" is documented on the Ninetyeast Ridge (Table 2). Also, the first "cherts" (opaline silica causing seismic "Horizon A" in all oceans) were encountered downhole in Zone P.11 in the following places: on the Ninetyeast Ridge at Site 216 and probably at Site 217 (an empty core barrel is unexplained, otherwise); on the Chagos-Laccadive Ridge at Sites 219 and 220; perhaps on and near the Mascarene Plateau, at Sites 237 and 236 where the highest cherts are recorded at a slightly lower level but where assemblages seem to be mixed and dating lacks precision; and on the Naturaliste Plateau at Site 264 where the chert occurs in the Guembelitria assemblages correlated with Zones P.10-P.11. There is, then, an oceanic event which must have considerable significance in space and time, especially since it correlates with a "cool interval" at Site 214 (where chert was not encountered).

Middle Eocene, Upper Part

There is widespread hiatus in the oceanic record in the Zone P.12-P.13 interval (Figure 6). Zone P.13 is identified clearly only at Site 214, Ninetyeast Ridge (Table 2). However, the evidence there for a "warm interval" at that time correlates excellently with two other phenomena. One is the virtually isochronous transgression, expressed mostly as shelf carbonate facies, over the continental margins around the northern and eastern Indian Ocean and regardless of latitude or stratotectonic situation. (The transgression is, it might be added, part of the classical Lutetian transgression which is world-wide.) The second happening was the establishment, in the north, of Nummulites-Discocyclina and Alveolina assemblages in these neritic carbonates (Figure 8), and an excursion southwards by Discocyclina to Site 253 on the Ninetyeast Ridge (McKelvey and Fleet, 1974, fig. 7) and to the Carnarvon Basin in Western Australia. In carbonates probably equivalent to Zone P.14, oceanic-type chert horizons are found in southern Australia (McGowran, 1973), well beyong the known migration limit of Discocyclina at that time; this is a distinctly younger record than the Zone P.11 "surface" in the Indian Ocean.

Late Eocene

Carbonates with larger foraminifera occur in many continental sections to the north (Nagappa, 1959; Adams, 1970; Figure 8). At this time, some of the elements in that fauna migrated south with a thinning out around the Australian margin: Discocyclina, Asterocyclina, Nummulites, Pellatispira in the Carnarvon Basin, Asterocyclina and the "tropical" dasyclad alga Neomeris on the southern margin, west of the Eucla Basin, Halkyardia and Linderina but no discocylinids in South Australia, and no larger foraminifera in neritic carbonates in the Otway Basin to the southeast. In the Indian Ocean, calcareous ooze accumulated again in Zones P.14-P.15, but there is then an unconformity in every documented section (Figure 6). Foraminiferal tests in Zones P.14-P.15 are badly affected, and the assemblages biased in composition, by dissolution.

Correlations and Interpretations

Pimm and Sclater (1974) suggest that the Paleocene-Early Eocene hiatus, "which affects only shallow depths on the Ninetyeast Ridge, may be an event related solely to local tectonism". But the hiatus occurs there in calcareous ooze as well as in neritic-type, glauconitic sediments, and is identified on the Chagos-Laccadive Ridge and on the Mascarene and Naturaliste Plateaus as well. Davies et al. (1975) summarize evidence for an early Tertiary hiatus in the Indian Ocean and southwest Pacific, and explain it in the same terms as the better-known Oligocene hiatus - by circulation patterns and changes in skeleton productivity in the plankton.

The present review seems to reveal two distinct and successional sets of data which require explanation:

1) In the Paleocene and Early Eocene, there is a stratigraphic sequence (Sloss, 1963) recorded in the Indian Ocean and around its northern and eastern margin without apparent regard for facies, latitude, or stratotectonic situation. This sequence is interrupted in the ocean, in ooze as in neritic sediment, by a remarkably isochronous unconformity across the Paleocene/Eocene boundary. Equally striking is the termination of the marginal sequence within the Lower Eocene.

2) In the Middle and Late Eocene there is a second stratigraphic sequence around the northern and eastern margin, identified as being virtually isochronous at its base (Zones P.12-P.13). In contrast, sediment in the ocean extends up to Zone P.11 but is sporadic thereafter. A further correlation is that a "cool interval" and chert horizons in the ocean (Zones P.10-P.11) match the widespread oceanic carbonate sediment (and marginal hiatus) whereas the later (Zone P.13) warm interval matches the marginal carbonate sequence and oceanic hiatus.

The earlier set of events shows some evidence of basic tectonic control (McGowran, 1977a). Thus, using the revised ages of early Tertiary geomagnetic anomalies (Sclater et al., 1974; see Figure 9) the synchronism and parallelism across the Paleocene/Eocene boundary in the Indian Ocean indicates a pause in the sinking of morphological/ tectonic elements in the Anomaly 22-24 interval, which is also the time of the probable first, pre-Himalayan collision north of India (Moore et al., 1974), of the commencement of seafloor spreading south of Australia (Weissel and Hayes, 1972), and of its cessation in the Tasman Sea (Hayes and Ringis, 1973). All these events are clustered within a stratigraphic sequence whose identification in all sedimentary and tectonic environments must be a reflection of oceanic and continental tectonism across at least three plates, even though the control pathway is not yet clear.

In the Middle-Upper Eocene the role of oceanic circulation in causing the observed stratigraphic pattern is rather more apparent, although the onset of a major, platewide transgression some 46 m.y. ago (Zones P.12-P.13; just before Anomaly 19) suggests a tectonic event at that time: it is tectonoeustatic, not glacio-eustatic. McGowran (1977a) has summarized evidence for the establishment then of oceanic conditions south of Australia, significantly before the circum-Antarctic current was established during the later Oligocene, as proposed by Kennett et al. (1975). Part of that evidence is the allochronous distribution of Horizon A cherts. Without commitment on the questions of the ultimate source of the silica, or of when it actually assumed its present habit in the sediment, it can be stated that (a) the enclosing sediment in the Indian Ocean correlates rather closely with evidence for a cool climatic interval including a northward excursion by extratropical species, and (b) that the horizon does not extend above Zone P.11, whereas in the Eucla Basin it is estimated to occur in Zone P.14. That is, there is reason to suggest significant deflection of oceanic water to the south of Australia by that time.

We can extend that line of reasoning further by considering the mutual correlation (Zone P.13) of the marginal transgression, the establishment of carbonate shelf facies and larger foraminiferal faunas, the southward migration by _Discocyclina_, and the "warm interval" on the Ninetyeast Ridge. All of these events are actually part of one event - the expansion of the tropical zone as Antarctic water becomes quarantined in the south, in contrast to the situation earlier in the Middle Eocene and, indeed, in the Paleocene and Maastrichtian when the neritic carbonate environments were available but not colonized by extant larger foraminifera (Figure 8). The tropical zone may have extended further, but episodically, in the Late Eocene, when _Asterocyclina_ and other pantropical elements reached southern Australia. That migrations to the south began much earlier on the western oceanic margin (Madagascar; Figure 8) is understandable if an anticlockwise circulation is assumed (McGowran, 1968).

Finally, there is the contrast in the later Middle and Late Eocene between a poor oceanic record and a prominent continental (neritic) record of biogenic carbonates. Berger (1973) suggested that the calcite compensation depth was relatively shallow in the Eocene, and that this may be due to the continental shelves acting as carbonate traps, causing shelf-basin fractionation and a compensating rise in the CCD. Although backtracking analysis (Berger, 1973) was not carried out here, it would seem clear that fractionation did not occur up to Zone P.11, but became a powerful factor thereafter. Thus, Berger's suggestion is supported. I see no such effect in the Maastrichtian or Paleocene - Early Eocene sequences, when there was substantial neritic carbonate accumulation but no quarantining of Antarctic water. The present southern oceanographic regime may be largely a Neogene phenomenon (Kennett et al., 1975) but its beginnings are perceived in the Middle Eocene.

Acknowledgements. Dr. Lawrence A. Frakes supplied the material from Site 264. Mrs. Joan Brumby typed the tables and the manuscript; Miss Milly Swan prepared the locality map and Mr. Richard Barrett did the photography.

References
Initial Reports of the Deep Sea Drilling Project
(Volumes cited in numerical order)

Burns, R. E., Andrews, J. E., van der Lingen, G. J., Churkin, M., Galehouse, J. S., Packham, G. H., Davies, T. A., Kennett, J. P., Dumitrica, P., Edwards, A. R., and Von Herzen, R. P., 1973, Initial Reports of the Deep Sea Drilling Project, 21, Washington (U. S. Government Printing Office).

von der Borch, C. C., Sclater, J. G., Gartner, S., Jr., Hekinian, R., Johnson, D. A., McGowran, B., Pimm, A. C., Thompson, R. W., Veevers, J. J., and Waterman, L. S., 1974, Initial Reports of the Deep Sea Drilling Project, 22, Washington, D. C. (U. S. Government Printing Office).

Whitmarsh, R. B., Weser, O. E., Ross, D. A., Ali, Syed, Boudreaux, J. E., Coleman, R., Fleisher, R. L., Girdler, R. W., Jipa, D., Kidd, R. B., Mallik, T. K., Manheim, F. T., Matter, A., Nigrini, C., Siddiquie, H. N., Stoffers, P., and Supko, P. R., 1974, Initial Reports of the Deep Sea Drilling Project, 23, Washington, D. C. (U. S. Government Printing Office).

Fisher, R. L., Bunce, E. T., Cernock, P. J., Clegg, D. C., Cronan, D. S., Damiani, V. V., Dmitriev, L. V., Kinsman, D. J. J., Roth, P. H., Thiede, J., and Vincent, E., 1974, Initial Reports of the Deep Sea Drilling Project, 24, Washington, D. C. (U. S. Government Printing Office).

Davies, T. A., Luyendyk, B. P., Rodolfo, K. S., Kempe, D. R. C., McKelvey, B. C., Leidy, R. D., Horvath, G. J., Hyndman, R. D., Thierstein, H. R., Boltovskoy, E., Doyle, P., 1974, Initial Reports of the Deep Sea Drilling Project, 26, Washington (U. S. Government Printing Office).

Veevers, J. J., Heirtzler, J. R., Bolli, H. M., Carter, A. N., Cook, P. J., Krasheninnikov, V. A., McKnight, B. K., Proto-Decima, F., Renz, G. W., Robinson, P. T., Rocker, K., and Thayer, P. A., 1974, Initial Reports of the Deep Sea Drilling Project, 27, Washington, D. C. (U. S. Govt. Printing Office).

Hayes, D. E., Frakes, L. A., Barrett, P., Burns, D. A., Pei-Hsin Chen, Ford, A. B., Kaneps, A. G., Kemp, E. M., McCollum, D. W., Peper, D. J. W., Well, R. E., and Webb, P. N., 1974, Initial Reports of the Deep Sea Drilling Project, 28, Washington, D. C. (U. S. Govt. Printing Office).

References Cited from the Initial Reports of the Deep Sea Drilling Project

Akers, W. H., Appendix V. Results of shore laboratory studies on Tertiary and Quaternary foraminifera from Leg 26, 26, 973-982, 1974a.

Akers, W. H., Foraminiferal range charts for Arabian Sea and Red Sea sites, Leg 23, 26, 1013-1050, 1974b.

Berggren, W. A., G. P. Lohmann, and R. Z. Poore, Shore laboratory report on Cenozoic planktonic foraminifera: Leg 22, 22, 635-656, 1974.

Fleisher, R. L., Cenozoic planktonic foraminifera and biostratigraphy, Arabian Sea Deep Sea Drilling Project, Leg 23A, 22, 1001-1072, 1974.

Gartner, S., Nannofossil biostratigraphy, Leg 22, Deep Sea Drilling Project, 22, 577-599, 1974.

Gartner, S., D. A. Johnson, and B. McGowran, Paleontology synthesis of deep sea drilling results from Leg 22 in the northeastern Indian Ocean, 22, 805-814, 1974.

Heiman, M. E., W. E. Frerichs, and E. Vincent, Paleogene planktonic foraminifera from the western tropical Indian Ocean, Deep Sea Drilling Project, Leg 24, 24, 851-857, 1974.

Johnson, D. A., Radiolaria from the eastern Indian Ocean, DSDP leg 22, 22, 521-575, 1974.

Kennett, J. P., R. E. Houtz, P. B. Andrews, A. R. Edwards, V. A. Gostin, M. Hajos, M. Hampton, D. G. Jenkins, S. V. Margolis, A. T. Ovenshine, and K. Perch-Nielsen, Cenozoic paleooceanography in the southwest Pacific Ocean, Antarctic glaciation, and the development of the circum-Antarctic current, in Kennett, J. P., Houtz, R. E. et al., 29, 1155-1169, 1974.

Krasheninnikov, V. A., Cretaceous and Paleogene planktonic foraminifera, 27, 663-672, 1974.

McGowran, B., Foraminifera, 22, 609-628, 1974.

McKelvey, B. C., and A. J. Fleet, Eocene basaltic pyroclastics at Site 253, Ninetyeast Ridge, 26, 553-565, 1974.

Moore, D. G., J. R. Curray, R. W. Raitt, and F. J. Emmel, Stratigraphic-seismic section correlations and implications to Bengal Fan history, 22, 403-412, 1974.

Olsson, R. K., Appendix V. Shore laboratory report on the foraminifera from Leg 21 sites, Deep Sea Drilling Project, 22, 865-881, 1974.

Pessagno, E. A., Jr., and F. Y. Michael, Mesozoic foraminifera, Leg 22, Site 217, 22, 629-634, 1974.

Sclater, J. G., R. D. Jarrard, B. McGowran, and S. Gartner, Jr., Comparison of the magnetic and biostratigraphic time scales since the Late Cretaceous, 22, 381-386, 1974.

Sclater, J. G., C. C. von der Borch, J. J. Veevers, R. Hekinian, R. W. Thompson, A. C. Pimm, B. McGowran, S. Gartner, Jr., and D. A. Johnson, Regional synthesis of the deep sea drilling results from Leg 22 in the eastern Indian Ocean, 22, 815-831, 1974.

Premoli Silva, I., and H. M. Bolli, Late Cretaceous to Eocene planktonic foraminifera and stratigraphy of Leg 15 sites in the Caribbean Sea, in Edgar, N. T., Saunders, J. B., et al., 15, 499-547, 1973.

Sigal, J., Comments on Leg 25 sites in relation to the Cretaceous and Paleogene biostratigraphy in the eastern and southeastern Africa coast and Madagascar regional setting, in Simpson, E. S. W. Schlich, R., et al., 25, 687-723, 1974.

Vincent, E., Cenozoic planktonic biostratigraphy and paleooceanography of the tropical western Indian Ocean, 24, 1111-1150, 1974.

Vincent, E., J. M. Gibson, and L. Brun, Paleocene and Early Eocene microfacies, benthonic foraminifera, and paleobathymetry of Deep Sea Drilling Project Sites 236 and 237, western Indian Ocean, 24, 859-885, 1974.

Other References Cited

Adams, C. G., A reconsideration of the East Indian Letter classification of the Tertiary, Bull. Brit. Mus. (Nat. Hist.) Geol., 19, 85-137, 1970.

Audley-Charles, M. G., and D. J. Carter, Palaeogeographical significance of some aspects of Palaeogene and early Neogene stratigraphy and tectonics of the Timor Sea region, Palaeogeogr., Palaeoclimatol., Palaeoecol., 11, 247-264, 1972.

Bain, J. H. C., and J. G. Binnekamp, The foraminifera and stratigraphy of the Chimbu Limestone, New Guinea, Australia, Bur. Mineral Resources, Geol. and Geophys., Bull. 139, 1-12, 1973.

Banerji, R. K., Foraminiferal biostratigraphy and geological evolution of the Thanjavur Sub-basin, India, J. Geol. Soc. India, 14, 257-274, 1973.

Belford, D. J., Paleocene planktonic foraminifera from Papua and New Guinea, Australia, Bur. Mineral Resources, Geol. and Geophys., Bull. 92, 1-33, 1967.

Berger, W. H., Cenozoic sedimentation in the eastern tropical Pacific, Bull. Geol. Soc. of Amer. 84, 1941-1954, 1973.

Berggren, W. A., A Cenozoic time-scale - some implications for regional geology and paleobiogeography, Lethaia, 5, 195-215, 1972.

Berggren, W. A., Late Paleocene-Early Eocene benthonic foraminiferal biostratigraphy and paleoecology of Rockall Bank, Micropaleontology, 20, 426-448, 1974a.

Berggren, W. A., Paleocene benthonic foraminiferal biostratigraphy, biogeography and palaeoecology of Libya and Mali, Micropaleontology, 20, 449-465, 1974b.

Bhatia, S. B., News report - India, Micropaleontology, 20, 373-384, 1974.

Blow, W. H., Late Middle Eocene to Recent planktonic foraminiferal biostratigraphy, in Bronnimann, P., and Renz, H. H., eds., Proceedings of the First International Conference on Planktonic Microfossils, Leiden, E. J. Brill, 1, 199-422, 1969.

Bolli, H. M., Zonation of Cretaceous to Pliocene marine sediments based on planktonic foraminifera, Bol. Informativo, 9, 3-32, 1966.

Condon, M. A., D. Johnstone, C. E. Prichard, and M. H. Johnstone, The Giralia and Marilla Anticlines, Northwest Division, Western Australia, Australia, Bur. Mineral Resources, Geol. and Geophys., Bull. 25, 1-86, pls. 1-7, text-figs. 1-14, 1956.

Cushman, J. A., Paleocene foraminifera from the Gulf Coastal region of the United States and adjacent areas, U. S. Geol. Surv. Prof. Paper 232, 1-75, 1951.

Datta, A. K., R. K. Banerji, and K. S. Soodan, A review of recent contributions to the Mesozoic and Cenozoic foraminiferal biostratigraphy of India, U. N. Economic Commission for Asia and Far East, Mineral Resources Development Series, 36, 108-117, 1970.

Davies, H. L., and I. E. Smith, Geology of eastern Papua, Bull. Geol. Soc. of Amer., 82, 3299-3312, 1971.

Davies, T. A., O. E. Weser, B. P. Luyendyk, and R. B. Kidd, Unconformities in the sediments of the Indian Ocean, Nature, 253, 15-19, 1975.

Dorreen, J. M., The western Gaj River section, Pakistan, and the Cretaceous-Tertiary boundary, Micropaleontology, 20, 178-193, 1974.

Fatmi, A. N., Lithostratigraphic units of the Kohar-Potwar Province, Indus Basin, Pakistan, Mem. Geol. Surv. Pakistan, 10, 80 pp., 1974.

Gansser, A., The Indian Ocean and the Himalayas. A geological interpretation, Eclog. Geol. Helv., 59, 831-848, 1966.

Govindan, A., Upper Cretaceous planktonic foraminifera from the Pondicherry area, south India, Micropaleontology, 18, 169-193, 1972.

Hayes, D. E., and J. Ringis, Sea floor spreading in the Tasman Sea, Nature, 243, 454-458, 1973.

Heirtzler, J. R., G. O. Dickson, E. M. Herron, W. C. Pitman, and X., Le Pichon, Marine magnetic anomalies, geomagnetic field reversals, and motions of the ocean floor and continents, J. Geophys. Res., 73, 2119-2136, 1968.

Jenkins, D. G., New Zealand Cenozoic planktonic foraminifera, N. Z. Geol. Surv., Paleont. Bull., 42, 1-288, 1971.

Khosla, S. C., Classification of the Lower Tertiary beds of Rajasthan, Bull. Indian Geologists' Assoc., 4, 54-60, 1972a.

Khosla, S. C., Ostracodes from the Eocene beds of Rajasthan, India, Micropaleontology, 18, 476-507, 1972b.

Lindsay, J. M., Cenozoic foraminifera and stratigraphy of the Adelaide Plains Sub-basin, South Australia, Geol. Surv. S. Aust., Bull., 42, 1-60, 1969.

Lindsay, J. M., and W. K. Harris, Fossiliferous marine and non-marine Cainozoic rocks from the eastern Eucla Basin, South Australia, South Aust. Dept. Mines, Mineral Resources Review 138, 29-42, 1975.

Luterbacher, H., and I. Premoli Silva, Biostratigraphia del limite Cretaceo-Terziario nell'appennino centrale, Rivista Ital. Paleont., 70, 67-128, 1964.

Lys, M., La Limite Crétacé-Tertiaire et l'Eocène inférieur dans le bassin de Majunga (Madagascar), Rept. 21st Internat. Geol. Congr., Section 5, 120-130, 1960.

McGowran, B., Foraminiferal evidence for the Paleocene age of the Kings Park Shale (Perth Basin, Western Australia), J. Roy. Soc. W. Aust., 47, 81-86, 1964.

McGowran, B., Two Paleocene foraminiferal faunas from the Wangerrip Group, Pebble Point coastal section, western Victoria, Proc. Roy. Victoria, 79, 9-74, 1965.

McGowran, B., Late Cretaceous and Early Tertiary correlations in the Indo-Pacific region, Geol. Soc. India, Memoir No. 2, 335-360, 1968.

McGowran, B., The role of planktonic foraminifera in the biostratigraphy of the Paleocene in Australia, U. N. Econmic Commission for Asia and Far East, Mineral Resources Development Series, No. 30, 94-104, 1969.

McGowran, B., Late Paleocene in the Otway Basin: biostratigraphy and age of key microfaunas, Trans. Roy. Soc. South Australia, 94, 1-14, 1970.

McGowran, B., Rifting and drift of Australia and the migration of mammals, Science, 180, 759-761, 1973.

McGowran, B., Paleogene foraminiferal stratigraphy on the Ninetyeast Ridge and in southern Australia, and some geohistorical implications, 1977, in press, a.

McGowran, B., Early Tertiary foraminiferal biostratigraphy in southern Australia: progress report, 1977, in press, b.

McGowran, B., and J. M. Lindsay, A Middle Eocene planktonic foraminiferal assemblage from the Eucla Basin, Geol. Surv. South Aust., Quart. Geol. Notes, 30, 2-10, 1969.

McGowran, B., J. M. Lindsay, and W. K. Harris, Attempted reconciliation of Tertiary biostratigraphic systems, Otway Basin, in Wopfner, H., and Douglas, J. G., Eds., The Otway Basin in southeast Australia, Geol. Survs. South Aust. and Victoria, Special Bulletin, 273-281, 1971.

McKenzie, D. P., and J. G. Sclater, The evolution of the Indian Ocean since the Late Cretaceous, Geophys. J. Roy. Astron. Soc., 25, 437-528, 1971.

Mohan, Madan, and K. S. Soodan, Middle Eocene planktonic foraminiferal zonation of Kutch, India, Micropaleontology, 16, 37-46, 1970.

Nagappa, Y., Foraminiferal biostratigraphy of the Cretaceous-Eocene succession in the India-Pakistan-Burma region, Micropaleontology, 5, 145-192, 1959.

Owen, M., Upper Cretaceous planktonic foraminifera from Papua New Guinea, Australia, Bureau Mineral Resources, Geol. and Geophys., Bull. 140, 47-65, 1973.

Palmieri, V., Occurrence of Danian at Port Moresby, Geol. Survey Queensland, 63, 1-8, 1971.

Pessagno, E. A., Upper Cretaceous planktonic foraminifera from the western Gulf Coastal Plain, Palaeontographica Americana, 5, no. 37, 445 pp., 1967.

Pimm, A. C., B. McGowran, and S. Gartner, Early sinking history of Ninetyeast Ridge, north-eastern Indian Ocean, Bull. Geol. Soc. Amer., 85, 1219-1224, 1974.

Pimm, A. C., and J. G. Sclater, Early Tertiary hiatuses in the north-eastern Indian Ocean, Nature, 252, 362-365, 1974.

Pitman, W. C., R. L. Larson, and E. M. Herron, Magnetic lineations of the oceans, Geol. Soc. America (Map), 1974.

Plummer, H. J., Foraminifera of the Midway Formation in Texas, Univ. Texas Bull. 2644, 1-206, 1927.

Powell, C. McA., and P. J. Conaghan, Plate tectonics and the Himalayas, Earth Planet. Sci. Letters, 20, 1-12, 1973.

Quilty, P. G., Cainozoic stratigraphy in the Perth area, Jour. Roy. Soc. W. Aust., 57, 16-31, 1974a.

Quilty, P. G., Tertiary stratigraphy of Western Australia, Jour. Geol. Soc. Aust., 21, 301-318, 1974b.

Rajagopalan, N., A re-study of the Pondicherry Formation, Geol. Soc. India, Mem. No. 2, 128-135, 1968.

Raju, D.S.N., Eocene-Oligocene planktonic foraminiferal biostratigraphy of Cauvery Basin, south India, Geol. Soc. India, Mem. No. 2, 286-299, 1968.

Raju, D.S.N., and D. K. Guha, Contributions on Cretaceous and Cenozoic Microfauna, paleoecology, stratigraphic classification and correlation in India, U. N. Econmic Commission for Asia and Far East, Mineral Resources Development Series, 36, 117-129, 1970.

Saito, T., and J. van Donk, Oxygen and carbon isotope measurements of Late Cretaceous and Early Tertiary foraminifera, Micropaleontology, 20, 152-177, 1974.

Samanta, B. K., The Eocene succession of Garo Hills, Assam, India, Geol. Mag. 105, 124-135, 1968.

Samanta, B. K., Eocene planktonic foraminifera from the Garo Hills, Assam, India, Micropaleontology, 15, 325-350, 1969.

Samanta, B. K., Upper Eocene planktonic foraminifera from the Kopili Formation, Mikir Hills, Assam, India, Contr. Cushman Found. Foram. Res., 21, 28-39, 1970a.

Samanta, B. K., Middle Eocene planktonic foraminifera from Lakhpat, Cutch, western India, Micropaleontology, 16, 185-215, 1970b.

Samanta, B. K., Planktonic foraminifera from the early Tertiary Pondicherry Formation of Madras, south India, Jour. Paleontology, 44, 605-641, 1970c.

Samanta, B. K., Early Tertiary stratigraphy of the area around Garampani, Mikir-North Cachar Hills, Assam, J. Geol. Soc. India, 12, 318-327, 1971.

Samanta, B. K., Planktonic foraminiferal biostratigraphy of the early Tertiary of the Rakhi Nala section, Sulaiman Range, West Pakistan, J. Geol. Soc. India, 13, 317-328, 1972.

Samanta, B. K., Planktonic foraminifera from the Paleocene-Eocene succession in the Rakhi Nala, Sulaiman Range, Pakistan, Bull. British Mus. Nat. Hist. (Geol.), 22, 421-482, 1973a.

Samanta, B. K., On the age of the Pondicherry Formation of Madras, south India, J. Geol. Soc. India, 14, 289-295, 1973b.

Samanta, B. K., Planktonic foraminiferal biostratigraphy of the Late Middle to Upper Eocene succession in Assam, eastern India, Bull. Indian Geologists' Assoc., 6, 99-126, 1973c.

Schmidt, R. R., and D.S.N. Raju, Globorotalia palmerae Cushman and

Bermudez and closely related species from the Lower Eocene, Cauvery Basin, south India, Proc. Koninkl. Nederl. Akad. Wetenschappen, ser. B., 76, 167-184, 1973.

Sclater, J. G., and R. L. Fisher, Evolution of the east central Indian Ocean, with emphasis on the tectonic setting of the Ninetyeast Ridge, Bull. Geol. Soc. of Amer., 85, 683-702, 1974.

Sigal, J., N. P. Singh, and M. Lys, Paleocene-Eocene boundary in the Jaisalmer area, India, Jour. Foram. Res., 1, 190-194, 1971.

Singh, N. P., Stratigraphy of the Eocene of Rajasthan, Indian Sci. Congr., 56th Session, Proc., 216-217 (Abstract), 1969.

Sloss, L. L., Sequences in the cratonic interior of North America, Bull. Geol. Soc. Amer., 74, 93-114, 1963.

Smith, C. C., and E. A. Pessagno, Jr., Planktonic foraminifera and stratigraphy of the Corsicana Formation (Maastrichtian), North Central Texas, Cushman Foundation Foraminiferal Research, Spec. Publ. No. 12, 1-68, 1973.

Thomas, B. M., and D. N. Smith, A summary of the petroleum geology of the Carnarvon Basin, Jour. Aust. Petroleum Exploration Assoc., 14, 66-76, 1974.

Tourmarkine, M., and H. M. Bolli, Evolution de Globorotalia cerroazulensis (Cole) dans l'Eocene Moyen et superieur de Possagno Italie), Rev. Micropaleontologie, 13, 131-145, 1970.

van Hinte, J., The Cretaceous time-scale and planktonic-foraminiferal zones, Proc. Koninkl. Nederl. Akad. Wetenschappen, ser. B, 75, 1-8, 1972.

Visser, W. A., and J. J. Hermes, Geological results of the exploration for oil in Netherlands New Guinea, Govt. Printing Office, The Hague, 265 pp., 1962.

Weissel, J. K., and D. E. Hayes, Magnetic anomalies in the southeast Indian Ocean, in Hayes, D. E., Ed., Antarctic oceanology II: the Australian-New Zealand sector, Washington, D. C., Amer. Geophys. Un., Antarctic Research Series, 19, 165-196, 1972.

Wright, C. A., Distribution of Cainozoic foraminiferids in the B.O.C. A. L. Scott Reef No. 1 Well, Aust. N. Z. Assoc. Advancement of Science, 45th Congress, Abstracts Section 3 (Geology), 94-95, 1973.

CHAPTER 19. OLIGOCENE PLANKTONIC FORAMINIFERAL ASSEMBLAGES FROM
DEEP SEA DRILLING PROJECT SITES IN THE INDIAN OCEAN

Robert L. Fleisher

Exxon Production Research Company
Houston, Texas 77001

Abstract. Planktonic foraminifera of Oligocene age were recovered at only thirteen of the forty-six Indian Ocean Deep Sea Drilling Project sites, all but one of which were situated on topographically high structures. Calcium carbonate dissolution is intense in all assemblages, which are often reduced to a few solution-resistant and typically long-ranging species. This preservation pattern suggests a shallow lysocline in the Indian Ocean throughout most of this interval.

Introduction

Planktonic foraminiferal associations of Oligocene age were recovered at thirteen of the forty-six Indian Ocean sites occupied during Deep Sea Drilling Project Legs 22 through 27. Each of the generally-recognized planktonic foraminiferal zones is represented in the sequence penetrated in at least one hole. With few exceptions, however, detailed reports on the Oligocene faunas are not available in either the Initial Reports volumes or subsequent publications. Range charts based on closely-spaced samples spanning the entire recovered Oligocene section, for example, have been constructed for only five sites (Boltovskoy, 1974; Fleisher, 1975). Thus a summary of Oligocene assemblages from other sites can be made only in the most general terms.

This treatment reflects the overall poor quality of the Oligocene Indian Ocean planktonic foraminiferal faunas. Unconformities, frequently representing long periods of time, can be documented within this interval at most sites, and their absence cannot be conclusively demonstrated at many others. A discussion of the paleoceanographic significance of these unconformities is beyond the range of this report, which will deal solely with the composition and biostratigraphy of the foraminiferal faunas. The stratigraphic discontinuities, however, substantially reduce the value of the planktonic foraminiferal successions in reconstructing the regional biostratigraphic sequence.

Intense calcium carbonate dissolution has drastically altered the species composition of the Oligocene faunas and substantially increased the difficulty involved in making foraminiferal correlations and age determinations. Oligocene diversities are typically low, in terms of both recognizable species and broad morphotypes, even in continental margin sequences where carbonate dissolution has been minimal. This low diversity has been further reduced by the virtually complete elimi-

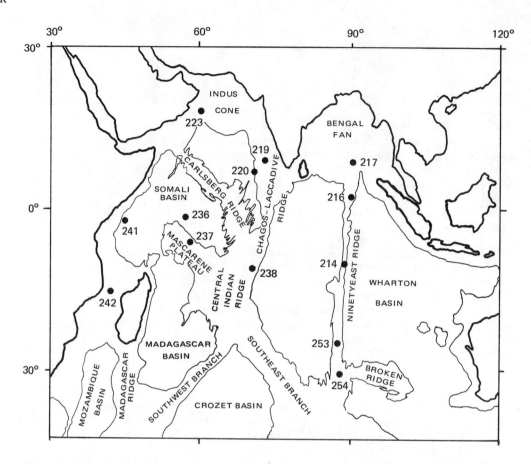

Fig. 1. Location of Deep Sea Drilling Project sites in the Indian Ocean at which Oligocene planktonic foraminiferal assemblages were recovered. Contour is on 4000 m water depth.

nation of the spinose species of Globigerina; at several of the sites, indeed, the faunas comprise little more than a handful of solution-resistant and typically long-ranging forms. The relatively poor preservation, together with the location of the thirteen sites, demonstrate the intensity of dissolution throughout the Indian Ocean during the Oligocene, in that the greatly reduced foraminiferal faunas were recovered from the intrabasinal sites most suited to the preservation of carbonate tests, that is, from high-standing topographic features. A shallow lysocline clearly persisted in the Indian Ocean throughout this interval.

Because detailed descriptions of the Oligocene foraminiferal populations have for the most part not been published, this paper is limited in scope to a summary of the results reported in the Initial Reports and supplemented by additional studies (Fleisher, 1975) of Sites 219, 220, and 237. The faunas are related to a zonal sequence modified slightly from that of Bolli (1966) and Jenkins and Orr (1972) for application to deep-ocean assemblages. The available data are too general to permit paleoecologic interpretations of any real significance to be drawn.

The taxonomic system employed here is somewhat at variance, particularly at the generic level, with general practice in the field. A thorough discussion of the rationale for this system is inappropriate

in this paper, but a brief section of taxonomic notes is
included to minimize confusion in terminology.

Oligocene Biostratigraphic Zonation Systems

The biostratigraphic studies of Deep Sea Drilling Project samples
document the difficulties involved in attempting to utilize the standard
interregional zonation systems for most deep-sea assemblages. Both the
Blow-Berggren (Blow, 1969; Berggren, 1972) and Bolli (1966) zonations
were constructed very largely from observations of land-based marine
sections deposited along continental margins, where dissolution has had
little effect on the faunas. The most readily dissolved species, in
general, are among those living in near-surface waters. These forms
are often abundant in well-preserved assemblages, and a number of the
zones in both biostratigraphic systems are defined on the basis of such
species. These zones are frequently unrecognizable in deep-sea sediments.

In all of the discussions in the Initial Reports, Oligocene Indian
Ocean planktonic foraminiferal assemblages have been related to the Blow-
Berggren zonal sequence. The results have not been entirely satisfactory,
primarily because several of the zone boundaries are based on solution-
susceptible species. Blow (1969) recognized his N.4/P.22 boundary,
thought to be coincident with the Miocene/Oligocene boundary, at the
lowest appearance of Globigerinoides primordius. This species, however,
is relatively susceptible to dissolution, and its appearances tend to
be sporadic near the base of its range even in fairly well-preserved
deep-ocean suites. More important, there is growing evidence (summarized by Meulenkamp, 1975) that G. primordius evolved within the Late
Oligocene. The problem of recognizing the Miocene/Oligocene boundary
is at present unresolved -- in this report, it is placed more or less
arbitrarily at the basal occurrence of "Turborotalia" kugleri -- but
clearly Zones N.4 and P.22 cannot be delineated in the sense intended by
Blow. Additionally, the absence in most deep-sea samples of both
Globigerina anguliofficinalis and G. angulisuturalis precludes the precise recognition of the P.21/P.20 boundary.

The zonation system proposed by Bolli (1966) is much more suitable for
application to these faunas, in that the zone boundaries are based on
the ranges of species particularly resistant to dissolution. Jenkins
and Orr (1971, 1972), proposing a zonation for DSDP samples from the
eastern tropical Pacific Ocean, adopted this system with only minor
changes. In effect, the Bolli zonation is employed here with modifications only in zone names (Figure 2). All but one of these changes merely serve to bring the names into line with the generic taxonomy used in
this report. However, the nominate species of Bolli's uppermost Oligocene zone, Globigerina ciperoensis, is generally absent. This interval,
between the extinction of "Turborotalia" opima opima and the initial
appearance of "T." kugleri, is here termed the Globoquadrina binaiensis
Zone. G. binaiensis, a form which first appears shortly above the base
of the interval, is both distinctive and relatively resistant to dissolution.

In virtually all of the DSDP samples considered here, Oligocene faunas
appear to be dominated by a relatively few species which could collectively be termed the resistant assemblage. Species of four genera
typically are prominent. Globoquadrina galavisi and G. tripartita are
found in most samples, with (within their ranges) G. binaiensis, G.
sellii, G. tapuriensis, and G. pseudovenezuelana usually somewhat less
abundant. "Turborotalia" kugleri, "T." opima nana, and "T." opima opima
are generally preserved selectively, as are Catapsydrax dissimilis s.l.
and C. unicavus s.l. Tenuitella angustiumbilicata and T. clemenciae are

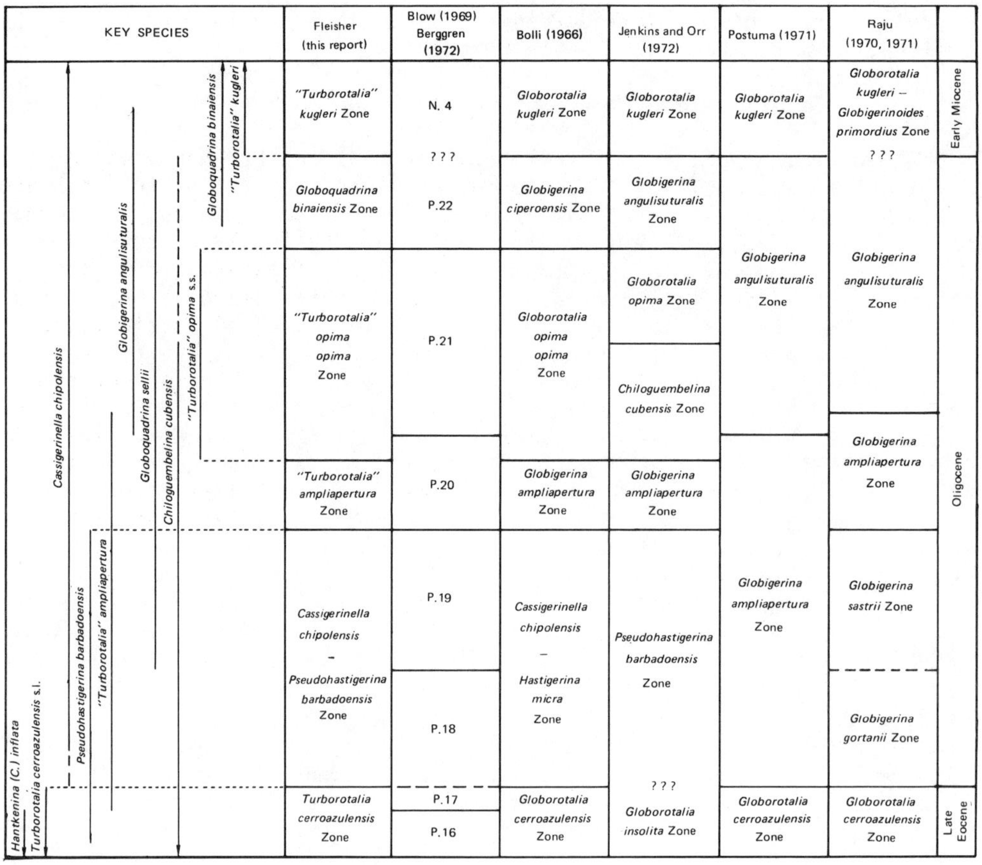

KEY SPECIES		Fleisher (this report)	Blow (1969) Berggren (1972)	Bolli (1966)	Jenkins and Orr (1972)	Postuma (1971)	Raju (1970, 1971)	
	"Turborotalia" kugleri *Globoquadrina binaiensis*	*"Turborotalia" kugleri* Zone	N. 4	*Globorotalia kugleri* Zone	*Globorotalia kugleri* Zone	*Globorotalia kugleri* Zone	*Globorotalia kugleri — Globigerinoides primordius* Zone	Early Miocene
			???				???	
Cassigerinella chipolensis *Globigerina angulisuturalis* *Globoquadrina sellii* *Chiloguembelina cubensis* *"Turborotalia" opima s.s.*		*Globoquadrina binaiensis* Zone	P.22	*Globigerina ciperoensis* Zone	*Globigerina angulisuturalis* Zone	*Globigerina angulisuturalis* Zone	*Globigerina angulisuturalis* Zone	Oligocene
		"Turborotalia" opima opima Zone	P.21	*Globorotalia opima opima* Zone	*Globorotalia opima* Zone			
					Chiloguembelina cubensis Zone		*Globigerina ampliapertura* Zone	
		"Turborotalia" ampliapertura Zone	P.20	*Globigerina ampliapertura* Zone	*Globigerina ampliapertura* Zone			
Hantkenina (C.) inflata *Turborotalia cerroazulensis s.l.* *Pseudohastigerina barbadoensis* *"Turborotalia" ampliapertura*		*Cassigerinella chipolensis*	P.19	*Cassigerinella chipolensis*	*Pseudohastigerina barbadoensis* Zone	*Globigerina ampliapertura* Zone	*Globigerina sastrii* Zone	
		Pseudohastigerina barbadoensis Zone	P.18	*Hastigerina micra* Zone			*Globigerina gortanii* Zone	
		Turborotalia cerroazulensis Zone	P.17	*Globorotalia cerroazulensis* Zone	??? *Globorotalia insolita* Zone	*Globorotalia cerroazulensis* Zone	*Globorotalia cerroazulensis* Zone	Late Eocene
			P.16					

Fig. 2. Correlation of Oligocene planktonic foraminiferal zonation systems. The zonation of Raju (1970, 1971) is based on stratigraphic sections in southeastern India. From Fleisher (1975).

usually present and often common, but are sometimes unreported because of their small size. In addition to these four groups, another resistant species, Neogloboquadrina siakensis, is often present in samples from the Upper Oligocene.

Eastern Indian Ocean Sites

Five drilling sites in the eastern Indian Ocean, all located on the crest of the Ninetyeast Ridge, penetrated sediments of Oligocene age containing planktonic foraminifera. They are considered here in order, from north to south.

Site 217: Oligocene sediments are present in the interval from Core 7 (230 m) through Core 9, Section 5 (314 m), but because coring was so discontinuous no significant conclusions can be drawn. No data is available concerning the fauna of Core 7, although foraminifera are apparently present. The assemblages of Core 8 are dominated by species of Globoquadrina, particularly G. tripartita, and are essentially uniform throughout the core. The presence as well of G. globularis, G. baroemoenensis, and G. venezuelana restricts the fauna to a Late Oligocene age. Berggren et al. (1974) assign the samples an age within P.21 ("Turborotalia" opima opima Zone), but the foraminiferal fauna could also be representative of a younger Globoquadrina binaiensis Zone age. Samples from Core 9, Sections 1 through 5 contain relatively common Pseudohastigerina sp. together with a resistant Oligocene assemblage.

Site 216: McGowran (1974) placed the top of the Oligocene at the lowest occurrence of Globigerinoides sp., at the base of Core 6 (177 m). The base of "T." kugleri, however, is somewhat lower (200 m), and the Miocene/Oligocene boundary is drawn at this horizon.

Preservation is particularly poor throughout the Oligocene at this site, from Core 9 (200 m) through Core 15 (257 m). Among the larger forms, only species of Catapsydrax and "Turborotalia" are present; Chiloguembelina sp. and Pseudohastigerina sp. are present in Core 14 (244 m) and perhaps slightly higher. In general, the Oligocene assemblages at Site 216 are so poorly preserved that no zone boundaries can be definitively recognized; even the base of the Oligocene cannot be located from the foraminiferal evidence.

Site 214: McGowran (1974) regarded the Oligocene section at this site as being virtually complete; at least, his documentation (p. 613) suggests that all of the Oligocene zones are represented. In the absence of detailed range charts, this stratigraphic situation cannot be verified, but his conclusion suggests that the entire Oligocene is represented by the interval from Core 24, Section 3 (223 m) to Core 27, Section 6 (257 m). The apparent compression of the section may reflect the presence of unconformities within the sequence; the virtual coincidence of the base of "Turborotalia" mendacis and the extinction of Chiloguembelina cubensis in Core 25 suggests that a portion of the Late Oligocene may be unrepresented.

As documented by Berggren et al. (1974), the diversity of Oligocene faunas at this site is extremely low; indeed, there is an apparent discrepancy between the two reports as to the nature and composition of the fauna. As reported by Berggren et al., the assemblages are dominated by resistant species of Globoquadrina and Catapsydrax. These authors place the top of the Eocene above Core 27, but the coincidence of base of Globoquadrina tapuriensis and the highest occurrences of of Turborotalia cerroazulensis and Globigerinatheka spp. at the base of Core 27 (McGowran, 1974) suggest that the Oligocene/Eocene boundary, apparently unconformable, should be placed there.

Site 253: Oligocene planktonic foraminifera were recovered from a twenty-seven meter interval between Core 10, Section 3 and Core 13, Section 1. The basal occurrence of "T." kugleri, at 88 m, is slightly below the Miocene/Oligocene boundary recognized by Boltovskoy (1974), in the absence of Globigerinoides spp., on the basis of somewhat ambiguous species ranges. The faunas are dominated by the typical solution-resistant assemblage, although relatively large bullate forms referred to Globigerina cryptomphala (Boltovskoy, 1974, Pl. 2) may be polyspecific. "T." opima opima is present from the middle of Core 11 through the middle of Core 12 (97 to 109 m). Oligocene faunas below this level include both Globoquadrina sellii and G. tapuriensis at most horizons, an association characteristic of the Cassigerinella chipolensis-Pseudohastigerina barbadoensis Zone.

Site 254: Boltovskoy (1974) assigned an Oligocene age only to the foraminifera recovered in Core 20 (167 to 176 m). In fact, the Miocene/Oligocene boundary is difficult to recognize. Both Globigerinoides primordius and "T." kugleri are absent, and typically Paleogene species such as Globoquadrina tripartita, G. galavisi (Globigerina yeguaensis of Boltovskoy) and "T." opima s.l. range several cores higher. The fauna throughout this interval, Cores 18 through 20 (148 to 176 m), comprises almost entirely the solution-resistant assemblage described above, although rare specimens assigned to Catapsydrax cf. africanus are also present.

Western Indian Ocean Sites

Site 223 (Owen Ridge): The Upper Oligocene, in Cores 28 through 30 (496 to 533 m), contains a heavily dissolved and poorly preserved (crushed) assemblage of solution-resistant forms. "T." kugleri was not observed in any recovered samples, and the topmost Oligocene may be missing. The "T." opima opima Zone, Late Oligocene, is represented in Cores 29 and 30 (520 to 533 m).

Within the top of Core 31 is a well-preserved Early Oligocene assemblage including Globoquadrina tapuriensis and Pseudohastigerina barbadoensis. The unconformable base of the Oligocene sequence is within Core 31.

Site 219 (Chagos-Laccadive Ridge): The entire Oligocene sequence recovered at this site, in Cores 15 and 16 (156 to 174 m), represents the Cassigerinella chipolensis - Pseudohastigerina barbadoensis Zone. Most of the dominant forms are resistant, long-ranging species, but several others, including Chiloguembelina cubensis, P. barbadoensis, Subbotina angiporoides, and "Turborotalia" ampliapertura, are present and sometimes common.

Site 220 (west flank, Chagos-Laccadive Ridge): Only the Late Oligocene is represented (Fleisher, 1974, 1975), in Cores 6 through 10 (93 to 168 m) at Site 220. Cores 6 through 9 (93 to 156 m) contain a reduced fauna indicative of the Globoquadrina binaiensis Zone, dominated by species of Globoquadrina (including G. binaiensis), Catapsydrax, "Turborotalia", and Tenuitella. Neogloboquadrina siakensis and Turborotalita primitiva are also present.

The fauna contained in the sediments recovered in Cores 9-CC and 10 is broadly similar, except that N. siakensis, Globoquadrina spp., and T. primitiva are essentially absent. "T." opima opima is found in several samples in this interval.

Site 241 (East African continental rise): A relatively diverse and well-preserved Late Oligocene fauna occurs in the lower section of Core 16 (457-458 m). The assemblage comprises, among other forms, Globigerina ciperoensis, Globoquadrina sellii, G. binaiensis, "Turborotalia" cf. kugleri (probably "T." mendacis or "T." pseudokugleri) and is assigned to the G. binaiensis Zone (see Zobel, in Simpson et al., 1974).

Site 236 (Somali Basin): Planktonic foraminifera are poorly developed throughout the Oligocene section recovered in Cores 20 through 27 (178 to 254 m). Precise data are available for only a few samples from this interval, but these contain almost exclusively long-ranging and resistant forms. The only planktonic zone recognizable is the "T." opima opima Zone in Core 21 through the upper part of Core 24 (Heiman et al., 1974).

Site 237 (Mascarene Plateau): The planktonic foraminiferal populations recovered from the short Oligocene interval, Core 21, Section 3, through Core 22 (191 to 200 m) are among the better preserved Indian Ocean assemblages of this age. As at other sites, resistant species of Globoquadrina, Catapsydrax, Neogloboquadrina, and "Turborotalia" are dominant, but species of Globigerina (primarily G. praebulloides, G. anguliofficinalis, and G. ouachitaensis), "Globigerina" ("G." woodi s.l.), and Protentella (P. cf. nicobarensis) are present, although rare, in many samples.

The sequence below the base of "T." kugleri, in Core 21, Section 2 (190 m), is assigned to the Globoquadrina binaiensis Zone, except for Core 22-CC, which contains "T." opima opima. The top samples in Core 23 contain Late Eocene assemblages including Turborotalia cerroazulensis s.l. (Fleisher, 1975).

Site 238 (Central Indian Ridge): Although a substantial Oligocene section was recovered at Site 238 (Cores 49 through 54, 450 to 506 m), the faunas are, in general, poorly preserved and non-diverse. "T." kugleri is present in Core 48, Section 2 (Heiman et al., 1974), but the base of its range cannot at present be located precisely. Cores 50 through 53 (462 to 500 m) contain very largely resistant forms, including "T." opima opima throughout and "T." pseudokugleri and Neogloboquadrina siakensis in Core 50 only. If the "T." ampliapertura Zone is represented, it is present only in the bottom few meters of Core 53. The lowest sediments in this hole, in Core 54, contain a relatively diverse fauna including Turborotalia pseudoampliapertura (Heiman et al., 1974) and Pseudohastigerina barbadoensis (Vincent, in Fisher et al., 1974), an Early Oligocene association.

Site 242 (Davie Ridge, Mozambique Channel): The information available on Oligocene planktonic foraminifera at this site is not sufficiently detailed to permit a significantly informative summary to be presented (Zobel, in Simpson et al., 1974). The base of "T." kugleri falls within Core 9, and the remainder of that core contains a fauna indicative of the Globoquadrina binaiensis Zone. "T." opima opima occurs in Cores 10 and 11 with Pseudohastigerina sp. and Chiloguembelina sp., but because nannofossil data indicate a Late Oligocene age, the latter two forms are assumed to be reworked. Cores 12 through 14 contain an Early Oligocene fauna of resistant species.

Taxonomic Notes

The prevalent systems of genus-level taxonomy applied to Cenozoic planktonic foraminifera, following in general the practices of Blow (1969), are based almost exclusively on considerations of gross test morphology. Approaches of this kind have the practical advantage of relative ease of application. Because of the broadly similar environmental pressures acting on all planktonic foraminifera, however, similar morphotypes have evolved independently in a number of separate and distinct lineages. Test form alone is not sufficiently sensitive, for the most part, to distinguish these essentially unrelated isomorphs at the generic level.

An alternative approach to classification can be based on primarily phyletic, rather than morphologic, criteria. The justification for this approach, and the techniques involved in applying it, have been discussed at considerable length elsewhere (Fleisher, 1974, 1975), and a detailed reiteration of these arguments is inappropriate here. However, the resultant application of several common generic names is at variance with general practice. For the sake of clarity, these usages require explanation.

1) Globigerina: This name is restricted to trochospirally coiled forms, typically but not universally with an umbilical aperture, with a spinose surface wall texture. The primary modern descendent is the type species, G. bulloides. Dominant Oligocene forms include species of the G. praebulloides, G. ouachitaensis, and G. ciperoensis-G. angulisuturalis groups. Several groups of globigeriniform species, however, are essentially unrelated to this stock, and are excluded from this genus. These are the species assigned to Subbotina and "Globigerina."

Included in Subbotina are coarsely-perforate and apparently non-spinose species descended from the Paleocene S. triloculinoides. These forms are restricted to the Paleogene, and like the primary Oligocene representative, S. angiporoides, are characterized by a low, frequently slit-like aperture bordered by a rope-like apertural rim.

"Globigerina" woodi is the dominant Late Oligocene and Miocene reppresentative of another group of species, similar in wall texture and test morphology to Subbotina but probably unrelated and largely Neogene in age. The lineage, which may have evolved from "Turborotalia" ampliapertura, includes such species as "G." druryi, "G." nepenthes, and "G." decoraperta. Although the lineage is phyletically independent from Globigerina s.s., there appears to be no available generic name which can properly be applied to it.

2) Globoquadrina: The type species, G. dehiscens, is the most distinctive species in one of a series of lineages which evolved from the G. galavisi-G. tripartita complex near the end of the Oligocene. The other descendent lineages are usually placed in this genus; the practice adopted here is to include the ancestral lineage, G. galavisi to G. binaiensis, as well. The phyletic relation to G. dehiscens is suffiently close that the distinctive morphological character common to most species of the genus, an umbilical aperture with a well-developed tooth, as also present in the G. galavisi group.

3) Neogloboquadrina: This name is applied to the phyletic sequence which culminated, in the Late Neogene, in the species N. acostaensis, N. humerosa, and N. dutertrei. N. siakensis, which evolved in the Late Oligocene, is the earliest species which can be assigned to this lineage with some degree of confidence.

4) "Turborotalia": It appears that no readily applicable generic name is available for the bulk of turborotaliform species. The type species of Turborotalia s.s., T. cerroazulensis, is a member of a lineage which became virtually extinct, with no apparent long-lived descendants, near the end of the Eocene. The sole survivor, the globigeriniform T. pseudoampliapertura, disappeared within the Early Oligocene.

For the present, therefore, all turborotaliform species which cannot be assigned on phyletic grounds to Neogloboquadrina, Turborotalia, or Tenuitella (small forms with matte-like, microperforate wall texture), are placed in "Turborotalia." Obviously, this grouping probably represents little more than a polyphyletic form-genus, but the taxonomy of these species will become clearer as their phylogenetic history is better understood.

"T." ampliapertura is included on the basis of its apparent evolutionary descent from "T." increbescens.

References

Berggren, W. A., A Cenozoic time-scale -- some implications for regional geology and paleobiogeography, Lethaia, 5, 195-215, 1972.

Berggren, W. A., G. P. Lohmann, and R. Z. Poore, Shore laboratory report on Cenozoic planktonic foraminifera: Leg 22, in von der Borch, C. C., Sclater, J. G., et al., Initial Reports of the Deep Sea Drilling Project, Washington, D. C. (U. S. Government Printing Office) 22, 635-655, 1974.

Blow, W. H., Late Middle Eocene to Recent planktonic foraminiferal biostratigraphy, in Proc. 1st Internat. Conf. on Planktonic Microfossils, edited by P. Brönnimann and H. H. Renz, E. J. Brill and Co., Leiden, 1, 199-422, 1969.

Bolli, H. M., Zonation of Cretaceous to Pliocene marine sediments based on planktonic foraminifera, Bol. Inform., Asoc. Venez. Geol., Min., Petrol., 9, no. 1, 3-32, 1966.

Boltovskoy, E., Neogene planktonic foraminifera of the Indian Ocean (DSDP, Leg 26), in Davies, T. A., Luyendyk, B. P., et al., Initial Reports of the Deep Sea Drilling Project, Washington, D. C. (U. S. Government Printing Office) 26, 675-742, 1974.

Fleisher, R. L., Cenozoic planktonic foraminifera and biostratigraphy, Arabian Sea, Deep Sea Drilling Project Leg 23A, in Whitmarsh, R. B.,

Weser, O. E., Ross, D. A., et al., Initial Reports of the Deep Sea Drilling Project, Washington (U. S. Government Printing Office) 23, 1001-1072, 1974.

Fleisher, R. L., Early Eocene to Early Miocene planktonic foraminiferal biostratigraphy of the western Indian Ocean, unpub. Ph.D. Dissert., Dept. Geol. Sci., Univ. So. Calif., Los Angeles, Cal., 1975.

Heiman, M. E., W. E. Frerichs, and E. Vincent, Paleogene planktonic foraminifera from the western tropical Indian Ocean, Deep Sea Drilling Project, Leg 24, in Fisher, R. L., Bunce, E. T., et al., Initial Reports of the Deep Sea Drilling Project, Washington (U. S. Government) 24, 851-858, 1974.

Jenkins, D. G., and W. N. Orr, Cenozoic planktonic foraminiferal zonation and the problem of test solution, Rev. Espan. Micropal., 3, 301-304, 1971.

Jenkins, D. G., and W. N. Orr, Planktonic foraminiferal biostratigraphy of the eastern Equatorial Pacific -- DSDP Leg 9, in Hollister, C. D., Ewing, J. I., et al., Initial Reports of the Deep Sea Drilling Project, Washington (U. S. Government Printing Office) 9, 1060-1193, 1972.

McGowran, B., Foraminifera, in von der Borch, C. C., Sclater, J. G., et al., Initial Reports of the Deep Sea Drilling Project, Washington (U. S. Government Printing Office) 22, 609-627, 1974.

Meulenkamp, J. E., Report of the Working Group on Micropaleontology, in Report on Activity of the R.C.M.N.S. Working Groups (1971-1975), edited by J. Senes, I.U.G.S. Regional Committee on Mediterranean Neogene Stratigraphy, Bratislava, 9-29, 1975.

Postuma, J. A., Manual of planktonic foraminifera, Elsevier Publ. Co., Amsterdam, The Netherlands, 1-420, 1971.

Raju, D. S. N., Zonal distribution of selected foraminifera in the Cretaceous and Cenozoic sediments of Cauvery Basin and some problems of Indian biostratigraphic classification, Centre Adv. Stud. Geol., Panjab Univ., Publ. 7, 85-110, 1970.

Raju, D. S. N., Upper Eocene to Early Miocene planktonic foraminifera from the subsurface sediments in Cauvery Basin, South India, Jb. Geol. B. A. Sond. 17, 7-68, 1971.

Vincent, E., Foraminifera (in Fisher, R. L., et al., Site 238): in Fisher, R. L., Bunce, E. T., et al., Initial Reports of the Deep Sea Drilling Reports, Washington (U. S. Government Printing Office) 24, p. 475, 1974.

Zobel, B., Foraminifera: Quaternary, Neogene, and Oligocene (in Schlich, R., et al., Site 241): in Simpson, E.S.W., Schlich, R., et al., Initial Reports of the Deep Sea Drilling Project, Washington, (U. S. Government Printing Office) 25, 98-99, 1974.

Zobel, B., Foraminifera: Quaternary, Neogene, and Oligocene (in Simpson, E.S.W., et al., Site 242): in Simpson, E.S.W., Schlich, R., et al., Initial Reports of the Deep Sea Drilling Project, Washington (U. S. Government Printing Office), 25, 149-150, 1974.

CHAPTER 20. INDIAN OCEAN NEOGENE PLANKTONIC FORAMINIFERAL BIO-
STRATIGRAPHY AND ITS PALEOCEANOGRAPHIC IMPLICATIONS

Edith Vincent

Scripps Institution of Oceanography
La Jolla, California, 92093

Abstract. The Neogene planktonic foraminiferal biostratigraphy of the Indian Ocean is discussed mainly from material of 16 DSDP sites between 12°N and 48°S, which were above the CCD throughout Neogene times, and where 4074 m of calcareous sediments were continuously cored. Foraminiferal faunas are common to abundant and well preserved in Upper Miocene through Pleistocene sediments at all these sites, except the deeper Somali Basin Site (236) and the shallow Gulf of Aden sites (231, 232 and 233), where the faunas are affected by significant dissolution. Comparison of Upper Neogene assemblages from the last four sites with well preserved faunas from the other sites provides data on the solution susceptibility of extinct species. Middle and Lower Miocene faunas are moderately to poorly preserved at all sites where they occur. Foraminiferal events are evaluated in relation to planktonic zonations of other fossil groups with special attention to paleontological events calibrated to the paleomagnetic-radiometric time scale. The tropical Neogene biostratigraphy is similar to that of the tropical Pacific. Further taxonomic reevaluation of the Globorotalia connoidea-G. inflata bioseries in sections of mid-latitude sites is needed for a more detailed biostratigraphic interpretation of these sites. Latitudinal control of appearances or disappearances of many planktonic foraminiferal species is well documented through the Neogene.

It was not possible to use the first appearance of Globorotalia truncatulinoides to delineate the Pliocene/Pleistocene boundary in the Indian Ocean because of the discrepancy between its level and the extinction level of Discoaster brouweri. The first G. truncatulinoides occurs well above (up to 70 m) the extinction level of D. brouweri in expanded sections whereas in slowly accumulating sequences it is coincident or slightly above this nannofloral event. There is, in general, good agreement among various fossil groups for the recognition of the Miocene/Pliocene boundary at low-latitude sites, but at none of the mid-latitude sites is this boundary clearly identified. The Oligocene/Miocene boundary is best estimated on a foraminiferal basis by the initial appearance level of Turborotalia kugleri.

Changes in faunal content and in rates of accumulation of calcareous sediments at these 16 sites, together with changes in the isotopic composition of foraminiferal tests at subantarctic Site 281 are tentatively related to global climatic changes associated with Antarctic glaciation. The two distinct regimes which occur today in the eastern and western Indian Ocean appear to have been in existence since the late Middle Miocene. Sedimentation was reduced in both Oligocene and Early Miocene throughout most of the Indian Ocean and resumed significantly only at the end of the Early Miocene. During the Early and Middle Miocene relatively warm surface-water temperatures (optimum for the development of the globoquadrinid group) prevailed in mid as well as low latitudes. Intensified bottom water circulation associated with the buildup of the East Antarctica ice sheet resulted in a significant decrease in sediment accumulation rates in the temperate and subtropical southeast Indian Ocean during the Late Miocene, but sedimentation was not reduced during that time either on the crest of the central and northern Ninetyeast Ridge, or in the western Indian Ocean, where sediments accumulated in the early Late Miocene (prior to 6.3 m.y.) at rates comparable to those of the Middle Miocene. Bottom currents, however, were apparently intense throughout the ocean in late Middle Miocene and earliest Late Miocene, as evidenced by the general telescoping of fossil zones at that time and by an interval of mixed, reworked foraminiferal faunas spanning the Middle/Late Miocene boundary throughout the northwestern Indian Ocean. The widespread presence of this interval with reworked sediments, which have been apparently winnowed during transport, may possibly be related to increased tectonic activity and circulation readjustments in the northwest Indian Ocean after the opening of the Gulf of Aden and the closure of the Indo-Tethyan seaway. A marked increase in sediment accumulation rates and in the abundance of siliceous fossils at all low-latitude sites in Late Miocene, at approximately 6.3 m.y., corresponded to the time of maximum Antarctic ice accumulation and of the onset of the modern equatorial circulation pattern.

Cooling during the Late Miocene moved the area of optimum conditions for globoquadrinids north of 25°S. The extinction of Globoquadrina dehiscens was paleoceanographically controlled, decreasing in age from south to north with progressive cooling. The uppermost Miocene (Kapitean) severe cooling recorded throughout the Pacific is not recognizable in the Indian Ocean from present data. There is faunal evidence in the temperate, subtropical and tropical areas for water temperatures cooler in Late than Early Pliocene. A substantial increase in velocity of bottom water associated with the Late Pliocene cooling, which removed Upper Pliocene to Quaternary sediments from the South Indian and South Australian Basins was felt throughout the Indian Ocean. Telescoping of Upper Pliocene sequences (especially pronounced in the uppermost Pliocene) occurs throughout the eastern ocean as far north as the Andaman-Nicobar region. In the western part of the ocean, disturbances of Upper Pliocene and lowermost Pleistocene sedimentary sequences are evidenced at many sites. Sites 214 and 216 on the crest of the central and northern Ninetyeast Ridge appear to remain in an area better protected from bottom currents than any other sites during the entire Neogene and, thus, provide the best undisturbed sequences for biostratigraphic investigations.

In the southern subtropical area, cool conditions prevailed in the earliest Pleistocene; a marked warming occurred between -1.42 and -1.18 m.y., followed by a relatively cool interval with small-scale temperature fluctuations up to -0.52 m.y. and from that time on a warming trend occurred. A warming trend is recorded as well in cores from the subantarctic southeast area during the last 0.8 m.y., with superimposed temperature fluctuations which were essentially synchronous with north hemisphere climatic changes. The 10°C surface isotherm did not shift north of 36°S during the Pleistocene.

Introduction

The purpose of this paper is to summarize the planktonic foraminiferal biostratigraphy of the Neogene (the interval of time from the beginning of the Miocene to the present: Hoernes, 1856; Berggren and Van Couvering, 1974) of the Indian Ocean mainly from results of the 56 sites drilled in this ocean during Legs 22 to 29 of the GLOMAR CHALLENGER (Fig. 1, Tables 1 and 2). The Indian Ocean has been much neglected compared to other oceanic areas. Scientific interest in this ocean developed primarily from the stimulus of the International Indian Ocean Expedition (IIOE) during the period 1962-1965. Before this time foraminiferal and biostratigraphic studies were very sparse, and more than one-half of all the papers on foraminifera from this ocean have been published in the last decade. A thorough review of foraminiferal research in the Indian Ocean up to 1971 was presented by Bandy, Lindenberg and Vincent (1972).

Broad-scale distribution patterns of planktonic foraminifera were developed from plankton tows by Beliaeva (1964), Boltovskoy (1969), Bé and Tolderlund (1971) and Bé and Hutson (in press), and from surface sediments by Beliaeva (1964) and Bé and Hutson (in press). These authors show that in the Indian Ocean, as in the other oceans, there is a latitudinal control of species distribution. Bé and Hutson recognized nine living assemblages which are ecologically and geographically meaningful but are reduced to five death assemblages in the sediments (Fig. 1). These authors pointed out that, despite the distortion and partial loss of ecological information during and after the sedimentation of foraminiferal tests, the faunal characteristics of the plankton ecosystem are sufficiently well preserved in the sediments for successful reconstructions of past environments.

A number of planktonic foraminiferal studies have been carried out from gravity and piston cores collected in the Indian Ocean (see references in Bandy, Lindenberg and Vincent, 1972, classified by geographical areas). These investigations, however, mainly involve sediments younger than Pliocene. Pre-Pleistocene Neogene planktonic foraminiferal faunas have been recovered in surface sediments in areas where older sediments are outcropping (e.g. Gulf of Aden: Ramsay and Funnell, 1969; Southwest Indian Ocean: Saito and Fray, 1964; Ubaldo, 1973; Vincent, 1976), and below the sediment surface more especially in deep-sea cores showing telescoped Quaternary sections or substantial stratigraphic hiatuses. Five cores which penetrated Pliocene and possibly latest Miocene sediments from the tropical and subtropical Indian Ocean were studied by Parker (1967). Although the paleomagnetic stratigraphy of a large number of cores covering most of the Indian Ocean has been established (Opdyke and Glass, 1969; Watkins and Kennett, 1972), detailed paleontologic investigations of these cores are scanty. These cores include two VEMA cores, V20-163 on the Ninetyeast Ridge (Hays et al., 1969; Gartner, 1973; Bandy, 1973, 1975) and V16-66 on the Southwest Indian Ridge (Ericson et al., 1963; Glass et al., 1967; Bandy et al., 1971), and a number of ELTANIN cores (Cruises 39 to 55; not reported on fig. 1) located south and southwest of Australia (Watkins et al., 1973; Kennett and Watkins, 1976; Vella et al., 1975; Williams and Johnson, 1974). All these cores contain substantial stratigraphic hiatuses.

Few data on Neogene planktonic biostratigraphy from lands bordering the Indian Ocean are available. These areas consist of fragments of

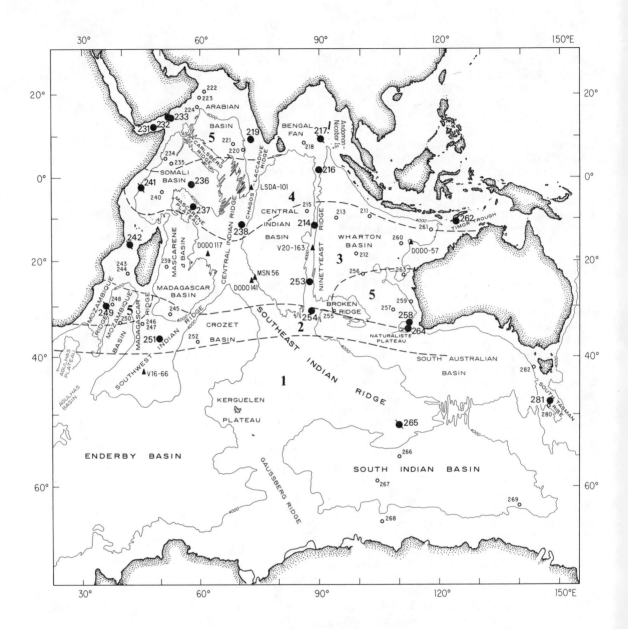

Fig. 1. DSDP sites in the Indian Ocean. Large black circles: sites valuable for Neogene planktonic biostratigraphy. Large black circles with a diagonal line: sites valuable for Neogene planktonic biostratigraphy, but where discontinuous coring or lack of available foraminiferal data precludes a detailed biostratigraphy. Small open circle: other sites. Black triangles: piston or gravity cores. Unbroken line: 4000 m bathymetric contour. Dashed line: boundary between surface fossil death - assemblages (from Be and Hutson, in press). Largest numbers refer to the latter assemblages: 1= Polar-Subpolar Assemblage; 2 = Transitional Assemblage; 3 = Tropical-Subtropical Assemblage; 4 = Tropical Assemblage; 5 = Tropical-Subtropical Boundary Current Assemblage.

TABLE 1. Location of DSDP Sites in the Indian Ocean and of piston and gravity cores studied for Neogene planktonic foraminifera older than Pleistocene (Cores from ELTANIN Cruises 39 to 55 in the southeast Indian Ocean, between 70°E and 120°E and between Antarctica and 30°S, have been omitted).

Leg	DSDP Site	Latitude	Longitude	Water Depth, m	Physiographic Province
22	211	9°46.53'S	102°41.95'E	5,535	Wharton Basin
	212	19°11.34'S	99°17.84'E	6,243	Wharton Basin
	213, 213A	10°12.71'S	93°53.77'E	5,611	Wharton Basin
	214	11°20.21'S	88°43.08'E	1,665	Ninetyeast Ridge
	215	8°07.30'S	86°47.50'E	5,319	Central Indian Basin
	216, 216A	1°27.73'N	90°12.48'E	2,247	Ninetyeast Ridge
	217, 217A	8°55.57'N	90°32.33'E	3,020	Ninetyeast Ridge
	218	8°00.42'N	86°16.97'E	3,759	Bengal Fan
23	219	9°01.75'N	72°52.67'E	1,764	Chagos Laccadive Ridge
	220	6°30.97'N	70°59.02'E	4,036	Arabian Basin
	221	7°58.18'N	68°24.37'E	4,650	Arabian Basin
	222	20°05.49'N	61°30.56'E	3,546	Arabian Basin
	223	18°44.98'N	60°07.78'E	3,633	Arabian Basin
	224	16°32.51'N	59°42.10'E	2,500	Arabian Basin
24	231	11°53.41'N	48°14.71'E	2,161	Gulf of Aden
	232, 232A	14°28.93'N	51°54.87'E	1,758	Gulf of Aden
	233, 233A	14°19.68!N	52°08.11'E	1,860	Gulf of Aden
	234, 234A	4°28.96'N	51°13.48'E	4,738	Somali Basin
	235	3°14.06'N	52°41.64'E	5,146	Somali Basin
	236	1°40.62'S	57°38.85'E	4,504	Somali Basin
	237	7°04.99'S	58°07.48'E	1,640	Mascarene Plateau
	238	11°09.21'S	70°31.56'E	2,844	Central Indian Ridge
25	239	21°17.67'S	51°40.73'E	4,971	Mascarene Basin
	240, 240A	3°29.28'S	50°03.42'E	5,082	Somali Basin
	241	2°22.24'S	44°40.77'E	4,054	Somali Basin
	242	15°50.65'S	41°49.23'E	2,275	Mozambique Channel
	243	22°54.49'S	41°23.99'E	3,879	Mozambique Channel
	244	22°55.87'S	41°25.98'E	3,768	Mozambique Channel
	245, 245A	31°32.02'S	52°18.11'E	4,857	Madagascar Basin
	246	33°37.21'S	45°09.60'E	1,030	Madagascar Ridge
	247	33°37.53'S	45°00.68'E	944	Madagascar Ridge
	248	29°31.78'S	37°28.48'E	4,994	Mozambique Basin
	249	29°56.99'S	36°04.62'E	2,088	Mozambique Ridge
26	250, 250A	33°27.74'S	39°22.15'E	5,119	Mozambique Basin
	251, 251A	36°30.26'S	49°29.08'E	3,489	S. E. Indian Ridge
	252	37°02.44'S	59°14.33'E	5,032	Crozet Basin
	253	24°52.65'S	87°21.97'E	1,962	Ninetyeast Ridge
	254	30°58.15'S	87°53.72'E	1,253	Ninetyeast Ridge
	255	31°07.87'S	93°43.72'E	1,144	Broken Ridge
	256	23°27.35'S	100°46.46'E	5,361	Wharton Basin
	257	30°59.16'S	108°20.99'E	5,278	Wharton Basin
	258, 258A	33°47.69'S	112°28.42'E	2,793	Naturaliste Plateau

TABLE 1. (continued)

Leg	DSDP Site	Latitude	Longitude	Water Depth, m	Physiographic Province
27	259	29°37.05'S	112°41.78'E	4,712	Eastern edge of Wharton Basin
	260	16°08.67'S	110°17.92'E	5,709	"
	261	12°56.83'S	117°53.56'E	5,687	"
	262	10°52.19'S	123°50.78'E	2,315	Timor Rrough
	263	23°19.43'S	110°58.81'E	5,065	Eastern edge of Wharton Basin
28	264	34°58.13'S	112°02.68'E	2,873	Naturaliste Plateau
	265	53°32.45'S	109°56.74'E	3,582	S.E. Indian Ridge
	266	56°24.13'S	110°06.70'E	4,173	South Indian Ridge
	267, 267A	59°15.74'S	104°29.30'E	4,564	South Indian Ridge
	267B	59°14.55'S	104°29.94'E	4,539	South Indian Basin
	268	63°56.99'S	105°09.34'E	3,544	Antarctic Cont. Rise
	269, 269A	61°40.57'S	140°04.21'E	4,285	South Indian Basin
29	280	48°57.44'S	147°14.08'E	4,191	South Tasman Rise
	281	47°59.84'S	147°45.85'E	1,601	South Tasman Rise
	282	42°14.76'S	143°29.18'E	4,217	South Tasman Rise
Piston and Gravity Cores					
DODO 57		15°40'S	112°44'E	3,660	NW Australian Cont. Rise
DODO 117P		18°21'S	62°04'E	3,398	Central Indian Ridge
DODO 141G		24°42'S	73°05'E	3,600	Central Indian Ridge
LSDA 101G		2°41'S	73°12'E	2,960	Chagos Laccadive Ridge
MSN 56P		23°56'S	73°53'E	3,700	Central Indian Ridge
V16-66		42°39'S	45°40'E	2,995	S. W. Indian Ridge
V20-163		17°12'S	88°41'E	2,706	Ninetyeast Ridge

Gondwanaland composed predominantly of igneous and metamorphic rocks dating back to Precambrian and bordered by a fringe of younger sediments. A few reports of Neogene foraminiferal faunas of shallow-water origin have been published (e.g. in southern Mozambique: Smitter, 1955). Deep-water Neogene marine facies, however, are well developed in the Andaman-Nicobar islands in the Bay of Bengal. Planktonic foraminifera, which are abundant and well preserved in the rocks of this area, were first described by Schwager (1866) and were recently studied in detail by Srinivasan and his co-workers (Srinivasan, 1969, 1975; Srinivasan and Srivastava, 1972, 1974, 1975; Srinivasan and Sharma, 1973.

An overall planktonic biostratigraphic scheme of the Neogene of the Indian Ocean cannot be established from these scattered studies. Only after deep-sea drilling throughout the entire ocean performed by the GLOMAR CHALLENGER during Phase III of the Deep Sea Drilling Project in the years 1972-1973 was it possible to construct such a scheme. Neogene sediments were penetrated at all 56 DSDP sites and a total of about 14,300 meters of Neogene was drilled. However, not all the Neogene sections can be used for calcareous plankton biostratigraphy

TABLE 2. References to special chapters in the Initial Reports of the Deep Sea Drilling Project dealing with Neogene planktonic foraminifera in the Indian Ocean.

Leg 22:	McGowran, 1974
	Berggren, Poore and Lohman, 1974*
	Berggren and Poore, 1974**
Leg 23A:	Fleisher, 1974
	Akers, 1974a*
Leg 24:	Vincnet, Frerichs and Heiman, 1974
Leg 25:	Zobel, 1974
Leg 26:	Boltovskoy, 1974a
	Boltovskoy, 1974b
	Ajers, 1974b*
Leg 27:	Rögl, 1974
	Olsson, 1974*
Leg 28:	Kaneps, 1975a
	Kennett, 1975*
Leg 29:	Jenkins, 1975

* Shore-based studies additional to studies made by shipboard scientists.
** Subsequent study not in the Initial Reports.

because 1) a number of sites are located below the calcite compensation depth (CCD), which occurs today at about 5000 meters in the tropical Indian Ocean shallowing gradually southwards to about 3900 m south of 50°S (Berger and Winterer, 1974; van Andel, 1975; Kolla et al., 1976), 2) displaced sediments disturb the biostratigraphic record at a number of sites, especially in the thick turbidite sequences of the Bengal Fan and Indus Cone, and 3) of insufficient coring and recovery of some calcareous sections. Calcareous sections at a number of sites, however, located on the many relatively shallow ridges or plateaus which characterize the Indian Ocean floor, where continuous coring with good recovery was achieved are particularly valuable (Fig. 1; Table 3).

Planktonic Zonations

The evolution of planktonic organisms through time has provided the framework for a system of planktonic zonations. Neogene planktonic foraminiferal zonations widely employed for tropical areas are those developed by Bolli (1957, 1966) and Bolli and Bermudez (1965). Blow (1969), following Banner and Blow (1965a), modified Bolli's original scheme using an abbreviated letter and number system of renaming Bolli's zones. Although Blow's scheme may be objectionable because of its inflexibility with respect to further subdivision, it is convenient because of its brevity and easiness as a mnemonic

TABLE 3. Summary of Neogene sections drilled at DSDP sites in the Indian Ocean. **Sites valuable for planktonic foraminiferal biostratigraphy whose detailed biostratigraphy is shown on Figures 3 to 18. *Sites valuable for planktonic foraminiferal biostratigraphy but whose value is deceased by discontinuous coring or lack of available foraminiferal data. Thickness: min. = minimum. Coring: C = continuous, D = discontinuous. Core recovery: G = good, M = moderately good, P - poor. Planktonic foraminifera: numbers refer to core numbers; abundance, A = abundant, C = common, R = rare; preservation, G = good, M = moderate, P = poor.

Physiographic Province	Site	Thickness Neogene Section, m	Coring and Recovery	Core Numbers	Stratigraphic Sequence Sediment Type	Underlying Section	Planktonic Foraminifera
Wharton Basin	211	304 min.	D,G	1-9	Pleistocene-Pliocene sands and siliceous oozes. Sands are turbidites from distal portion of Nicobar fan.	Undated brown clay which overlies a Cretaceous calcareous ooze.	R in sandy horizons; Miocene or younger.
	213	88 min.	C,G	1-10 1A	Apparently continuous Mid. Miocene-Pleistocene sequence. 66 m of Up. Mio-Pleist. siliceous ooze overlying 22 m of Mid.-Up. Mio. zeolitic clay.	Undated zeolitic brown clay which overlies Eocene nanno ooze.	Barren
	212	288.5 min.	D,M	1-15	Mid. Miocene to Pliocene. Brown clay interbedded with thick calcareous units displaced from shallow water.	Undated zeolitic clay which overlies Eocene nanno chalk.	Small forms well sorted in the calcareous units; Miocene with mixing.
	256	125.5 min.	D,G	1-5	Pleistocene-Pliocene / Brown detrital clay	Cretaceous brown detrital clay.	Mostly barren.
	257	?85.5	D,G	1-3	Undated / Brown detrital clay		
Eastern edge of Wharton Basin	261	104.5 min.	D,P	1-2	Pleistocene radiolarian clay.	Cretaceous zeolitic clay.	Mostly barren
				3-4	Lo. Plio.-Up. Mio. displaced nanno ooze.		A, M; reworking and mixing.
	260	101 min.	D,M	1-3	Nanno foram and rad ooze with clay; mostly turbidite.	Oligocene nanno clay	1 (Pleist.) and 2 (Lo. Plio.): A, M, with reworking; 3: mixed Cretaceous, Paleogene and Mid. Mio.
	263	100 min.	D,M-G	1-3	Pleistocene to Up. Pliocene turbidite foram nanno ooze.	Paleocene nanno clay	A, G-M; mixing
	259	8 min.	D,G-P	1	Pleistocene nanno ooze.	Eocene Zeolitic clay	A, G (dominated by G. inflata)
				2-3(?)	Undated zeolitic clay.		Barren
Timor Sea	262**	442	C,G	1-47	Up. Pliocene-Pleistocene. 414 m of clay rich nanno ooze (rad rich in upper 261 m) overlying 28 m of shallow dolomitic mud and calcarenite.	Bottom hole	1-44 (Pleisto.): A,G; 45-47 (Plio.): R, M-P.
Ninetyeast Ridge	217*	210	D,G	1-8 1A-4A	Virtually continuous Lo. Miocene-Pleistocene sequences. Nanno ooze and chalk	Oligocene nanno ooze and chalk	
	216**	196.5	Up. Mio.-Pleis.:D,G; Mid.-Lo. Mio.:C,G	1-8 1A-6A			Up. Mio.-Pleist.: A-G; Lo.-Mid. Mio.: C, M-P
	214**	221	C,G	1-24			
	253**	86.5	C,G	1-10			
	254**	167	C,G	1-19			
Bengal Fan	218	773	D,M	1-27	Mid. Miocene-Pleistocene turbidites. Silts with nanno ooze layers.	Bottom hole	R, P; mixing.
Central Indian Basin	215	74	C,G	1-8	Up. Miocene-Pleistocene almost continuous sequence (short gap in Up. Mio.). 64 m of siliceous ooze overlying 10 m of clay.	Undated zeolitic brown clay which overlies Eocene nanno ooze.	Barren
Broken Ridge	255	61	D,P	1-6	Lo. Miocene-Pleistocene possibly continuous sequence. Nanno foram ooze.	Eocene cherty gravel.	Mid. Mio.-Pleist.: A,G; Lo. Mio.: R,P; common mixing; Oligo. reworking in lower part.

NEOGENE PLANKTONIC FORAMINIFERA

Physiographic Province	Site	Thickness Neogene Section, m	Coring and Recovery	Core Numbers	Stratigraphic Sequence Sediment Type	Underlying Section	Planktonic Foraminifera
Naturaliste Plateau	258**	114	C,M-G	1-4 1A-8A	Virtually continuous Up. Miocene-Pleistocene sequence. Nanno ooze.	Cretaceous silicified chalk and limestone.	A, G
	264**	29.5	C,G	1-2 1A-2A	Uppermost Miocene-Pleistocene continuous sequence. Foram nanno ooze	Eocene nanno ooze and chalk	A, G (P in basal few meters)
South Tasman Rise	282	55	D,M	1-5	6 m of Pleist. nanno and foram oozes overlying 7 m of Up. Miocene nanno ooze, in turn overlying 42 m of Lo. Miocene detrital silty clay nanno ooze.	Oligocene detrital silty and clayey nanno ooze.	C-R, M-P
	281**	121.5	C,G	1-13	Virtually continuous sequence Lo. Miocene/Pleistocene. Nanno and foram oozes.	Oligocene glauconitic sand.	1-5 (Pleist.-Plio.): A,G-M; 6-13 (Mio.): A-M.
	280	6	D,G	1	1 m of Pleist. foram-nanno ooze overlying 5 m of Lo. Plio. to Up. Mio. siliceous nanno ooze and detrital clay.	Oligocene silty diatom ooze.	R, M
Southeast Indian Ridge	265*	445	D,G	1-17	370 m of Pleist.-Lo. Plio. diatom ooze overlying 75 m of Up.-Mid. Mio. diatom bearing nanno ooze.	Basalt.	R, M to barren. Very low diversity.
South Indian Basin	266	370	D,G	1-22	148 m Pleist.-Lo. Plio. diatom ooze overlying 105 m of Up.-Lo. Mio. mixed nanno ooze, nanno clay, diatom clay and diatom ooze, in turn overlying 117 m of Lo.-Mid. Mio. to Lo. Mio. nanno chalk and claystone.	Basalt.	Mostly barren. Sporadically R, P. Very low diversity (no more than 2 species).
	267	110	D,M	1-3 1A-3A	110 m of Pleist. to Lo. Mio. silty clays and diatom ooze.	Oligocene nanno ooze and chalk.	Barren
	267B	310	D,G	1B-9B	310 m of Lo. Plio.-Lo. to Mid. Mio. diatomaceous silty clays.	Eocene to Oligocène chalk.	
	269	426 min.	D,M	1-11 1A	426 m of Pleist.-Lo. to Mid. Mio. diatomaceous clays and silts, largely turbidites. (Cherts between 320-426 m).	Undated sequence with same lithology.	
Antarctic Continental Rise	268	237 min.	D,M	1-10	160 m of Pleist.-Plio. clay, silty clay, sand and diatom ooze overlying 77 m of Miocene clay, silty clay and clayey nanno ooze.	Oligocene to Lo. Miocene (?) clays, silts and chert.	1 (Pleist.): A, M; 8 (Lo. Mio.): R, M; very low diversity (2 species only); other cores: barren.
Chagos Laccadive Ridge	219**	137 min.	Up. Mio.-Pleist.:C,G; Lo.-Mid. Mio: D,G	1-14	Virtually continuous Lo. Mio. Pleistocene sequence. Nanno ooze and chalk.	Oligocene nanno ooze and chalk.	A,G
Arabian Basin	220	45 min.	C,M	1-5	Mid. Mio. to Pleist. nanno ooze and detrital clay with turbidite foram sand layers.	Oligocene nanno ooze and chalk.	R, P
	221	155 min.	D,M	1-15	128 m of Pleist.-Up. Plio. silt with graded sand layers (distal margin of Indus Cone) overlying 27 m of unfossiliferous brown clay (Plio-Mio.?)	Oligocene brown clay passing down to nanno ooze.	Almost barren.
	222	1300	D,M-G	1-36	Pleist.-Up. Mio. carbonate-rich detrital silty clay with sand and silt layers. Distal margin of Indus Cone.	Bottom hole.	R, P
	223	496	D,G	1-27	Continuous Lo. Mio.-Pleist. sequence (except a hiatus in Lo. Mio.). Silty and siliceous nanno chalk and ooze. Several chalk breccias.	Oligocene nanno chalk.	R, M
	224	325	D,P	1-4	Lo. Mio.-Pleist. Clayey silt nanno ooze and chalk.	Oligocene detrital clayey silt and claystone.	1 (Pleist.): C, G; 2 (Up. Mio.): R to barren; 3-5 (Lo.-Mid. Mio): barren.
Gulf of Aden	231**	566.5	C,G	1-61	Mid. Mio.-Pleist. (Continuous sequences, Nanno ooze with occasional sand layers)	Basalt	Pleist.: A,M Mio.-Plio.: C,M-P
	232**	434	C,G	1-19 1A-30A	Up. Mio.-Pleist.	Bottom hole	
	233**	234.5	Plio.-Pleist. C,G; Up. Mio.:C,P	1-19 1A-7A	Up. Plio.-Pleist.	Diabase	

Physiographic Province	Site	Thickness Neogene Section,m	Coring and Recovery	Core Numbers	Stratigraphic Sequence Sediment Type	Underlying Section	Planktonic Foraminifera
Somali Basin	234	161.5	D,G-M	1-9	10 m of Plio.-Up. Mio. nanno clay to nanno ooze overlying 151.5 m of Lo.-Mid. Mio. clay to nanno clay.	Oligocene clay.	1 (Plio.-Up.Mio.): R, P; 2-9 (Lo.-Mid. Mio.): barren.
	235	500 min.	D,G-M	1-15	45 m of Pleist. nanno ooze overlying 455 m of Mid. Mio.-Plio. aternations of nanno clay and nanno ooze.	Undated brown clay.	R, P. Layers with A, G, well-sorted minute forms (probably, distal turbidites).
	240	187	D,G-M	1-5 1A-3A	Lo. Mio.-Pleist. sequence mixed with Lo. Eocene below 165 m. Nanno-rad ooze interbedded with clays and silts.	Lo. Eocene nanno ooze.	Upper 2 m (Pleist.): C,G. Below: R, P. Mixing.
	241*	457	D,G	1-16	Apparently continuous Lo. Mio.-Pleist. sequence. Clayey nanno ooze.	Oligocene clayey nanno ooze.	Pleist.-Plio: C, G-M; Mio.: R, M-P.
	236**	187	C,G	1-20	Virtually continuous Lo. Mio.-Pleist. sequence. 19 m of Pleist. nanno and rad oozes overlying 120.5 m of Up.-Mid. Mio. nanno and foram oozes, in turn overlying 47.5 m of Mid.-Lo. Mio. clay with layers of nanno clay and ooze.	Oligocene nanno ooze to chalk.	A-C, M-P; better preserved in coarse-grained ooze than in fine-grained ooze.
Mascarene Plateau	237**	192.5	C,G	1-21	Virtually continuous Lo. Mio.-Pleist. sequences. Nanno ooze	Oligocene nanno ooze.	Pleist.-Up. Mio.: A, G; Mid.-Lo. Mio.: A-C, M-P.
Central Indian Ridge	238**	454.5	C,G	1-49			
Mascarene Basin	239	215	D,G	1-12	Pleist.-Lo. Mio. sequence. Distal turbidite of silt and clay with varying amount of nanno ooze.	Oligocene with same lithology.	R, M-P; mixing. 3 cc and below: very small-sized forms.
Madagascar Basin	245	35	D,G	1 1A	Mid. Miocene? Brown silt-rich clay.	Eocene clay and ooze.	Almost barren.
Madagascar Ridge	246	125	D,P	1-4	Probably continuous Lo. Mio.-Pleist. sequence. Foram ooze.	Lo. Eocene shelly calcareous sand.	A, G
	247	No penetration					
Mozambique Basin	248	314	D,P	1-10	(Lo. Mio.?)-Mid. Mio. to Pleist. sequence. Terrigenous turbidite with various amounts of clay, silt and sand.	Lo. Eocene silty clay.	R, P
	250	691	D,G	1-3 1A-21A	116 m of Pleist.-uppermost Plio. nanno ooze and detrital silty clay overlying 575 m of Up. Plio.-Lo. Mio. detrital clays.	Cretaceous detrital clay.	1-4A (Pleist.-Up. Plio): C, G; 5A-8A (Plio.-Up. Mio.): R, M-P; 9A-21A: mostly barren.
Mozambique Ridge	249*	178	C,G	1-16	2 m of Pleist.-Plio. overlying 176 m of Up.-Mid. Miocene. Nanno ooze.	Upper Cretaceous nanno chalk.	A, G
Mozambique Channel	243	No penetration					
	244						
	242*	485	D,G	1-8	Apparently continuous Lo. Mio.-Pleist. sequence. Clayey nanno ooze.	Oligocene clayey nanno chalk.	Up. Mio.: A,G; Mid.-Lo. Mio.: R, M.
Southwest Indian Ridge	251**	489.5	Pleist.-Plio.: C,G; Up.Mio.-up. Mid. Mio.: D,G; lo. Mid. Mio.:C,M	1-10 1A-30A	Virtually continuous Mid. Mio.-Pleist. sequence. 240 m of nanno ooze overlying 249.5 m of nanno chalk.	Basalt	Pleist.-Up. Mio: A,M; Mid.-Lo. Mio: A-L, M-P.
Crozet Basin	252	247	D,G	1-7	(Mid. Mio.?)-Up. Mio.-Pleist. sequence. Radiolarian detrital clay.	Bottom hole.	Barren.

TABLE 3. Continued

device and was preferred by most authors of the Indian Ocean DSDP Initial Reports.

Brönnimann and Resig (1971) noted that the ranges of certain biochronologically important taxa differ in the tropical Pacific from those established by Blow (1969) using data from other areas and proposed a number of modifications in Neogene zones, most of which are supported by data obtained in the tropical Indian Ocean (Fleisher, 1974). Difficulties in recognizing Blow's zonal boundaries have often been found because of the lack or rarity of the nominate zonal taxa.

The absence of an index species in a faunal assemblage may result from a geographic location outside the latitudinal range of the taxon or from elimination by selective dissolution. Bolli's and Blow's zonations were constructed mainly from observations of land-based marine sections deposited in tropical areas along continental margins where dissolution has had little effect on the faunas. Solution susceptibilities for modern planktonic foraminifera at oceanic depths are fairly well known (Berger, 1970), and it may prove possible to use this information to provide data on the test solution of extinct species and assemblages (Berger and von Rad, 1972). In this respect, a comparison of late Neogene foraminiferal assemblages from the Gulf of Aden sites (231, 232, and 233) and Site 236 in the Somali Basin, where faunas are affected by significant dissolution, with assemblages from other low-latitude sites with well-preserved faunas will be very informative. Solution is pronounced in Early Miocene sediments throughout the Indian Ocean.

An additional difficulty in using Blow's zonation results from taxonomic discrimination. Some zonal boundaries are defined by taxa distinguished by rather fine morphologic criteria. Consequently differing taxonomic interpretations of various workers may lead to miscorrelations. This possibility of differing interpretations has been kept in mind in attempting this synthesis based on preliminary investigations by many different workers, and the correlations suggested in this paper may be modified after re-evaluation of equivocal taxa and important lineages.

A number of authors have proposed "revised" or new zonations. Zonal schemes developed in tropical areas include those of Jenkins and Orr (1971, 1972), Krasheninnikov (1971), and Kaneps (1973) for the Pacific, Bolli (1970), Bolli and Premoli Silva (1973), and Lamb and Beard (1972) for the Caribbean and Gulf of Mexico, Srinivasan and Srivastava (1974, 1975) for the Pliocene of the Andaman-Nicobar Islands in the Bay of Bengal. The latter scheme, discussed below, is non-applicable to other parts of the Indian Ocean.

Berggren (1973) has proposed a new zonation for the Pliocene interval for which the Blow sequence is particularly susceptible to modification. His sixfold Pliocene zonation is suitable for the Atlantic but is not entirely applicable to other oceanic areas. Rather than developing a formal zonation which is not applicable worldwide, Parker (1973, 1974) preferred to use informal "spatial" intervals to subdivide Pliocene cores from the tropical Atlantic.

Despite the difficulties encountered in applying Blow's zonation in certain areas of the Indian Ocean, none of the authors of the Initial Reports felt it would be appropriate to erect new local zonal schemes at such an early stage of investigation. In a number of instances when the nominate taxa of Blow's zones were rare or absent, zones were assigned based on the rest of the faunal assemblage. In many cases larger units than single zones were used and several zones were combined.

Blow's zonation is generally not applicable in temperate regions. Foraminiferal zonations constructed for mid-latitude regions include those of Cati et al. (1968) and Cita (1973) for the Mediterranean area, Jenkins (1966, 1967a, 1971), Kennett (1973) and Kennett and Vella (1975) for the temperate regions of New Zealand and southwestern Pacific, Kennett (1973) for that part of the southwestern Pacific intermediate between temperate and subtropical areas, and Poore and Berggren (1975) for the North Atlantic. Conflicting correlations between New Zealand and Mediterranean stratigraphies remain to be solved. Correlations between temperate and tropical areas are

MAGNETIC EPOCH	TIME m.y.	EPOCH		PLANK FORAM ZONES Blow (1969)	CALCAREOUS NANNOPLANTON ZONES				RADIOLARIAN ZONES Riedel and Sanfilippo (1971)	DIATOM ZONES Schrader (1974)
					Martini (1971)		Gartner (1974)	Bukry (1973a)		
BRUNHES		PLEISTOCENE		N23	NN21	*E. huxleyi*	*E. huxleyi*	*E. huxleyi*	"QUATERNARY"	TID 1
	1				NN20	*G. oceanica*	*G. oceanica*	*G. caribbeanica*		
MATUYAMA				N22	NN19	*P. lacunosa*	*P. lacunosa*	*E. annula* *C. macintyrei*		
	2		LATE		NN18	*D. brouweri*	*D. brouweri*	*D. pentaradiatus*	*P. prismatium*	TID 7
GAUSS		PLIOCENE		N21	NN17	*D. pentaradiatus*				
	3				NN16	*D. surculus*	*D. surculus*	*D. tamalis*		TID 8
				N20	NN15	*R. pseudoumbilica*	*R. pseudoumbilica*	*D. asymmetricus*	*S. pentas*	
GILBERT	4		EARLY	N19	NN14	*D. asymmetricus*	*D. asymmetricus*	*S. neoabies*		
					NN13	*C. rugosus*	*C. rugosus*	*C. rugosus*		
	5			N18				*C. acutus*		TID 12
					NN12	*C. tricorniculatus*	*C. tricorniculatus*	*T. rugosus*	*S. peregrina*	TID 13
	6						*D. quinqueramus*	*C. primus*		TID 15
	7		LATE	N17				*D. berggrenii*		TID 16
	8				NN11	*D. quinqueramus*	*D. neohamatus*	*D. neoerectus*	*O. penultimus*	
	9	MIOCENE		N16				*D. bellus*		
	10								*O. antepenultimus*	
					NN10	*D. calcaris*				TID 21
11	11			N15	NN9	*D. hamatus*	*D. hamatus*	*D. hamatus*	*C. petterssoni*	
12	12			N14	NN8	*C. coalitus*	*C. coalitus*	*C. coalitus*		
13			MIDDLE	N13	NN7	*D. kugleri*	*D. exilis*	*D. kugleri*		
14	13			N12	NN6	*D. exilis*		*C. miopelagicus*		
				N11 N10					*D. alata*	
15	14			N9	NN5	*S. heteromorphus*		*S. heteromorphus*		
	15			N8			*S. heteromorphus*			
16	16			N7	NN4	*H. ampliaperta*		*H. ampliaperta*		
17	17		EARLY						*C. costata*	
18	18			N6	NN3	*S. belemnos*	*S. belemnos*	*S. belemnos*		
19	19			N5	NN2	*D. druggi*	*T. carinatus*	*D. druggi*	*C. virginis*	UNZONED
20	20									
	21									
	22							*D. deflandrei*		
21	23			N4	NN1	*T. carinatus*		*C. abisectus*	*L. elongata*	
	24									
22	25	OLIGO.	LATE	P22	NP25	*S. ciperoensis*	*S. ciperoensis*	*S. ciperoensis*	*D. ateuchus*	

especially difficult because of the latitudinal control on the distribution of planktonic faunas. Sites located at the transition between these two areas are very valuable to provide insight into their biostratigraphic correlations. Such sites are uncommon and in this respect the calcareous sections recovered in the mid-latitudes of the Indian Ocean during Legs 25 and 26, as well as at Site 264 (Leg 28), are particularly significant.

Jenkins temperate zonal scheme was used by this author to subdivide the Neogene sections at Sites 280, 281 and 282 just south of Australia, close to the present-day subtropical convergence (Jenkins, 1975). Jenkins made some modifications of his earlier zonation (Jenkins, 1971). His former Globorotalia miozea sphericomiozea Zone (equivalent to the Kapitean New Zealand Stage) became the Globorotalia miozea conomiozea Zone and his former Pliocene-Pleistocene Globorotalia inflata Zone was subdivided into three zones (in ascending order): the Globorotalia puncticulata, Globorotalia inflata and Globorotalia truncatulinoides Zones. The threefold planktonic foraminiferal zonation for the Pliocene-Pleistocene of the Antarctic region developed by Kennett (1970) and Keany and Kennett (1972) was recognized at Site 265 in the Antarctic region of the Indian Ocean (Kennett, 1975).

"Foraminiferal events" (preferred here to the word "datum"), i.e. first and last occurrences of selected species, were evaluated at various sites and are the main basis for the biostratigraphic discussion below. Multiple zonations provide the mean to evaluate the reliability of a biostratigraphic event in one group of organisms in the light of events observed in other groups. It is important, therefore, to compare planktonic foraminiferal events to those of other groups. The simultaneous occurrence at a number of Indian Ocean low-latitude sites (Sites 214, 232, 236, 237 and 238) of calcareous and siliceous microfossils is of special interest. These sequences permit direct comparison and correlation of zonation schemes based on the two microfossil types.

The calcareous nannoplankton zonation developed by Gartner (1969, 1974) was used by this author during Leg 22, whereas the zonal scheme constructed by Martini (1971) was used by Müller (1974), Thierstein (1974) and Proto Decima (1974) during DSDP Legs 25, 26 and 27 respectively, as well as by Burns (1975a) at Site 264 of Leg 28. The zonation developed by Bukry (1973a) was utilized by this author for his shore-based study of material recovered during Legs 22 to 25. Because synchronous zones of different systems may have different zone names and, on the other hand, zones from different systems may

---

Fig. 2. Time relationship and correlation of Neogene planktonic microfossil zones. The correlation of Blow's (1969) planktonic foraminiferal zones with those of Martini (1971) for calcareous nannoplankton and the assignment of absolute ages to the zone boundaries are from Berggren (1973) and Van Couvering and Berggren (in press). The calcareous nannoplankton zonal scheme of Bukry (1973a) and Gartner (1969, 1974) are correlated to Martini's zonation. The ages recognized for the radiolarian zone boundaries are those proposed by Theyer and Hammond (1974a, b). The boundaries between the tropical Indian Ocean diatom zones (TID) of Schrader (1974) are placed according to their relationships to other fossil zonations at DSDP Sites 213, 215 and 238 (Vincent, 1974b). The paleomagnetic time-scale is from Van Couvering and Berggren (in press) and Theyer and Hammond (1974a, b).

bear the same name but not be synchronous, confusion may result for the non-specialist. For this reason, various calcareous nannoplankton zonal schemes referred to in the biostratigraphic discussion below are reproduced (Figure 2) showing the correlations of Bukry's and Gartner's zonations correlated with that of Martini. Martini's zonation is correlated with the foraminiferal zonation and the absolute time-scale according to Berggren (1973), Berggren and Van Couvering (1974) and Van Couvering and Berggren (in press). Roth (1974) applied his own zonation to the calcareous nannoplankton sequences obtained during Leg 24. This author uses the same zonal markers as Bukry but does not utilize subzones and his zones bear some name differences (see Fig. 3 in Vincent, 1974b). Data provided by both Bukry and Roth have been combined for subdivision of Leg 24 sites in this paper but the nomenclature of Bukry's zonation is utilized. Boudreaux (1974) did not specify which criteria he was using in his subdivision of Leg 23 sequences.

Martini's and Bukry's zonal schemes proved applicable in both low- and mid-latitude regions of the Indian Ocean and provide a very valuable tool for indirect comparison of low- and mid-latitude foraminiferal assemblages. In contrast, high-latitude calcareous nannoplankton assemblages recovered in the Indian Ocean during Legs 28 and 29 showed a low diversity and did not allow a high-resolution biostratigraphy. At Subantarctic Sites 280, 281 and 282, Edwards and Perch-Nielsen (1975) applied biostratigraphic units which they correlated with New Zealand Stages, and at Antarctic sites 265 through 269 Burns (1975b) did not subdivide the nannofossil sequences into zones.

The Neogene radiolarian zonation of Riedel and Sanfilippo (1971), which has been calibrated to the paleomagnetic time-scale (Theyer and Hammond, 1974a,b), was used by various workers at all low-latitude sites of the Indian Ocean containing siliceous microfossils. For this reason this zonal scheme is reproduced in this paper (Fig. 2) rather than the slightly modified new version given later (Riedel and Sanfilippo, in press). Calcareous sections from mid-latitude sites do not contain sufficient radiolarian faunas to allow a comparison between calcareous and siliceous zonations. High-latitude sites yielded significant radiolarian sequences which appear to be similar to the Antarctic zonation of Hays (1965) and Hays and Opdyke (1967). The sequences at subantarctic sites 280, 281 and 282 were subdivided by Petrushevskaya (1975) into tentative zones marked by letters, whereas Chen (1975) adapted and modified Hay's zonation at antarctic sites 265 through 269.

Schrader (1974) developed a zonation for the late Neogene diatom sequences recovered at low-latitude sites 213, 215 and 238. Schrader's sequence of tropical Indian Ocean Diatom (TID) zones is similar to that of Burckle (1972) for the equatorial Pacific, which has been calibrated to the paleomagnetic stratigraphy sequence (Hays et al., 1969; Burckle, 1972; Saito et al., 1975), but his TID zones are shorter and hence more numerous. The relationship of the TID zones at Sites 213, 215 and 238 with the radiolarian zonation recognized at these three sites and the calcareous plankton zonation of Site 238 has been discussed by Vincent (1974b). This author also compared the TID zonation with the NPD (North Pacific Ocean Diatom) zonation defined by Schrader (1973) in the northeast Pacific and with the correlative calcareous plankton zones of Ingle (1973) and Bukry (1973c). The biostratigraphic position of a number of the TID zone boundaries varies from site to site. The locations of some of these boundaries, however, are consistent when compared to the sequences of other microfossil groups and the TID zonal boundaries which appear most reliable are indicated in Figure 2. Diatoms were recovered in abundance at high-latitude sites. Diatom sequences were not zoned at subantarctic sites 280 to 282 but antarctic sites 265 through 269 were subdivided by McCollum (1975) according to the high-latitude zonation which he developed during Leg 28. The latter author discussed the correlation of his zonation with the paleomagnetic stratigraphy established in piston cores from the southern ocean and with other zonations constructed by other workers for various parts of the Pacific.

It was pointed out by Leg 28 biostratigraphers in their Site Reports of Sites 266 and 267 that although both calcareous and siliceous microfossils occur together in these sections, precise age determinations on the basis of any single fossil group or on combinations of various groups are uncertain. This uncertainty probably results from our present lack of knowledge concerning high-latitude planktonic biostratigraphy. In the biostratigraphic section given below, foraminiferal events are evaluated with respect to their relationships with the calcareous nannoplankton and radiolarian zonations at low- and mid-latitude sites but comparison between the various fossil groups is not undertaken for high-latitude sites.

One of the most remarkable advances in stratigraphy in recent years has been the correlation and calibration of planktonic biostratigraphy with the paleomagnetic reversal scale (Berggren et al., 1967; Hays and Opdyke, 1967; Hays et al., 1969; Gartner, 1969, 1973; Hays 1970; Bandy et al., 1971; Burckle, 1972; Berggren, 1973; Berggren and Van Couvering, 1974; Kennett and Watkins, 1974; Ryan et al., 1974; Theyer and Hammond, 1974a,b; Bandy, 1975; Saito et al., 1975; Van Couvering and Berggren, in press). Unfortunately direct correlation of calcareous plankton with the paleomagnetic reversal scale has been possible so far only in the late Neogene and its calibration to the paleomagnetic time-scale for the early Neogene is done indirectly by way of its relationship to radiolarian zones. Calibration of the paleomagnetic reversal scale with the radiometric time-scale, based upon direct calibration to potassium-argon dates back to about 5 m.y. (Cox, 1969), and from 5 to 20 m.y. upon calibration of biostratigraphic events to K-Ar dates and suggested correlations between the sea-floor magnetic anomalies and paleomagnetic epochs, provides the framework for the construction of a Neogene time-scale. The time-scales developed by Berggren (1973) for the Pliocene-Pleistocene, Berggren and Van Couvering (1974) for the Late Miocene-Pliocene, and Van Couvering and Berggren (in press) for the Early and Middle Miocene, were used in constructing Figure 2.

Planktonic Foraminiferal Biostratigraphy

Introduction

Among the 56 sites drilled in the Indian Ocean (Tables 1 and 3; Fig. 1), those which yielded calcareous sections and where sufficient coring and recovery was achieved are used to discuss the Neogene planktonic foraminiferal biostratigraphy of the Indian Ocean. These calcareous sections are located 1) in the open-ocean, between about 1600 and 3000 meters water depths, on the major shallow-ridges and plateaus, i.e. the Chagos-Laccadive Ridge (Site 219), the Mascarene Plateau (Site 237), the Central Indian Ridge (Site 238), the Southwest Indian Ridge (Site 251), the Mozambique Ridge (Site 249), the Ninetyeast Ridge (Sites 217, 216, 214, 253, 254), the Naturaliste

Plateau (Sites 258, 264), the South Tasman Rise (Site 281), and the Southeast Indian Ridge (Site 265); 2) very close to continental margins, between 1750 and 2300 meters water depth, where hemipelagic sedimentation resulted in expanded sections, i.e. in the Gulf of Aden (Sites 231, 232, 233), in the Mozambique Channel (Site 242), and in the Timor Trough (Site 262); and 3) in the Somali Basin between 4000 and 4500 meters water depth (Sites 241 and 236). Sites from 1) and 2) have been located above the CCD throughout Neogene time, whereas Site 236 lay close to the CCD during Early and Middle Miocene. times when the paleodepth at the site was from about 4200 to 4400 m. Most of these sites yielded well-preserved Upper Miocene to Pleistocene foraminiferal assemblages (except in the Gulf of Aden where calcium carbonate dissolution is pronounced at shallow depths), but rather poorly preserved Middle and Lower Miocene assemblages.

The zonation of various fossil groups as recognized by authors of the Initial Reports at 16 of these sites (mostly continuously cored) is plotted against the stratigraphic column on Figures 3 to 18. Selected paleontologic events, i.e top and bottom of species ranges taken from authors range charts (rare occurrences at either end or outside of a species range, thought to be due to displacement, have been in some cases omitted) are reported on these figures. The biostratigraphy of the remaining 5 sites (217, 241, 242, 249 and 265) cannot be evaluated with precision because intermittent uncored intervals preclude the placement of paleontological events within a reasonable degree of uncertainty, or because foraminiferal range charts were not available. For purpose of comparison, data from Parker (1967) for DODO and MONSOON cores, from various workers for the paleomagnetically dated cores V16-66 and V20-163, and from Srinivasan and Srivastava (1975) for land sections for the Andaman-Nicobar Islands are presented on Figures 19, 20 and 21 respectively.

Bé and Hutson's (in press) study of fossil assemblages in surface sediments of the Indian Ocean (Fig. 1) indicate that there are five dominant faunal assemblages which are from south to north: 1) the Polar-Subpolar Assemblage dominated by Neogloboquadrina pachyderma, and to a lesser extent by Globigerina bulloides and Turborotalita quinqueloba; 2) the Transitional Assemblage dominated by Globorotalia inflata associated with smaller percentages of G. truncatulinoides; 3) the Tropical-Subtropical Assemblage dominated by Globigerinoides ruber and to a lesser degree by Globigerinita glutinata, Globigerinoides sacculifer, Globigerinella siphonifera and Globigerina rubescens; 4) the Tropical Assemblage dominated by Globorotalia menardii with important contribution of Neogloboquadrina dutertrei, G. sacculifer, Pulleniatina obliquiloculata, Globorotalia tumida, and to a lesser extent of G. siphonifera; and 5) the Tropical-Subtropical Boundary-Current Assemblage dominated by Globigerinita glutinata with a significant proportion of Globigerina bulloides and to a lesser extent of Globorotalia menardii.

A comparison between the faunal composition of the highest samples recovered at DSDP Sites with Bé and Hutson's faunal assemblages can be done only in very general terms in the absence of detailed quantitative studies at DSDP Sites. However, semi-quantitative studies at many sites show the latitudinal dependence of the recovered faunal assemblages. Low-latitude sites appear to be dominated by G. menardii associated with other tropical species, whereas mid-latitude sites are dominated by G. inflata and high-latitude Antarctic Site 265 by N. pachyderma. For convenience the various sites are discussed below according to these three subdivisions, low-, mid-, and high-latitude sites, even though they do not exactly correspond to distinct faunal provinces (see Fig. 1). Low-latitude sites (from 20°N to 20°S include those drilled during Leg 22, 23, 24 and Site 262 (Leg 27); mid-latitude sites, (from 20°S to 40°S) those drilled during Leg 26 and Site 264 (Leg 28), and high-latitude sites (south of 40°S, south of the Subtropical Convergence) those drilled during Legs 28 and 29. Keeping all sites of one Leg studied by the same author in the same category is convenient for the sake of consistency in comparing data. As a result Site 214 (Leg 22) and 253 (Leg 26) on the Ninetyeast Ridge are treated separately with low and mid-latitudes sites respectively, even though they both belong to the same tropical-subtropical fossil assemblage province. Site 214 however, is close to the boundary with the tropical province, whereas Sites 253 is close to the boundary with the transitional province.

An important restriction for biostratigraphic study from DSDP material is the amount of vertical displacement of the sediments in the core barrel, especially for the non-indurated upper part of the sedimentary column. Cores obtained by deep-sea drilling techniques provide sediments more disturbed than in gravity and piston cores but represent nevertheless a large amount of valuable material unreachable otherwise. Variations in the detail of the studies published in the Initial Reports (which because of time limitation are often preliminary), in sampling intervals, as well as in taxonomic concepts from author to author often make comparison difficult. Sampling interval is a limiting factor especially for evaluation of paleontologic events. Sampling interval vary from study to study, but is usually close to 1.5 meters (each 9-m core is routinely cut into six sections; one sample is usually collected from each section). An interval of 1.5 meters with a typical average rate of accumulation of 10 to 20 m/m.y. for late Neogene calcareous oozes represents a time span of 150,000 to 75,000 yrs. and for a low rate of accumulation of 3 m/m.y., as typical for Early Miocene sediments at many sites of the Indian Ocean, 500,000 yrs. (almost 1 m.y. in some instances).

All these restrictions are kept in mind in attempting this synthesis study, which does not claim to go beyond the degree of resolution of DSDP Initial Reports. Correlations undertaken here retain a degree of uncertainty, which will doubtless decrease with further detailed analyses. Nevertheless, the general scheme of the Neogene planktonic biostratigraphy of the Indian Ocean was obtained, which points out which sections are valuable for further study and where additional data need to be obtained during future drilling programs.

The foraminiferal generic assignments used in this report mainly follow those of Fleisher (1974). However, the generic assignment Neogloboquadrina is made here for continuosa, acostaensis, humerosa, dutertrei and pachyderma species.

Pleistocene

Paleoclimatic Changes

Planktonic foraminifera, which are good temperature indicators, have been used extensively as a tool for Quaternary marine stratigraphy. The deep-sea record provides the best means of evaluating the paleoclimatic changes of this era and temperature fluctuations have been inferred from downcore variations in foraminiferal faunal composition. A quantitative paleoclimatic approach recently developed by Imbrie and Kipp (1971) in which faunal and floral characteristics are related to oceanographic parameters is now being used in the study of Pleistocene deep-sea cores of the entire Indian Ocean and has led to the reconstruction of sea-surface conditions in the world ocean 18,000 yrs ago (CLIMAP Project members, 1976). However, published Quaternary paleoclimatic studies of the Indian Ocean are still scanty. These reports have been compiled by Bandy, Lindenberg and Vincent (1972), who classified them by geographical area. They are based mainly on variations in the composition of the planktonic foraminiferal faunas established from analysis of either a few selected species or of the total fauna.

The Holocene/Pleistocene boundary (at approximately 10,000 radiocarbon years, at a temperature approximately midway between the maximum "cold" of the last glacial and the "warm" of the post-glacial according to the INQUA Holocene Commission) has been identified in a number of piston and gravity cores from various parts of the Indian Ocean by faunal changes, radiometric dating and oxygen isotopic studies. The boundary occurs from 15 to 40 cm below the sea floor in calcareous cores from tropical, subtropical and temperate areas of the open ocean (Conolly, 1967; Frerichs, 1968; Oba, 1969; Vincent, 1972, 1976; Thierstein et al., in press) indicating an average accumulation rate of 1.5 to 4 cm/1000 yrs for Holocene calcareous oozes. The boundary falls much lower in sections located very close to continental margins in areas where Holocene sediments accumulated much faster from about 20 cm/1000 yrs in the Somali Basin and the Timor Sea (Olausson et al., 1971), up to 40 cm/1000 yrs or more in the Arabian Sea (Schott et al., 1970; Zobel, 1973), and up to almost 100 cm/1000 yrs in the Gulf of Aden where this high value results from a large supply of aeolian matter (Olausson and Olsson, 1969).

A Holocene age for core tops is by no means certain in DSDP material. As mentioned earlier a comparison between the faunal composition of the highest recovered samples and the surface sediment fossil-assemblages described by Bé and Hutson (in press) can be done only in very general terms. Differences in faunal composition between modern assemblages and assemblages of colder Pleistocene intervals are not important enough to be detected without detailed quantitative study. It is, therefore, difficult to recognize if Holocene sediments are present. However, because the top of the sedimentary section is often not recovered with deep-sea drilling techniques, it is probable that the upper 40 cm are largely missing and that at sites where calcareous oozes accumulated at an average normal rate of 2 to 4 cm/1000 yrs, Holocene sediments are not represented. Therefore, although a Holocene age was tentatively assigned by several authors at a few DSDP sites, no attempt is made in this paper to differentiate Holocene from Pleistocene and the entire Quaternary sequences are attributed to the Pleistocene.

Rögl (1974) tentatively assigned a Holocene age to the uppermost core of Site 262 in the Timor Sea on the basis of a discontinuity in the distribution of Globorotalia crassaformis subspp. and G. truncatulinoides. At Site 258 on the Naturaliste Plateau, Boltovskoy (1974a), following the criterion used by Parker (1973) in Atlantic cores, assigned a tentative Holocene age to the upper 8.5 meters of sediments containing pink-walled Globigerina rubescens and Globigerinoides ruber. On the basis of sedimentation rates, however, this age assignment is very unlikely and pink G. rubescens and G. ruber first appeared in the Indian Ocean prior to the 10,000 years Holocene/Pleistocene boundary. In the western Indian Ocean pink G. rubescens and G. ruber occur commonly in the upper 3 meters of the Gulf Aden sites, in the upper 3 meters at Site 237 and the upper 3 meters at Site 238, and were not found lower in these sections (Vincent et al., 1974).

Factors causing red-pigmentation of planktonic foraminifera are not yet known and occurrences of pigmented G. ruber have been discussed by various authors (Bé and Hamlin, 1967; Emiliani, 1969; Orr, 1969; Bronniman and Resig, 1971; Jenkins and Orr, 1972). The pink variety of G. ruber has been found in the living plankton in the Atlantic Ocean only (Bé and Hamlin, 1967; Boltovskoy, 1968; Bé and Tolderlund, 1971) but neither in the Pacific (Bradshaw, 1959) nor in the Indian Ocean (Bé and Hutson, in press). It occurs in the Atlantic in the surface sediments as well as in the living plankton (Orr, 1969; Boltovskoy, 1968; Parker, 1973). In the Pacific it was found in upper Pleistocene sediments but not in surface sediment assemblages except in sporadic occurrences of probable reworked specimens (Bronniman and Resig, 1971). Pink G. ruber is not present in the surface sediments of the Indian Ocean (Bé and Hutson, in press) and was not observed by the author in Holocene cores from the subtropical Indian Ocean (Vincent, 1976). Thus, the presence of pink G. ruber in DSDP core-tops supports the suggestion that the top of the sedimentary sequence is absent. A layer of pink Globigerinoides ruber appears to occur in the Indian Ocean, as in the Pacific, in upper Pleistocene, an occurrence which may be stratigraphically very useful.

Boltovskoy (1974b) analyzed Pleistocene samples of subtropical Site 253 on the southern Ninetyeast Ridge at 10-cm intervals and constructed a paleoclimatic curve based on the ratio between the tropical elements of the Globorotalia menardii complex and the temperate form G. inflata (Fig. 14b). This author placed the Holocene/Pleistocene boundary at 30 cm at a well-pronounced downward increase in Globorotalia inflata, which, assuming that the entire Holocene section is recovered, would indicate an average accumulation rate of 3 cm/1,000 yrs for Holocene sediments. This value would agree with the findings of Conolly (1967) and Vincent (1972, 1976) in deep-sea gravity cores located at similar latitudes in the eastern and western margins of the Indian Ocean. These authors observed a marked downward increase in G. inflata at a level varying between 15 to 40 cm below the sea-floor dated by radiocarbon at about -10,000 yrs B. P. However, even if the top of the sedimentary section was entirely recovered at Site 253, the condensing of the entire Pleistocene at this site makes a 10,000 yr age assignment for the 30 cm-horizon unlikely. The Pleistocene sequence is 9 m thick which indicates an average accumulation rate of 0.5 cm/1000 yrs for the entire Pleistocene. The latter value is supported by the occurrence of the nannoplankton species Pseudoemiliania lacunosa up to a least 2.55 m in the section (Thierstein, 1974). This species, known to have become extinct at approximately -458,000 yrs (Thierstein et al., in press) indicates a minimum age of -458,000 yrs for the 2.55 m horizon, and an average accumulation rate of about 0.5 cm/1000 yrs for both Pleistocene intervals above and below this horizon.

Boltovskoy placed the "Pliocene/Pleistocene boundary" at Site 253 at 5.9 meters at the level of equal abundance of Globorotalia truncatulinoides and G. tosaensis and of an abrupt change in the G. menardii/ G. inflata ratio. However, as discussed later, this level is higher than the boundary, which on the basis of nannofossil data and on the oldest occurrence of G. truncatulinoides lies at 9 meters at the bottom of Core 1 (Fig. 14a). It appears, thus, that the entire curve constructed by Boltovskoy covers only the Quaternary section. Assuming a constant sedimentation rate of 0.5 cm/1000 yrs, ages have been assigned to the climatic changes recorded at Site 253 (Fig. 14b). These age assignments, however, remain tentative without further calibration of the sequence to the time-scale and without further correlation with the calcareous nannoplankton zonation which needs to be refined with more closely spaced samples. Furthermore, restrictions in the use of the ratio G. menardii/G. inflata as a paleoclimatic tool should be kept in mind.

Variations in the abundance of G. inflata, a species characteristic of the temperate water mass, are a good indicator of shifting of this water mass through time, and have indeed proved so in subtropical regions of the Indian Ocean (Conolly, 1967; Vincent, 1976). On the other hand the tropical species Globoratalia menardii has been proved to be of great value for Pleistocene paleoclimatic studies in the Indian Ocean (Stubbings, 1939; Conolly, 1967; Oba, 1969; Olausson et al., 1971), where increase in its abundance delineates warmer intervals. This faunal criterion, however, was found not applicable for Pleistocene paleoclimatic investigations in parts of the Indian Ocean especially in areas marginal to the habitat of G. menardii. As shown by Vincent (1976) and Morin et al. (1970), variations in abundance of the G. menardii complex in the subtropical southwestern Indian Ocean show an inverse relation to temperature. Doubts have also been expressed (Morin et al., 1970) on the significance of the variations in abundance of the G. menardii complex reported by Ericson and Wollin (1970) in cores from the southeastern Pacific. In the area of Site 253, increases in abundance of G. menardii may possibly correlate to increases in temperature, as it apparently does at about the same latitude in deep sea cores off western Australia (Conolly, 1967). However, before drawing paleoclimatic conclusions, it would be preferable to evaluate changes in the relative abundance of Globorotalia inflata in respect to the whole planktonic fauna and to construct ratios using tropical species other than G. menardii (e.g. Neogloboquadrina dutertrei and/or Pulleniatina obliquiloculata). As Lidz (1966) pointed out, climatic curves based on the abundance of several species give better agreement with the oxygen isotope sequence than curves based on a single species.

Quantitative foraminiferal faunal analyses for climatic interpretation were conducted by Zobel (1974) in the expanded Quaternary section of Site 241 in the western Somali Basin. Radiocarbon dates were obtained for three levels of Core 1. No significant change in the faunal composition was observed above and below the presumed 10,000 yrs level. The relative abundances of diagnostic species remain uniform throughout the upper 3 meters. Furthermore, assuming that the radiocarbon ages are correct, there would have been an average sediment accumulation rate of 56 cm/1000 yrs during the Holocene, and assuming a constant rate during the Quaternary, about 1000 m of sediments would have accumulated during that period. This extremely high rate is unlikely; the thickness of the Quaternary section at Site 241 is approximately 70 meters. In view of these observations, Zobel concluded that the upper section of Site 241 contains mixed sediments.

No other quantitative faunal study has been carried out from DSDP Indian Ocean Quaternary sediments. However, variations in the abundance of cooler and warmer-water species are evidenced by the semiquantitative analyses reported on the range-charts of various authors.

Foraminiferal Events Calibrated to the Paleomagnetic and/or the Oxygen Isotope Time Scale.

A few isotope studies of planktonic foraminifera from Indian Ocean cores have shown an oxygen isotope record comparable to the Atlantic and Pacific records for the latest Quaternary back to about -500,000 years (oxygen isotope stage 13) (Oba, 1969; Vincent and Shackleton, in press; Thierstein et al., in press). No direct calibration of planktonic events to the paleomagnetic time scale for the entire Quaternary has yet been achieved in the Indian Ocean. All or part of the Brunhes and Matuyama intervals are absent in Core V20-163 and in the Eltanin cores south of Australia, and the lower part of the Matuyama (including the Olduvai event) is missing in Core V16-66.

Foraminiferal events known to occur within the Pleistocene and which have been calibrated to the paleomagnetic and/or the oxygen isotope time-scales in the Pacific Ocean are in order of increasing age: 1) the extinction of Globoquadrina pseudofoliata at approximately -0.23 m.y. (mid-oxygen isotope stage 7; Thompson and Saito, 1974; Thompson, 1976); 2) the extinction of Globorotalia tosaensis at about -0.6 m.y. (within oxygen isotope stage 16, Thompson, 1976); 3) a marked upward decrease in the abundance of Sphaeroidinella dehiscens at about -0.7 m.y. (at the Brunhes/Matuyama paleomagnetic epochs boundary; Hays et al., 1969); 4) the first evolutionary appearance of Pulleniatina finalis at about -1.4 m.y. (midway between the Jaramillo and Olduvai paleomagnetic events; Hays et al. 1969). All these events were not identified in all Pleistocene sequences of the Indian Ocean but were recorded in a number of sections. They are indicated in Figures 3 to 18, as well as the levels of appearance of the nannofossil Emiliania huxleyi and of the extinction of Pseudoemiliania lacunosa. These two nannofossil events have been consistently recorded in late oxygen isotope Stage 8 and mid-oxygen isotope Stage 12 respectively in cores from all oceans from 54°N to 43°S (Thierstein et al., in press). Their ages are thus estimated to be -0.27 m.y. and -0.458 m.y. respectively, using the age-scale proposed by Shackleton and Opdyke (1973) for oxygen isotope stages, on the assumption of a uniform sedimentation rate throughout the Brunhes Epoch. The synchroneity of these two biostratigraphic datums in tropical to subpolar areas makes them excellent markers for Pleistocene stratigraphy. The radiolarian species Axoprunum angelinum (= Stylatractus universus), known to have become extinct in various parts of the world very shortly after the extinction of Pseudoemiliania lacunosa at the oxygen isotope Stages 11/12 boundary (Hays and Shackleton, 1975), is also an excellent Pleistocene biostratigraphic marker. This radiolarian species was not reported in Indian Ocean DSDP sections except at Site 262 in the Timor Sea where Renz (1974) reported its occurrence as ranging up to the top of the recovered sequence. The species illustrated by Renz (op. cited, pl. 14), however, is probably not S. universus because the intermediate size spines on the spherical shell, which are diagnostic of the species (Kling, 1973), are not present (Kling, personal communication).

Globoquadrina pseudofoliata was noted only at Site 219 in the Arabian Sea where it last occurs in a biostratigraphic position lower than in the Pacific, below the extinction level of Pseudoemiliania lacunosa. G. pseudofoliata was reported by Parker (1967) in the Pliocene of Core DODO 117P but does not range in this core into the Pleistocene. The last occurrence of Globorotalia tosaensis has been reported at various sites inconsistently above or below the extinction level of P. lacunosa and at the Pliocene/Pleistocene boundary in DODO cores. The rarity of this species in most sections restricts its use as a biostratigraphic marker. Pulleniatina finalis was not separated from P. obliquiloculata s.l. at most sites except at Sites 214 and 219. It first occurs at these two sites in the lower part of the Pleistocene section, as in Pacific cores. Contrary to the sequential order of events recorded in the Pacific, however, the last occurrence of G. tosaenis was reported below the first occurrence of P. finalis at Site 219. No data are available for Site 214 on the upper range

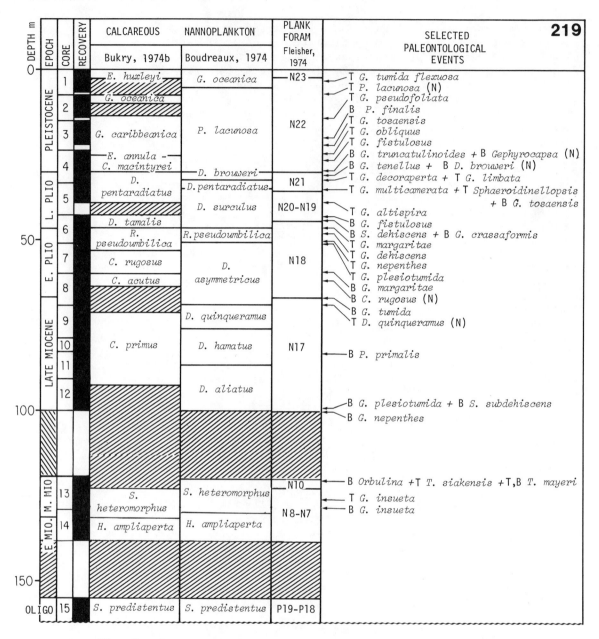

Fig. 3. Neogene planktonic biostratigraphy at Site 219. Solid pattern indicates recovered interval. Hachured pattern indicates intervals of undetermined age. Within paleontological events (N) = nannofossil, (R) = radiolarian, other events are foraminiferal; T = top of range, B = base of range. Fossil zone boundaries are reproduced from various authors. Epoch boundaries are located according to reinterpretation in this paper. (Symbols apply to Figures 3 through 18.)

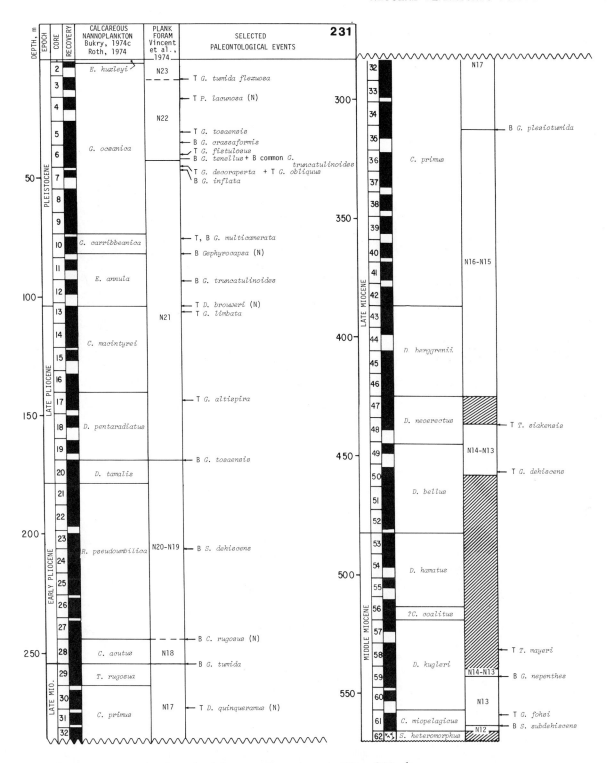

Fig. 4. Neogene planktonic biostratigraphy at Site 231 (see Figure 3 for symbols).

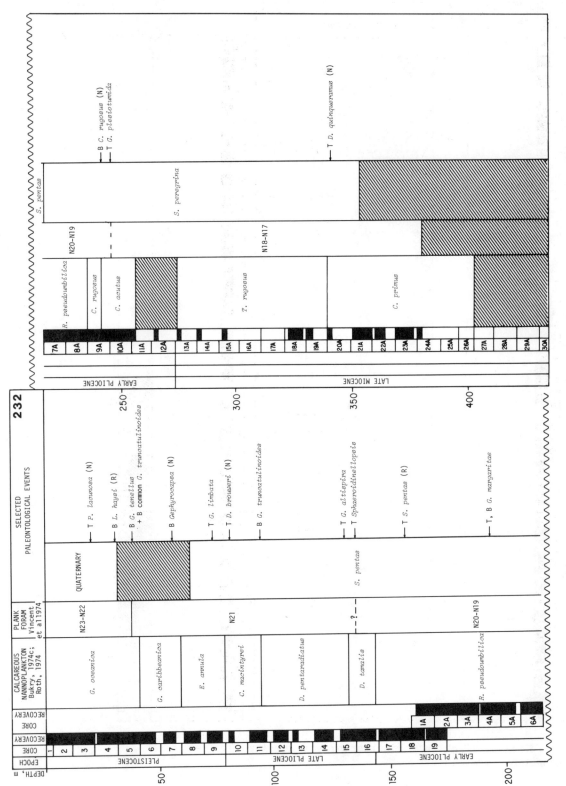

Fig. 5. Neogene planktonic biostratigraphy at Site 232 (see Figure 3 for symbols).

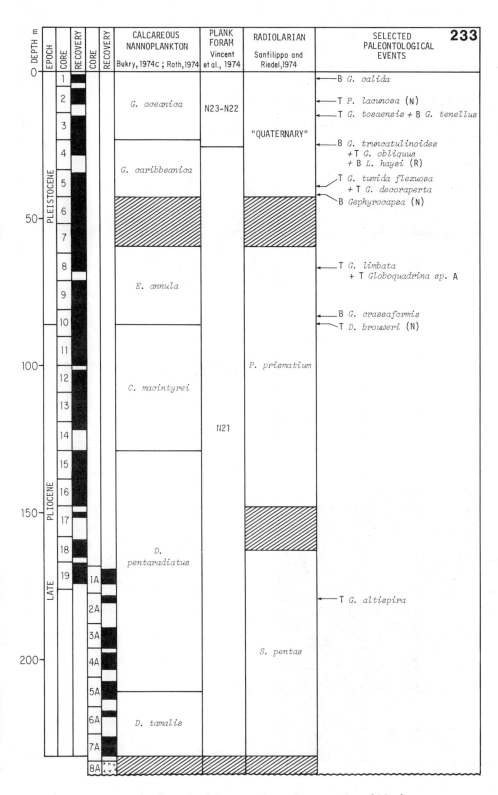

Fig. 6. Neogene planktonic biostratigraphy at Site 233 (see Figure 3 for symbols).

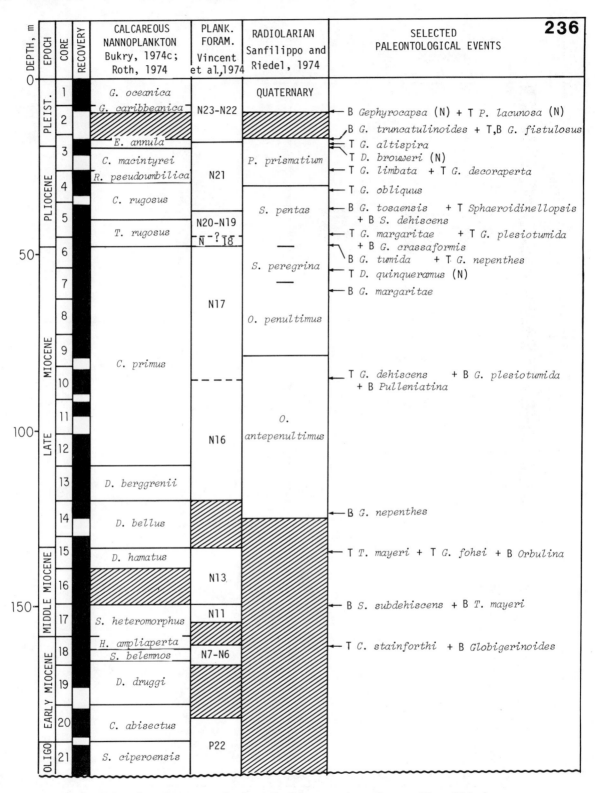

Fig. 7. Neogene planktonic biostratigraphy at Site 236 (see Figure 3 for symbols).

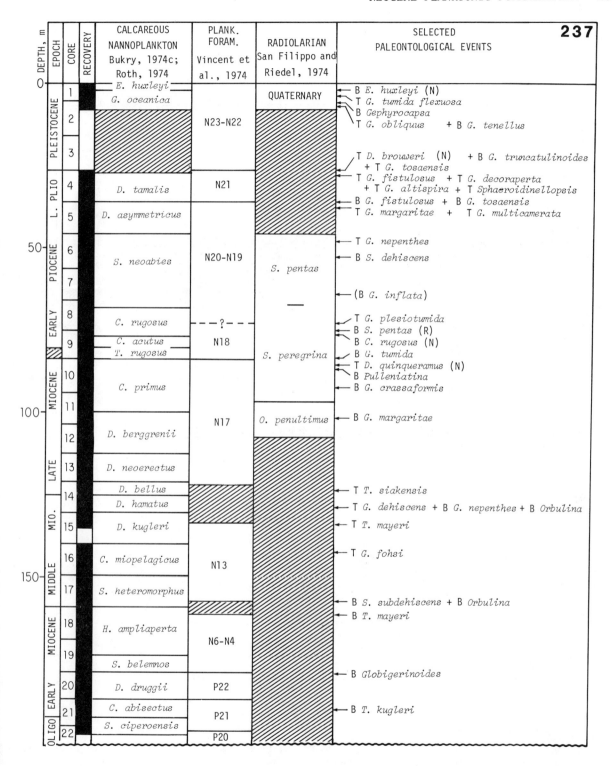

Fig. 8. Neogene planktonic biostratigraphy at Site 237 (see Figure 3 for symbols). The abnormally low range of G. inflata needs reevaluation.

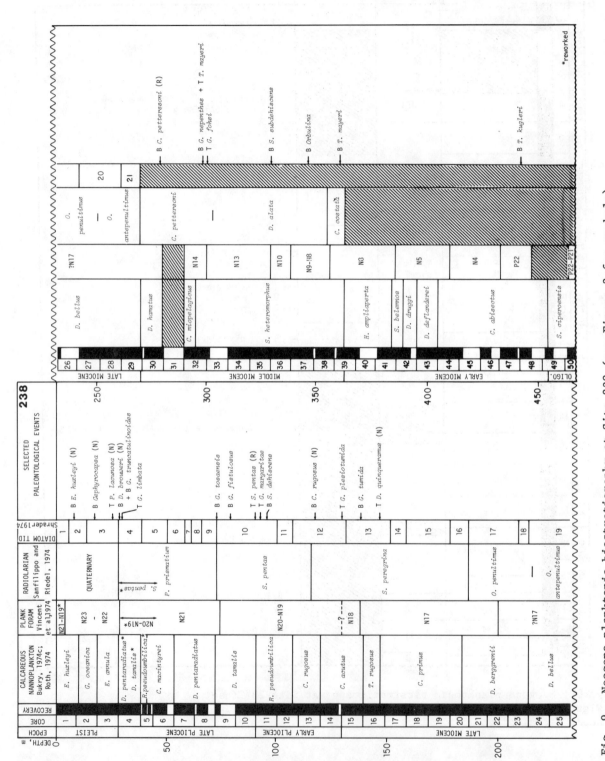

Fig. 9. Neogene planktonic biostratigraphy at Site 238 (see Figure 3 for symbols).

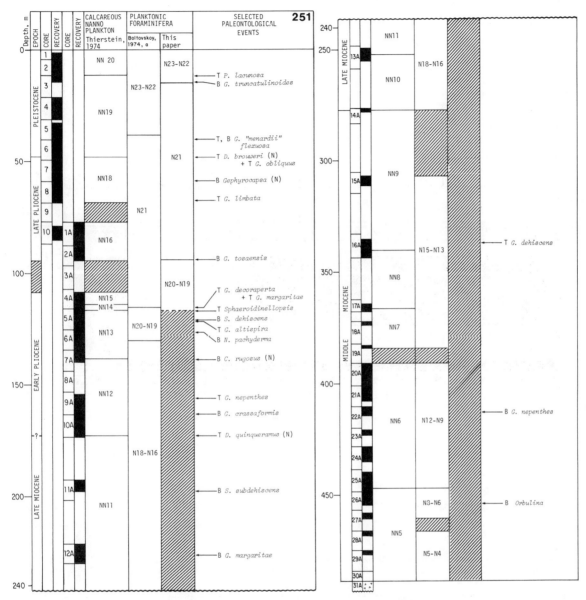

Fig. 10. Neogene planktonic biostratigraphy at Site 251 (see Figure 3 for symbols).

of G. tosaensis which ranges up to the highest sample analyzed by Berggren et al. (1974). It appears, thus, that without further detailed study and/or additional data a consistent succession of Pleistocene foraminiferal events calibrated to the paleomagnetic and oxygen isotope time-scales in the Pacific cannot be recognized in the Indian Ocean.

Extinction of Globorotalia tumida flexuosa

Flexuose forms referred to G. menardii as well as to G. tumida have been reported in Indian Ocean sections. However, because of the taxonomic ambiguity attached to these forms (Ericson and Wollin,

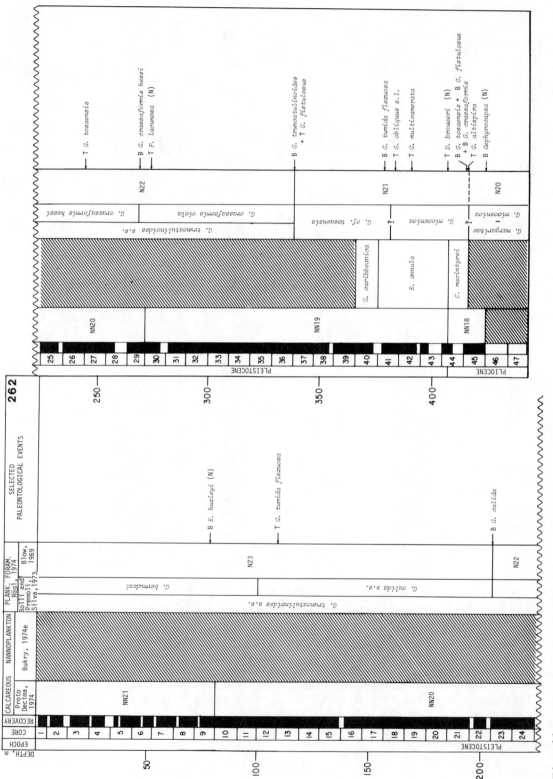

Fig. 11. Neogene planktonic biostratigraphy at Site 262 (see Figure 3 for symbols).

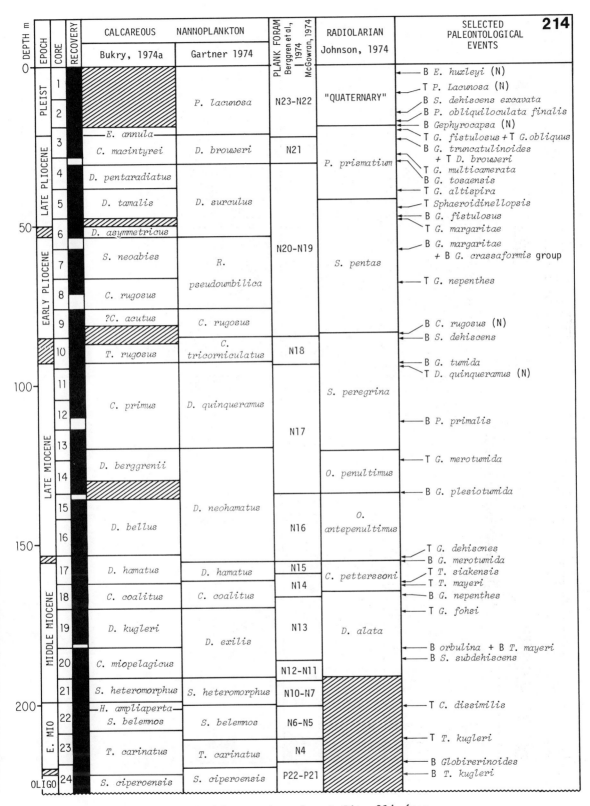

Fig. 12. Neogene planktonic biostratigraphy at Site 214 (see Figure 3 for symbols).

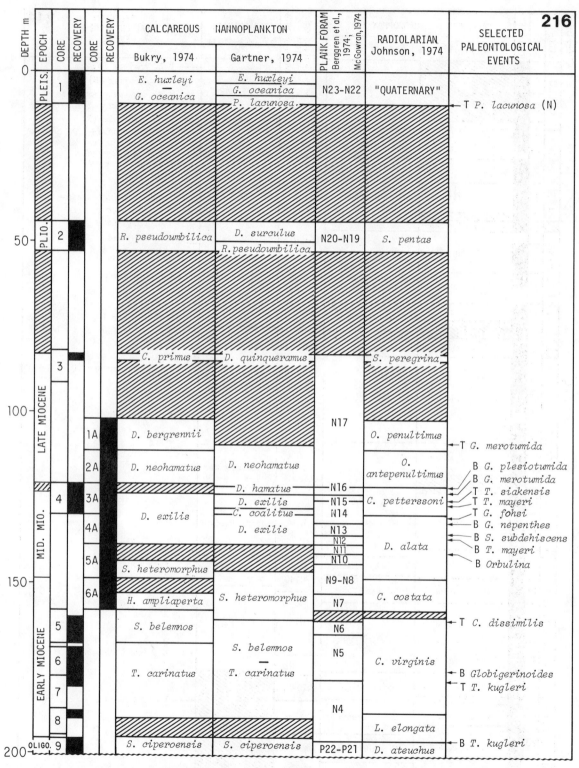

Fig. 13. Neogene planktonic biostratigraphy at Site 216 (see Figure 3 for symbols).

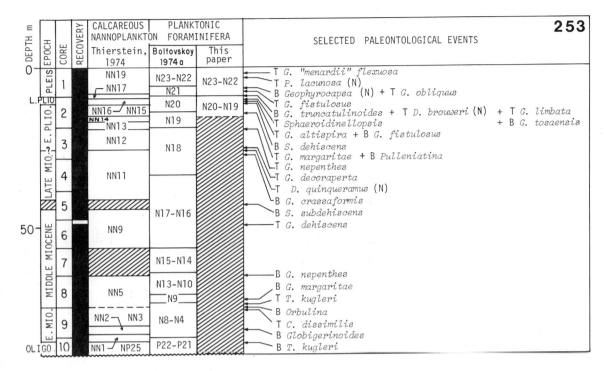

Fig. 14a. Neogene planktonic biostratigraphy at Site 253 (see Figure 3 for symbols).

1956; Ericson et al., 1961, 1964; Srinivasan et al., 1974; Adelseck, 1975) various authors may have referred similar forms to different taxa. Forms referred to G. tumida flexuosa were found in noticeable abundance in Pleistocene horizons of tropical sections at Sites 262, 237, 219, 231 and 233 and were probably included with G. tumida at other tropical sites where they were not reported.

The youngest occurrences of this subspecies were recorded in the upper part of these sequences in the interval comprised between the first occurrence of Emiliania huxleyi and the last occurrence of Pseudoemiliania lacunosa (within Zone NN 20 of Martini and the Gephyrocapsa oceanica Zone of Bukry), at a level which approximates the first appearance of Globigerinella adamsi at both Sites 219 and 231. Globorotalia tumida flexuosa may, thus, have become extinct earlier in the tropical Indian Ocean, prior to the initial appearance of E. huxleyi, than in the Atlantic-Caribbean. In the latter area the extinction level of G. tumida flexuosa occurs stratigraphically higher than the first appearance of E. huxleyi (DSDP Sites 147, 148, 149 and 154: Bolli and Premoli Silva, 1973; Hay and Beaudry, 1973). It occurs at the top of oxygen isotope Stage 5 and of Zone X of Ericson et al. at about -80,000 years (Emiliani, 1966; Ericson et al., 1961; Ericson and Wollin, 1968; Srinivasan et al., 1974).

An early extinction of G. tumida flexuosa in the Indian Ocean is supported by Oba's (1967, 1969) findings. This author observed the occurrence in noticeable abundance of "G. menardii flexuosa" (following Ericson's nomenclature) in the lower part of three cores from the central equatorial Indian Ocean. From the oxygen isotope curve constructed for two of these cores, it appears that this horizon is certainly older than oxygen isotope stage 5 and occurs in what appears to be oxygen isotope stage 7 or 8 (Oba, 1969, his figs. 26 and 27).

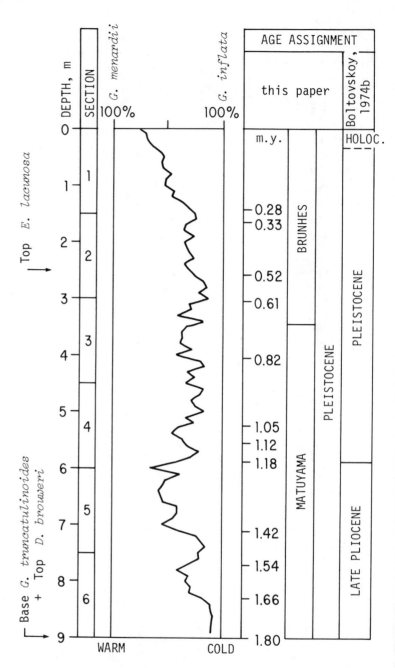

Fig. 14b. Climatic curve obtained from Globorotalia menardii/G. inflata ratios in Core 1 at Site 253 (from Boltovskoy, 1974b).

No younger occurrence of flexuose forms of the G. menardii-tumida group was reported by Oba in these three cores, even though two of them contain an apparently continuous Pleistocene-Holocene sequence.

In the southwest Pacific G. tumida flexuosa disappears as in the Indian Ocean at a level older than the base of the E. huxleyi nannofossil zone (Burns, Andrews et al., 1973). It is not clear, however, from published data when this subspecies became extinct (if it did) from other areas of the tropical Pacific. Bronniman et al. (1971) noted its disappearance at a level approximating the extinction level

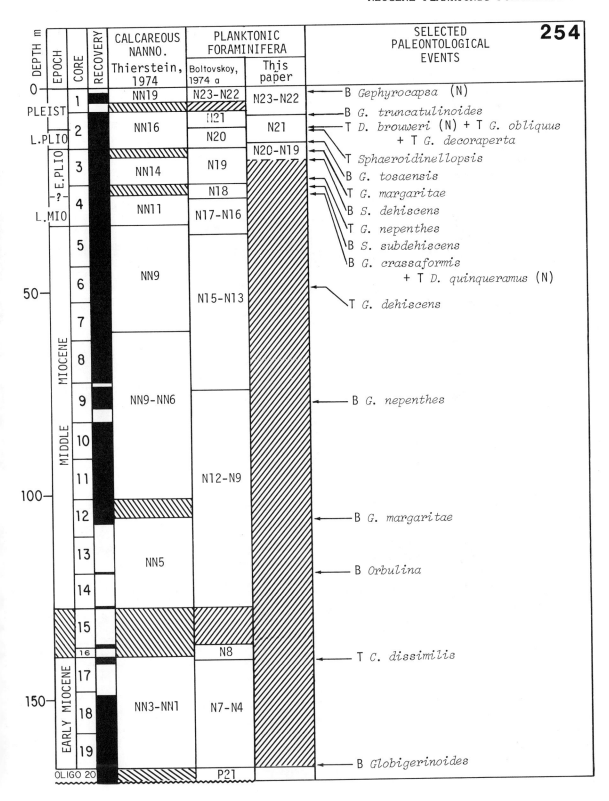

Fig. 15. Neogene planktonic biostratigraphy at Site 254 (see Figure 3 for symbols).

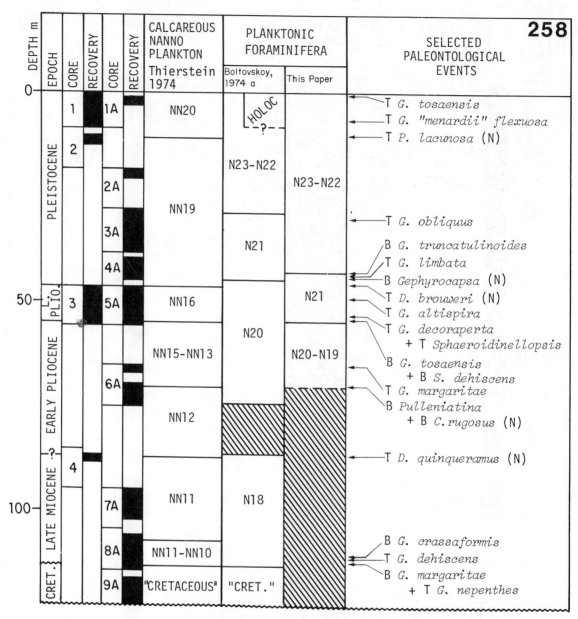

Fig. 16. Neogene planktonic biostratigraphy at Site 258 (see Figure 3 for symbols).

of *Pseudoemiliania lacunosa* whereas Jenkins and Orr (1972) reported its range up to the top of the Pleistocene. Parker (1967), who pointed out that a gradation exists from the flexuose to the typical form of *G. tumida* did not separate these two forms neither did Hays et al. (1969).

No flexuose forms attributed to *G. menardii* were noted at the low-latitude Indian Ocean sites even in the northern part of this ocean where these forms have been observed living (Bé and McIntyre, 1970). Their absence in the highest samples recovered from DSDP drilling supports the suggestion that these core-tops do not represent surface sediment. Forms referred to *G. menardii* forma *flexuosa* by Boltovskoy

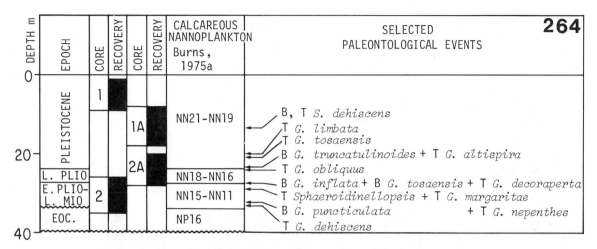

Fig. 17. Neogene planktonic biostratigraphy at Site 264 (see Figure 3 for symbols).

(1974a) however, were reported by this author with rare and scattered occurrences in Pliocene and Pleistocene sediments at mid-latitude Sites 251, 253 and 258. The specimen illustrated by Boltovskoy (his pl. 9, fig. 12) appears to have a morphology intermediate between G. menardii and G. tumida and it is probable that the forms called G. menardii forma flexuosa by Boltovskoy are part of the group referred to G. tumida flexuosa by other workers of DSDP sections. An early extinction of "G. menardii forma flexuosa" is recorded at Sites 251 and 258; this form last occurs slightly above the extinction level of P. lacunosa at Site 258 and well below it at the higher latitude Site 251. "G. menardii forma flexuosa" last occurs at 40 cm sub-bottom depth at Site 253. According to the age estimates based on sedimentation rates as postulated above for this site, an age of -80,000 yrs would be assigned to the 40 cm-horizon. This age appears much younger than the age of the extinction of flexuose forms in all other Indian Ocean sections and suggests that the top of the sedimentary sequence was not entirely recovered at Site 253.

The range of G. tumida flexuosa is discontinuous. It does not extend lower than the lower Pleistocene at mid-latitude sites as well as at the low-latitude Site 262, whereas it extends further down into the Lower Pliocene at other low-latitude sites.

Pleistocene Zonations

Blow (1969) proposed a late Quaternary Globigerina calida calida/ Sphaeroidinella dehiscens excavata Assemblage Zone (N 23), to which these species, as well as Globigerinella adamsi and Hastigerinella rhumbleri (= H. digitata) are restricted. Brönnimann and Resig (1971) have discussed the problems involved in recognizing this stratigraphic unit in the western Pacific. Most workers did not differentiate Zones N 23 and N 22 in the Pleistocene sections of the Indian Ocean. The subspecies S. dehiscens excavata was differentiated from S. dehiscens s.l. only at Site 214 on the Ninetyeast Ridge where it first occurs slightly above the appearance of Pulleniatina finalis. The lower range of G. calida was used by Rögl (1974) to place the N23/N22 zonal boundary at Site 262 in the Timor Sea in the middle of the Pleistocene section. This species, however, appears earlier in the northwestern Indian Ocean at Sites 219 and 231, just below the

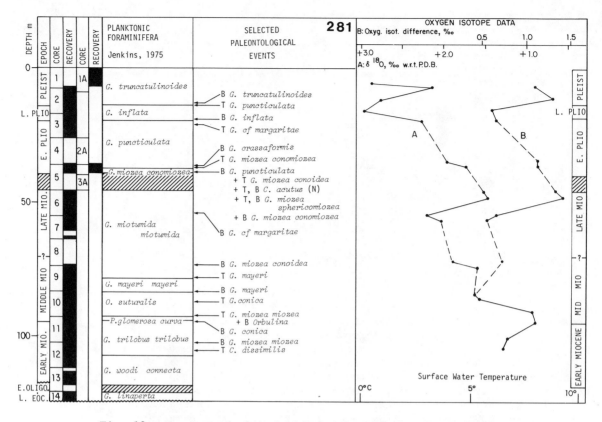

Fig. 18. Neogene planktonic biostratigraphy at Site 281 (see Figure 3 for symbols). Oxygen isotope data are from Shackleton and Kennett (1975a). Curve A: oxygen isotope composition of planktonic foraminifera; deviation per mil from PDB. Curve B: difference in oxygen isotope composition between planktonic and benthonic foraminifera. Surface water temperatures are obtained from Curve A for the interval below Core 10 (deposited before the accumulation of the Antarctic ice sheet) and from Curve B above to correct for changes in ocean isotopic composition as ice sheets accumulated (assuming that the bottom temperature at the site would have remained essentially unchanged from the time the Antarctic ice started to accumulate).

base of the Pleistocene, where Fleisher (1974) and Vincent (1974a) tentatively placed the base of N23 at the first occurrence of Globigerinella adamsi.

The initial appearance of Globorotalia crassaformis hessi was used by Rögl to subdivide the lower Pleistocene of Site 262 in a manner similar to the lower Pleistocene subdivision of Bolli and Premoli Silva (1973) in the Caribbean. It appears, however, that this subspecies first appears in the Timor Sea (just above the extinction level of Pseudoemiliania lacunosa at Site 262) later than in the Caribbean (significantly below this nannofossil event at Sites 148 and 154A (Bolli and Premoli Silva, 1973)). This subspecies was not differentiated from G. crassaformis s.l. at other Indian Ocean sites.

Coiling Changes

Among the Pleistocene planktonic foraminiferal events recorded at Site 262 in the Timor Sea there are distinct and frequent changes in coiling direction in various globorotaliids (Globorotalia menardii, G. tumida, G. truncatulinoides, G. tosaensis and G. crassaformis) in the lower part of the section between 290 and 413.5 meters (Cores 32 to 44) and in Pullen atina obliquiloculata between 337.5 and 413.5 meters (Cores 37 to 44), Rögl, 1974, his fig. 2). Variations in the coiling direction in the species Globorotalia tumida and the genus Pulleniatina (whose modern populations are exclusively sinistrally and dextrally coiled respectively) have provided excellent means of stratigraphic correlations over wide areas in both the Pacific and Atlantic basins as shown by Hays et al. (1969) and Saito et al. (1975) in their studies of paleomagnetically dated cores.

A sharp left-to-right coiling change in Pulleniatina occurring at about -3.6 m.y. (just above the Gilbert C event) was observed by these authors in deep-sea cores from all oceans. In Pacific cores this genus shows numerous coiling changes in the late Pliocene and early Pleistocene and exhibits a final switch from left to right at about -0.9 m.y. (Jaramillo event). In Atlantic cores Pulleniatina is generally absent from the interval between -3.6 and -2.3 m.y. (Gauss and lower Matuyama intervals) and its pattern of coiling change after its reappearance in this basin no longer follows the Pacific pattern. In the Indian Ocean core V20-163, a sharp coiling change from left to right occurs in Pulleniatina just above the Gilbert C event, as in cores from other oceans, but the coiling pattern of the genus in late Pliocene and Pleistocene cannot be investigated in this core because these series are lacking. As seen at various DSDP low-latitude sites, Pulleniatina is present in the Indian Ocean throughout the Pliocene and its coiling pattern probably follows the Pacific pattern. Therefore, the final switch from left to right-coiling recorded at 337.5 meters at Site 262 is probably correlative with the -0.9 m.y. youngest switch recorded in the Pacific. The location of this event at Site 262, about midway between the Pliocene/Pleistocene boundary and the extinction level of Pseudoemiliania lacunosa, supports this age assignment. Coiling-direction patterns in Pulleniatina were not investigated in other sections of the Indian Ocean.

At the three Gulf of Aden sites, a horizon with dextrally coiled Globorotalia tumida was found in the lower part of the Pleistocene sections, below the base of the common occurrence of Globorotalia truncatulinoides. This horizon is possibly correlative with the interval between Cores 35 and 32 at Site 262 in the Timor Sea in which Rögl (1974) reported that populations of the G. menardii-tumida group are dominantly dextral. The presence of right-coiling G. tumida near the Pliocene/Pleistocene boundary was also noted by Kierstead et al. (1969) in equatorial Pacific cores.

A distinct change in coiling preference from left to right in Globorotalia crassaformis occurs at 60 m at Site 262 in the Timor Sea. This coiling change occurs simultaneously in the various subspecies of G. crassaformis which remain dextral to the top of the section, and might possibly correlate with a similar change recorded in the central equatorial Indian Ocean. In three cores from the latter area, Oba (1969) noted an upward change in coiling preference in G. crassaformis from left to right, which appears to occur within oxygen isotope stage 3 (80 cm in Core IC-6, top of Core Ka-18, 70 cm in Core IC-5: Oba, 1969, his figures 16 and 27). Preceding this youngest coiling change from left to right, two right coiling peaks of very short duration were noted by Oba. These two peaks appear to occur during oxygen isotope Stage 5 and near the Stages 5/6 boundary respectively (Oba, 1969, his figures 16, 26 and 27). Similar peaks were not reported by Rögl at Site 262. Coiling changes in G.

crassaformis have proved very useful for correlations between cores in a restricted area, as shown by Oba in the central equatorial Indian Ocean and by various workers in other areas as well. Correlations on this basis over distant areas, however, are hazardous because of the limited extent of righ-and left-coiling provinces of modern populations of this species. Coiling changes seen in cores probably reflect the shifting of these provinces. However, the significance of these provinces is not well understood and the boundaries between them is not well known. Areal distribution of dextral and sinistral provinces of modern Globorotalia crassaformis and Globorotalia truncatulinoides (see for example Ericson et al., 1964; Bandy, 1960; Parker, 1971, Parker and Berger, 1971; Bé and Tolderlund, 1971; Bé and Hutson, in press; Thiede, 1971) appear to show some relationship with surface-water temperature distribution, however the parallelism is not well defined and factors other than temperature probably are involved.

In a number of cores from the southwestern Indian Ocean, located in a province of sinistral modern populations of G. truncatulinoides, Vincent (1976) noted a downward change in coiling direction in this species dated by radiocarbon at about -6,000 yrs. This change was recorded in some, but not all, cores from other parts of the Indian Ocean (A. Bé, personal communication).

Pliocene/Pleistocene Boundary

Paleontological Events Associated with the Pliocene/Pleistocene Boundary

A number of paleontological events which are linked with the Olduvai normal paleomagnetic event (Gilsa event of Cox, 1969), approximately 1.6- 1.8 m.y. old, have been used as criteria for recognizing the Pliocene/Pleistocene boundary because some of these events are near the Pliocene/Pleistocene boundary stratotype. Paleontological events which have been found associated with the Olduvai event in tropical deep-sea cores include the extinctions of the nannofossil genus Discoaster (in the form of D. broweri) and the radiolarian Pterocanium prismatium, and the initial appearances of the foraminiferal species Globorotalia truncatulinoides and the diatom Pseudoeunotia doliolus (Ericson et al., 1963; Banner and Blow, 1965a; Berggren et al., 1967; Glass et al., 1967; Burckle, 1969; Hays and Berggren, 1971; Berggren, 1973; Berggren and Van Couvering, 1974; Briskin and Berggren, 1975; Saito et al., 1975; Berggren in press). These events, however, are not simultaneous. Hays et al. (1969), Burckle (1969) and Saito et al. (1975) noted that in the equatorial Pacific D. broweri became extinct at the base of the Olduvai event and P. doliolus appeared during the Olduvai event, while P. prismatium became extinct just after the Olduvai event and G. truncatulinoides appeared significantly later in the later part of the reversed interval between the Olduvai and the Jaramillo events.

The extinction of D. broweri, which is related to the base of the Calabrian (Bandy and Wilcoxon, 1970; Takayama, 1970), has been repeatedly recorded in the Olduvai event in low- and mid-latitude Atlantic and Pacific cores (Berggren et al., 1967; Glass et al., 1967; Phillips et al., 1968; Ericson and Wollin, 1968; Hays et al., 1969; Briskin and Berggren, 1975). The lowest occurrence of G. truncatulinoides, which has been reported near the base of the Calabrian is often associated with the extinction level of D. broweri (or a drastic reduction in abundance) in tropical and subtropical

deep-sea sediments of the open ocean, but has also been found well
above (e.g. in the Mediterranean: Cita et al., 1973) or below it
(e.g. in the Gulf of Mexico: Beard, 1969; in the southwest Pacific
at DSDP Sites 203, 206, 208 and 209: Burns, Andrews et al., 1973).
Furthermore, the evolutionary development of G. truncatulinoides from
its ancestor G. tosaensis has been a subject of controversy and the
speciation of the two forms appears to be a complex phenomenon.
Berggren et al. (1967) and Berggren (1968) described the evolutionary
transition from Globorotalia tosaensis to G. truncatulinoides in the
Olduvai event of a North Atlantic core but Parker (1973) who restudied
the same core found that Berggren's G. tosaensis is actually referable
to Blow's G. crassaformis ronda. In other Atlantic cores Ericson
and Wollin (1968) reported the appearance of "abundant" G. truncatulinoides at the base of the Olduvai event and Saito et al. (1975)
found the base of "common" G. truncatulinoides within the Olduvai,
the species ranging down in rare occurrence to below this paleomagnetic event. In the southwest Pacific Kennett (1973) described
a fairly simple gradation from G. tosaensis to G. truncatulinoides
with steadily increasing abundance of the morphotype of G. truncatulinoides at DSDP Site 206 but at nearby Sites 208 and 209 the inter-
relationship between the two forms is more complex and their relative
abundance fluctuates significantly within the interval of their overlap. Alternations of G. tosaensis and G. truncatulinoides have been
described in a Late Pliocene-early Pleistocene core in the southeast
Pacific and shown to be related to paleooceanographic oscillations
(Kennett and Getzeinauer, 1969).

It appears, thus, that the G. truncatulinoides first appearance
"datum" should be used with great caution, especially at different
latitudes in various water masses. Difficulty in using this "datum"
for the recognition of the Pliocene/Pleistocene boundary and differing
interpretations of this event (placed by some workers at the first
common occurrence of the species and by others at the first appearance
of isolated specimens) is exemplified by Indian Ocean DSDP data.

In Indian Ocean sediments the first occurrence of G. truncatulinoides
is often recorded at a level identical or very close to the extinction
levels of Discoaster brouweri and Pterocanium prismatium. Discrepancies between these events, however, have been noted at a number
of sites where G. truncatulinoides first appears significantly higher
in the sections that the extinction level of D. brouweri. Because of
these discrepancies, which may result either from ecological control
or from the difficulty of identification of the G. truncatulinoides
datum, and because siliceous fossils are present only in a few
sections, for the sake of uniformity the Pliocene/Pleistocene
boundary is arbitrarily placed in this report on nannofossil data.
The boundary is placed at the NN 19/NN 18 zonal boundary of Martini
(1971) and the Emiliania annula/Cyclococcolithina macintyrei subzonal
boundary of Bukry, both defined by the extinction level of Discoaster
brouweri (or rather an abrupt decrease in abundance to avoid reworked
specimens in Pleistocene, Bukry, 1973a).

Another nannofloral event associated with the Pliocene/Pleistocene
boundary is the appearance of the genus Gephyrocapsa which appears
near or just above the boundary in the open oceans (McIntyre et al.,
1967; Sachs and Skinner, 1973). The first appearance of this genus,
however, has been noted in upper Pliocene sediments of late Neogene
sections of Italy (Rio, 1974). The levels of disappearance of
Discoaster brouweri and appearance of the genus Gephyrocapsa are
reported on Figures 3 to 18. These two events are usually recorded
near each other, the base of Gephyrocapsa being usually slightly

above the disappearance of G. brouweri. In the hemipelagic expanded sections of Sites 251 and 262, however, Gephyrocapsa appears about 10 meters below the extinction level of D. brouweri. At the latter site in the Timor Sea Proto Decima (1974) noted an interesting evolutionary sequence of Gephyrocapsa from Core 45, CC to the top of the section. Higher evolved Gephyrocapsa species appear in Core 36 (where Rögl, 1974, reported the first appearance of G. truncatulinoides), only primitive forms being present below that level.

Other foraminiferal events known to occur near the Pliocene/Pleistocene boundary in tropical and subtropical regions of the Pacific and Atlantic oceans are the extinction levels of Globigerinoides fistulosus and Globigerinoides obliquus (e.g. Parker, 1967, 1973; Hays et al., 1969; Brönnimann and Resig, 1971; Jenkins and Orr, 1972; Berggren, 1973; Briskin and Berggren, 1975; Saito et al., 1975) and the appearance level of Globigerinoides tenellus (Parker, 1967, 1973). Dextrally coiled Globorotalia limbata range up almost to the Pliocene/Pleistocene boundary in cores from the Atlantic (Parker, 1973) and the Pacific (Kaneps, 1973). Parker (1973) discussed the taxonomic problems of G. limbata and follows Blow's (1969) interpretation. This species is ancestral to Globorotalia multicamerata that Parker differs from G. limbata only when specimens with eight or more chambers in the final whorl occur. Kaneps (1973) followed the same species concept. Parker pointed out that G. limbata has been probably included with G. multicamerata by many authors including herself (Parker, 1967) in her study of Indo-Pacific cores.

The upper range of Pulleniatina primalis has been found to approximate the Pliocene/Pleistocene boundary (Hays et al., 1969; Saito et al., 1975). The extinction level of Globigerina decoraperta is useful for late Pliocene-early Pleistocene biostratigraphy, although, as pointed out by Parker (1967) the use of this species as a stratigraphic marker is restricted because of its nondescript character and because other species seem to include pseudomorphs of it.

All the above discussed species which appear to be important for latest Pliocene to earliest Pleistocene biostratigraphy, have been reported in DSDP Indian Ocean sections. Their levels of appearance or disappearance are reported on Figure 3 to 18 and their sequential order at the various sites will be discussed, first at low-latitude and second at mid-latitude sites.

Two other species of importance for latest Pliocene biostratigraphy are Globorotalia exilis and Globorotalia miocenica which are known to have become extinct at, or shortly before, the Pliocene/Pleistocene boundary in Atlantic sections (Berggren, 1973; Berggren and Amdurer, 1973; Parker, 1973; Briskin and Berggren, 1975; Saito et al., 1975). These two species, however, are confined to the Atlantic. Kaneps (1973) argues that they evolved independently in the Atlantic from G. limbata (G. exilis evolved via G. multicamerata and G. pertenuis) after complete emergence of the Isthmus of Panama. The forms referred to Globorotalia exilis in the Pacific by Jenkins and Orr (1972, their pl. 23) are probably referable to G. limbata. Parker (1973) pointed out that specimens identified by Blow (1969) and by Jenkins and Orr (1972) as "G. miocenica" from tropical Pacific sections are presumably referable to her "Globorotalia pseudomiocenica", a species very close to Globorotalia merotumeda and possibly a senior synonym of that species.

Only a few references were made to G. exilis and G. miocenica in the Indian Ocean sections, all of which are questionable. McGowran (1974) reported the top range of "G. exilis" in the basal Pleistocene of Site 214. However, Berggren et al. (1974) did not report this species in their study of the same section. Forms referred to G. cf G. exilis were noted by Berggren et al. (1974) in a Late Miocene interval at Site 216 and by Rögl (1974) and Olsson (1974) in isolated occurrences in the lower part of the Pleistocene section at Site 262. Rögl reported that these forms are left-coiling. Boltovskoy (1974a) reported the common occurrence in the Miocene and Pliocene sediments recovered during Leg 26 of forms which he tentatively ascribed to "? G. miocenica". He illustrated these forms and pointed out that they have characteristics, such as for example a smaller number of chambers in final whorl and a highly vaulted last chamber on the ventral side, which distinguish this species from G. miocenica. From an examination of Boltovskoy's illustrations (his Plate 12, Figs. 1-3) it appears that his forms may possibly be related to Globorotalia conomiozea and more specifically to the subspecies G. conomiozea subconomiozea described by Bandy (1975) from the Indian Ocean subtropical core V20-163. The statigraphic range reported by Boltovskoy for "? G. miocenica" supports this suggestion, which however, remains tentative pending the comparison of Boltovskoy's specimens with Bandy's. Akers (1974b) referred to G. cf G. miocenica sinistral forms from the Miocene assemblages of Site 254 which are within the range of the "? G. miocenica" of Boltovskoy. Akers also reported three isolated occurrences of dextral specimens referred to "G. miocenica" in the Miocene of Site 253 and in the Pleistocene of Site 251. Forms referred to "G. miocenica s.l." by Zobel (1974) at mid-latitude sites of Leg 25 may include forms that Boltovskoy called "? G. miocenica" and also some other members of the G. conoidea-G inflata lineage which, although common at these latitudes, were not identified by this author.

Pliocene/Pleistocene Boundary at Low-Latitude Sites of the Indian Ocean

The progressive development of G. truncatulinoides from G. tosaensis was described by Rögl (1974) in the expanded section of Site 262 in the Timor Sea. Rögl reported the occurrence of transitional forms in Cores 39-37, the separation between the two species being nearly impossible in Core 37, Section 3 where many specimens have a fully developed keel, except for the last chamber. Specimens become fully keeled in Core 36, CC, at 337.5 m where Rögl placed the base of Zone N 22 of Blow and of the G. truncatulinoides s.s. Zone of Bolli and Premoli Silva. This horizon, however, is stratigraphically higher than the first occurrence of G. truncatulinoides as reported in the same section by Olsson (1974) in Core 40, Section 5 at 372 m, and well above the sharp decrease in abundance of Discoaster brouweri reported by both Proto Decima (1974) and Bukry (1974e) in Core 44 at 407 m.

Calcareous plankton data from Site 262 show the difficulty involved in the exact location of the Pliocene/Pleistocene boundary, depending as it does on which criterion is used. Unfortunately, a comparison of calcareous to siliceous plankton data was not possible at this site because siliceous fossils occur commonly only in the upper 250 meters of the section above Core 27. The discrepancy between Rögl's and Olsson's locations of the first occurrence of G. truncatulinoides exemplifies the lack of agreement between various workers in the interpretation of this event.

In pelagic sections of the open Indian Ocean where the sedimentation rate is low (Sites 214, 219, 236, 237 and 238) forms transitional between G. tosaensis and G. truncatulinoides were reported in a short interval. At these sites the appearance level of G. truncatulinoides was delineated at an horizon which approximates the extinction level of Discoaster brouweri. In contrast in the Gulf of Aden, where rapid hemipelagic sedimentation has produced an expanded stratigraphic section (Sites 231, 232 and 233), the interval showing transitional specimens is very thick (up to 95 meters at Site 231, between Cores 3 and 12) and G. truncatulinoides with a fully developed imperforate keel occurs as rare and isolated specimens for an interval up to 60 meters below the level at which it becomes common. Because Vincent et al (1974) could not determine whether the presence of these single specimens was due to downhole contamination or whether the initial appearance of G. truncatulinoides occurs lower in the sections than the level where it becomes common, they placed the N 22/N 21 zonal boundary at the latter level. This horizon proved easy to recognize and correlates within the Gulf of Aden, but is significantly higher in the sections (up to 60 m) than the extinction level of Discoaster.

It is very probable that the level of first common occurrence of G. truncatulinoides in the Gulf of Aden is correlative with the level of first apparent evolutionary appearance in the Timor Sea as shown by indirect age estimates of these levels. G. truncatulinoides first appears in the Timor Sea (Rögl's level of evolutionary appearance, not Olsson's level of first occurrence) contemporaneously with a change in coiling of Pulleniatina. As mentioned above, this horizon is probably correlative with a similar coiling change in the Pacific, paleomagnetically dated at about -0.9 m.y. Assuming a constant sediment accumulation rate in the Gulf of Aden during the Pleistocene the base of common G. truncatulinoides in this basin is estimated to be about 0.8 to 0.9 m.y. old. The duration of approximately 1.0 m.y. between this event and the extinction of Discoaster brouweri is represented only by about 2 meters of sediments in deep-sea pelagic calcareous sequences typically accumulating at about 2 cm/1000 yrs (a thickness which might be not more than the sampling interval used for DSDP Initial Reports).

The extinction level of the radiolarian Pterocanium prismatium, which in the open-ocean sections coincides with the extinction level of Discoaster brouweri (Site 238) or is slightly above it (Sites 214 and 236), occurs in the expanded hemipelagic sections of the Gulf of Aden (Sites 232 and 233) midway between the levels of extinction of D. brouweri and of common occurrence of Globorotalia truncatulinoides.

The age estimate for the extinction of P. prismatium in the tropical to subtropical Indian Ocean, based on an assumed constant sedimentation rate in the Gulf of Aden and a -1.8 m.y. age for the extinction of D. brouweri, is approximately -1.3 m.y. The radiolarian species Lamprocyrtis haysi first appeared slightly later in the Gulf of Aden and in other areas of the western tropical Indian Ocean (Sites 235 and 238; Sanfilippo and Riedel, 1974) at about -0.8 - 0.9 m.y., almost contemporaneously with the first appearance of common G. truncatulinoides in the Gulf of Aden. The base of the new Quaternary radiolarian zone of Riedel and Sanfilippo (in press), the Lamprocyrtis haysi Zone, defined by the evolutionary appearance of this species appears, thus, to occur later in time than the Pliocene/Pleistocene boundary.

The diatom species Pseudoeunotia doliolus was found to be of no value in recognizing the Pliocene/Pleistocene boundary in the tropical Indian Ocean because its level of first appearance (which defined the TID 5/TID 6 zonal boundary) varies with respect to the extinction of the radiolarian Pterocanium prismatium. It first appears 4 meters above the last occurrence of P. prismatium at Site 213, but well below it (11 and 22 meters, respectively) at Sites 215 and 238, being well below the extinction level of Discoaster brouweri at the latter site (Shrader, 1974; Johnson, 1974; Sanfilippo and Riedel, 1974).

In pelagic sections of the open ocean the extinctions of Globigerinoides fistulosus and Globigerinoides obliquus were recorded at the same level (or very close to each other) very slightly above the Pliocene/Pleistocene boundary at Sites 214, 219 and 238. In the expanded sections of the Timor Sea and the Gulf of Aden, these two foraminiferal events occur well above the boundary. At Site 262 G. fistulosus disappears later than G. obliquus simultaneously with the appearance of Globorotalia truncatulinoides (as reported by Rögl, 1974). In the Gulf of Aden G. obliquus and G. fistulosus are partially eliminated by dissolution; their uppermost occurrences, however, are recorded at the level where G. truncatulinoides becomes common. In the DODO cores studied by Parker (1967) G. fistulosus last occurs in the uppermost Pliocene and G. obliquus at a lower level in the late Pliocene. G. tenellus first appears at the Pliocene/Pleistocene boundary at Sites 219 and 238 and well above the boundary at the three Gulf of Aden sites at the level where G. truncatulinoides becomes common. This species was not reported at Sites 214 and 262 in the eastern Indian Ocean. In DODO cores it is present only in the Pleistocene.

At the western Indian Ocean sites drilled during Legs 23 and 24 Globorotalia limbata and G. multicamerata were identified by Fleisher (1974) and Vincent et al. (1974) (who compared their specimens) according to Parker's (1973) criterion. Right-coiling G. limbata, a common component of the Pliocene fauna, disappears simultaneously with Globigerina decoraperta, very slightly below the Pliocene/Pleistocene boundary at Sites 219, 236, 237 and 238, and G. multicamerata, which is rare, disappears slightly below that level. In the Gulf of Aden the last occurrence of right-coiling G. limbata approximates very closely the Pliocene/Pleistocene boundary at Sites 231 and 233 and is slightly above at Site 232. G. multicamerata was not reported in the Gulf of Aden except as a sole occurrence in the lower part of the Pleistocene section at Site 231. The last occurrence of Globigerina decoraperta is well above the Pliocene/Pleistocene boundary at the Gulf of Aden sites, slightly below the level where Globorotalia truncatulinoides becomes common. At the eastern low-latitude sites of the Indian Ocean G. limbata was not reported. Forms referred to G. multicamerata by Rögl (1974) and Berggren et al. (1974) disappear very slightly below the Pliocene/Pleistocene boundary at Site 214 and slightly above it at Site 262. It is possible that they (reported by Rögl as being right coiling at Site 262) include the forms like those referred to as G. limbata by by Fleisher (1974) and Vincent et al. (1974) in the western part of the ocean. Parker (1973) pointed out that she included G. limbata with G. multicamerata in her 1967 study of the Indian Ocean cores. Her youngest "G. multicamerata" occurs high in the upper Pliocene of Cored DODO 57P and DODO 117P (the part of the latter core attributed to the mid-Pliocene Zone N 20 by Parker is reassigned to the late Pliocene Zone N 21, see below). Globigerina decoraperta was not reported at the two eastern Sites 214 and 262. In Parker's cores this species was noted only in DODO 57P where it last occurs in latest Pliocene.

The subspecies Pulleniatina obliquiloculata praecursor was identified in low-latitude sections only at Site 219 (Fleisher,

1974) where it became extinct simultaneously with Globorotalia limbata and Globigerina decoraperta at the Pliocene/Pleistocene boundary. Fleisher reported the last occurrence of P. primalis at this site much lower, in early Pliocene. It is probable that the forms attributed to P. obliquiloculata praecursor by Fleisher were included with P. primalis by other authors as they were by Parker (1967), who included "P. obliquiloculata praecursor" with P. primalis because it has no "linear suture". The progressive transition of "P. obliquiloculata primalis" to "P. obliquiloculata obliquiloculata" was noted by Rögl at Site 262, in Core 44, slightly above the Pliocene/Pleistocene boundary. P. primalis was reported in the late Pliocene at Site 214 by Berggren et al (1974) as well as in Cores DODO 117P and DODO 57P by Parker (1967). At sites drilled on Leg 24, Vincent et al. (1974) included P. primalis with P. obliquiloculata s.l.

In the Gulf of Aden, Vincent et al. (1974) reported at all three sites 231, 232 and 233 the sole occurrence of Globoquadrina sp. A, a form morphologically intermediate between Globoquadrina and Pulleniatina, simultaneously with the disappearance of Globorotalia limbata at what closely approximates the Pliocene/Pleistocene boundary. Vincent et al. suggested that this morphotype may be an abnormal form developed in response to special oceanographic conditions in the Gulf of Aden at this time.

Pliocene/Pleistocene Boundary at Mid- and High-Latitude Sites of the Indian Ocean

In mid-latitudes sections, Boltovskoy (1974a) placed the N 22/N 21 zonal boundary at a level where the percentages of Globorotalia tosaensis and G. truncatulinoides reach 50%, and placed the Pliocene/Pleistocene boundary on the basis of this criterion. At the eastern Sites 258, 254 and 253 this horizon is significantly above the Pliocene/Pleistocene boundary based on nannofossil data, whereas the first occurrence of G. truncatulinoides approximates the boundary so defined (Figs. 14, 15, 16). In the expanded sections of the western Sites 250 and 251 the first occurrence of G. truncatulinoides reported by Boltovskoy is well above the Pliocene/Pleistocene boundary defined by the nannofossils. Akers (1974b) placed the level of first occurrence of G. truncatulinoides 13 meters lower than did Boltovskoy at Site 251. The discrepancy between these interpretations of the initial appearance of G. truncatulinoides illustrates again the difficulty of its identification.

No calcareous nannoplankton and siliceous fossil data are available at subantarctic Site 281 to substantiate the Pliocene/Pleistocene boundary which was placed at the first occurrence of G. truncatulinoides. At Antarctic Site 265 the Pliocene/Pleistocene boundary is defined in Core 10 by the extinction level of the radiolarian Clathrocylcas bicornis known to occur at the base of the Olduvai event in paleomagnetically dated cores (Hays and Opdyke, 1967). This horizon falls below the interval (Cores 5 to 9) assigned to the nannofossil Zone NN 19 (Site Report in Hayes et al., 1975, p. 62-64) and within the Globorotalia puncticulata Zone of Kennett (1970) assigned by him to the interval between Cores 7 to 13 (the Pliocene/Pleistocene boundary was erroneously reported as lying between Cores 4 and 5 in the site report in Hayes et al., 1975, p. 54). G. truncatulinoides, which has been shown to migrate into the Antarctic region only 300,000 years ago (Kennett, 1970) first appears much higher in Core 3 within nannofossil Zone NN 20 (assigned to

Cores 2 through 4, Site Report in Hayes et al., 1975, p. 60-61).

Globigerinoides fistulosus and G. obliquus last occur slightly above the Pliocene/Pleistocene boundary at the subtropical Site 253 as they do in low latitude sections. The highest occurrence of G. obliquus is also slightly above the boundary at Site 254, but at the three higher latitude Sites 258, 264 and 251 is at or below the boundary. G. fistulosus was not noted at sites south of Site 253.

The extinction level of Globorotalia limbata approximates the Pliocene/Pleistocene boundary in mid-latitudes as well as in low-latitudes. It coincides with the boundary at Site 253, is very slightly above at Site 258 and 264 and slightly below at Site 251. Pliocene G. limbata are dextrally coiled, as in low-latitude assemblages. G. limbata was identified in mid-latitude assemblages by Boltovskoy (1974a) at sites of Leg 26 and by Kaneps (1975b) at Site 264, following Parker's (1973) criterion and, therefore, in the same manner as Fleisher and Vincent et al., (1974) in tropical sections. Illustrations of the taxon from these various authors compare well. Akers (1974b), who studied the same sequences as Boltovskoy, did not report G. limbata and probably referred this taxon to "dextral G. cultrata", which according to Akers ranges up into the uppermost Pliocene at Site 258. Kennett did not report G. limbata at Site 264. G. multicamerata was not reported at mid-latitude sites.

According to Boltovskoy (1974a) the youngest occurrence of Globigerina decoraperta approximates the Pliocene/Pleistocene boundary at Site 254, as at lower latitudes, but is stratigraphically much lower in the Pliocene at Sites 251, 253 and 258. Kennett (1975) also noted the last occurrence of this species at Site 264 within the Pliocene. A discrepancy occurs between the ranges of this taxon as reported by Boltovskoy (1974a) and Akers (1974b). Akers did not report G. decoraperta at Sites 251 and 253 and noted its occurrence at Sites 254 and 258 with a much greater abundance in the Pliocene assemblages than reported by Boltovskoy. Also, according to Akers G. decoraperta ranges up into the Pleistocene at Site 258.

Pulleniatina is absent at Sites 251 and 254 and rare at Sites 253 and 258. At the two latter sites, however, events within the Pulleniatina group cannot be evaluated because of the apparent confusion in the taxonomic interpretation of various species and subspecies of this genus. At Site 253 Boltovskoy refers to "P. primalis" forms ranging up to the Pliocene/Pleistocene boundary, to be succeeded by "P. obliquiloculata praecursor" in the Pleistocene (Akers reported only the presence of Pleistocene "P. obliquiloculata" at this site). At Site 258 Boltovskoy did not report "P. primalis" but reported the simultaneous occurrence of "P. obliquiloculata s.s." and "P. obliquiloculata praecursor" throughout the Pliocene and Pleistocene (Akers reported only scattered rare occurrences of Pleistocene "P. primalis" at this site). Kaneps noted "P. praecursor" just above the Pliocene/Pleistocene boundary at Site 264. Globorotalia limbata, Globigerina decoraperta and Pulleniatina were not reported by Jenkins (1975) at high latitude Site 281.

Pliocene

Blow's Zonation

The N 21/N20 zonal boundary of Blow (1969) which approximates the Upper/Lower Pliocene boundary is defined by the initial appearance of Globorotalia tosaensis. The successive evolutionary development of _G. tosaensis_ from _G. crassaformis_ was observed in the subtropical Indian Ocean in the lower part of the Gauss interval of Core V 20-163 (Fig. 20). Typical _G. tosaensis_ appears first in the core just below the Mammoth event, very slightly above the first appearance of _Globigerinoides fistulosus_, and below the extinction levels of _Sphaeroidinellopsis_ spp. and _Globoquadrina altispira_ (Hays et al., 1969; Bandy, 1975). Nannofossil data from this core indicate that this foraminiferal event occurs within Zone NN 15 (Gartner, 1973).

At none of the DSDP Indian Ocean sites is _G. tosaensis_ a common component of the fauna (except in the lower Pleistocene assemblages of Site 253 on the southern Ninetyeast Ridge). In low-latitude sections all authors draw the base of Zone N 21 at the first occurrence of _G. tosaensis_, except at Site 232 in the Gulf of Aden where because of the scarcity of the species the base of this zone was tentatively placed by Vincent al. (1974) on other faunal criteria. In mid-latitude sections Boltovskoy (1974a) draw the base of N 21 at Sites 258 and 254 at a level stratigraphically higher than the lowermost _G. tosaensis_ and did not report the presence of this species at Site 251, although it was reported by Akers (1974b). At Site 253 the Upper Pliocene is very condensed; _G. tosaensis_ first appears at the same level as _G. truncatulinoides_ at the base of the Pleistocene section and Zone N 21, therefore, is not represented. At all sites the first occurrence of _G. tosaensis_ is stratigraphically higher with respect to the nannofossil zonation than it is in Core V20-163. It lies at the NN 16/NN 15 zonal boundary at Sites 237, 258 and 264, at the NN 17/NN 16 zonal boundary at Sites 214 and 219, and within Zone NN 16 at other sites. Its level is usually above the first occurence of _Globigerinoides fistulosus_ and is inconsistent with respect to the extinction levels of _Globoquadrina altispira_ and _Sphaeroidinellopsis_ spp., being either coincidental with these two events, below, or above. _G. tosaensis_ was not reported at high-latitude Site 281.

No attempt was made by most authors to differentiate Zones N 20 and N 19. In addition to the taxonomic difficulty in identifying the index species _Turborotalia pseudopima_, whose initial appearance was reported by Blow (1969) to mark the boundary between these two zones, this species apparently evolved earlier than first reported by Blow (Brönnimann and Resig, 1971). _T. pseudopima_ was identified only at low-latitude Sites 214 and 219 in the upper half of these Pliocene sections. At the mid-latitude Site 251, Boltovskoy (1974a) reported forms referred to _Globorotalia_ cf. _G. pseudopima_ as ranging throughout the Upper Miocene and Pliocene.

The Upper/Lower Pliocene boundary is placed in this report on nannofossil data at the base of NN 16.

Foraminiferal Events Calibrated to the Paleomagnetic Time Scale

Planktonic foraminiferal events other than the ones used by Blow for his zonation appear to be more useful for the subdivision of the Pliocene. A number of these events have been calibrated to the paleomagnetic scale in tropical and subtropical deep-sea cores of various oceans (Hays et al., 1969; Saito et al., 1975; Bandy, 1973, 1975). These events include by order of increasing age: 1) the extinction of _Globoquadrina altispira_ and _Sphaeroidinellopsis_ spp. accompanied by an abrupt upward increase in abundance of _Sphaeroidinella dehiscens_ at about -2.9 m.y. (at the top of the Mammoth event); 2) the extinction of _Globorotalia margaritae_ at about -3.4 m.y. (at the Gauss/Gilbert boundary); 3) a sharp left- to right-coiling change

in Pulleniatina at about -3.6 m.y. (just above the Gilbert A event);
4) the extinction of Globigerina nepenthes at about -3.7 m.y. (at the top of the Gilbert A event); 5) the initial appearance of Sphaeroidinella dehiscens at about -4.8 m.y. (midway between Epoch 5 and Gilbert C2 event; 6) the initial appearance of Globorotalia tumida at about -4.9 m.y. (shortly after Epoch 5); and 7) the extinction of Globoquadrina dehiscens at about -5.0 m.y. (just after Epoch 5). The latter event coincides with the Miocene/Pliocene boundary which has been estimated to be 5.0 to 5.1 m.y. old (Berggren, 1973; Berggren and Van Couvering, 1974; Saito et al., 1975). The four youngest events mentioned above have been observed in paleomagnetically dated cores from all oceans (Pacific: V24-59 and RC12-66; Indian: V20-163; Atlantic: RC11-252) whereas the three oldest ones have been calibrated in one Pacific core only (RC12-66). The stratigraphic position of these events at the various Indian Ocean sites (except for coiling changes in Pulleniatina, which were not investigated), their sequential order, and their relationships with paleontological events of of other fossil groups also calibrated to the paleomagnetic scale are discussed below.

Extinction of Globoquadrina altispira. The extinction level of Globoquadrina altispira occurs within the Discoaster surculus Zone of Gartner and of Martini (NN 16) and approximates the Discoaster tamalis/D. pentaradiatus subzonal boundary of Bukry at the low latitude Sites 214, 219, 232 and 238. Its position falls slightly higher with respect to the nannofossil zonation at Sites 231 and 233 within the Discoaster pentaradiatus of Bukry. The G. altispira extinction level approximates or is slightly below the extinction level of the radiolarian species Stichocorys peregrina at Sites 214, 233 and 238 in good agreement with the paleomagnetic age of the latter event dated at about -2.6 m.y. (Theyer and Hammond, 1974a). At Site 237 the top of G. altispira lies just below an unrecovered interval which occurs at the Pliocene/Pleistocene boundary and precludes the interpretation of latest Pliocene events, but its position with respect to the nannofossil and radiolarian zonations is in agreement with its position at the other sites. The occurrence of reworked G. altispira was observed just below the Pliocene/Pleistocene boundary at Sites 236 and 238. The reworking of lower Pliocene sediments into upper Pliocene (reworking was observed in other fossil groups as well, Fig. 9) is relatively easy to identify in the somewhat expanded section of Site 238 on the Central Indian Ridge where the reworked and in situ Early Pliocene faunas (with G. altispira) are distinctly separated by an interval of Late Pliocene sediments (without G. altispira). However, such is not the case at Site 236 in the Somali Basin where the greatly condensed Pliocene section makes interpretation of the mixed faunas difficult. It is probable that the top of the continuous and common occurrence of G. altispira in the lower part of Core 4 (Fig. 7) represents the real extinction level of this species at the latter site. At Site 262 in the Timor Sea, G. altispira occurs only in the lowermost part of the section. The top of its short range occurs 20 cm below the base of Globorotalia tosaensis and Globigerinoides fistulosus. The occurrence of these three foraminiferal events only slightly below the Pliocene/Pleistocene boundary suggests a telescoping of the latest Pliocene at this site, a suggestion supported by the apparent condensing of latest Pliocene of other eastern Indian Ocean sections (Sites 253, 258, 264 and Core DODO 57P). The interval in which these three foraminiferal events occur, however, was attributed to the latest

Pliocene nannofossil Zone NN 18 by Proto Decima (1974), a zonal assignment which does not agree with the mid-Pliocene age indicated by the foraminifera. The discrepancy between the nannofossil and foraminiferal data may result from sediment mixing. Shallow water conditions at Site 262 at the time of deposition of the lower portion of the sequence make the latter poorly suited for planktonic biostratigraphic investigation of the Pliocene.

In mid-latitude sections G. altispira is not as common as in those of low latitudes. Its level of extinction at the subtropical to temperate Sites 253 and 258 occurs within nannofossil Zone NN 16, as at tropical sites, and within an undifferentiated NN18-NN 16 interval at Site 264. At the temperate Site 251 G. altispira last occurs in a much lower stratigraphic position within the lowest Pliocene nannofossil zone NN 13. Boltovskoy did not report the presence of G. altispira at Site 254. Akers, however, reported this species at the latter site and placed the top of its range in the Middle Miocene within an interval attributed to the nannofossil zones NN 9 to NN 6. G. altispira was not reported at high-latitude Site 281.

The position of the first occurrence of Globorotalia tosaensis with respect to the extinction level of G. altispira varies from site to site. It is above the latter at Sites 214, 219 and 232, and below it at Sites 231, 233, 236 and 237. The two levels coincide at Site 238 as well as in Core DODO 57P.

Extinction of Sphaeroidinellopsis spp. In low-latitude sections the highest occurrence of Sphaeroidinellopsis spp. coincides with the top of G. altispira at Sites 232, 237 and 238, as well as in Core DODO 57P, but is slightly higher at Site 219 and slightly lower at Site 214. In the Gulf of Aden occurrences of Sphaeroidinellopsis spp. are very rare, probably because of the elimination of these forms by the pronounced amount of dissolution in this basin. As a result Sphaeroidinellopsis spp. were not found at Site 233 and their last occurrence at Site 231 is much lower than elsewhere, in the Upper Miocene. At the mid-latitude Sites 253, 254 and 258, the highest occurrence of Sphaeroidinellopsis spp. occupies a position relative to the nannofossil zonation similar to that in low-latitudes (within NN 16) and is slightly lower at Site 264 (just below the NN 16/NN 15 boundary). This event, however, is much lower at the higher latitude Site 251 where it falls within the Lower Pliocene zone NN 13. Sphaeroidinellopsis was not reported at high-latitude Site 281. The top range of Sphaeroidinellopsis is almost coincidental with the top of G. altispira at Site 253 but below the latter at Sites 258 and 264.

First Appearance of Globigerinoides fistulosus. The initial appearance of Globigerinoides fistulosus occurs in the lower Gauss just below the Mammoth event in the Indian Ocean Core V20-163 and at the top of this event in the Atlantic Core RC11-252. It occurs higher in the Pacific Core V24-59 where the species is probably eliminated in the lower part of its range by dissolution. The initial appearance of G. fistulosus was used as a mid-Pliocene marker in the tropical Pacific by Jenkins and Orr (1972) as well as in the tropical Atlantic by Parker (1973). This foraminiferal event is consistently recorded below the extinction of Globoquadrina altispira and above the extinction of Globorotalia margaritae at Sites 214, 219, 237, 238 and 253 in an interval spanning the Discoaster tamalis nannofossil subzone of Bukry (= early NN 16). This position is in good agreement with data from Core V20-163. The initial occurrences of Globigerinoides fistulosus and Globorotalia tosaensis are simultaneous at 237, while G. tosaensis first appears later than G. fistulosus at Sites 214, 219, 253 and in Cores DODO 57P and 117P. G. fistulosus is presumably eliminated by dissolution from the Pliocene sediments of the Gulf of Aden and at Site 236 in the Somali Basin where it is found in rare occurrences only in the uppermost Pliocene. This species was not reported at sites south of Site 253, which are probably outside of its latitudinal range.

Extinction of Globorotalia margaritae. Globorotalia margaritae is a rare component of low-latitude faunal assemblages of the Indian Ocean. Its extinction level, however, appears to be consistent relative to other fossil zonations and typically occurs close to the level where it has been recorded in paleomagnetically dated cores. It occurs within the Reticulofenestra pseudoumbilica nannofossil Zone of Bukry (in its upper part, i.e. in the Discoaster asymmetricus subzone where this subzone was differentiated) at Sites 214, 219, 237 and 238 and approximates the extinction level of the radiolarian species Spongaster pentas at Sites 214 and 238 (slightly above the latter at Site 214 and just below it at Site 238). The extinction level of S. pentas has been recorded in paleomagnetically dated cores of the equatorial Pacific at about -3.9 m.y. in the upper Gilbert Epoch about midway between the top of Gilbert Event A and the base of the Gauss (Theyer and Hammond, 1974a). In the Gulf of Aden G. margaritae was eliminated by dissolution and reported only once at Site 232 within the Reticulofenestra pseudoumbilica Zone of Bukry and within the lower Spongaster pentas Zone not far below the top range of S. pentas. The low stratigraphic occurrence of Globrotalia margaritae at Site 236 probably results from elimination by dissolution in its upper range. G. margaritae was not reported by Parker (1967) in DODO cores and was reported at Site 262 in the Timor Sea only as rare, probably reworked, specimens in the lower Pleistocene (Rögl, 1974).

In mid-latitude sections G. margaritae is a rare element of the faunal assemblages (except at the subtropical Site 253 where it is common) and disappears at the same biostratigraphical level relative to the nannoplankton zonation as in low-latitudes. Its upper occurrences are within NN 15 at Sites 253, 251 and 264, and within undifferentiated intervals between NN 16 and NN 14 and between NN 15 and NN 13 at Sites 254 and 258 respectively. At the latter site it disappears slightly below the initial appearance of Globorotalia inflata. At high-latitude Site 281, rare forms referred to G. cf. G. margaritae by Jenkins last occur slightly below the appearance of G. inflata.

Extinction of Globigerina nepenthes. Globigerina nepenthes is often a common component of the Middle Miocene faunas in the Indian Ocean but becomes rare in the upper part of its range. At the low-latitude sites the uppermost occurrence of this species was noted in the lower Reticulofenestra pseudoumbilica Zone of Bukry at Sites 219 and 237 and at the base of this zone at Site 214. These positions are in fairly good agreement with data from paleomagnetically dated cores. G. nepenthes disappears much lower at Sites 238 and 236, in lowermost Pliocene. The low disappearance at the deeper Site 236 probably results from dissolution. The species was apparently totally eliminated by dissolution from the Pliocene sediments of the Gulf of Aden. At the mid-latitude sites, G. nepenthes disappears lower in the sections than at low-latitudes, in the upper Miocene (except at Site 264). It last occurs in basal NN 13 at Site 253, within an unzoned interval between NN 14 and NN 11 at Site 254, within NN 12 at Site 251 and within an undifferentiated NN 11 - NN 10 interval at Site 258. G. nepenthes, however, was noted by Kennett (1975) and Kaneps (1975b) in the lower Pliocene at Site 264 where it disappears almost simultaneously with Globorotalia margaritae. It was not reported from high-latitude Site 281.

Initial Appearance of Sphaeroidinella dehiscens. The first appearance of the genus Sphaeroidinella is marked by individuals exhibiting a small, dorsal, sutural, supplementary aperture(s). The first appearance of these primitive forms (i.e. Sphaeroidinella dehiscens forma immatura Cushman) defines the base of the Early Pliocene Zone N 19 of Blow and has been referred to as the "Sphaeroidinella dehiscens Datum" by Blow (1969), Berggren (1969, 1973) and Hays et al., (1969). The first occurrence of S. dehiscens forma immatura is recorded in the basal Pliocene in the Mediterranean area, 40 feet above the base of the Trubi Marl in Sicily (Blow, 1969, p. 418) and within the basal Pliocene Sphaeroidinellopsis Acme Zone (e.g. Cita, 1973; Mazzola, 1971). This event, which has been paleomagnetically dated at -4.8 m.y. (Saito et al., 1975; Berggren, 1973), is therefore slightly younger than the Miocene/Pliocene boundary. It has, however, been equated to the "Miocene/Pliocene boundary" by many workers (e.g. in the Indian Ocean: Parker, 1967; Srinivasan and Srivastava, 1974, 1975).

Berggren and Poore (1974) in a study subsequent to the DSDP Leg 22 Initial Reports have examined in detail the S. dehiscens bioseries in the sediments of Site 214 on the Ninetyeast Ridge. They observed the evolution of Sphaeroidinellopsis subdehiscens to Sphaeroidinella dehiscens forma immatura by the development of a discrete, supplementary aperture on the spiral side within Core 10. This level approximates the Spongaster pentas/Stichocorys peregrina radiolarian zonal boundary, the Ceratolithus rugosus/C. tricornulatus (NN 13/NN 12) zonal boundary of Gartner and the Ceratolithus acutus/Triquetrorhabdulus rugosus subzonal boundary of Bukry.

A gradual increase in the size of Sphaeroidinella dehiscens, as well as a gradual enlargement of the secondary aperture, took place during the Early Pliocene (Blow, 1969). The aperture finally is large, extensive and fringed by a marked everted flange which coalesces with the apertural rim around the primary aperture. This morphologic step (which represents the "Sphaeroidinella dehiscens Datum" of Bandy, 1963, 1964) has been paleomagnetically dated -3.0 m.y., slightly prior to the extinction of Sphaeroidinellopsis spp. at -2.0 m.y. (Hays et al., 1969). Berggren and Poore observed the gradual development of Sphaeroidinella dehiscens forma immatura to typical S. dehiscens exhibiting a well developed flanged supplementary aperture in the Upper Pliocene sediments of Site 214 in the interval from Cores 9 to 6 and reported the appearance of the typical forms in Core 5. This level is slightly above the top of G. margaritae and slightly below the top of Sphaeroidinellopsis spp. (noted in Core 5, Section 4 by McGowran, 1974). It also lies in the upper Spongaster pentas radiolarian Zone, shortly above the last occurrence of S. pentas (paleomagnetically dated in the equatorial Pacific at -3.9 m.y. by Theyer and Hammond, 1974a), and near the Discoaster surculus/Reticulofenestra pseudoumbilica (= NN 15/NN 16) zonal boundary of Gartner. These relationships are in agreement with data from the equatorial Pacific.

The evolutionary development of S. dehiscens from S. dehiscens forma

immatura was not observed at other sites of the Indian Ocean and the
first appearance of S. dehiscens was reported at various biostratigraphic levels. At the tropical western Indian Ocean sites (Legs 23
and 24) the lowest specimens of S. dehiscens were significantly higher
in the sections with respect to other paleontological events than at
Site 214. Because of the delayed appearance of this zonal marker,
Vincent et al. (1974) tentatively placed the base of N 19 at the top
of the Globorotalia plesiotumida range in the sections recovered during
Leg 24. The high stratigraphic position of the first appearance of S.
dehiscens in the Gulf of Aden probably is due to solution of calcium
carbonate lower in the section. The same is true possibly at the
deeper Site 236 in the Somali Basin also but not at the relatively
shallower Sites 237 and 219 on the Mascarene Plateau and Chagos
Laccadive Ridge respectively. The base of S. dehiscens is slightly
below the top of Globigerina nepenthes at Site 237, slightly above
the top of Globorotalia margaritae at Site 219 and synchronous with
the top of Sphaeroidinellopsis spp. at Site 236. From the sequence
of paleontologic events at these sites, discussed above, it appears
that S. dehiscens first occurs in the western tropical Indian Ocean
3.0 to 3.7 m.y. ago. It is of interest that Bandy (1973, 1975), in
a careful restudy of the Indian Ocean Core V20-163 (initially studied
by Hays et al., 1969 and Saito et al., 1975), reported that primitive
S. dehiscens first appear in this core within the upper Gilbert at
about -3.5 m.y. (Fig. 20; a few scattered specimens of fully developed
S. dehiscens were found by Bandy below that level only in contaminated
samples from the outer margin of the core which had been carefully
separated from the center portion: Bandy, personal communication).
Variations in time of the initial appearance of S. dehiscens in
tropical regions of the Pacific Ocean were pointed out by Theyer and
Hammond (1974a). In the latter area at DSDP Sites 62, 71, 72 and
possibly 63, the base of foraminifera Zone N 19 correlates with a
level above the Spongaster pentas first appearance datum (about -4.6
m.y.) but at Site 73 the base of Zone N 19 falls below it (Bronniman
et al., 1971; Tracey et al., 1971). Discrepancies in the stratigraphic
position of the "Sphaeroidinella dehiscens datum" may result from the
difficulty involved in the recognition of S. dehiscens forma immatura,
as pointed out by Bronnimann and Resig(1971), Berggren (1973), Parker
(1973) et al. However, the possibility that this level is timetransgressive within tropical regions of the oceans should not be
excluded. This species, according to the present-day occurrence in
surface sediments, is one of the most limited latitudinally, as is
also true in the Pacific plankton (Bradshaw, 1959).

A delayed first appearance of S. dehiscens in subtropical to temperate latitudes of the Pacific has been reported in the south part
of this ocean (Kennett, 1973) as well as in the north (Vincent, 1975).
This appears to be also the case for the subtropical to temperate
area of the Indian Ocean. Sphaeroidinella dehiscens, which is a
minor constituent of the Pliocene fauna in this area, first appears
above the extinction level of Globorotalia margaritae at Sites 253
and 258 (within nannofossil Zone NN 16) and slightly below it at
Site 254 (within nannofossil Zone NN 14), and as high as in the
Pleistocene at Site 264. Surprisingly it appears lower at the
higher-latitude Site 251 within the early Pliocene nannofossil Zone
NN 13. No S. dehiscens was noted at high-latitude Site 281.

The two oldest Pliocene foraminiferal events calibrated to the
paleomagnetic time-scale in the equatorial Pacific, i.e. the initial
appearance of Globorotalia tumida and the disappearance of Globoquadrina dehiscens, both of which approximate the Miocene/Pliocene

boundary at about -5.0 m.y., will be discussed later in the discussion of this boundary.

Age Assignment of Cores DODO 117P and MSN 56P

It is suggested that the bottom Core of DODO 117P, located on the Central Indian Ridge is Upper Pliocene above the extinction level of G. altispira and Sphaeroidinellopsis (Fig. 19), and not Lower Pliocene as indicated by Parker (1967), and that the stratigraphically high first occurrence of G. tosaensis in this core, as at a number of DSDP sites, does not delineate the base of Zone N 21. This age assignment is supported by the presence of G. fistulosus throughout the bottom part of the core and by the absence of G. altispira and Sphaeroidinellopsis throughout the section, even though the faunal assemblages are tropical to subtropical in nature, as evidenced by the common occurrence of Spheroidinella dehiscens and G. fistulosus (samples from the bottom part of the core were reexamined by the author). It is also suggested that the interval of Core MSN 56P between 129 and 202 cm, which contains G. fistulosus and is above the last occurrence of G. altispira, is Upper and not Lower Pliocene (Fig. 19).

Andaman-Nicobar Islands

The late Neogene biostratigraphic sequence of the Andaman-Nicobar Islands (Fig. 21) in the Bay of Bengal, described by Srinivasan and Srivastava (1975), may be reevaluated in view of the foraminiferal events discussed above. Globigerina nepenthes, Globorotalia margaritae and Globoquadrina altispira disappear simultaneously in the Andaman-Nicobar Islands, whereas in continuous sequences elsewhere they disappear in sequential order. It is therefore suggested that a telescoping of part of the Pliocene occurs in the Andaman region, as well as in many other areas of the Indian Ocean (especially in the eastern part), which curtails the upper range of these species. In the Andaman-Nicobar Islands, Sphaeroidinellopsis spp. range upward for a short distance above the three other species. Although Sphaeroidinellopsis spp. typically disappear at the same level as G. altispira in tropical regions, they have been observed to range slightly higher in the Arabian Sea (Site 219) in the northern Indian Ocean, at the same latitude as the Andaman region. It appears, thus, that the Globorotalia multicamerata-Pulleniatina obliquiloculata Zone of Srinivasan and Srivastava above the range of G. altispira should be attributed a Late Pliocene age (N 21) rather than the Early Pliocene age (N 20) assigned by these authors. The concurrent range of Globigerinoides fistulosus, Globigerina nepenthes and Globorotalia margaritae in the interval spanning the Globigerina nepenthes, Globorotalia tumida flexuosa and Sphaeroidinella dehiscens Zones of Srinivasan and Srivastava (equated by these authors to the Lower Pliocene Zone N 19) is puzzling. The first occurrence of G. fistulosus has been shown to be a reliable biostratigraphic marker within the latitudinal range of the species, being sometimes artificially too high as a result of solution. If the earliest G. fistulosus in the Andaman-Nicobar Islands is not displaced, it would thus indicate a maximum age of about -3.1 m.y. (age of the initial appearance of this species just below the Mammoth event in Core V20-163) for the level which marks the boundary between the S. dehiscens and the Globorotalia tumida-Sphaeroidinellopsis subdehiscens Zones of Srinivasan and Srivastava (which the latter authors equated to the

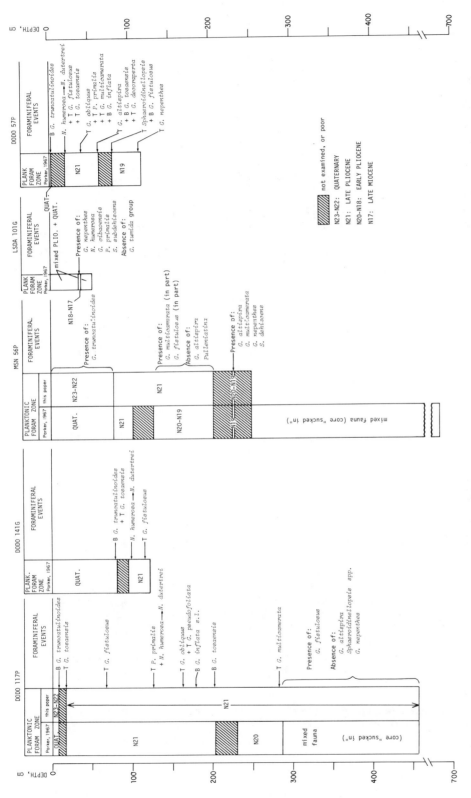

Fig. 19. Planktonic foraminiferal biostratigraphy of DODO, MONSOON and LUISIAD cores (data from Parker, 1967).

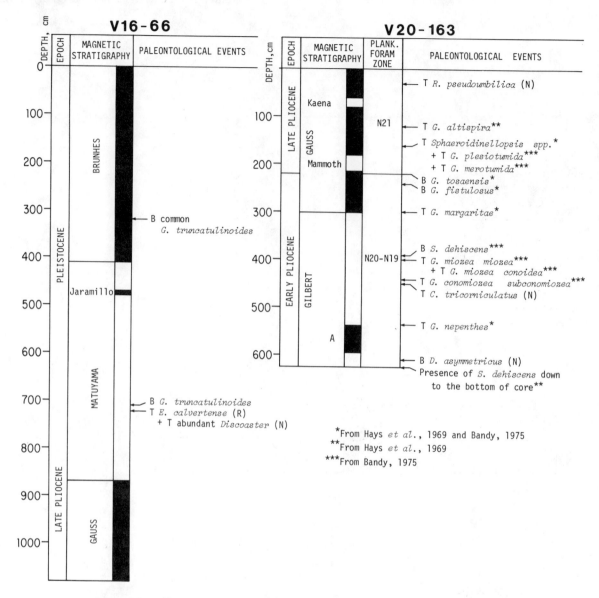

Fig. 20. Planktonic biostratigraphy of Cores V16-66 and V20-163. The magnetic stratigraphy is from Glass et al. (1967) for Core V16-66 and from Opdyke and Glass (1969) for Core V20-163. Paleontological data for Core V16-66 are from Ericson et al. (1963), Hays (1965), Glass et al. (1967), Berggren (1967) and Bandy et al. (1971). Nannofossil data for Core V20-163 are from Gartner (1973). (N): nannofossil; (R): radiolarian; other paleontological events are foraminiferal. T: top of range; B: bottom of range.

N 19/N 18 zonal boundary and to the "Miocene/Pliocene boundary"). This mid-Pliocene age would not be incompatible with the first appearance of Shaeroidinella dehiscens at this level since this event appears to be delayed in other areas of the Indian Ocean (including the Arabian Sea at the same latitude of the Indian Ocean), but is contradicted by the presence of G. nepenthes and G. margaritae which, if these species are in place, indicate an age older than

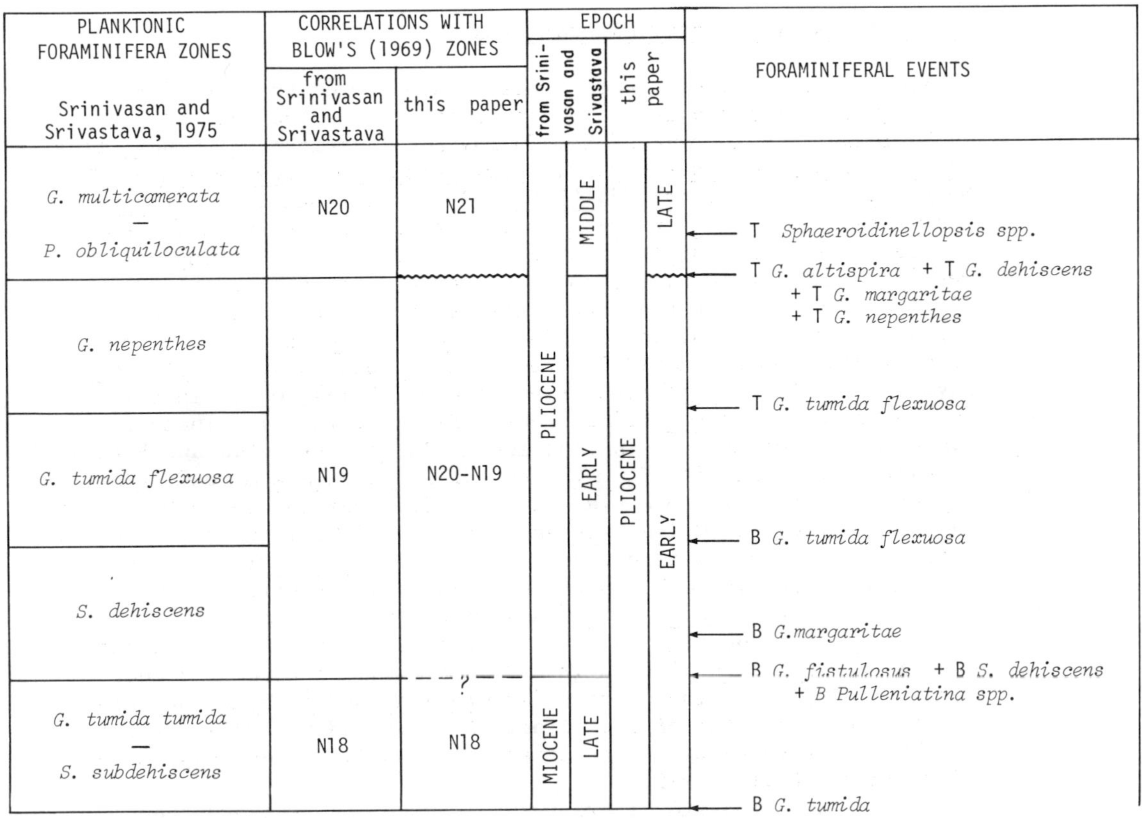

PLANKTONIC FORAMINIFERA ZONES Srinivasan and Srivastava, 1975	CORRELATIONS WITH BLOW'S (1969) ZONES		EPOCH		FORAMINIFERAL EVENTS
	from Srinivasan and Srivastava	this paper	from Srinivasan and Srivastava	this paper	
G. multicamerata — P. obliquiloculata	N20	N21	PLIOCENE MIDDLE	LATE	← T Sphaeroidinellopsis spp. ← T G. altispira + T G. dehiscens + T G. margaritae + T G. nepenthes
G. nepenthes	N19	N20-N19	PLIOCENE EARLY	PLIOCENE EARLY	← T G. tumida flexuosa ← B G. tumida flexuosa
G. tumida flexuosa					
S. dehiscens					← B G. margaritae ← B G. fistulosus + B S. dehiscens + B Pulleniatina spp.
G. tumida tumida — S. subdehiscens	N18	N18	MIOCENE	LATE	← B G. tumida

Fig. 21. Upper Neogene planktonic foraminiferal biostratigraphy of the Andaman-Nicobar Islands (data from Srinivasan and Srivastava, 1975). No indication of sediment thickness. T: top of range, B: bottom of range (see text for discussion).

about -3.7 m.y. Thus, it is probable that reworking of Pliocene sediments in the Andaman-Nicobar sections artificially modifies the range of some of these species.

The interval spanning the Globorotalia tumida flexuosa Zone of Srinivasan and Srivastava in the Andaman Nicobar Islands may possibly correlate with an interval including common G. tumida flexuosa in the lower Pliocene sediments of the Gulf of Aden (Site 232) and of the Arabian Sea (Site 219) in the western part of the ocean, at about the same latitude as the Andaman region. This horizon, which lies in the Reticulofenestra pseudoumbilica nannofossil zone of Bukry at both Sites 232 and 219, is slightly below the extinction level of Globorotalia margaritae and coincides with the extinction level of Globigerina nepenthes at the latter site. At Site 232 the age of the G. tumida flexuosa interval cannot be estimated with respect to the extinction levels of G. nepenthes and G. margaritae because these two taxa are eliminated from the Gulf of Aden by dissolution. It lies, however, just below the extinction horizon of the radiolarian species Spongaster pentas, which has been paleomagnetically dated at about -3.9 m.y. in the equatorial Pacific (Theyer and Hammond, 1974a), and which approximates the last occurrence level of G. margaritae at various sites of the Indian Ocean.

Other Low-Latitude Foraminiferal Events

Pliocene foraminiferal events which have proved to have biostratigraphic importance in low-latitude regions include the evolutionary development of the following bioseries: 1) Globigerina decoraperta -- G. rubescens, 2) Neogloboquadrina acostaensis -- N. humerosa -- N. dutertrei, 3) Globoquadrina venezuelana -- G. conglomerata, and 4) Pulleniatina spp. The progressive transitions between these taxa and their time of appearance in the tropical Pacific and Atlantic have been discussed by Parker (1967, 1973, 1974) who pointed out the great difficulty of pinpointing the exact appearance of the evolving species. Globigerina rubescens developed from G. decoraperta in mid-Pliocene. Neogloboquadrina acostaensis gave rise to N. humerosa in Early Pliocene (N 18), which in turn developed into N. dutertrei in latest Pliocene (N 21). These two evolutionary series are contemporaneous in both Atlantic and Pacific oceans. The transition from Globoquadrina venezuelana to G. conglomerata apparently took place in late Pliocene and early Quaternary in the Pacific, but G. venezuelana disappeared from the Atlantic, where G. conglomerata is now absent. Pulleniatina primalis gave rise in the Pacific to P. spectabilis in earliest Pliocene and shortly later to P. obliquiloculata in early Pliocene (N 19): P. spectabilis did not migrate to the Atlantic after its development, whereas P. obliquiloculata migrated apparently in Late Pliocene. Both the first appearances of P. spectabilis and N. humerosa were observed near the top of the Gilbert C event and in the lower part of this event respectively in equatorial Pacific cores (Hays et al., 1969; Saito et al., 1975).

It was found impossible, from present data, to evaluate the biostratigraphic significance of these events in the Indian Ocean. Such an attempt will be possible only after reexamination of these bioseries at various sites, with a uniform taxonomic interpretation. The appearance levels of these taxa as reported by various authors, are not represented on Figures 3 to 18. The different taxonomic concepts of workers are demonstrated by the discrepancy in ranges of forms referred to the same taxon (even in the same section by two different authors). Discrepancy in taxonomic interpretations of Pulleniatina primalis and P. obliquiloculata at various sites has been mentioned earlier. P. spectabilis, however, is a form easily identifiable by its planoconvex text and broad pseudocarina, and has not been reported in the Indian Ocean, either at DSDP sites or in DODO and V20-163 cores.

Globorotalia conoidea - G. inflata and G. crassaformis Group

The dominant components of upper Neogene temperate planktonic foraminiferal faunas in the world oceans are globorotaliids, which comprise the Globorotalia conoidea - G. inflata and G. crassaformis plexus. Kennett (1966, 1967, 1973), Kennett and Watkins (1974) and Kennett and Vella (1975) reported the evolution of G. conoidea to G. conomiozea, to G. puncticulata (including its early form G. puncticulata sphericomiozea), to G. inflata in temperate areas of New Zealand and of the southwest Pacific. Kennett and Watkins (1974) calibrated the various evolutionary events in that part of the lineage from G. conoidea to G. puncticulata to the paleomagnetic stratigraphy of two New Zealand sections. Occurrences of members of the G. conoidea - G. inflata bioseries in the Atlantic have been discussed by Berggren and Amdurer (1973) and Poore and Berggren

(1974, 1975) and in the North Pacific by Vincent (1975). These
authors related events in this foraminiferal group to other fossil
zonations and discussed their relationship with the paleomagnetic
time-scale as well as the biostratigraphic correlations of the
various areas with the Mediterranean and New Zealand regions.

The evolutionary appearance of G. puncticulata (including G.
puncticulata sphericomiozea) which typically occurs in temperate
areas at, or just below the NN 13/NN 12 nannofossil zonal boundary
is about -4.3 to -4.5 m.y. old, i.e. slightly younger than the
Miocene/Pliocene boundary. The earliest occurrence of G. puncticulata
has been erroneously equated however to the "Miocene/Pliocene
boundary" by a number of workers (e.g. Atlantic: Parker, 1973; North
Pacific: Ingle, 1973; South Pacific: Kennett, 1973, and traditionally
by New Zealand workers).

Globorotalia conomiozea has been reported to disappear in the
Pacific and Atlantic Oceans very shortly after the initial appearance
of its descendant G. puncticulata in the earliest Pliocene. G.
conomiozea survived longer in the Indian Ocean in the form of the
subspecies G. conomiozea subconomiozea which extinction is recorded
in Core V20-163 in the upper Gilbert at about -3.5 m.y. (Bandy, 1973,
1975).

The evolutionary transition of Globorotalia inflata from G.
puncticulata has been observed in the southwest Pacific (Kennett,
1973) and in the North Atlantic (Poore and Berggren, 1974; Berggren
and Amdurer, 1973) at, or just below the NN 16/NN 15 nannofossil
zonal boundary. This horizon, about -3.0 m.y. old is slightly older
than the extinctions levels of Sphaeroidinellopsis spp. and Globoquadrina altispira in tropical areas, and slightly younger than the
evolutionary appearance of Globorotalia tosaensis. Cita (1973)
observations of rare G. tosaensis together with forms transitional
between G. puncticulata and G. inflata in the Mediterranean also
show that the initial occurrence of G. tosaensis precedes the
evolutionary appearance of G. inflata. Globorotalia tosaensis, however, typically appears in most temperate sections slightly higher
than the lowermost G. inflata, and the latter species, a common
component of temperate assemblages appears to be a much more reliable mid-Pliocene biostratigraphic marker at mid-latitudes.

The first occurrence of Globorotalia crassaformis predates the
Gilbert A Event in the Indian Ocean Core V20-163 (Saito et al.,
1975), indicating that this species is at least older than -4.0 m.y.
There is evidence that it developed at least as early as G. puncticulata and possibly earlier. Although in mid-latitudes sections of
various parts of the world the first occurrence of G. crassaformis
is typically recorded above the first occurrence of G. puncticulata,
in the same sequential order as in the New Zealand sections described by Kennett and Watkins (1974), in a number of instances the
difference is minimal and the two events are often coincidental
(e.g. at Site 111 in the North Atlantic, Poore and Berggren, 1974;
at Site 310 in the North Pacific, Vincent, 1975; and in the lower
Tabianian Italian type section, Saito et al., 1975). Blow (1969, p.
421) considers that G. crassaformis developed earlier, from Globorotalia subscitula, in latest Miocene within N 17 and Kaneps (personal
communication) observed the transition of forms illustrated as G.
subscitula by Blow (1969) to G. crassaformis in the upper Miocene
sediments of one core from the Rio Grande Rise.

In low-latitudes as well as in high-latitudes G. inflata and G.
crassaformis first appear later than in mid-latitudes. In tropical
areas their first occurrences have been recorded at various levels

in the Upper Pliocene and the Pleistocene. G. inflata is not usually present in equatorial regions. Both species migrated in Antarctic regions at about -0.7 m.y. at the Brunhes/Matuyama boundary (Kennett, 1970; Keany and Kennett, 1972).

At the low-latitude Indian Ocean sites, G. inflata and G. crassaformis were found in rare occurrences in the Pliocene and the Pleistocene, being common only in some Pleistocene horizons in the Gulf of Aden, possibly reflecting intervals of cooling in this area. G. inflata first occurs high in the Gulf of Aden sections in the Pleistocene, but surprisingly low at Site 237 on the Mascarene Plateau in the Lower Pliocene. The species was not reported at other low-latitude sites. The surprisingly low range of G. inflata at tropical Site 237 should be reevaluated; it might possibly result from artificially displaced sediments or the species may have been misidentified for one of its forerunners.

Globorotalia crassaformis was noted at most low-latitude sites of the Indian Ocean. A late first occurrence of this species was reported in the Pleistocene in the Gulf of Aden, in the mid-Pliocene at Site 219 in the Arabian Sea (above the disappearance of Globorotalia margaritae and below the appearance of Globigerinoides fistulosus), and in the upper Lower Pliocene at Site 214 on the Ninetyeast Ridge (below the disappearance of Globorotalia margaritae). G. crassaformis first appears slightly above the Miocene/Pliocene boundary at Site 236 in the Somali Basin as in many areas of the Pacific and Atlantic, but low at Site 237 on the Mascarene Plateau below the Miocene/Pliocene boundary in the Ceratolithus primus Zone of Bukry.

Globorotalia puncticulata and G. conomiozea were not reported at any of the low-latitude Indian Ocean sites. Rare forms referred to G. conoidea by Vincent et al. (1974) were noted in the Late Miocene assemblages of Site 236 in the Somali Basin and forms assigned to G. miozea by Berggren et al. (1974) in the Miocene and Pliocene faunas of Site 214 on the Ninetyeast Ridge.

At the mid-latitude Site 264 on the Naturaliste Plateau Kennett (1975) reported the first occurrence of Globorotalia inflata and last occurrence of G. puncticulata simultaneously with the base of Globorotalia tosaensis, in lowermost Core 2A. Kaneps' (1975a) observations at the same site are in rather good agreement with Kennett's data. Kaneps reported the last occurrence of G. puncticulata slightly higher than did Kennett in Core 2A, Section 3, and the lowest occurrence of G. inflata, higher than Kennett did in Core 2A, Section 4. But he reported G. triangula (a subspecies of G. inflata which Parker, 1967 referred to as G. inflata variant, preceding typical G. inflata in Core DODO 117P; also called G. sp. 1 by Phleger et al., 1953 and Ericson et al., 1964) as low as the bottom of Core 2A. The base of G. inflata (including G. inflata triangula) is just slightly above the extinction levels of Globorotalia margaritae and Globigerina nepenthes, reported by both Kennett and Kaneps in the middle of Core 2, and coincides with the NN 16/NN 15 nannofossil zonal boundary. It appears, thus, that G. inflata first appears in the southeast Indian Ocean at the same biostratigraphic level as in the southwest Pacific and the Atlantic.

Earlier events in the G. conoidea - G. inflata lineage cannot be evaluated at Site 264 because of the discrepancies in Kennett's and Kaneps' reports and because of the condensing of the Lower Pliocene section which overlies unconformably Eocene sediments in lower Core 2. Kaneps (1975a, b) identified forerunners of G. puncticulata which he referred to as "G. miozea conoidea", "G. sphericomiozea", and "G. conomiozea" and reported that the evolutionary sequence is well developed in Core 2. Kennett, however, did not report any of these forms and noted only the abundant occurrence of a form he referred to as G. cf. G. conoidea in Core 2, Section 3.

It appears that the Neogene section overlying Eocene sediments at Site 264 is almost entirely Pliocene, as shown by the presence of the Pliocene marker G. puncticulata as far down as Core 2, Section 4, 26-28 cm. However, the basal Neogene sample just above the unconformity (Core 2, Section 4, 75-77 cm) analyzed by Kaneps (1975a, b) does not contain G. puncticulata but includes primitive forms of G. crassaformis transitional with G. subscitula (Kaneps, 1975a, b, personal communication). The nannofossil zonal assignment of this horizon indicates a Late Miocene to Early Pliocene age (NN 13- NN 9: Site Report in Hayes et al, 1975, p. 28-29). It appears, thus, that the sediments directly overlying the Eocene at Site 264 are at least as old as earliest Pliocene (pre G. puncticulata) and probably latest Miocene. That the earliest Neogene sediments on the Naturaliste Plateau are Late Miocene is shown by the NN 11 nannofloral assemblage directly overlying Cretaceous at nearby Site 258 (Thierstein, 1974).

At subantarctic Site 281 on the South Tasman Rise, Globorotalia inflata first appears in Core 3, Section 2 shortly below the disappearance of G. puncticulata (in Core 2, Section 5) and just slightly above the last occurrence of forms referred to Globorotalia cf G. margaritae (Jenkins, 1975). The sequential order of these three foraminiferal events agrees with data from elsewhere, and the level of appearance of G. inflata can probably be taken here as the Upper/Lower Pliocene boundary at about -3.0 m.y. However, the relationship of this event to the nannofossil and radiolarian zonations at this site cannot be compared with the relationships at other sites because of the differences in mid- and high-latitude floral and faunal assemblages. The base of G. crassaformis and the base of G. puncticulata in Core 3A, Section 1 and Core 3 ACC, respectively are separated only by a very short distance and the last Globorotalia conomiozea occurs in between. These three events occur in the same sequential order as in the temperate southwest Pacific sites 207 and 284 (Kennett, 1973; Jenkins, 1975).

Among the various taxa of the G. conoidea - G. inflata lineage, Boltovskoy (1974a) reported "G. conoidea" and "G. inflata" at the mid-latitude sites of Leg 26. It is very probable, however, that intermediate members of the bioseries which are present in the southeast Indian Ocean have been included with either "G. conoidea" or "G. inflata" by Boltovskoy. This assumption is supported by the inconsistency in the ranges of these two forms from site to site with respect to the extinction level of Globorotalia margaritae and to the nannofossil zonation. The forms referred to "? Globorotalia miocenica" by Boltovskoy might possibly, as mentioned earlier, be assigned to G. conomiozea subconomiozea. This possibility is supported by the Upper Pliocene last occurrences of these forms at Sites 251, 253 and 254 below the extinction level of Globorotalia margaritae and in the nannofossil zonal interval NN 13- NN 14. Among taxa of the G. conoidea-G. inflata bioseries, Akers (1974b) reported only "G. inflata" at the sites studied by Boltovskoy, and Zobel (1974) reported only "G. inflata" and "G. puncticulata" at Leg 25 sites. The latter author may also have included members of the lineage in the forms she referred to "G. miocenica s.l.". It appears, therefore, that this important lineage needs to be reevaluated at all mid-latitude sites of the Indian Ocean.

The first occurrences of Globorotalia crassaformis reported by Boltovskoy (1974a) in Leg 26 mid-latitude sections fall in nannofossil Zone NN 12 at Site 251, at the boundary between NN 11 and an overlying unzoned interval below NN 14 at Site 254, in uppermost NN11 at Site 253, and in an undifferentiated NN 11 - NN 10 interval at Site 258. It is not possible to tell precisely whether the earliest G. crassaformis is above or below the Miocene/Pliocene boundary (straddled by NN 12) at Site 251 and 254, but it appears to be definitely below at Site 253 and 258 (as at low-latitude Site 237 on the Mascarene Plateau. Because of the difficulty in drawing taxonomic boundaries between G. crassaformis and other related forms, the taxonomic status of these early G. crassaformis should doubtless be reexamined before drawing definite conclusions as to their age. It is probable, however, that at mid-latitudes of the Indian Ocean G. crassaformis first appeared in the Upper Miocene as suggested by Blow and Kaneps.

Miocene/Pliocene Boundary

Paleontological Events Associated with the Miocene/Pliocene Boundary

The type Miocene/Pliocene boundary occurs between the Late Miocene Messinian and the Early Pliocene Tabianian (equals Zanclean of Sicily) stages of Italy. Biostratigraphic evidence which can be directly correlated with the base of the Pliocene Italian type sections has been summarized by Berggren (1973) and Berggren and Van Couvering (1974) who placed the Miocene/Pliocene boundary at approximately 5 m.y. in the lowermost reversed interval of the Gilbert paleomagnetic Epoch. A similar conclusion was reached by Saito et al. (1975) who suggested the use of the Gilbert/Epoch 5 paleomagnetic boundary (estimated to be 5.1 m.y. old by Talwani et al., 1971) as the criterion for delineating the Miocene/Pliocene boundary outside the Mediterranean region.

The planktonic foraminiferal events closest to this paleomagnetic boundary in the equatorial Pacific Core RC12-66 are the extinction of Globoquadrina dehiscens just after Epoch 5 (at about -5.10 m.y.) and the slightly younger evolutionary development of Globorotalia tumida from its ancestor G. plesiotumida (equals base of foraminifera Zone N 18) shortly after Epoch 5 (at about -4.9 m.y.) (Saito et al., 1975). The calcareous nannoplankton species Ceratolithus acutus* first appears in Core RC12-66 simultaneously with the earliest G. tumida and apparently ranges no higher than the Gilbert B event in Core V24-59 (Gartner, 1973). This species which is most common in lower Pliocene sediments occurs from about 6 m to 48 m above the base of the Trubi Formation at the newly proposed Zanclean type section of Capo Rosello in Sicily (Cita and Gartner, 1973; Gartner and Bukry in the middle part of Zone NN 12 of Martini. Two cal-

---

* Ceratolithus acutus Gartner and Bukry, 1974, has been previously referred to Ceratolithus sp. by Gartner (1973) and incorrectly to C. amplificus by many authors. Gartner and Bukry (1974, 1975) discussed in detail the taxonomy of this species. C. acutus was erroneously put in synonymy with C. armatus Müller, 1974, by Vincent (1975, p. 1136). C. armatus, at present known only from the Indian Ocean DSDP Site 242 in the Lower Pliocene part of Core 4, appears to be a locally successful derivation of the more widespread C. acutus (Gartner and Bukry, 1975).

careous nannofossil events which bracket the Miocene/Pliocene boundary in Pacific equatorial cores are the first appearance of Ceratolithus rugosus in the Gilbert C event (at about -4.5 m.y.) and the last occurrence of Discoaster quinqueramus near the top of the reversed event in normal Epoch 5 (at about 5.6 m.y.) (Gartner, 1973).

No radiolarian species evolved or became extinct at the end of Epoch 5 in paleomagnetically dated cores from the equatorial Pacific. The first appearance of Pterocanium prismatium, however, (and thus, the Spongaster pentas/Stichocorys peregrina zonal boundary as defined by Riedel and Sanfilippo,(1971) lies shortly above, in the Gilbert C event at about -4.5 m.y. (Theyer and Hammond, 1974a; Saito, et al., 1975). Theyer and Hammond pointed out that P. prismatium and Spongaster pentas first appearances are nearly coincidental in Core M70-39 in the Gilbert C event, and because of the restricted distribution of P. prismatium proposed that the S. pentas appearance datum be used to define the base of the S. pentas Zone. This proposal was followed by Riedel and Sanfilippo (in press) in their recently revised radiolarian zonation. Riedel et al. (1974) demonstrated that the radiolarian fauna at one level of the Capo Rosello section of the Trubi Formation belongs in the Stichocorys peregrina Zone, and Sanfilippo et al. (1973) reported a similar radiolarian fauna in the stratotype Tabianian.

Extinction Level of Globoquadrina dehiscens

The last occurrence of Globoquadrina dehiscens, which approximates the Miocene/Pliocene boundary in the equatorial Pacific Core RC12-66 has also been reported in the uppermost Miocene in various tropical areas of the Pacific and Atlantic (e.g. Parker, 1967, 1973). Berggren (1973) considered the extinction of G. dehiscens as a reliable datum for distinguishing the Miocene/Pliocene boundary. This species, however, has been shown to be of no use in delineating the boundary in some tropical areas where it last occurs lower in the Miocene (e.g.: in Upper Miocene in Blake Plateau cores, Kaneps, 1970; in upper Middle Miocene in the Bodjonegoro-1 well, Bolli, 1964, 1966; in lower Upper Miocene in the Caribbean and Gulf of Mexico, Lamb and Beard, 1972). In temperate areas, Kennett and Watkins (1974) showed that in the Upper Miocene the last appearance of G. dehiscens in New Zealand decreased in age from south to north in response to progressive cooling. Vincent (1975) observed a similar latitudinal effect in the North Pacific where the disappearance of G. dehiscens decreases in age from north to south in association with a Late Miocene cooling.

Srinivasan and Srivastava (1975) and Kennett and Srinivasan (1975) showed that G. dehiscens occurs within the Lower Pliocene in ELTANIN Core 35-12 in the southeast Indian Ocean in the Great Australian Bight and in the Andaman-Nicobar Islands in the northern Indian Ocean. Kennett and Srinivasan pointed out that the species has been reported in Lower Pliocene sediments of the subtropical South Atlantic as well (DSDP, Site 15; Blow, 1970), and suggested that G. dehiscens was entirely eliminated from the Pacific Ocean before the beginning of the Pliocene but that in the Indian Ocean and South Atlantic Ocean (at least in some areas) it lingered on, finally becoming extinct towards the end of the Early Pliocene (although, as discussed above, the age of the Pliocene level at which G. dehiscens disappears in the Andaman-Nicobar Islands, simultaneously with Globoquadrina altispira, Globigerina nepenthes and Globorotalia margaritae (Fig. 21) , cannot be precisely defined).

Results from Indian Ocean DSDP Sites confirm Kennett and Srinivasan's findings. Globoquadrina dehiscens disappears in the upper Lower Pliocene in the Arabian Sea (Site 219) at the same latitude as the Andaman-Nicobar region, and near the Middle/Upper Miocene boundary in the Gulf of Aden and between the equator and 12°S at Sites 236, 237, 238 and 214. (At the latter site it disappears essentially in upper Core 17 where McGowran, 1974, p. 612, placed the top of its range and where Berggren et al., 1974 reported the top of its consistent and common occurrence, although the latter authors noted two rare isolated occurrences in Upper Miocene). The disappearance level of G. dehiscens at latitudes intermediate between the Andaman region and the Equator cannot be specified because of the discontinuous coring performed at Sites 216 and 217. The species, however, occurs commonly in the Upper Miocene sediments at these two sites. G. dehiscens was not reported in DODO and V20-163 cores. It disappears in the Middle Miocene (within nannofossil Zone NN 9) at mid-latitude Sites 251, 253 and 254 and in Lower Miocene at subantarctic Site 281. G. dehiscens occurs in lowermost Neogene sediments (Upper Miocene or lowest Pliocene) just above the unconformity on the Naturaliste Plateau (in the lowest Neogene sample at Site 264, Kaneps, 1975b; in lower Core 8A at Site 258, Boltovskoy, 1974) and disappears, therefore later in this area than at the same latitude farther west. That the species survived longer close to Australia than at the same latitude of the open ocean to the west is also shown by its Lower Pliocene occurrence in the Great Australian Bight.

Thus, it can be concluded that the disappearance of Globoquadrina dehiscens in the Indian Ocean, which shows a narrow latitudinal dependence, is paleocenographically controlled and is of no use for delineating the Miocene/Pliocene boundary in this ocean.

Miocene/Pliocene Boundary at Low-Latitude Sites of the Indian Ocean

The transition of Globorotalia tumida from its ancestor G. plesiotumida is gradational and the taxonomic boundary is difficult to draw. Some uncertainty, therefore, may remain in the recognition of the earliest forms of G. tumida by various authors. Berggren and Poore (1974), in their study subsequent to the Initial Reports, reevaluated this transition at Site 214. They reported the earliest occurrence of G. tumida in Core 10, Section 6 and on this basis drew the Miocene/Pliocene boundary between Cores 10 and 11. This horizon is significantly below the Ceratolithus acutus Subzone tentatively recognized by Bukry in Core 9, Section 2, and closely approximates the Triquetrorhabdulus rugosus/Ceratolithus primus subzonal boundary of Bukry and the Ceratolithus tricorniculatus/Discoaster quinqueramus zonal boundary of Gartner (= NN 12/NN 11), both placed by these authors between Sections 5 and 6 of Core 10 (Bukry, 1974a, p. 602; Gartner, 1974, p. 596). The last occurrence of D. quinqueramus which defines these two floral boundaries is noted in lowermost Core 10 by Gartner et al. (1974, p. 811) but in Core 12, Section 1 by Gartner (1974, p. 585), and Berggren and Poore used the latter position to place the NN 12/NN 11 boundary between Cores 11 and 12. It appears, thus, that the placement of basal NN 12 by Berggren and Poore in the stratigraphic column at Site 214 is too low and that the initial appearance of G. tumida in this section occupies a position relative to nannofossils slightly lower than in equatorial Pacific cores, near the extinction level of Discoaster quinqueramus. Calcareous faunal and floral data at other low-latitude sites of the Indian Ocean support this finding.

At Sites 219, 236 and 237 the first occurrence of G. tumida reported by Fleisher (1974) and Vincent et al. (1974) nearly coincides with the last occurrence of Discoaster quinqueramus and is a few meters below the interval assigned to the C. acutus Subzone. In the more expanded sections of low-latitude Sites 231 and 238, however, the lower range of G. tumida nearly coincides with the base of the C. acutus Subzone. Because Vincent et al. (1974) used a large sampling interval in these two thick sequences, this event should be reevaluated at the two latter sites after the examination of more closely spaced samples (a study currently in progress by the author). At Site 232 Vincent et al. (1974) could not separate G. tumida from G. plesiotumida because transitional forms were found throughout the lower 200 meters and they combined the foraminiferal zones N 18 and N 17. The C. acutus Subzone, however, was recognized at this site in Cores 9A and 10A at about 25 meters below the first appearance of the radiolarian species Pterocanium prismatium.

The first occurrence of P. prismatium delineates the Spongaster pentas/Stichocorys peregrina radiolarian zonal boundary in Core 9, CC at Site 214, about 9 meters above the base of Globorotalia tumida. This level coincides with the N 18/N 19 foraminiferal zonal boundary, well defined by the evolutionary appearance of Sphaeroidinella dehiscens (Berggren and Poore, 1974), and with the Ceratolithus tricorniculatus/C. rugosus nannofossil zonal boundary of Gartner (= NN 12/NN 13). S. pentas first appears shortly above in Core 9, Section 1 in the C. rugosus Zone of Gartner (= NN 13).

The Spongaster pentas/Stichocorys peregrina zonal boundary was delineated at Site 238 at the first occurrence of Pterocanium prismatium in Core 13, 20 meters above the first Globorotalia tumida and within the C. rugosus Subzone of Bukry (= NN 13), but was not recognized at Sites 236 and 237 where the Spongaster pentas and the Stichocorys peregrina Zones were combined. As a result of displacement or downward reworking, the first occurrences of S. pentas are lower at both Sites 232 and 238 than its known evolutionary appearance elsewhere; this species ranges below the base of Pterocanium prismatium and overlaps Ommatartus antepenultimus at Site 232, and ranges lower than Acrobotrys tritubus at Site 238. At Site 237, however, S. pentas seems to appear at its known level of evolutionary appearance in a biostratigraphic position similar to Site 214. It appears in low Core 8, and 10 meters above Globorotalia tumida, within the C. rugosus Subzone of Bukry (= NN 13), and coincides with the N 18/N 19 foraminiferal boundary tentatively placed by Vincent et al. (1974) at the uppermost occurrence of G. plesiotumida.

In summary, it appears that at low-latitude sites of the Indian Ocean the initial appearance of Globorotalia tumida is typically shortly prior to the appearance of Ceratolithus acutus and coincides, or is just above, the extinction level of Discoaster quinqueramus in a slightly lower biostratigraphic position relative to the calcareous nannoplankton than in the equatorial Pacific. In the latter area, both G. tumida and C. acutus appear simultaneously. This slight discrepancy should be reevaluated after reexamination of the Globorotalia plesiotumida-G. tumida bioseries at various sites and may not be significant, considering the degree of biostratigraphic resolution which can be obtained from DSDP cores. The Miocene/Pliocene boundary at low-latitude sites is drawn in this paper as a hachured area between the first occurrences of G. tumida and C. acutus. The Spongaster pentas/Stichocorys peregrina radiolarian boundary occurs shortly above the Miocene/Pliocene boundary so delineated, and typically falls at the base, or in, nannofossil zone NN 13.

The TID 12/TID 13 diatom boundary of Shrader (1974) is defined by the first appearance of Nitzschia jouseae, an event which also defines the base of the N. jouseae Zone of Burckle (1972) and correlates with the Gilbert C_1 event, about -4.3 m.y. This diatom boundary should therefore occur just above the S. pentas/S. peregrina radiolarian zonal boundary. The TID 12/TID 13 boundary occupies this position at Sites 213 and 215 on either side of the Ninetyeast Ridge where it falls just above the radiolarian zonal boundary (Johnson, 1974; Shrader, 1974), but is below the latter at Site 238. The base of TID 12 is at Site 238 in upper Core 15 slightly above the base of N 18 and the Ceratolithus acutus Zone, and thus slightly above the Miocene/Pliocene boundary. The TID 12/TID 13 is equivalent to the NPD 10/NPD 11 diatom boundary (North Pacific Ocean Diatom) which also approximates the base of N 18 at DSDP Site 173 in the northeast Pacific (Shrader, 1973; Ingle, 1973; Bukry, 1973b).

Thus, it can be concluded that there is in general a good agreement among various fossil groups for the recognition of the Miocene/Pliocene boundary at low-latitude sites of the Indian Ocean.

Miocene/Pliocene Boundary at Mid- and High-Latitude Sites of the Indian Ocean

Recognition of the Miocene/Pliocene boundary in mid- and high-latitude regions outside the latitudinal range of G. tumida is not easily done on a foraminiferal basis in the absence of this index species. Mid-latitude calcareous sections located at the transition between tropical and temperate areas which yield mixed tropical and temperate foraminiferal assemblages are very valuable. Faunas from such sections include members of the Globorotalia plesiotumida - G. tumida as well as the Globorotalia conoidea - G. inflata, bioseries and provide a link for Upper Miocene-Lower Pliocene biostratigraphic correlations between tropical and temperate areas. Although a number of such calcareous sequences were recovered in the Indian Ocean during Leg 25, 26 and 28, it is not possible from the present data to investigate the Miocene/Pliocene boundary at these sites because of the taxonomic confusion involved in the interpretation of these two important foraminiferal groups by various workers.

The calcareous nannoplankton index species Ceratolithus acutus is usually found in mid- as well as in low-latitudes and provides a good marker for recognition of the boundary in both regions. Unfortunately Bukry's zonation was not applied to the calcareous sequences recovered after Leg 25. Neither Thierstein (1974) or Burns (1975a) differentiated the C. acutus interval within the NN 12 zonal interval of Martini at the mid-latitude sites of Leg 26 and at Site 264 (Leg 28) respectively. During Leg 26, Thierstein drew the "Miocene/Pliocene boundary" at the base of NN 13 (defined by the initial appearance of Ceratolithus rugosus) but, as mentioned earlier, this event is younger than the Miocene/Pliocene boundary. Boltovskoy (1974a) tentatively placed the "Miocene/Pliocene boundary" at an even younger level at the base of the common occurrence of "G. inflata" and G. crassaformis. Both the latter author and Akers (1974b) did not separate the Lower Pliocene marker Globorotalia puncticulata from its ancestors or descendants and did not report the presence of G. tumida. In the absence of these two foraminiferal markers, of the nannoplankton marker C. acutus, and of siliceous fossils, the Miocene/Pliocene boundary is best estimated at the Leg 26 Sites (Figs. 10, 14-16) at the last Discoaster quinqueramus.

At Site 264 on the Naturaliste Plateau Kennett (1975) reported the rare, but consistent, occurrence of G. tumida throughout the Neogene section (down to Core 2, Section 3), whereas Kaneps noted its presence in only one sample. As mentioned above, the lowermost Neogene section (below the lowest sample analyzed by Kennett) may possibly include the Miocene/Pliocene boundary, in view of the absence of G. puncticulata, the presence of primitive G. crassaformis (Kaneps, 1975a, b), and the nannofossil zonal assignment.

The "Miocene/Pliocene boundary" in New Zealand has been traditionally placed at the upper limit of the Kapitean Stage (at the boundary between the Kapitean and Opoitian stages) marked by the first appearance of the non-keeled members of the Globorotalia conoidea-G. inflata bioseries, i.e. G. puncticulata sphericomiozea, the earlier form of G. puncticulata (variously called G. inflata by Hornibrook, 1958, and Jenkins, 1967; G. crassaformis by Kennett, 1966, 1967; G. puncticulata by Hornibrook and Edwards, 1971, Kennett and Watkins, 1974, Kennett, 1973, and Collen and Vella, 1973; and G. puncticulata sphericomiozea by Kennett and Watkins, 1974). This event, however, is slightly higher than the Miocene/Pliocene boundary as evidenced by the data summarized above, and at least part of the Kapitean is Pliocene.

There are conflicting data on the age of the base of the Kapitean, which has been assigned a Pliocene age of about -4.7 m.y. by Kennett and Watkins (1974) from their paleomagnetic study, but which appears to be older than the Miocene/Pliocene boundary from faunal correlations with the Late Miocene Messinian Stage of Italy. The base of the Kapitean is defined by the evolutionary appearance of Globorotalia conomiozea. The Kapitean index species G. conomiozea occurs in upper Miocene Italian sections and its initial appearance has been recently proposed as a marker for the recognition of the lower boundary of the Messinian (Tortonian/Messinian boundary) (D'Onofrio et al., 1975). Biostratigraphic data from other areas also provide evidence that G. conomiozea straddles the Miocene/Pliocene boundary (e.g. in Japan: Ikebe et al., 1972; in the North Atlantic at DSDP Site 116: Poore and Berggren, 1975; in the South Atlantic on the flank of the Rio Grande Rise: Berggren, in Van Couvering et al., 1976, in prep., personal communication).

At the subantarctic Site 281, Globorotalia tumida is not present and Jenkins (1975, p. 460) placed the Miocene/Pliocene boundary as a dashed line within his Globorotalia miozea conomiozea Zone (defined by the total range of this species; thus its base equates the base of the Kapitean). Unfortunately, the presence of an unrecovered interval prevents the recognition of the thickness of this zone at Site 281 and there is some confusion in Jenkins' range chart because he artificially expanded the G. miozea conomiozea zone by misplacing Core 3A between Cores 3 and 4 instead of between Cores 4 and 5 (see Fig. 18). G. miozea conomiozea, G. miozea sphericomiozea (according to Jenkins nomenclature) and G. puncticulata first occur at the same level in the first sample above the unrecovered interval (Core 5, CC and Core 3A, CC). The occurrence of one specimen of Ceratolithus acutus (referred to "C. amplificus") reported from shipboard examinations (Kennett, Houtz et al., 1975, p. 280) in Core 5, CC supports the placement of the Miocene/Pliocene boundary near this horizon. The position in the stratigraphic column of the samples recovered with the core catchers of Cores 5 and 3A is equivocal. These two cores cut the interval between 36 m and 45.5 m subbottom depth but only Sections 1 and 2 were recovered. The core catcher sample is arbitrarily placed on Figure 18 at the base of

Section 2 at 39 m and the Miocene/Pliocene boundary is drawn below this horizon as a hatched area throughout the unrecovered interval.

The Miocene/Pliocene boundary at high-latitude antarctic Site 265 was based on siliceous fossils. A foraminiferal age assignment or zonation of the Upper Miocene-Lower Pliocene interval is not feasible at this site. Assemblages contain abundant Globigerina woodi and are below the first occurrence of Neogloboquadrina pachyderma, which possibly first appeared in Antarctic waters later than its first occurrence in temperate regions to the north, in the Late Miocene (Kennett, 1975).

Late Miocene

Globorotalia merotumida-plesiotumida-tumida Bioseries

Planktonic foraminifera from the Globorotalia merotumida-plesiotumida-tumida bioseries are of primary importance for subdivision of the Upper Miocene in tropical areas. Based on these taxa Banner and Blow (1965a) and Blow (1969) defined the following Upper Miocene foraminiferal zonal boundaries: N 18/N 17 (Miocene/Pliocene boundary) at the first evolutionary appearance of G. tumida, N 17/N 16 at the first evolutionary appearance of G. plesiotumida, and N 16/N 15 at the first occurrence of G. merotumida simultaneously with the initial appearance of Turborotalia acostaensis. A discussion of the characteristics of members of this lineage has been given by Banner and Blow (1965b). However, taxonomic boundaries within this series of intergrading forms prove very difficult to draw and might be interpreted differently by various workers. This appears to be partly the case for Indian Ocean planktonic foraminiferal studies, as shown by the discrepancy in reported ranges of some of these taxa in relation to other fossil groups.

Berggren and Poore (1974), in their reevaluation of this lineage at Site 214 on Ninetyeast Ridge, illustrated their concept of the taxa and discussed the characteristics useful in separating the three forms. They observed between Core 17 to 10 gradual, but distinct, changes within a group of related forms which leads to the evolutionary development of G. tumida. These authors reported that forms assigned to G. merotumida range through Cores 17 (upper part) to 14 and appear to lose their identity in the interval of Cores 13 to 12. They placed the initial appearance near the base of Core 14 although transitional forms to G. merotumida occur over the interval of Cores 16 and 15. Morphologic intergradation between G. merotumida and G. plesiotumida is seen as high as Core 13. Within the interval of Cores 11 and 10, the morphologic transition occurs relatively rapidly in the upper part of Core 11 and the lower part of Core 10, and the first true G. tumida occurs at the base of Core 10 whereas G. plesiotumida disappears just above in lower Core 10. These various foraminiferal events (which have a stratigraphic position different than first estimated by Berggren et al., 1974, and McGowran, 1974, in the Initial Reports) are shown on Figure 12. The extinction level of G. merotumida falls within N 17, very slightly below the Stichocorys peregrina/Ommatartus penultimus radiolarian zonal boundary, the Discoaster berggrenii/Ceratolithus primus nannofossil subzonal boundary of Bukry and the Discoaster neohamatus/D. quinqueramus zonal boundary of Gartner (there is an excellent agreement between Bukry's and Gartner's data for the placement in Core 13 of the two latter nannofossil boundaries, marked by the first occurrence of non-birefringent ceratoliths, including Ceratolithus primus).

The initial appearance of G. plesiotumida coincides with the Ommatartus antepenultimus/O. penultimus radiolarian zonal boundary between Cores 14 and 15 and falls within the D. neohamatus Zone of Gartner and between the Discoaster berggrenii and D. bellus Subzones of Bukry. The first occurrence of G. merotumida coincides in uppermost Core 17 with the Cannartus petterssoni/O. antepenultimus radiolarian boundary and with the Discoaster hamatus/D. neohamatus boundary of Gartner (equivalent to the NN 10/NN 9 boundary of Martini). The equivalent nannofossil boundary of Bukry, the D. hamatus/D. bellus is placed very slightly higher between Cores 16 and 17. It is also at this level (Core 16, CC) that Neogloboquadrina acostaensis first appears.

The Upper/Middle Miocene boundary (which equates the boundary between the Serravalian and Tortonian Stages) has been discussed by Van Couvering and Berggren (in press). These authors place the boundary at the base of paleomagnetic Epoch 10,* which approximates the C. petterssoni/O. antepenultimus radiolarian boundary (Theyer and Hammond, 1974a) and lies within foraminiferal Zone N 15 and nannofossil Zone NN 9. There is, therefore, good agreement between various fossil groups at Site 214 for the placement of the Upper/Middle Miocene boundary in uppermost Core 17. The position of the N 16/ N 15 and the D. hamatus/D. neohamatus (= NN 9/NN 10) boundaries in coincidence with the C. petterssonni/O. antepenultimus boundary, instead of being slightly above, does not represent a biostratigraphically significant difference considering the condensing of the section and the sampling interval involved.

Taxa of the G. merotumida-plesiotumida-tumida group have been identified and utilized by Berggren et al. (1974), Fleisher (1974) and Vincent et al. (1974) to subdivide the Upper Miocene foraminiferal sequences at other low-latitude sites of the Indian Ocean. It is possible, however, that subsequent reevaluation of this bioseries will somewhat modify the placement of the various events in the stratigraphic column.

At Site 216 on the northern Ninetyeast Ridge, discontinuous coring of the upper part of the sedimentary sequence precludes the recognition of the latest events in the G. plesiotumida-G. tumida bioseries (i.e. base of G. tumida and top of G. plesiotumida). The last occurrence of G. merotumida was recorded as at Site 214 in N 17, in the O. penultimus radiolarian Zone, and in the Discoaster neohamatus Zone of Gartner and the Discoaster berggrenii Subzone of Bukry. The C. petterssonni/O. antepenultimus radiolarian boundary delineates the Upper/Middle Miocene boundary between Cores 2A and 3A. This level coincides with the D. hamatus/D. neohamatus nannofossil boundary of Gartner as at Site 214. The base of N 17 (lowest G. plesiotumida), however, placed just below that level in Core 3A, Section 1 appears slightly too low relative to the two other fossil groups. Zone N 16 is very condensed, its base being marked by the simultaneous first occurrence of G. merotumida and Turborotalia acostaensis in Core 3A, Section 2. Zone N 15, if present, is extremely short because the N 14 zonal index Turborotalia siakensis ranges up as high as Core 4, Section 3.

The great condensing of N 16 and N 15 at Site 216, as well as of the Discoaster hamatus Zone (identified only in Core 4, Section 1), and the absence of the Catinaster coalitus Zone between the D. hamatus and D. exilis zones in upper Core 4 indicate a telescoping of the

---

* Ryan et al. (1974), however, place this boundary in the lower part of paleomagnetic Epoch 11.

section near the Upper/Middle Miocene boundary. As a result, a certain amount of mixing and reworking may occur in this part of the sequence as shown by the presence of an horizon (Core 3A, CC) assigned to the C. coalitus Zone by Gartner (1974) in the middle of the D. exilis Zone.

In the tropical sections of the western Indian Ocean (Sites 219, 231, 236, 237 and 238) the initial appearance of G. tumida was noted in a position relative to other fossil groups similar to the sequence at Site 214 (as discussed above with the Miocene/Pliocene boundary). The last occurrence of Globorotalia plesiotumida was reported higher in the sections however, within the lowest Pliocene at Sites 232, 236, 237 and 238 and in upper Lower Pliocene at Site 219. It was noted below the Miocene/Pliocene boundary at Site 231 in the Gulf of Aden where elimination of the species by dissolution higher up may be responsible for this low position. The inconsistency in the range of forms referred to G. merotumida (not reported on Figures 3 to 9) from site to site suggests discrepancies in the taxonomic interpretation of these forms. Zonal assignments of the interval spanning Zones N 16 and N 15 were further complicated by common reworking of older sediments into the lower part of upper Miocene sequences.

An interval of upper and middle Miocene mixed foraminiferal faunas spanning the Upper/Middle Miocene boundary was found at all tropical western Indian Ocean sites. The co-occurrence of Upper and Middle Miocene assemblages leads various workers to assign different ages to the mixed intervals. For example, at Site 219 Akers (1974a) assigned a Middle Miocene age to the entire interval spanning Cores 9 through 12, whereas Fleisher (1974) attributed a Late Miocene (N 17) age to this interval. As pointed out by the latter author the combined presence of Upper Miocene and less common Middle Miocene faunas should be attributed to reworking rather than downhole contamination. The latter would be a reasonable explanation only if the Upper Miocene species were present in just a few samples at the tops and bottoms of cores. Instead, they are consistently present throughout the recovered interval and dominate the combined faunas. This likely explanation is supported by the presence at this site of a probable unconformity directly below the mixed interval. The probable unconformity, which lies in the non-cored interval between Cores 12 and 13 (between N 17 and N 10) represents the age to which the reworked specimens are referred. It is thus probable that nearby contemporaneous outcrops were available as suitable sources for the reworked fauna.

At other western tropical Indian Ocean sites it cannot be determined on a foraminiferal basis whether an unconformity spans the Upper/Middle Miocene boundary or whether the mixed faunas represent a condensed lowermost Upper Miocene and uppermost Middle Miocene interval. An almost complete calcareous nannoplankton sequence, however, is present at all these sites and all nannofloral zones were recognized with the exception of the Catinaster coalitus Zone (only questionably identified at Site 231). It appears thus that this fossil group shows much less reworking than did the foraminiferal, suggesting that the reworked sediments have been winnowed during transport. Because of the lesser disturbance in the nannofloral sequences the latter were used for placing the Upper/Middle Miocene boundary, drawn at the Discoaster bellus/D. hamatus zonal boundary of Bukry. At Site 238 on the Central Indian Ridge this boundary approximates the Cannartus petterssoni/Omnatartus antepenultimus radiolarian boundary, as at Site 214, and the base of

TID 21. The NPD level equivalent to the base of TID 21 (NPD 18/NPD 18, however, is within the Middle Miocene in the northeast Pacific).

The Upper/Middle Miocene boundary placed on nannofossil data at Site 231 in the Gulf of Aden falls between Cores 52 and 53 within an interval (Cores 58 through 48) assigned to the Middle Miocene foraminiferal zones N 14 - N 13. In this interval the preservation of planktonic foraminifera is very poor and the assemblages contain only Middle Miocene elements (common Turborotalia siakensis) without Upper Miocene components. It is possible that intensive dissolution eliminated the Upper Miocene fauna from this interval preventing the recognition of sediment mixing.

At mid-latitude sites the Upper/Middle Miocene boundary is placed in this report on nannofossil data as in low-latitude sections (at the NN 9/NN 10 boundary). Globorotalia merotumida and G. plesiotumida were not included in Boltovskoy's range charts for the mid-latitude Sites of Leg 26 because of the uncertainty regarding the taxonomic separation of these two species (Boltovskoy, 1974a). Akers (1974b), however, identified G. plesiotumida at these sites. He noted the last occurrence of this species in NN 12 at Site 253, in a position similar to low-latitude sites, but lower in the Upper Miocene (within NN 11) at higher latitude Sites 251 and 258. The base of G. plesiotumida was reported by Akers in Middle Miocene (NN 9) at a level much lower than its known evolutionary appearance, which suggests that Akers included forerunners of G. plesiotumida with this species. At subantarctic Site 281 the Upper/Middle Miocene boundary is not identified. It falls within the Globorotalia miotumida miotumida Zone of Jenkins (1975).

Other Foraminiferal Events

The first occurrence of Globorotalia margaritae in the Mediterranean area coincides with the Miocene/Pliocene boundary (Cita, 1973) and this species has been widely used by many workers as a Pliocene marker. In the equatorial Pacific, however, this species first appears in the Upper Miocene during Event A of paleomagnetic Epoch 5 at about -5.5 m.y. (Saito et al., 1975). In the Indian Ocean the first occurrence of this species has been reported at various biostratigraphic levels. It was noted in the Lower Pliocene in the Andaman-Nicobar Islands and at Sites 219 and 214, in the Upper Miocene at Sites 236, 237, 238, 251 and 258, and as low as in the Middle Miocene at Sites 253 and 254. G. margaritae is present in the Lower Pliocene at Site 264 on the southern Naturaliste Plateau but is absent in the lowermost Neogene sample of this site just above the unconformity. At subantarctic Site 281, forms referred to G. cf. G. margaritae by Jenkins (1975) range down into the Upper Miocene Globorotalia miotumida miotumida Zone.

Although the differences in biostratigraphic levels at which G. margaritae first appears at various latitudes may have paleooceanographic significance, no conclusion can be drawn before reexamination of the taxon. As pointed out by Berggren and Ul Haq (1976), G. margaritae can be easily misidentified with the morphologically similar (and probably ancestrally related) G. cibaoensis. In any case the rarity of G. margaritae in the Indian Ocean decreases its biostratigraphic value in this ocean.

Pulleniatina primalis first appeared in the equatorial Pacific during the early part of paleomagnetic Epoch 5 at about -6.0 m.y. (Saito et al., 1975). The first occurrence of this species in low-latitude sections of the Indian Ocean at Sites 214, 237 and 238 shortly above the Ommatartus penultimus/Stichocorys peregrina radiolarian boundary (dated approximately -6.2 m.y. by Theyer and Hammond, 1974a) and within the Ceratolithus primus Zone of Bukry is in good agreement with this data. At Site 219 and 236 P. primalis first occurs within the C. primus Zone of Bukry as well. This level cannot be compared at Site 236 with the radiolarian zonation which could not be well recognized at this site. The Ommatartus antepenultimus/O. penultimus boundary placed at Site 236 in Core 9 (above the lower P. primalis) occupies a position relative to the calcareous nannoplankton higher than at other sites. P. primalis first occurs in the Andaman-Nicobar Islands in a biostratigraphic position higher than in other low-latitude sections, in the Lower Pliocene (Srinivasan and Srivastava, 1975). At mid-latitude sites the first appearance of Pulleniatina is delayed. Species of this genus first occur in the Lower Pliocene at Site 253, in the Upper Pliocene at Site 258, in the Pleistocene at Site 264, and were not noted at Sites 251, 254 and 281.

A multitude of opinions exist as to the taxonomic and stratigraphic position of Neogloboquadrina pachyderma. It has been suggested that this species evolved form N. continuosa (Jenkins, 1971; Bandy, 1972b). This event, termed by Bandy the "Turborotalia pachyderma datum", has been recognized in circum North Pacific sections in Middle Miocene near the foramniferal zonal boundary N 12/N 13 (Bandy 1972a, b; Bandy et al., 1969; Bandy and Ingle, 1970: Parker, 1964; Ingle, 1967, 1973; Ikebe et al., 1972). In the southwest Pacific, however, the first occurrence of N. pachyderma has been observed in the Upper Miocene (Kennett, 1973; in the Globigerina nepenthes Zone of this author and in the Discoaster bellus nannofossil subzone of Bukry at Sites 206 and 207). In the Indian Ocean this species first occurs at subantarctic Site 281 in the Upper Miocene, as in the southwest Pacific, in the Globorotalia miotumida miotumida Zone of Jenkins (1975). It does not appear, however, below the Pliocene at Antarctic Site 265 and at mid-latitude sites of the southwest Indian Ocean (in the Lower Pliocene nannofossil zone NN 13 and in the Upper Pliocene zone NN 16 at Sites 251 and 250, respectively). At Site 264 on the Naturaliste Plateau, N. pachyderma is not present in the lowermost part of the Neogene section in Core 2, Section 4 and first appears in Core 2, Section 3 (in the nannofossil interval NN 13-NN 11).

N. pachyderma was not reported at other mid-latitude sites of Leg 26, nor at low-latitude sites of the eastern Indian Ocean. It was noted, however, at low-latitude sites of the western part of the ocean, in the Arabian Sea (Site 219), the Gulf of Aden (Sited 231, 232 and 233) and the Somali Basin (236) from Upper Miocene (N 17) to Pleistocene. It is also present in Pleistocene sediments on the Mascarene Plateau (Site 237).

Middle and Early Miocene

Late Middle Miocene

The N 15/N 14 zonal boundary, marked by the extinction level of Turborotalia siakensis, was identified at Site 214 only. It falls within the Cannartus pettersoni (radiolarian) Zone and within the Discoaster hamatus Zone of both Gartner and Bukry (= NN 9), in agreement with the estimate of the position of this fossil boundary relative to other fossil zonations by Berggren and Van Couvering (1974).

The biostratigraphic position of the N 14/ N 13 zonal boundary (the "Globigerina nepenthes datum") has been discussed by Van Couvering and Berggren (in press) and is somewhat ambiguous. In the tropical Pacific, Brönnimann et al. (1971, p. 1742) report its position within the C. petterssoni Zone, but Hays et al. (1972) place it somewhat lower in the upper part of the Dorcadospyris alata Zone. In the latter case the range of G. nepenthes overlaps that of Globorotalia fohsi lobata for several meters (Jenkins and Orr, 1972) at the top of the D. alata Zone. The overlap of these two foraminiferal species was also noted in the North Atlantic at DSDP Site 11 by Berggren and Amdurer (1973). In the Caribbean and in Java, on the other hand, Bolli (1957, 1966) and Blow (1959, 1969) recorded a gap between the last appearance of G. fohsi s.l. and the initial appearance of G. nepenthes.

In the Indian Ocean Srinivasan (1975) observed the progressive development of Globigerina nepenthes from G. druryi in sediments from the Middle Miocene Hut Bay Formation of the Andaman Islands as well as the stratigraphic overlap within this formation of G. nepenthes and G. fohsi lobata and also G. fohsi robusta. On Ninetyeast Ridge G. nepenthes first appears very slightly below the C. petterssoni/D. alta radiolarian boundary at both Sites 214 and 216. G. nepenthes and G. fohsi overlap for a short interval at Site 216 but do not overlap at Site 214. At Site 214 the N 14/N 13 boundary is slightly above the Catinaster coalitus/Discoaster kugleri and the C. coalitus/Discoaster exilis nannofossil boundaries of Bukry and Gartner respectively (= NN 7/ NN 8), in agreement with the estimate of Van Couvering and Berggren (in press) for the relative position of these calcareous foraminiferal and floral boundaries. At Site 216, however the foraminiferal boundary has a lower position relative to the nannofossil sequence; it falls within the D. exilis Zone of both Bukry and Gartner (= NN 7 + NN 6).

At Site 238 on the Central Indian Ridge the first occurrence of G. nepenthes marks the N 14/ N 13 boundary slightly below the first occurrence of the radiolarian C. petterssoni, as at Site 214 and 216, but lies in a lower stratigraphic position relative to the nannofossil zonation, in the Sphenolithus heteromorphus Zone of Bukry (= NN 5). Members of the G. fohsi bioseries do not overlap G. nepenthes at this site. In the Gulf of Aden at Site 231 the N 14/ N 13 boundary could not be determined because foraminiferal species known to become extinct within Zone N 13 (e.g. Globigerinoides subquadratus, Turborotalia mayeri) are present above the first appearance of forms referable to G. nepenthes. For this reason Vincent et al. (1974) attributed the upper part of Core 59 at Site 231 to undifferentiated N 14 - N 15. The lowermost forms referable to G. nepenthes occur as far down as Core 59, Section 2. This level, which may possibly represent the "G. nepenthes datum", falls within the Discoaster kugleri subzone of Bukry (= NN 7), in a position relative to the nannofossil zonation similar to that of Site 216. Contrary to the situation at the latter site, however, members of the G. fohsi group do not overlap G. nepenthes at Site 231.

In mid-latitude sections G. nepenthes first occurs within the nannofossil Zone NN 6 at the western Site 251, and within undifferentiated intervals NN 9 to NN 5 and NN9 to NN 6 at the eastern Sites 253 and 254 respectively. G. fohsi was not noted at the three latter sites. However, the forerunners of this species, G. peripheronda and G. peripheroacuta, were found and the first occurrence of G. nepenthes approximates that of G. peripheroacuta at all three sites.

The N 13/N 12 boundary marked by the first appearance of _Sphaeroidinellopsis subdehiscens_ was recognized at Site 214 and 216 on Ninetyeast Ridge and in the Gulf of Aden at Site 231 just above basement. The boundary falls within the _Discoaster exilis_ nannofossil Zone of both Gartner and Bukry, being within the lower part of Bukry's Zone (the _Coccolithus miopelagicus_ Subzone, = NN 6) at both Sites 214 and 231. This position relative to the nannofossil zonation is stratigraphically lower than the position estimated by Van Couvering and Berggren (in press) (i.e. within NN 7). The N 13/N 12 boundary falls at both Sites 214 and 216 within the _D. alata_ radiolarian zone. It is not known if the lowest occurrence of _S. subdehiscens_ at Sites 238, 236 and 237 in the western Indian Ocean delineates the base of Zone N 13 because the latter is stratigraphically immediately supra-adjacent to Zones N 10, N 11 and N 6 respectively. It is possible, however, that intermediate zones were not identified because of their short duration combined with a too large sampling interval. The position of the first occurrence of _S. subdehiscens_ relative to the nannofossil zonation is lower at the three later sites than at the eastern Sites 214 and 216 and falls within, or at the base of the _Spenolithus heteromorphus_ Zone (= NN 5). It falls, however, within the _D. alata_ radiolarian zone at Site 238 as it does at Sites 214 and 216.

At mid-latitude sites _S. subdehiscens_ cannot be used as a Middle Miocene marker. It first appears much higher than at low-latitude sites, within the Upper Miocene nannofossil Zone NN 11 at Sites 251, 253 and 254 and is not present neither at Site 258 on the Naturaliste Plateau nor at subantarctic Site 281. A foraminiferal event useful in mid-latitude areas is the initial appearance of _Globorotalia conoidea_ estimated by Berggren and Amdurer (1973) to occur in the upper Middle Miocene at a level approximately 11.9 m.y. old. This event takes place slightly later than the extinction of _Turborotalia mayeri_, a species which disappear within N 14 at low-latitude Sites 214, 216 and 238. _G. conoidea_ first occurs shortly above the disappearance of _T. mayeri_ at Subantarctic Site 281. Biostratigraphic correlations, however, cannot be undertaken on the basis of these two foraminiferal events between Site 281 and the mid-latitude sites of Leg 26. At the latter sites, Boltovskoy (1974a) did not report _T. mayeri_ and probably included forerunners of _G. conoidea_ with this taxon, as suggested by the very low range of forms he referred to "_G. conoidea_" (within nannofossil Zone NN 5 or lower).

Early Middle Miocene and Early Miocene

Zonal subdivisions of strata of early Middle Miocene (N 12 to N 9) and of Early Miocene (N 8 to N 4) age were found difficult at all low and mid-latitude sites. In most instances zonal boundaries could not be located and zones were often combined into wider intervals. Except for the expanded sequences of Sites 238, 251 and 254 this part of the Neogene section is very condensed at all Indian Ocean sites where the average rate of sediment accumulation is very low (about 3 to 5 m/m.y.) as in preceding Oligocene times. Mixed faunas and signs of significant dissolution were commonly reported. It is noteworthy that at a number of sites (especially in the western Indian Ocean at Site 238 and at sites of Leg 25, as well as at Site 253 on the southern Ninetyeast Ridge) planktonic foraminiferal assemblages contain a great number of extremely small-sized individuals usually unidentifiable (Vincent et al., 1974; Zobel, personal communication; Boltovskoy (1974a). These considerations and the short time spans of

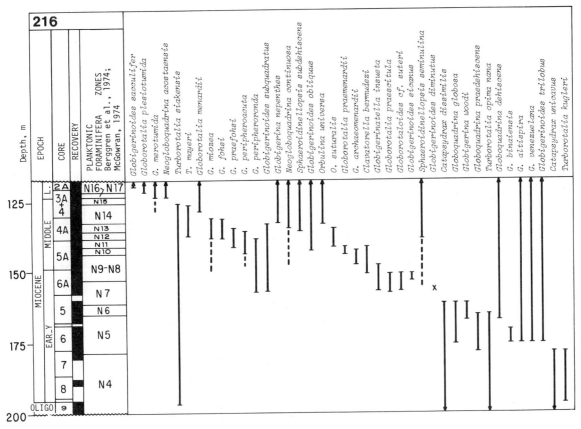

Fig. 22. Ranges of Lower and Middle Miocene planktonic foraminifera at Site 216 (data from Berggren et al., 1974).

many of the early Neogene zones precluded a detailed biostratigraphic investigation of this interval, which may be obtained with further study of more closely spaced samples.

The only low-latitude site where lower Neogene zonal boundaries were well identified is Site 216 on the northern Ninetyeast Ridge. Although condensed, the series appears to be continuous and faunal assemblages do not show mixing. A range chart of planktonic foraminifera at this site constructed from Berggren et al. (1974) data is presented in Figure 22. This site is especially valuable because the co-occurrence of both calcareous and siliceous fossils throughout the lower Neogene sediments permits a direct comparison of the two fossil groups for this interval.

The successive appearance of members of the short-lived Globorotalia peripheronda-G. fohsi lineage (Bolli, 1950, 1967) which marks the zonal boundaries for the interval N 10 to N 12 (Blow, 1969) is well recorded at Site 216. The N 12/N 11 (first appearance of G. fohsi, N 11/N 10 (first appearance of G. praefohsi), and N 10/N 9 (first appearance of G. peripheroacuta) boundaries all fall within the D. alata radiolarian Zone. Subspecies of G. fohsi were not differentiated by Berggren et al. (1974) at this site. In the nearby Andaman region, however, Srinivasan (1975) noted the evolutionary development of G. fohsi robusta from G. fohsi lobata in the Hut Bay Formation of Little Andaman Island (above the initial appearance of G. nepenthes as mentioned earlier). In other islands of this region, where strata equivalent to the upper part of Zone N 10 and lower part of Zone N 11 are present, Srinivasan and Srivastava (1972) reported the evolutionary development of G. praefohsi from G. peripheroacuta.

At other low-latitude sites of the Indian Ocean most members of the G. peripheroronda-G. fohsi bioseries were reported but they usually occurred together in mixed assemblages, and their biostratigraphic succession as known elsewhere is not represented. At mid-latitude sites (251, 253 and 254) the only members of the bioseries present are G. peripheronda and G. peripheroacuta and the latter species disappears lower in the sections than its forerunner G. peripheronda. None of the members of the bioseries occur at subantarctic Site 281.

The Sphenolithus heteromorphus/Discoaster exilis nannofossil boundary of both Bukry and Gartner (= NN 5/NN 6) cannot be precisely correlated with the equivalent foraminiferal level at Site 216 because of a too wide spacing of nannofossil data in this interval, but probably lies within N 11 as shown by Van Couvering and Berggren (in press). This nannofossil zonal boundary occupies a similar position relatively to the foraminiferal zonation at Site 214 (within an undifferentiated N 12 - N 11 interval), but a higher position at the western Sites 238 and 237 (within N 13 and N 14 respectively.

The development of Orbulina suturalis from Praeorbulina glomerosa curva is a widely noted and important foraminiferal event called the "Orbulina datum", which delineates the base of Zone N 9 and the Early/Middle Miocene boundary. Controversies have resulted from various estimates of the biochronological and radiochronological age of this event in different parts of the world. The controversy on the radiometric age of the "Orbulina datum" still lives on. It has been recently discussed and summarized by Van Couvering and Berggren (in press) who accept a 15 m.y. date for this event, which is used in this report (see Fig. 2)*. The "Orbulina datum" has been found to occur within the lowermost part of nannofossil Zone NN 5 and very slightly above the Calocycletta costata/D. alata radiolarian zonal boundary (Van Couvering and Berggren, in press; Theyer and Hammond 1974a; Ryan et al., 1974; see Figure 2). Most Indian Ocean sections are of no use for investigation of the biostratigraphic significance of the "Orbulina datum" because calcium carbonate dissolution appears to be pronounced at most sites in this part of the stratigraphic column and Orbulina spp, which are highly susceptible to solution, are probably eliminated in the lower part of their range. As a result the base of Zone N 9 could not be determined at any of the low- and mid-latitude sites. Elimination of orbulinids and praeorbulinids by dissolution is shown by the high stratigraphic first occurrence of the former and the rarity of the latter (noted at Sites 216, 219, 238, 251 and 253), but at none of these sites was the successive development of various subspecies of praeorbulinids observed. Orbulina spp first occur at various Lower Miocene levels (within NN 6 at Site 214), or as high as in the Middle Miocene (within nannofossil Zone NN 9 at Sites 236 and 237). At subantarctic Site 281, however, the appearance of O. suturalis succeeding to the disappearance of P. glomerosa curva was noted by Jenkins (1975) who placed the Early/Middle Miocene boundary at the lowest occurrence of

---

* These authors (personal communication) have more recently revised their age estimates and now use the dates of 15.5 m.y. for the "Orbulina datum" and 16 m.y. for the "Globigerinoides sicanus datum", while Ryan et al. (1974) use a date of 16 m.y. for the "Orbulina datum".

the latter species (the base of his P. glomerosa curva Zone). At all other sites of the Indian Ocean the Early/Middle Miocene boundary is placed, in this paper, at the Sphenolithus heteromorphus/Helicopontosphaera ampliaperta zonal boundary of Bukry. This boundary closely approximates the D. alata/C. costata radiolarian boundary at Sites 216 and 238 and usually falls within undifferentiated foraminiferal intervals covering Zones N 10 to N 7.

The N 8/N 7 foraminiferal boundary, marked by the initial appearance of Globigerinoides sicanus, was identified only at Site 216. It falls within the H. ampliaperta (nannofossil) Zone of Bukry (= NN 4) and the C. costata (radiolarian) Zone in agreement with the conclusion of Van Couvering and Berggren (in press) for the biostratigraphic position of this boundary. Bronnimann and Resig (1971) noted that in the tropical Pacific the range of G. sicanus is greater than that ascribed to it by Blow (1969). The absence of Catapsydrax stainforthi (which according to Blow, 1969, becomes extinct near the top of N 7) at all Indian Ocean sites (except 236) precludes verification of the observed overlap of these two species. C. stainforthi was noted only at Site 236 (in Core 18, Section 4), but G. sicanus was not reported at this site, where Zone N 8 was not identified. Fleisher (1974), accepting Bronnimann and Resig's(1971) conclusions that Zones N 7 and N 8 cannot be distinguished, referred the interval from Core 13, Section 2 to the bottom of Core 14 at Site 219 in the Arabian Sea to undifferentiated N 8 - N 7, although G. sicanus ranges down to the base of this interval (Fig. 3).

At mid-latitude sites G. sicanus was noted at Sites 253 and 254. Its lowermost occurrence at these two sites, however, cannot be taken to delineate the base of Zone N 8 because it is biostratigraphically too high at Site 253 (within nannofossil Zone NN 5) and too low at Site 254 (within an undifferentiated nannofossil interval NN 3 - NN 1 and below the last occurrence of Catapsydrax dissimilis). G. sicanus was not reported at the higher latitude Sites 251 and 281.

The N 7/N 6 foraminiferal boundary marked by the extinction of C. dissimilis was recognized at Sites 214 and 216 on Ninetyeast Ridge. It is closely coincident with the H. ampliaperta/Sphenolithus belemnos nannofossil boundary of Bukry (= NN 4/NN 3) at both sites and to the Calocyclatta costata/C. virginis radiolarian boundary at Site 214.

Berggren et al. (1974) placed the N6/N 5 boundary between Sections 3 and 4 of Core 5 at Site 216 (Figs. 13 and 22) although Globigerinatella insueta (whose first appearance defines this boundary) first occurs higher in the section in Core 6A, Section 5. The high appearance of this species at Site 216 may possibly result from dissolution of lower specimens. The N 5 zonal assignment to the lower part of Core 5 and to Core 6 by Berggren et al. (1974) at the latter site is based on the presence of Globoquadrina praedehiscens throughout this interval together with Globigerina binaensis in Core 6. The N 6/N 5 boundary, thus, falls within the nannofossil Sphenolithus belemnos Zone of Bukry (= NN 3) and in the upper C. virginis radiolarian Zone.

The N 5/N 4 boundary defined by the extinction of Globorotalia kugleri was recognized at Sites 214, 216 and 238. It falls within the Triquetrorhabdulus carinatus nannofossil zone of Bukry (which encompasses NN 1 and NN 2) at all three sites, being within its lowest subzone (the Cyclicargolithus abisectus subzone, = NN 1) at Site 238 and within the C. virginis (radiolarian) Zone at Site 216.

The position of the three foraminiferal boundaries N 7/N 6, N 6/N 5 and N 5/N 4 relative to the nannofossil zonation is in low-latitude sections of the Indian Ocean in agreement with Van Couvering and

Berggren (in press) estimates. The relationships of the C. costata/
C. virginis and C. virginis/Lychnocanoma elongata radiolarian boundaries with calcareous plankton have been discussed by Theyer and
Hammond (1974b). They pointed out that whether the C. costata/C.
virginis radiolarian boundary falls above, below or coincides with
the N 7/N 6 foraminiferal boundary in the tropical Pacific is unclear, as all three situations have been encountered (Hays et al.,
1972; Winterer et al., 1971; Tracey et al., 1971). In the Indian
Ocean at Site 216 the foraminiferal and radiolarian boundaries are
almost coincident. Theyer and Hammond (1974b) showed that the C.
virginis/L. elongata boundary occurs in the lowermost Neogene (just
below Paleomagnetic Epochs 20/21 boundary at about -22.6 m.y.).
They noted that in the Pacific its precise correlation with the
foraminiferal and nannofossil zones is difficult to establish but
that it probably correlates with the upper part of foraminiferal
Zone N 4 or lower part of Zone N 5, and the upper part of nannofossil Zone NN 1 (possibly lower NN 2). In the Indian Ocean at
Site 216 the C. virginis/L. elongata radiolarian boundary falls
within foraminiferal Zone N 4 and within the Triquetrorhabdulus
carinatus nannofossil Zone of Bukry (which encompasses NN 1 and NN 2).

Oligocene/Miocene Boundary

The base of the Aquitanian stratotype defines the base of the
Miocene (George et al., 1969) and thus of the Neogene. Blow (1969)
correlated the base of foraminiferal Zone N 4, marked by the evolutionary appearance of Globigerinoides primordius (the "Globigerinoides datum") with the base of the Aquitanian. Following Blow many
workers have used the Globigerinoides datum to place the base of the
Miocene. The Aquitanian stratotype, however, rests on a transgressive unconformity and in more complete stratigraphic sections
in the Aquitanian Basin the Globigerinoides datum occurs well below
the level of the base of the stratotype (Eames, 1970; Anglada, 1971;
Scott, 1972). In Puerto Rico, G. primordius has been observed to
appear below Globorotalia kugleri (Seiglie, 1973). Van Couvering
and Berggren (in press) and Theyer and Hammond (1974b) cite other
evidence showing that the Globigerinoides datum occurs well within
the Upper Oligocene nannofossil Zone NP 25. These authors revised
the biostratigraphic position of the Oligocene/Miocene boundary and
proposed an approximate correlation of this boundary with lowermost
nannofossil Zone NN 1, the middle of foraminiferal Zone N 4, and a
level slightly above the Dorcadospyris ateuchus/L. elongata radiolarian boundary. The boundary occurs in the lower part of paleomagnetic Epoch 21 and has an estimated age of -23 to -24 m.y.

For practical purposes the best approximation is therefore to place
the Oliogcene/Miocene boundary at the NN 1/NP 25 nannofossil boundary which has been used in this report to draw the base of the
Neogene. In the Initial Reports the Oligocene/Miocene boundary was
placed by various workers on different criteria. It was drawn by
Vincent et al. (1974) at the western tropical Sites 236, 237 and 238
on the basis of nannofossil data (higher than in the present report
at the base of the Discoaster deflandrei Subzone of Bukry and the
base of the Triquetrorhabdulus carinatus Zone of Roth, a level equivalent to mid-NN 1). At the eastern low-latitude Sites 214 and 216,
Berggren et al. (1974) drew arbitrarily the Oligocene/Miocene boundary
at the initial occurrence of Turborotalia kugleri. This criterion
was also followed by Fleisher (this volume).

It appears that on a foraminiferal basis the Oligocene/Miocene is best approximated in the Indian Ocean by the initial appearance of T. kugleri, which closely coincides at the low-latitude sites menioned above (i.e. 214, 216, 237 and 238; T. kugleri was not noted at Site 236) with the base of the T. carinatus Zone of Bukry (= base NN 1) as well as at mid-latitude Site 253 (T. kugleri was not reported at mid-latitude Site 254). The boundary between Zones N 4 and P 22 cannot be delineated in the Indian Ocean by the "Globigerinoides datum". Significant dissolution was observed at all sites in this part of the stratigraphic column and specimens of this genus, which are highly susceptible to dissolution, are very probably eliminated near the base of their range. As a result, the Oligocene/Miocene boundary based on nannofossil data falls within intervals assigned to Zone P 22.

Paleoceanography and Sediment Accumulation Rates

During the Neogene, changes in the global climate and oceanic circulation due to glaciation strongly influenced the sedimentary record. Neogene stratigraphy, thus, cannot be dissociated from Neogene climatology, which may provide one of the best means for compiling chronology of this time interval. Paleocenaographic studies require detailed quantitative faunal analyses, which have not been undertaken in DSDP Indian Ocean Initial Reports. From semi-quantitative analyses, however, some faunal changes are apparent which, together with changes in the sediment accumulation rates and changes in the isotopic composition of foraminiferal tests at subantarctic Site 281, can be related to climatic and paleoceanographic variations in other oceanic areas. In attempting this correlative study no account is taken for latitudinal movement of sites which, according to reconstructions of plate-motion in the Indian Ocean (McKenzie and Sclater, 1971) has been slight throughout the Neogene, especially for the western sites.

Isotopic and Foraminiferal Changes

Stable isotopic analyses made by Shackleton and Kennett (1975a,b) on benthonic and planktonic foraminifera from DSDP sites of the subantarctic sector of the southeast Indian Ocean and adjacent southwest Pacific have been interpreted by these authors in terms of glacially induced isotopic changes in the ocean. The Paleogene sections analyzed are from Sites 277 (Campbell Plateau) and 279 (Macquarie Ridge) and the Neogene sections those from Sites 279, 281 (South Tasman Rise) and 284 (Challenger Plateau). Shackleton and Kennett have shown that the present day temperature structure of oceanic deep waters was probably established at the beginning of the Oligocene. During the Oligocene, glaciers on the Antarctic continent probably descended to sea level and there was sea-ice production; but if an ice-sheet was present it must have been much smaller than that of the present day. After a period of climatic stability throughout the Oligocene there was an Early Miocene rise in surface temperature of over 2°C followed by a cooling and then a second similar temperature rise at the beginning of the Middle Miocene. The latter warming is recorded at Site 281 in Core 11 (Fig. 18). The isotopic transition between Cores 10 (Middle Miocene) and 7 (Upper Miocene) is interpreted as representing the relatively rapid accumulation of an isotopically light East Antarctic continental ice sheet. From this time on surface temperature estimation is complicated by the change in oceanic isotopic composition, and Shackleton and Kennett estimated surface temperatures at Site 281 (Curve B on Figure 18) by the surface-to-bottom gradient. It appears from their estimate that after the accumulation of the East Antarctic ice-sheet, the surface temperature at Site 281 (upper Core 6) rose during the Late Miocene to values similar to those at the beginning of the Middle Miocene.

The uppermost Miocene and Pliocene isotopic records at Site 281 show a general cooling trend between Cores 6 and 3. However, big gaps in sampling prevent the recognition of fluctuations during this interval. A more complete record was obtained at nearby Pacific Site 284. This section was subdivided by Jenkins (1975), using the same zonation as at Site 281, and was correlated to New Zealand stages by Kennett and Vella (1975). The latter authors examined quantitative changes in the planktonic foraminiferal fauna at Site 284 which suggest substantial climatic fluctuations, a suggestion strongly supported by the oxygen isotopic record. An intense cooling is recorded during the interval which corresponds to the Kapitean and lower Opoitian stages of New Zealand (with a brief warming episode over the Kapitean/Opoitian boundary) that reflects substantially greater dimensions of the East Antarctic ice-sheet than at the present day. The age of this severe cooling (accompanied by a regression facies in New Zealand) has been the subject of controversies. Kennett and Watkins (1974) correlated the Kapitean to the lower Gilbert paleomagnetic Epoch between -4.7 m.y. and -4.3 m.y., which would thus indicate an Early Pliocene age (see Fig. 2) for this stage. This age assignment (followed by Kennett and Vella, 1975; Shackleton and Kennett, 1975a,b; Kennett et al., 1975; Kemp et al., 1975; Hayes and Frakes, 1975) contradicts the correlation of the Kapitean with the Late Miocene intense cooling observed throughout the North Pacific and with the Late Messinian Stage of Italy (latest Miocene), during which the Mediterranean dried up. Dessication of the Mediterranean at that time is thought to be linked to the glacio-eustatic lowering of sea level caused by the increase of Antarctic ice volume associated with the Kapitean. Recent reevaluations of the Kapitean age assignment (Ryan et al., 1974; Van Couvering et al., 1976), however, place the base of this stage within the Messinian, thus establishing the synchroneity during the latest Miocene of the North and South Pacific cooling events and the glacio-eustatic regressions observed in various parts of the world. Ryan et al. (1970) assigned an age of approximately -6 to -4.3 m.y. to the Kapitean whereas Van Couvering et al. (1976) assign an age of -5.5 to -5.1 m.y. Using the latter age estimates permits the correlation of the brief warming events recorded at the Kapitean/Opoitian boundary in the Southwest Pacific and in uppermost Zone N 17 in the North Pacific (Ingle, 1973; Vincent, 1975, her Fig. 8). The Kapitean (bounded by the levels of initial appearances of Globorotalia conomiozea and G. puncticulata respectively, thus equivalent to the lower G. miozea conomiozea Zone of Jenkins, 1975) corresponds to an unrecovered interval at Site 281 lying between Cores 5 and 6, as indicated by the first occurrences of G. conomiozea and G. puncticulata (in Cores 5,CC and 3,CC) just above this interval (Fig. 18). Thus, it is not possible to recognize any Kapitean excursion on the isotopic curves obtained at this site.

Following the Kapitean-lower Opoitian cooling a prolonged warm interval is recorded at Site 284 during the rest of the Opoitian (Early Pliocene and earliest Late Pliocene) interrupted by a brief cooling (at about -3.4 - 3.5 m.y.) that reflects a lesser advance of the Antarctic ice sheet. Another substantial cooling occurs during the Late Pliocene (from about -2.6 to -2.15 m.y.) associated

with the Waipipian Stage, during which marked changes in the oxygen
isotopes record the accumulation of a substantial northern-hemisphere
ice sheet. These paleoclimatic trends are similar to those determined
for the New Zealand marine sequences (Kennett, 1967, 1968; Devereux
et al., 1970; Kennett et al., 1971; Kennett and Watkins, 1974) and
the North Pacific (Bandy, 1972a; Bandy and Ingle, 1970; Ingle, 1967,
1973, 1975; Hays et al., 1969; Kent et al., 1971; Echols, 1973;
Vincent, 1975).

The cool surface-temperature value obtained at Site 281 in upper
Core 3, slightly above the first appearance of Globorotalia inflata,
is probably associated with the Late Pliocene (Waipipian) cooling.
The paleoclimatic faunal tools used by Kennett and Vella (1975) at
nearby Site 284, located just north of the Subtropical Convergence,
should be applied at Site 281, situated just south of it, in order
to define detailed climatic oscillations at the latter site. Kennett
and Vella utilized as paleothermometers the frequency variations in
Neogloboquadrina pachyderma (this cool-water species being more abundant in cooler intervals) and those (reciporocal) in Globigerina woodi,
as well as changes in the ratio of abundances of the cooler-water form
N. pachyderma to the warmer-water form Globigerina falconensis and of
the cooler-water form Globigerina bulloides to the warmer-water form
G. falconensis. All these species were noted by Jenkins (1975) as
occurring at Site 281, but without indication of their frequency. It
is reported, however, in the site report (Kennett, Houtz et al., 1974,
Chapter 8) that there is some evidence in the Lower Miocene to Pliocene foraminiferal sequence that the Subtropical Convergence oscillated
north and south over the area (Op. cit., p. 277) and also that a significant warming occurred in the Early Miocene and an obvious cooling
in the Late Miocene and earliest Pliocene (Op. cit., p. 271).

Site 264 on the southern Naturaliste Plateau lies at the boundary
between the transitional and subtropical faunal provinces (Fig. 1) and
is ideally located to record latitudinal shifts of these provinces
through time. Unfortunately, the great condensation of the Neogene
section at this site (Fig. 17) decreases its value. Semi-quantitative
data from Kennett (1975) show that here N. pachyderma occurs commonly
throughout the Neogene sequence, without significant variations in
its abundance. Minor changes in the frequency of this species in
the Pleistocene section, however, may possibly reflect climatic
variations. A lower abundance of N. pachyderma in the mid-Pleistocene
sequence (Core 1A, Section 3) may indicate a relatively warmer interval,
whereas its higher abundance in the uppermost Pleistocene sample (Core
1, Section 3) may reflect a cooler interval. G. bulloides and G. falconensis are also common components of the fauna at Site 264 and a
marked decrease in the ratio of G. bulloides to G. falconensis in
Core 2A, Section 6 at the level of first appearance of Globorotalia
inflata may possibly reflect a warm interval preceding the Late Pliocene cooling. Data from both Kennett and Kaneps (1975b) at this site
show that the warm-water form Globorotalia menardii is present throughout the Lower Pliocene section, is absent in the Upper Pliocene (except for one single occurrence in Core 2A, Section 4) and occurs intermittently in the Pleistocene. This trend suggests conditions of deposition cooler for the Upper Pliocene sequence of the southern
Naturaliste Plateau than for the Lower Pliocene.

No trend in the abundance of G. menardii, however, was observed
in the Pliocene sequence of Site 258 on the northern Naturaliste
Plateau where this species was found (Boltovskoy, 1974a) as rare to
very rare in isolated occurrences throughout the entire Pliocene and
uppermost Miocene (Core 5A, Section 1 through Core 8, Section 3). G.

menardii, however, occurs at this site throughout lower Core 8A (being common in Section 5) in the lowest part of the Upper Miocene interval overlying the Cretaceous, possibly indicating warmer conditions for this part of the Upper Miocene (equated to an undifferentiated NN 11-NN 10 nannofossil interval, see Fig. 16). *G. menardii* occurs consistently again at Site 258 throughout the Pleistocene sequence. Variations in its abundance accompanied by simultaneous changes in the abundance of the warm-water species *Sphaeroidinella dehiscens* and the cooler-water species *Globorotalia truncatulinoides* probably reflect temperature changes. Frequency variations in these species indicate a general trend toward warmer conditions for the lower Pleistocene section (Core 3A and below, with rare continuous occurrences of *G. menardii*, rare to very rare *S. dehiscens* and very rare isolated occurrences of *G. truncatulinoides*), and cooler conditions for the upper Pleistocene section (Core 2A and above, without *S. dehiscens*, with very rare isolated occurrences of *G. menardii* and continuous occurrences of rare to common *G. truncatulinoides*).

At other mid-latitude sites the only discernible trend in abundance fluctuations of *G. menardii* is in the Pleistocene sequence of Site 253 in which this species is abundant to common and was used by Boltovskoy (1974b) to construct a paleoclimatic curve (Fig. 14b). In the remaining mid-latitude sequences, *G. menardii* was noted with very rare isolated occurrences in the Pliocene of Site 253, in the uppermost Pleistocene of Site 254, and throughout the Upper Miocene to Pleistocene of Site 251. *G. menardii* is a dominant component of the Upper Miocene to Pleistocene faunas at low-latitude sites, and trends in its frequency variations in these sections cannot be recognized without further detailed quantitative studies. There is, however, an apparent decrease in the relative abundance of this species in the uppermost Miocene, Pliocene and lowermost Pleistocene at Sites 219 (Arabian Sea) and 237 (Mascarene Plateau), which may reflect a general cooling during this interval in the northwest Indian Ocean.

Indications of paleoclimatic trends at the mid-latitude sites of Leg 26 cannot be obtained from variations in ratios of *G. bulloides* to *G. falconensis* because Boltovskoy (1974a) did not include forms belonging to this group in his range-charts, in view of the uncertainty regarding their taxonomy (op. cit. p. 677). Boltovskoy reported that in these sections *N. pachyderma* is present only at the southernmost Site 251. This species is here very rare with isolated occurrences in the Lower Pliocene (Cores 6A to 4A), rare with consistent occurrences in the Upper Pliocene equivalent to nannofossil Zone NN 16 (Cores 10, 1A and 2A), virtually absent (except in one horizon) in the uppermost Pliocene and lowermost Pleistocene (Cores 5 to 8), and again rare with consistent occurrences in the rest of the Pleistocene (Cores 4 to 1). These frequency oscillations in *N. pachyderma* probably reflect paleoclimatic changes and seem to indicate surface temperatures at Site 251 which are cooler during most of the Late Pliocene than during the Early Pliocene.

At low-latitude sites of the eastern Indian Ocean none of the cool-water forms *G. bulloides*, *N. pachyderma* and *G. inflata* were found, whereas these species were noted at the low-latitude sites of the western part of the ocean. *G. bulloides* and *N. pachyderma* are rare but occur consistently in the Upper Miocene to Pleistocene sequences of the Somali Basin (236), Arabian Sea (219), and Gulf of Aden (231, 232, 233) probably reflecting the presence of upwelling in these basins. Rare isolated occurrences of *G. inflata* were noted in the Upper Pliocene and Pleistocene on the Mascarene Plateau (237) and in the Gulf of Aden (231).

The warm-water forms, resistant to solution, that compose the globoquadrinid group are abundant to common in the Lower and Middle Miocene sections of the Indian Ocean in mid- as well as in low-latitudes. They are present in the Lower Miocene only at subantarctic Site 281. At mid-latitude sites these forms decrease markedly in abundance in the Upper Miocene, are rare to very rare through the Upper Miocene and Pliocene and are absent from the Pleistocene (except in the lowermost Pleistocene at Site 254). At low-latitude sites, on the other hand, they remain common throughout the entire Neogene but occur with lower frequencies in Upper Pliocene and lower Pleistocene of most sections. It, thus, appears that relatively warm conditions (optimum for the development of globoquadrinids) prevailed in the mid- as well as in the low-latitudes of the Indian Ocean during the Early and Middle Miocene and that cooling starting in the Late Miocene displaced the zone of optimum conditions for this warm-water group north of 25°S (the latitude of Site 253). The group is restricted to low-latitudes in the Pleistocene (as Globoquadrina conglomerata). At these latitudes a decrease in abundance in the Upper Pliocene and lower Pleistocene reflects a time of cooling. As mentioned earlier the extinction of Globoquadrina dehiscens in the Indian Ocean appears to be paleocenaographically controlled and occurs progressively later from south to north in response to progressive cooling during the Middle Miocene to the Early Pliocene. It is noteworthy that this species lived longer in the area of the Naturaliste Plateau (Sites 258, 264) where it survived until the latest Miocene, than at lower latitudes of the open ocean to the west (Sites 253, 254) where it became extinct in the Middle Miocene as it did at temperate Site 251. Although it is difficult to explain from present-day distribution of surface isotherms (Wyrtki, 1971) that temperatures were warmer near southwest Australia than to the west, these conditions appear to be also evidenced by: 1) the last occurrence of Globoquadrina altispira on the Naturaliste Plateau at the same Upper Pliocene level as in the tropical region, whereas at Site 254 to the west this species disappears at a much lower stratigraphic position similar to that of temperate Site 251; and 2) by the presence of warm-water pulleniatinids in sections of the Naturaliste Plateau, while these forms are absent at Site 254 to the west as well as at temperate Site 251.

Antarctic sites from the southern Indian Ocean (Leg 28) show evidence of a gradual cooling throughout the Miocene (Kemp, et al., 1975; Hayes and Frakes, 1975). The distribution of the first ice-rafted detritus and the boundary between a calcareous biofacies and an overlying siliceous biofacies show a pronounced diachronism growing progressively younger towards the north. The change in facies occurs in the interval Late Oligocene to Early Miocene at the southern Sites 267 and 268, in the Middle Miocene at Site 266 and in Late Miocene at Site 265 (Table 3). The earliest ice-rafting is evident from the Middle Miocene at Site 267, the Late Miocene at Site 266 and is not apparent at Site 265. Ciesielski and Weaver (1974) constructed paleotemperature curves from quantitative silicoflagellate studies at Site 266 as well as in nearby paleomagnetically dated piston cores. Their findings imply that after the glacial maximum of the Late Miocene-earliest Pliocene circumpolar surface water temperature increased significantly from about -4.2 to -3.95 m.y., suggesting the retreat of the western Antarctic ice sheet at this time with its reestablishment during a succeeding climatic cooling in the late Gilbert.

A series of paleoclimatic cycles is indicated in the section corresponding to the Matuyama and Brunhes paleomagnetic Epochs (uppermost Pliocene and Pleistocene) in Antarctic and Subantarctic deep-sea cores from the southeast Indian Ocean (Keany and Kennett, 1972; Theyer, 1972; Bandy and Casey (1973) and from the Pacific (Hays, 1965, 1967; Kennett, 1970; Keany and Kennett, 1972; Bandy et al., 1971). The relative amplitude of the cycles is somewhat lower in the Matuyama indicating more uniform southern ocean conditions during the Matuyama than the Brunhes. Different interpretations, however, have been given by various investigators. Most workers have claimed that much colder climates prevailed during the Brunhes than during the late Matuyama, whereas Kennett (1970) and Keany and Kennett (1972) find evidence that the Brunhes, although a time of erratic high-intensity climatic cycles, had a generally warmer temperature than the Matuyama.

Isotopic and faunal data at Site 284 indicate rather stable paleoglacial conditions in the early Matuyama (Mangapanian) when the northern-hemisphere ice sheet would have been about half the size of those which accumulated during the Brunhes Epoch. One major interglacial and one major glacial episode are recorded in the lower Pleistocene (Hautawan) at Site 284 (below a mid-Pleistocene disconformity). The documentation of a major glaciation in the earliest Pleistocene is of significance. Kennett et al. (1971) defined two brief cooling episodes near the Pliocene/Pleistocene boundary in marine sediments of New Zealand. It is probable that associated with these coolings there was an increase in the erosional efficiency of bottom waters through the South Indian Basin during the latest Pliocene and earliest Pleistocene. This increase in erosion is evidenced by a major sedimentary hiatus representing the interval of time between approximately -1.3 to -2.2 m.y., accompanied by the occurrence of micromanganese nodules in piston cores from this basin (Weaver and McCollum, 1974). Pleistocene surface temperature at Site 281 could be estimated from oxygen isotopic analysis from only two samples (Fig. 18). The lowest one in Core 2, Section 3, very slightly above the Pliocene/Pleistocene boundary, indicates a relatively warm surface temperature and may possibly correspond to the interval between the two coolings recorded in New Zealand near the boundary. However, this suggestion remains tentative in view of the scarcity of data and lack of a precise age determination. The uppermost sample analyzed at Site 281 indicates a cooling trend throughout the interval represented by Core 2.

ELTANIN piston cores from the subantarctic southeast Indian Ocean, located between the Subtropical Convergence (39°S) and the Australasian-Subantarctic front (48°S), between about 90°E and 115°E, have been quantitatively studied by Vella et al. (1975) and Douglas and Johnson (1974) for their planktonic foraminiferal content which reflects shifts of these two oceanographic features during the last one million years. Vella et al. using ratios of G. bulloides to G. inflata and of left- to right-coiling N. pachyderma as paleothermometers, showed that a warming trend with superimposed temperature fluctuations occurred during the last 0.8 m.y. These fluctuations correspond closely in age to northern-hemisphere glacial and interglacial phases, according to the glaciation dates given by Berggren and Van Couvering (1974). It appears that in the southeast Indian Ocean a period of severe cooling between -1.0 and -0.65 m.y. corresponds in age to the first major northern hemisphere glaciation, which occurs in the interval between the Jaramillo and the Brunhes/Matuyama boundary. Subsequent coolings occur at -0.55, -0.17, and -0.04 m.y.

The correlation of small-scale paleoclimatic fluctuations requires

extremely accurate stratigraphy, which is indeed beyond the degree
of biostratigraphic precision obtained in DSDP sections. Furthermore, owing to the lack of quantitative faunal analyses in DSDP
reports rough trends can be detected, but not detailed climatic
variations. However, the paleoclimatic curve constructed by Boltov-
sky (1974b) from a study of the section at 10 cm interval at Site
253 in the southern subtropical region can be analyzed for detailed
Pleistocene temperature fluctuations in this area (Fig. 14b). Ages
have been tentatively assigned to the temperature changes indicated
by Boltovskoy's curve on the basis of a constant sedimentation rate
of 0.5 m/m.y. throughout the section. Although this assumption is
supported by the last occurrence of Pseudoemiliania lacunosa at 2.5
meters, these age assignments are approximate because varying intensities of calcium carbonate solution during warmer and colder
periods probably resulted in intermittent changes in the sedimentation rates. Age assignments are especially questionable in the
uppermost part of the section because of uncertainty in the recovery
of the top sediments. Furthermore, restrictions in the reliability
of temperature changes obtained from only one paleoclimatic tool
(ratio of G. menardii to G. inflata) should be kept in mind.

The coolest temperature recorded in the Pleistocene sequence of
Site 253 in its lowermost part between 9 and 8.3 m (-1.86 to -1.66
m.y.), may possibly correspond to the lowermost Pleistocene glacial
episode recorded in the southwest Pacific at Site 284 and in New
Zealand. The well-marked warm interval between 7.1 and 5.9 m
(-1.42 to -1.18 m.y.) may possibly correspond to the succeeding
interglacial episode recorded in the lower Pleistocene at Site 284.
Following this warm interval, a brief cooling and a brief warming
are recorded from 5.9 to 5.59 m (-1.18 to -1.12 m.y.) and from 5.59
to 5.25 m (-1.12 to 1.05 m.y.) respectively. The cooling recorded
at 5.25 m marks the beginning of a long relatively cool interval (up
to 2.59 m) with small-scale temperature fluctuations. The length
of time represented by this interval appears to be from -1.05 to
-0.52 m.y. and, thus, encompasses the time of severe cooling
recorded farther south (Vella et al., 1975) which correlates with
the first northern hemisphere glaciation. It corresponds in age to
late Matuyama and early Brunhes time, the boundary between these two
magnetic epochs based on sedimentation rate estimates lying at
3.45 m in the section. From 2.59 m to the top of the sedimentary
sequence a warming trend is observed (with a cool episode between
1.65 and 1.40 m), and the warmest temperature indicated by the entire
curve is at the top. It appears, thus, that a warming trend occurred
in the southern subtropical Indian Ocean during about the last
-0.6 m.y., that is, through most of the Brunhes epoch as was observed further south in the subantarctic region.

The presence of G. menardii only in the upper half of the Pleistocene section (upper 7.4 m) at Site 254, located south of Site 253
at the boundary between the subtropical and temperate provinces
(Fig. 1), may reflect a similar upper Pleistocene warming trend.
However, such is not the case in sections of the Naturaliste Plateau
where the faunal trends mentioned earlier indicate cooler conditions
for the upper half of the Pleistocene sequence at Site 258 (upper
28.5 m; Core 2A and above) and for the upper part of Core 1 at
Site 264.

At temperate Site 251, Neogloboquadrina pachyderma occurs only in
the upper part of the Pleistocene above 30.5 m. In this expanded
section the age of this level is estimated to be 1.1 m.y. old,
assuming a constant Pleistocene accumulation rate (supported by the

stratigraphic position of the last P. lacunosa). This age corresponds to the well-marked change in temperature from warmer below to cooler above recorded at Site 253 (Fig. 14b). A peak at 5.4 m (Core 2, Section 3) in abundance of N. pahcyderma is estimated to be about 0.2 m.y. old and corresponds to a cooler excursion on the curve of Site 253. N. pachyderma decreases from this level to the top of the sequence. It appears, thus, that rough trends in temperature change in the temperate area indicated by frequency variations in N. pachyderma are in general agreement with those defined in the subtropical region. It should be mentioned that N. pachyderma remains right coiling throughout the Pleistocene at Site 251 and 264 (Boltovskoy, 1974a; Kennett, 1975). Living right-coiling populations of this species in the Indian Ocean inhabit waters with surface temperature between 10°C and 18°C (Bé and Hutson, in press), which implies that the 10°C surface isotherm did not shift north of the latitude of those sites during the Pleistocene.

In tropical sections of the western Indian Ocean, the upper part of the Pleistocene at Sites 219 (upper 15 m), 237 (upper 3.7 m), and 238 (upper 23.5 m) appears to contain faunal assemblages of a generally warmer type than the lower part as evidenced by a higher relative abundance of G. menardii, Globoquadrina conglomerata and Pulleniatina obliquiloculata, a lower frequency of Globorotalia truncatulinoides and the absence of G. inflata. No attempt is made to assign an age to these not very well-defined faunal changes in slowly accumulating sections of the open ocean. However, in the expanded hemipelagic tropical sections of the Gulf of Aden and of the Timor Sea age assignments can be attempted, on the basis of average accumulation rates, to relatively well-defined intervals of cooling or warming.

At Site 262 in the Timor Sea, an interval between 150 and 270 m (Cores 17 to 29) contains a greater abundance of the warm-water forms P. obliquiloculata, Globigerinoides conglobatus, G. sacculifer, Globorotalia menardii, Neogloboquadrina dutertrei and Globigerina rubescens than does the rest of the Pleistocene section (Akers, 1974b) and probably corresponds to a time of warming dated at -0.25 to -0.45 m.y. This interval occurs just above the last occurrence of P. lacunosa and appears to correlate with the relatively warmer excursion on the climatic curve of Site 253 between -0.33 and -0.52 m.y. At Site 231 in the Gulf of Aden, an interval between 17 and 33.5 m (Core 4 to Core 5, Section 5) in which G. inflata is rare to common, G. menardii decreases in abundance and G. truncatulinoides increases in abundance probably reflects a time of cooling dated -0.45 to -0.71 m.y. This interval occurs just below the last occurrence of P. lacunosa and appears to correspond to the cooler interval of the climatic curve of Site 253 between -0.52 and -0.61 m.y. It also probably correlates with the cooling defined in the subantarctic area at -0.55 m.y. by Vella et al. (1975) and would, thus, correlate with oxygen isotope Stage 12, according to age estimates of Shackleton and Opdyke (1973).

Sediment Accumulation Rates

Average rates of accumulation in the Neogene series at the sites represented on Figures 3 to 19 are indicated on Table 4. None of the rates were corrected for effect of compaction. Rates were averaged for each interval without accounting for periods of reduced or increased sedimentation or of breaks in deposition within the intervals. However, significant short term fluctuations in sedi-

TABLE 4. Average sediment accumulation rates in m/m.y. at selected sites of the Indian Ocean. Subbottom depths of Series boundaries are derived from Figures 3 to 19. When the boundaries occur in an interval of undetermined age, the boundary was placed arbitrarily in the middle of the undetermined interval. Whenever the Series boundaries could not be identified directly from the paleontological events indicated on the left, a best estimate (indicated between brackets) was obtained from other events approximately equivalent. At Site 264 the base of the Neogene section, which may possibly include the uppermost Miocene, is questionably placed at the Miocene/Pliocene boundary.

mentation rates may occur, as exemplified in Table 4 for two specific series (Pleistocene and Upper Miocene) when reliable paleontologic datums could be used. Sediment accumulation depends upon the balance between the rate of sediment supply and removal, both controlled by a variety of processes. A detailed discussion of sediment accumulation rates throughout the entire Indian Ocean is beyond the scope of this report. The complexity of the physiography and of the tectonic activity of the Indian Ocean, both controlling factors in sediment distribution, makes the reconstruction of the sedimentary history of this basin particularly difficult. However, keeping in mind the danger of regionalizing the local sedimentation regimes observed at each site, some general trends in carbonate accumulation which seem of particular interest for their paleoceanographic implications are apparent.

All sequences considered here have been above the CCD throughout Neogene times and consist essentially of calcareous oozes and chalks. Clay, however, is common in the Lower to Middle Miocene interval of the deeper Somali Basin Site 236 which probably lay close to the CCD during that time. Pelagic sedimentation prevailed at all sites except those in the Gulf of Aden (231, 232, 233) and the Timor Sea (262) where hemipelagic sedimentation resulted in expanded sections which accumulated much faster than those in the open ocean. Siliceous biogenic components make up a significant amount of the sediments in the upper Neogene of low-latitude sections, now located in the equatorial belt of high productivity. The rather abrupt increase in biogenic silica in these sections occurs in the Late Miocene (as well as at other low-latitude deep-basin sites located below the CCD).

All sites, but Site 236, are located between 1600 and 3000 m water depth and, as far as bottom circulation is concerned, are under the influence of intermediate- and deep-water masses. Site 236 in the Somali Basin, at a water depth of 4500 m, is under the influence of Antarctic Bottom Water (AABW). At the present time cold, "aggressive" AABW forms by sinking from the surface along the continental slope of Antarctica. There is no evidence, however, for the formation of bottom water along this continent in the Indian Ocean (Wyrtki, 1973). AABW penetrates into the Indian Ocean both from the west and east and extends north. Its various paths detected from data on bottom-water potential temperature, turbidity, current indications and Antarctic diatom transport, have been described by Kolla et al. (1976). AABW of Weddell Sea origin flows northwards from the Agulhas Basin into the Mozambique Basin (see Fig. 1) and from the Enderby Basin into the Crozet Basin. From the latter basin it flows through the Southwest Indian Ridge, at about 29-26°S and 60-64°E, into the Madagascar, Mascarene, Somali and Arabian Basins. East of the Kerguelen Plateau AABW formed along the Adelie Coast and Ross Sea flows west in the South Indian Basin toward the Kerguelen Plateau, then parallel to it and finally returns eastward hugging the Southeast Indian Ridge. Some of this water flows through the Ridge at about 120-125°E into the South Australian Basin and subsequently between Broken Ridge and the Naturaliste Plateau into the Wharton Basin.

The deep water of the Indian Ocean originates from the North Atlantic Deep Water (NADW) which enters the Indian Ocean to the south of Africa at a depth between 2500 and 3200 m. Together with the bottom water it forms a rather uniform water mass of great volume, the circumpolar water, which flows east in the lower parts of the Circumpolar Current. It also spreads north into the Indian Ocean where it can be followed to the north of Madagascar and into the Wharton Basin (Wyrtki, 1973).

Intermediate water masses include the North Indian Intermediate Water (NIIW), of Persian Gulf, Red Sea and Arabian sea origin, which spreads throughout the northern Indian Ocean, and the Antarctic Intermediate Water (AIW), formed in the Antarctic Convergence at about 50°S by sinking of surface waters, which moves north and mixes in part north of the equator with NIIW. It appears that most of the water ascending from the bottom and the deep water in the northern part of the ocean returns south mixed with NIIW, especially along the western side of the ocean.

It, thus, appears that in most of the Indian Ocean, erosion by bottom currents is largely controlled by the circum Antarctic circulation which, when increased would enhance bottom current effects at deep as well as at intermediate depths. It also appears that the eastern and western Indian Ocean, east and west of the Central Indian Basin, are under two distinct water circulation regimes. Bottom and deep waters spread northward in both areas but originate in two different regions on either side of the Kerguelen Plateau-Gaussberg Ridge. The southward flow of intermediate water is more intense in the western area. Two distinct eastern and western circulation patterns must have been in existence throughout the Neogene when India was moving northward.

Recent erosional effects related to flows of AABW in the western Indian Ocean have been delineated: 1) in the Crozet Basin by manganese nodule formation and sediment reworking (Kolla et al., 1976), and by an extensive hiatus between Quaternary and Eocene sediments (Hekinian et al., 1969); 2) in the northwest corner of the Madagascar Basin by the presence of only a veneer of Quaternary sediments at the top of Core DODO 117 (Fig. 19) located at the base of the Southwest Indian Ridge flank; and 3) in the Mascarene Basin by an erosional surface at the top of a core (Leclaire, 1974). Current action of AABW in the Somali Basin apparently have been in effect throughout the Neogene, as evidenced by the alternance of fine- and coarse-grained ooze layers throughout the sedimentary sequence at Site 236. The fine-grained layers, which contain more poorly preserved planktonic foraminifera, are probably autochtonous, whereas the coarse-grained layers with better preserved planktonic foraminifera may be derived from transported sediments (Vincent, 1974b). This suggestion is supported by the abundance of reworked older elements (radiolarians and foraminifera, apparently winnowed) associated with the coarse-grained layers and by variations in the relative abundance of benthonic foraminiferal species, which vary from 1% to 10%.

Erosional effects of AABW are widespread in the southeast Indian Ocean where Watkins and Kennett (1972) and Kennett and Watkins (1976) have shown the existence of extensive areas of erosion or non-deposition associated with manganese-nodule pavements in the South Australian Basin and the South Indian Basin. In deep parts of these two basins sediments ranging in age from Quaternary to Pliocene have been systematically removed. Kennett and Watkins related this major regional sedimentary disconformity to a substantial increase in the velocity of AABW, associated with the Late Pliocene cooling. Apparently two major pulses of bottom-water activity have occurred since that time. One of these includes part of the late Gauss and the early Matuyama (-2.5 to -1.5 m.y.) and the other occurred within the Brunhes.

Strong erosive bottom-current activity at intermediate depths has occurred in various parts of the Indian Ocean. Geologically old outcrops and manganese pavements on flanks of the Agulhas Plateau (Saito and Fray, 1964) result from intense erosive action of NADW

(Kolla et al., 1976). Scouring effects of the southward Mozambique Agulhas Current on the African shelf and slope have been evidenced in the southern area of the Mozambique Channel (Vincent, 1973, 1976). From an analysis of water movement in the southwest Indian Ocean (Vincent, 1976), it appears that circulation at intermediate depths in this area is complex as a result of the encounter of the AIW from the south and NIIW from the north. AIW tends to penetrate into the Mozambique Channel on its western side, whereas NIIW tends to extend south of the channel on its eastern side. Vincent showed that the Mozambique Ridge is swept by a large counterclockwise swirl of AIW which prevents sedimentation on the Ridge, in the area covered with manganese nodules lying under the central part of the gyre. The intense erosive effect of this system appears to have been in action for a long time as evidenced by the lacuna of most of the Pliocene and Pleistocene further north on the Ridge at Site 249 in an area which at the time is under the influence of this swirl. At the latter site a veneer of uppermost Pleistocene (Emiliania huxleyi Zone) unconformably overlies lowermost Pliocene (Ceratolithus rugosus Zone of Bukry) in Core 1, Section 2 (Bukry, 1974d; Müller, 1974). Cores raised from the flanks of the Mozambique Ridge yielded mixed foraminiferal faunas of Late Cretaceous, Eocene, Miocene and recent age assumed to be derived from submarine erosion of formations outcropping at shallower depths on the Ridge (Saito and Fray, 1964).

In the eastern Indian Ocean erosion over much of the crest and flanks of the Kerguelen Plateau to depths of 1700 m has been attributed to high current velocities associated with the eastward flow of Antarctic Circumpolar Current (Kennett and Watkins, 1976). The Naturaliste Plateau, which appears to be under the influence of Circumpolar Deep Water admixed with NADW (Wyrtki, 1971), has recently experienced strong current activity as indicated by the analysis of bottom photographs (Kennett and Watkins, 1976) and by the occurrence of a reversed magnetic interval in Core RC8-55 indicating that sediments from at least the entire Brunhes interval are missing (Opdyke and Glass, 1969).

Bottom-current activity resulting in sea-floor erosion in the Indian Ocean at intermediate as well as at deep depths has prevailed for a considerable time during the Cenozoic possibly since at least the Late Eocene to Early Oligocene when the development of the AABW started. The production and introduction of this water mass northward in the ocean is probably responsible for the existence of major hiatuses and/or reduced sedimentation centered in Oligocene sections in many areas (Davies, et al. 1975), a phenomenon observed also in the western Pacific in the Tasman-Coral Sea between Australia and New Zealand (Kennett et al., 1972). In this respect it is noteworthy that the only thick calcareous Oligocene sequence (about 160 m) drilled in the Indian Ocean lies at Site 242 in the Mozambique Channel on the Davie Ridge, a locale protected from the northward flow of deep water by the Mozambique Channel sill. This site, which is located in a subtropical area at a latitude midway between the latitudes of Sites 214 and 253 (see Fig. 1), where apparently low-energy currents prevailed throughout Oligocene-Neogene times and where hemipelagic sedimentation produced an expanded, apparently continuous section, would offer a unique opportunity for biostratigraphic and paleocenographic studies; unfortunately coring at this site was widely scattered.

By Late Oligocene time the South Tasman Rise had moved north far enough to allow the initiation of a Circum Antarctic Current. The development of an active circumpolar deep-sea circulation at this time resulted in a highly erosive western boundary current system in the Pacific to the east of New Zealand and probably decreased the important northward flowing bottom currents in the Tasman-Coral Sea, thus, allowing reinitiation of sediment deposition in the latter area (Kennett, et al., 1974). A similar decrease of northward bottom currents, diminishing erosive effects, probably occurred in the Indian Ocean. Sedimentation, however, appears to have resumed significantly only at the end of the Early Miocene. During the Early Miocene preservation of planktonic foraminifera remained poor, as in the Oligocene, and the sequence is still very condensed. Sediments were not deposited on the Naturaliste Plateau and accumulated in other areas with low rates of about 2 to 5 m/m.y. A higher rate of approximately 10 m/m.y. occurred at Site 238 on the Central Indian Ridge. At this site, as well as at Site 251 on the Southwest Indian Ridge, sediments accumulated throughout the Neogene usually two to three times more rapidly than at other sites. Accumulation rate changes at these two sites, however, show trends similar to those derived at other sites. The increased depositional activity at Sites 238 and 251 possibly reflects penecontemporaneous reworking at these sites, a result of their location close to the ridge crest and to an active transform fault.

In the Middle Miocene accumulation rates increase to values of about 6 to 10 m/m.y. (and to values of 20 to 47 m/m.y. at Sites 238 and 251 respectively), with the exception, however, of Site 254 on the southern Ninetyeast Ridge where an exceptionally high rate of 22 m/m.y. is obtained. There was still no deposition on the Naturaliste Plateau, but sedimentation resumed at Site 249 on the Mozambique Ridge following a Cretaceous-Lower Miocene gap.

Sediment accumulation rates decrease again in the Late Miocene in the southeast Indian Ocean (including southern Ninetyeast Ridge: Sites 253 and 254), with overall values of 1 to 5 m/m.y. Most of the Upper Miocene (except the uppermost part) is still missing on the southern Naturaliste Plateau (Site 264) but sedimentation resumed earlier in the Late Miocene on the northern part of the Plateau (Site 258). This reduction in sedimentation in the southeast Indian Ocean is possibly related to an increase in velocity of bottom currents flowing northward from the South Indian Basin during the time of initiation and build up of the East Antarctic Ice Sheet. A suggestion supported by a gap in sedimentation in Late Miocene at Antarctic Site 265 on the flank of the Southeast Indian Ridge.

By contrast to the southeastern part of the ocean, the overall Upper Miocene accumulation rates of the western part and of central and northern Ninetyeast Ridge (Sites 214 and 216) increase to values of 8 to 15 m/m.y. (with high values of about 25 m/m.y. at Sites 238 and 251 and of 41 m/m.y. for the hemipelagic section of Site 231 in the Gulf of Aden). It appears, however, that during the early Late Miocene (from -10.5 to -6.3 m.y.) rates remain of the same order of magnitude as during the Middle Miocene and increase abruptly at about -6.3 m.y. The age of this increase was identified in low-latitude sections by the position of the Ommatartus penultimus/Stichocorys peregrina radiolarian boundary (paleomagnetically dated within Paleomagnetic Epoch 6 by Theyer and Hammond, 1974a) and/or the contemporaneous Discoaster berggrenii/Ceratolithus primus nannofossil boundary of Bukry (both these nannofossil and radiolarian zonal boundaries coincide at Sites 237, 238 and 214, as well as in the North Pacific at Site 310). It is probable that bottom currents moving north from the Agulhas and Enderby Basin did not reduce sedimentation in the western Indian Ocean during the time of East Antarctic ice build up as much as did those originating from the South Indian Basin. Erosive effects from the latter were particularly strong in the southeast Indian Ocean (including the area of the southern Ninetyeast Ridge) but were probably attenuated to the north and did not affect as much the central and northern part of the Ninetyeast Ridge. Bottom currents, however, were apparently intense throughout the ocean near the Middle/Late Miocene boundary, as evidenced by the general condensing of fossil zones at that time (even on the central and northern Ninetyeast Ridge) and by an interval of mixed and reworked foraminiferal faunas spanning the Middle/Late Miocene boundary throughout the western Indian Ocean. The latter area may have been subject to more intense tectonic activity near the end of the Middle Miocene when the Arabian and African plates were elevated above sea level (Buchbinder and Gvirtzman, 1976). One possible consequence of such tectonic activity would be to expose outcrops to the reworking and winnowing action of bottom currents. A significant change in water circulation, which would have affected the western part of the ocean more than its eastern part, probably also occurred at this time when the closure of the connection of the Indian Ocean with the Tethys and Paratethys through the "Mesopotamian Trough" took place (Buchbinder and Gviertzman, 1976) and when the Gulf of Aden opened.

Sediment accumulation increased abruptly at all low-latitude sites at -6.3 m.y. about the time of "maximum ice development". This increase in accumulation appears to be associated with the onset of the modern surface equatorial circulation pattern, characterized by a high productivity belt in the equatorial region and upwelling in the northwestern basins (Somali Basin, Arabian Sea, Gulf of Aden), as shown by the occurrence of abundant biogenic silica in the sediments from that time on. It has been suggested that the establishment of the modern equatorial current system at that time was made possible by the movement of the tip of India north of the Equator (Vincent, 1974b; Leclaire, 1974). It also was probably favored by the equatorial communication with Pacific surface waters at the same time (Edwards, 1975, his Fig. 4). The age assignment of -6.3 m.y. for the change in sedimentation regime is based on the paleomagnetic age of the O. penultimus/S. peregrina boundary and should be supported by additional paleomagnetic data to determine whether it does indeed correspond to the time of increase in Antarctic ice. Dating of intermediate paleontologic datums within the Late Miocene by further paleomagnetic studies would also allow the determination of detailed fluctuations in accumulation rates within this epoch and possibly link them with paleoceanographic and paleoclimatic fluctuations. Variations in accumulation rates within the Late Miocene cannot be recognized at mid-latitude sites from present paleontological data.

In the southeast Indian Ocean, following the reduced Late Miocene sedimentation, accumulation rates increased in the Early Pliocene with average values of 7 to 16 m/m.y. but remain very low on the southern Naturaliste Plateau (Site 264). Rates decrease to about half of their Early Pliocene values in the Late Pliocene at Sites 281, 253 and 258, a decrease probably related to the intensified bottom circulation associated with the Late Pliocene cooling which removed most of the Upper Pliocene to Quaternary sediments in the South Australian and South Indian Basins. Telescoping of the Upper Pliocene is especially pronounced in its upper part as indicated by the absence or marked condensation of fossil zones in that part of the sequence at all southeast Indian Ocean sites (Figs. 14 to 18).

The effects of this increased bottom circulation are felt far north in the east Indian Ocean on both sides of the Wharton Basin. The entire uppermost Pliocene and Pleistocene is missing in Core V20-163 on the eastern flank of the Ninetyeast Ridge (Fig. 20) and these series accumulated at an extremely slow rate of less than 0.5 m/m.y. on the NW Australian continental rise (DODO 57). The erosive effects of Upper Pliocene intensified circulation are apparently felt as far north as in the Timor Sea and the Andaman-Nicobar region, where telescoping of Upper Pliocene sequences is suggested by curtailment of foraminiferal species ranges.

Pleistocene sedimentation rates in the southeast Indian Ocean vary from 8 m/m.y. on the South Tasman Rise (281) to 5 m/m.y. on the southern Ninetyeast Ridge (253, 254) and to higher values of 13 and 26 m/m.y. on the southern and northern parts of the Naturaliste Plateau respectively (264 and 258). The consistently higher rate at Site 258 is possibly due to the physiographic position of the former site in an area better protected from the northward bottom water flow. The stratigraphic position of the last occurrence of P. lacunosa at Sites 253, 254 and 258 indicate a similar average rate of accumulation for the Pleistocene sediments above and below this level, whereas at Site 262 in the Timor Sea the uppermost Pleistocene, above the last P. lacunosa, accumulated six times as fast as Pleistocene sediments below this level. The extremely high rate of accumulation of approximately 600 m/m.y. for the uppermost Pleistocene at the latter site is supported by diatom data (Jouse and Kazarina, 1974).

At Site 214 on the crest of the central part of the Ninetyeast Ridge average accumulation rates have values for the Lower and Upper Pliocene similar to those of the uppermost Miocene and are slightly lower for the Pleistocene. There is no evidence of faunal mixing or zonal condensing in the late Neogene at this site, which, because of its location at the top of the Ridge, may have been protected from the northward flow of deep water. Furthermore, this site is under the influence of intermediate water circulation less active in the eastern part than the western part of the ocean. Site 214, thus, appears to provide the most suitable section for biostratigraphic investigation of the upper Neogene in the tropical Indian Ocean (Sites 216 and 217 farther north on the crest of the ridge were discontinuously cored).

In the western Indian Ocean, following the accelerated sedimentation of the latest Miocene, accumulation rates remain high throughout the rest of the Neogene, as on the crest of the central and northern Ninetyeast Ridge, but there is evidence of disturbance in the sedimentary sequences at many sites. Average accumulation rates for the entire Pliocene and Pleistocene are 14 to 16 m/m.y. on the Chagos Laccadive Ridge (219) and the Mascarene Plateau (237), 28 to 35 m/m.y. on the Central Indian Ridge (238) and Southwest Indian Ridge (251), and about 51 m/m.y. in the Gulf of Aden. The average accumulation rate for the same stratigraphic interval is reduced (9 m/m.y.) at deeper Site 236 in the Somali Basin, where biostratigraphic mixing is very pronounced throughout the Pliocene, especially in its uppermost part, very likely as a result of intensified AABW activity. Disturbance of the sedimentary sequence also occurs in the Upper Pliocene at Site 237, where the rate of accumulation is reduced and in the Upper Pliocene and Pleistocene sequences at Site 238. At the latter site, a large section of Lower Pliocene slumped into uppermost Pliocene sediments (see Fig. 9) and is responsible for the high Upper Pliocene accumulation rate value. This interval of intense reworking is overlain by lower Pleistocene (with very low sedimentation rates) which is in turn overlain by uppermost Pleistocene (with rapid rates of sedimentation). The high accumulation rate of this latter interval again results from the influx of a large amount of reworked older sediments at the top of the section (see Fig. 9), a phenomenon also observed farther north at the top of Core LSDA 101 (see Fig. 19). Although the sedimentary sequence at Site 219 is apparently continuous without evidence of faunal mixing or sedimentary disturbance and without significant changes in the average accumulation rates throughout the Pliocene-Pleistocene, the possibility of an unconformity near, or spanning, the Pliocene/Pleistocene boundary was suggested by Fleisher (1974), in view of the slight condensing of uppermost Pliocene floral and faunal zones and the presence of glauconite grains in the highest Pliocene samples.

In the southwest Indian Ocean most of the Pliocene and Pleistocene are missing at Site 249 on the Mozambique Ridge, whereas at Site 251 on the Southwest Indian Ridge these series are apparently continuous, with an increased rate for the Upper Pliocene. There is no apparent disturbance near the Pliocene/Pleistocene boundary at the latter site. Further south on the ridge, however, at the site of Core V16-66 a hiatus occurs at this boundary, which removed part of the Matuyama including the Olduvai event, as shown by radiolarian, nannofossil and foraminiferal data (Fig. 20). Other cores paleomagnetically studied from the flank of the Southwest Indian Ridge (V16-60 and V16-69; Opdyke and Glass, 1969) contain hiatuses in the Gauss-Brunhes interval as shown by the absence of entire magnetic events. The stratigraphy of these cores, however, cannot be interpreted in the absence of available biostratigraphic data.

Summary and Conclusions

1). During DSDP Legs 22 through 29 about 14,300 meters of Neogene sediments were drilled at 56 sites in the Indian Ocean. Material from 16 of these sites, where approximately 4074 meters of calcareous sediments were continuously cored with good recovery is used to discuss the Neogene planktonic foraminiferal biostratigraphy. These are located: 1) in the open ocean between 1600 and 3000 m on the shallow ridges which characterize the Indian Ocean floor, including 5 low-latitude sites (from 20°N to 20°S) in the tropical region (Sites 214, 216, 219, 237 and 238), 5 mid-latitude sites in the subtropical to temperate region (Sites 251, 253, 254, 258 and 264), and one at 48°S in the subantarctic region (Site 281); 2) very close to continental margins between 1750 and 2300 m in the Gulf of Aden (Sites 231, 232 and 233) and the Timor Sea (Site 262); and 3) in the Somali Basin at 4500 m (Site 236). All these sites were above the CCD throughout Neogene time, except Site 236 which lay close to the CCD during Early and Middle Miocene times when the paleodepth at the site was about 4200 to 4400 m.

Foraminiferal faunas are common to abundant and well preserved in Upper Miocene through Pleistoce sediments at all these sites, except the deeper Somali Basin Site (236) and the shallow Gulf of Aden sites (231, 232 and 233), where the faunas are affected by significant dissolution. Comparison of Upper Neogene assemblages from the last four sites with well preserved faunas from the other sites provides data on the solution susceptibility of extinct species. Middle and Lower Miocene faunas are moderately to poorly preserved at all sites where they occur.

Planktonic foraminiferal data from piston and gravity cores throughout the Indian Ocean, including two paleomagnetically dated cores (V20-163 and V16-66), as well as from land sections exposed in the Andaman-Nicobar Islands are compared to those from the DSDP sequences. Five additional DSDP sites (low-latitude Sites 217, 241, 242, mid-latitude Site 249, and Antarctic Site 265) provide calcareous sequences valuable for Neogene biostratigraphy, but intermittent coring or lack of available data precludes a detailed biostratigraphy at these sites.

2). Although Blow's (1969) zonation was applied at all low-latitude sites, restrictions in its use were often encountered because of the lack or rarity of the nominate zonal taxa or because of difficulties in taxonomic discrimination. It was found generally unapplicable at mid-latitude sites. "Foraminiferal events" (first and last occurrences of selected species) are evaluated in relation to planktonic zonations of other fossil groups with special attention to paleontological events calibrated to the paleomagnetic-radiometric time scale.

The tropical Indian Ocean Neogene foraminiferal biostratigraphy is similar to that of the tropical Pacific. The latitudinal control of appearances or disappearances of many planktonic foraminiferal species is well evidenced in Neogene Indian Ocean sequences. It restricts their use for correlation between temperate and tropical areas but, on the other hand, make them very useful for paleoceanographic studies if they are calibrated to an independent time scale. Co-occurrences of common calcareous and siliceous microfossils in many Upper Miocene to Pleistocene low-latitude sequences of the Indian Ocean are of special interest, as they permit direct comparison and correlation of zonal schemes based on the two microfossil types. Calcareous sections from mid-latitude sites do not contain radiolaria to allow a comparison between calcareous and siliceous zonations. However, calcareous nannoplankton zonal schemes, which proved applicable in both low- and mid-latitude regions of the Indian Ocean, provide a very valuable tool for indirect correlation of low- and mid-latitude foraminiferal assemblages. Sites located at the transition between temperate and tropical areas are very valuable because they provide insight into the biostratigraphic correlation of these two areas. In this respect, a number of calcareous sections in the mid-latitudes of the Indian Ocean are particularly significant. However, becaue of the differing taxonomic interpretation of various workers, further reevaluation of several important bioseries (such as the Globorotalia merotumida-G. tumida and the G. conoidera-G. inflata lineages) is needed for a more detailed biostratigraphic interpretation of these mid-latitude sites.

3). The Holocene/Pleistocene boundary has been recognized in a number of piston and gravity cores in the Indian Ocean by means of planktonic faunal changes, radiometric dating and oxygen isotope stratigraphy, but was not identified in DSDP sections. The top of the sedimentary sequence appears to be unrecovered at many DSDP sites, as shown by discrepancies between known living foraminiferal assemblages and core-top assemblages. For example, Globorotalia menardii forma flexuosa, which has been observed in the northern part of the Indian Ocean, is not present in the sediments drilled in this area, and pink specimens of Globigerinoides ruber are common in a number of DSDP core-tops but are not present in living assemblages. A layer of pink G. ruber occurs in the Indian Ocean, as in the Pacific, in the Upper Pleistocene, an occurrence which provide a useful stratigraphic tool. The succession of Pleistocene foraminiferal events calibrated to the paleomagnetic and oxygen isotope time-scales in the Pacific cannot be recognized in the Indian Ocean from the present data. Globorotalia tumida flexuosa appears to have become extinct earlier in the tropical Indian Ocean (prior to the initial appearance of the nannofossil species Emiliania huxleyi) than in the Atlantic-Caribbean region.

Changes in coiling direction of Pulleniatina are recorded in the Lower Pleistocene in the Timor Sea (Site 262). The final switch from left- to right-coiling, recorded at 337.5 m in this expanded section, is probably correlative with the youngest switch recorded in Pacific paleomagnetically dated cores at -0.9 m.y. An horizon with dextrally coiled Globorotalia tumida occurs in the lower part of the Pleistocene section in the expanded sequences of the Gulf of Aden and Timor Sea. A distinct change in coiling preference from left to right in Globorotalia crassaformis, which occurs in the upper Pleistocene sequence of the Timor Sea, possibly correlates with a similar change recorded in piston cores of the central equatorial Indian Ocean which appears to occur within oxygen isotope Stage 3. A change in coiling in Globorotalia truncatulinoides was observed in a number of gravity cores in the southwest Indian Ocean, dated by radiocarbon at -6,000 yrs.

4). It was not possible to use the first appearance of Globorotalia truncatulinoides to delineate the Pliocene/Pleistocene boundary because of the difficulty involved in the recognition of this event and because of the discrepancy between its level and the extinction level of Discoaster brouweri. For the sake of consistency, the latter level is used for placing the Pliocene/Pleistocene boundary. The first G. truncatulinoides occurs well above (30 to 70 m) the extinction level of D. brouweri in the expanded sections of the Timor Sea (Site 262) and the southwest Indian Ocean (Sites 250 and 251), whereas in slowly accumulating open-ocean sequences it is coincident or slightly above this nannofloral event. In the expanded hemipelagic sections of the Gulf of Aden the first "common" occurrence of G. truncatulinoides is well above (up to 65 meters) the extinction of D. brouweri, but rare isolated specimens occur as low as the latter. It is very probable that the level of first common occurrence of G. truncatulinoides in the Gulf of Aden is correlative with the level of its initial appearance in the Timor Sea, as shown by associated foraminiferal events (extinctions of Globigerinoides fistulosus and G. obliquus) and by indirect age estimates of these levels (about -0.9 m.y.).

In addition to the first occurrence, or base of common occurrence of G. truncatulinoides, the extinction levels of Globigerinoides fistulosus, G. obliquus, and dextrally coiled Globorotalia limbata appear to be the most useful foraminiferal events for latest Pliocene-earliest Pleistocene biostratigraphy of the tropical to subtropical Indian Ocean. G. fistulosus and G. obliquus became extinct later than the Pliocene/Pleistocene boundary based on nannofossil data, whereas G. limbata typically disappears at or slightly below the boundary. G. limbata, a species more resistant to solution than Globigerinoides obliquus and G. fistulosus, is common in Upper Pliocene tropical and subtropical assemblages as far south as the latitude of Site 253 (25°S), and is rare in Upper Pliocene assemblages at mid-latitude sites south of Site 253. The extinction of Globigerina decoraperta and the transitional development of Pulleniatina obliquiloculata from P. primalis took place near the Pliocene/Pleistocene boundary but the taxonomic difficulty involved in the identification of these species restricts their biostratigraphic use.

The sequential order (by order of increasing age) of 1) the extinction of Discoaster brouweri, 2) the extinction of the radiolarian species Pterocanium prismatium, and 3) the appearance of Globorotalia truncatulinoides in the tropical and subtropical Indian Ocean is identical to the sequential order of these events in the equatorial Pacific. The diatom species Pseudoeunotia doliolus, known to evolve at the Pliocene/Pleistocene boundary in the equatorial Pacific, first occurs in the tropical Indian Ocean either above or below the boundary as identified by other fossil groups.

5). Among the four mid-Pliocene foraminiferal events recorded in paleomagnetically dated cores as occurring near the NN 15/NN 16 nannofossil zonal boundary, which is used in this report for placing the Upper/Lower Pliocene boundary (i.e. extinctions of Globoquadrina altispira and Sphaeroidinellopsis spp. and initial appearances of Globorotalia tosaensis and Globigerinoides fistulosus) the extinctions of G. altispira and Sphaeroidinellopsis spp. provide reliable biostratigraphic tools for the tropical and subtropical Indian Ocean, as shown by their consistent relationship at various sites to the nannofossil and radiolarian zonations. Both these species are common in the area of the Indian Ocean and are solution-resistant. G. altispira, however, is the more resistant as shown by the partial elimination of Sphaeroidinellopsis in the Gulf of Aden. It is also more abundant in the faunal assemblages and thus appears to be a better biostratigraphic marker. The first appearance of G. fistulosus, shortly below the extinction of G. altispira and Sphaeroidinellopsis spp., is also a reliable event for tropical and subtropical regions, but this species is significantly affected by solution as shown by its total elimination from lower Upper Pliocene sediments of the Gulf of Aden and at Site 236. These three foraminiferal events, however, cannot be used as biostratigraphic markers for the temperate areas of the Indian Ocean where G. altispira and Sphaeroidinellopsis disappeared much lower in the Lower Pliocene and G. fistulosus is absent.

The scarcity of Globorotalia tosaensis at all Indian Ocean DSDP sites restricts its use as a biostratigraphic marker and its first occurrence is usually higher than its known evolutionary appearance. The rarity of Globorotalia margaritae at all sites also restricts its use as a biostratigraphic marker. The stratigraphic position of its extinction level, however, in the Lower Pliocene at both low- and mid-latitude is in fairly good agreement with data from paleomagnetically dated cores. Globigerina nepenthes is rare in the upper part of its range at all sites and its extinction level occupies a stratigraphic position in agreement with data from paleomagnetically dated cores only at a few low-latitude sites. This species usually disappears lower, in the Upper Miocene in mid-latitudes. Both G. margaritae and G. nepenthes appear to be susceptible to solution and are mostly eliminated by dissolution from the Gulf of Aden and at Site 236.

The simultaneous disappearance of G. nepenthes, G. margaritae and G. altispira in the sequence of the Andaman-Nicobar Islands in the Bay of Bengal suggests that a telescoping of part of the Pliocene occurs in this region, which curtails the upper range of these species. A lower Pliocene interval with common Globorotalia tumida flexuosa in the tropical sections of the northern Indian Ocean between about 9°N and 14°N (in the Andaman-Nicobar Islands, the Arabian Sea and the Gulf of Aden) provides a useful correlative tool.

The evolution of Sphaeroidinella dehiscens from S. dehiscens forma immatura was observed in the lowermost Pliocene at Site 214 but not at other sites, where the first S. dehiscens occurs at various biostratigraphic levels significantly higher in the sections.

The initial occurrence of Globorotalia inflata in the southeast Indian Ocean on the southern Naturaliste Plateau (Site 264) and on the South Tasman Rise (Site 281) appears to be the most reliable mid-Pliocene biostratigraphic marker at these latitudes. It appears, as in the Atlantic and southwest Pacific, slightly above the last G. margaritae occurrence and this appearance coincides with the NN 15/ NN 16 nannofossil boundary. Earlier events in the lineage of Globorotalia conoidea - G. conomiozea - G. puncticulata - G. inflata can be evaluated neither at Site 264 (because of the great condensation of the uppermost Miocene to Lower Pliocene section which overlies unconformably Eocene sediments and because of discrepancies in the reports of various authors) nor at Site 281 (where a wide unrecovered interval encompasses the uppermost Miocene). Members of this bioseries dominate the Upper Neogene faunal assemblages at all other Indian Ocean mid-latitude sites but cannot be used from present data for biostratigraphy because of the confusion in their taxonomic interpretation. Globorotalia conomiozea survived longer in the Indian Ocean than in the Pacific and Atlantic oceans as the subspecies G. conomiozea subconomiozea whose extinction is recorded in the paleomagnetically dated Core V20-163 in the Upper Gilbert at about -3.5 m.y.

The taxonomic interpretation of Globorotalia crassaformis should be reevaluated before drawing definite conclusions as to the age of its initial occurrence in the Indian Ocean. It is probable, however, that at mid-latitudes this species first appears in the uppermost Miocene where it apparently developed from G. subscitula.

6). The extinction of Globoquadrina dehiscens cannot be used to recognize the Miocene/Pliocene boundary in the Indian Ocean. It appears to be paleoceanographically controlled, decreasing in age from south to north, and shows a narrow latitudinal dependence. There is, in general, good agreement among various fossil groups for the recognition of the Miocene/Pliocene boundary at low-latitude sites of the Indian Ocean. Globorotalia tumida, however, first appears typically shortly below the first occurrence of the nannofossil species Ceratolithus acutus, and the Miocene/Pliocene boundary is drawn in these sections as a hachured area between these two events. At none of the mid-latitude sites of the Indian Ocean is the Miocene/Pliocene boundary clearly identified. Further taxonomic reevaluation of the G. conoidea - G. inflata and the G. plesiotumida - G. tumida bioseries at these sites, as well as the discrimination of C. acutus which is present at both low- and mid-latitudes, will shed light on the problem of the recognition of this boundary. The Miocene/Pliocene boundary at mid-latitude sites is best estimated (in this paper) at the last occurrence of the nannofossil Discoaster quinqueramus.

7). Detailed taxonomic investigation of the Globorotalia merotumida-plesiotumida-tumida bioseries at Site 214 on Ninetyeast Ridge allowed the calibration of the various Upper Miocene events within the interval of this lineage to other fossil zonations. Reliable stratigraphic correlations with other sites based on members of this group cannot be undertaken before taxonomic reexamination of these taxa. Pulleniatina first appears in low-latitude sections in the Upper Miocene in a biostratigraphic position similar to that of the equatorial Pacific but appears much higher at mid-latitude sites, at various levels of the Pliocene and Pleistocene, or is totally absent. Forms referred to Globorotalia margaritae first occur at various biostratigraphic levels from the Middle Miocene to Lower Pliocene. Neogloboquadrina pachyderma first occurs in the Upper Miocene at subantarctic Site 281 on the South Tasman Rise, as in the southwest Pacific, but does not appear below the Pliocene at mid-latitude sites and at Antarctic Site 265.

8). There is a telescoping of the sedimentary sequences near the Upper/Middle Miocene boundary at most sites of the Indian Ocean. Good agreement between various fossil groups for identifying this boundary was found only at Site 214 on the Ninetyeast Ridge. At all tropical western Indian Ocean sites an interval of mixed lower Upper Miocene and upper Middle Miocene foraminiferal assemblages occurs, whereas an almost complete, apparently undisturbed calcareous nannoplankton sequence is present. The latter is, therefore, used for placing the Upper/Middle Miocene boundary. In the Upper Miocene of the Gulf of Aden the nannofossil zonal boundaries occupy a stratigraphic position consistently lower than the apparent foraminiferal zonal boundaries. The stratigraphic position of the latter are artificially displaced because of partial elimination of foraminiferal zonal key species by dissolution.

9). Foraminiferal zonal subdivisions of the Middle and Lower Miocene were found difficult to detect at most low- and mid-latitude sites because of the presence of condensed sections, mixed faunas and the significant amount of dissolution in this interval. The zonal sequence from N 13 to N 15 was identified only at a few low-latitude sites and that of N 4 of N 12 only at Site 216 on Ninetyeast Ridge. The latter site is especially valuable because, although condensed, the series appears to be continuous, and foraminiferal assemblages do not show mixing. In addition, the co-occurrence of both calcareous and siliceous fossils throughout the lower Neogene sediments permits a direct comparison of the two fossil groups in this interval.

Globigerina nepenthes usually overlaps the range of Globorotalia fohsi s.l., except at Sites 214, 231 and 238. The gap between the occurrences of these two species at the three latter sites, as well as in some other tropical provinces of the world, may possibly result from selective solution. The first appearance of Sphaeroidinellopsis subdehiscens cannot be used as a mid-Middle Miocene marker in mid-latitudes because it is delayed into the Upper Miocene. Recognition of the initial appearance of Globorotalia conoidea in mid-latitude sections, after further reexamination of this taxon and its forerunners, will bring a useful link for correlations of low- and mid-latitude sequences. The successive appearance of members of the Globorotalia peripheronda-G. fohsi bioseries was recognized at Site 216 and in the Andaman-Nicobar Islands while at other low-latitude sites most members of this group occur together in mixed assemblages. At mid-latitude sites the only members of the bioseries present are G. peripheroronda and G. peripheroacuta. Most Indian Ocean sections are of no use for investigation of the biostratigraphic significance of the "Orbulina datum" because this species, highly susceptible to solution, usually is eliminated in the lower part of its range. The Lower/Middle Miocene boundary, therefore, is placed on nannofossil data.

The foraminiferal zonal boundaries of the Lower and Middle Miocene occupy a consistently lower stratigraphic position in the western tropical Indian Ocean than in the eastern region where good agreement was found between the two fossil groups with what has been observed elsewhere. This discrepancy possibly reflects disturbance and reworking of the foraminiferal sequence in the western area.

10). The Oligocene/Miocene boundary, placed according to recent revisions of its biostratigraphic position at the base of nannofossil Zone NN 1, is best estimated on a foraminiferal basis by the initial appearance of Turborotalia kugleri.

11). An attempt has been made to relate changes in faunal content and in rates of accumulation of calcareous sediments at all sites together with changes in the isotopic composition of foraminiferal tests at subantarctic Site 281 to global climatic changes associated with Antarctic glaciation. Recent erosional effects as evidenced in various parts of the Indian Ocean, and widespread gaps and/or reduced sedimentation in the geological record at deep as well as at intermediate depths involve the entire oceanwide circulation, which appears to be largely controlled by the Antarctic circulation. The two distinct regimes which occur today in the eastern and western Indian Ocean appear to have been in existence since the late Middle Miocene. Following the Oligocene epoch of major hiatus and/or reduced sedimentation in most areas as a result of the northward introduction of Antarctic Bottom Water, sedimentation remained reduced in the Early Miocene only to resume significantly at the end of the Early Miocene, when the development of an active circumpolar deep-sea circulation probably decreased northward bottom currents. During the Early and Middle Miocene relatively warm surface water tempertures (optimum for the development of the globoquadrinid group) prevailed in mid as well as low latitudes.

Gradual East Antarctic ice build-up from mid-Middle Miocene to latest Miocene is indicated in the subantarctic area by the oxygen isotopic record at Site 281 and in the Antarctic region by the distribution of ice-rafted detritus. Associated intensified water circulation resulted in a significant decrease in sediment accumulation rates in the temperate and southern subtropical southeast Indian Ocean during the Late Miocene. Erosive effects during that time were not as strong either on the crest of the central and northern Ninetyeast Ridge, or in the western Indian Ocean, where sediments accumulated in the early Late Miocene (prior to 6.3 m.y.) at rates of the same order of magnitude as in the Middle Miocene. Bottom currents, however, were apparently intense throughout the ocean in late Middle Miocene and earliest Late Miocene as evidenced by the general telescoping of fossil zones at that time and by an interval of mixed foraminiferal faunas spanning the Middle/Late Miocene boundary throughout the northwestern Indian Ocean. The mixed faunas consist of common upper Middle Miocene species reworked into lower Upper Miocene assemblages whereas in this interval the calcareous nannoplankton sequence is complete and apparently undisturbed. It thus appears that this fossil group shows much less reworking than did the foraminiferal, suggesting that the reworked sediments have been winnowed during transport. Increased tectonic activity in the northwestern Indian Ocean in late Middle Miocene may have provided suitable outcrops for the reworking and winnowing action of bottom currents. A circulation readjustment must have taken place in this area at about this time after the closure of the Indo-Tethyan-Parathethyan seaway through the Persian Gulf and the opening of the Gulf of Aden. A marked increase in sediment accumulation rates and in the abundance of siliceous fossils at all low-latitude sites in Late Miocene, at approximately -6.3 m.y., corresponds to the time of maximum Antarctic ice accumulation and the onset of the modern equatorial circulation pattern.

Cooling during the Late Miocene moved the area of optimum conditions for globoquadrinids north of 25°S. The extinction of _Globoquadrina dehiscens_ was paleoceanographically controlled, decreasing in age from south to north with progressive cooling. The uppermost Miocene (Kapitean) severe cooling recorded throughout the Pacific is not recognizable in the Indian Ocean from available studies because of a lack of: 1) isotopic data at Site 281 from this interval which corresponds to a gap in recovery; 2) detailed quantitative faunal analyses;

and 3) accurate biostratigraphic correlations near the Miocene/Pliocene boundary between low and mid latitudes. There is faunal evidence, however, in the temperate, subtropical and tropical areas that water temperatures were cooler in Late than in Early Pliocene. The Late Pliocene cooling was associated with a substantial increase in velocity of bottom water which removed Upper Pliocene to Quaternary sediments from the South Indian and South Australian Basins and was felt throughout the Indian Ocean. Telescoping of Upper Pliocene sequences (especially pronounced in the uppermost Pliocene) occurs throughout the eastern ocean as far north as the Andaman-Nicobar region. In the western part of the ocean, disturbances of Upper Pliocene and lowermost Pleistocene sedimentary sequences are evidenced at many sites either by condensing, faunal mixing, reworking, or slumping. Sites 214 and 216 on the crest of the central and northern Ninetyeast Ridge appear to be located throughout the Neogene in an area better protected from bottom currents than any other sites and, thus, provide the best undisturbed sequences for biostratigraphic investigations.

Cores from the subantarctic southeast Indian Ocean record a warming trend during the last 0.8 m.y. with superimposed temperature fluctuations, which were essentially synchronous with north hemisphere climatic changes. Similar trends are indicated in the southern subtropical area at Site 253 by ratios of Globorotalia menardii to G. inflata and age estimates based on sediment accumulation rates. In the latter area, cool conditions prevailed in the earliest Pleistocene; a marked warming occurred between -1.42 and -1.18 m.y., followed by a relatively cool interval with small-scale temperature fluctuations up to -0.52 m.y. From that time on a warming trend occurred. The age estimate, based on sedimentation rates, of several Pleistocene climatic changes evidenced by faunal variations in a few expanded sections of the temperate and tropical areas are in general agreement with the climatic fluctuations recorded at Site 253. It is noteworthy that the time of first common occurrence of Globorotalia truncatulinoides in the Gulf of Aden and of the youngest switch in coiling direction of Pulleniatina in the Timor Sea, estimated to be -0.9 m.y., corresponds to a generally cooler interval (from -1.05 to -0.82 m.y.) at Site 253. It is also a time of cooling climax in the subantarctic southeast Indian Ocean, as well as in the North Pacific. The 10°C surface isotherm did not shift north of 36°S during the Pleistocene.

Acknowledgments. The author is greatly indebted to William A. Berggren who critically reviewed the manuscript and to Frances Parker who made valuable editorial comments and suggestions. This work was supported by the Oceanography Section, National Science Foundation, Grant OCE 76-84029.

References

Adelseck, C. G., Living Globorobalia menardii (d'Orbigny) forma neoflexuosa from the eastern tropical Pacific Ocean, Deep-Sea Res., 22, 689-691, 1975.

Anglada, R., Sur la position du datum à Globigerinoides (Foraminiferida) la zone N4 (Blow 1967) et la limite Oligo-Miocène en Méditerranée, C. R. Acad. Sci., Paris, 272, 1067-1070, 1971.

Akers, W. H., Foraminiferal range charts for Arabian Sea and Red Sea sites, leg 23, in Davies, T. A., Luyendyk, B. P., et al., Initial Reports of the Deep Sea Drilling Project, 26, Washington (U. S. Government Printing Office),1013-1050, 1974a.

Akers, W. H., Results of shore laboratory studies on Tertiary and

Quaternary foraminifera from Leg 26, in Davies, T. A., Luyendyk, B. P., et al., Initial Reports of the Deep Sea Drilling Project, 26, Washington (U. S. Government Printing Office), 973-982, 1974b.

Bandy, O. L., The geologic significance of coiling ratios in the foraminifer Globigerina pachyderma (Ehrenberg), J. Paleontology, 34, 671-681, 1960.

Bandy, O. L., Miocene-Pliocene boundary in the Philippines as related to Late Tertiary stratigraphy of deep-sea sediments, Science, 142, 1290-1292, 1963.

Bandy, O. L., Cenozoic planktonic foraminiferal zonation, Micropaleont., 10, 1-17, 1964.

Bandy, O. L., Neogene planktonic foraminiferal zones, California and some geologic implications, Palaeography, Palaeoclimat. Palaeoecol., 12, 131, 1972a.

Bandy, O. L., Origin and development of Globorotalia (Turborotalia) pachyderma (Ehrenberg), Micropaleontology, 18, 294, 1972b.

Bandy, O. L., Chronology and paleoenvironmental trends, Late Miocene-Early Pliocene, western Mediterranean, in Messinian events in the Mediterranean, Drooger, C. W., ed., Konink. Nederl. Akad. Van Wetenschappen, Amsterdam, 21-25, 1973.

Bandy, O. L., Messinian evaporite deposition and the Miocene/Pliocene boundary, Pasquasia-Capodarso sections, Sicily, in Late Neogene Epoch Boundaries, Saito, T. and Burckle, L. H., eds., Micropaleontology, Spec. Publ. No. 1, 49-63, 1975.

Bandy, O. L., and R. E. Casey, Paleotemperature characteristics, Matuyama-Brunhes sequence in the Southern Ocean, 9th INQUA Congress, Abstracts, Christchurch, New Zealand, 10-11, 1973.

Bandy, O. L., and J. C. Ingle, Jr., Neogene planktonic events and radiometric scale, California, in Bandy, O. L., ed., Radiometric dating and paleontologic zonation, Geol. Soc. Am. Spec. Paper 125, 131, 1970.

Bandy, O. L., and J. A. Wilcoxon, The Pliocene-Pleistocene boundary, Italy and California, Geol. Soc. Amer. Bull., 81, 2939-2948, 1970.

Bandy, O. L., E. A. Butler, and R. C. Wright, Alaskan Upper Miocene marine glacial deposits and the Turborotalia pachyderma datum plane, Science, 166, 607, 1969.

Bandy, O. L., R. E. Casey, and R. C. Wright, Late Neogene planktonic zonation, magnetic reversals, and radiometric dates, Antarctic to the Tropics, Amer. Geophys. Un., Antarct. Res. Ser. (Antarctic Oceanology, 1) 15, 1-26, 1971.

Bandy, O. L., H. G. Lindenberg, and E. Vincent, History of research, Indian Ocean foraminifera, Jour. Mar. Biol. Assoc. India, 13, 86-105, 1972.

Banner, F. T., and W. H. Blow, Progress in the planktonic foraminiferal biostratigraphy of the Neogene, Nature, 208, 1164-1166, 1965a.

Banner, F. T., and W. H. Blow, Two new taxa of the Globorotaliinae (Globigerinacae Foraminifera) assisting determination of the late Miocene/middle Miocene boundary, Nature, 207, 1351-1354, 1965b.

Bé, A. W. H., and W. H. Hamlin, Ecology of recent planktonic foraminifera, Pt. 3, Distribution in the North Atlantic during the summer of 1962, Micropaleont. 13, 87-106, 1967.

Bé, A. W. H., and W. H. Hutson, Ecologic and biogeographic patterns of planktonic foraminiferal life and fossil assemblages, in Symposium "Marine Plankton and Sediments", 3rd Planktonic Conference, Kiel, 1974, in press.

Bé, A. W. H., and A. McIntyre, Globorotalia menardii flexuosa (Koch): an "extinct" foraminiferal subspecies living in the Indian Ocean, Deep-Sea Res., 17, 595-601, 1970.

Bé, A. W. H., and D. S. Tolderlund, Distribution and ecology of living planktonic foraminifera in surface waters of the Atlantic and Indian Oceans, in Funnell, B. M. and Riedel, W. R., eds., Micropaleontology of the Oceans, Cambridge University Press, 105-149, 1971.

Beard, J. H., Pleistocene paleotemperature record based on planktonic foraminifera, Gulf of Mexico, Gulf Coast Assoc. Geol. Soc. Trans. 19, 535-553, 1969.

Beliaeva, N. V., Raspredelenie planktonnykh foraminifer v vodakh i na dnie Indiiskogo Okeana. (Distribution of planktonic foraminifera in the water and on the floor of the Indian Ocean), Trudy Inst., Okeanol., Akad. NAUK SSSR, 68, 12-83, 1964.

Berger, W. H., Planktonic foraminifera: selective solution and the lysocline, Marine Geol., 8, 111-138, 1970.

Berger, W. H., and U. von Rad, Cretaceous and Cenozoic sediments from the Atlantic Ocean floor, in Hayes, D. E., Pimm, A. C., et al., Initial Reports of the Deep Sea Drilling Project, 14, Washington, (U. S. Government Printing Office), 787-954, 1972.

Berger, W. H., and E. L. Winterer, Plate stratigraphy and the fluctuating carbonate line, in Pelagic sediments on land and under the sea, Hsü, K. J. and Jenkyns, H. C., eds., Spec. Publ. Int. Assn. Sediment. No. 1, 11-48, 1974.

Berggren, W. A., Micropaleontology and the Pliocene/Pleistocene boundary in a deep-sea core from the south-central North Atlantic, Ciorn. Ceol., ser. 2, 35, 291-312, 1968.

Berggren, W. A., Rates of evolution of some Cenozoic planktonic Foraminifera, Micropaleont. 15, 351-365, 1969.

Berggren, W. A., The Pliocene time-scale: calibration of planktonic foraminiferal and calcareous nannoplankton zones, Nature, 243, 391-397, 1973.

Berggren, W. A., The Pliocene/Pleistocene boundary in deep sea sediments: 1975, Gior. Geol., in press.

Berggren, W. A., and Amdurer, M., Late Paleogene (Oliogocene) and Neogene planktonic foraminiferal biostratigraphy of the Atlantic Ocean (lat. 30°N to lat. 30°S), Riv. Ital. Paleontol. 79, 337, 1973.

Berggren, W. A., and B. U. Haq, The Andalusian stage (Late Miocene): biostratigraphy, biochronology and paleoecology, Palaeogeog., Palaeoclimat. Palaeoecol., 20, 67-129, 1976.

Berggren, W. A., and R. Z. Poore, Late Miocene-Early Pliocene planktonic foraminiferal biochronology: Globorotalia tumida and Sphaeroidinella dehiscens lineages. Riv. Ital. Paleontol., 80, 689-698, 1974.

Berggren, W. A., and J. A. Van Couvering, The late Neogene: Biostratigraphy, biochronology and paleoclimatology of the last 15 million years in marine and continental sediments, Palaeogeogr., Palaeoecol., Palaeoclimatol., 16, 1-216, 1974.

Berggren, W. A., J. D. Phillips, A. Bertels, and D. Wall, Late Pliocene-Pleistocene stratigraphy in deep-sea cores from the south central North Atlantic, Nature, 216, 253-254, 1967.

Berggren, W. A., G. P. Lohmann, and R. Z. Poore, Shore laboratory report on Cenozoic planktonic foraminifera: Leg 22, in von der Borch, C. C., Sclater, J. G., et al., Initial Reports of the Deep Sea Drilling Project, 22, Washington (U. S. Government Printing Office), 635-655, 1974.

Blow, W. H., Age, correlation, and biostratigraphy of the upper Tocuyo (San Lorenzo) and Pozon formations, eastern Falcon, Venezuela, Bull. Amer. Pal., 39, 67-252, 1959.

Blow, W. H., Late Middle Eocene to Recent planktonic foraminiferal biostratigraphy, in Brönnimann, P. and Renz, H. H., eds., Proc.

Int. Conf. Planktonic Microfossils, 1st., Geneva, 1967, 1. E. J. Brill, Leiden, 199-421, 1969.

Blow, W. H., Deep Sea Drilling Project, Leg 3, foraminifera from selected samples, in Maxwell, A. E., et al., Initial Reports of the Deep Sea Drilling Project, 3, Washington (U. S. Government Printing Office), 629-661, 1970.

Bolli, H. M., The direction of coiling in the evolution of some Globorotaliidae, Cushman Found. Foram. Res., Contr. 1, 82-89, 1950.

Bolli, H. M., Planktonic foraminifera from the Oliogocene-Miocene Cipero and Lengua formations of Trinidad, B.W.I., U. S. Nat. Mus, 215, 97-123, 1957.

Bolli, H. M., Observations on the stratigraphic distribution of some warm water planktonic Foraminifera in the young Miocene to Recent, Eclogae Geol. Helv., 57, 541-552, 1964.

Bolli, H. M., Zonation of Cretaceous to Pliocene marine sediments based on planktonic foraminifera, Bol. Inf. Asoc. Venez. Geol. Min. Pet., 8, 119-149, 1966.

Bolli, H. M., The subspecies of Globorotalia fohsi Cushman and Ellisor and the zones based on them, Micropaleontology, 13, 502-512, 1967.

Bolli, The foraminifera of Sites 23-31, Leg 4, in Bader, R. G. et al., Initial Reports of the Deep Sea Drilling Project, 4, Washington (U. S. Government Printing Office), 577-643, 1970.

Bolli, H. M., and P. J. Bermudez, Zonation based on planktonic foraminifera of middle Miocene to Pliocene warm-water sediments, Bol. Inf. Asoc. Venez. Geol. Min. Pet., 8, 119-149, 1965.

Bolli, H. M., and I. Premoli-Silva, Oliogocene to recent planktonic foraminifera and stratigraphy of the Leg 15 sites in the Caribbean Sea, in, Edgar, N. T., Saunders, J. B., et al., Initial Reports of the Deep Sea Drilling Project, 15, Washington (U. S. Government Printing Office), 475-497, 1973.

Boltovskoy, E., Living planktonic foraminifera of the Eastern part of the Tropical Atlantic, Rev. Micropaleontology, 11, 85, 1968.

Boltovskoy, E., Living planktonic foraminifera at the 90°E meridian from the equator to the Antarctic, Micropaleontology, 15, 237-255, 1969.

Boltovskoy, E., Neogene planktonic foraminifera of the Indian Ocean (DSDP, Leg 26), in Davies, T. A., Luyendyk, B. P., et al., Initial Reports of the Deep Sea Drilling Project, 26, Washington (U. S. Government Printing Office), 675-741, 1974a.

Boltovskoy, E., Late Pliocene and Quaternary paleoclimatic changes, Indian Ocean, DSDP, Leg 26, in Davies, T. A., Luyendyk, B. P., et al., Initial Reports of the Deep Sea Drilling Project, 26, Washington (U. S. Government Printing Office), 743-744, 1974b.

Boudreaux, J. E., Calcareous nannoplankton ranges, Deep Sea Drilling Project Leg 23, in Whitmarsh, R. B., Weser, O. E., Ross, D. A., et al., Initial Reports of the Deep Sea Drilling Project, 23, Washington (U. S. Government Printing Office), 1073-1090, 1974.

Bradshaw, J. S., Ecology of living planktonic foraminifera in the North and equatorial Pacific Ocean, Contr. Cushman Found. Foram. Res., 10, 25-64, 1959.

Briskin, M., and W. A. Berggren, Pleistocene stratigraphy and quantitative paleooceanography of tropical North Atlantic core V16-205, in Late Neogene Epoch Boundaries, Saito, T. and Burckle, L. H., eds., Micropaleontol., Spec. Publ. No. 1, 167-198, 1975.

Brönnimann, P., and J. Resig, A Neogene globigerinacean biochronologic time-scale of the Southwestern Pacific, in Winterer, E. L., et al., Initial Reports of the Deep Sea Drilling Project, 7, Washington (U. S. Government Printing Office), 1235-1469, 1971.

Brönnimann, P., E. Martini, J. Resig, W. R. Riedel, A. Sanfilippo, and T. Worsley, Biostratigraphic synthesis - Late Oliogocene and Neogene of the western tropical Pacific, in Winterer, E. L., et al., Initial Reports of the Deep Sea Drilling Project, 7, Washington (U. S. Government Printing Office), 1723-1745, 1971.

Buchbinder, B., and G. Gvirtzman, The breakup of the Tethys Ocean into the Mediterranean Sea, the Red Sea, and the Mesopotamian Basin during the Miocene: a sequence of fault movements and dessication events, Abstr., 1st Congr. Pacific Neog. Stratigr., Tokyo, 32-35, 1976.

Bukry, D., Low-latitude coccolith biostratigraphic zonation, in Edgar, N. T., Saunders, J. B., et al., Initial Reports of the Deep Sea Drilling Project, 15, Washington (U. S. Government Printing Office), 685-703, 1973a.

Bukry, D., Coccolith and silicoflagellate stratigraphy, Deep Sea Drilling Project, Leg 18, Eastern North Pacific, in Kulm, L. D., von Huene, R., et al., Initial Reports of the Deep Sea Drilling Project, 18, Washington (U. S. Government Printing Office), 817-832, 1973b.

Bukry, D., Coccolith and silicoflagellate stratigraphy, Eastern Indian Ocean, Deep-Sea Drilling Project, Leg 22, in von der Borch, C. C., Sclater, J. G., et al., Initial Reports of the Deep Sea Drilling Project, 22, Washington (U. S. Government Printing Office), 601-607, 1974a.

Bukry, D., Coccolith stratigraphy, Arabian and Red Seas, Deep-Sea Drilling Project, Leg 23, in Whitmarsh, R. B., Weser, O. E., Ross, D. A., et al., Initial Reports of the Deep Sea Drilling Project, 23, Washington (U. S. Government Printing Office), 1091-1093, 1974b.

Bukry, D., Coccolith zonation of cores from the western Indian Ocean and the Gulf of Aden, Deep Sea Drilling Project, Leg 24, in Fisher, R. L., Bunce, E. T., et al., Initial Reports of the Deep Sea Drilling Project, 24, Washington (U. S. Government Printing Office), 995-996, 1974c.

Bukry, D., Phytoplankton stratigraphy, offshore East Africa, Deep Sea Drilling Project, Leg 25, in Simpson, E. S., Schlich, R., et al., Initial Reports of the Deep Sea Drilling Project, 25, Washington (U. S. Government Printing Office), 635-646, 1974d.

Bukry, D., Coccolith stratigraphy, offshore Western Australia, Deep Sea Drilling Project, Leg 27, in Veevers, J. J., Heirtzler, J. R., et al., Initial Reports of the Deep Sea Drilling Project, 27, Washington (U. S. Government Printing Office), 623-630, 1974e.

Burckle, L. H., Late Cenozoic planktonic diatom zones in Equatorial Pacific sediments, Geol. Soc. Am. Abstracts with programs for 1969, Part 7, 24, 1969.

Burckle, Late Cenozoic planktonic diatom zones from the Eastern Equatorial Pacific, Beih. Nova Hedwigia-Heft, 39, 217-246, 1972.

Burns, D. A., Nannofossils: in Site 264, Hayes, D. E., Frakes, L. A., et al., Initial Reports of the Deep Sea Drilling Project, 28, Washington (U. S. Government Printing Office), 25-26, 1975a.

Burns, D. A., Nannofossil biostratigraphy for Antarctic sediments, Leg 28, Deep Sea Drilling Project, in Hayes, D. E., Frakes, L. A., et al., Initial Reports of the Deep Sea Drilling Project, 28, Washington (U. S. Government Printing Office), 589-598, 1975b.

Burns, R. E., J. E. Andrews, et al., Initial Reports of the Deep Sea Drilling Project, 21, Washington (U. S. Government Printing Office), 931 pp., 1973.

Cati, F., M. L. Colalongo, U. Crescenti, S. D'Onofrio, A. Pomesano Cherchi, G. Salvatorini, S. Sartoni, I. Premoli-Silva, C. F. Wezel, V. Bertoline, G. Bizon, H. M. Bolli, A. M. Cati Borsetti, L. Dondi, H. Feinberg, D. G. Jenkins, E. Perconig, M. Sampo, and R. Sprovieri,

Biostratigrafia del Neogene mediterraneo basata sui foraminiferi planctonici, Bol. Soc. Geol. Ital., 87, 491-503, 1968.

Chen, Pei-Hsu, Antarctic radiolaria, in Hayes, D. E. Frakes, L. A., et al., Initial Reports of the Deep Sea Drilling Project, 28, Washington (U. S. Government Printing Office), 437-513, 1975.

Ciesielski, P. F., and F. M. Weaver, Early Pliocene temperature changes in the Antarctic Seas, Geology, 2, 511-515, 1974.

Cita, M. B., Pliocene biostratigraphy and chronostratigraphy, in Ryan, W.B.F., Hsü, K. J., et al., Initial Reports of the Deep Sea Drilling Project, 13, Washington (U. S. Government Printing Office), 1343-1380, 1973.

Cita, M. B., and S. Gartner, The stratotype Zanclean foraminiferal and nannofossil biostratigraphy, Riv. Italiana Paleontologia e Stratigrafia, 79, 503-558, 1973.

Cita, M. B., M. A. Chierici, R. Ciampo, M. Zei Montcharmont, S. D'Onofrio, W.B.F. Ryan, and R. Scorziello, The Quaternary record in the Tyrrhenian and Ionian Basins of the Mediterranean, in Ryan, W.B.F., Hsü, K. J., et al., Initial Reports of the Deep Sea Drilling Project, 13, Washington (U. S. Government Printing Office), 1263-1339, 1973.

CLIMAP Project Members, The surface of the Ice-Age Earth, Science, 191, 1131-1137, 1976.

Collen, J. D., and P. Vella, Pliocene planktonic foraminifera, southern North Island, New Zealand, J. Foram. Res., 3, p. 13, 1973.

Conolly, J. R., Postglacial-glacial change in climate in the Indian Ocean, Nature, 214, 873-875, 1967.

Cox, A., Geomagnetic reversals, Science, 163, 237-245, 1969.

Davies, T. A., O. E. Weser, B. P. Luyendyk, and R. B. Kidd, Unconformities in the sediments of the Indian Ocean, Nature, 253, 15-19, 1975.

Devereux, I., C. H. Hendy, and P. Vella, Pliocene and early Pleistocene sea temperature fluctuations, Mangaopori Stream, New Zealand, Earth Planet. Sci. Lett., 8, 163-168, 1970.

Douglas, F. W., and W. C. Johnson, III, Late Pleistocene paleotemperature model for the southern Indian Ocean, Antarct. J., 9, 269-261, 1974.

Eames, F. E., Some thoughts on the Neogene/Paleogene boundary, Palaeogeogr. Palaeoecol., Palaeoclimatol, 8, 37-48, 1970.

Echols, R. J., Foraminifera, Leg 19, Deep Sea Drilling Project, in Creager, J. S., Scholl, D. S., et al., Initial Reports of the Deep Sea Drilling Project, 19, Washington (U. S. Government Printing Office), 721-735, 1973.

Edwards, A. R., Southwest Pacific Cenozoic paleogeography and an integrated Neogene paleocirculation model, in Andrews, J. E., Packham, G., et al., 1975, Initial Reports of the Deep Sea Drilling Project, 30, Washington (U. S. Government Printing Office), 667-684, 1975.

Edwards, A. R., and K. Perch-Nielsen, Calcareous nannofossils from the southern Southwest Pacific, Deep Sea Drilling Project, Leg 29, in Kennett, J. P., Houtz, R. E., et al., Initial Reports of the Deep Sea Drilling Project, 20, Washington (U. S. Government Printing Office), 469-539, 1975.

Emiliani, C., Isotopic paleotemperatures, Science, 154, 851-856, 1966.

Emiliani, A new paleontology, Micropaleontology, 15, 265, 1969.

Ericson, D. B., and G. Wollin, Correlation of six cores from the equatorial Atlantic and the Caribbean, Deep Sea Res., 3, 104-125, 1956.

Ericson, D. B., and G. Wollin, Pleistocene climates and chronology in deep-sea sediments, Science, 162, 1227-1234, 1968.

Ericson, D. B., and G. Wollin, Pleistocene climates in the Atlantic and Pacific Oceans: A comparison based on deep-sea sediments, Science, 167, 1483-1485, 1970.

Ericson, D. B., and J. Wollin, Coiling direction of Globorotalia truncatulinoides in deep-sea cores, Deep-Sea Res., 2, 152-158, 1954.

Ericson, D. B., M. Ewing, G. Wollin, and B. C. Heezen, Atlantic deep-sea sediment cores, Geol. Soc. Amer. Bull., 72, 193-286, 1961.

Ericson, D. B., M. Ewing, G. Wollin, and B. C. Heezen, Pliocene-Pleistocene boundary in deep-sea sediments, Science, 139, 727-737, 1963.

Ericson, D. B., M. Ewing, G. Wollin, and B. C. Heezen, The Pleistocene Epoch in deep-sea sediments, Science, 146, 723-732, 1964.

Frerichs, W. E., Pleistocene-Recent boundary and Wisconsin glacial biostratigraphy in the northern Indian Ocean, Science, 159, 1456-1458, 1968.

Fleisher, R. L., Cenozoic planktonic foraminifera and biostratigraphy. Arabian Sea, Deep Sea Drilling Project, Leg 23A, in Whitmarsh, R.B., Weser, O. E., Ross, D. A., et al., Initial Reports of the Deep Sea Drilling Project, 23, Washington (U. S. Government Printing Office), 1001-1074, 1974.

Gartner, Jr., S., Correlation of Neogene planktonic foraminifera and calcareous nannofossil zones, Trans. Gulf Coast Assoc. Geol. Soc., 19, 585-599, 1969.

Gartner, Jr., S., Absolute chronology of the late Neogene nannofossil succession in the equatorial Pacific, Geol. Soc. Am. Bull., 84, 2021-2034, 1973.

Gartner, Jr., S., Nannofossil biostratigraphy, Leg 22, Deep Sea Drilling Project, in von der Borch, C. C., Sclater, J. G., et al., Initial Reports of the Deep Sea Drilling Project, 22, Washington (U. S. Government Printing Office), 577-600, 1974.

Gartner, S., and D. Bukry, Ceratolithus acutus Gartner and Bukry n. sp. and Ceratolithus amplifucus Bukry and Percival-nomenclatural clarification: Tulane Studies, Geology and Paleontology, 11, 115-118, 1974.

Gartner, S., and D. Bukry, Morphology and phylogeny of the coccolithophycean family Ceratolithacae, Jour. Res. U. S. Geol. Survey, 3, 451-465, 1975.

Gartner, S., D. A. Johnson, and B. McGowran, Paleontology synthesis of Deep-Sea Drilling results from Leg 22 in the northeastern Indian Ocean, in von der Borch, C. C., Sclater, J. G., et al., Initial Reports of the Deep Sea Drilling Project, 22, Washington (U. S. Government Printing Office), 805-813, 1974.

George, T. N., Chairman, and Members of the Stratigraphy Committee of the Geological Society of London, Recommendations on stratigraphical usage, Proc. Geol. Soc. London, 1956, 139-166, 1969.

Glass, B., D. B. Ericson, B. C. Heezen, N. D. Opdyke, and J. A. Glass, Geomagnetic reversals and Pleistocene chronology, Nature, 216, 437-442, 1967.

Hay, W. W., and F. M. Beaudry, Calcareous nannofossils: Leg 15, Deep Sea Drilling Project, in Edgar, N. T., Saunders, J. B., et al., Initial Reports of the Deep Sea Drilling Project, 15, Washington (U. S. Government Printing Office), 625-683, 1973.

Hayes, D. E., and L. A. Frakes, General synthesis, Deep Sea Drilling Project, Leg 28, in Hayes, D. E., Frakes, L. A., et al., Initial Reports of the Deep Sea Drilling Project, 28, Washington (U. S. Government Printing Office), 919-942, 1975.

Hayes, D. E., L. A. Frakes, et al., Initial Reports of the Deep Sea Drilling Project, 28, Washington (U. S. Government Printing Office), 1917 pp., 1975.

Hays, J. D., Radiolaria and late Tertiary and Quaternary history of Antarctic seas: Biology of the Antarctic Sea II, Am. Geophys. Un., Antarct. Res., 5, 125-184, 1965.

Hays, J. D., Quaternary sediments of the Antarctic Ocean, in Progress in Oceanography, Sears, M., ed., 4, Pergamon, Oxford, 117-131, 1967.

Hays, J. D., The stratigraphy and evolutionary trends of Radiolaria in North Pacific deep-sea sediments, Geol. Soc. Am. Mem., 126, 185-218, 1970.

Hays, J. D., and W. A. Berggren, Quaternary boundaries, in Funnell, B. M. and Riedel, W. R., eds., Micropalaeontology of Oceans, Cambridge Univ. Press, 669-691, 1971.

Hays, J. D., and N. D. Opdyke, Antarctic Radiolaria magnetic reversals and climatic change, Science, 158, 1001-1011, 1967.

Hays, J. D., and N. J. Shackleton, The extinction of Axoprunum angelinum, a time-synchronous biostratigraphic level, Geol. Soc. Amer., Abstracts with programs, 7, 1106-1107, 1975.

Hays, J. D., T. Saito, N. D. Opdyke, and L. H. Burckle, Pliocene-Pleistocene sediments of the equatorial Pacific - their paleomagnetic, biostratigraphic and climatic record, Geol. Soc. Am. Bull., 80, 1481-1514, 1969.

Hays, J. D., et al., Initial Reports of the Deep Sea Drilling Project, 9, Washington (U. S. Government Printing Office), 1205 pp., 1972.

Hekinian, R., F. Mélières, W. D. Nesteroff, and L. H. Burckle, Une importante lacune dans les dépôts profonds au Nord de Crozet (Indien) sud), Bull. Assoc. Fr. Et. Gr. Prof. Océan, 6, 29-35, 1969.

Hoernes, M., Die fossilen Mollusken der Tertiar-Beckens von Wien, Abh. K. K. Geol. Reichsanstalt, 3, 1-733, 1856.

Hornibrook, N. de B., New Zealand Upper Cretaceous and Tertiary foraminiferal zones and some overseas correlations, Micropaleont., 4, 25-38, 1958.

Hornibrook, N. de B., and A. R. Edwards, Integrated planktonic foraminiferal and calcareous nannoplankton datum levels in the New Zealand Cenozoic, in Farinacci, A., ed., Plankt. Conf. Proc. 2nd, Rome, 1970, 649-657, 1971.

Ikebe, N., Y. Takayanagi, M. Chiji, and K. Chinzei, Neogene biostratigraphy and radiometric time scale of Japan - an attempt at international correlation, Pacific Geol., 4, 39-78, 1972.

Imbrie, J., and N. Kipp, A new micropaleontological method for quantitative paleoclimatology: application to a late Pleistocene Caribbean core, in Turekian, K., ed., Late Cenozoic Glacial Ages, New Haven, Conn., Yale Univ. Press, 73-181, 1971.

Ingle, J. C., Jr., Foraminiferal biofacies and the Miocene-Pliocene boundary in southern California, Am. Paleontol. Bull., 52, 217, 1967.

Ingle, J. C., Jr., Neogene foraminifera from the Northeastern Pacific Ocean, Leg 18, Deep Sea Drilling Project, in Kulm, L. D., von Huene, R., et al., Initial Reports of the Deep Sea Drilling Project, 18, 517-568, 1973.

Ingle, J. C., Jr., Pleistocene and Pliocene foraminifera from the Sea of Japan, Leg 31, Deep Sea Drilling Project, in Karig, D. E. Ingle, J. C., Jr., et al., 1975, Initial Reports of the Deep Sea Drilling Project, 31, Washington (U. S. Government Printing Office), 693-701, 1975.

Jenkins, D. G., Planktonic foraminiferal zones and new taxa from the Danian to lower Miocene to New Zealand, New Zealand J. Geol. Geophy., 8, p. 1088, 1966.

Jenkins, D. G., Planktonic foraminiferal zones and new taxa from the lower Miocene to the Pleistocene of New Zealand, New Zealand J. Geol. Geophys., 10, 1064, 1967.

Jenkins, D. G., New Zealand Cenozoic planktonic Foraminifera, New Zealand Geol. Surv. Paleont. Bull., 42, 278, 1971.

Jenkins, D. G., Cenozoic planktonic foraminiferal biostratigraphy of the Southwestern Pacific and Tasman Sea, Deep Sea Drilling Project Leg 29, in Kennett, J. P., Houtz, R. E., et al., Initial Reports of the Deep Sea Drilling Project, 29, Washington (U. S. Government Printing Office), 449-467, 1975.

Jenkins, D. G., and W. N. Orr, Cenozoic planktonic foraminiferal zonation and the problem of test solution, Rev. Espan. Micropal., 3, 301-304, 1971.

Jenkins, D. G., and W. N. Orr, Planktonic foraminiferal biostratigraphy, of the eastern Equatorial Pacific, in Hays, J. D., Initial Reports of the Deep Sea Drilling Project, 9, Washington (U. S. Government Printing Office), 1060-1193, 1972.

Johnson, D. A., Radiolaria from the Eastern Indian Ocean, Deep Sea Drilling Project, Leg 22, in von der Borch, C. C., Sclater, J. G., et al., Initial Reports of the Deep Sea Drilling Project, 22, Washington (U. S. Government Printing Office), 521-575, 1974.

Jouse, A. P., and G. H. Kazarina, Pleistocene diatoms from Site 262, Leg 27, Deep Sea Drilling Project, in Veevers, J. J., Heirtzler, J. R., et al., Initial Reports of the Deep Sea Drilling Project, 27, Washington (U. S. Government Printing Office), 925-945, 1974.

Kaneps, A., Late Neogene biostratigraphy (planktonic foraminifera), biogeography, and depositional history, Ph.D. thesis, Columbia University, New York, 1970.

Kaneps, A., Cenozoic planktonic foraminifera from the eastern equatorial Pacific Ocean, in van Andel, T. H., Heath, G. R., et al., Initial Reports of the Deep Sea Drilling Project, 16, Washington (U. S. Government Printing Office), 713-746, 1973.

Kaneps, A., Foraminifera: In Site 264, Hayes, D. E., Frakes, L. A., et al., Initial Reports of the Deep Sea Drilling Project, 28, Washington (U. S. Government Printing Office), 23-25, 1975a.

Kaneps, A., Cenozoic planktonic foraminifera from Antarctic deep-sea sediments, Leg 28, Deep Sea Drilling Project, in Hayes, D. E., Frakes, L. A., et al., Initial Reports of the Deep Sea Drilling Project, 28, Washington (U. S. Government Printing Office), 573-583, 1975b.

Keany, J., and J. P. Kennett, Pliocene-early Pleistocene paleoclimatic history recorded in Antarctic-subantarctic deep-sea cores, Deep Sea Res., 19, 529-548, 1972.

Kemp, E. M., L. A. Frakes, and D. E. Hayes, Paleoclimatic significance of diachronous biogenic biofacies, Leg 28, Deep Sea Drilling Project, in Hayes, D. E., Frakes, L. A., et al., Initial Reports of the Deep Sea Drilling Project, 28, Washington (U. S. Government Printing Office), 909-917, 1975.

Kennett, J. P., The Globorotalia crassaformis bioseries in north Westland and Marlborough, New Zealand, Micropaleontology, 12, 235, 1966.

Kennett, J. P., Recognition and correlation of the Kapitean Stage (upper Miocene, New Zealand), New Zealand J. Geol. Geophys., 10, 143, 1967.

Kennett, J. P., Paleo-oceanographic aspects of the foraminifera zonation in the upper Miocene-lower Pliocene of New Zealand, Gior. Geol., 35, 143, 1968.

Kennett, J. P., Pleistocene paleoclimates and foraminiferal biostratigraphy in subantarctic deep-sea cores, Deep-Sea Res., 17, 124-140, 1970.

Kennett, J. P., Middle and late Cenozoic planktonic foraminiferal biostratigraphy of the southwest Pacific-Deep-Sea Drilling Project Leg 21, in Burns, R. E., Andrews, J. E., et al., Initial Reports of the Deep Sea Drilling Project, 21, Washington (U. S. Government Printing Office), 575-640, 1973.

Kennett, J. P., Neogene planktonic foraminiferal stratigraphy in Deep Sea Drilling Sites, Southeast Indian Ocean, in Hayes, D. E., Frakes, L. A., et al., Initial Reports of the Deep Sea Drilling Project, 28, 705-708, 1975.

Kennett, J. P., and K. R. Geitzenauer, The Pliocene-Pleistocene boundary in a South Pacific deep-sea core, Nature, 224, 899, 1969.

Kennett, J. P., and M. S. Srinivasan, Stratigraphic occurrences of the Miocene planktonic foraminifer Globoquadrina dehiscens in Early Pliocene sediments of the Indian Ocean, Rev. Espan. Micropal., 7, 5-14, 1975.

Kennett, J. P., and P. Vella, Late Cenozoic planktonic foraminifera and paleoceanography at Deep-Sea Drilling Project Site 284 in the cool subtropical South Pacific, in Kennett, J. P., Houtz, R. E., et al., Initial Reports of the Deep Sea Drilling Project, 29, Washington (U. S. Government Printing Office), 769-799, 1975.

Kennett, J. P., and N. D. Watkins, Late Miocene-Early Pliocene paleomagnetic stratigraphy, paleoclimatology, and biostratigraphy in New Zealand, Bull. Geol. Soc. Am., 85, 1385-1398, 1974.

Kennett, J. P., and N. D. Watkins, Regional deep-sea dynamic processes recorded by late Cenozoic sediments of the southeastern Indian Ocean, Geol. Soc. Amer. Bull., 87, 321-339, 1976.

Kennett, J. P., N. D. Watkins, and P. Vella, Paleomagnetic chronology of Pliocene-Early Pleistocene climates and the Plio-Pleistocene boundary in New Zealand, Science, 171, 276-279, 1971.

Kennett, J. P., R. E. Burns, J. E. Andrews, J. Churkin, Jr., T. A. Davies, P. Dumitrica, A. R. Edwards, J. S. Galehouse, G. H. Packham, and G. J. Van der Lingen, Australian-Antarctic continental drift, palaeocirculation changes and Oligocene deep-sea erosion, Nature, Phys. Sci., 239, 51-55, 1972.

Kennett, J. P., R. E. Houtz, P. B. Andrews, A. R. Edwards, V. A. Gostin, M. Hajos, M. A. Hampton, D. G. Jenkins, S. V. Margolis, A. T. Ovenshine, and K. Perch-Nielsen, Development of the Circum-Antarctic Current, Science, 186, 144-147, 1974.

Kennett, J. P., and R. E. Houtz, et al., Initial Reports of the Deep Sea Drilling Project, 29, Washington (U. S. Government Printing Office), 1197 pp., 1974.

Kennett, J. P., P. B. Andrews, A. R. Edwards, V. A. Gostin, M. Hajós, M. A. Hampton, D. G. Jenkins, S. V. Margolis, A. T. Owenshine, and K. Perch-Nielsen, Cenozoic paleoceanography in the Southwest Pacific Ocean, Antarctic glaciation, and the development of the Circumantarctic current, in Kennett, J. P. Houtz, R. E., et al., Initial Reports of the Deep Sea Drilling Project, 29, Washington (U. S. Government Printing Office), 1155-1169, 1975.

Kent, D., N. D. Opdyke, and M. Ewing, Climate change in the North Pacific using ice-rafted detritus as a climatic indicator, Geol. Soc. Am. Bull., 82, 2741-2754, 1971.

Kierstead, C. H., R. R. D., Leidy, R. L. Fleisher, and A. Boersma, Neogene zonation of tropical Pacific cores, Int. Conf. Plankt. Microfossils, 1st Proc.,Brönnimann, P., and Renz, H. H., eds., Leiden (Brill), 2, 328-338, 1969.

Kling, S. A., Radiolaria from the Eastern North Pacific, Deep-Sea Drilling Project, Leg 18, in Kulm, L. D., von Huene, R., et al., Initial Reports of the Deep Sea Drilling Project, 18, Washington (U. S. Government Printing Office), 617, 1973.

Kolla, V., L. Sullivan, S. S. Streeter, and M. G. Langseth, Spreading of Antarctic bottom water and its effects on the floor of the Indian Ocean inferred from bottom-water potential temperature, turbidity, and sea-floor photography, Marine Geology, 21, 171-189, 1976.

Krasheninnikov, V., Cenozoic foraminifera, in Fisher, A. G., et al., Initial Reports of the Deep Sea Drilling Project, 6, Washington (U. S. Government Printing Office), 1055-1068, 1971.

Lamb, J. L., and J. H. Beard, Late Neogene planktonic foraminifers in the Caribbean, Gulf of Mexico and Italian stratotypes, Univ. Kans. Paleontol. Contrib., Art. 57 (Protozoa 8), 67 pp., 1972.

Leclaire, L., Late Cretaceous and Cenozoic pelagic deposits - Paleoenvironment and paleoceanography of the central western Indian Ocean, in Simpson, E. S. W., Schlich, R., et al., Initial Reports of the Deep Sea Drilling Project, 25, Washington (U. S. Government Printing Office), 481-513, 1974.

Lidz, L., Deep-sea Pleistocene biostratigraphy, Science, 154, 1448-1452, 1966.

Martini, E., Standard Tertiary and Quaternary calcareous nannoplankton zonation, in Farinacci, A., ed., Proc. Planktonic Conf., 2nd, Rome, 1970, 739-785, 1971.

Mazzola, G., Les Foraminifères planctoniques du Mio-Pliocène de l'Algérie nord-occidentale, in Farinacci, A., ed., Proc. Plankt. Confer., 2nd Roma 1970, 787-812, 1971.

McCollum, D. W., Diatom stratigraphy of the Southern Ocean, in Hayes D. E., Frakes, L. A., et al., Initial Reports of the Deep Sea Drilling Project, 28, Washington (U. S. Government Printing Office), 515-571, 1975.

McGowran, B., Foraminifera, in von der Borch, C. C., Sclater, J. G., et al., Initial Reports of the Deep Sea Drilling Project, 22, Washington (U. S. Government Printing Office), 609-627, 1974.

McIntyre, A., A. W. H. Bé, and R. Preikstas, Coccoliths and the Pliocene-Pleistocene boundary, Progress in Oceanography, 4, 3-25, Oxford, 1967.

McKenzie, D., and J. G. Sclater, The evolution of the Indian Ocean since the Late Cretaceous, Geophys. J. Roy. Astron. Soc., 25, 437-528, 1971.

Morin, R. W., F. Theyer, and E. Vincent, Pleistocene climates in the Atlantic and Pacific Oceans: A reevaluated comparison based on deep-sea sediments, Science, 169, 365-366, 1970.

Müller, C., Calcareous nannoplankton, Leg 25 (Western Indian Ocean), in Simpson, E.S.W., Schlich, R., et al., Initial Reports of the Deep Sea Drilling Project, 25, Washington (U. S. Government Printing Office), 579-633, 1974.

Oba, T., Planktonic foraminifera from the deep-sea cores of the Indian Ocean, Tohoku Univ., Sci. Rep., 2nd ser. (Geol.), 38, 193-219, 1967.

Oba, T., Biostratigraphy and isotopic paleotemperature of some deep-sea cores from the Indian Ocean, Tohoku Univ., Sci. Rep., 2nd ser. (Geol.), 41, 129-195, 1969.

Olausson, E., and I. U. Olsson, Varve stratigraphy in a core from the Gulf of Aden, Palaeogeog., Palaeoclimatol., Palaeoecol., 6, 87-103, 1969.

Olausson, E., B. U. Haq, G. B. Gunvor, and I. U. Olsson, Evidence in Indian Ocean cores of Late Pleistocene changes in oceanic and atmospheric circulation, Geologiska Foreningens Stockholm Forhandlingar, 93, 51-84, 1971.

Olsson, R. K., Shore laboratory report on the foraminifera from Leg 27 sites, Deep-Sea Drilling Project, in Veevers, J. J., Heirtzler, J. R., et al., Initial Reports of the Deep Sea Drilling Project, 27, Washington (U. S. Government Printing Office), 967-975, 1974.

Onofrio, S. d', L. Gianelli, S. Iaccarino, E. Morlotti, M. Romeo, G. Salvatorini, M. Sampo, and R. Sprovieri, Planktonic foraminifera of the Upper Miocene from some Italian sections and the problem of the lower boundary of the Messinian, Boll. Soc. Paleontol. Italiana, 14, 177-196, 1975.

Opdyke, N. D., and B. P. Glass, The paleomagnetism of sediment cores from the Indian Ocean, Deep-Sea Res., 16, 249-261, 1969.

Orr, W. N., Variation and distribution of Globigerinoides ruber in the Gulf of Mexico, Micropaleontology, 15, 373, 1969.

Parker, F. L., Foraminifera from the experimental MOHOLE drilling near Guadalupe Island, Mexico, J. Paleontol., 38, 617, 1964.

Parker, F. L., Late Tertiary biostratigraphy (planktonic Foraminifera) of tropical Indo-Pacific deep-sea cores, Am. Paleontol. Bull., 52, 115-203, 1967.

Parker, F. L., Distribution of planktonic foraminifera in recent deep-sea sediment, in The Micropaleontology of Oceans, Funnell, B. M. and Riedel, W. R., eds., Cambridge Univ. Press, 289-307, 1971.

Parker, F. L., Late Cenozoic biostratigraphy (planktonic foraminifera) of tropical Atlantic deep-sea sections, Rev. espan. Micropal., 5, 253-289, 1973.

Parker, F. L., Upper Neogene biostratigraphy (planktonic foraminifera) of DSDP sites 139 and 141, J. Foram. Res., 4, 9-15, 1974.

Parker, F. L., and W. H. Berger, Faunal and solution patterns of planktonic foraminifera in surface sediments of the South Pacific, Deep-Sea Res., 18, 73, 1971.

Petrushevskaya, M. G., Cenozoic Radiolarians of the Antarctic, Leg 29, Deep-Sea Drilling Project, in Kennett, J. P., Houtz, R. E., et al., Initial Reports of the Deep Sea Drilling Project, 29, Washington (U. S. Government Printing Office), 541-675, 1975.

Phillips, J. D., W. A. Berggren, A. Bertels, and D. Wall, Paleomagnetic stratigraphy and micropaleontology of three deep-sea cores from the central North Atlantic Ocean, Earth Planet. Sci. Lett., 4, 118-130, 1968.

Phleger, F. B, F. L. Parker, and J. F. Peirson, North Atlantic foraminifera, Swedish Deep-Sea Exped. Repts., 7, 3-122, 1953.

Poore, R. Z., and W. A. Berggren, Pliocene biostratigraphy of the Labrador Sea: Calcareous plankton, J. Foram. Res., 4, 91-108, 1974.

Poore, R. Z., and W. A. Berggren, Late Cenozoic planktonic foraminiferal biostratigraphy and paleoclimatology of the northeastern Atlantic: DSDP site 116, J. Foram. Res., 5, 270-293, 1975.

Proto Decima, F., Leg 27 calcareous nannoplankton, in Veevers, J. J., Heirtzler, J. R., et al., Initial Reports of the Deep Sea Drilling Project, 27, Washington (U. S. Government Printing Office), 589-621, 1974.

Ramsay, A. T. A., and B. M. Funnell, Upper Tertiary microfossils from the Alula-Fartak Trench, Gulf of Aden, Deep-Sea Res., 16, 25-43, 1969.

Renz, G. W., Radiolaria from Leg 27 of the Deep Sea Drilling Project, in Veevers, J. J., Heirtzler, J. R., et al., Initial Reports of the Deep Sea Drilling Project, 27, Washington (U. S. Government Printing Office), 769-841, 1974.

Riedel, W. R., and A. Sanfilippo, Cenozoic Radiolaria from the western tropical Pacific, Leg 7, in Winterer, E. L., et al., Initial Reports of the Deep Sea Drilling Project, 7, Washington (U. S. Government Printing Office), 1529-1672, 1971.

Riedel, W. R., and A. Sanfilippo, Stratigraphy and evolution of tropical Cenozoic Radiolarians, in Symposium "Marine Plankton and Sediments", 3rd Planktonic Conference, Kiel, 1974, in press.

Riedel, W. R., A. Sanfilippo, and M. B. Cita, Radiolarians from the stratotype Zanclean (lower Pliocene Sicily), Riv. Ital. Paleont. 80, 699-734, 1974.

Rio, D., Remarks on late Pliocene-Early Pleistocene calcareous nannofossile, stratigraphy in Italy, Ateneo Parmense, acta nat., 10, 409-449, 1974.

Rögl, F., The evolution of the Globorotalia truncatulinoides and Globorotalia crassaformis group in the Pliocene and Pleistocene of the Timor Trough, Deep Sea Drilling Project, Leg 27, Site 262, in Veevers, J. J., Heirtzler, J. R., et al., Initial Reports of the Deep Sea Drilling Project, 27, Washington (U. S. Government Printing Office), 743-768, 1974.

Roth, P. H., Calcareous nannofossils from the Northwestern Indian Ocean, Leg 24, Deep Sea Drilling Project, in Fisher, R. L., Bunce, E. T., et al., Initial Reports of the Deep Sea Drilling Project, 24, Washington (U. S. Government Printing Office), 969-994, 1974.

Ryan, W. B. F., M. B. Cita, Rawsson, M. Dreyfus, L. H. Burckle, and T. Saito, A paleomagnetic assignment of Neogene stage boundaries and the development of isochronous datum planes between the Mediterranean, the Pacific and Indian oceans in order to investigate the response of the world ocean to the Mediterranean "salinity crisis", Riv. Ital. Paleontol., 80, 631-688, 1974.

Sachs, J. B., H. C. Skinner, Late Pliocene-Early Pleistocene nannofossil stratigraphy in the North Central Gulf Coast Area, Proc. of Symp. on Calcareous Nannofoss. Gulf Coast Sect. SEPM, 94-125, Houston, 1973.

Saito, T., and C. Fray, Cretaceous and Tertiary sediments from the southwestern Indian Ocean, Spec. Pap. Geol. Soc. Am., 82, 171-172, 1964.

Saito, T., L. H. Burckle, and J. D. Hays, Late Miocene to Pleistocene biostratigraphy of equatorial Pacific sediments, in Saito, T., and Burckle, L. H., eds., Late Neogene Epoch Boundaries, Micropaleontol. Spec. Publ. No. 1, (Micropaleontology Press, New York), 226-244, 1975.

Sanfilippo, A., and W. R. Riedel, Radiolaria from the west-central Indian Ocean and Gulf of Aden, Deep Sea Drilling Project, Leg 24, in Fisher, R. L., Bunce, E. T., et al., Initial Reports of the Deep Sea Drilling Project, 24, Washington (U. S. Government Printing Office), 997-1035, 1974.

Sanfilippo, A., L. H. Burckle, E. Martini, and W. R. Riedel, Radiolarians, diatoms, silicoflagellates and calcareous nannofossils in the Mediterranean Neogene, Micropaleont., 19, 209-234, 1973.

Schott, W., U. von Stackelberg, F. J. Eckhardt, B. Mattiat, J. Peters and B. Zobel, Geologische Untersuchungen an Sedimenten des indisch-pakistanischen Kontinentalrandes (Arabisches Meer): Geol. Rundschau, 60, 264-275, 1970.

Schrader, H. J., Cenozoic diatoms from the Northeast Pacific, Leg 18, in Kulm, L. D., von Huene, R., et al., Initial Reports of the Deep Sea Drilling Project, 18, Washington (U. S. Government Printing Office), 673-798, 1973.

Schrader, H. J., Cenozoic marine planktonic diatom stratigraphy the tropical Indian Ocean, in Fisher, R. L., Bunce, E. T., et al. Initial Reports of the Deep Sea Drilling Project, 24, Washington (U. S. Government Printing Office), 887-967, 1974.

Schwager, C., Fossile foraminiferen von Kar Nicobar. Novara Exped. 1857-1859, Wien, 2, Geol. Theil 2, 187-268, 1866.

Scott, G. H., Phyletic trends for trans-Atlantic lower Neogene Globigerinoides, Rev. Esp. Micropal., 3, 283-292, 1972.

Seiglie, G. A., Revision of Mid-Tertiary stratigraphy of southwestern Puerto Rico, Am. Assoc. Petrol. Geol. Bull., 57, 405-406, 1973.

Shackleton, N. J., and J. P. Kennett, Paleotemperature history of the Cenozoic and the initiation of Antarctic glaciation: Oxygen and carbon isotope analyses in Deep Sea Drilling Project sites 277, 279, and 281, in Kennett, J. P., Houtz, R. E., et al., Initial Reports of the Deep Sea Drilling Project, 29, Washington (U. S. Government Printing Office), 742-755, 1975a.

Shackleton, N. J., and J. P. Kennett, Late Cenozoic oxygen and carbon isotopic changes at Deep Sea Drilling Project site 284: Implications for glacial history of the Northern Hemisphere and Antarctic, in Kennett, J. P., Houtz, R. E., et al., Initial Reports of the Deep Sea Drilling Project, 29, Washington (U. S. Government Printing Office), 801-807, 1975b.

Shackleton, N. J., and N. D. Opdyke, Oxygen isotope and palaeomagnetic stratigraphy of Equatorial Pacific Core V28-238: oxygen isotope temperature and ice volumes on a 10^5 year and 10^6 year scale, Quaternary Res., 3, 39-55, 1973.

Smitter, Y. H., A foraminiferal fauna from the Tertiary sediments of southern Mozambique, Paleontol. Africana, 3, 109-118, 1955.

Srinivasan, M. S., Miocene foraminifera from Hut Bay, Little Andaman Island, Bay of Bengal, Cushman Found. Foram. Res., Contr., 20, 102-105, 1969.

Srinivasan, M. S., Middle Miocene planktonic foraminifera from the Hut Bay formation, Little Andaman Island, Bay of Bengal, Micropaleontology, 21, 133-150, 1975.

Srinivasan, M. S., and V. Sharma, Stratigraphy and microfauna of Car Nicobar Island, Bay of Bengal, Geol. Soc. India, 14, 1-11, 1973.

Srinivasan, M. S., and S. S. Srivastava, Foraminifera from Nancowry and Kamorta islands, Bay of Bengal, and some aspects of Early Neogene orogeny, Earth Sci. India. Bull., 1, 27-35, 1972.

Srinivasan, M. S., and S. S. Srivastava, Sphaeroidinella dehiscens datum and Miocene-Pliocene boundary, Amer. Assoc. Petr. Geol. Bull., 58, 304-311, 1974.

Srinivasan, M. S., and S. S. Srivastava, Late Neogene biostratigraphy and planktonic foraminifera of Andaman-Nicobar Islands, Bay of Bengal, in Late Neogene Epoch Boundaries, Saito, T., and Burckle, L. H., eds., Micropaleontol., Spec. Publ., No. 1, 124-161 (Micropaleontology Press, New York), 1975.

Srinivasan, M. S., J. P. Kennett, and A. W. H. Bé, Globorotalia menardii neoflexuosa new subspecies from the northern Indian Ocean, Deep-Sea Res., 21, 321-324, 1974.

Stubbings, H. G., Stratification of biological remains in marine deposits, J. Murray Exp. 1933-34, Sci. Rept., 3, 159-192, 1939.

Takayama, T., The Pliocene-Pleistocene boundary in the Lamont core V21-98 and at le Castella, southern Italy, Jour. Mar. Geol., 7, 70-77, 1970.

Talwani, M., C. C. Windisch, and M. G. Langseth, Jr., Reykjanes Ridge crest: a detailed geophysical study, J. Geophys. Res., 76, 473-517, 1971.

Theyer, F., Late Neogene paleomagnetic and planktonic zonation, southeast Indian Ocean-Tasman Basin, Univ. Southern California, unpublished doctoral dissertation, 200 pp., 1972.

Theyer, F., Globorotalia truncatulinoides datum plane: evidence for a Gauss (Pliocene) age in subantarctic cores, Nature Phys. Sci., 142-145, 1973a.

Theyer, F., Reply, Nature Phys. Sci., 244, 47-48, 1973b.

Theyer, F., and S. R. Hammond, Paleomagnetic polarity sequence and radiolarian zones, Brunhes to Polarity epoch 20, Earth Planet. Sci. Lett., 22, 307-319, 1974a.

Theyer, F., and S. R. Hammond, Cenozoic magnetic time-scale in deep-sea cores: Completion of the Neogene, Geology, 487-492, 1974b.

Thiede, J., Variations in coiling ratios of Holocene planktonic foraminifera, Deep-Sea Res., 18, 823-831, 1971.

Thierstein, H. R., Calcareous nannoplankton - Leg 26, Deep Sea Drilling Project, in Davies, T. A., Luyendyk, B. P., et al., Initial Reports of the Deep Sea Drilling Project, 26, Washington (U. S. Government Printing Office), 619-667, 1974.

Thierstein, H. R., K. R. Geitzenauer, B. Molfino, and N. J. Shackleton, Global synchroneity of late Quaternary coccolith datums: validation by oxygen isotopes, Geology, in press.

Thompson, P. R., Planktonic foraminiferal dissolution and the progress towards a Pleistocene equatorial Pacific transfer function, J. Foram. Res., 6, 208-227, 1976.

Thompson, P. R., and T. Saito, Pacific Pleistocene sediments: planktonic foraminifera dissolution cycles and geochronology, Geology, 333-336, 1974.

Tracey, J. I., Jr., et al., Initial Reports of the Deep Sea Drilling Project, 8, Washington (U. S. Government Printing Office), 1037 pp., 1971.

Ubaldo, M. L., Etude des foraminifères de sondages du canal de Mozambique (Océan Indien), Mem. Junta Investig. Ultamar, 62, 278 pp., 1973.

van Andel, T. H., Mesozoic/Cenozoic calcite compensation depth and the global distribution of calcareous sediments, Earth Planet. Sci. Lett., 26, 187-194, 1975.

Van Couvering, J. A., and W. A. Berggren, The biostratigraphical basis of the Neogene time-scale, in Hazel, J. E. and Kauffman, Ed., eds., New Concepts in Biostratigraphy, Pal. Soc. Am. Spec. Publ., in press.

Van Couvering, J. A., R. E. Drake, E. Aguirre, and G. H. Curtis, The terminal Miocene event, Marine Micropaleont., 1, 263-286, 1976.

Vella, P., B. B. Ellwood, and N. D. Watkins, Surface-water temperature changes in the Southern Ocean southwest of Australia during the last one million years, Quaternary Studies, Suggate, R. P. and Cresswell, M. M., eds., Roy. Soc. N. Z., Wellington, 297-309, 1975.

Vincent, E., Climatic change at the Pleistocene-Holocene boundary in Southwestern Indian Ocean, Palaeoecology of Africa, the surrounding islands and Antarctica, van Zinderen Bakker, Sr., E. M., ed., Balkema, Cape Town, 6, 45-54, 1972.

Vincent, E., Sediment distribution in the Southwestern Indian Ocean, Amer. Assoc. Petrol. Geol. Bull., 57, p. 810, 1973.

Vincent, E., Foraminifera: In Site 231, Fisher, R. L., Bunce, E. T., et al., Initial Reports of the Deep Sea Drilling Project, 24, Washington (U. S. Government Printing Office), 22-24, 1974a.

Vincent, E., Cenozoic planktonic biostratigraphy and paleoceanography of the tropical western Indian Ocean, in Fisher, R. L., Bunce, E.T., et al., Initial Reports of the Deep Sea Drilling Project, 24, Washington (U. S. Government Printing Office), 1111-1150, 1974b.

Vincent, E., Neogene planktonic foraminifera from the central North Pacific, Leg 32, Deep Sea Drilling Project, in Larson, R. L., Moberly, R., et al., Initial Reports of the Deep Sea Drilling Project, 32, Washington (U. S. Government Printing Office), 765-801, 1975.

Vincent, E., Planktonic foraminifera, sediments and oceanography of the late Quaternary, Southwest Indian Ocean, Allan Hancock Monog. Marine Biol., 9, 235 pp., 1976.

Vincent, E., and N. J. Shackleton, Agulhas current temperature distribution delineated by oxygen isotope analysis of foraminifera in surface sediments, Orville L. Bandy memorial volume, in press.

Vincent, E., W. E. Frerichs, and M. E. Heiman, Neogene planktonic foraminifera from the Gulf of Aden and the western tropical Indian Ocean, Deep Sea Drilling Project, Leg 24, in Fisher, R. L., Bunce, E. T., et al., Initial Reports of the Deep Sea Drilling Project, 24, Washington (U. S. Government Printing Office), 827-849, 1974.

Watkins, N. D., and J. P. Kennett, Regional sedimentary disconformities and upper Cenozoic changes in bottom water velocities between Australia and Antarctica, in Hayes, D. E., ed., Antarctic Res. Ser. 19, Antarctic Oceanology II, The Australian-New Zealand Sector, 273-293, 1972.

Watkins, N. D., and P. Vella, Palaeomagnetism and the Globorotalia truncatulinoides datum in the Tasman Sea and Southern Ocean, Nature, Phys. Sci., 244, 45-47, 1973.

Weaver, F. M., and D. W. McCollum, Sedimentary hiatus in the South Indian Basin, Antarct. J., 9, 250-251, 1974.

Williams, D. F., and W. C. Johnson II, Late Pleistocene paleotemperature model for the Southern Indian Ocean, U. S. Ant. Jour., 9, 260-261, 1974.

Winterer, E. L., et al., Initial Reports of the Deep Sea Drilling Project, 7, Washington (U. S. Government Printing Office), 1757 pp., 1971.

Wyrtki, K., Oceanographic atlas of the International Indian Ocean Expedition, Washington, 531 pp., 1971.

Wyrtki, K., Physical oceanography of the Indian Ocean, in The Biology of the Indian Ocean, Zeitzschel, B., ed., Springer Verlag, N. Y., 18-36, 1973.

Zobel, B., Biostratigraphische Untersuchungen an Sedimenten des indisch, pakistanischen Kontinentalrandes (Arabisches Meer), Meteor Forsch. - Ergebnisse, Reihe C., 12, 9-73, 1973.

Zobel, B., Quaternary and Neogene foraminifera: Biostratigraphy, in Simpson, E. S. W., Schlich, R., et al., Initial Reports of the Deep Sea Drilling Project, 25, Washington (U. S. Government Printing Office), 573-578, 1974.

CHAPTER 21. SYNTHESIS OF THE CRETACEOUS BENTHONIC FORAMINIFERA RECOVERED BY THE DEEP SEA DRILLING PROJECT IN THE INDIAN OCEAN

Viera Scheibnerova

Geological and Mining Museum, Geological Survey of New South Wales
Department of Mines, Sydney, Australia

Abstract. In 1971 to 1973 eight legs (22-29) comprising 62 sites crossed various parts of the Indian Ocean. Of these, four legs (24, 25, 26 and 27) recovered Cretaceous in 18 sites. Seventeen sites penetrated Late Cretaceous horizons, and 8 sites Early Cretaceous horizons and one also entered the Jurassic. Where the oldest sediments included Jurassic, the lowest Early Cretaceous (Neocomian) was also represented. In other sites, the oldest Cretaceous sediments were of Middle or Late Albian or upper Late Cretaceous age. The Cretaceous profiles recovered include hiatuses before and after the Albian.

Except for one site (249) which is off the East African margin, all Late Cretaceous sites are in the Eastern Indian Ocean, in a rather restricted area (about 29°N to 35°S and 100°E to 120°E) off the western and northwestern Australian continental shelf.

Late Cretaceous sediments were recovered much more widely; from 5 sites in the western Indian Ocean, ranging from about 10°N to 35°S and 12 sites in the Eastern Indian Ocean from the northern part of the Ninetyeast Ridge to the Naturaliste Plateau (about 10°N to 35°S).

Stratigraphically, the recovered sections are all fragmentary, ranging in extent from fractions of a stage (Site 235, 250) to one to three stages at best. This is in part due to incomplete coring programs and poor recovery, but mainly because over wide areas in the Indian Ocean there are considerable intervals of the Cretaceous missing altogether, such as is the case in the southern Indian Ocean (Legs 26 and 27) where the Early Cretaceous is relatively well developed but most of the Late Cretaceous (and also the Tertiary) is missing.

Due to the distribution pattern of the Cretaceous and incomplete recoveries of what is present the scope of this study is considerably restricted.

The most complete recoveries within the Early Cretaceous have been made in the Albian, and to a lesser degree in the Aptian and older in Legs 26 and 27, where faunas and floras, including the benthonic foraminifera, have been studied in considerable detail.

It can be seen that the Late Cretaceous occurrences in the Indian Ocean are spread over a much wider area, but these records are also still only fragmentary. This general pattern of incomplete vertical distribution holds also for the Cretaceous on adjacent landmasses, in western Australia in particular (see Belford 1958). The most complete sections come from Site 217 with an apparently continuous Maastrichtian-Campanian-Santonian section and from Site 258 with a Cenomanian to Santonian section, which continues into the Albian and older Cretaceous.

Cretaceous sediments recovered by the DSDP contain both benthonic and planktonic foraminifera, although the numbers of these vary greatly. Many samples are barren in both pre-Albian, Albian and post-Albian deposits.

Benthonic foraminifera from both Early and Late Cretaceous deposits recovered by DSDP in the Indian Ocean correlate closely with coeval faunas on adjacent landmasses especially in South Africa, the Indian Peninsula and Australia. The more southerly the locality, the closer is the correlation with the Great Australian Basin.

The most important Cretaceous genera of foraminifera recovered by the DSDP in the Indian Ocean are: Lingulogavelinella, Orithostella, Gavelinella, Pseudopatellinella, Discorbis, Pseudolamarckina, Osangularia, Gubkinella, Eponides, Praebulimina, Neobulimina, Aragonia, Globorotalites, Labrospira, Textularia, Recurvoides, Praetrochamminoides, Plectorecurvoides, Praecystammina und Pseudobolivina. Many represent good stratigraphic markers.

The composition and nature of the microfaunas, and changes of them, proved to be good general indicators of Cretaceous bathymetry in the Indian Ocean.

Data on stratigraphic zonation in S-N and E-W distribution are restricted to the W and NW margin of Australia.

The most thorough palaeontological reports are those of Legs 26 and 27. They present general information, and taxonomic descriptions and illustrations of complete or nearly complete foraminiferal faunas. Both the Early and the Late Cretaceous benthonic foraminifera were described and figured only from Leg 27. In the others only the planktonic taxa received attention.

Late Jurassic and Early Cretaceous Foraminifera (Leg 27)

Benthonic species of pre-Albian sediments are represented by primitive agglutinated foraminifera studies in Leg 27 by Bartenstein (1974) and Kuznetsova (1974). Bartenstein (op. cit.) described species of the following genera: Bathysiphon, Kalamopsis, Hyperammina, Hippocrepina, Proteonina and Sorosphaera. He compared some of the foraminiferal associations from sites 259 and 261 (leg 27) with those of certain flysch deposits, e.g. of the Bavarian flysch as described by Pflaumann (1967).

Kuznetsova (1974) described selected Late Jurassic - Early Cretaceous foraminifera from Site 261 (Leg 27). She characterized these as showing the following features:
1) All foraminifera are benthonic with agglutinated forms predominant.
2) Preservation is very good especially that of the agglutinated forms.
3) Assemblages are highly diverse, with over 90 species present
   (Kuznetsova op. cit. attributed them to 44 genera from 12 families).
4) Many assemblages are rich in number of individual specimens.
5) The absence of "index species" made age determinations based on
   this fauna difficult.

Foraminiferal elements characteristic of the Callovian and Oxfordian such as Ophthalmidiidae, Ceratobuliminidae and Trocholina are missing, while the Stephanolithon bigoti nannoplankton species, also characteristic for Callovian and Oxfordian, does occur. The increase in numbers of agglutinated forms upwards in the Jurassic - Early Cretaceous section was explained by Kuznetsova (1974) by a gradual decrease in temperature. This illustrates the important fact that at the close of the Jurassic and beginnning of the Early Cretaceous a climatic differentiation existed, and in this area a relatively cooler (Austral) climate reigned. Though this is in good agreement with the paleogeographic and palaeobioprovincial development of the area as outlined on the basis of foraminifera (Scheibnerová 1970, 1971a), it contradicts the conclusions based on the interpretation of the molluscan faunas from New Zealand and other parts of the Austral Biogeoprovince (as defined by Scheibnerová (1970, 1971a)) by Stevens (1965 and elsewhere). Stevens does not distinguish the cooler Austral bioprovince in the Jurassic, as his belemnite and other molluscan faunas allegedly show quite strong Tethyan affinity at that time. These faunas show some differences when compared with the tropical ones, but this is on a quantitative rather than a qualitative level.

The scarcity of calcareous foraminifera was partly explained by Kuznetsova (1974, p. 674) as indicating abyssal deposition well below the lysocline. However, this conclusion is opposed by the present author on the grounds of the similarity of these faunas with their coeval counterparts on the Australian continent (especially in the Early Cretaceous, pre-Albian deposits) and the shallow to extremely shallow (in the range of 200 down to perhaps 50 m) nature of coeval sediments recovered in the Indian Ocean elsewhere (Site 263, Leg 27).

In the author's opinion there is no convincing evidence for great depth before the beginning of the Late Cretaceous. The scarcity of calcareous forms can be attributed to low temperatures, extremely shallow depth and solution of calcitic tests in the sediment (after burial in the sediment) due to reducing conditions (see also Exon, 1972, Scheibnerová 1974a).

Foraminifera of Aptian and Younger Cretaceous Deposits in the Indian Ocean (Legs 22, 25, 26 and 27)

During Leg 22 Late Cretaceous sediments were penetrated in Sites 211, 212, 216 and 217. The oldest sediment at Site 211, between diabase and basalt, contain a "fairly diverse assemblage of small benthonic specimens but no planktonics" (McGowran 1974, p. 625). The generic composition of this assemblage is: Spiroplectammina, Marssonella, Gaudryina, Glomospira, Bathysiphon, Praebulimina, Reussella, Nuttallides, Pullenia, Gyroidinoides, Anomalinoides, Cibicides and Angulogavelinella. The fauna was determined as Late Senonian to Maastrichtian. Although McGowran (1974) admits that most of the above genera are known from shelf deposits, he favours "... a depth great enough for planktonic to have been removed by selective solution."

The lowest carbonate unit (lower part of cores 29-35) at Site 212 contains according to McGowran "... abundant planktonics and some calcareous benthonics, uniformly minute in size." Planktonics are represented by a "Globigerinelloides - Hedbergella - Heterohelix assemblage with no juvenile Globotruncana." The age of this assemblage was indicated as Campanian - Maastrichtian based on cf. Globotruncanella citae. In core 36 again, an assemblage of non-calcareous forms (Pelosina, Ammodiscus, Glomospira, Bathysiphon, Reophax, Haplophragmoides, cf. Adercotryma was interpreted as indicating a deep water origin.

The lowest sample at Site 216 is Late Maastrichtian, based on the presence of Globotruncanella mayaroensis (core 24cc). The associated benthonics include Stensioeina, Angulogavelinella and Cibicides indicating a shelf facies. Below the base of G. mayaroensis in Site 216 a generalized age of Campanian - Maastrichtian was given by McGowran (1974) and the assemblages characterized by low number of planktonics and sparse, but diverse benthonics including Gaudryina, Marssonella, Spiroplectammina, Lenticulina, Nodosaria, Marginulina, polymorphinids,

Gyroidinoides, Angulogavelinella, Cibicides, Praebulimina and Allomorphina. A shallow-marine environment was indicated by McGowran (1974) for these foraminiferal faunas supported by the presence of echinodermal remains, and molluscan fragments including Inoceramus.

Foraminiferal faunas of Site 217 closely parallel those of the Toolonga Calcilutite, Korojon Calcarenite and Miria Marl and contain numerous benthonic foraminiferal species as described by Belford (1960). Only generic determinations were made on the benthonic foraminifera.

Of eight sites drilled during Leg 25, only three holes penetrated sediments older than Tertiary. Of these only one (Site 249) penetrated pre-Albian deposits. Here, above the basalt, Neocomian and Barremian ages were determined by Sigal (1974). The only microfossils mentioned by him from these sediments are ostracodes, which correlate quite well with those known to occur in deposits of similar age in Madagascar (e.g. in the Majunga Basin, studied by Grekoff, 1963). In this site, the pre-Albian deposits are overlain by the Albian-Vraconian fossil-bearing beds.

Late Cretaceous sediments were recovered in Sites 239 (Early Campanian), 241 (?Turonian, Campanian) and 249 (Early Campanian, Early Masstrichtian). Both benthonic and planktonic foraminifera were studied, but only a few planktonic forms were illustrated from Site 249 (Sigal 1974, plate 2). Vraconian sediments in Site 249 are overlain by Late (not terminal) Cretaceous beds yielding both benthonic and planktonic species.

A few tens of centimeters of Late Cretaceous sediments at Site 239 contain an unusual microfaunal assemblage composed of benthonic species alone. No details, however, were given by Sigal (1974, p. 694). The Late Cretaceous of Site 241 contains a "...sparse, poor and exclusively benthonic microfauna assemblage" (Sigal op. cit. p. 693).

In general, Early Cretaceous sequences recovered on DSDP Leg 25 were found quite different from those known to occur on adjacent land. More studies are necessary to elucidate this problem. However, from general notes of Sigal (op. cit.) it is quite clear that the Late Cretaceous foraminiferal fauna is predominantly benthonic, and that the planktonic assemblages were less diversified than the coeval tropical associations. Most probably, conclusions drawn for other DSDP legs are also valid for Leg 25 (see especially Herb 1974). It is unlikely, therefore, that benthonic taxa of Leg 25 would be completely different from those of other DSDP legs in the Indian Ocean and from those occurring on adjacent land.

Leg 26 Cretaceous sediments were recovered from Broken Ridge (Site 255), Wharton Basin (Sites 256 and 257) and from the Naturaliste Plateau (Site 258). In Site 255 the oldest sediments are Santonian, in Site 256 Late Albian. All these ages were obtained from planktonic foraminifera (Herb, 1974) and nannoplankton (Thierstein, 1974). The age determinations based on planktonic foraminifera in Site 258 differ from those based on nannoplankton.

The oldest Cretaceous deposits recovered in Leg 26 are Albian. They were penetrated in Sites 256, 257 and 258. Their age was determined by nannofossils (Thierstein, 1974) and planktonic foraminifera (Herb, 1974). In Site 256 Thierstein (1975) recorded the Late Albian Eifellithus turriseiffeli Zone in the basal sediments, while in Site 257 the oldest recovered sediments belong to the Middle Albian (Praediscosphaera cretacea zone.

According to Herb's (1974) planktonic foraminiferal studies, the oldest sediments recovered in Site 256 contain Schackoina cf. cenomana and are Late Albian based on strong affinities with the Albian of Sites 257 and 258.

In Site 257 the oldest sediments are assigned to the Middle Albian based on a "primitive" arrangement of the "supplementary" apertures of forms assigned to the genus Ticinella.

The lowermost interval in Site 258, between Core 17, Section 5 and sample 20cc shows a restricted assemblage of small-sized planktonic forms including Globigerinelloides casseyi indicating most probably Late Albian or Vraconian. However, Herb also noted that "...A Middle Albian age for cores 19 and 20, as determined by the nannoplankton, cannot be excluded." For this same interval in Site 258 Thierstein identified Middle Albian Praediscophaera cretacea Zone from the oldest recorded sediments.

The present author studied benthonic foraminiferal faunas of Sites 256, 257 and especially 258. Benthonic foraminiferal species of the basal deposits in these three sites differ in that Lingulogavelinella frankei occurs only in the basal sediments in Site 258. The first specimens of L. frankei occur in samples 258-20cc and are quite rare, typical forms were found in sample 14-1 (125-127). The age of these samples ranges from Middle Albian to Middle Cenomanian based on nannofossil determinations by Thierstein (1974) or Albian to Cenomanian based on nannofossils as determined by Bukry (1974). Planktonic foraminiferal ages by Herb (1974) range from Late Albian (sample 20cc) to Cenomanian (sample 14-1 (125-127). Herb states that "A lowermost interval, between Core 17, Section 5 and Core 20cc, shows very restricted assemblages of small-sized forms... This assemblage indicates most probably Upper Albian or Vraconian. However, a Middle Albian age for Cores 19 and 20, as determined by the nannoplankton, cannot be excluded." The benthonic foraminiferal species of the above interval, however, are the same and do not indicate any differences in age within this interval (i.e. Core 17 Section 5 to Core 20cc). Species additional to the ones occurring in the above interval, such as Rotaliatina asiatica Vasilenko, Charltonina sp. and a problematic form resembling a globotruncanid (Scheibnerova, 1977) appear for the first time above the interval in sample 258-16-1 (bulk) and persist into the sample 258-13cc (bottom). Rotaliatina asiatica was described as an Early Cenomanian species from southern Turkmenia and Mangyshlak, where it represents the Early Cenomanian R. asiatica Zone (Vasilenko, 1961). If this determination is correct, sample 258-16-1 (bulk) and the whole of core 15 would represent Early Cenomanian.

In sample 258-14cc Pseudopatellinella howchini occurs for the first time and continues up to sample 13cc inclusive. This species was originally described from the upper part of the Cretaceous section in the South Australian portion of the Great Australian Basin (Ludbrook, 1966). Planktonic foraminiferal and nannoplankton evidence from the Indian Ocean indicate that it is a Late Albian-Cenomanian species. This fact is in good agreement with the stratigraphic interpretation of the marine Cretaceous sequences of the Great Australian Basin as proposed by the present author, who in many places interprets the upper part of the sequence as including also an early part of the early Cenomanian.

Another important species, Tappanina laciniosa, occurs first in sample 258-14cc. This species is known from the Late Cenomanian and Early and Middle Turonian Greenhorn Formation of the Great Plains of the United States. The vertical sequence of the above species strongly indicates the necessity to shift the Albian/Cenomanian boundary in Site 258 downwards. The shift of the Albian/Cenomanian boundary down to sample 258-16-1 (bulk) or at least down to sample 258-15cc by the present author is in a good agreement with the vertical distribution of the above species and also with that of Schackoina cenomana in Site 258.

In samples between 258-16-1 (bulk) and 258-14cc a peculiar form occurs, which is like a double-keeled globotruncanid of a very small size. At this stage a more precise determination is impossible; however, this form may be of considerable importance.

Benthonic foraminiferal assemblage of the interval between 16-1 (bulk) (or at least 15cc) up to 258-13cc inclusive are quite distinct. They differ slightly from those in the underlying sequences and are markedly different from those above; they are here interpreted as Cenomanian in age. Planktonic foraminiferal faunas in this interval are quite rare and of low specific diversity, lacking most of the typical Cenomanian species known in the Tethyan Biogeoprovince (see also Herb, 1974). However, Herb described a fragment of Rotalipora reicheli from one sample (258-14cc) and Schackoina cenomana (in samples 258-15cc, 258-14cc and 258-14-1(125-127)); nevertheless he left in doubt the exact position of the Albian/Cenomanian boundary, as well as a more precise determination of Albian sequences. This is understandable considering the low specific diversity of planktonic foraminiferal faunas.

It is of importance that Lingulogavelinella frankei was found to occur throughout the Cenomanian; however, in the uppermost Cenomanian it was rare. While most of the species are known to occur in coeval sediments on adjacent land, Albian/Cenomanian deposits of Leg 26 are most similar to those of the Great Australian Basin in their specific composition, contrasting with those of Leg 27 which show more affinities with South Africa, peninsular India and Western Australia.

Site 258 of Leg 26 contains the most complete Cretaceous sequence. Sequences above Cenomanian include Turonian, Coniacian, Santonian and Campanian. Benthonic foraminiferal assemblages from several samples (cores 12-5) were studied by the present author. They contain many species known to occur in coeval sediments in Western Australia (Belford, 1960, or as yet undescribed).

Of special interest are some agglutinated forms found in a few samples from Site 256, especially Recurvoides sp. and Labrospira pacifica Krasheninnikov. L. pacifica was described from the Late Cretaceous of the Pacific Ocean (Krasheninnikov, 1973) and also from Leg 27 in the Indian Ocean (Krasheninnikov, 1974). Recurvoides sp. resembles the species pseudosymmetricus Krasheninnikov and is identical with the specimens found in coeval sediments in Western Australia and some boreholes drilled off Western Australia.

The oldest Cretaceous sediments recovered by the DSDP in Leg 27 are those of Sites 259, 260 and 261 (Bartenstein, 1974 and Kuznetsova, 1974), and 263 (Scheibnerova, 1974) ranging from Late Jurassic to Early Cretaceous. The Early Cretaceous faunas of Leg 27 are mostly agglutinated with nodosariids being the dominant calcareous forms.

An interesting discrepancy exists regarding the age of Cretaceous sequence in Site 263. While foraminifera, spores and pollen and dinoflagellates (Wismann and Williams, 1974), nannoplankton in part (Bukry, 1974) and megafossils in part (Speden, 1974) indicate Early Cretaceous (pre-Middle Albian), nannofossils as studied by Proto Decima (1974) are assigned for the entire section to Late Albian.

Foraminiferal faunas of Site 263 are different from those of Sites 259 and 260. However, they correlate closely with those of the adjacent Canning Basin and the oldest forms of the Great Australian Basin. The Aptian or older age based on these foraminifera is well documented, although the fauna is mostly agglutinated. It is composed of species which are absent from post-Aptian sediments in Sites 259 and 260, which could be explained by a substantially different ecology. However, in the Cretaceous section of the Great Australian Basin there is a sharp difference between the Aptian (or older) and Albian foraminiferal faunas. Moreover, some of the Trochammina and Textularis species determined by Kuznetsova from undoubtedly Early Cretaceous (pre-Albian) deposits in Site 261 resemble the species of the lower portions of Site 263. Recently, a few specimens of Gubkinella strongly resembling G. californica Dailey (Dailey, 1970) (in sample 263-8cc) and an ostracode similar to Cytheropteron sp. from the Neocomian of Leg 25 (Sigal, 1974) were found in Site 263 (sample 263-23cc). These forms also support an older age for the sediments in question.

If the younger age determination based on nannofossils (Proto Decima, 1974) were correct, all Roma faunas of the Great Australian Basin would be Late Albian. This appears doubtful because the age of the Roma fauna is determined as Aptian by ammonite and pelecypod faunas (Day, 1969, 1974).

The summary of the palaeontology published in the Initial Reports (Bolli, 1974) mentions only Proto Decima's younger age in 263. Speden (1974) determined the age of ?Aucellina sp. from Site 263-17cc as ?Neocomian - Turonian. The ammonite fragment of Stevens (1974) determined by him as ?cf Prohysteroceras (Goodhallites) richardsi Whitehouse was extremely poor. The only belemnite guard, ?Parahibolites sp. indet., found in 263 (Core 26-18-5(15)) is better preserved, although as mentioned by Stevens, many of its diagnostic features are missing.

Another important piece of evidence for the pre-Albian age of the lower part of the Cretaceous section of 263 is provided by Veevers and Johnstone (1974) who correlate the lower part of the Cretaceous section of 263 with similar sequences in Western Australia: "... a direct link between the continental margin and oceanic Early Cretaceous sequences is provided by the seismic profile between Pendock No. 1 Well and Site 263 "... "The lithological similarities are striking ... The Conclusion is inescapable that the Lower Cretaceous sequence at Site 263 is part of the Winning Group, and, though now in an oceanic setting (the base of Site 263 is 5811 m below sea level), was deposited in a fairly shallow environment." This statement is supported by the foraminifera, spore-pollen and dinoflagellates and Bukry's (1974) nannofossil determinations.

There are slight differences in age determinations based on different animal groups in Sites 259 and 260 (in part). At Site 259 in Cores 11-17 benthonic foraminifera indicate Late Albian (Scheibnerova, 1974b) planktonic foraminifera: Albian (Krasheninnikov, 1974) and the nannoplankton: Middle Albian. These age discrepancies are only minor, but they deserve attention, especially for the stratigraphic value of the Austral benthonic foraminifera, which is progressively being established.

The benthonic foraminifera of Sites 259 and 260 studied by the present author showed some interesting features. Not only do they include taxa known to occur in adjacent areas (Indian Peninsula, South Africa, Australia), but instead of Lingulogavelinella frankei being present in its typical form it occurs as cf. frankei. This absence of the typical form was explained by either different stratigraphic position of the beds or the different, in this case deeper water, nature of the environment yielding this form. Scheibnerova (1974b) discussed the difference in age and concludes that if the beds with atypical frankei from Site 259 were undoubtedly Late Albian, than the typical frankei could be a Cenomanian species. However, bearing in mind the Middle Albian age determination based on the nannofossils of the beds in question, the absence of the typical frankei would be explained by implying a Middle Albian age for those beds. Such an explanation is favoured by the occurrence of typical frankei in Leg 26, Site 258 (see also below).

However, the relatively deeper water nature of sediments in Site 259 (Leg 27) can still be maintained.

Specimens referred by Scheibnerova (1974b) to ?Lingulogavelinella sp. from Site 259 are identical with forms referred by Krasheninnikov (1974) to Globigerinelloides gyroidinaeformis Moullade. It is the present author's opinion that these are not planktonic forms. They in fact were referred to Globigerinelloides only tentatively (Moullade, 1966). Nevertheless, their stratigraphic value may be considerable as they have not been reported from sediments with typical frankei in Site 258, but from the Early and Middle Albian of Northern Europe. This again would support an age not younger than Middle Albian for the beds in question. This conclusion is further supported by the similarity of a foraminiferal assemblage from Leg 26, Site 257 to those discussed above and determined as Middle Albian.

Late Cretaceous sediments are completely absent from Sites 259 and 263. They occur, however, in much reduced thickness in Site 260, Core 6 (Gascoyne Abyssal Plain) and Site 261, Cores 5-8 (Argo Abyssal Plain). These sediments contain an assumedly autochthonous fauna of characteristic agglutinated foraminifera. The brown clays of Site 260 also contain reworked planktonic foraminifera of Albian (?), Cenomanian and Late Turonian - Coniacian ages.

Two assemblages of agglutinated foraminifera were distinguished by Krasheninnikov (1974) including 21 genera:
1) An upper assemblage with Praecystammina globigerinaeformis and
2) a lower assemblage with Haplophragmium lueckei.

Since the upper assemblage contains reworked planktonic foraminifera, the youngest of which indicate Late Turonian-Coniacian age, the age of sediments containing the upper assemblage of agglutinated foraminifera was regarded as not older (within the Late Cretaceous) than upper Turonian-Coniacian. Since the upper and lower agglutinated assemblages have some species in common, the age of the lower assemblage, which does not contain planktonic foraminifera, was considered by Krasheninnikov (1974) to be upper Cretaceous also, although slightly older than the upper assemblage. The upper assemblage with Praecystammina globigerinaeformis has also been recognized in the Pacific Ocean (Leg 20, Krasheninnikov, 1973).

These agglutinated foraminifera are characterized by their small dimensions, and a very-fine-grained, homogeneous thin wall with a smooth surface, these features supposedly reflecting deep water conditions.

Some of them (Recurvoides, Labrospira pacifica, Glomospira charoides) have been identified by the present author in Late Cretaceous sediments in Leg 26, Site 256, Cores 7cc 9-1 (120-122) in samples without planktonic foraminifera. These Late Cretaceous assemblages differ markedly from the coeval assemblages recovered in Leg 26, Site 258 in which Herb (1974) determined many planktonic taxa of non-tropical nature.

Cretaceous Palaeobathymetry of the Indian Ocean

Except for Gubkinella, pre-Albian sediments recovered by the DSDP in the Indian Ocean have not yielded planktonic foraminifera. These faunas are mainly composed of agglutinated and calcareous benthonic forms. The agglutinating component predominates, the calcareous when present, often shows signs of dissolution.

The partial dissolution and/or absence of calcareous (both benthonic and planktonic) tests may be explained in two ways:
1) Dissolution due to excessive depth below the lysocline or
2) dissolution in sediment due to reducing, low oxygen, low pH

conditions in a relatively shallow to extremely shallow, stagnant environment.

It is of importance to know which kind of dissolution took place, and at what time, as this has a major impact on a palaeogeographic reconstruction of the Cretaceous Indian Ocean.

As far as the benthonic foraminifera are concerned, they are composed of two types of assemblages:
1) Those known to occur on adjacent land, in sediments deposited in epicontinental, shelf seas, which were relatively to extremely shallow, and
2) those composed of taxa so far unknown in sediments from land areas.

All pre-Albian, Albian and Cenomanian, and pre-Santonian deposits recovered by the DSDP in the Indian Ocean contain taxa of agglutinated and calcareous foraminifera known to occur in coeval sediments on land. There is little doubt about their relatively shallow to extremely shallow depth of deposition. To imply any great depth to these foraminiferal species would be difficult to explain.

The mere absence of calcareous (including planktonic) foraminifera from these sediments does not justify an excessive depth, because the same restriction in numbers of calcareous forms has been observed in similar deposits on land. Similar dissolution phenomena and a reduction in numbers of planktonic foraminifera have been observed in the Great Australian Basin. Besides secondary reasons for the dissolution, primary factors must be implied, in this case shallow depth, cool (lower than optimum) temperatures and restricted open marine connections. All three factors, but especially the shallow depth and cool temperatures must have had considerable affect in the newly opened Indian Ocean, especially in those parts adjacent to the surrounding continents and to the south. As a logical consequence of these primary conditions the existence of environments with shallow stagnant waters must also be expected, and is supported by a striking lithological similarity with land sections (see Veevers and Johnstone 1974, Scheibnerová, 1974b). It is the present author's opinion that great depths equivalent to present situation below the lysocline, can be ruled out for the Early and early Late Cretaceous Indian Ocean. Foraminiferal faunas of these deposits show features of the assemblages of the inner and outer shelf and upper (middle) slope. Assemblages typical of the lower slope and abyssal zones have not been recovered in these sediments. It is the author's opinion that such environments had not developed in the Indian and Atlantic Oceans in the early stages of their opening.

On the other hand, in the Late Cretaceous brown zeolitic clays recovered in the Gascoyne and Argo Abyssal Plains Krasheninnikov (1974) found and described in detail assumed autochthonous characteristic agglutinated forms (see below), which were previously described in coeval sediments recovered by the DSDP in the Pacific Ocean (Leg 20), but which differ strongly from those in contemporaneous Cretaceous marine sediments on land. This, together with the lithology (brown zeolitic clays) of sediments yielding these faunas shows their deepwater nature. The calcareous component is completely missing or substantially reduced, therefore the depth of this environment was presumably below the lysocline at a considerable depth. Here, the lack of coeval planktonic foraminiferal species can indeed be attributed to dissolution due to excessive depth.

Some authors (Bartenstein, 1974, Kuznetsova, 1974) compared some of the Early Cretaceous benthonic assemblages recovered by the DSDP in the Indian Ocean with those of the Alpine-Carpathian flysch. However, this again does not necessarily imply great depth; similar associations in the flysch deposits could have been carried down by turbidity currents from shallow environments.

It is the author's opinion that elsewhere in sediments not containing the above characteristic agglutinated forms no excessive depth can be deduced, even if traces of dissolution are present. Low specific diversity and small numbers of planktonic individuals here can successfully be attributed to non-tropical, below optimum temperatures, which prevailed in the southern and central parts of the Cretaceous Indian Ocean. This conclusion is strongly supported by comparison of the recovered benthonic and planktonic foraminiferal species with those of the rest of the Austral Biogeoprovince; similar comparisons also show many affinities with its northern counterparts, the Boreal Biogeoprovince.

Stratigraphic Importance of Aptian/Albian and Cenomanian Benthonic Foraminifera as Recovered by the DSDP in the Indian Ocean

Data on the Austral Cretaceous foraminifera were meagre until recently. However, in the last five years sufficient information has been published, which shows that during the Cretaceous extensive marine deposition occurred especially during the Aptian/Albian and Cenomanian, followed by hiatuses after the Cenomanian. Deposits of the Turonian, Santonian/Campanian and Maastrichtian are also known. These sediments yield benthonic foraminifera composed of genera and species correlating very well with those known from coeval Boreal sediments in the Northern Hemisphere. Therefore, they were interpreted as assemblages of relatively cool environment when compared with the coeval tropical environments (Scheibnerová, 1970). At the same time, these foraminiferal assemblages as a rule show features of quite shallow to extremely shallow depth (in the range of 100-200m) judging from their typical occurrences in Cretaceous sediments on land. This shallowness is especially typical of Early and early Late Cretaceous sediments not only on land, but also in the sediments recovered by the DSDP in the Indian Ocean.

Taxonomically important Austral foraminiferal genera are Lingulogavelinella, Orithostella, Pseudopatelinella, Gavelinella, Charltonina, Reinholdella, Hoeglundina, Pseudolamarckina, Discorbis, Neobulimina, Praebulimina, Tappanina, Valvulineria, Coryphostoma, Textularis, Trochammina, and Rotaliatina. The stratigraphic value of some of them has been studied over the past five years by the present author, based mainly on material from the Great Australian Basin in Central and Eastern Australia. The DSDP material from the Indian Ocean is of special interest, because of its proximity to the known land sections and because of the associated, sometimes intermittent, planktonic foraminifera in relative abundance. This enables at least some correlation with the existing Cretaceous tropical zonation.

Although there remain many problems to be solved regarding the stratigraphic value of benthonic foraminifera in the Austral Biogeoprovince, especially in terms of European stages, it is becoming clear that the stratigraphically most diagnostic genera are Lingulogavelinella and Orithostella; also important are Gavelinella, Discorbis, Charltonina, Rotaliatina, Pseudopatellinella, Textularia, Trochammina, Verneuilina, Neobulimina, Praebulimina, Coryphostoma and Tappanina. However, it has to be kept in mind that the species of the above genera are facies controlled, and different genera and species have to be used as stratigraphic markers for different facies. Although it is often neglected, a similar facies control also applies for the planktonic zonation, especially in the Late Cretaceous (for more detail see Scheibnerová, 1974). It is also evident that most benthonic species are not confined to a short time span. Stratigraphic investigation should therefore also focus on datum levels (first occurrence in particular). One of the most typical examples for this is the case of L. frankei (sensu Scheibnerová, 1971c) which according to recent data by the present author covers the time span Late Albian to Late Cenomanian. The fixing of the Albian/Cenomanian boundary in the Austral Biogeoprovince therefore is extremely difficult when using Lingulogavelinella or other benthonic foraminifera.

The first occurrence of Lingulogavelinella albiensis albiensis (Malapris) on the other hand clearly defines the Aptian/Albian boundary especially in the Great Australian Basin where it is common (Scheibnerová, 1974a).

Orithostella species appear first in the Albian; many species occur also in the Cenomanian. They seem to be good stratigraphic markers especially in and around the area of the Indian Ocean (for more detail see Scheibnerová 1974b and Narayanan and Scheibnerová (in press).

Tappanina laciniosa found recently in Site 258 seems to define the Late Cenomanian and Early Turonian in the Indian Ocean.

The Pseudopatellinella howchni first occurrence seems to be a good indicator for the Late Albian and the base of the (late) Cenomanian in Site 258 (Leg 26). This is of interest, because this species was described by Ludbrook (1966) from the upper parts of the marine Cretaceous section in the South Australian portion of the Great Australian Basin. Discorbis sp. occurs extensively in the Albian and Cenomanian in the Austral Biogeoprovince (for more detail see Lambert, Malumian and Scheibnerová, in prep.).

Permission to publish this paper was given by the Undersecretary New South Wales Department of Mines.

References

Bartenstein, H., Upper Jurassic - Lower Cretaceous primitive arenaceous foraminifera from DSDP sites 259 and 261, Eastern Indian Ocean, in Veevers, J. J., Heirtzler, J. R. et al., 1974, Initial Reports of the Deep Sea Drilling Project, Washington (U. S. Government Printing Office), 27, 683-695, figure 1-2, 1974.

Belford, D. J., Upper Cretaceous foraminifera from the Toolonga Calcilutite and Gingin Chalk, Western Australia, Australia Bur. Min. Res., Geol. Geophys. Bull., 57, 1-198, pls. 1-35, text-figures 1-15, appendix 1-2, 1960.

Bolli, H. M., Synthesis of the Leg 27 biostratigraphy and paleontology, in Veevers, J. J., Heirtzler, J. R., et al., 1974, Initial Reports of the Deep Sea Drilling Project, Washington (U. S. Government Printing Office) 27, 993-999, figures 1-5, 1974.

Bukry, D., Cretaceous and Paleogene coccoliths stratigraphy, Deep Sea Drilling Project, Leg 26, in Davies, T. A., Luyendyk, B. P., et al., Initial Reports of the Deep Sea Drilling Project, Washington (U. S. Government Printing Office), 26, 669-673, 1974.

Dailey, D. H., Some new Cretaceous foraminifera from the Budden Canyon Formation, northwestern Sacramento Valley, California, Contr. Cushman Fdn. Foramin. Res., 21, pt. 3, 100-111, 1970.

Day, R. W., The Lower Cretaceous of the Great Artesian Basin, in Stratigraphy and Palaeontology, Essays in honour of Dorothy Hill, edited by K. S. W. Campbell, Australian National University Press, Canberra, 140-173, 1969.

Day, R. W., Aptian Ammonites from the Eromanga and Surat Basins, Queensland, G. S. Qld, Publ. no. 360, Palaeontological Papers, no. 34, 1-19, 1974.

Eicher, D. L., and P. Worstell, Cenomanian and Turonian foraminifera from the Great Plains, United States, Micropaleontology, 16, no. 3, 269-324, 1970.

Exon, N., Sedimentation in the outer Flensburg Fjörd area (Baltic Sea) since the last glaciation, Meyriana, 22, 5-62, 1972.

Grekoff, N., Contribution a l'etude des Ostracodes du Mesozoic moyen (Bathnien - Valanginien) due bassin de Majunga, Madagascar, Rev. Inst. Franc. Petro, 18, 1709-1762, 1963.

Herb, R., Cretaceous planktonic foraminifera from the eastern Indian Ocean, in Davies, T. A., Luyendyk, B. P., et al., Initial Reports of the Deep Sea Drilling Project, Washington (U. S. Government Printing Office) 26, 619-667, 1974.

Krasheninnikov, V. A., Cretaceous benthonic foraminifera, Leg 20, Deep Sea Drilling Project, in Heezen, B. C., MacGregor, I. D., Initial Reports of the Deep Sea Drilling Project, Washington (U. S. Government Printing Office) 20, 205-219, 1973.

Krasheninnikov, V. A., Upper Cretaceous benthonic agglutinated foraminifera, Leg 27 of the Deep Sea Drilling Project, in Veevers, J. J. Heirtzler, J. R., et al., Initial Reports of the Deep Sea Drilling Project, Washington (U. S. Government Printing Office) 27, 631-661, 1974.

Kuznetsova, K. I., Distribution of benthonic foraminifera in Upper Jurassic and Lower Cretaceous deposits at Site 261, DSDP Leg 27, in the eastern Indian Ocean, in Veevers, J. J., Heirtzler, J. R., et al., Initial Reports of the Deep Sea Drilling Project, Washington (U. S. Government Printing Office) 27, 673-681, 1974.

Ludbrook, N. H., Cretaceous biostratigraphy of the Great Artesian Basin in South Australia, Geol. Soc. S. Australia Bull., 40, 1-225, 1966.

McGowran, B., Foraminifera, in von der Borch, C. C., Sclater, J. G., et al., Initial Reports of the Deep Sea Drilling Project, Washington (U. S. Government Printing Office) 22, 609-627, 1974.

Moullade, M., Etude stratigraphique et micropaleontologique du Cretace in rieur de la "Fosse Vocontienne", Doc. Labor. Geol. Fac. Sci. Lyon, 15, 1-369, 1966.

Narayanan, V., and V. Scheibnerova, Lingulogavelinella and Orithostella (Foraminifera) from the Utatur group of the Trichinopoly Cretaceous, South (peninsular) India, Revista espanola de micropaleontologia, 1977 (in press)

Pessagno, E. A., and F. Y. Michael, Results of the shore laboratory studies on Mesozoic planktonic foraminifera from Leg 26, Sites 255, 256, 257 and 258, in Davies, T. A., Luyendyk, B. P., et al., Initial Reports of the Deep Sea Drilling Project, Washington (U. S. Government Printing Office) 26, 969-972, 1974.

Pflaumann, U., Zur Oekologie des bayersschen Flysches auf Grund der Mikrofossilfuehrung, Geol. Rundsch., 56, 1, 1967.

Proto Decima, F., Leg 27 calcareous nannoplankton, in Veevers, J. J., Heirtzler, J. R., Initial Reports of the Deep Sea Drilling Project, Washington (U. S. Government Printing Office) 27, 589-621, 1974.

Scheibnerova, V., Some notes on the palaeoecology and palaeogeography of the Great Artesian Basin during the Cretaceous, Search, 1, no. 3, 125-126, 1970.

Scheibnerova, V., Foraminifera and their Mesozoic Biogeoprovinces, Geol. Surv. N.S.W. Rec., 13, no. 3, 135-174, 1971a.

Scheibnerova, V., Palaeoecology and Palaeogeography of the Cretaceous deposits of the Great Artesian Basin, Geol. Surv. N.S.W. Rec., 13, no. 1, 1-48, 1971b.

Scheibnerova, V., Lingulogavelinella (foraminifera) in the Cretaceous of the Great Artesian Basin, Australia, Micropaleontology, 17, no. 1, 109-116, 1971c.

Scheibnerova, V.,The ecology of Scutuloris and other important genera from the Early Cretaceous of the Great Artesian Basin, Rev. espan. Micropal. 6, no. 2, 229-255, 1974a.

Scheibnerova, V., Aptian-Albian benthonic foraminifera from DSDP Leg 27, Sites 259, 260 and 263, Eastern Indian Ocean, in Veevers, J. J., Heirtzler, J. R., Initial Reports of the Deep Sea Drilling Project, Washington (U. S. Government Printing Office) 27, 697-741, 1974b.

Scheibnerova, V., Cretaceous foraminifera of the Great Austral Basin, Mem. Geol. Surv. N.S.W., 1977 (in press).

Sigal, J., Comments on Leg 27 sites in relation to the Cretaceous and Paleogene stratigraphy in the eastern and southeastern Africa coast and Madagascar regional setting, in Simpson, E.S.W., Schlich, R., et al., Initial Reports of the Deep Sea Drilling, Washington (U. S. Government Printing Office) 25, 687-723, 1974.

Speden, I. G., Cretaceous bivalvia from cores, Leg 27, in Veevers, J. J., Heirtzler, J. R., Initial Reports of the Deep Sea Drilling Project, Washington (U. S. Government Printing Office) 27, 977-981, 1974.

Stevens, G. R., Faunal realms in Jurassic and Cretaceous belemnites, Geol. Mag., 102, 175, 1965.

Stevens, G. R., Cretaceous belemnites, in Atlas of palaeobiogeography, edited by A. Hallam, Amsterdam (Elsevier), 385, 1973.

Stevens, G. R., Leg 27 Cephalopoda, in Veevers, J. J., Heirtzler, J. R. et al., Initial Reports of the Deep Sea Drilling Project, Washington (U. S. Government Printing Office) 27, 983-989, 1974.

Thierstein, H. R., Calcareous nannoplankton - Leg 26, Deep Sea Drilling Project, in Davies, T. A., Luyendyk, B. P., et al., Initial Reports of the Deep Sea Drilling Project, Washington (U. S. Government Printing Office) 26, 619-667, 1974.

Vasilenko, V. P., Foraminifery verkhnego mela poluostrova Mangyshlaka, Trudy VNIGRI, no. 171, 1-487, 1961.

Veevers, J. J., and M. H. Johnstone, Comparative stratigraphy and structure of the Western Australian margin and the adjacent deep ocean floor, in Veevers, J. J., Heirtzler, J. R., et al., Initial Reports of the Deep Sea Drilling Project, Washington (U. S. Government Printing Office) 27, 571-586, 1974.

Wisemann, J. F., and A. J. Williams, Palynological investigation of samples from sites 259-261 and 263, Leg 27, in Veevers, J. J., Heirtzler, J. R., et al., Initial Reports of the Deep Sea Drilling Project, Washington (U. S. Government Printing Office) 27, 915-924, 1974.

CHAPTER 22. NEOGENE DEEP WATER BENTHONIC FORAMINIFERA OF THE INDIAN
           OCEAN

E. Boltovskoy

Museo Argentino de Ciencias Naturales "B. Rivadavia" and Consejo
Nacional de Investigaciones Cientificas y Tecnicas, Argentina

Abstract. Neogene - Recent benthonic foraminifera were studied in 499 samples from 15 sites which were drilled in the Indian Ocean during legs 23, 25, 26 and 27 of the Deep Sea Drilling Project. These sites were located at depths ranging from 1,030 - 5,709m. 306 species were identified, none of them new. Nine species were put in nomenclatura aperta. The unicameral calcareous species belonging to the genera Lagena, Oolina, Fissurina and Parafissurina were not studied. The stratigraphic subdivisions of the sites examined in this study had been previously established by the paleontological staffs of the legs mentioned above by means of planktonic foraminifera and other microfossils. The present study of benthonic foraminifera revealed that the number of cores which contain displaced or/and reworked fauna is much higher than previously thought; only five of the 15 sites have not yielded redeposited specimens. Four of these five sites are located far from the ocean margin. From 10 sites containing redeposited specimens eight should be considered as with heterogeneous material throughout their whole length and two as partially with redeposited specimens. These data cast some doubt on several conclusions made by previous authors who did not realize that sequences at several sites were rather strongly mixed with redeposited material and thus probably are not completely reliable.
  On the basis of the data obtained exclusively from the sites which have sediments in situ the following conclusions were drawn: 1) Only Miocene guide fossils ("Bulava indica"*, Bulimina macilenta, B. miolaevis, B. jarvisi, Planulina marialana gigas, and Siphogenerina vesca) could be established. 2) The calcareous: agglutinated species ratio increased with an increase in depth from the 1,000 - 2,000 m range to the 3,500 - 4,000 m range. 3) Qualitative changes in the benthonic foraminiferal faunas as related to the depth changes were insignificant. However, it was observed that Orthomorphina and Osangularia preferred shallower depths and Nonion was found mainly at greater depths. 4) It is believed that the Indian Ocean has not suffered any drastic bathymetric changes since Miocene time.

---

* "Bulava indica"(a very good Middle and Late Miocene indicator) is a
  fossil of uncertain origin and therefore this name should be written
  for the present in quotation marks. It is not clear if it is a new
  benthonic taxon or a fragment of some planktonic foraminifera.

Introduction

So far in the Indian Ocean the Deep Sea Drilling Project has had six legs, 22 through 27. All of them took place in 1972 and only planktonic foraminifera were studied in the material recovered. Benthonic foraminiferal species were cited sporadically and not by all the authors. Therefore, I decided to dedicate a special study to the benthonic forms of the Neogene deposits drilled during these legs.

Material and Methods

The main material for this report consisted of the samples collected during Leg 26 (in which I participated) and examined by me for their planktonic foraminiferal content (Boltovskoy, 1974). At that time I studied almost 900 samples from 9 sites. For the present paper only selected material of six sites was studied because several sites and cores appeared to be barren or contained dissolved, damaged or redeposited planktonic foraminifera.

In addition, I received samples from three other legs which took place in the Indian Ocean. Below is a table showing all the sites studied, their location and number of samples provided from each of them.

Table 1

LOCATION OF SITES

Leg	Site	Latitude	Longitude	Depth (m)	Number of samples
	219	09°01.75'N	72°52.67'E	1,764 m	28
23	220	06°30.97'N	70°59.02'E	4,043 m	9
23	223	18°44.98'N	60°07.78'E	3,633 m	4
25	239	21°17.67'S	51°40.73'E	4,971 m	21
25	242	15°50.65'S	41°49.23'E	2,275 m	15
25	246	33°37.21'S	45°09.60'E	1,030 m	3
26	250	33°27.74'S	39°22.15'E	5,119 m	65
26	251	36°30.26'S	49°29.08'E	3,489 m	60
26	253	24°52.65'S	87°21.97'E	1,962 m	76
26	254	30°58.15'S	87°53.72'E	1,253 m	75
26	255	31°07.87'S	93°43.72'E	1,144 m	18
26	258	33°47.69'S	112°28.42'E	2,793 m	25
27	260	16°08.67'S	110°17.92'E	5,709 m	6
27	262	10°52.19'S	123°50.78'E	2,298 m	86
27	263	23°19.43'S	110°57.81'E	5,065 m	6
				Total number of samples	497

Aims of this Study

The geological age of all the samples studied was previously determined by means of planktonic foraminifera and/or other microfossils used by the scientists of each leg. Benthonic foraminifera, and especially those of great depths, are less changeable than planktonic ones. In addition, the quantitative and qualitative content of benthonic assemblages, as well as the morphological appearance of the tests, depend on the character of the bottom, depth and other ecological parameters. They are far worse for age determination than planktonic species. Therefore, if I found, for instance, a typical Pliocene benthonic indicator in Miocene sediments, I did not change the previously established age of the sample, but considered that the range of this benthonic species was previously interpreted erroneously. Nor did I try to correlate the sediments of the sites studied with those from other areas by means of benthonic foraminifera. Such correlation should be done on the basis of personal studies of different faunas and I did not have these faunas at my disposal. However, I was able to compare faunas of the legs cited.

The main objectives of this report were as follows: 1) To identify taxonomically the benthonic foraminiferal species which compose the faunas found in the Neogene deposits of the sites studied and 2) To establish: a) the vertical ranges of the most common and typical benthonic foraminiferal species, b) the assemblages typical of different geological times, c) those benthonic species which can serve as guide fossils, and d) which parts of which cores contain resedimented material.

Displaced Fauna

The study of the benthonic foraminifera of the sites where planktonic forms have been studied is very important because such studies may yield new information which can be useful in the reinterpretation of previous conclusions based on planktonic studies. This is because planktonic foraminifera can give clear evidence that the material is redeposited only in two cases: a) if planktonic faunas of given age contains specimens of a quite different age, or b) if planktonic specimens show evidence of having been transported. If, however, the transported faunas is more or less contemporaneous with that in situ, or if the tests were not damaged in transport, it cannot be stated with certainty that a given sample has displaced specimens. In such cases benthonic specimens are of particular importance.

There are several benthonic genera and species which dwell solely in the shallow water of the shelf zone. As all the cores studied were taken at great depths (between 1,030 and 5,709 m) the occurrence of these species in a sample is proof that the sample contains displaced material.

In the majority of cases the redeposited benthonic specimens found were small. Some of them were even extremely small. The presence of small specimens in a redeposited fauna is understandable. Small shells are transported much easier than are large ones. They even can be transported long distances by surface currents after being lifted from the bottom by heavy storms. Moreover, foraminiferal shells can be transported from the littoral zone to zones of great depths by icebergs and floating algae (Boltovskoy and Lena, 1969). However, all these factors are insignificant compared to post-mortem transport carried out by turbidity currents and submarine landslides.

The following species found in the material studied should be considered as typical shallow water ones and consequently as obviously displaced: Amphistegina lessonii, Asterigerinata pacifica, Cancris sagra,

Cymbaloporetta bradyi, Discorbis parkerae, Discorbinella biconcava, Elphidium advenum, E. complanatum, E. craticulatum, E. crispum, E. discoidale, E. sp., Glabratella sp., Guttulina regina, Miliolinella lutea, M. subrotunda, Patellina corrugata, Peneroplis pertusus, Planorbulina mediterranensis, Quinqueloculina seminulum, Rotalia beccarii koeboeensis, R. margaritifera, R. schroeteriana, Spiroloculina antillarum.

There are several other species which probably prefer a shelf environment. Since, however, their habitat is still not definitely determined, I prefer not to include them in the above list.

The study of benthonic foraminifera revealed that obviously displaced specimens are very often present in the material of the sites under study (see Table 2).

Table 3 shows the distribution and quantity of redeposited specimens at Site 262. Although this site does not contain the highest percentage of samples with displaced specimens, I chose it because the displaced fauna found there is qualitatively and quantitatively rich and many specimens are large, thick-walled and strongly worn.

An analysis of Tables 2 and 3 leads to a rather unhappy conclusion.

Ten sites of the fifteen studied here appeared to contain displaced fauna. In eight of these ten sites the displaced fauna is distributed in all the sequences from Pleistocene down to Lower Miocene. At the two sites displaced faunas were observed only in parts of their length.

Of the remaining five sites, which did not contain displaced faunas, four sites (251, 253, 254, 255) are located far from the shore and only one (220) relatively near the shore.

Inevitably the problem arises. If such a high proportion of samples with heterogeneous material was obtained from the Indian Ocean, we can anticipate similar situations in the other oceans. And if such is the case what errors could be created on the basis of studies of material presumably uncontaminated, but in reality with displaced or/and reworked elements.

It seems to me that the main conclusion which can be drawn from this observation is that in studies of deep-sea cores an examination of planktonic fossils only is insufficient to obtain good results. A study of benthonic foraminifera should be included as well, especially in case of sites located on the peripheral parts of the oceans.

As mentioned above, no displaced benthonic specimens were found at the sites located far away from the continents and islands (Sites 251, 253, 254, 255). However, in some of them (Sites 253, 254) several samples were contaminated with redeposited planktonic foraminifera (Boltovskoy, 1974, figs. 4,5). Evidently this contamination occurred either during the drilling or during the core treatment aboard the GLOMAR CHALLENGER. But this problem is out of the scope of the present report.

Relationship Between Fauna and Depth

To establish the relationship between the qualitative characteristics of the fauna and depth, only sites that did not yield redeposited specimens were taken into account. They were divided, according to their depth, into two groups: a) those located in the abyssal zone, between c. 3,5000 and 4,000 m (Sites 220 and 251), and b) those located in the bathyal zone, between c. 1,000 and 2,000 m (Sites 253, 254 and 255). A comparison of the faunas found in these two zones shows that there is only a very insignificant difference in their faunas. Almost all the species were found either in both zones or if only in one of them then as isolated specimens. Thus, it cannot be stated that they prefer a particular depth zone. However, an exception, can be made for the following genera. Nonion was consistently more numerous at great depths.

Table 2

Sites containing redeposited benthonic foraminifera

SITE	SEQUENCE WITH REDEPOSITED SPECIMENS	CHARACTER OF REDEPOSITED FAUNA AND REFERENCE
219	Entire sequence	Reworked planktonic foraminifera primarily early Middle Miocene in age (Fleisher, 1974)
223	Entire sequence	Shallow water displaced benthonic species: several Elphidium, Quinqueloculina seminulum, Rotalia, etc. (this study)
239	Entire sequence	Small-sized displaced benthonic shallow water species mainly Elphidium and Miliolinella (this study)
242	Pliocene only	Displaced specimens of Elphidium sp. (this study)
246	Entire sequence	Shallow water displaced benthonic foraminifera (Discorbinella biconcava, Elphidium sp. (this study)
250	Entire sequence except Quaternary	Displaced littoral benthonic foraminifera (Elphidium crispum, Amphistegina lessonii, Glabratella sp. and others) (Boltovskoy, 1974, and this study)
258	Entire sequence	Displaced shallow water benthonic foraminifera mainly small-sized Elphidium (this study)
260	Entire sequence	Reworked benthonic foraminiferal species of older ages an shallow water benthonic specimens (Bolli, 1974, and this study)
262	Entire sequence	Shallow water species are rare in the Quaternary but common in the Pliocene (mainly large displaced Rotalia (Bolli, 1974, and this study)
263	Entire sequence	Redeposited shallow water benthonic species: Elphidium advenum, E. crispum, Miliolinella subrotunda, Rotalia beccarii koeboensis and others (Bolli, 1974, and this study)

Orthomorphina and Osangularia apparently preferred depth of 1,000 - 2,000 m. At these same depths numerous specimens of Planulina marialana gigas were also found. This species was absent at greater depths.

Using the data from the above cited sites, the relationship between the calcareous and agglutinated species at different depths was established. The sites located in the abyssal zone contained 103 calcareous and 9 agglutinated species. The sites located in the bathyal zone contained 122 calcareous and 17 agglutinated benthonic foraminifera. As the number of sites in each group was different, to compare directly their species diversity is not possible. However, we can compare the calcareous: agglutinated ratio of their faunas. This ratio is 11 at deeper sites and 7 at shallower sites. Thus, it appears that in the Indian Ocean during Neogene-Quaternary time an increase in depth is correlated with an increase in the proportion of calcareous species in the assemblages. Certainly this increase was only above the critical depth where the dissolution of calcareous benthonic foraminifera becomes

Table 3
Redeposited shallow water benthonic foraminiferal species in samples at Site 262.

AGE	CORE	SECTION	SAMPLE	Amphistegina lessonii	Cancris sagra	Cymbaloporetta bradyi	Discorbinella biconcava	Discorbis parkera	Elphidium advenum	Elphidium complanatum	Elphidium craticulatum	Elphidium crispum	Elphidium discoidale	Elphidium sp.	Guttulina regina	Miliolinella lutea	Miliolinella subrotunda	Patellina corrugata	Peneroplis pertusus	Planorbulina mediterranensis	Quinqueloculina seminulum	Rotalia beccarii koeboeensis	Rotalia margaritifera	Rotalia schroeteriana
QUATERNARY	1	3	140-142										1					1						
	1		CC										2			1							
	2	3	75-77	1								1												
	4		CC		1							1						1						
	5	4	10-12										3											
	5		CC	1																	1			
	6	4	15-16	2								1												
	8	4	10-12					2																
	8		CC	3						1														
	9	4	10-12	2				1								1								
	9		CC	1																				
	10	4	10-12								1	1												
	11	4	10-12	1																	7			
	12	4	10-12			4			2				4	1							3			
	12		CC																		1			
	13	4	10-12										3											
	14	3	140-142										2		1									
	15	3	140-142							1			1											
	16	3	125-127	4		2					2													
	16		CC	5					1															
	17	3	110-112										2											
	17		CC				2						2	1										
	18	3	140-142					1					2											
	18		CC							1			2											
	19		CC	1	1								1			1								
	20		CC										3											
	21	3	140-142	1	1								1											
	21		CC						1			1	10											
	22	3	140-142										3											
	22		CC	1						1			2											
	23	3	140-142									2	2								1			
	24	3	140-142										2											
	25		CC										1											
	26	3	140-142										1											
	27	3	140-142									1	1											
	29		CC					1				1												
	34		CC							1											10			
	34	4	10-12										2											
	35		CC									2									4			
	36	3	140-142						1	1			2											
PLIOCENE	36		CC																		1			
	37		CC					1													5			
	38		CC			1		1				1												
	39	4	10-12			2															2			
	39		CC			1															2			
	40	3	140-142									1									1			
	40		CC			1		1				1									4			
	41	4	140-142					1													2	3		
	41		CC	1				1				1												
	42	3	140-142		2																			
	42		CC			1						1									4			
	43	2	140-142									1									1			
	43		CC		1		2	3													1	3		
	44	3	140-142	1		1		2				1									2	6	2	
	44		CC	1					2					1				1			12	16		
	45	4	10-12	2					22		3	1										6	15	
	45		CC	2					10														10	
	46		CC	2					16			1											10	

very intense. About 4,000 - 5,000 m appears to be the critical depth for planktonic foraminiferal shells in medium latitudes. The critical depth is somewhat greater for benthonic foraminifera as they are more resistant to dissolution.

Brief Paleontological Description of the Sites

The stratigraphic subdivisions used in this report were those established by the paleontological staffs of previous legs. The following papers on the distribution of the planktonic foraminifera at the sites were utilized by me: Sites 219, 220, 223 (Fleisher, 1974); Sites 239, 242, 246 (Zobel, 1974); Sites 250, 251, 253, 254, 255, 258 (Boltovskoy, 1974); Sites 260, 262, 263 (Bolli, 1974).

Site 219

This site is located in the Arabian Sea relatively near the southwestern shore of the Indian Subcontinent and is associated with the Chagos-Laccadive Ridge.
A previous study determined the presence of the following sequences at this site: Quaternary, Pliocene, Upper, Middle and Lower Miocene. However, according to Fleisher (1974, p. 1006), "Reworked specimens are present sporadically below core 2, but appear constantly below core 5 and commonly below core 10. Most species represented are of Early Middle Miocene age". Then Fleisher adds (loc. cit.) that "occasional specimens ... are probably somewhat older". Since cores 1 and 2 mentioned contain uppermost Quaternary sediments, it turns out that practically the whole sequence is contaminated.
The most common species found practically throughout the whole length of this site are: Bulimina rostrata, Cassidulina subglobosa subglobosa, Cibicides bradyi, C. wuellerstorfi, Epistominella exigua, Gyroidina soldanii, Osangularia culter, Stilostomella cf S. annulifera, S. ex gr. S. lepidula and Uvigerina proboscidea, s.l.
I did not find any shallow water benthonic foraminiferal species in my 28 samples.

Site 220

This site was also drilled in the Arabian Sea, somewhat SSE of the foregoing one.
According to Fleisher (1974), it contains a highly dissolved Pleistocene fauna and below it planktonic assemblages so poorly preserved that his assignment of them to the Pliocene and Upper Miocene was carried out with great reservation. The occurrence of "Bulava indica", in Core 5 indicates that the standpoint of Fleisher that that part of site is Miocene in age is correct.
The benthonic foraminifera are rather large, and are rich qualitatively and quantitatively with presumably no displaced specimens. The most common species are: Epistominella exigua, Eponides bradyi, Oridorsalis umbonatus, some Stilostomella and Uvigerina.

Site 223

I had at my disposal only four samples of the Quaternary unit at this site. All of them contained displaced shallow water benthonic specimens (Elphidium advenum, E. crispum, Quinqueloculina seminulum, Cancris sagra, Rotalia sp., etc.). The most common species, presumably in situ, were Cibicides wuellerstorfi and Uvigerina dirupta.

Site 239

This site is located in the Abyssal Plain of the Western Mascarene Basin between Madagascar and the Island of Reunion. It was drilled at a rather great depth (4971 m) and therefore the age determination of the sequence encountered was done mainly by means of nannofossils. The following stratigraphic units in the Neogene sequence were determined: Quaternary, Pliocene, Upper, Middle and Lower Miocene.

The benthonic foraminiferal fauna encountered at this site is small-sized. An appreciable number of specimens were found only in the Quaternary sediments. In the Pliocene sediments many specimens were too small to be identified and several Miocene samples were either barren or contained specimens so small that they were practically unidentifiable. However, in all samples where determinable assemblages were found, there were always specimens of several displaced shallow water species (mainly Elphidium and Miliolinella species).

Site 242

This site was drilled on the eastern flank of the Davie Ridge which is located between Africa and Madagascar. The following units were previously recorded at this site: Quaternary, Pliocene, Upper, Middle and Lower Miocene.

The most frequent and typical species of this site are: Cassidulina subglobosa subglobosa, Epistominella exigua, Oridorsalis umbonatus, Pullenia osloensis and various Stilostomella. Vertical changes in their distribution were not evident. In the upper part of the site (Pleistocene - Upper Miocene) Bolivina globulosa is numerous.

Site 246

Site 246 is located near the crest of the Madagascar Ridge, about 480 miles south of Madagascar. The Neogene core material ranged in age from Quaternary to Lower Miocene. Unfortunately I had at my disposal only three samples, one Pliocene, one Middle Miocene and one Lower Miocene. The last one contained a strongly attacked and reworked fauna. The Middle Miocene sample was very poor qualitatively and quantitatively; the most abundant elements in it were: Uvigerina porrecta and Stilostomella cf. S. annulifera. The Pliocene sample was also poor and contained rare specimens of Discorbinella biconcava which very probably were redeposited from a shallow water environment.

Site 250

This is one of the deepest sites (5119 m) and consequently its foraminiferal fauna is strongly chemically attacked. Many samples are completely barren while the fauna in the others is quantitatively very poor and, in addition, very small-sized. The following strata were previously identified at this site: Quaternary, Upper Pliocene, Upper Miocene and Lower Miocene. In the Lower Miocene sediments some evidently reworked Paleogene planktonic foraminifera and probably displaced benthonic foraminifera of Lower Miocene in age were found (Boltovskoy, 1974).

The most common benthonic foraminifera recorded at Site 250 are: Cassidulina subglobosa subglobosa, Epistominella exigua and Pullenia subcarinata quinqueloba, but even these species were represented by a quite reduced number of specimens mainly in Upper Pliocene and Quaternary deposits.

Only in the Quaternary sequence no displaced shallow water specimens were found. In the Upper Pliocene sediments, however, were encountered Amphistegina lessonii and Miliolinella subrotunda and in the Lower Miocene other shallow water species, as for instance, Elphidium crispum, Spiroloculina antillarum and Glabratella sp.

Site 251

This is an interesting site because it contains a complete sequence of Neogene deposits (Quaternary, Upper Pliocene, Middle and Lower Pliocene not separated, Upper Miocene, Middle Miocene and Lower Miocene) presumably without interruption.

The benthonic foraminifera are well preserved and are represented by many species, the most common of which are: Cassidulina subglobosa subglobosa, Cibicides wuellerstorfi, Eggerella bradyi, Eponides bradyi, Gyroidina soldanii, Nonion affine, N. pompilioides, Oridorsalis umbonatus, Pullenia bulloides and P. subcarinata quinqueloba. These species occurred rather frequently throughout the whole site. Uvigerina dirupta is numerous in the Quaternary deposits, rare in the Pliocene and Upper and Middle Miocene ones and absent in the Lower Miocene sediments. Stilostomella ex gr. S. lepidula is, on the contrary, missing in the Quaternary deposits but present (although in small numbers but persistent) in the underlying Neogene strata.

There are no vertical changes of importance with respect to the species content of this site. No displaced specimens were found.

Site 253

This is also a valuable site as it has a good sequence of Neogene deposits which could be subdivided into smaller units. It is believed that the sequences are uninterrupted.

The benthonic foraminiferal fauna of this site is rich, of normal size, not chemically attacked and there are no specimens which can be considered as redeposited.

The most common species found in all sequences and which make up the main part of the assemblages are: Cibicides kullenbergi, Oridorsalis umbonatus and Stilostomella ex gr. S. lepidula. In smaller quantities of specimens the following were also rather consistently present. Eggerella bradyi, Osangularia culter, several Pleurostomella, Stilostomella subspinosa and Uvigerina proboscidea, s.l.

Cassidulina subglobosa subglobosa and Cibicides wuellerstorfi are abundant or common in the Pleistocene-Middle Miocene sequences and missing in the Lower Miocene one.

Although the planktonic foraminiferal fauna in the Miocene sediments of this site turned out to be very different from that of the Oligocene (Boltovskoy, 1974, fig. 4), no evidence of a fauna change was observed in the benthonic foraminiferal assemblages. The majority of the benthonic species, especially the most abundant ones, cross the Miocene-Oligocene boundary without showing any change in their morphology or abundance.

Several species of Site 253 have a rather short vertical range and are sufficiently numerous and typical to be considered as index fossils (for the Miocene sequence, e.g. "Bulava indica", Bulimina macilenta, B. jarvisi, Planulina marialana gigas, Rectuvigerina royoi).

Site 254

This site, like Site 253 is also located on the Ninetyeast Ridge, but almost 6° further south.

The tests of foraminifera are very well preserved and by means of planktonic ones the Neogene deposits were subdivided into Quaternary, Upper, Middle and Lower Pliocene and Miocene. The limits between these subdivisions (especially the Miocene/Oligocene limit) were well expressed and located.

The following benthonic foraminifera were found rather consistently throughout the sequence: Cibicides kullenbergi, C. wuellerstorfi, Karreriella bradyi, Laticarinina pauperata, Oridorsalis umbonatus, Stilostomella ex gr. S. lepidula and Uvigerina proboscidea, s.l.

The following species were not found in the Pleistocene but occurred frequently and consistently in all the strata below it: Orthomorphina aff. antillea, and Stilostomella cf. S. annulifera.

It was not possible to locate any limit between strata of different ages based on the vertical distribution of benthonic foraminifera.

Site 255

This site is located on the submarine mountain called Broken Ridge. It was drilled down to 108.5 m below the sea bottom, but the recovery was very poor and only a small number of samples was at my disposal for study.

Quaternary, Pliocene, Upper, Middle and Lower Miocene sediments were previously identified at this site. Below the Lower Miocene deposits a strongly contaminated, very small-sized fauna of unknown age was encountered (Boltovskoy, 1974).

The benthonic foraminiferal fauna at Site 255 is very similar to those faunas at Site 253 and 254. This is explained by the fact that all these sites are located near each other and at approximately the same depth.

Site 255 does not contain any displaced benthonic specimens.

Site 258

Site 258 is situated on the Naturaliste Plateau near SW Australia. On the basis of previous studies the following Neogene sequences were determined: Quaternary, Upper, Middle and Lower Pliocene and Upper Miocene.

The range chart of planktonic foraminifera was considered to be reasonably reliable with the exception of the vertical distribution of Globigerinoides sicanus de Stefani. This species is well known as a Miocene indicator. However, at Site 258 it was found as high as Middle Pliocene. Since the specimens found did not look reworked and no other reworked planktonic species were observed in the same strata, it was believed that the specimens of G. sicanus are in situ and thus the upper limit of this species should be extended to include the Lower and Middle Pliocene (Boltovskoy, 1974).

However, the present study of the benthonic foraminifera revealed that that conclusion most probably was wrong. Many samples at Site 258 contained small specimens of various Elphidium sp. Practically the whole sequence at this site should be considered as containing displaced (and/or reworked) elements. This probably explains the strange occurrence of Globigerinoides sicanus in the Pliocene sediments. It also makes occurrence data of other species questionable.

The most common benthonic foraminifera found at Site 258 are: Cassidulina subglobosa subglobosa, Cibicides kullenbergi, C. wuellerstorfi, Trifarina bradyi, Uvigerina dirupta.

Site 260

This site is the deepest site studied in this report (5,702 m). It is located in the southern part of the Gascoyne Abyssal Plain. Neogene strata are almost non-existent in this site, and only relatively thin sequences of the Quaternary, Lower Pliocene and Middle Miocene sediments were discovered. However, all of them contained abundant redeposited material and fauna was poor.

I had at my disposal only six samples from this site. They do not contain displaced shallow water benthonic foraminifera, but many specimens were evidently redeposited from the older sequences (e.g. Cassidulina carandelli, C. cuneata, C. subglobosa horizontalis and others). Thus, the whole site should be considered as heterogeneous in age.

Site 262

Site 262 was drilled in the western part of the Timor Trough. The sediments comprise very thick Quaternary (337.5 m) and less thick Pliocene (100 m) sediments. It was considered the most complete and fossiliferous tropical marine sequence of this age ever encountered by the Deep Sea Drilling Project (Bolli, 1974). With the exception of the lowermost part, all the sequences are rich in planktonic foraminifera and were interpreted as deep water deposits. According to Bolli (1974), an euxinic bottom milieu is indicated for almost all the Quaternary at this site. In the lowermost part (Core 45 and downwards) rapid shallowing took place and the environment is interpreted as being a typical nearshore zone with wave action, strong turbulent water and currents.

The benthonic foraminifera found at this site are qualitatively very rich and quantitatively in many samples abundant. The most common deep-water species found at this site are: Bulimina aculeata, B. inflata, Bolivina robusta, Cassidulina laevigata, C. subglobosa subglobosa, Cibicides wuellerstorfi, Nonion affine, Uvigerina dirupta, U. proboscidea. However, it contains also rather frequent displaced shallow water foraminifera as, e.g. Elphidium crispum, E. complanatum, E. advenum, E. sp., Miliolinella lutea, M. subrotunda, Amphistegina lessonii, Patellina corrugata, Quinqueloculina seminulum, Rotalia beccarii koeboeensis, R. margaritifera, R. schroeteriana and some others (see Table 3), as well as several reworked species mainly typical of Lower and Middle Miocene ages (Casidulina carandelli, C. cuneata, C. subglobosa horizontalis, Textularia bermudezi).

Site 263

Site 263 is on the eastern edge of the Cuvier Abyssal Plain. Of the Neogene sediments only the Quaternary and probable Upper Pliocene sequences were identified by means of planktonic foraminifera and nannoplankton. According to this previous study (Bolli, 1974), the Quaternary sequence contains reworked Pliocene species and the Upper Pliocene fauna is heterogeneous and thus it is possible that the Upper Pliocene sequence is in reality Quaternary in age.

I had at my disposal six samples from this site. All of them contained specimens of typical benthonic shallow water species, e.g. Elphidium advenum, E. crispum, Miliolinella subrotunda, Rotalia beccarii koeboeensis and others.

Benthonic Foraminiferal Species as Stratigraphic Indicators

As stated above, the entire sequence of only five sites (220, 251, 253, 254, 255) contained no displaced and/or reworked specimens. This statement is based on both the present study and previous investigations made by means of planktonic foraminifera and nannoplankton. The displaced elements found in previous studies at Site 254 are evidently the result of contamination which took place during the drilling process.

In addition, the following intervals at two other sites also did not contain displaced faunas: the Quaternary and Miocene at Site 242 and the Quaternary at Site 250. Thus, the species distribution in these sequences also can be considered absolutely reliable.

To be on the safe side, the conclusions with respect to the vertical distribution of benthonic foraminiferal species and their suitability for use as guide fossils are drawn exclusively on the basis of data obtained from the sequences which contained no indication of redeposited faunas.

The large majority of benthonic foraminifera were vertically distributed throughout the whole Neogene-Quaternary sequence. Those species which occurred consistently and more or less frequently are: Anomalina globulosa, Astrononion umbilicatulum, Bolivina pusilla, Bulimina inflata, B. rostrata, Cassidulina subglobosa subglobosa, Cibicides bradyi, C. kullenbergi, Eggerella bradyi, Epistominella exigua, Gyroidina lamarckiana, G. soldanii, Karreriella bradyi, Laticarinina pauperata, Nonion affine, N. pompilioides, Oridorsalis umbonatus, Orthomorphina aff. O. antillea, Pullenia bulloides, P. osloensis, P. subcarinata quinqueloba, Pyrgo murrhina, Stilostomella cf. S. annulifera, S. ex gr. S. lepidula and Uvigerina proboscidea, s.l.

The following species also occurred rather consistently but their lowermost appearance in the material studied was recorded in the Middle Miocene sediments: Bolivina globulosa, Cassidella bradyi, Gyroidina umbonata, Uvigerina dirupta and U. hirsuta.

Several species were limited to a particular sequence; some of these species can be used as guide fossils and are discussed below.

Quaternary

Several foraminifera were found only in Quaternary sediments especially at Site 262. However, all of them, except Hyalinea balthica, are also well known in older sediments in other areas. As for Hyalinea balthica, this species was accepted by the Seventh Congress of INQUA (Denver, Colorado, USA, 1965) as the guide fossil for the commencement of the Quaternary, and was considered to be solely Quaternary. However, subsequent studies have shown that this is not the case. Hyalinea balthica can now be considered as a Pleistocene guide species only for the Mediterranean area because it has been found in deposits of other ages (e.g. according to Bandy, 1968, in the Philippine Islands even in Miocene strata). In my material I found isolated and small-sized specimens of Hyalinea balthica at four sites (239, 258, 262, 263) in the Quaternary sediments, but at two sites (258, 262) this species was recorded in the Pliocene strata too. It should be mentioned that all these sites contain numerous displaced specimens. Thus, it is not clear whether this species can be interpreted as an indicator of the Quaternary in the Indian Ocean.

Pliocene

Only two species, Pyrulina extensa and Uvigerina longistriata, were found restricted to the Pliocene sediments. Both occur sporadically and in very small numbers. They cannot be considered as Pliocene guide specimens because Pyrulina extensa is well known in Recent ocean (Cushman, 1923) and Uvigerina longistriata was described from Miocene (Perconig, 1955).

Table 4

Range chart of selected species in the L. Pliocene - L. Miocene sequence

Species	Lower Pliocene	Miocene			Age and areas of initial description
		U	M	L	
Bolivinopsis cubensis		x	x	x	Paleocene and Lower Eocene, Cuba
Bulava indica		x	x		U. and M. Miocene, Indian Ocean
Bulimina macilenta		x	x	x	Eocene, California, USA
Bulimina jarvisi			x	x	U. Oligocene, Eocene, Trinidad, W.I.
Cassidulina carandelli		x	x	x	L. Miocene, Spain
Cassidulina cuneata		x	x	x	U. Miocene-U. Oligocene, New Zealand
Cassidulina subglobosa horizontalis		x	x	x	L. Miocene-U. Oligocene, Venezuela
Planulina marialana gigas		x	x	x	Oligomiocene, Cuba
Pleurostomella obtusa	x	x	x	x	L. Cretaceous, France
Rectuvigerina royoi	x	x	x	x	Pliocene, Venezuela
Saracenaria latifrons jamaicensis		x	x	x	Miocene, Jamaica
Siphogenerina vesca			x	x	L. Miocene, New Zealand
Stilostomella tuckerae		x	x	x	U. Oligocene, Cuba
Uvigerina miozea		x	x	x	L. Miocene, New Zealand

Miocene

The number of species which were found to be limited to the Miocene sequence is larger. Table 4 shows their range in the deep water Neogene sequence of the Indian Ocean. Two specimens are included which range up into the Lower Pliocene. Those which occurred in the Miocene sequence at a very few sites, and then only as isolated specimens, are not included.

An analysis of Table 4 shows that the ranges of selected species found in the present material do not differ very much from the stratigraphic data given in the initial descriptions of the species. However, Pleurostomella obtusa is an exception. It was described in the L. Cretaceous of France and found in the Miocene and even the L. Pliocene in the Indian Ocean.

The following species are frequent, easily identified and can be considered as the best Miocene deep water Indian Ocean guide fossils: "Bulava indica", Bulimina macilenta, B. miolaevis, B. jarvisi, Planulina marialana gigas, Rectuvigerina royoi and Siphogenerina vesca. Rectuvigerina royoi occurred in small numbers in the lowermost Pliocene sediments as well.

Main Conclusions

1) At all the sites (except 262) the benthonic foraminiferal fauna of the Neogene-Quaternary sequence did not show any change which could be interpreted as indicating a dramatic change in depth. Thus, it is concluded that since the Early Miocene the Indian Ocean has not undergone bathymetric alteration of any importance. A single drastic faunal change was observed at Site 262 where, from Core 45 (Pliocene) downward rapid shallowing is indicated. However, this site is located quite near Timor Island and the geological events which created the transition from shallow to the deeper water in the Pliocene are of local importance only.

2) The contamination of sites by redeposited fauna (of shallow water origin and/or from older strata) is much higher than previous studies of these sites, based on planktonic foraminifera and other microplankton, indicated. Of 15 sites studied only 5 turned out to contain no displaced specimens. Eight contained heterogeneous material throughout practically their entire length. Two contained displaced elements only partially. These data are rather alarming in terms of the reliability of conclusions drawn by previous studies which unknowingly were based on questionable material. It is highly recommended that future paleontological and stratigraphic studies of the sites include not only an examination of planktonic foraminifera but of the benthonic ones as well.

3) The large majority of benthonic foraminifera recorded in this report were found throughout the whole Neogene-Quaternary sequence and thus are not useful as guide fossils. There are practically no guide fossils among benthonic foraminifera in the sites studied either for Quaternary or Pliocene sediments. The following species were found solely in the Miocene sequence and can be considered as indicators of this system (at least in the area under study) because they are constant, sufficiently frequent and easy to determine: "Bulava indica", Bulimina macilenta, B. miolaevis, B. jarvisi, Planulina marialana gigas and Siphogenerina vesca. These conclusions are based on the data obtained from sites in which all the sediments showed no signs of redeposited faunas.

Acknowledgments. I am indebted to Drs. R. L. Fleisher (Houston, Texas, U.S.A.), B. Zobel (Hannover, Germany) and H. M. Bolli (Zürich, Switzerland) for their helpful suggestion with respect to the most interesting cores of Legs 22, 25 and 27, respectively, in which the cited researchers participated.

I am not less grateful to the Curatorial Staff of the Deep Sea Drilling Project who provided me with the material requested.

I wish to express my sincere thanks also to Mr. I. Riobo de Magaldi and Miss A. M. Leverone for their technical help in preparing this report.

References to Benthonic Foraminifera Cited in Text

Unfortunately a limited space has not allowed to give the complete list of all the species found. Only those are listed below which were cited in the text. However, in view of great confusions existing now in the benthonic foraminiferal taxonomy, I consider that it is necessary to give for all these species the references by whom they were firstly described.

Amphistegina lessonii d'Orbigny, 1826, Ann. Sci. Nat., ser. 1, v. 7, p. 304, no. 3, pl. 17, figs. 1-4 (A. quoii in description of plate); Mod. 98.
Anomalina globulosa Chapman & Parr, 1937, Australasian Antarct. Exp., C, v. 1, p. 117, pl. 9, fig. 27.

Asterigerinata pacifica Uchio, 1960, Cushman Found. Foram. Res., Sp.
  Publ. 5, p. 67, pl. 10, figs. 26-31.
Astrononion umbilicatulum Uchio, 1952, Japan. Assoc. Petr. Technolog.,
  Journ., v. 17, no. 1, p. 36, text fig. 1.
Bolivina globulosa Cushman, 1933, Cushman Lab. Foram. Res., Contr.,
  v. 9, p. 80, pl. 8, fig. 9.
Bolivina pusilla Schwager, 1866, Novara Exp., Geol., v. 2, p. 254, pl. 7,
  fig. 101.
Bolivinopsis cubensis (Cushman & Bermúdez) = Spirolectoides cubensis
  Cushman & Bermúdez, 1937, Cushman Lab. Foram. Res., Contr., v. 13,
  pt. 1, p. 13, pl. 1, figs. 44 and 45.
"Bulava indica" Boltovskoy, Rev. Espan. Micropal., v. 8, no. 2, p. 301-
  303, 1977.
Bulimina aculeata d'Orbigny, 1826, Ann. Sci. Nat., ser. 1, v. 7,
  p. 269, no. 7.
Bulimina inflata Seguenza, 1862, Accad. Gioenia Sci. Nat. Catania,
  Atti, ser. 2, v. 18, p. 109, pl. 1, fig. 10.
Bulimina macilenta Cushman & Parker = Bulimina denticulata Cushman &
  Parker, 1936, Cushman Lab. Foram. Res., Contr., v. 12, p. 42, pl. 7,
  figs. 7,8; Bulimina macilenta Cushman & Parker, new name, 1939,
  ibidem, v. 15, p. 93.
Bulimina miolaevis Finlay, 1940, Roy. Soc. New Zealand, Trans. Proc.,
  v. 69, pt. 1, p. 454, pl. 64, figs. 70,71.
Bulimina rostrata Brady, 1884, Challenger Exp., Zool., v. 9, p. 408,
  pl. 51, figs. 14,15.
Bulimina jarvisi Cushman & Parker, 1936, Cushman Lab. Foram. Res.,
  Contr., v. 12, pt. 2, p. 39, pl. 7, fig. 1.
Cancris sagra (d'Orbigny) = Rotalia sagra d'Orbigny, 1839, in de la
  Sagra, Hist. Phys. Nat. Pol. Cuba, Foram., p. 91, pl. 5, figs. 14,15.
Cassidella bradyi (Cushman) = Virgulina bradyi Cushman, 1922, U. S.
  Nat. Mus., Bull. 104, p. 115, pl. 24, fig. 1.
Cassidulina carandelli Colom, 1943, R. Soc. Espanola Hist. Nat., Bol.,
  v. 41, p. 324, pl. 23, figs. 65-67.
Cassidulina cuneata Finlay, 1940, Roy. Soc. New Zealand, Trans. Proc.,
  v. 69, pt. 1, p. 456, pl. 63, figs. 62-66.
Cassidulina laevigata d'Orbigny, 1826, Ann. Sci. Nat., ser. 1, v. 7,
  p. 282, no. 1, pl. 15, figs. 4,5.
Cassidulina subglobosa horizontalis Cushman & Renz, 1941, Cushman
  Lab. Foram. Res., Contr., v. 17, p. 26, pl. 4, fig. 8.
Cassidulina subglobosa subglobosa Brady = Cassidulina subglobosa
  Brady, 1884, Challenger Exp., Zool., v. 9, p. 430, pl. 54, fig. 17.
Cibicides bradyi (Trauth) = Truncatulina dutemplei Brady, 1884,
  Challenger Exp., Zool., v. 9, p. 665, pl. 95, fig. 5 - Truncatulina
  bradyi Trauth, 1918, Denkschr. K. Akad. Wiss. Wien, Math.-Nat. Kl.,
  v. 95, p. 235.
Cibicides kullenbergi Parker, 1953, in Phleger, Parker & Peirson, 1953,
  Swedish Deep Sea Exp., Repts., v. 1, p. 49, pl. 11, figs. 7,8.
Cibicides wuellerstorfi (Schwager) = Anomalina wuellerstorfi Schwager,
  1866, Novara Exp., Geol., v. 2, p. 258, pl. 7, figs. 105-107.
Cymbaloporetta bradyi (Cushman) = Cymbalopora poeyi (d'Orbigny) var.
  bradyi Cushman, 1915, U. S. Nat. Mus., Bull. 71, pt. 5, p. 25, pl.
  10, fig. 2; pl. 14, fig. 2.
Discorbinella biconcava (Parker & Jones) = Discorbina biconcava Parker
  & Jones, 1862, in Carpenter, Introd. Stud. Foram., p. 201, pl. 19,
  fig. 10.
Discorbis parkerae Natland, 1950, Geol. Soc. Amer., Mem., v. 43,
  p. 27, pl. 6, fig. 11.

Eggerella bradyi (Cushman) = Verneuilina bradyi Cushman, 1911, U. S.
   Nat. Mus., Bull. 71, pt. 2, p. 54, text fig. 87.
Elphidium advenum (Cushman) = Polystomella advena Cushman, 1922,
   Carnegie Inst. Washington, Publ. 311, p. 56, pl. 9, figs. 11, 12.
Elphidium complanatum (d'Orbigny) = Polystomella complanata d'Orbigny,
   1839, in Barker, Webb & Berthelot, Hist. Nat. Iles Canaries, v. 2,
   no. 2, Foram., p. 129, pl. 2, figs. 35, 36.
Elphidium craticulatum (Fichtel & Moll) = Nautilus craticulatus
   Fichtel & Moll, 1798, Test. Microsc. p. 51, pl. 5, figs. h-k.
Elphidium crispum (Linne) = Nautilus crispus Linne, 1767, Syst. Nat.,
   13th ed., p. 1162.
Elphidium discoidale (d'Orbigny) = Polystomella discoidale d'Orbigny,
   1839, in de la Sagra, Hist. Phys. Pol. Nat. Cuba, Foram., p. 56,
   pl. 6, figs. 23, 24.
Epistominella exigua (Brady) = Pulvinulina exigua Brady, 1884, Challenger Exp., Zool., v. 9, p. 696, pl. 10, figs. 13,14.
Eponides bradyi Earland = Truncatulina pygmaea Hantken in Brady, 1884,
   Challenger Exp., Zool., v. 9, p. 666, pl. 95, figs. 9, 10 = Eponides
   bradyi Earland, 1934, Discovery Repts., v. 10 (1935), p. 187,
   pl. 8, figs. 36-38.
Guttulina regina (Brady, Parker & Jones) = Polymorphina regina Brady,
   Parker & Jones, 1870, Trans. Linn. Soc., v. 27, p. 241, pl. 41,
   fig. 32.
Gyroidina lamarckiana (d'Orbigny) = Rotalia lamarckiana d'Orbigny,
   1839, in Barker, Webb & Berthelot, Hist. Nat. Iles Canaries, v. 2
   no. 2, Foram., p. 131, pl. 2, figs. 13-15.
Gyroidina soldanii d'Orbigny, 1826, Ann. Sci. Nat., ser. 1, v. 7,
   p. 276, no. 5; Mod. 36.
Gyroidina umbonata (Silvestri) = Rotalia soldanii d'Orbigny var.
   umbonata Silvestri, 1898, Accad. Pont. Nuovi Lincei, Mem., v. 15
   p. 329, pl. 6, fig. 14.
Hyalinea balthica (Schroeter) = Nautilus balthicus, 1783,
   Einleitung, v. 1, p. 20, pl. 1, fig. 2.
Karreriella bradyi (Cushman) = Gaudryina bradyi Cushman, 1911, U. S.
   Nat. Mus., Bull. 71, pt. 2, p. 67, text fig. 107.
Laticarinina pauperata (Parker & Jones) = Pulvinulina repanda var.
   menardii subvar. pauperata Parker & Jones, 1865, Roy. Soc. London,
   Philos. Trans., v. 155, p. 395, pl. 16, figs. 50,51.
Miliolinella lutea (d'Orbigny) = Triloculina lutea d'Orbigny, 1839,
   Voy. Amér. Mérid., v. 5, no. 5, p. 70, pl. 9, figs. 6-8.
Miliolinella subrotunda (Montagu) - Vermiculum subrotundum Montagu,
   1803, Test. Britan., p. 521.
Nonion affine (Reuss) = Nonionina affinis Reuss, 1851, Deutsch. Geol.
   Ges., Zeitschr., v. 3, p. 72, pl. 5, fig. 32.
Nonion pompilioides (Fichtel & Moll) = Nautilus pompilioides Fichtel
   & Moll, 1798, Test. Microsc., p. 31, pl. 2, figs. a-c.
Oridorsalis umbonatus (Reuss) = Rotalina umbonata Reuss, 1851, Deutsch.
   Geol. Ges., Zeitschr., v. 3, p. 75, pl. 5, fig. 35.
Orthomorphina aff. O. antillea (Cushman) = Nodosaria antillea Cushman
   1923, U. S. Nat. Mus., Bull. 104, pt. 4, p. 91, pl. 14, fig. 9.
Osangularia culter (Parker & Jones) = Planorbulina culter Parker &
   Jones, 1865, Roy. Soc. London, Philos. Trans., v. 155, p. 421,
   pl. 19, fig. 1.
Patellina corrugata Williamson, 1858, Rec. Foram. Gr. Britain, p. 46,
   pl. 3, figs. 86-89.
Peneroplis pertusus (Forskål) = Nautilus pertusus Forskål, 1775, Descr.
   Anim., p. 125, no. 65.
Planorbulina mediterranensis d'Orbigny, 1826, Ann. Sci. Nat., ser. 1,
   v. 7, p. 280, no. 2, pl. 14, figs. 4-6; Mod. 79.

Planulina marialana gigas Keijzer, 1945, Utrecht Univ. Geogr. Geol.
    Meded., Physiogr.-Geol. Reeks, ser. 2, no. 6, p. 206, pl. 5, fig. 77.
Pleurostomella obtusa Berthelin, 1880, Soc. Géol. Franc, Mém., sér. 3,
    v. 1, no. 5, p. 29, pl. 1, fig. 9.
Pullenia bulloides (d'Orbigny) = Nonionina bulloides d'Orbigny, 1846,
    Foram. Foss. Vienne, p. 107, p. 5, figs. 9,10.
Pullenia osloensis Feyling-Hanssen = Pullenia quinqueloba (Reuss) subsp. minuta Feyling-Hanssen, 1954, Norsk. Geol. Tidsskr., v. 33,
    no. 1-2, p. 133, pl. 2, fig. 3 (emend. P. osloensis Feyling-Hanssen,
    1954, ibidem, no. 3-4, p. 194).
Pullenia subcarinata quinqueloba (Reuss) = Nonionina quinqueloba
    Reuss, 1851, Deutsch. Geol.Ges.,Zeitschr.,v.3,p.71, pl.5, fig. 31.
Pyrgo murrhina (Schwager) = Biloculina murrhina Schwager, 1866, Novara
    Exp., Geol., v. 2, p. 203, pl. 4, fig. 15.
Pyrulina extensa (Cushman) = Polymorphina longicollis Brady, 1881,
    Quart. Journ. Micr. Sci., v. 21, p. 64; Challenger Exp., Zool.,
    v. 9, p. 572, pl. 73, figs. 18, 19 = Polymorphina extensa Cushman,
    1923, U. S. Nat. Mus., Bull. 104, pt. 4, p. 156, pl. 41, figs. 7,8.
Quinqueloculina seminulum (Linné) = Serpula seminulum Linné, Syst.
    Nat., 13th ed., p. 1264, no. 791.
Rectuvigerina royoi Bermúdez & Fuenmayor, 1963, in Bermúdez & Seiglie,
    1963, Inst. Ocean. Univ. Oriente, Bol., v. 2, no. 2, p. 144, pl. 18,
    fig. 9.
Rotalia beccarii koeboeensis Leroy, 1939, Natuurk, Tijdschr. Nederl.-
    Ind., v. 99, no. 6, p. 255, pl. 6, figs. 13-15.
Rotalia margaritifera (Brady) = Truncatulina margaritifera Brady,
    1881, Quart. Journ. Micr. Sci., v. 21, p. 66; 1884, Challenger
    Exp., Zool., v. 9, p. 667, pl. 96, fig. 2.
Rotalia schroeteriana Parker & Jones, 1862, in Carpenter, Introd. Stud.
    Foram., p. 213, pl. 13, figs. 7-9.
Saracenaria latifrons jamaicensis Cushman & Todd, 1945, Cushman Lab.
    Foram. Res., Sp. Publ. 15, p. 32, pl. 5, fig. 7.
Siphogenerina vesca Finlay, 1939, Roy. Soc. New Zealand, Trans. Proc.,
    v. 69, pt. 1, p. 109, pl. 13, figs. 46,47.
Spiroloculina antillarum d'Orbigny, 1839, in de la Sagra, Hist. Phys.
    Pol. Nat. Cuba, Foram., p. 166, pl. 9, figs. 3,4.
Stilostomella cf. S. annulifera (Cushman & Bermúdez) = Ellipsonodo
    saria annulifera Cushman & Bermúdez, 1936, Cushman Lab. Foram. Res.
    Contr., v. 12, no. 2, p. 28, pl. 5, figs. 8,9.
Stilostomella ex gr. S. lepidula (Schwager) = Nodosaria lepidula
    Schwager, 1866, Novara Exp., Geol., v. 2, p. 210, pl. 5, figs. 27,28.
Stilostomella subspinosa (Cushman) = Ellipsonodosaria subspinosa
    Cushman, 1943, Cushman Lab. Foram. Res., Contr., v. 19, pt. 4,
    p. 92, pl. 16, figs. 6,7.
Stilostomella tuckerae (Hadley) = Ellipsonodosaria tuckerae Hadley,
    1934, Bull. Amer. Pal., v. 20, no. 70A, p. 21, pl. 3, figs. 1,2.
Textularia bermudezi Cushman & Todd, 1945, Cushman Lab. Foram. Res.,
    Sp. Publ. 15, p. 3, pl. 1, fig. 7.
Trifarina bradyi Cushman, 1923, U. S. Nat. Mus., Bull. 104, pt. 4,
    p. 99, pl. 22, figs. 3-9.
Uvigerina dirupta Todd = Uvigerina peregrina Cushman var. dirupta
    Todd, 1948, in Cushman & McCulloch, Hancock Pacif. Exp., v. 6,
    no. 5, p. 267, pl. 34, fig. 3.
Uvigerina hispida Schwager, 1866, Novara Exp., Geol., v. 2, p. 249
    pl. 17, fig. 95.
Uvigerina longistriata Perconig, 1955, Serviz. Geol. Ital., Boll.,
    v. 77, fasc. 2, 3, p. 182, pl. 2, figs. 1-4.
Uvigerina miozea Finlay, 1939, Roy. Soc. New Zealand, Trans. Proc.,
    v. 69, pt. 1, p. 102, pl. 12, figs. 12-14.

Uvigerina porrecta Brady, 1879, Quart. Journ. Micr. Sci., v. 19, n.s., p. 60, pl. 8, figs. 15, 16.
Uvigerina proboscidea Schwager, sensu lato, 1866, Novara Exp., Geol. v. 2, p. 250, pl. 7, fig. 96.

References

Bandy, O. L., Paleoclimatology and Neogene planktonic foraminiferal zonation. Proc. IV Sess., Bologna 1967, Giorn. Geol. (2), 35, fasc. 2, p. 277-290, 1968.

Bolli, H. M., Synthesis of the leg 27 biostratigraphy and paleontology, in Veevers, J. J., Heirtzler, J. R., et al., Initial Reports of the Deep Sea Drilling Project, Washington (U. S. Government Printing Office) 27, 993-999, 1974.

Boltovskoy, E., Neogene planktonic foraminifera of the Indian Ocean (DSDP, leg 26), in Davies, T. A., Luyendyk, B. P., et al., Initial Reports of the Deep Sea Drilling Project, Washington (U. S. Government Printing Office) 26, 675-741, 1974.

Boltovskoy, E. and H. Lena, Los epibiontes de Macrocystic flotante como indicadores hidrológicos, Neotrópica, 15, no. 48, 135-137, 1969.

Cushman, J. A., The foraminifera of the Atlantic Ocean, Pt. 4, Lagenidae, U. S. Nat. Mus., Bull. 104, 1-228, 1923.

Fleisher, R. L., Cenozoic planktonic foraminifera and biostratigraphy, Arabian Sea, Deep Sea Drilling Project, leg 23A, in Whitmarsh, R. B., Weser, O. E., et al., Initial Reports of the Deep Sea Drilling Project, Washington (U. S. Government Printing Office), 23, 1001-1072, 1974.

Perconig, E., Due nuove specie di Uvigerina del Neogene della Pianura Padana, Serviz. Geol. Ital., Boll., 77, fasc. 2, 3, 181-191, 1955.

Zobel, B., Quaternary and Neogene foraminifera: biostratigraphy, in Simpson, E.S.W., Schlich, R., et al., Initial Reports of the Deep Sea Drilling Project, Washington (U. S. Government Printing Office) 25, 573-578, 1974.